To my sisters, Linda and Chris, and to my parents, Lillian and William

ABOUT THE AUTHOR

William J. Palm III is Professor of Mechanical Engineering and Applied Mechanics at the University of Rhode Island. In 1966 he received a B.S. from Loyola College in Baltimore, and in 1971 a Ph.D. in Mechanical Engineering and Astronautical Sciences from Northwestern University in Evanston, Illinois.

During his 33 years as a faculty member, he has taught 19 courses. One of these is a freshman MATLAB course, which he helped develop. He has authored eight textbooks dealing with modeling and simulation, system dynamics, control systems, and MATLAB. These include *System Dynamics* (McGraw-Hill, 2005). He wrote a chapter on control systems in the *Mechanical Engineers' Handbook* (M. Kutz, ed., Wiley, 1999), and was a special contributor to the fifth editions of *Statics* and *Dynamics,* both by J. L. Meriam and L. G. Kraige (Wiley, 2002).

Professor Palm's research and industrial experience are in control systems, robotics, vibrations, and system modeling. He was the Director of the Robotics Research Center at the University of Rhode Island from 1985 to 1993, and is the coholder of a patent for a robot hand. He served as Acting Department Chair from 2002 to 2003. His industrial experience is in automated manufacturing; modeling and simulation of naval systems, including underwater vehicles and tracking systems; and design of control systems for underwater-vehicle engine-test facilities.

format compact	get rid of empty lines
format short	fix 4 digit
format long	fix 14 digit
format short e	sci 4 digit
format long e	sci 15 digit
format short g	float 5 digit
format long g	float 15 digit
format bank	fix 2 digit
format loose	opposite of compact

CONTENTS

PREFACE

Formerly used mainly by specialists in signal processing and numerical analysis, MATLAB* in recent years has achieved widespread and enthusiastic acceptance throughout the engineering community. Many engineering schools now require a course based entirely or in part on MATLAB early in the curriculum. MATLAB is programmable and has the same logical, relational, conditional, and loop structures as other programming languages, such as Fortran, C, BASIC, and Pascal. Thus it can be used to teach programming principles. In most schools a MATLAB course has replaced the traditional Fortran course, and MATLAB is the principal computational tool used throughout the curriculum. In some technical specialties, such as signal processing and control systems, it is the standard software package for analysis and design.

The popularity of MATLAB is partly due to its long history, and thus it is well developed and well tested. People trust its answers. Its popularity is also due to its user interface, which provides an easy-to-use interactive environment that includes extensive numerical computation and visualization capabilities. Its compactness is a big advantage. For example, you can solve a set of many linear algebraic equations with just three lines of code, a feat that is impossible with traditional programming languages. MATLAB is also extensible; currently more than 20 "toolboxes" in various application areas can be used with MATLAB to add new commands and capabilities.

MATLAB is available for MS Windows and Macintosh personal computers and for other operating systems. It is compatible across all these platforms, which enables users to share their programs, insights, and ideas.

TEXT OBJECTIVES AND PREREQUISITES

This text is intended as a stand-alone introduction to MATLAB. It can be used in an introductory course, as a self-study text, or as a supplementary text. The text's material is based on the author's experience in teaching a required two-credit semester course devoted to MATLAB for engineering freshmen. In addition, the text can serve as a reference for later use. The text's many tables, and its referencing system in an appendix and at the end of each chapter, have been designed with this purpose in mind.

A secondary objective is to introduce and reinforce the use of problem-solving methodology as practiced by the engineering profession in general and as applied to the use of computers to solve problems in particular. This methodology is introduced in Chapter 1.

---

*MATLAB is a registered trademark of The MathWorks, Inc.

The reader is assumed to have some knowledge of algebra and trigonometry; knowledge of calculus is not required for the first seven chapters. Some knowledge of high school chemistry and physics, primarily simple electrical circuits, and basic statics and dynamics is required to understand some of the examples.

TEXT ORGANIZATION

This text is an update to the author's previous text.* In addition to providing new material based on MATLAB 7, the text incorporates the many suggestions made by reviewers and other users.

The text consists of 10 chapters. The first chapter gives an overview of MATLAB features, including its windows and menu structures. It also introduces the problem-solving methodology. Chapter 2 introduces the concept of an array, which is the fundamental data element in MATLAB, and describes how to use numeric arrays, cell arrays, and structure arrays for basic mathematical operations.

Chapter 3 discusses the use of functions and files. MATLAB has an extensive number of built-in math functions, and users can define their own functions and save them as a file for reuse.

Chapter 4 treats programming with MATLAB and covers relational and logical operators, conditional statements, for and while loops, and the switch structure. A major application of the chapter's material is in simulation, to which a section is devoted.

Chapter 5 treats two- and three-dimensional plotting. It first establishes standards for professional-looking, useful plots. In the author's experience beginning students are not aware of these standards, so they are emphasized. The chapter then covers MATLAB commands for producing different types of plots and for controlling their appearance. Function discovery, which uses data plots to discover a mathematical description of the data, is a common application of plotting, and a separate section is devoted to this topic. The chapter also treats polynomial and multiple linear regression as part of its modeling coverage.

Chapter 6 covers the solution of linear algebraic equations, which arise in applications in all fields of engineering. "Hand" solution methods are reviewed first. This review has proved helpful to many students in the author's classes. This coverage also establishes the terminology and some important concepts that are required to use the computer methods properly. The chapter then shows how to use MATLAB to solve systems of linear equations that have a unique solution. The use of MATLAB with underdetermined and overdetermined systems is covered in two optional sections.

Chapter 7 reviews basic statistics and probability and shows how to use MATLAB to generate histograms, perform calculations with the normal distribution, and create random number simulations. The chapter concludes with linear

---

* *Introduction to MATLAB 6 for Engineers,* McGraw-Hill, New York, 2000.

and cubic-spline interpolation. This chapter can be skipped if necessary. None of the following chapters depend on it.

Chapter 8 covers numerical methods for calculus and differential equations. Analytical methods are reviewed to provide a foundation for understanding and interpreting the numerical methods. Numerical integration and differentiation methods are treated. Ordinary differential equation solvers in the core MATLAB program are covered, as well as the linear-system solvers in the Control System toolbox.

Chapter 9 introduces Simulink,* which is a graphical interface for building simulations of dynamic systems. The coverage of Simulink has been expanded to a separate chapter in light of its growing popularity, as evidenced by recent workshops held by various professional organizations such as the ASEE. This chapter need not be covered to read Chapter 10.

Chapter 10 covers symbolic methods for manipulating algebraic expressions and for solving algebraic and transcendental equations, calculus, differential equations, and matrix algebra problems. The calculus applications include integration and differentiation, optimization, Taylor series, series evaluation, and limits. Laplace transform methods for solving differential equations are also introduced. This chapter requires the use of the Symbolic Math toolbox or the Student Edition of MATLAB.

Appendix A contains a guide to the commands and functions introduced in the text. Appendix B is an introduction to producing animation and sound with MATLAB. While not essential to learning MATLAB, these features are helpful for generating student interest. Appendix C summarizes functions for creating formatted output. Appendix D is a list of references. Appendix E, which is available on the text's website, contains some suggestions for course projects and is based on the author's experience in teaching a freshman MATLAB course. Answers to selected problems and an index appear at the end of the text.

All figures, tables, equations, and exercises have been numbered according to their chapter and section. For example, Figure 3.4–2 is the second figure in Chapter 3, Section 4. This system is designed to help the reader locate these items. The end-of-chapter problems are the exception to this numbering system. They are numbered 1, 2, 3, and so on to avoid confusion with the in-chapter exercises.

The first four chapters constitute a course in the essentials of MATLAB. The remaining six chapters are independent of each other, and may be covered in any order, or may be omitted if necessary. These chapters provide additional coverage and examples of plotting and model building, linear algebraic equations, probability and statistics, calculus and differential equations, Simulink, and symbolic processing, respectively.

---

*Simulink is a registered trademark of The MathWorks, Inc.

SPECIAL REFERENCE FEATURES

The text has the following special features, which have been designed to enhance its usefulness as a reference.

- Throughout each of the chapters, numerous tables summarize the commands and functions as they are introduced.
- At the end of each chapter is a guide to tables in that chapter. These master tables will help the reader find descriptions of specific MATLAB commands.
- Appendix A is a complete summary of all the commands and functions described in the text, grouped by category, along with the number of the page on which they are introduced.
- At the end of each chapter is a list of the key terms introduced in the chapter, with the page number referenced.
- Key terms have been placed in the margin or in section headings where they are introduced.
- The index has four sections: a listing of symbols, an alphabetical list of MATLAB commands and functions, a list of Simulink blocks, and an alphabetical list of topics.

PEDAGOGICAL AIDS

The following pedagogical aids have been included:

- Each chapter begins with an overview.
- **Test Your Understanding** exercises appear throughout the chapters near the relevant text. These relatively straightforward exercises allow readers to assess their grasp of the material as soon as it is covered. In most cases the answer to the exercise is given with the exercise. Students should work these exercises as they are encountered.
- Each chapter ends with numerous problems, grouped according to the relevant section.
- Each chapter contains numerous practical examples. The major examples are numbered.
- Each chapter has a summary section that reviews the chapter's objectives.
- Answers to many end-of-chapter problems appear at the end of the text. These problems are denoted by an asterisk next to their number (for example, **15***).

Two features have been included to motivate the student toward MATLAB and the engineering profession:

- Most of the examples and the problems deal with engineering applications. These are drawn from a variety of engineering fields and show realistic applications of MATLAB. A guide to these examples appears on the inside front cover.

- The facing page of each chapter contains a photograph of a *recent* engineering achievement that illustrates the challenging and interesting opportunities that await engineers in the 21st century. A description of the achievement, its related engineering disciplines, and a discussion of how MATLAB can be applied in those disciplines accompanies each photo.

An Instructor's Manual is available online for instructors who have adopted this text for a course. This manual contains the complete solutions to all the **Test Your Understanding** exercises and to all the chapter problems. The text website (at http://www.mhhe.com/palm) also has downloadable files containing the major programs in the text, PowerPoint slides keyed to the text, and suggestions for projects.

ACKNOWLEDGMENTS

Many individuals are due credit for this text. Working with faculty at the University of Rhode Island in developing and teaching a freshman course based on MATLAB has greatly influenced this text. Email from many users contained useful suggestions. The following people, as well as several anonymous reviewers, patiently reviewed the manuscript and suggested many helpful corrections and additions.

Steven Ciccarelli, *Rochester Institute of Technology*
Dwight Davy, *Case Western Reserve University*
Yueh-Jaw Lin, *The University of Akron*
Armando Rodriquez, *Arizona State University*
Thomas Sullivan, *Carnegie Mellon University*
Daniel Valentine, *Clarkson University*
Elizabeth Wyler, *Thomas Nelson Community College*

The MathWorks, Inc. has always been very supportive of educational publishing. I especially want to thank Naomi Fernandes of The MathWorks, Inc. for her help. Carlise Paulson, Michaela Graham, Michelle Flomenhoft, and Peggy Lucas of McGraw-Hill efficiently handled the manuscript reviews and guided the text through production.

My sisters, Linda and Chris, and my mother Lillian, have always been there, cheering my efforts. My father was always there for support before he passed away. Finally, I want to thank my wife, Mary Louise, and my children, Aileene, Bill, and Andy, for their understanding and support of this project.

William J. Palm III
Kingston, Rhode Island
April, 2004

Introduction to MATLAB 7 for Engineers

Photo courtesy of NASA Jet Propulsion Laboratory.

Engineering in the 21st Century...

Remote Exploration

It will be many years before humans can travel to other planets. In the meantime, unmanned probes have been rapidly increasing our knowledge of the universe. Their use will increase in the future as our technology develops to make them more reliable and more versatile. Better sensors are expected for imaging and other data collection. Improved robotic devices will make these probes more autonomous, and more capable of interacting with their environment, instead of just observing it.

NASA's planetary rover *Sojourner* landed on Mars on July 4, 1997, and excited people on Earth while they watched it successfully explore the Martian surface to determine wheel-soil interactions, to analyze rocks and soil, and to return images of the lander for damage assessment. Then in early 2004, two improved rovers, *Spirit* and *Opportunity,* landed on opposite sides of the planet. In one of the major discoveries of the 21st century, they obtained strong evidence that water once existed on Mars in significant amounts.

About the size of a golf cart, the new rovers have six wheels, each with its own motors. They have a top speed of 5 centimeters per second on flat hard ground and can travel up to about 100 meters per day. Needing 100 watts to move, they obtain power from solar arrays that generate 140 watts during a four-hour window each day. The sophisticated temperature control system must not only protect against nighttime temperatures of -96°C, but must also prevent the rover from overheating.

The robotic arm has three joints (shoulder, elbow, and wrist), driven by five motors, and it has a reach of 90 centimeters. The arm carries four tools and instruments for geological studies. Nine cameras provide hazard avoidance, navigation, and panoramic views. The on-board computer has 128 MB of DRAM and coordinates all the subsystems including communications.

All engineering disciplines were involved with the rovers' design and launch. The MATLAB Neural Network, Signal Processing, Image Processing, PDE, and various control system toolboxes are well suited to assist designers of probes and autonomous vehicles like the Mars rovers. ■

CHAPTER 1

An Overview of MATLAB®*

OUTLINE

This is the most important chapter in the book. By the time you have finished this chapter, you will be able to use MATLAB to solve many kinds of engineering problems. Section 1.1 provides a "quick-start" introduction to MATLAB as an interactive calculator. Section 1.2 covers the main menus and toolbar. Section 1.3 gives an overview of MATLAB, and directs the reader to the appropriate chapter where more detailed information is available. Section 1.4 discusses how to create, edit, and save MATLAB programs. Section 1.5 introduces the extensive MATLAB Help System. Section 1.6 treats the use of conditional statements and loops. Section 1.7 discusses methodologies for approaching engineering problems, with particular emphasis on a methodology to use with computer software such as MATLAB. A number of practice problems are given at the end of the chapter.

---

*MATLAB is a registered trademark of The MathWorks, Inc.

How to Use This Book

The book's chapter organization is flexible enough to accommodate a variety of users. However, it is important to cover at least the first four chapters, in that order. Chapter 2 covers *arrays,* which are the basic building blocks in MATLAB. Chapter 3 covers file usage, functions built into MATLAB, and user-defined functions. Chapter 4 covers programming using relational and logical operators, conditional statements, and loops.

Use Section 1.3 to determine those MATLAB features for which you want more detailed information. This section will guide you to the appropriate chapter.

Chapters 5 through 10 are independent chapters that can be covered in any order, or can be omitted. They contain in-depth discussions of how to use MATLAB to solve several common types of engineering problems. Chapter 5 covers two- and three-dimensional plots in more detail, and shows how to use plots to build mathematical models from data. Chapter 6 treats the solution of linear algebraic equations, including cases having nonunique solutions. Chapter 7 covers probability, statistics, and interpolation applications. Chapter 8 introduces numerical methods for calculus and ordinary differential equations. Chapter 9 covers Simulink®,* which is a graphical user interface for solving differential equation models. Chapter 10 covers symbolic processing in MATLAB, with applications to algebra, calculus, differential equations, linear algebra, and transforms.

Reference and Learning Aids

The book has been designed as a reference as well as a learning tool. The special features useful for these purposes are as follows.

- Throughout each chapter margin notes identify where new terms are introduced.
- Throughout each chapter short Test Your Understanding exercises appear. Where appropriate, answers immediately follow the exercise so you can measure your mastery of the material.
- Homework exercises conclude each chapter. These usually require more effort than the Test Your Understanding exercises.
- Each chapter contains tables summarizing the MATLAB commands introduced in that chapter.
- At the end of each chapter is:
  - A summary guide to the commands covered in that chapter,
  - A summary of what you should be able to do after completing that chapter, and
  - A list of key terms you should know.
- Appendix A contains tables of MATLAB commands, grouped by category, with the appropriate page references.

---

*Simulink is a registered trademark of The MathWorks, Inc.

- Two indexes are included. The first is an index of MATLAB commands and symbols; the second is an index of topics.

Software Updates and Accuracy

Software publishers can release software updates faster than book publishers can release new editions. This text documents the pre-release version of MATLAB 7 as of the spring of 2004. There will be additional updates, numbered 7.1, 7.2, and so forth, that will change some of the program's features. The best way to protect yourself against obsolete information is to check the "What's New?" file provided with the program, and to learn how to use the extensive MATLAB Help System, which is covered in Section 1.5.

MATLAB and Related Software

MATLAB is both a computer programming language and a software environment for using that language effectively. It is maintained and sold by The MathWorks, Inc., of Natick, Massachusetts, and is available for MS Windows and other computer systems. The MATLAB interactive environment allows you to manage variables, import and export data, perform calculations, generate plots, and develop and manage files for use with MATLAB. The language was originally developed in the 1970s for applications involving matrices, linear algebra, and numerical analysis (the name MATLAB stands for "Matrix Laboratory"). Thus the language's numerical routines have been well-tested and improved through many years of use, and its capabilities have been greatly expanded.

MATLAB has a number of add-on software modules, called *toolboxes,* that perform more specialized computations. They can be purchased separately, but all run under the core MATLAB program. Toolboxes deal with applications such as image and signal processing, financial analysis, control systems design, and fuzzy logic. An up-to-date list can be found at The MathWorks website, which is discussed later in this chapter. This text uses material from the core MATLAB program, from two of the toolboxes (the Control Systems toolbox, in Chapter 8, and the Symbolic Math toolbox, in Chapter 10), and from Simulink (in Chapter 9). All of the examples and problems in the first seven chapters can be done with the core MATLAB program.

On MS Windows systems MATLAB 7 requires Windows XP or Windows NT to run. The Student Edition of MATLAB contains the core MATLAB program, some commands from two toolboxes (the Signal Processing toolbox and the Symbolic Math toolbox), and the Simulink program. The Simulink program is based on MATLAB, and requires MATLAB to run.

This book does not explain how to install MATLAB. If you purchased it for your own computer, the installation is easily done with the instructions that come with the software. If you will be using MATLAB in a computer lab, it will have been installed for you.

In the next section we introduce MATLAB by means of some simple sessions to illustrate its interactive nature, basic syntax, and features.

1.1 MATLAB Interactive Sessions

We now show how to start MATLAB, how to make some basic calculations, and how to exit MATLAB.

Conventions

In this text we use `typewriter font` to represent MATLAB commands, any text that you type in the computer, and any MATLAB responses that appear on the screen, for example, `y = 6*x`. Variables in normal mathematics text appear in italics; for example, $y = 6x$. We use boldface type for three purposes: to represent vectors and matrices in normal mathematics text (for example, $\mathbf{Ax} = \mathbf{b}$), to represent a key on the keyboard (for example, **Enter**), and to represent the name of a screen menu or an item that appears in such a menu (for example, **File**). It is assumed that you press the **Enter** key after you type a command. We do not show this action with a separate symbol.

Starting MATLAB

DESKTOP

To start MATLAB on a MS Windows system, double-click on the MATLAB icon. You will then see the MATLAB *Desktop*. The Desktop manages the Command window and a Help Browser, as well as other tools. The default appearance of the Desktop is shown in Figure 1.1–1. Three windows appear. These are the

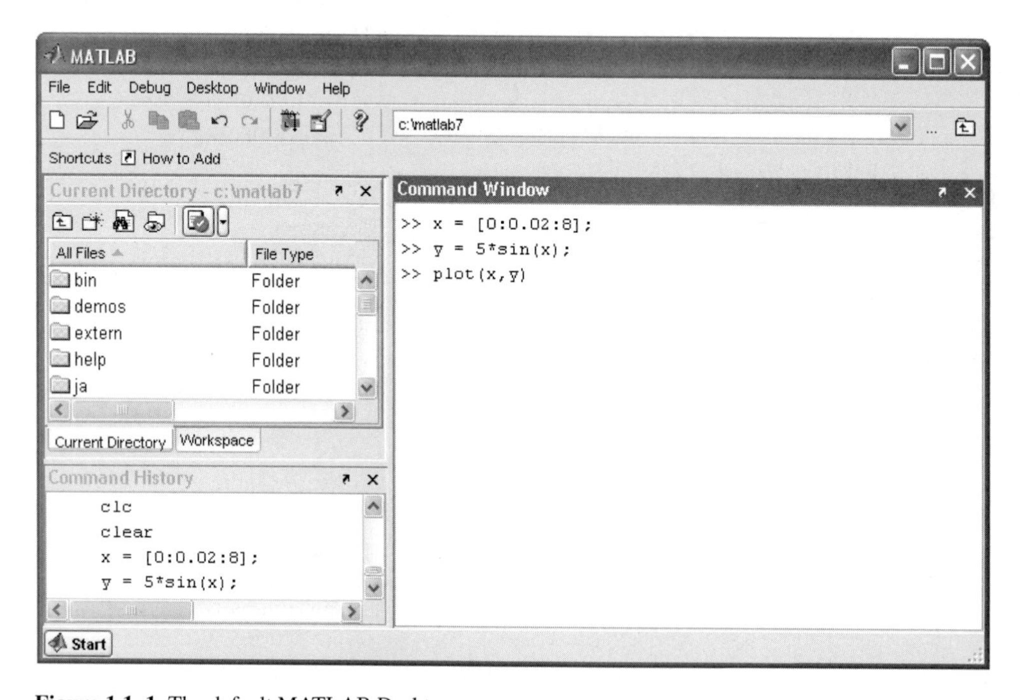

Figure 1.1–1 The default MATLAB Desktop.

Command window, the Command History window, and the Current Directory window. Across the top of the Desktop are a row of menu names, and a row of icons called the *toolbar*. To the right of the toolbar is a box showing the directory where MATLAB looks for and saves files. We will describe the menus, toolbar, and directories later in this chapter.

You use the Command window to communicate with the MATLAB program, by typing instructions of various types called *commands, functions,* and *statements*. Later we will discuss the differences between these types, but for now, to simplify the discussion, we will call the instructions by the generic name *commands*. MATLAB displays the prompt (>>) to indicate that it is ready to receive instructions. Before giving MATLAB instructions, make sure the cursor is located just after the prompt. If it is not, use the mouse to move the cursor. The prompt in the Student Edition looks like `EDU>>`. We will use the normal prompt symbol >> to illustrate commands in this text.

COMMAND WINDOW

Three other windows appear in the default Desktop. The Current Directory window is much like a file manager window; you can use it to access files. Double-clicking on a file name with the extension .m will open that file in the MATLAB Editor. The Editor is discussed in Section 1.4.

Underneath the Current Directory window is the Workspace window. To activate it, click on its tab at the bottom of the Current Directory window. The Workspace window displays the variables created in the Command window. Double-click on a variable name to open the Array Editor, which is discussed in Chapter 2.

The fourth window in the default Desktop is the Command History window. This window shows all the previous keystrokes you entered in the Command window. It is useful for keeping track of what you typed. You can click on a keystroke and drag it to the Command window or the Editor. Double-clicking on a keystroke executes it in the Command window.

You can alter the appearance of the Desktop if you wish. For example, to eliminate a window, just click on its Close-window button (×) in its upper right-hand corner. To undock, or separate the window from the Desktop, click on the button containing an arrow. You can manipulate other windows in the same way. To restore the default configuration, click on the **View** menu, then click on **Desktop Layout,** and select **Default.**

Entering Commands and Expressions

To see how simple it is to use MATLAB, try entering a few commands on your computer. If you make a typing mistake, just press the **Enter** key until you get the prompt, and then retype the line. Or, because MATLAB retains your previous keystrokes in a command file, you can use the up-arrow key (↑) to scroll back through the commands. Press the key once to see the previous entry, twice to see the entry before that, and so on. Use the down-arrow key (↓) to scroll forward through the commands. When you find the line you want, you can edit it using the left- and right-arrow keys (← and →), and the **Backspace** key, and the

Delete key. Press the **Enter** key to execute the command. This technique enables you to correct typing mistakes quickly.

Note that you can see your previous keystrokes displayed in the Command History window. You can copy a line from this window to the Command window by highlighting the line with the mouse, holding down the left mouse button, and dragging the line to the Command window.

Make sure the cursor is at the prompt in the Command window. To divide 8 by 10, type `8/10` and press **Enter** (the symbol `/` is the MATLAB symbol for division). Your entry and the MATLAB response looks like the following on the screen (we call this interaction between you and MATLAB an *interactive session,* or simply *a session*).

SESSION

```
>> 8/10
ans =
     0.8000
```

MATLAB uses high precision for its computations, but by default it usually displays its results using four decimal places. This is called the *short* format. This default can be changed by using the `format` command, which is discussed later in this section. MATLAB uses the notation `e` to represent exponentiation to a power of 10; for example, MATLAB displays the number 5.316×10^2 as `5.316e+02`.

VARIABLE

MATLAB has assigned the answer to a variable called `ans`, which is an abbreviation for *answer*. A *variable* in MATLAB is a symbol used to contain a value. You can use the variable `ans` for further calculations; for example, using the MATLAB symbol for multiplication (`*`):

```
>> 5*ans
ans =
     4
```

Note that the variable `ans` now has the value 4.

You can use variables to write mathematical expressions. We will soon see why this is an advantage. You can assign the result to a variable of your own choosing, say `r`, as follows:

```
>> r=8/10
r =
   0.8000
```

Spaces in the line improve its readability; for example, you can put a space before and after the = sign if you want. MATLAB ignores these spaces when making its calculations.

If you now type `r` at the prompt, you will see

```
>> r
r =
   0.8000
```

thus verifying that the variable `r` has the value 0.8. You can use this variable in further calculations. For example,

```
>> s=20*r
s =
   16
```

When we do not specify a variable name for a result, MATLAB uses the symbol `ans` as a temporary variable containing the most recent answer.

MATLAB has hundreds of functions available. One of these is the *square root* function, `sqrt`. A pair of parentheses is used after the function's name to enclose the value—called the function's *argument*—that is operated on by the function. For example, to compute the square root of 9, you type `sqrt(9)`. We will see more MATLAB functions in this chapter; an extensive list of mathematical functions is given in Chapter 3. Other types of functions are covered throughout the text.

ARGUMENT

Order of Precedence

SCALAR

A *scalar* is a single number. A *scalar variable* is a variable that contains a single number. MATLAB uses the symbols `+ - * / ^` for addition, subtraction, multiplication, division, and exponentiation (power) of scalars. These are listed in Table 1.1–1. For example, typing `x = 8 + 3*5` returns the answer `x = 23`. Typing `2^3-10` returns the answer `ans = -2`. The *forward slash* (/) represents *right division,* which is the normal division operator familiar to you. Typing `15/3` returns the result `ans = 5`.

MATLAB has another division operator, called *left division,* which is denoted by the *backslash* (\). The left division operator is useful for solving sets of linear algebraic equations, as we will see in Section 1.3. A good way to remember the difference between the right and left division operators is to note that the slash slants toward the denominator. For example, $7/2 = 2\backslash 7 = 3.5$.

PRECEDENCE

The mathematical operations represented by the symbols `+ - * / \`, and `^` follow a set of rules called *precedence*. Mathematical expressions are evaluated starting from the left, with the exponentiation operation having the highest order of precedence, followed by multiplication and division with equal precedence, followed by addition and subtraction with equal precedence. Parentheses can be

Table 1.1–1 Scalar arithmetic operations

Symbol	Operation	MATLAB form
^	exponentiation: a^b	`a^b`
*	multiplication: ab	`a*b`
/	right division: $a/b = \frac{a}{b}$	`a/b`
\	left division: $a\backslash b = \frac{b}{a}$	`a\ b`
+	addition: $a + b$	`a+b`
-	subtraction: $a - b$	`a-b`

Table 1.1–2 Order of precedence

Precedence	Operation
First	Parentheses, evaluated starting with the innermost pair.
Second	Exponentiation, evaluated from left to right.
Third	Multiplication and division with equal precedence, evaluated from left to right.
Fourth	Addition and subtraction with equal precedence, evaluated from left to right.

used to alter this order. Evaluation begins with the innermost pair of parentheses, and proceeds outward. Table 1.1–2 summarizes these rules. For example, note the effect of precedence on the following session.

```
>> 8 + 3*5
ans =
     23
>> 8 + (3*5)
ans =
     23
>>(8 + 3)*5
ans =
     55
>>4^2-12- 8/4*2
ans =
     0
>>4^2-12- 8/(4*2)
ans =
     3
>> 3*4^2 + 5
ans =
     53
>>(3*4)^2 + 5
ans =
     149
>>27^(1/3) + 32^(0.2)
ans =
     5
>>27^(1/3) + 32^0.2
ans =
     5
>>27^1/3 + 32^0.2
ans =
     11
```

To avoid mistakes, you should feel free to insert parentheses wherever you are unsure of the effect precedence will have on the calculation. Use of parentheses also improves the readability of your MATLAB expressions. For example, parentheses are not needed in the expression `8 + (3*5)`, but they make clear our intention to multiply 3 by 5 before adding 8 to the result.

Test Your Understanding

T1.1–1 Use MATLAB to compute the following expressions.

a. $6\frac{10}{13} + \frac{18}{5(7)} + 5(9^2)$

b. $6(35^{1/4}) + 14^{0.35}$

(Answers: *a.* 410.1297 *b.* 17.1123.)

The Assignment Operator

The = sign in MATLAB is the called the *assignment* or *replacement* operator. It works differently than the equals sign you know from mathematics. When you type `x=3`, you tell MATLAB to assign the value 3 to the variable `x`. This usage is no different than in mathematics. However, in MATLAB we can also type something like this: `x = x + 2`. This tells MATLAB to add 2 to the current value of `x`, and to replace the current value of `x` with this new value. If `x` originally had the value 3, its new value would be 5. This usage of the = operator is different from its use in mathematics. For example, the mathematics equation $x = x + 2$ is invalid because it implies that $0 = 2$ (subtract x from both sides of the equation to see this).

It is important to understand this difference between the MATLAB operator = and the equals sign of mathematics. The variable on the *left-hand* side of the = operator is replaced by the value generated by the *right-hand* side. Therefore, one variable, and only one variable, must be on the left-hand side of the = operator. Thus in MATLAB you cannot type `6 = x`. Another consequence of this restriction is that you cannot write in MATLAB expressions like the following:

```
>>x+2=20
```

The corresponding equation $x + 2 = 20$ is acceptable in algebra, and has the solution $x = 18$, but MATLAB cannot solve such an equation without additional commands (these commands are available in the Symbolic Math toolbox, which is described in Chapter 10).

Another restriction is that the right-hand side of the = operator must have a computable value. For example, if the variable `y` has not been assigned a value, then the following will generate an error message in MATLAB.

```
>>x = 5 + y
```

In addition to assigning known values to variables, the assignment operator is very useful for assigning values that are not known ahead of time, or for changing the value of a variable by using a prescribed procedure. The following example shows how this is done.

EXAMPLE 1.1–1 Volume of a Circular Cylinder

The volume of a circular cylinder of height h and radius r is given by $V = \pi r^2 h$. A particular cylindrical tank is 15 m tall and has a radius of 8 m. We want to construct another cylindrical tank with a volume 20 percent greater but having the same height. How large must its radius be?

■ Solution

First solve the cylinder equation for the radius r. This gives

$$r = \sqrt{\frac{V}{\pi h}}$$

The session is shown below. First we assign values to the variables `r` and `h` representing the radius and height. Then we compute the volume of the original cylinder, and increase the volume by 20 percent. Finally we solve for the required radius. For this problem we can use the MATLAB built-in constant `pi`.

```
>>r = 8;
>>h = 15;
>>V  = pi*r^2*h;
>>V = V + 0.2*V;
>>r = sqrt(V/(pi*h))
ans =
      r = 8.7636
```

Thus the new cylinder must have a radius of 8.7636 m. Note that the original values of the variables `r` and `V` are replaced with the new values. This is acceptable as long as we do not wish to use the original values again. Note how precedence applies to the line `V = pi*r^2*h;`. It is equivalent to `V = pi*(r^2)*h;`.

Variable Names

WORKSPACE

The term *workspace* refers to the names and values of any variables in use in the current work session. Variable names must begin with a letter and must contain less than 32 characters; the rest of the name can contain letters, digits, and underscore characters. MATLAB is case-sensitive. Thus the following names represent five different variables: `speed`, `Speed`, `SPEED`, `Speed_1`, and `Speed_2`.

Managing the Work Session

Table 1.1–3 summarizes some commands and special symbols for managing the work session. A semicolon at the end of a line suppresses printing the results to the screen. If a semicolon is not put at the end of a line, MATLAB displays the results of the line on the screen. Even if you suppress the display with the semicolon, MATLAB still retains the variable's value.

You can put several commands on the same line if you separate them with a comma—if you want to see the results of the previous command—or semicolon

Table 1.1–3 Commands for managing the work session

Command	Description
`clc`	Clears the Command window.
`clear`	Removes all variables from memory.
`clear var1 var2`	Removes the variables `var1` and `var2` from memory.
`exist('name')`	Determines if a file or variable exists having the name 'name'.
`quit`	Stops MATLAB.
`who`	Lists the variables currently in memory.
`whos`	Lists the current variables and sizes, and indicates if they have imaginary parts.
`:`	Colon; generates an array having regularly spaced elements.
`,`	Comma; separates elements of an array.
`;`	Semicolon; suppresses screen printing; also denotes a new row in an array.
`...`	Ellipsis; continues a line.

if you want to suppress the display. For example,

```
>>x=2;y=6+x,x=y+7
y =
    8
x =
    15
```

Note that the first value of `x` was not displayed. Note also that the value of `x` changed from 2 to 15.

If you need to type a long line, you can use an *ellipsis,* by typing three periods, to delay execution. For example,

```
>>NumberOfApples = 10; NumberOfOranges = 25;
>>NumberOfPears = 12;
>>FruitPurchased = NumberOfApples + NumberOfOranges  ...
+NumberOfPears
FruitPurchased =
      47
```

Use the arrow, tab, and control keys to recall, edit, and reuse functions and variables you typed earlier. For example, suppose you mistakenly enter the line

```
>>volume = 1 + sqr(5)
```

MATLAB responds with the error message

```
Undefined function or variable 'sqr'.
```

because you misspelled `sqrt`. Instead of retyping the entire line, press the up-arrow key (↑) once to display the previously typed line. Press the left-arrow key (←) several times to move the cursor and add the missing `t`, then press **Enter.** Repeated use of the up-arrow key recalls lines typed earlier.

You can use the *smart recall* feature to recall a previously typed function or variable whose first few characters you specify. For example, after you have entered the line starting with `volume`, typing `vol` and pressing the up-arrow key (↑) once recalls the last-typed line that starts with the function or variable whose name begins with `vol`. This feature is case-sensitive.

You can use the *tab completion* feature to reduce the amount of typing. MATLAB automatically completes the name of a function, variable, or file if you type the first few letters of the name and press the **Tab** key. If the name is unique, it is automatically completed. For example, in the session listed earlier, if you type `Fruit` and press **Tab,** MATLAB completes the name and displays `FruitPurchased`. Press **Enter** to display the value of the variable, or continue editing to create a new executable line that uses the variable `FruitPurchased`.

If there is more than one name that starts with the letters you typed, MATLAB displays nothing. In this case press the **Tab** key again to see a list of the possibilities.

The up-arrow (↑) and down-arrow (↓) keys move up and down through the previously typed lines one *line* at a time. Similarly, the left-arrow (←) and rightarrow (→) keys move left and right through a line one *character* at a time. To move through one *word* at a time, press **Ctrl** and → simultaneously to move to the *right;* press **Ctrl** and ← simultaneously to move to the *left*. Press **Home** to move to the beginning of a line; press **End** to move to the end of a line.

Press **Del** to delete the character at the cursor; press **Backspace** to delete the character before the cursor. Press **Esc** to clear the entire line; press **Ctrl** and **k** simultaneously to delete (*kill*) to the end of the line.

MATLAB retains the last value of a variable until you quit MATLAB or clear its value. Overlooking this fact commonly causes errors in MATLAB. For example, you might prefer to use the variable `x` in a number of different calculations. If you forget to enter the correct value for `x`, MATLAB uses the last value, and you get an incorrect result. You can use the `clear` function to remove the values of *all* variables from memory, or you can use the form `clear var1 var2` to clear the variables named `var1` and `var2`. The effect of the `clc` command is different; it clears the Command window of everything in the window display, but the values of the variables remain.

You can type the name of a variable and press **Enter** to see its current value. If the variable does not have a value (i.e., if it does not exist), you see an error message. You can also use the `exist` function. Type `exist('x')` to see if the variable `x` is in use. If a 1 is returned, the variable exists; a 0 indicates that it does not exist. The `who` function lists the names of all the variables in memory, but does not give their values. The form `who var1 var2` restricts the display to the variables specified. The wildcard character `*` can be used to display variables that match a pattern. For instance, `who A*` finds all variables in the current workspace that start with A. The `whos` function lists the variable names and their sizes, and indicates whether or not they have nonzero imaginary parts.

Table 1.1–4 Special variables and constants

Command	Description
ans	Temporary variable containing the most recent answer.
eps	Specifies the accuracy of floating point precision.
i,j	The imaginary unit $\sqrt{-1}$.
Inf	Infinity.
NaN	Indicates an undefined numerical result.
pi	The number π.

The difference between a function and a command or a statement is that functions have their arguments enclosed in parentheses. Commands, such as `clear`, need not have arguments, but if they do, they are not enclosed in parentheses; for example, `clear x`. Statements cannot have arguments; for example, `clc` and `quit` are statements.

You can quit MATLAB by typing `quit`. On MS Windows systems you can also click on the **File** menu, and then click on **Exit MATLAB.**

Predefined Constants

MATLAB has several predefined special constants, such as the built-in constant `pi` we used in Example 1.1–1. Table 1.1–4 lists them. The symbol `Inf` stands for ∞, which in practice means a number so large that MATLAB cannot represent it. For example, typing `5/0` generates the answer `Inf`. The symbol `NaN` stands for "not a number." It indicates an undefined numerical result such as that obtained by typing `0/0`. The symbol `eps` is the smallest number which, when added to 1 by the computer, creates a number greater than 1. We use it as an indicator of the accuracy of computations.

The symbols `i` and `j` denote the imaginary unit, where $i = j = \sqrt{-1}$. We use them to create and represent complex numbers, such as `x = 5 + 8i`.

Try not to use the names of special constants as variable names. Although MATLAB allows you to assign a different value to these constants, it is not good practice to do so.

Complex Number Operations

MATLAB handles complex number algebra automatically. For example, the number $c_1 = 1 - 2i$ is entered as follows: `c1 = 1-2i`.

Caution: Note that an asterisk is not needed between `i` or `j` and a number, although it is required with a variable, such as `c2 = 5 - i*c1`. This convention can cause errors if you are not careful. For example, the expressions `y = 7/2*i` and `x = 7/2i` give two different results: $y = (7/2)i = 3.5i$ and $x = 7/(2i) = -3.5i$.

Addition, subtraction, multiplication, and division of complex numbers are easily done. For example,

```
>>s = 3+7i;w = 5-9i;
>>w+s
ans =
     8.0000 - 2.0000i
>>w*s
ans =
     78.0000 + 8.0000i
>>w/s
ans =
     -0.8276 - 1.0690i
```

Complex conjugates have the same real part but imaginary parts of opposite sign; for example, $-3 + 7i$ and $-3 - 7i$ are complex conjugates. The product of two conjugates is the sum of the squares of the real and imaginary parts; for example,

```
>>(-3 + 7i)*(-3 - 7i)
ans =
     58
```

because $\sqrt{3^2 + 7^2} = 58$. More complex number functions are discussed in Chapter 3.

Test Your Understanding

T1.1–2 Given $x = -5 + 9i$ and $y = 6 - 2i$, use MATLAB to show that $x + y = 1 + 7i$, $xy = -12 + 64i$, and $x/y = -1.2 + 1.1i$.

Formatting Commands

The `format` command controls how numbers appear on the screen. Table 1.1–5 gives the variants of this command. MATLAB uses many significant figures in its

Table 1.1–5 Numeric display formats

Command	Description and example
`format short`	Four decimal digits (the default); 13.6745.
`format long`	16 digits; 17.27484029463547.
`format short e`	Five digits (four decimals) plus exponent; 6.3792e+03.
`format long e`	16 digits (15 decimals) plus exponent; 6.379243784781294e−04.
`format bank`	Two decimal digits; 126.73.
`format +`	Positive, negative, or zero; +.
`format rat`	Rational approximation; 43/7.
`format compact`	Suppresses some line feeds.
`format loose`	Resets to less compact display mode.

calculations, but we rarely need to see all of them. The default MATLAB display format is the `short` format, which uses four decimal digits. You can display more by typing `format long`, which gives 16 digits. To return to the default mode, type `format short`.

You can force the output to be in scientific notation by typing `format short e`, or `format long e`, where `e` stands for the number 10. Thus the output `6.3792e+03` stands for the number 6.3792×10^3. The output `6.3792e-03` stands for the number 6.3792×10^{-3}. Note that in this context `e` does *not* represent the number e, which is the base of the natural logarithm. Here `e` stands for "exponent." It is a poor choice of notation, but MATLAB follows conventional computer programming standards that were established many years ago.

Use `format bank` only for monetary calculations; it does not recognize imaginary parts.

1.2 Menus and the Toolbar

The Desktop manages the Command window and other MATLAB tools. The default appearance of the Desktop is shown in Figure 1.1–1. Besides the Command window, the default Desktop includes three other windows, the Command History, Current Directory, and Workspace windows, which we discussed in the previous section. Across the top of the Desktop are a row of menu names, and a row of icons called the *toolbar*. To the right of the toolbar is a box showing the *current directory,* where MATLAB looks for files. We now describe the menus and the toolbar.

CURRENT DIRECTORY

Other windows appear in a MATLAB session, depending on what you do. For example, a graphics window containing a plot appears when you use the plotting functions; an editor window, called the Editor/Debugger, appears for use in creating program files. Each window type has its own menu bar, with one or more menus, at the top. *Thus the menu bar will change as you change windows.* To activate, or select, a menu, click on it. Each menu has several items. Click on an item to select it. *Keep in mind that menus are context-sensitive. Thus their contents change, depending on which features you are currently using.*

The Desktop Menus

Most of your interaction will be in the Command window. When the Command window is active, the default MATLAB 7 Desktop (shown in Figure 1.1–1) has six menus: **File, Edit, Debug, Desktop, Window,** and **Help.** Note that these menus change depending on what window is active. Every item on a menu can be selected with the menu open either by clicking on the item or by typing its underlined letter. Some items can be selected without the menu being open by using the shortcut key listed to the right of the item. Those items followed by three dots (`...`) open a submenu or another window containing a dialog box.

The three most useful menus are the **File, Edit,** and **Help** menus. The **Help** menu is described in Section 1.5. The **File** menu in MATLAB 7 contains the following items, which perform the indicated actions when you select them.

The File Menu in MATLAB 7

New Opens a dialog box that allows you to create a new program file, called an M-file, using a text editor called the Editor/Debugger, or a new Figure or Model file (a file type used by Simulink).

Open... Opens a dialog box that allows you to select a file for editing.

Close Command Window Closes the Command window.

Import Data... Starts the Import Wizard which enables you to import data easily.

Save Workspace As... Opens a dialog box that enables you save a file.

Set Path... Opens a dialog box that enables you to set the MATLAB search path.

Preferences... Opens a dialog box that enables you to set preferences for such items as fonts, colors, tab spacing, and so forth.

Print... Opens a dialog box that enables you to print all of the Command window.

Print Selection... Opens a dialog box that enables you to print selected portions of the Command window.

File List Contains a list of previously used files, in order of most recently used.

Exit MATLAB Closes MATLAB.

The **Edit** menu contains the following items.

The Edit Menu in MATLAB 7

Undo Reverses the previous editing operation.

Redo Reverses the previous Undo operation.

Cut Removes the selected text and stores it for pasting later.

Copy Copies the selected text for pasting later, without removing it.

Paste Inserts any text on the clipboard at the current location of the cursor.

Paste Special... Inserts the contents of the clipboard into the workspace as one or more variables.

Select All Highlights all text in the Command window.

Delete Clears the variable highlighted in the Workspace Browser.

Find... Finds and replaces phrases.

Find Files... Finds files.

Clear Command Window Removes all text from the Command window.

Clear Command History Removes all text from the Command History window.

Clear Workspace Removes the values of all variables from the workspace.

You can use the **Copy** and **Paste** selections to copy and paste commands appearing on the Command window. However, an easier way is to use the up-arrow key to scroll through the previous commands, and press **Enter** when you see the command you want to retrieve.

Use the **Debug** menu to access the Debugger, which is discussed in Chapter 4. Use the **Desktop** menu to control the configuration of the Desktop and to display toolbars. The **Window** menu has one or more items, depending on what you have done thus far in your session. Click on the name of a window that appears on the menu to open it. For example, if you have created a plot and not closed its window, the plot window will appear on this menu as **Figure 1.** However, there are other ways to move between windows (such as pressing the **Alt** and **Tab** keys simultaneously).

The toolbar, which is below the menu bar, provides buttons as shortcuts to some of the features on the menus. Clicking on the button is equivalent to clicking on the menu, then clicking on the menu item; thus the button eliminates one click of the mouse. The first seven buttons from the left correspond to the **New M-File, Open File, Cut, Copy, Paste, Undo,** and **Redo.** The eighth button activates Simulink. The ninth button activates the GUIDE Quick Start, which is used to create and edit graphical user interfaces (GUIs). The tenth button (the one with the question mark) accesses the Help System.

1.3 Computing with MATLAB

This section provides an overview of the computational capabilities of MATLAB, and points out where in the book these capabilities are discussed in more detail. The following chapters also provide numerous self-help exercises and examples of how these features can be used to solve engineering problems.

Arrays (Chapter 2)

MATLAB has hundreds of functions, which we will discuss throughout the text. For example, to compute sin x, where x has a value in radians, you type `sin(x)`. To compute cos x, type `cos(x)`. The exponential function e^x is computed from `exp(x)`. The natural logarithm, ln x, is computed by typing `log(x)`. (Note the spelling difference between mathematics text, ln, and MATLAB syntax, `log`.) You compute the base 10 logarithm by typing `log10(x)`. The inverse sine, or arcsine, is obtained by typing `asin(x)`. It returns an answer in radians, not degrees.

One of the strengths of MATLAB is its ability to handle collections of numbers, called *arrays,* as if they were a single variable. A numerical array is an ordered collection of numbers (a set of numbers arranged in a specific order). An example of an array variable is one that contains the numbers 0, 1, 3, and 6, in that order. We can use square brackets to define the variable `x` to contain this collection by typing `x = [0, 1, 3, 6]`. The elements of the array must be separated by commas or spaces. Note that the variable `y` defined as `y = [6, 3, 1, 0]` is not the same as `x` because the order is different.

We can add the two arrays `x` and `y` to produce another array `z` by typing the single line `z = x + y`. To compute `z`, MATLAB adds all the corresponding numbers in x and y to produce z. The resulting array z contains the numbers 6, 4, 4, 6. In most other programming languages, this operation requires more than one command. Because of this capability for handling arrays, MATLAB programs can be very short. Thus they are easier to create, read, and document.

You need not type all the numbers in the array if they are regularly spaced. Instead, you type the first number and the last number, with the spacing in the middle, separated by colons. For example, the numbers 0, 0.1, 0.2, . . . , 10 can be assigned to the variable u by typing `u = [0:0.1:10]`.

To compute $w = 5 \sin u$ for $u = 0, 0.1, 0.2, \ldots, 10$, the session is:

```
>>u = [0:0.1:10];
>>w = 5*sin(u);
```

The single line, `w = 5*sin(u)`, computed the formula $w = 5 \sin u$ 101 times, once for each value in the array u, to produce an array z that has 101 values. This illustrates some of the power of MATLAB to perform many calculations with just a few commands.

Because you typed a semicolon at the end of each line in the above session, MATLAB does not display the results on the screen. The values are stored in the variables u and w if you need them. You can see all the u values by typing u after the prompt or, for example, you can see the seventh value by typing `u(7)`. You can see the w values the same way. The number 7 is called an *array index*, because it points to a particular element in the array.

ARRAY INDEX

```
>>u(7)
ans =
     0.6000
>>w(7)
ans =
     2.8232
```

You can use the `length` function to determine how many values are in an array. For example, continue the previous session as follows:

```
>>m = length(w)
m =
   101
```

Arrays that display on the screen as a single row of numbers with more than one column are called *row arrays*. You can create *column* arrays, which have more than one row, by using a semicolon to separate the rows.

Polynomial Roots

We can describe a polynomial in MATLAB with an array whose elements are the polynomial's coefficients, *starting with the coefficient of the highest power of x*. For example, the polynomial $4x^3 - 8x^2 + 7x - 5$ would be represented by the array `[4,-8,7,-5]`. The *roots* of the polynomial $f(x)$ are the values of x such that $f(x) = 0$. Polynomial roots can be found with the `roots(a)` function, where a is the polynomial's coefficient array. The result is a *column* array that contains the polynomial's roots. For example, to find the roots of $x^3 - 7x^2 + 40x - 34 = 0$, the session is

```
>>a = [1,-7,40,-34];
>>roots(a)
ans =
      3.0000 + 5.000i
      3.0000 - 5.000i
      1.0000
```

The roots are $x = 1$ and $x = 3 \pm 5i$. The two commands could have been combined into the single command `roots([1,-7,40,-34])`.

Roots of functions other than polynomials can be obtained with the `fzero` function, covered in Chapter 3.

Useful applications of arrays are discussed in more detail in Chapter 2.

Test Your Understanding

T1.3–1 Use MATLAB to determine how many elements are in the array `[cos(0):0.02:log10(100)]`. Use MATLAB to determine the 25th element. (Answer: 51 elements and 1.48.)

T1.3–2 Use MATLAB to find the roots of the polynomial $290 - 11x + 6x^2 + x^3$. (Answer: $x = -10, 2 \pm 5i$.)

Built-in and User-Defined Functions (Chapter 3)

We have seen several of the functions built in to MATLAB, such as the `sqrt` and the `sin` functions. Table 1.3–1 lists some of the commonly used functions. Chapter 3 gives extensive coverage of the built-in functions.

Table 1.3–1 Some commonly used mathematical functions

Function	MATLAB syntax[1]
e^x	`exp(x)`
$\sqrt{x}$	`sqrt(x)`
$\ln x$	`log(x)`
$\log_{10} x$	`log10(x)`
$\cos x$	`cos(x)`
$\sin x$	`sin(x)`
$\tan x$	`tan(x)`
$\cos^{-1} x$	`acos(x)`
$\sin^{-1} x$	`asin(x)`
$\tan^{-1} x$	`atan(x)`

[1]The MATLAB trigonometric functions use radian measure.

MATLAB users can create their own functions for their special needs. Creation of user-defined functions is covered in Chapter 3.

Working with Files

MATLAB uses several types of files that enable you to save programs, data, and session results. As we will see in Section 1.4, MATLAB function files and program files are saved with the extension `.m`, and thus are called M-files.

MAT-FILES

MAT-files have the extension `.mat` and are used to save the names and values of variables created during a MATLAB session.

ASCII FILES

ASCII files are files written in a specific format designed to make them usable to a wide variety of software. The ASCII abbreviation stands for American Standard Code for Information Interchange. M-files are ASCII files that are written in the MATLAB language. Because they are ASCII files, M-files can be created using just about any word processor—generically called a text editor—because the ASCII file format is the basic format that all word processing programs can recognize and create. M-files are machine independent. MAT-files are *binary* files, not ASCII files. Binary files are generally readable only by the software that created them, so you cannot read a MAT-file with a word processor. In general, transferring binary files between machine types (MS Windows and Macintosh, for example) is not easy. However, MAT-files contain a machine signature that allows them to be transferred. They can also be manipulated by programs external to MATLAB. Binary files provide more compact storage than ASCII files.

DATA FILE

The third type of file we will be using is a data file, specifically an ASCII data file, that is, one created according to the ASCII format. You may need to use MATLAB to analyze data stored in such a file created by a spreadsheet program, a word processor, or a laboratory data acquisition system or in a file you share with someone else.

Saving and Retrieving Your Workspace Variables

If you want to stop using MATLAB but continue the session at a later time, you must use the `save` and `load` commands. Typing `save` causes MATLAB to save the workspace variables, that is, the variable names, their sizes, and their values, in a binary file called `matlab.mat`, which MATLAB can read. To retrieve your workspace variables, type `load`. You can then continue your session as before. Of course, if you exited MATLAB after using the `save` command, you cannot recover your keystrokes or the MATLAB responses. To save the workspace variables in another file named `filename.mat`, type `save filename`. To load the workspace variables, type `load filename`; this loads all the workspace variables from the file `filename.mat`. If the saved MAT-file `filename` contains the variables `A`, `B`, and `C`, then loading the file `filename` places these variables back into the workspace. If the variables already exist in the workspace, they are overwritten with the values of the variables from the file `filename`.

To load the workspace variables, the filename must have the extension `.mat` or no extension at all. If the file name does not have an extension, MATLAB assumes that it is `.mat`.

To save just some of your variables, say, `var1` and `var2`, in the file `filename.mat`, type `save filename var1 var2`. You need not type the variable names to retrieve them; just type `load filename`.

You can save the variables in ASCII single-precision (eight digits) format by typing `save filename -ASCII`. To save the variables in ASCII double-precision (16 digits) format, type `save filename -double`. ASCII files containing single-precision data are recognizable by their use of the `E` format to represent numbers. For example, the number 1.249×10^2 is represented as `1.249E+002`. ASCII files containing double-precision data use the `D` format; for example, `1.249D+002`. As an alternative to the `save` function, you can select **Save Data** from the **File** menu in the Command window. You can also save variables from the Workspace Browser.

Directories and Search Path It is important to know the location of the files you use with MATLAB. File location frequently causes problems for beginners. Suppose you use MATLAB on your home computer and save a file to a removable disk, as discussed later in this section. If you bring that disk to use with MATLAB on another computer, say, in a school's computer lab, you must make sure that MATLAB knows how to find your files. Files are stored in *directories,* called *folders* on some computer systems. Directories can have subdirectories below them. For example, suppose MATLAB was installed on drive c: in the directory `c:\matlab`. Then the `toolbox` directory is a subdirectory under the directory `c:\matlab`, and `symbolic` is a subdirectory under the `toolbox` directory. The *path* tells us and MATLAB how to find a particular file. For example, the file `solve.m` is a function in the Symbolic Math toolbox. The path to this file is `c:\matlab\toolbox\symbolic`. The full name of a file consists of its path and its name, for example, `c:\matlab\toolbox\symbolic\solve.m`.

PATH

Working with Removable Disks In Section 1.4 you will learn how to create and save M-files. Suppose you have saved the file `problem1.m` in the directory `\homework` on a disk, which you insert in drive a:. The path for this file is `a:\homework`. As MATLAB is normally installed, when you type `problem1`,

1. MATLAB first checks to see if `problem1` is a variable and if so, displays its value.
2. If not, MATLAB then checks to see if `problem1` is one of its own commands, and executes it if it is.
3. If not, MATLAB then looks in the current directory for a file named `problem1.m` and executes `problem1` if it finds it.
4. If not, MATLAB then searches the directories in its search path, in order, for `problem1.m` and then executes it if found.

You can display the MATLAB search path by typing `path`. If `problem1` is on the disk only and if directory a: is not in the search path, MATLAB will not find the file and will generate an error message, unless you tell it where to look. You can do this by typing `cd a:\homework`, which stands for "change directory to a:\homework." This will change the current directory to `a:\homework` and

Table 1.3–2 System, directory, and file commands

Command	Description
`addpath dirname`	Adds the directory `dirname` to the search path.
`cd dirname`	Changes the current directory to `dirname`.
`dir`	Lists all files in the current directory.
`dir dirname`	Lists all the files in the directory `dirname`.
`path`	Displays the MATLAB search path.
`pathtool`	Starts the Set Path tool.
`pwd`	Displays the current directory.
`rmpath dirname`	Removes the directory `dirname` from the search path.
`what`	Lists the MATLAB-specific files found in the current working directory. Most data files and other non-MATLAB files are not listed. Use `dir` to get a list of all files.
`what dirname`	Lists the MATLAB-specific files in directory `dirname`.

force MATLAB to look in that directory to find your file. The general syntax of this command is `cd dirname`, where `dirname` is the full path to the directory. The *main* directory on the disk is a:, so if your file is in the main directory, be sure to include the colon, and type `cd a:`.

An alternative to this procedure is to copy your file to a directory on the hard drive that is in the search path. However, there are several pitfalls with this approach: (1) if you change the file during your session, you might forget to copy the revised file back to your disk; (2) the hard drive becomes cluttered (this is a problem in public computer labs, and you might not be permitted to save your file on the hard drive); (3) the file might be deleted or overwritten if MATLAB is reinstalled; and (4) someone else can access your work!

You can determine the current directory (the one where MATLAB looks for your file) by typing `pwd`. To see a list of all the files in the current directory, type `dir`. To see the files in the directory `dirname`, type `dir dirname`.

The `what` command displays a list of the MATLAB-specific files in the current directory. The `what dirname` command does the same for the directory `dirname`.

You can add a directory to the search path by using the `addpath` command. To remove a directory from the search path, use the `rmpath` command. The Set Path tool is a graphical interface for working with files and directories. Type `pathtool` to start the browser. To save the path settings, click on **Save** in the tool. To restore the default search path, click on **Default** in the browser.

These commands are summarized in Table 1.3–2.

Decision-Making Programs in MATLAB (Chapter 4)

The usefulness of MATLAB greatly increases with its ability to use decision-making functions in its programs. These functions enable you to write programs whose operations depend on the results of calculations made by the program. MATLAB also can use loops to perform calculations repeatedly, a specified number of times, or until some condition is satisfied. This allows us to solve problems of great complexity or problems requiring numerous calculations.

Section 1.6 gives an introduction to these topics. Chapter 4 covers them in greater detail.

Plotting with MATLAB (Chapter 5)

MATLAB contains many powerful functions for easily creating plots of several different types, such as rectilinear, logarithmic, surface, and contour plots. As a simple example, let us plot the function $y = \sin 2x$ for $0 \leq x \leq 10$. We choose to use an increment of 0.01 to generate a large number of `x` values in order to produce a smooth curve. The function `plot(x,y)` generates a plot with the `x` values on the horizontal axis (the abscissa) and the `y` values on the vertical axis (the ordinate). The session is:

```
>>x = [0:0.02:8];
>>y = 5*sin(x);
>>plot(x,y),xlabel('x'),ylabel('y')
```

GRAPHICS WINDOW

The plot appears on the screen in a *graphics window,* named **Figure No. 1,** as shown in Figure 1.3–1. The `xlabel` function places the text in single quotes as a label on the horizontal axis. The `ylabel` function performs a similar function

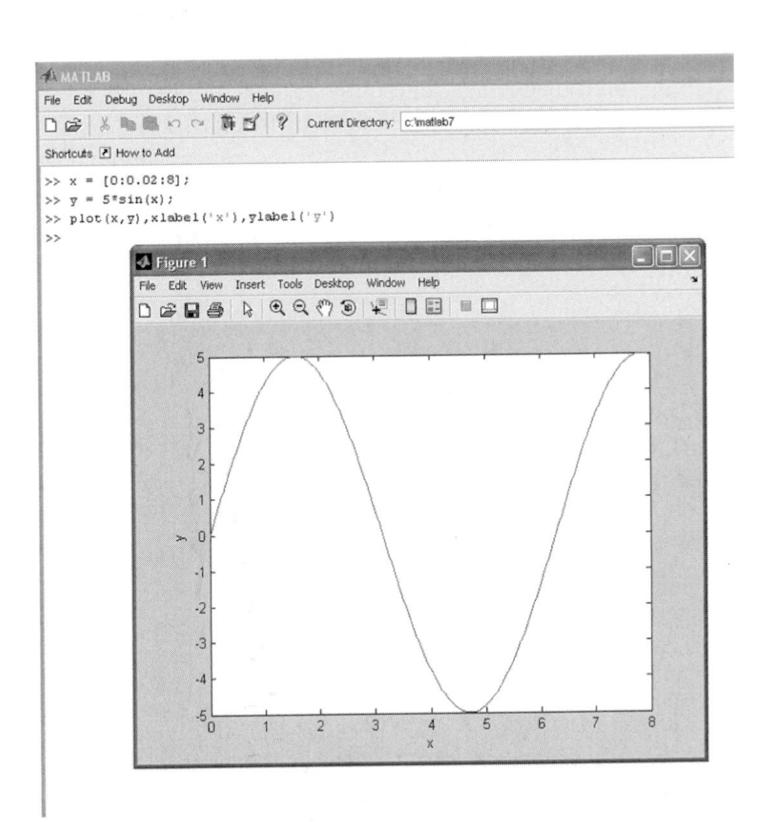

Figure 1.3–1 A graphics window showing a plot.

SM

for the vertical axis. When the plot command is successfully executed, a graphics window automatically appears. If a hard copy of the plot is desired, the plot can be printed by selecting **Print** from the **File** menu on the graphics window. The window can be closed by selecting **Close** on the **File** menu in the graphics window. You will then be returned to the prompt in the Command window.

Other useful plotting functions are `title` and `gtext`. These functions place text on the plot. Both accept text within parentheses and single quotes, as with the `xlabel` function. The `title` function places the text at the top of the plot; the `gtext` function places the text at the point on the plot where the cursor is located when you click the left mouse button.

OVERLAY PLOT

You can create multiple plots—called *overlay* plots—by including another set or sets of values in the `plot` function. For example, to plot the functions $y = 2\sqrt{x}$ and $z = 4 \sin 3x$ for $0 \leq x \leq 5$ on the same plot, the session is

```
>>x = [0:0.01:5];
>>y = 2*sqrt(x);
>>z = 4*sin(3*x);
>>plot(x,y,x,z),xlabel('x'),gtext('y'),gtext('z')
```

After the plot appears on the screen, the program waits for you to position the cursor and click the mouse button, once for each `gtext` function used.

Although MATLAB displays different colors for each curve, if you are going to print the plot on a black-and-white printer, you should label each curve so that you know which curve represents y and which curve represents z. One way of doing this is to use the `gtext` function to place the labels y and z next to the appropriate curves, as shown in the above session. Another way is to use the `legend` function, which is discussed in Chapter 5.

The plotting functions `xlabel`, `ylabel`, `title`, and `gtext` must be placed after the `plot` function and separated by commas.

You can also distinguish curves from one another by using different line types for each curve. For example, to plot the z curve using a dashed line, replace the `plot(x,y,x,z)` function in the above session with `plot(x,y,x,z,'--')`. Other line types can be used. These are discussed in Chapter 5.

In the above example, we had many values in the arrays to be plotted, and thus the curve plotted in Figure 1.3–1 is smooth. When plotting functions, you should always use arrays that have several hundred points so they will plot as continuous curves. This should be done because functions are defined at an infinite number of points. You should *never* plot a function using data markers or a small number of points.

Sometimes it is useful or necessary to obtain the coordinates of a point on a plotted curve. The function `ginput` can be used for this purpose. Place it at the end of all the plot and plot formatting statements, so that the plot will be in its final form. The command `[x,y] = ginput(n)` gets n points and returns the x and y coordinates in the vectors `x` and `y`, which have a length n. Position the cursor using a mouse, and press the mouse button. The returned coordinates have the same scale as the coordinates on the plot.

Table 1.3–3 Some MATLAB plotting commands

Command	Description
`[x,y] = ginput(n)`	Enables the mouse to get n points from a plot, and returns the x and y coordinates in the vectors `x` and `y`, which have a length n.
`grid`	Puts grid lines on the plot.
`gtext('text')`	Enables placement of text with the mouse.
`plot(x,y)`	Generates a plot of the array `y` versus the array `x` on rectilinear axes.
`title('text')`	Puts text in a title at the top of the plot.
`xlabel('text')`	Adds a text label to the horizontal axis (the abscissa).
`ylabel('text')`	Adds a text label to the vertical axis (the ordinate).

DATA MARKER

In cases where you are plotting *data,* as opposed to functions, you should use *data markers* to plot each data point (unless there are very many data points). To mark each point with a plus sign +, the required syntax for the `plot` function is `plot(x,y,'+')`. You can connect the data points with lines if you wish. In that case, you must plot the data twice, once with a data marker, and once without a marker.

For example, suppose the data for the independent variable is `x = [15:2:23]` with units of seconds, and the dependent variable values are `y = [20, 50, 60, 90, 70]` with units of volts. To plot the data with plus signs use the following session:

```
>>x = [15:2:23];
>>y = [20, 50, 60, 90, 70];
>>plot(x,y,'+',x,y),xlabel('x (seconds)'),ylabel('y (volts)')
```

Never forget to label your plots with the units of measurement! Other data markers are available. These are discussed in Chapter 5.

Table 1.3–3 summarizes these plotting commands. The `grid` command puts grid lines on the plot. We will discuss other plotting functions, and the Plot Editor, in Chapter 5. The chapter also discusses how to use plots to develop mathematical models from data. This process is called *function discovery and regression.*

Test Your Understanding

T1.3–3 Use MATLAB to plot the function $s = 2\sin(3t+2) + \sqrt{5t+1}$ over the interval $0 \le t \le 5$. Put a title on the plot, and properly label the axes. The variable s represents speed in feet per second; the variable t represents time in seconds.

T1.3–4 Use MATLAB to plot the functions $y = 4\sqrt{6x+1}$ and $z = 5e^{0.3x} - 2x$ over the interval $0 \le x \le 1.5$. Properly label the plot and each curve. The variables y and z represent force in newtons; the variable x represents distance in meters.

Linear Algebraic Equations (Chapter 6)

You can use the left division operator (\) in MATLAB to solve sets of linear algebraic equations. For example, consider the set

$$6x + 12y + 4z = 70$$
$$7x - 2y + 3z = 5$$
$$2x + 8y - 9z = 64$$

To solve such sets in MATLAB you must create two arrays; we will call them `A` and `B`. The array `A` has as many rows as there are equations, and as many columns as there are variables. The rows of `A` must contain the coefficients of x, y, and z in that order. In this example, the first row of `A` must be 6, 12, 4; the second row must be 7, −2, 3, and the third row must be 2, 8, −9. The array `B` contains the constants on the right-hand side of the equation; it has one column and as many rows as there are equations. In this example, the first row of `B` is 70, the second is 5, and the third is 64. The solution is obtained by typing `A\B`. The session is

```
>>A = [6,12,4;7,-2,3;2,8,-9];
>>B = [70;5;64];
>>Solution = A\B
Solution =
     3
     5
    -2
```

The solution is $x = 3$, $y = 5$, and $z = -2$.

This method works fine when the equation set has a unique solution. To learn how to deal with problems having a nonunique solution (or perhaps no solution at all!), see Chapter 6.

Test Your Understanding

T1.3–5 Use MATLAB to solve the following set of equations.

$$6x - 4y + 8z = 112$$
$$-5x - 3y + 7z = 75$$
$$14x + 9y - 5z = -67$$

(Answer: $x = 2$, $y = -5$, $z = 10$.)

Statistics (Chapter 7)

MATLAB has a number of useful functions for performing statistical calculations and other types of data manipulation. For example, you can compute the mean (the average) of a set of values stored in the array `x` by typing `mean(x)`. The standard deviation is obtained by typing `std(x)`. Chapter 7 covers these topics, as well as methods for obtaining several types of specialized histogram plots used in statistical analysis. The chapter also describes methods for developing simulations based on random number generation and methods for interpolating data.

Numerical Calculus, Differential Equations, and Simulink (Chapters 8 and 9)

Given a set of x and y values, MATLAB can numerically compute the derivative dy/dx and the integral $\int y\,dx$. In addition, MATLAB can numerically solve differential equations, which are equations involving derivatives; for example,

$$\frac{dy}{dx} + 5y^2 = 3\sin 8x$$

where the desired solution is $y(x)$. Chapter 8 deals with these methods.

Chapter 9 treats Simulink, which is a graphical user interface, built on top of MATLAB, for solving differential equations.

Symbolic Processing (Chapter 10)

Given a function $y(x)$, MATLAB can be used to obtain the derivative dy/dx and the integral $\int y\,dx$ in *symbolic* form, that is, as a formula instead of as a set of numerical values. This can be done with the Symbolic Math toolbox. In addition, this toolbox can be used to symbolically solve many types of algebraic, transcendental, and differential equations. Chapter 10 covers these methods.

1.4 Script Files and the Editor/Debugger

You can perform operations in MATLAB in two ways:

1. In the interactive mode, in which all commands are entered directly in the Command window, or
2. By running a MATLAB program stored in *script* file. This type of file contains MATLAB commands, so running it is equivalent to typing all the commands—one at a time—at the Command window prompt. You can run the file by typing its name at the Command window prompt.

Using the interactive mode is similar to using a calculator, but is convenient only for simpler problems. When the problem requires many commands, a repeated set of commands, or has arrays with many elements, the interactive mode is inconvenient. Fortunately, MATLAB allows you to write your own programs to avoid this difficulty. You write and save MATLAB programs in M-files, which have the extension `.m`; for example, `program1.m`.

MATLAB uses two types of M-files: *script* files and *function* files. You can use the Editor/Debugger built into MATLAB to create M-files. A *script file* contains a sequence of MATLAB commands, and is useful when you need to use many commands or arrays with many elements. Because they contain commands, script files are sometimes called *command* files. You execute a script file at the Command window prompt by typing its name without the extension `.m`.

Another type of M-file is a *function file,* which is useful when you need to repeat the operation of a set of commands. You can create your own function files. We discuss them in Chapter 3.

Script files may contain any valid MATLAB commands or functions, including user-written functions. When you type the name of a script file at the Command window prompt, you get the same result as if you had typed at the Command window prompt all the commands stored in the script file, one at a time. When you type the name of the script file, we say that you are "running the file" or "executing the file." The values of variables produced by running a script file are available in the workspace; thus, we say the variables created by a script file are *global* variables.

GLOBAL VARIABLE

Creating and Using a Script File

COMMENT

The symbol % designates a *comment,* which is not executed by MATLAB. Comments are not that useful for interactive sessions, and are used mainly in script files for the purpose of documenting the file. The comment symbol may be put anywhere in the line. MATLAB ignores everything to the right of the % symbol. For example, consider the following session.

```
>>% This is a comment.
>>x = 2+3 % So is this.
x =
   5
```

Note that the portion of the line before the % sign is executed to compute x.

Here is a simple example that illustrates how to create, save, and run a script file, using the Editor/Debugger built into MATLAB. However, you may use another text editor to create the file. The sample file is shown below. It computes the sine of the square root of several numbers and displays the results on the screen.

```
% Program example1.m
% This program computes the sine of
% the square root and displays the result.
x = sqrt([5:2:13]);
y = sin(x)
```

To create this new M-file (in the MS Windows environment), in the Command window select **New** from the **File** menu, then select **M-file.** You will then see a new edit window. This is the Editor/Debugger window as shown in Figure 1.4–1. Type in the file as shown above. You can use the keyboard and the **Edit** menu in the Editor/Debugger as you would in most word processors to create and edit the file. When finished, select **Save** from the **File** menu in the Editor/Debugger. In the dialog box that appears, replace the default name provided (usually named `Untitled`) with the name `example1`, and click on **Save.** The Editor/Debugger will automatically provide the extension `.m` and save the file in the MATLAB current directory, which for now we will assume to be on the hard drive.

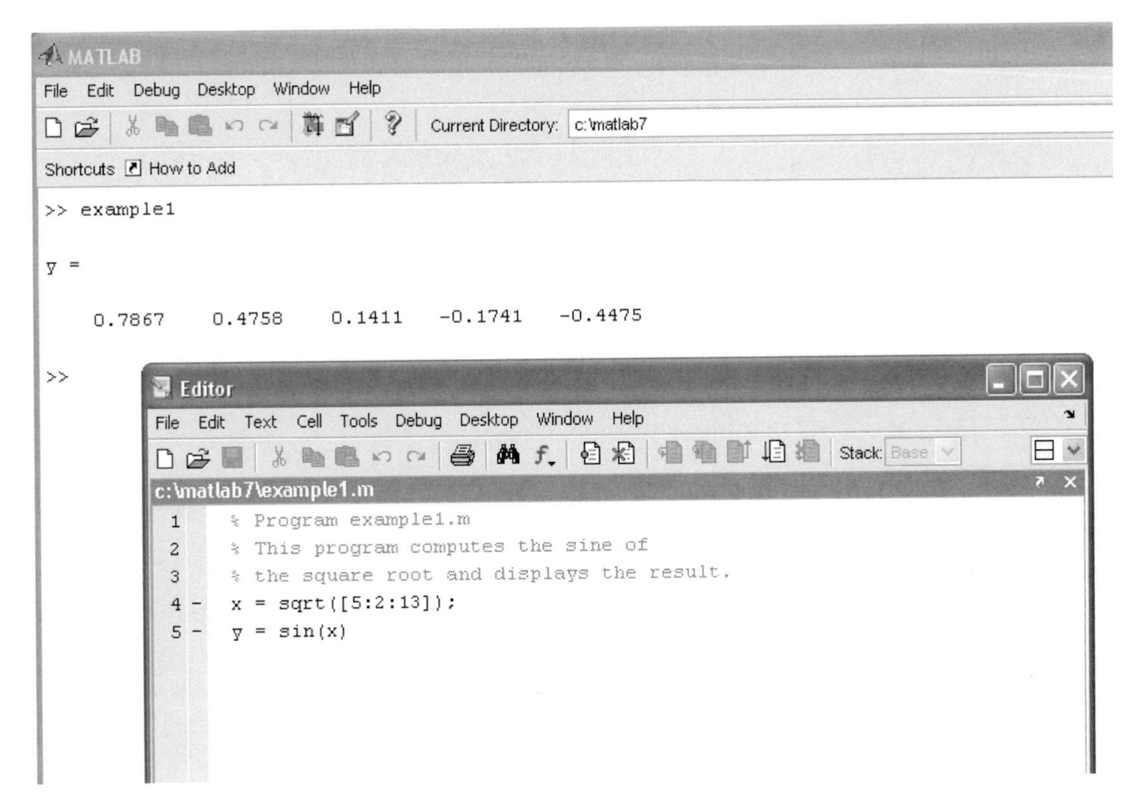

Figure 1.4–1 The MATLAB Command window with the Editor/Debugger open.

Once the file has been saved, in the MATLAB Command window type the script file's name `example1` to execute the program. You should see the result displayed in the Command window. The session looks like the following.

```
>>example1
y =
   0.7867   0.4758   0.1411   0.1741   -0.4475
```

Figure 1.4–1 shows a MATLAB MS Windows screen containing the resulting Command window display and the Editor/Debugger opened to display the script file.

Effective Use of Script Files

Create script files to avoid the need to retype commonly used procedures. The above file, `example1`, implements a very simple procedure, for which we ordinarily would not create a script file. However, it illustrates how a script file is useful. For example, to change the numbers evaluated from `[3:2:11]` to `[2:5:27]`, simply edit the corresponding line and save the file again (a common oversight is to forget to resave the file after making changes to it).

Here are some other things to keep in mind when using script files:

1. The name of a script file must follow the MATLAB convention for naming variables; that is, the name must begin with a letter, and may include digits and the underscore character, up to 31 characters.
2. Recall that typing a variable's name at the Command window prompt causes MATLAB to display the value of that variable. Thus, do not give a script file the same name as a variable it computes because MATLAB will not be able to execute that script file more than once, unless you clear the variable.
3. Do not give a script file the same name as a MATLAB command or function. You can check to see if a command, function or file name already exists by using the `exist` command. For example, to see if a variable `example1` already exists, type `exist('example1');` this will return a 0 if the variable does not exist, and a 1 if it does. To see if an M-file `example1.m` already exists, type `exist ('example1.m','file')` *before* creating the file; this will return a 0 if the file does not exist, and a 2 if it does. Finally, to see if a built-in function `example1` already exists, type `exist('example1', 'builtin')` *before* creating the file; this will return a 0 if the built-in function does not exist, and a 5 if it does.
4. As in interactive mode, all variables created by a script file are global variables, which means that their values are available in the basic workspace. You can type `who` to see what variables are present.
5. Variables created by a function file are *local* to that function, which means that their values are not available outside the function. All the variables in a script file are global. Thus, if you do not need to have access to all the variables in a script file, consider using a function file instead. This will avoid "cluttering" the workspace with variable names, and will reduce memory requirements. Creation of user-defined function files is discussed in Chapter 3.

LOCAL VARIABLE

6. You can use the `type` command to view an M-file without opening it with a text editor. For example, to view the file `example1`, the command is `type example1`.

Note that not all functions supplied with MATLAB are "built-in" functions. For example, the function `mean.m` is supplied but is not a built-in function. The command `exist ('mean.m', 'file')` will return a 2, but the command `exist ('mean', 'builtin')` will return a 0. You may think of built-in functions as primitives that form the basis for other MATLAB functions. You cannot view the entire file of a built-in function in a text editor, only the comments.

Effective Use of the Command and Editor/Debugger Windows

Here are some tips on using the Command and Editor/Debugger windows effectively.

1. You can use the mouse to resize and move windows so they can be viewed simultaneously. Or you can dock the Editor/Debugger window inside the

Desktop by selecting **Dock** from the view menu of the Editor/Debugger. To activate a window, click on it.

2. If the Editor/Debugger is not docked, use the **Alt-Tab** key combination to switch back and forth quickly between the Editor/Debugger window and the Command window. In the Command window, use the up-arrow key to retrieve the previously typed script-file name, and press **Enter** to execute the script file. This technique allows you to check and correct your program quickly. After making changes in the script file, be sure to save it before switching to the Command window.
3. You can use the Editor/Debugger as a basic word processor to write a short report that includes your script file, results, and discussion, perhaps to present your solution to one of the chapter problems. First use the mouse to highlight the results shown in the Command window, then copy and paste them to the Editor/Debugger window above or below your script file (use **Copy** and **Paste** on the **Edit** menu). Then, to save space, delete any extra blank lines, and perhaps the prompt symbol. Type your name and any other required information, add any discussion you wish, and print the report from the Editor/Debugger window, or save it and import it into the word processor of your choice (change the file name or its extension if you intend to use the script file again!).

Debugging Script Files

Debugging a program is the process of finding and removing the "bugs," or errors, in a program. Such errors usually fall into one of the following categories.

1. Syntax errors such as omitting a parenthesis or comma, or spelling a command name incorrectly. MATLAB usually detects the more obvious errors and displays a message describing the error and its location.
2. Errors due to an incorrect mathematical procedure, called *runtime errors*. They do not necessarily occur every time the program is executed; their occurrence often depends on the particular input data. A common example is division by zero.

MATLAB error messages usually allow you to find syntax errors. However, run-time errors are more difficult to locate. To locate such an error, try the following:

1. Always test your program with a simple version of the problem, whose answers can be checked by hand calculations.
2. Display any intermediate calculations by removing semicolons at the end of statements.
3. Use the debugging features of the Editor/Debugger, which are covered in Chapter 4. However, one advantage of MATLAB is that it requires relatively simple programs to accomplish many types of tasks. Thus you probably will not need to use the Debugger for many of the problems encountered in this text.

Programming Style

Comments may be put anywhere in the script file. However, it is important to note that the first comment line before any executable statement is the line searched by the `lookfor` command, discussed later in this chapter. Therefore, if you intend to use the script file in the future, consider putting key words that describe the script file in this first line (called the H1 line).

A suggested structure for a script file is the following.

1. *Comments section* In this section put comment statements to give:
   a. The name of the program and any key words in the first line.
   b. The date created, and the creators' names in the second line.
   c. The definitions of the variable names for every input and output variable. Divide this section into at least two subsections, one for input data, and one for output data. A third, optional section may include definitions of variables used in the calculations. *Be sure to include the units of measurement for all input and all output variables!*
   d. The name of every user-defined function called by the program.
2. *Input section* In this section put the input data and/or the input functions that enable data to be entered. Include comments where appropriate for documentation.
3. *Calculation section* Put the calculations in this section. Include comments where appropriate for documentation.
4. *Output section* In this section put the functions necessary to deliver the output in whatever form required. For example, this section might contain functions for displaying the output on the screen. Include comments where appropriate for documentation.

The programs in this text usually omit the comments in order to save space. These comments are not necessary here because the text discussion associated with the program provides the required documentation, and because we all know who wrote these programs!

Documenting Units of Measurement

We emphasize again that you must document the units of measurement for all the input and all the output variables. More than one dramatic failure of an engineering system has been traced to the misunderstanding of the units used for the input and output variables of the program used to design the system. Table 1.4–1 lists common units and their abbreviations. The foot-pound-second system (FPS) is also called the U.S. Customary System and the British Engineering System. SI (Système Internationale) is the international metric system.

Using Script Files to Store Data

You might have applications which require you to access the same set of data frequently. If so, you can store the data in an array within a script file. An example is

Table 1.4–1 SI and FPS units

	Unit name and abbreviation	
Quantity	**SI unit**	**FPS unit**
Time	second (s)	second (sec)
Length	meter (m)	foot (ft)
Force	newton (N)	pound (lb)
Mass	kilogram (kg)	slug
Energy	joule (J)	foot-pound (ft-lb), Btu (= 778 ft-lb)
Power	watt (W)	ft-lb/sec, horsepower (hp)
Temperature	degrees Celsius (°C), kelvin (K)	degrees Fahrenheit (°F), degrees Rankine (°R)

a set of daily temperature measurements at a particular location, which are needed from time to time for calculations. As a short example, consider the following script file, whose name is `mydata.m`. The array `temp_F` contains temperatures in degrees Fahrenheit.

```
% File mydata.m: Stores temperature data.
% Stores the array temp_F,
% which contains temperatures in degrees Fahrenheit.
temp_F = [72, 68, 75, 77, 83, 79]
```

A session to access this data from the Command window, and convert the temperatures to degrees Celsius, is

```
>>mydata
temp_F =
    72   68   75   77   83 79
>>temp_C = 5*(temp_F - 32)/9
temp_C =
    22.2222  20.0000  23.8889  25.0000   28.3333  26.1111
```

Thus 68° Fahrenheit corresponds to 20° Celsius.

Test Your Understanding

T1.4–1 Create, save, and run a script file that solves the following set of equations for given values of a, b, and c. Check your file for the case $a = 112$, $b = 75$, and $c = -67$. The answers for this case are $x = 2$, $y = -5$, $z = 10$.

$$6x - 4y + 8z = a$$

$$-5x - 3y + 7z = b$$

$$14x + 9y - 5z = c$$

Table 1.4–2 Input/output commands

Command	Description
`disp(A)`	Displays the contents, but not the name, of the array `A`.
`disp('text')`	Displays the text string enclosed within single quotes.
`format`	Controls the screen's output display format (see Table 1.1–5).
`fprintf`	Performs formatted writes to the screen or to a file (see Appendix C).
`x = input('text')`	Displays the text in quotes, waits for user input from the keyboard, and stores the value in `x`.
`x = input('text','s')`	Displays the text in quotes, waits for user input from the keyboard, and stores the input as a string in `x`.
`k=menu('title','option1','option2',...`	Displays a menu whose title is in the string variable `'title'`, and whose choices are `'option1'`, `'option2'`, and so on.

Controlling Input and Output

MATLAB provides several useful commands for obtaining input from the user and for formatting the output (the results obtained by executing the MATLAB commands). Table 1.4–2 summarizes these commands. The methods presented in this section are particularly useful with script files.

You already know how to determine the current value of any variable by typing its name and pressing **Enter** at the command prompt. However, this method, which is useful in the interactive mode, is not useful for script files. The `disp` function (short for "display") can be used instead. Its syntax is `disp(A)`, where `A` represents a MATLAB variable name. Thus typing `disp(Speed)` causes the value of the variable `Speed` to appear on the screen, but not the variable's name.

The `disp` function can also display text. You enclose the text within single quotes. For example, the command `disp('The predicted speed is:')` causes the message to appear on the screen. This command can be used with the first form of the `disp` function in a script file as follows (assuming the value of `Speed` is 63):

```
disp('The predicted speed is:')
disp(Speed)
```

When the file is run, these lines produce the following on the screen:

```
The predicted speed is:
   63
```

User Input

The `input` function displays text on the screen, waits for the user to enter something from the keyboard, and then stores the input in the specified variable. For example, the command `x = input('Please enter the value` of x:') causes the message to appear on the screen. If you type 5 and press **Enter,** the variable x will have the value 5.

A *string variable* is composed of text (alphanumeric characters). If you want to store a text input as a string variable, use the other form of the input command. For example, the command Calendar = input('Enter the day of the week:','s') prompts you to enter the day of the week. If you type Wednesday, this text will be stored in the string variable Calendar.

STRING VARIABLE

Use the menu function to generate a menu of choices for user input. Its syntax is

```
k = menu('title','option1','option2',...)
```

The function displays the menu whose title is in the string variable 'title', and whose choices are string variables 'option1', 'option2', and so on. The returned value of k is 1, 2, ... depending on whether you click on the button for option1, option2, and so forth. For example, the following script uses a menu to select the data marker for a graph, assuming that the vectors x and y already exist.

```
k = menu('Choose a data marker','o','*','x');
type = ['o','*','x'];
plot(x,y,x,y,type(k))
```

Test Your Understanding

T1.4–2 The surface area A of a sphere depends on its radius r as follows: $A = 4\pi r^2$. Write a script file that prompts the user to enter a radius, computes the surface area, and displays the result.

Example of a Script File

The following is a simple example of a script file that shows the preferred program style. The speed v of a falling object dropped with no initial velocity is given as a function of time t by $v = gt$, where g is the acceleration due to gravity. In SI units, $g = 9.81$ m/s^2. We want to compute and plot v as a function of t for $0 \leq t \leq t_f$, where t_f is the final time entered by the user. The script file is the following.

```
% Program Falling_Speed.m: Plots speed of a falling object.
% Created on March 1, 2004 by W. Palm III
%
% Input Variable:
% tf = final time (in seconds)
%
% Output Variables:
% t = array of times at which speed is computed (seconds)
% v = array of speeds (meters/second)
```

```
%
% Parameter Value:
g = 9.81; % Acceleration in SI units
%
% Input section:
tf = input('Enter the final time in seconds:');
%
% Calculation section:
dt = tf/500;
t = [0:dt:tf]; % Creates an array of 501 time values.
v = g*t;
%
% Output section:
plot(t,v),xlabel('Time (seconds)'),ylabel('Speed (meters/second)')
```

After creating this file, you save it with the name `Falling_Speed.m`. To run it, you type `Falling_Speed` (without the `.m`) in the Command window at the prompt. You will then be asked to enter a value for t_f. After you enter a value and press **Enter,** you will see the plot on the screen.

1.5 The MATLAB Help System

If all the MATLAB documentation were printed, it would fill a volume many times the size of this book. Therefore, it is impossible for us to describe all of the details of MATLAB. This book gives you in-depth coverage of the basic MATLAB language and an overview of the available tools to alert you to their existence. It gives you all the material you need to do the homework problems. If you need information about a topic covered in this book, remember to use the following special features that were designed as reference aids.

- Throughout each chapter margin notes identify where key terms are introduced.
- Each chapter contains tables summarizing the MATLAB commands introduced in that chapter.
- At the end of each chapter is a summary guide to the commands covered in that chapter.
- Appendix A contains tables of MATLAB commands, grouped by category, with the appropriate page references.
- There are two indexes. The first lists MATLAB commands and symbols, while the second lists topics.

To explore the more advanced features of MATLAB not covered in this book, you will need to know how to use effectively the MATLAB Help System. MATLAB has these options to get help for using MathWorks products.

1. *Help Browser* This graphical user interface helps you find information and view online documentation for your MathWorks products.
2. *Help Functions* The functions `help`, `lookfor`, and `doc` can be used to display syntax information for a specified function.
3. *Other Resources* For additional help, you can run demos, contact technical support, search documentation for other MathWorks products, view a list of other books, and participate in a newsgroup.

The Help Browser

The Help Browser enables you to search and view documentation for MATLAB and your other MathWorks products. To open the Help Browser, select MATLAB **Help** from the **Help** menu, or click the question mark button in the toolbar. The Help Browser contains two window "panes": the Help Navigator pane on the left and the Display pane on the right (see Figure 1.5–1). The Help Navigator contains four tabs:

- *Contents:* a contents listing tab,
- *Index:* a global index tab,
- *Search:* a search tab having a find function and full text search features, and
- *Demos:* a bookmarking tab to start built-in demonstrations.

Use the tabs in the Help Navigator to find documentation. You view documentation in the Display pane. To adjust the relative width of the two panes, drag the separator bar between them. To close the Help Navigator pane, click the close box (x) in the

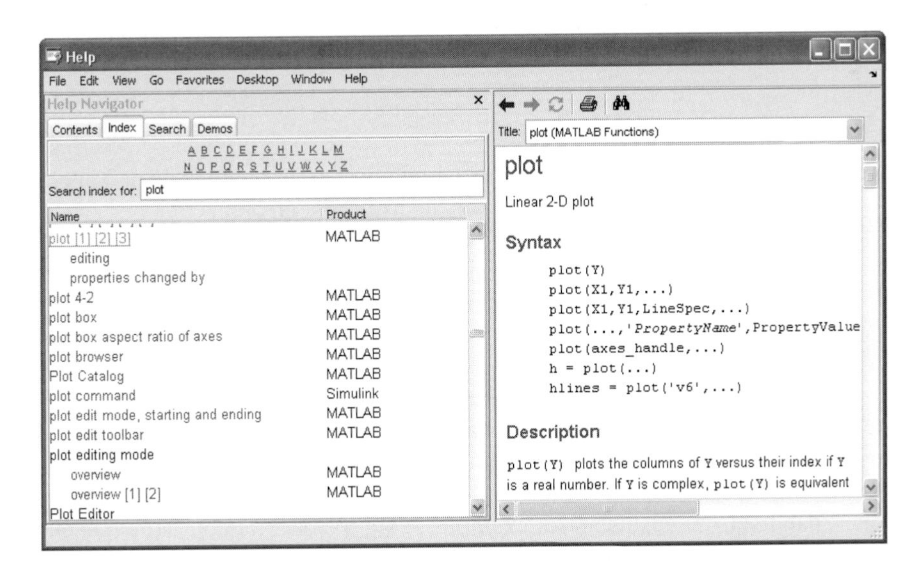

Figure 1.5–1 The MATLAB Help Browser.

pane's upper right corner. This is useful once you have found the documentation you wanted because it provides more screen space for the documentation itself. To open the Help Navigator pane from the display pane, click on **Help Navigator** in the **View** menu.

Viewing Documentation

After finding documentation with the Help Navigator, view the documentation in the Display pane. While viewing a page of documentation, you can:

- Scroll to see contents not currently visible in the window.
- View the previous or next page in the document by clicking the left or right arrow at the top of the page.
- View the previous or next item in the index by clicking the left or right arrow at the bottom of the page.
- Reload a page by clicking the Refresh button (circular arrows) in the Help Browser toolbar.
- Find a phrase in the currently displayed page by clicking on the binoculars icon and typing it in the **Find what:** box in the Help Browser toolbar and pressing the **Enter** key.
- Add that page to your list of favorite documents by clicking **Add to Favorites** in the Help Browser toolbar.

The box above the Display pane contains the title of the help page currently displayed in the Display pane. Click on the arrow to the right of the box to see a list of the help pages you previously accessed. Then click on the name of a page, and it appears in the Display pane.

Using the Contents Tab

Click the **Contents** tab in the Help Navigator to list the titles and table of contents for all product documentation. To expand the listing for an item, click the + to the left of the item. To collapse the listings for an item, click the - to the left of the item, or double-click the item. Click on an item to select it. The first page of that document appears in the Display pane. Double-clicking an item in the contents listing expands the listing for that item and shows the first page of that document in the Display pane.

The Contents pane is synchronized with the Display pane. By default, the item selected in the Contents pane always matches the documentation appearing in the Display pane. Thus, the contents tree is synchronized with the displayed document. This feature is useful if you access documentation with a method other than the Contents pane, for example, using the back button in the Display pane. With synchronization, you always know to what section the displayed page belongs.

Using the Index Tab

Click the **Index** tab in the Help Navigator pane to find specific index entries (keywords) from all of your MathWorks documentation. Type a word or words in the "Search index for" box. As you type, the index highlights the matching entries. Scroll down in the Help Navigator pane to see more matching entries. Click on an entry to display the corresponding page. If you do not find a matching index entry or if the corresponding page does not contain the information you seek, try a less specific topic by using only part of the wording, or use the **Search** tab.

Using the Search Tab

Click the **Search** tab in the Help Navigator pane to find all MATLAB documents containing a specified phrase. Type the phrase in the "Search for" box. Then click the **Go** button. The list of documents and the heading under which the phrase is found in that document then appear in the Help Navigator pane. Select an entry from the list of results to view that document in the Display pane.

Using the Favorites Menu

Click on the **Favorites** menu to add a page to the Favorites list, or to view a list of documents you previously designated as favorites. Select an entry and that document then appears in the Display pane. To remove a document from the list of favorites, right-click the document in the favorites list and select **Delete** from the pop-up menu. To designate a document page as a favorite, you can either:

- Click **Add to Favorites** in the **Help Browser** toolbar while that document is open in the Display pane, or
- Right-click the document name listed under the **Contents** tab and click the **Add to Favorites** button, or
- Right-click the document name in the **Help Browser** search results list and click the **Add to Favorites** button.

Help Functions

Three MATLAB functions can be used for accessing online information about MATLAB functions:

- `help funcname` Displays in the Command window a description of the specified function `funcname`.
- `lookfor topic` Displays in the Command window a brief description for all functions whose description includes the specified key word `topic`.
- `doc funcname` Opens the Help Browser to the reference page for the specified function `funcname`, providing a description, additional remarks, and examples.

The `help` Function The `help` function is the most basic way to determine the syntax and behavior of a particular function. Information is displayed directly in the Command window. For example, typing `help log10` in the Command window produces the following display:

```
LOG10  Common (base 10) logarithm.
  LOG10(X) is the base 10 logarithm of the elements of X.
  Complex results are produced if X is not positive.

  See also LOG, LOG2, EXP, LOGM.
```

Note that the display describes what the function does, warns about any unexpected results if nonstandard argument values are used, and directs the user to other related functions.

All the MATLAB functions are organized into logical groups, upon which the MATLAB directory structure is based. For instance, all elementary mathematical functions such as `log10` reside in the `elfun` directory, and the polynomial functions reside in the `polyfun` directory. To list the names of all the functions in that directory, with a brief description of each, type `help polyfun`. If you are unsure of what directory to search, type `help` to obtain a list of all the directories, with a description of the function category each represents. Throughout this text we point out the appropriate directory name so that you can get more information if you need it.

Typing `helpwin topic` displays the help text for the specified `topic` inside the Desktop Help Browser window. Links are created to functions referenced in the "See Also" line of the help text. You can also access the Help window by selecting the **Help** option under the **Help** menu, or by clicking the question mark button on the toolbar.

The `lookfor` Function The `lookfor` function allows you to search for functions on the basis of a key word. It searches through the first line of help text, known as the H1 line, for each MATLAB function, and returns all the H1 lines containing a specified key word. For example, MATLAB does not have a function named `sine`. So the response from `help sine` is

```
sine.m not found
```

However, typing `lookfor sine` produces over a dozen matches, depending on which toolboxes you have installed. For example, you will see

```
ACOS      Inverse cosine.
ACOSH     Inverse hyperbolic cosine.
ASIN      Inverse sine.
ASINH     Inverse hyperbolic sine.
COS       Cosine.
COSH      Hyperbolic cosine.
SIN       Sine.
SINH      Hyperbolic sine.
...
```

Table 1.5–1 MATLAB Help functions

Function	Use
`doc`	Displays the start page of the documentation in the Help Browser.
`doc function`	Displays the documentation for the MATLAB function `function`.
`doc toolbox/function`	Displays the documentation for the specified toolbox function.
`doc toolbox`	Displays the documentation road map page for the specified toolbox.
`help`	Displays a list all the function directories, with a description of the function category each represents.
`help function`	Displays in the Command window a description of the specified function `function`.
`helpwin topic`	Displays the help text for the specified `topic` inside the desktop Help Browser window.
`lookfor topic`	Displays in the Command window a brief description for all functions whose description includes the specified key word `topic`.
`type filename`	Displays the M-file `filename` without opening it with a text editor.

From this list you can find the correct name for the sine function. Note that all words containing sine are returned, such as cosine. Adding `-all` to the `lookfor` function searches the entire help entry, not just the H1 line.

The `doc` Function Typing `doc` displays the start page of the documentation in the Help Browser. Typing `doc function` displays the documentation for the MATLAB function `function`. Typing `doc toolbox/function` displays the documentation for the specified toolbox function. Typing `doc toolbox` displays the documentation road map page for the specified toolbox.

Table 1.5–1 summarizes the MATLAB Help functions.

The MathWorks Website

If your computer is connected to the Internet, you can access The MathWorks, Inc., the home of MATLAB. You can use electronic mail to ask questions, make suggestions, and report possible bugs. You can also use a solution search engine at The MathWorks website to query an up-to-date database of technical support information. The website address is http://www.mathworks.com.

1.6 Programming in MATLAB

MATLAB has *relational operators, conditional statements,* and *loops*. Relational operators are used to make comparisons. The conditional statements allow us to write programs that make decisions. A loop is a structure for repeating a calculation a number of times. Conditional statements and loops are best used in a script file, rather than in an interactive session.

Table 1.6–1 Relational operators

Relational operator	Meaning
<	Less than.
<=	Less than or equal to.
>	Greater than.
>=	Greater than or equal to.
==	Equal to.
~=	Not equal to.

In this section we will limit our applications of relational operators, conditional statements, and loops to the use of *scalar* variables. In Chapter 4 we will show how to use *arrays* to make decisions and computations in loops.

Relational Operators

MATLAB has six *relational* operators to make comparisons between arrays. These operators are shown in Table 1.6–1. Note that the "equal to" operator consists of two = signs, not a single = sign as you might expect. The single = sign is the *assignment* operator in MATLAB. The result of a comparison using the relational operators is a *logical* value, which is either a 0 (if the comparison is *false*) or a 1 (if the comparison is *true*), and the result can be used as a *logical variable*, which is a variable containing logical values.

LOGICAL VARIABLE

For example, if `x = 2` and `y = 5`, typing `z = x < y` returns the value `z = 1`, because `x` is less than `y`. Typing `u = x==y` returns the value `u = 0` because `x` does not equal `y`. To make the statements more readable, we can group the operations using parentheses. For example, `z = (x < y)` and `u = (x==y)`. For the operators consisting of two symbols, there cannot be a space between them.

The relational operators compare arrays on an element-by-element basis. The arrays must have the same dimension. The only exception occurs when we compare an array to a scalar. In that case, all the elements of the array are compared to the scalar. For example, suppose that `x = [6,3,9]` and `y = [14,2,9]`. The following MATLAB session shows some examples.

```
>> z = (x < y)
z =
   1 0 0
>>z = (x > y)
z =
   0 1 0
>>z = (x ~= y)
z =
   1 1 0
>>z = (x == y)
z =
   0 0 1
```

```
>>z = (x > 8)
z =
   0 0 1
```

We can also use the relational operators to address arrays. For example, with `x = [6,3,9,11]` and `y = [14,2,9,13]`, typing `z = x(x<y)` finds all the elements in `x` that are less than the corresponding elements in `y`. The result is the array `z = [6, 11]`.

The `find` Function We can use the `find` function to create decision-making programs, especially when we combine it with the relational operators. The function `find(x)` computes an array containing the indices of the *nonzero* elements of the numeric array `x`. For example, consider the session

```
>>x = [-2,0,4];
>>y = find(x)
y =
   1    3
```

The resulting array `y = [1, 3]` indicates that the first and third elements of `x` are nonzero. Note that the `find` function returns the *indices,* not the *values*. In the following session, note the difference between the result obtained by `x(x<y)` and the result obtained by `find(x<y)`.

```
>>x = [6,3,9,11];y = [14,2,9,13];
>>values = x(x<y)
values =
     6    11
>>how_many = length(values)
how_many =
     2
>>indices = find(x<y)
indices =
     1  4
```

Thus two values in the array `x` are less than the *corresponding* values in the array `y`. They are the first and fourth values, 6 and 11. To find out how many, we could also have typed `length(indices)`.

In the above example, there were only a few numbers in the arrays `x` and `y`, and thus we could have obtained the answers by visual inspection. However, these MATLAB methods are useful either where there is so much data that visual inspection would be time-consuming, or where the values are generated by a program.

Analysis of Temperature Data — EXAMPLE 1.6–1

The arrays `temp_A` and `temp_B` given in the table contain the water temperature in degrees Fahrenheit of two ponds measured at noon for 10 days. Determine how many days the temperature of pond A was above 60°. On what days did this occur? Determine the temperature of pond A on the days when it was greater than or equal to the temperature of pond B.

Day	1	2	3	4	5	6	7	8	9	10
Temperature in pond A	55	62	60	61	63	65	62	59	58	56
Temperature in pond B	54	59	62	64	68	68	62	59	57	53

■ Solution

The session is

```
>>A = [55, 62, 60, 61, 63, 65, 62, 59, 58, 56];
>>B = [54, 59, 62, 64, 68, 68, 62, 59, 57, 53];
>>when = find(A>60)
when =
     2  4  5  6  7
>>how_many1 = length(when)
how_many1 =
     5
>>above = A(A>=B)
above =
     55  62  62  59   58  56
>>how_many2 = length(above)
ans =
     6
```

The temperature of pond A was above 60° on five days: days 2, 4, 5, 6, and 7. The temperature of pond A was above the temperature of pond B on six days. On those days its temperature was 55, 62, 62, 59, 58, and 56° respectively.

Test Your Understanding

T1.6–1 Suppose that `x = [-9,-6,0,2,5]` and `y = [-10,-6, 2,4,6]`. What is the result of the following operations? Determine the answers by hand, and then use MATLAB to check your answers.

a. `z = (x < y)`
b. `z = (x > y)`
c. `z = (x ~= y)`
d. `z = (x == y)`
e. `z = (x > 2)`

T1.6–2 Suppose that `x = [-4,-1,0,2,10]` and `y = [-5,-2, 2,5,9]`. Use MATLAB to find the values and the indices of the elements in `x` that are greater than the corresponding elements in `y`.

Conditional Statements

The MATLAB *conditional* statements allow us to write programs that make decisions. Conditional statements contain one or more of the `if`, `else`, and `elseif` statements. The `end` statement denotes the end of a conditional statement. These conditional statements read somewhat like their English language equivalents. For example, suppose that `x` is a scalar, and that we want to compute $y = \sqrt{x}$ only if $x \geq 0$. In English, we could specify this as: If x is greater than or equal to zero, compute y from $y = \sqrt{x}$, otherwise, do nothing. The `if` statement in the following script file accomplishes this in MATLAB, assuming that the variable `x` already has a scalar value.

```
if x >= 0
   y = sqrt(x)
end
```

If `x` is negative, the program takes no action.

When more than one action must occur as a result of a decision, we can use the `else` and `elseif` statements. The statements after the `else` are executed if all the preceding `if` and `elseif` expressions are false. The general form of the `if` statement is

```
if expression
   commands
elseif expression
   commands
else
   commands
end
```

The `else` and `elseif` statements may be omitted if not required.

Suppose that we want to compute y from $y = \sqrt{x}$ for $x \geq 0$, and $y = -\sqrt{-x}$ for $x < 0$. The following statements will calculate y, assuming that the variable `x` already has a scalar value.

```
if x >= 0
   y = sqrt(x)
else
   y = -sqrt(-x)
end
```

As another example, suppose that we want to compute y such that

$$y = \begin{cases} 15\sqrt{4x} + 10 & \text{if } x \geq 9 \\ 10x + 10 & \text{if } 0 \leq x < 9 \\ 10 & \text{if } x < 0 \end{cases}$$

The following statements will compute y, assuming that the variable x already has a scalar value.

```
if x >= 9
   y = 15*sqrt(4x) + 10
elseif x >= 0
   y = 10*x + 10
else
   y = 10
end
```

Note that the elseif statement does not require a separate end statement.

When the variable being tested (x in the previous session) is an array rather than a scalar, the if-elseif-else-end structure can give unexpected results if not used carefully. This is discussed in Chapter 4.

Loops

A *loop* is a structure for repeating a calculation a number of times. Each repetition of the loop is a *pass*. There are two types of explicit loops in MATLAB: the for loop, used when the number of passes is known ahead of time, and the while loop, used when the looping process must terminate when a specified condition is satisfied, and thus the number of passes is not known in advance.

A simple example of a for loop is

```
m = 0;
x(1) = 10;
for k = 2:3:11
   m = m+1;
   x(m+1) = x(m) + k^2;
end
```

The *loop variable* k is initially assigned the value 2. During each successive pass through the loop k is incremented by 3, and x is calculated until k exceeds 11. Thus, k takes on the values 2, 5, 8, 11. The variable m indicates the index of the array x. The program then continues to execute any statements following the end statement. When the loop is finished the array x will have the values x(1)=14, x(2)=39, x(3)=103, x(4)=224. The name of the loop variable need not be k.

The while loop is used in cases where the looping process must terminate when a specified condition is satisfied, and thus the number of passes is not known in advance. A simple example of a while loop is

```
x = 5;
k = 0;
while x < 25
   k = k + 1;
```

```
    y(k) = 3*x;
    x = 2*x-1;
end
```

The loop variable x is initially assigned the value 5, and it keeps this value until the statement x = 2*x - 1 is encountered the first time. Its value then changes to 9. Before each pass through the loop, x is checked to see if its value is less than 25. If so, the pass is made. If not, the loop is skipped and the program continues to execute any statements following the end statement. The variable x takes on the values 9, 17, and 33 within the loop. The resulting array y contains the values y(1) = 15, y(2) = 27, y(3) = 51.

Plotting with a for Loop — EXAMPLE 1.6–2

Write a script file to plot the function:

$$y = \begin{cases} 15\sqrt{4x} + 10 & x >= 9 \\ 10x + 10 & 0 \leq x \leq 9 \\ 10 & x < 0 \end{cases}$$

for $-5 \leq x \leq 30$.

■ Solution

We choose a spacing $dx = 35/300$ to obtain 301 points, which is sufficient to obtain a smooth plot. The script file is the following:

```
dx = 35/300;
x = [-5:dx:30];
for k = 1:length(x)
    if x(k) >= 9
      y(k) = 15*sqrt(4*x(k)) + 10;
    elseif x(k) >= 0
      y(k) = 10*x(k) + 10;
    else
      y(k) = 10;
    end
end
plot(x,y),xlabel('x'),ylabel('y')
```

Note that we must use the index k to refer to x within the loop, as x(k).

Series Calculation with a for Loop — EXAMPLE 1.6–3

Write a script file to compute the sum of the first 15 terms in the series $5k^2 - 2k$, $k = 1, 2, 3, \ldots, 15$.

■ Solution

Because we know how many times we must evaluate the expression $5k^2 - 2k$, we can use a `for` loop. The script file is the following:

```
total = 0;
for k = 1:15
    total = 5*k^2 - 2*k + total;
end
disp('The sum for 15 terms is:')
disp(total)
```

The answer is 5960.

Note that it might be tempting to use the variable `sum`, instead of `total` to represent the series sum. However, `sum` is a built-in MATLAB function, so although this particular program would work with `total` replaced by `sum`, it is good practice to avoid using built-in function names as variables. Before creating the program, you can check to see if `total` is used by MATLAB by typing `exist('total')`. The result is `ans = 0`, which means that the variable `total` does not exist and the name is not used by MATLAB.

EXAMPLE 1.6–4 Series Calculation with a `while` Loop

Write a script file to determine how many terms are required for the sum of the series $5k^2 - 2k$, $k = 1, 2, 3, \ldots$ to exceed 10,000. What is the sum for this many terms?

■ Solution

Because we do not know how many times we must evaluate the expression $5k^2 - 2k$, we use a `while` loop. The script file is the following:

```
total = 0;
k = 0;
while total < 1e+4
   k = k + 1;
   total = 5*k^2 - 2*k + total;
end
disp('The number of terms is:')
disp(k)
disp('The sum is:')
disp(total)
```

The sum is 10,203 after 18 terms.

EXAMPLE 1.6–5 Growth of a Bank Account

Determine how long it will take to accumulate at least \$10,000 in a bank account if you deposit \$500 initially and \$500 at the end of each year, if the account pays 5 percent annual interest.

■ **Solution**

Because we do not know how many years it will take, a `while` loop should be used. The script file is the following.

```
amount = 500;
k=0;
while amount < 10000
  k = k+1;
  amount = amount*1.05 + 500;
end
amount
k
```

The final results are `amount = 1.0789e+004`, or \$10,789, and `k = 14`, or 14 years.

The Editor/Debugger is capable of automatically indenting to improve the readability of a file. For example, `if`, `else`, `elseif`, `for`, and `while` structures do not require indenting, but doing so enables the reader to identify the structure more easily. The Editor/Debugger automatically indents the lines after `if`, `else`, `elseif`, `for`, and `while` statements when you press the **Enter** key. It continues to indent until the corresponding `end` statement is reached. It also uses *syntax highlighting* to identify key statements by displaying them in different colors.

Table 1.6–2 summarizes these statements. Chapter 4 covers these topics in greater depth, and also introduces logical (Boolean) operators and string variables. More information on script files, function files, and the Editor/Debugger is given in Chapter 3.

Test Your Understanding

T1.6–3 Write a script file using conditional statements to evaluate the following function, assuming that the scalar variable `x` has a value. The function is $y = \sqrt{x^2+1}$ for $x < 0$, $y = 3x + 1$ for $0 \leq x < 10$, and $y = 9\sin(5x - 50) + 31$ for $x \geq 10$. Use your file to evaluate y for $x = -5$, $x = 5$, and $x = 15$, and check the results by hand.

T1.6–4 Use a `for` loop to determine the sum of the first 20 terms in the series $3k^2$, $k = 1, 2, 3, \ldots 20$. (Answer: 8610.)

T1.6–5 Use a `while` loop to determine how many terms in the series $3k^2$, $k = 1, 2, 3, \ldots$ are required for the sum of the terms to exceed 2000. What is the sum for this number of terms? (Answer: 13 terms, with a sum of 2457.)

Table 1.6–2 Some MATLAB programming statements

Command	Description
`else`	Delineates an alternate block of commands.
`elseif`	Conditionally executes an alternate block of commands.
`end`	Terminates `for`, `while`, and `if` statements.
`find(x)`	Computes an array containing the indices of the nonzero elements of the array `x`.
`for`	Repeats commands a specified number of times.
`if`	Executes commands conditionally.
`while`	Repeats commands an indefinite number of times.

1.7 Problem-Solving Methodologies

Designing new engineering devices and systems requires a variety of problem-solving skills. (This variety is what keeps engineering from becoming boring!) When solving a problem, it is important to plan your actions ahead of time. You can waste many hours by plunging into the problem without a plan of attack. Here we present a plan of attack, or *methodology,* for solving engineering problems in general. Because solving engineering problems often requires a computer solution and because the examples and exercises in this text require you to develop a computer solution (using MATLAB), we also discuss a methodology for solving computer problems in particular.

Steps in Engineering Problem Solving

MODEL

Table 1.7–1 summarizes the methodology that has been tried and tested by the engineering profession for many years. These steps describe a general problem-solving procedure. Simplifying the problem sufficiently and applying the appropriate fundamental principles is called *modeling,* and the resulting mathematical description is called a *mathematical model,* or just a *model*. When the modeling is finished, we need to solve the mathematical model to obtain the required answer. If the model is highly detailed, we might need to solve it with a computer program. Most of the examples and exercises in this text require you to develop a computer solution (using MATLAB) to problems for which the model has already been developed. Thus we will not always need to use all the steps shown in Table 1.7–1. More discussion of engineering problem solving can be found in [Eide, 1998].*

Example of Problem Solving

Consider the following simple example of the steps involved in problem solving. Suppose you work for a company that produces packaging. You are told that a new packaging material can protect a package when dropped, provided that the

*References appear in Appendix C.

Table 1.7–1 Steps in engineering problem solving

1.	Understand the purpose of the problem.
2.	Collect the known information. Realize that some of it might later be found unnecessary.
3.	Determine what information you must find.
4.	Simplify the problem only enough to obtain the required information. State any assumptions you make.
5.	Draw a sketch and label any necessary variables.
6.	Determine which fundamental principles are applicable.
7.	Think generally about your proposed solution approach and consider other approaches before proceeding with the details.
8.	Label each step in the solution process.
9.	If you solve the problem with a program, hand check the results using a simple version of the problem. Checking the dimensions and units and printing the results of intermediate steps in the calculation sequence can uncover mistakes.
10.	Perform a "reality check" on your answer. Does it make sense? Estimate the range of the expected result and compare it with your answer. Do not state the answer with greater precision than is justified by any of the following: (a) The precision of the given information. (b) The simplifying assumptions. (c) The requirements of the problem. Interpret the mathematics. If the mathematics produces multiple answers, do not discard some of them without considering what they mean. The mathematics might be trying to tell you something, and you might miss an opportunity to discover more about the problem.

package hits the ground at less than 25 ft/sec. The package's total weight is 20 lb, and it is rectangular with dimensions of 12 by 12 by 8 in. You must determine whether the packaging material provides enough protection when the package is carried by delivery persons.

The steps in the solution are as follows:

1. *Understand the purpose of the problem.* The implication here is that the packaging is intended to protect against being dropped while the delivery person is carrying it. It is not intended to protect against the package falling off a moving delivery truck. In practice, you should make sure that the person giving you this assignment is making the same assumption. Poor communication is the cause of many errors!
2. *Collect the known information.* The known information is the package's weight, dimensions, and maximum allowable impact speed.
3. *Determine what information you must find.* Although not explicitly stated, you need to determine the maximum height from which the package can be

dropped without damage. You need to find a relationship between the speed of impact and the height at which the package is dropped.

4. *Simplify the problem only enough to obtain the required information. State any assumptions you make.* The following assumptions will simplify the problem and are consistent with the problem statement as we understand it:
   a. The package is dropped from rest with no vertical or horizontal velocity.
   b. The package does not tumble (as it might when dropped from a moving truck). The given dimensions indicate that the package is not thin and thus will not "flutter" as it falls.
   c. The effect of air drag is negligible.
   d. The greatest height the delivery person could drop the package from is 6 ft (and thus we ignore the existence of a delivery person 8 ft tall!).
   e. The acceleration g due to gravity is constant (because the distance dropped is only 6 ft).
5. *Draw a sketch and label any necessary variables.* Figure 1.7–1 is a sketch of the situation, showing the height h of the package, its mass m, its speed v, and the acceleration due to gravity g.
6. *Determine which fundamental principles are applicable.* Because this problem involves a mass in motion, we can apply Newton's laws. From physics we know that the following relations result from Newton's laws and the basic kinematics of an object falling a short distance under the influence of gravity, with no air drag or initial velocity:
   a. Height versus time to impact t_i: $h = \frac{1}{2} g t_i^2$.
   b. Impact speed v_i versus time to impact: $v_i = g t_i$.
   c. Conservation of mechanical energy: $mgh = \frac{1}{2} m v_i^2$.

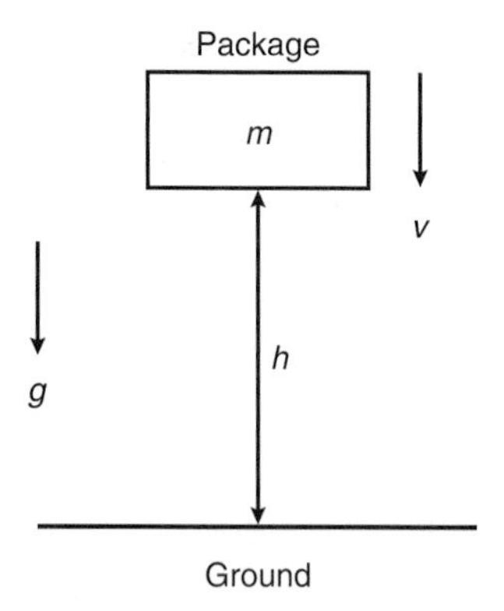

Figure 1.7–1 Sketch of the dropped-package problem.

7. *Think generally about your proposed solution approach and consider other approaches before proceeding with the details.* We could solve the second equation for t_i and substitute the result into the first equation to obtain the relation between h and v_i. This approach would also allow us to find the time to drop t_i. However, this method involves more work than necessary because we need not find the value of t_i. The most efficient approach is to solve the third relation for h.

$$h = \frac{1}{2}\frac{v_i^2}{g} \tag{1.7–1}$$

Notice that the mass m cancels out of the equation. The mathematics just told us something! It told us that the mass does not affect the relation between the impact speed and the height dropped. Thus we do not need the weight of the package to solve the problem.

8. *Label each step in the solution process.* This problem is so simple that there are only a few steps to label:
   a. Basic principle: conservation of mechanical energy

$$h = \frac{1}{2}\frac{v_i^2}{g}$$

   b. Determine the value of the constant g: $g = 32.2$ ft/sec^2.
   c. Use the given information to perform the calculation and round off the result consistent with the precision of the given information:

$$h = \frac{1}{2}\frac{25^2}{32.2} = 9.7 \text{ ft}$$

Because this text is about MATLAB, we might as well use it to do this simple calculation. The session looks like this:

```
>>g=32.2;
>>vi=25;
>>h=vi^2/(2*g)
h =
   9.7050
```

9. *Check the dimensions and units.* This check proceeds as follows, using (1.7–1),

$$[\text{ft}] = \left[\frac{1}{2}\right]\frac{[\text{ft/sec}]^2}{[\text{ft/sec}^2]} = \frac{[\text{ft}]^2}{[\text{sec}]^2}\frac{[\text{sec}]^2}{[\text{ft}]} = [\text{ft}]$$

which is correct.

10. *Perform a reality check and precision check on the answer.* If the computed height were negative, we would know that we did something wrong. If it were very large, we might be suspicious. However, the computed height of 9.7 ft does not seem unreasonable.

Table 1.7–2 Steps for developing a computer solution

1. State the problem concisely.
2. Specify the data to be used by the program. This is the "input."
3. Specify the information to be generated by the program. This is the "output."
4. Work through the solution steps by hand or with a calculator; use a simpler set of data if necessary.
5. Write and run the program.
6. Check the output of the program with your hand solution.
7. Run the program with your input data and perform a reality check on the output.
8. If you will use the program as a general tool in the future, test it by running it for a range of reasonable data values; perform a reality check on the results.

If we had used a more accurate value for g, say $g = 32.17$, then we would be justified in rounding the result to $h = 9.71$. However, given the need to be conservative here, we probably should round the answer *down* to the nearest foot. So we probably should report that the package will not be damaged if it is dropped from a height of less than 9 ft.

The mathematics told us that the package mass does not affect the answer. The mathematics did not produce multiple answers here. However, many problems involve the solution of polynomials with more than one root; in such cases we must carefully examine the significance of each.

Steps for Obtaining a Computer Solution

If you use a program such as MATLAB to solve a problem, follow the steps shown in Table 1.7–2. More discussion of modeling and computer solutions can be found in [Starfield, 1990] and [Jayaraman, 1991].

MATLAB is useful for doing numerous complicated calculations and then automatically generating a plot of the results. The following example illustrates the procedure for developing and testing such a program.

EXAMPLE 1.7–1 Piston Motion

Figure 1.7–2a shows a piston, connecting rod, and crank for an internal combustion engine. When combustion occurs, it pushes the piston down. This motion causes the connecting rod to turn the crank, which causes the crankshaft to rotate. We want to develop a MATLAB program to compute and plot the distance d traveled by the piston as a function of the angle A, for given values of the lengths L_1 and L_2. Such a plot would help the engineers designing the engine to select appropriate values for the lengths L_1 and L_2.

We are told that typical values for these lengths are $L_1 = 1$ ft and $L_2 = 0.5$ ft. Because the mechanism's motion is symmetrical about $A = 0$, we need consider only angles in the range $0 \leq A \leq 180°$. Figure 1.7–2b shows the geometry of the motion. From this figure we can use trigonometry to write the following expression for d:

$$d = L_1 \cos B + L_2 \cos A \tag{1.7–2}$$

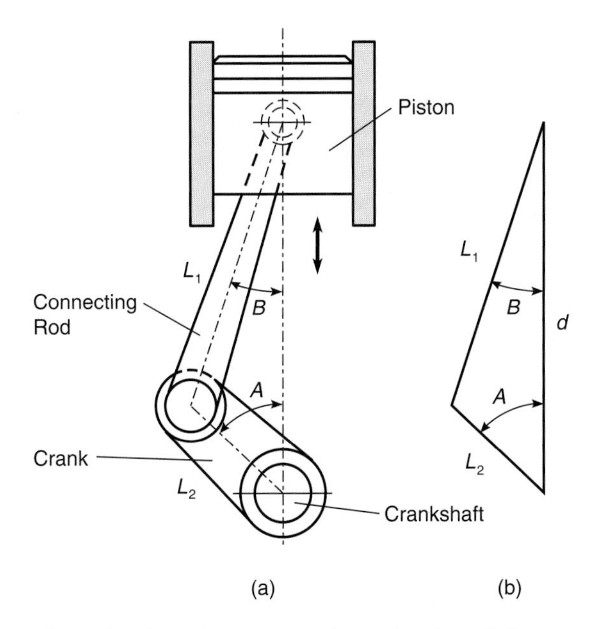

Figure 1.7–2 A piston, connecting rod, and crank for an internal combustion engine.

Thus to compute d given the lengths L_1 and L_2 and the angle A, we must first determine the angle B. We can do so using the law of sines, as follows:

$$\frac{\sin A}{L_1} = \frac{\sin B}{L_2}$$

Solve this for B:

$$\sin B = \frac{L_2 \sin A}{L_1}$$

$$B = \sin^{-1}\left(\frac{L_2 \sin A}{L_1}\right) \tag{1.7–3}$$

Equations (1.7–2) and (1.7–3) form the basis of our calculations. Develop and test a MATLAB program to plot d versus A.

■ Solution

Here are the steps in the solution, following those listed in Table 1.7–2.

1. *State the problem concisely.* Use equations (1.7–2) and (1.7–3) to compute d; use enough values of A in the range $0 \leq A \leq 180°$ to generate an adequate (smooth) plot.
2. *Specify the input data to be used by the program.* The lengths L_1 and L_2 and the angle A are given.

3. *Specify the output to be generated by the program.* A plot of d versus A is the required output.

4. *Work through the solution steps by hand or with a calculator.* You could have made an error in deriving the trigonometric formulas, so you should check them for several cases. You can check for these errors by using a ruler and protractor to make a scale drawing of the triangle for several values of the angle A, measure the length d, and compare it to the calculated values. Then you can use these results to check the output of the program.

   Which values of A should you use for the checks? Because the triangle "collapses" when $A = 0$ and $A = 180°$, you should check these cases. The results are $d = L_1 - L_2$ for $A = 0$, and $d = L_1 + L_2$ for $A = 180°$. The case $A = 90°$ is also easily checked by hand, using the Pythagorean theorem; for this case $d = \sqrt{L_1^2 - L_2^2}$. You should also check one angle in the quadrant $0 < A < 90°$ and one in the quadrant $90° < A < 180°$. The following table shows the results of these calculations using the given typical values: $L_1 = 1$, $L_2 = 0.5$ ft.

A (degrees)	**d (ft)**
0	1.5
60	1.15
90	0.87
120	0.65
180	0.5

5. *Write and run the program.* The following MATLAB session uses the values $L_1 = 1$, $L_2 = 0.5$ ft.

```
>>L_1 = 1;
>>L_2 = 0.5;
>>R = L_2/L_1;
>>A_d = [0:0.5:180];
>>A_r = A_d*(pi/180);
>>B = asin(R*sin(A_r));
>>d = L_1*cos(B)+L_2*cos(A_r);
>>plot(A_d,d),xlabel('A (degrees)'), ...
ylabel('d (feet)'),grid
```

   Note the use of the underscore (_) in variable names to make the names more meaningful. The variable `A_d` represents the angle A in degrees. Line 4 creates an array of numbers 0, 0.5, 1, 1.5, . . . , 180. Line 5 converts these degree values to radians and assigns the values to the variable `A_r`. This conversion is necessary because MATLAB trigonometric functions use radians, not degrees. (A common oversight is to use degrees.) MATLAB provides the built-in constant `pi` to use for π. Line 6 uses the inverse sine function `asin`.

   The `plot` command requires the label and grid commands to be on the same line, separated by commas. The line-continuation operator, called an *ellipsis,* consists of three periods. This operator enables you to continue typing the line after you press **Enter.** Otherwise, if you continued typing without using the ellipsis, you would not see the entire line on the screen. Note that the prompt is not visible when you press **Enter** after the ellipsis.

The `grid` command puts grid lines on the plot so that you can read values from the plot more easily. The resulting plot appears in Figure 1.7–3.

6. *Check the output of the program with your hand solution.* Read the values from the plot corresponding to the values of `A` given in the preceding table. You can use the `ginput` function to read values from the plot. The values should agree with each other and they do.
7. *Run the program and perform a "reality check" on the output.* You might suspect an error if the plot showed abrupt changes or discontinuities. However, the plot is smooth and shows that d behaves as expected. It decreases smoothly from its maximum at $A = 0$ to its minimum at $A = 180°$.
8. *Test the program for a range of reasonable input values.* Test the program using various values for L_1 and L_2 and examine the resulting plots to see whether they are reasonable. Something you might try on your own is to see what happens if $L_1 \leq L_2$. Should the mechanism work the same way it does when $L_1 > L_2$? What does your intuition tell you to expect from the mechanism? What does the program predict?

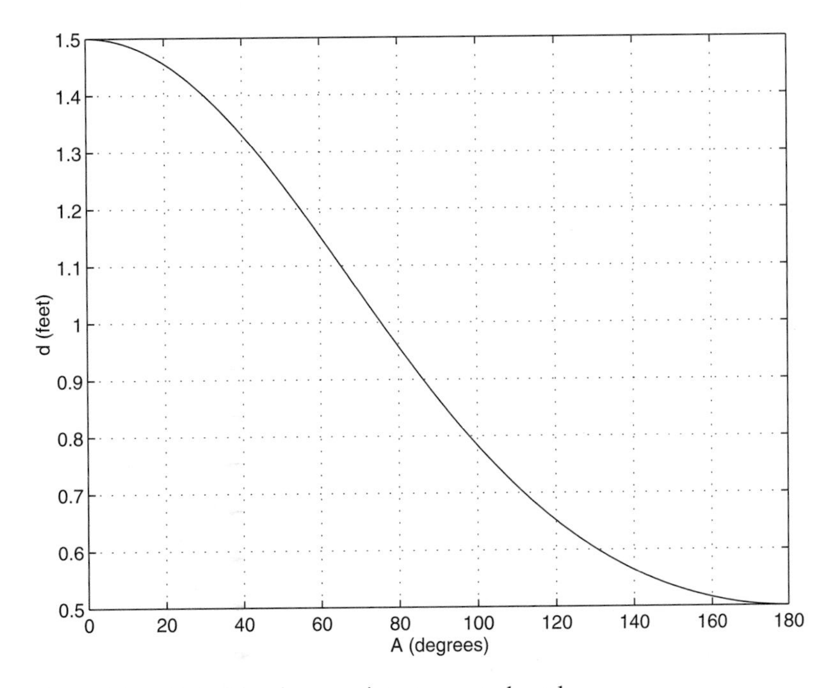

Figure 1.7–3 Plot of the piston motion versus crank angle.

Besides using degrees instead of radians, a common mistake in programming is forgetting to put the multiplication symbol (`*`) between two variables. This is easy to overlook because we are used to writing xy in algebra, rather than $x*y$. Forgetting to enclose function arguments in parentheses, as `sin 2`, is another common mistake, instead of typing `sin(2)`.

1.8 Summary

You should now be familiar with basic operations in MATLAB. These include

- Starting and exiting MATLAB,
- Computing simple mathematical expressions, and
- Managing variables.

You should also be familiar with the MATLAB menu and toolbar system.

The chapter gives an overview of the various types of problems MATLAB can solve. These include

- Using arrays and polynomials,
- Using relational operators,
- Creating plots,
- Solving linear algebraic equations,
- Creating M-file programs,
- Using conditional statements, and
- Using loops.

The following chapters give more details on these topics. You can also obtain more information using the MATLAB Help System.

You should be familiar with the methodology used for problem solving, and the specific methodology used for solving problems by computer, as discussed in Section 1.7.

Table 1.8–1 is a guide to the MATLAB commands and features introduced in this chapter.

Table 1.8–1 Guide to commands and features introduced in this chapter

Scalar arithmetic operations	Table 1.1–1
Order of precedence	Table 1.1–2
Commands for managing the work session	Table 1.1–3
Special variables and constants	Table 1.1–4
Some commonly used mathematical functions	Table 1.3–1
System, directory, and file commands	Table 1.3–2
Some MATLAB plotting commands	Table 1.3–3
Input/output commands	Table 1.4–2
MATLAB Help functions	Table 1.5–1
Relational operators	Table 1.6–1
Some MATLAB programming statements	Table 1.6–2

Key Terms with Page References

Problems

Answers to problems marked with an asterisk are given at the end of the text.

Section 1.1

1. Make sure you know how to start and quit a MATLAB session. Use MATLAB to make the following calculations, using the values: $x = 10$, $y = 3$. Check the results using a calculator.

a. $u = x + y$ *b.* $v = xy$ *c.* $w = x/y$
d. $z = \sin x$ *e.* $r = 8 \sin y$ *f.* $s = 5 \sin (2y)$

2.* Suppose that $x = 2$ and $y = 5$. Use MATLAB to compute the following.

a. $\dfrac{yx^3}{x - y}$ *b.* $\dfrac{3x}{2y}$ *c.* $\dfrac{3}{2}xy$ *d.* $\dfrac{x^5}{x^5 - 1}$

3. Suppose that $x = 3$ and $y = 4$. Use MATLAB to compute the following, and check the results with a calculator.

a. $\left(1 - \dfrac{1}{x^5}\right)^{-1}$ *b.* $3\pi x^2$ *c.* $\dfrac{3y}{4x - 8}$ *d.* $\dfrac{4(y - 5)}{3x - 6}$

4. Evaluate the following expressions in MATLAB for the given value of x. Check your answers by hand.

a. $y = 6x^3 + \dfrac{4}{x}$, $x = 2$ *b.* $y = \dfrac{x}{4}3$, $x = 8$

c. $y = \dfrac{(4x)^2}{25}$, $x = 10$ *d.* $y = 2\dfrac{\sin x}{5}$, $x = 2$

e. $y = 7(x^{1/3}) + 4x^{0.58}$, $x = 20$

5. Assuming that the variables a, b, c, d, and f are scalars, write MATLAB statements to compute and display the following expressions. Test your statements for the values $a = 1.12$, $b = 2.34$, $c = 0.72$, $d = 0.81$, $f = 19.83$.

$$x = 1 + \frac{a}{b} + \frac{c}{f^2} \qquad s = \frac{b-a}{d-c}$$

$$r = \frac{1}{\frac{1}{a} + \frac{1}{b} + \frac{1}{c} + \frac{1}{d}} \qquad y = ab\frac{1}{c}\frac{f^2}{2}$$

6. Use MATLAB to calculate

a. $\frac{3}{4}(6)(7^2) + \frac{4^5}{7^3 - 145}$ *b.* $\frac{48.2(55) - 9^3}{53 + 14^2}$

c. $\frac{27^2}{4} + \frac{319^{4/5}}{5} + 60(14)^{-3}$

Check your answers with a calculator.

7. The volume of a sphere is given by $V = 4\pi r^3/3$, where r is the radius. Use MATLAB to compute the radius of a sphere having a volume 30 percent greater than that of a sphere of radius 5 ft.

8.* Suppose that $x = -7 - 5i$ and $y = 4 + 3i$. Use MATLAB to compute

a. $x + y$ *b.* xy *c.* x/y

9. Use MATLAB to compute the following. Check your answers by hand.

a. $(3 + 6i)(-7 - 9i)$ *b.* $\frac{5 + 4i}{5 - 4i}$

c. $\frac{3}{2}i$ *d.* $\frac{3}{2i}$

10. Evaluate the following expressions in MATLAB, for the values $x = 5 + 8i$, $y = -6 + 7i$. Check your answers by hand.

a. $u = x + y$ *b.* $v = xy$ *c.* $w = x/y$

d. $z = e^x$ *e.* $r = \sqrt{y}$ *f.* $s = xy^2$

11. Engineers often need to estimate the pressure exerted by a gas in a container. The *ideal gas law* provides one way of making the estimate. The law is

$$P = \frac{nRT}{V}$$

More accurate estimates can be made with the *van der Waals* equation:

$$P = \frac{nRT}{V - nb} - \frac{an^2}{V^2}$$

where the term nb is a correction for the volume of the molecules, and the term an^2/V^2 is a correction for molecular attractions. The values of a and b depend on the type of gas. The gas constant is R, the *absolute* temperature is T, the gas volume is V, and the number of gas molecules is indicated by n. If $n = 1$ mol of an ideal gas were confined to a volume of $V = 22.41$ L at 0°C (273.2 K), it would exert a pressure of 1 atmosphere. In these units, $R = 0.08206$.

For chlorine (Cl_2), $a = 6.49$ and $b = 0.0562$. Compare the pressure estimates given by the ideal gas law and the van der Waals equation for 1 mol of Cl_2 in 22.41 L at 273.2 K. What is the main cause of the difference in the two pressure estimates: the molecular volume or the molecular attractions?

12. The *ideal gas law* relates the pressure P, volume V, absolute temperature T, and amount of gas n. The law is

$$P = \frac{nRT}{V}$$

where R is the gas constant.

An engineer must design a large natural gas storage tank to be expandable to maintain the pressure constant at 2.2 atmospheres. In December when the temperature is 4°F (−15°C), the volume of gas in the tank is 28,500 ft^3. What will the volume of the same quantity of gas be in July when the temperature is 88°F (31°C)? (Hint: Use the fact that n, R, and P are constant in this problem. Note also that K = °C + 273.2.)

Section 1.3

13. Suppose x takes on the values $x = 1, 1.2, 1.4, \ldots, 5$. Use MATLAB to compute the array `y` that results from the function $y = 7\sin(4x)$. Use MATLAB to determine how many elements are in the array `y`, and the value of the third element in the array `y`.

14. Use MATLAB to determine how many elements are in the array `[sin(-pi/2):0.05:cos(0)]`. Use MATLAB to determine the 10th element.

15. Use MATLAB to calculate

a. $e^{(-2.1)^3} + 3.47\log(14) + \sqrt[4]{287}$

b. $(3.4)^7\log(14) + \sqrt[4]{287}$

c. $\cos^2\left(\dfrac{4.12\pi}{6}\right)$

d. $\cos\left(\dfrac{4.12\pi}{6}\right)^2$

Check your answers with a calculator.

16. Use MATLAB to calculate

a. $6\pi \tan^{-1}(12.5) + 4$ *b.* $5 \tan[3 \sin^{-1}(13/5)]$

c. $5 \ln(7)$ *d.* $5 \log(7)$

Check your answers with a calculator.

17. The Richter scale is a measure of the intensity of an earthquake. The energy E (in joules) released by the quake is related to the magnitude M on the Richter scale as follows.

$$E = 10^{4.4} 10^{1.5M}$$

How much more energy is released by a magnitude 7.3 quake than a 5.5 quake?

18.* Use MATLAB to find the roots of $13x^3 + 182x^2 - 184x + 2503 = 0$.

19. Use MATLAB to find the roots of the polynomial $36x^3 + 12x^2 - 5x + 10$.

20. Determine which search path MATLAB uses on your computer. If you use a lab computer as well as a home computer, compare the two search paths. Where will MATLAB look for a user-created M-file on each computer?

21. Use MATLAB to plot the function $T = 6 \ln t - 7e^{0.2t}$ over the interval $1 \le t \le 3$. Put a title on the plot and properly label the axes. The variable T represents temperature in degrees Celsius; the variable t represents time in minutes.

22. Use MATLAB to plot the functions $u = 2 \log_{10}(60x + 1)$ and $v = 3 \cos(6x)$ over the interval $0 \le x \le 2$. Properly label the plot and each curve. The variables u and v represent speed in miles per hour; the variable x represents distance in miles.

23. The Fourier series is a series representation of a periodic function in terms of sines and cosines. The Fourier series representation of the function

$$f(x) = \begin{cases} 1 & 0 < x < \pi \\ -1 & -\pi < x < 0 \end{cases}$$

is

$$\frac{4}{\pi}\left(\frac{\sin x}{1} + \frac{\sin 3x}{3} + \frac{\sin 5x}{5} + \frac{\sin 7x}{7} + \cdots\right)$$

Plot on the same graph the function $f(x)$ and its series representation using the four terms shown.

24. A *cycloid* is the curve described by a point P on the circumference of a circular wheel of radius r rolling along the x axis. The curve is described in parametric form by the equations

$$x = r(\phi - \sin \phi)$$
$$y = r(1 - \cos \phi)$$

Use these equations to plot the cycloid for $r = 10$ inches and $0 \le \phi \le 4\pi$.

25. Use MATLAB to solve the following set of equations.

$$7x + 14y - 6z = 95$$
$$12x - 5y + 9z = -50$$
$$-5x + 7y + 15z = 145$$

26. It is known that the function $y = ax^3 + bx^2 + cx + d$ passes through the following (x, y) points: $(-1, 8)$, $(0, 4)$, $(1, 10)$, and $(2, 68)$. Use the MATLAB left division operator / to compute the coefficients a, b, c, and d by writing and solving four linear equations in terms of the four unknowns a, b, c, and d.

Section 1.4

27. Create, save, and run a script file that solves the following set of equations for given values of a, b, and c. Check your file for the case $a = 95$, $b = -50$, and $c = 145$.

$$7x + 14y - 6z = a$$
$$12x - 5y + 9z = b$$
$$-5x + 7y + 15z = c$$

28. A fence around a field is shaped as shown in Figure P28. It consists of a rectangle of length L and width W, and a right triangle that is symmetrical about the central horizontal axis of the rectangle. Suppose the width W is known (in meters), and the enclosed area A is known (in square meters). Write a MATLAB script file in terms of the given variables W and A to determine the length L required so that the enclosed area is A. Also determine the total length of fence required. Test your script for the values $W = 6$ m and $A = 80$ m^2.

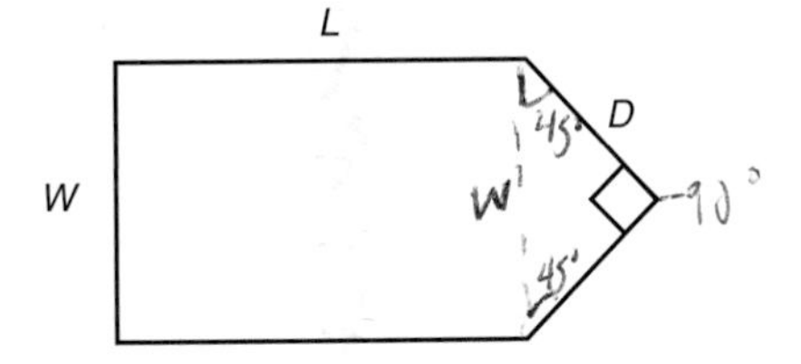

Figure P28

29. The four-sided figure shown in Figure P29 consists of two triangles having a common side a. The law of cosines for the top triangle states that

$$a^2 = b_1^2 + c_1^2 - 2b_1c_1 \cos A_1$$

and a similar equation can be written for the bottom triangle. Develop a procedure for computing the length of side c_2 if you are given the lengths of sides b_1, b_2, and c_1, and the angles A_1 and A_2 in degrees. Write a script file to implement this procedure. Test your script using the following values: $b_1 = 180$ m, $b_2 = 165$ m, $c_1 = 115$ m, $A_1 = 120°$ and $A_2 = 100°$.

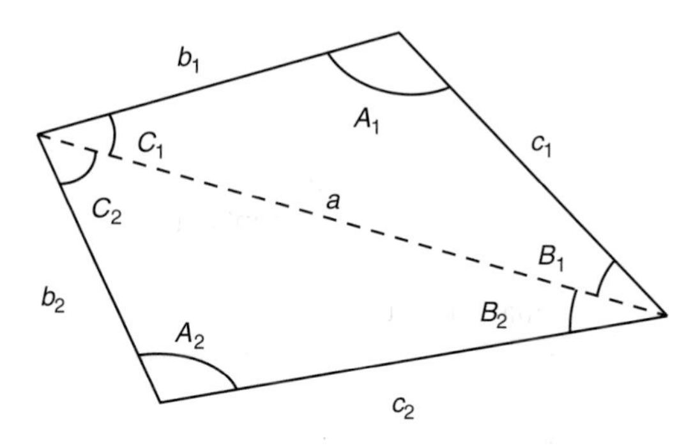

Figure P29

30. A fenced enclosure consists of a rectangle of length L and width $2R$, and a semicircle of radius R, as shown in Figure P30. The enclosure is to be built to have an area A of 1600 ft^2. The cost of the fence is \$40 per foot for the curved portion, and \$30 per foot for the straight sides. Use a plot to determine with a resolution of 0.01 ft the values of R and L required to minimize the total cost of the fence. Also compute the minimum cost, and use a plot of the cost versus R to analyze the sensitivity of the cost to a ± 20 percent change in R from its optimum value.

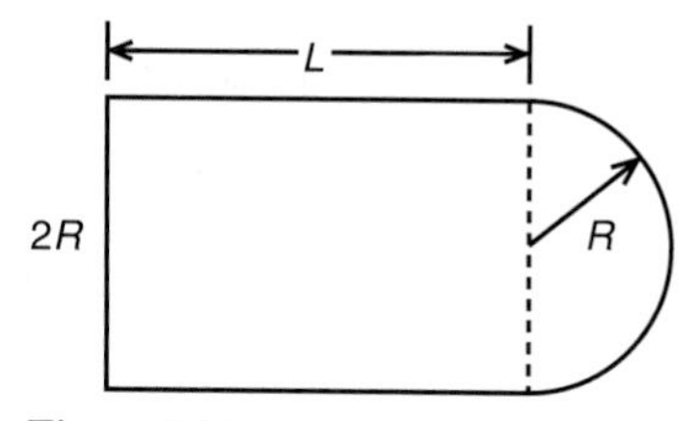

Figure P30

Section 1.5

31. Use the MATLAB help facilities to find information about the following topics and symbols: plot, label, cos, cosine, :, and *.

32. Use the MATLAB help facilities to determine what happens if you use the `sqrt` function with a negative argument.

33. Use the MATLAB help facilities to determine what happens if you use the `exp` function with an imaginary argument.

Section 1.6

34. Suppose that `x = [-15,-8,9,8,5]` and `y = [-20,12,-4,8,9]`. What is the result of the following operations? Determine the answers by hand, and then use MATLAB to check your answers.

a. `z = (x < y)`
b. `z = (x > y)`
c. `z = (x ~= y)`

d. `z = (x == y)`
e. `z = (y > -4)`

35. Suppose that `x = [-15,-8,9,8,5]` and `y = [-20,12,-4,8,9]`. Use MATLAB to find the values and the indices of the elements in `x` that are greater than the corresponding elements in `y`.

36. Write a script file using conditional statements to evaluate the following function, assuming that the scalar variable `x` has a value. The function is $y = e^{x+1}$ for $x < -1$, $y = 2 + \cos(\pi x)$ for $-1 \leq x < 5$, and $y = 10(x - 5) + 1$ for $x \geq 5$. Use your file to evaluate y for $x = -5$, $x = 3$, and $x = 15$, and check the results by hand.

37. Use a `for` loop to plot the function given in Problem 36 over the interval $-2 \leq x \leq 6$. Properly label the plot. The variable y represents height in kilometers, and the variable x represents time in seconds.

38. Plot the function $y = 10(1 - e^{-x/4})$ over the interval $0 \leq x \leq x_{max}$, using a `while` loop to determine the value of x_{max} such that $y(x_{max}) = 9.8$. Properly label the plot. The variable y represents force in newtons, and the variable x represents time in seconds.

39. Use a `for` loop to determine the sum of the first 10 terms in the series $5k^3$, $k = 1, 2, 3, \ldots 10$.

40. Use a `while` loop to determine how many terms in the series 2^k, $k = 1, 2, 3, \ldots,$ are required for the sum of the terms to exceed 2000. What is the sum for this number of terms?

41. One bank pays 5.5 percent annual interest, while a second bank pays 4.5 percent annual interest. Determine how much longer it will take to accumulate at least $50,000 in the second bank account if you deposit $1000 initially, and $1000 at the end of each year.

Section 1.7

42. *a.* With what initial speed must you throw a ball vertically for it to reach a height of 20 ft? The ball weighs 1 lb. How does your answer change if the ball weighs 2 lb?

b. Suppose you wanted to throw a steel bar vertically to a height of 20 ft. The bar weighs 2 lb. How much initial speed must the bar have to reach this height? Discuss how the length of the bar affects your answer.

43. Consider the motion of the piston discussed in Example 1.7–1. The piston *stroke* is the total distance moved by the piston as the crank angle varies from 0° to 180°.

a. How does the piston stroke depend on L_1 and L_2?

b. Suppose $L_2 = 0.5$ ft. Use MATLAB to plot the piston motion versus crank angle for two cases: $L_1 = 0.6$ ft and $L_1 = 1.4$ ft. Compare each plot with the plot shown in Figure 1.7–3. Discuss how the shape of the plot depends on the value of L_1.

Photo courtesy of Stratosphere Corporation.

Engineering in the 21st Century...

Innovative Construction

We tend to remember the great civilizations of the past in part by their public works, such as the Egyptian pyramids and the medieval cathedrals of Europe, which were technically challenging to create. Perhaps it is in our nature to "push the limits," and we admire others who do so. The challenge of innovative construction continues today. As space in our cities becomes scarce, many urban planners prefer to build vertically rather than horizontally. The newest tall buildings push the limits of our abilities, not only in structural design but also in areas that we might not think of, such as elevator design and operation, aerodynamics, and construction techniques. The photo above shows the 1149-ft-high Las Vegas Stratosphere Tower, the tallest observation tower in the United States. It required many innovative techniques in its assembly. The construction crane shown in use is 400 ft tall.

Designers of buildings, bridges, and other structures will use new technologies and new materials, some based on nature's designs. Pound for pound, spider silk is stronger than steel, and structural engineers hope to use cables of synthetic spider silk fibers to build earthquake-resistant suspension bridges. *Smart* structures, which can detect impending failure from cracks and fatigue, are now close to reality, as are *active* structures that incorporate powered devices to counteract wind and other forces. The MATLAB Financial toolbox is useful for financial evaluation of large construction projects, and the MATLAB Partial Differential Equation toolbox can be used for structural design. ■

CHAPTER 2

Numeric, Cell, and Structure Arrays

OUTLINE

The MATLAB sessions in Chapter 1 used scalar arithmetic to acquaint you with the MATLAB Command window, its Figure window, and a text editor window. In this chapter we begin to explore MATLAB commands in more depth. One of the strengths of MATLAB is the capability to handle collections of numbers, called *arrays,* as if they were a single variable. For example, when we add two arrays `A` and `B` by typing the single command `C = A + B`, MATLAB automatically adds all the corresponding numbers in `A` and `B` to produce `C`. In most other programming languages, this operation requires more than one command. The array-handling feature means that MATLAB programs can be very short. Thus they are easy to create, read, and document.

The MATLAB array capabilities make it a natural choice for engineering problems that require a set of data analyzed. If you have been using a spreadsheet for data analysis, you may find that MATLAB is an easier and more powerful tool for such work.

The array is the basic building block in MATLAB. We explain how to create, address, and edit arrays and how to use an important array operation, called *element-by-element* operation, to perform addition, subtraction, multiplication, division, and exponentiation to solve practical problems. We then introduce *matrix* operations, which are performed differently than element-by-element operations, and have their own applications. You probably have performed algebra with polynomials by hand. We explain how to use arrays in MATLAB to do polynomial algebra and root finding.

The following classes of arrays are now available in MATLAB 7:

Array						
numeric	character	logical	cell	structure	function handle	Java

So far we have used only numeric arrays, which are arrays containing only numeric values. Within the numeric class are the subclasses *single* (single precision), *double* (double precision), *int8*, *int16*, and *int32* (signed 8-bit, 16-bit, and 32-bit integers), and *uint8*, *uint16*, and *uint32* (unsigned 8-bit, 16-bit, and 32-bit integers). A character array is an array containing strings. The elements of logical arrays are "true" or "false," which, although represented by the symbols 1 and 0, are not numeric quantities. We will study the logical arrays in Chapter 4. Cell arrays and structure arrays are covered in Sections 2.6 and 2.7 of this chapter. Function handles are treated in Chapter 3. The Java class is not covered in this text.

We introduce two new data structures: *cell* arrays and *structure* arrays. These data structures enable one array to store different types of data (for example, string data, real numbers, and complex numbers). With cell arrays you can access such data by its location, but with structure arrays you can access it by name also. This feature enables you to create and use databases having different types of information (for example, a list of people's names, their addresses, and their phone numbers). We introduce these structures in Sections 2.6 and 2.7.

2.1 Arrays

We can represent the location of a point in three-dimensional space by three Cartesian coordinates x, y, and z. As shown in Figure 2.1–1, these three coordinates specify the *vector* **p**. (In mathematical text we often use boldface type to indicate vectors.) The set of *unit vectors* **i**, **j**, **k**, whose lengths are 1 and whose directions coincide with the x, y, and z axes, respectively, can be used to express the vector mathematically as follows: $\mathbf{p} = x\mathbf{i} + y\mathbf{j} + z\mathbf{k}$. The unit vectors enable us to associate the vector components x, y, z with the proper coordinate axes; therefore, when we write $\mathbf{p} = 5\mathbf{i} + 7\mathbf{j} + 2\mathbf{k}$, we know that the x, y, and z coordinates of the vector are 5, 7, and 2, respectively. We can also write the components in a specific order, separate them with a space, and identify the group with brackets, as follows: [5 7 2]. As long as we agree that the vector components will be written in the order x, y, z, we can use this notation instead of the unit-vector notation. In fact, MATLAB uses this style for vector notation. MATLAB allows us to separate

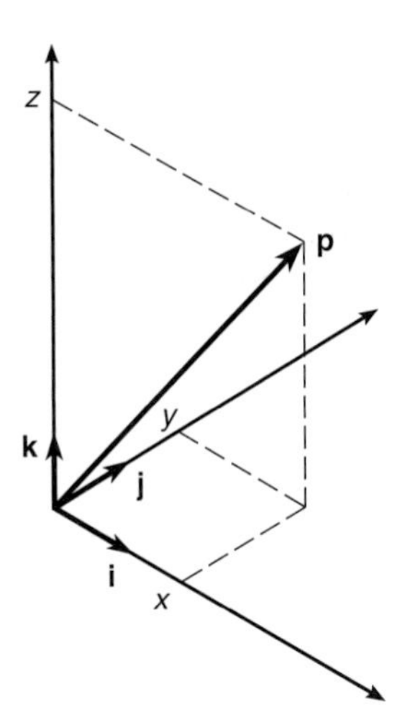

Figure 2.1–1 Specification of a position vector using Cartesian coordinates.

the components with commas for improved readability if we desire so that the equivalent way of writing the preceding vector is [5, 7, 2]. This expression is a *row* vector, which is a horizontal arrangement of the elements.

ROW VECTOR

We can also express the vector as a *column* vector, which has a vertical arrangement, as follows:

COLUMN VECTOR

$$\begin{bmatrix} 5 \\ 7 \\ 2 \end{bmatrix}$$

A vector can have only one column, or only one row. Thus, a vector is a special case of an array. In general, arrays can have more than one column and more than one row.

Creating Vectors in MATLAB

Although a position vector cannot have more than three components, the concept of a vector can be generalized to any number of components. In MATLAB a vector is simply a list of scalars, whose order of appearance in the list might be significant, as it is when specifying xyz coordinates. As another example, suppose we measure the temperature of an object once every hour. We can represent the measurements as a vector, and the 10th element in the list is the temperature measured at the 10th hour.

To create a row vector in MATLAB, you simply type the elements inside a pair of square brackets, separating the elements with a space or a comma. Brackets are required for arrays in some cases, but not all. To improve readability, we will always use them. The choice between a space or comma is a matter of personal preference, although the chance of an error is less if you use a comma. (You can also use a comma followed by a space for maximum readability.)

TRANSPOSE

To create a column vector, you can separate the elements by semicolons; alternatively, you can create a row vector and then use the *transpose* notation ('), which converts a row vector into a column vector, or vice versa. For example:

```
>>g = [3;7;9]
g =
    3
    7
    9
>>g = [3,7,9]'
g =
    3
    7
    9
```

The third way to create a column vector is to type a left bracket ([ ) and the first element, press **Enter,** type the second element, press **Enter,** and so on until you type the last element followed by a right bracket ( ]) and **Enter.** On the screen this sequence looks like

```
>>g = [3
7
9]
g =
    3
    7
    9
```

Note that MATLAB displays row vectors horizontally and column vectors vertically.

You can create vectors by "appending" one vector to another. For example, to create the row vector `u` whose first three columns contain the values of `r = [2,4,20]` and whose fourth, fifth, and sixth columns contain the values of `w = [9,-6,3]`, you type `u = [r,w]`. The result is the vector `u = [2,4,20,9,-6,3]`.

The colon operator (:) easily generates a large vector of regularly spaced elements. Typing

```
>>x = [m:q:n]
```

creates a vector `x` of values with a spacing `q`. The first value is `m`. The last value is `n` if `m - n` is an integer multiple of `q`. If not, the last value is less than `n`. For example, typing `x = [0:2:8]` creates the vector `x = [0,2,4,6,8]`, whereas typing `x = [0:2:7]` creates the vector `x = [0,2,4,6]`. To create a row vector `z` consisting of the values from 5 to 8 in steps of 0.1, you type `z = [5:0.1:8]`. If the increment `q` is omitted, it is presumed to be 1. Thus `y = [-3:2]` produces the vector `y = [-3,-2,-1,0,1,2]`.

The increment `q` can be negative. In this case `m` should be greater than `n`. For example, `u = [10:-2:4]` produces the vector `[10,8,6,4]`.

The `linspace` command also creates a linearly spaced row vector, but instead you specify the number of values rather than the increment. The syntax is `linspace(x1,x2,n)`, where `x1` and `x2` are the lower and upper limits and `n` is the number of points. For example, `linspace(5,8,31)` is equivalent to `[5:0.1:8]`. If `n` is omitted, the spacing is 1.

The `logspace` command creates an array of logarithmically spaced elements. Its syntax is `logspace(a,b,n)`, where `n` is the number of points between 10^a and 10^b. For example, `x = logspace(-1,1,4)` produces the vector `x = [0.1000, 0.4642, 2.1544, 10.000]`. If `n` is omitted, the number of points defaults to 50.

Two-Dimensional Arrays

An array is a collection of scalars arranged in a logical structure. An array may have a single row; if so, it is called a row vector. A column vector has a single column. An array can have multiple rows, multiple columns, or both. Such a two-dimensional array is called a *matrix*. In mathematical text we often use boldface type to indicate vectors and matrices, for example, **A**. If possible, vectors are usually denoted by lowercase letters and matrices by uppercase letters. An example of a matrix having three rows and two columns is

MATRIX

$$\mathbf{M} = \begin{bmatrix} 2 & 5 \\ -3 & 4 \\ -7 & 1 \end{bmatrix}$$

A matrix should not be confused with a determinant, which also has rows and columns but can be reduced to a *single* number. Two parallel lines usually denote a determinant; square brackets usually denote matrices.

ARRAY SIZE

We refer to the *size* of an array by the number of rows and the number of columns. For example, an array with 3 rows and 2 columns is said to be a 3×2 array. *The number of rows is always stated first!* A row vector has a size of $1 \times n$ where n is the number of columns. A column vector has a size of $n \times 1$ where n is the number of rows.

We sometimes represent a matrix **A** as $[a_{ij}]$ to indicate its elements a_{ij}. The subscripts i and j—called *indices*—indicate the row and column location of the element a_{ij}. *The row number must always come first!* For example, the element a_{32} is in row 3, column 2. Two matrices **A** and **B** are equal if they have the same size and if all their corresponding elements are equal; that is, $a_{ij} = b_{ij}$ for every value of i and j.

Creating Matrices

The most direct way to create a matrix is to type the matrix row by row, separating the elements in a given row with spaces or commas and separating the rows with semicolons. For example, typing

```
>>A = [2,4,10;16,3,7];
```

creates the following matrix:

$$\mathbf{A} = \begin{bmatrix} 2 & 4 & 10 \\ 16 & 3 & 7 \end{bmatrix}$$

Remember, spaces or commas separate elements in different *columns,* whereas semicolons separate elements in different *rows*.

If the matrix has many elements, you can press **Enter** and continue typing on the next line. MATLAB knows you are finished entering the matrix when you type the closing bracket (`]`).

You can also create a matrix from row or column vectors. A row vector `r` can be appended to a matrix `A` if `r` and `A` have the same number of columns. The command `B = [A r]` appends the row vector `r` to the matrix `A`. This increases the number of columns in `A`. Use a semicolon to increase the number of rows. The command `B = [A;r]` appends the row vector `r` to the matrix `A`. Note the difference between the results given by `[a b]` and `[a;b]` in the following session:

```
>>a = [1,3,5]
a =
   1  3  5
>>b = [7,9,11]
b =
   7  9  11
>>c = [a b];
c =
   1  3  5  7  9  11
>>D = [a;b]
D =
   1  3  5
   7  9  11
```

You need not use symbols to create a new array. For example, typing `D = [[1,3,5];[7,9,11]];` produces the same result as typing `D = [a;b]`.

Matrices and the Transpose Operation

The transpose operation interchanges the rows and columns. In mathematics text we denote this operation by the superscript T. For an $m \times n$ matrix $\mathbf{A}$ with m rows and n columns, $\mathbf{A}^T$ (read "A transpose") is an $n \times m$ matrix. For example, if

$$\mathbf{A} = \begin{bmatrix} -2 & 6 \\ -3 & 5 \end{bmatrix}$$

then

$$\mathbf{A}^T = \begin{bmatrix} -2 & -3 \\ 6 & 5 \end{bmatrix}$$

If $\mathbf{A}^T = \mathbf{A}$, the matrix **A** is *symmetric*. Only a square matrix can be symmetric, but not every square matrix is symmetric.

If the array contains complex elements, the transpose operator (`'`) produces the *complex conjugate transpose;* that is, the resulting elements are the complex conjugates of the original array's transposed elements. Alternatively, you can use the *dot transpose* operator (`.'`) to transpose the array without producing complex conjugate elements, for example, `A.'`. If all the elements are real, the operators `'` and `.'` give the same result.

Array Addressing

Array indices are the row and column numbers of an element in an array and are used to keep track of the array's elements. For example, the notation `v(5)` refers to the fifth element in the vector `v`, and `A(2,3)` refers to the element in row 2, column 3 in the matrix `A`. *The row number is always listed first!* This notation enables you to correct entries in an array without retyping the entire array. For example, to change the element in row 1, column 3 of a matrix **D** to 6, you can type `D(1,3) = 6`.

The colon operator selects individual elements, rows, columns, or "subarrays" of arrays. Here are some examples:

- `v(:)` represents all the row or column elements of the vector `v`.
- `v(2:5)` represents the second through fifth elements; that is `v(2)`, `v(3)`, `v(4)`, `v(5)`.
- `A(:,3)` denotes all the elements in the third column of the matrix `A`.
- `A(:,2:5)` denotes all the elements in the second through fifth columns of `A`.
- `A(2:3,1:3)` denotes all the elements in the second and third rows that are also in the first through third columns.

You can use array indices to extract a smaller array from another array. For example, if you create the array **B**

$$\mathbf{B} = \begin{bmatrix} 2 & 4 & 10 & 13 \\ 16 & 3 & 7 & 18 \\ 8 & 4 & 9 & 25 \\ 3 & 12 & 15 & 17 \end{bmatrix} \tag{2.1–1}$$

by typing

```
>>B = [2,4,10,13;16,3,7,18;8,4,9,25;3,12,15,17];
```

and then type

```
>>C = B(2:3,1:3);
```

you can produce the following array:

$$\mathbf{C} = \begin{bmatrix} 16 & 3 & 7 \\ 8 & 4 & 9 \end{bmatrix}$$

EMPTY ARRAY

The *empty* or *null* array contains no elements and is expressed as `[]`. Rows and columns can be deleted by setting the selected row or column equal to the null array. This step causes the original matrix to collapse to a smaller one. For example, `A(3,:) = []` deletes the third row in `A`, while `A(:,2:4) = []` deletes the second through fourth columns in `A`. Finally, `A([1 4],:) = []` deletes the first and fourth rows of `A`.

Suppose we type `A = [6,9,4;1,5,7]` to define the following matrix:

$$\mathbf{A} = \begin{bmatrix} 6 & 9 & 4 \\ 1 & 5 & 7 \end{bmatrix}$$

Typing `A(1,5) = 3` changes the matrix to

$$\mathbf{A} = \begin{bmatrix} 6 & 9 & 4 & 0 & 3 \\ 1 & 5 & 7 & 0 & 0 \end{bmatrix}$$

Because **A** did not have five columns, its size is automatically expanded to accept the new element in column 5. MATLAB adds zeros to fill out the remaining elements.

MATLAB does not accept negative or zero indices, but you can use negative increments with the colon operator. For example, typing `B = A(:,5:-1:1)` reverses the order of the columns in **A** and produces

$$\mathbf{B} = \begin{bmatrix} 3 & 0 & 4 & 9 & 6 \\ 0 & 0 & 7 & 5 & 1 \end{bmatrix}$$

Suppose that `C = [-4,12,3,5,8]`. Then typing `B(2,:) = C` replaces row 2 of `B` with `C`. Thus **B** becomes

$$\mathbf{B} = \begin{bmatrix} 3 & 0 & 4 & 9 & 6 \\ -4 & 12 & 3 & 5 & 8 \end{bmatrix}$$

Suppose that `D = [3,8,5;2,-6,9]`. Then typing `E = D([2,2,2],:)` repeats row 2 of `D` three times to obtain

$$\mathbf{E} = \begin{bmatrix} 2 & -6 & 9 \\ 2 & -6 & 9 \\ 2 & -6 & 9 \end{bmatrix}$$

Using `clear` to Avoid Errors

You can use the `clear` command to protect yourself from accidentally reusing an array that has the wrong dimension. Even if you set new values for an array, some previous values might still remain. For example, suppose you had previously used the 2×2 array `A = [2, 5; 6, 9]`, and you then create the 5×1 arrays `x = [1:5]'` and `y = [2:6]'`. Suppose you now redefine `A` so that its columns will be x and y. If you then type A(:,1) = x to create the first column, MATLAB displays an error message telling you that the number of rows in A and x must be the same. MATLAB thinks A should be a 2 × 2 matrix because A was previously defined to have only two rows and its values remain in memory. The clear command wipes A and all other variables from memory and avoids this error. To clear A only, type clear A before typing A(:,1) = x.

Some Useful Array Functions

MATLAB has many functions for working with arrays (see Table 2.1–1). Here is a summary of some of the more commonly used functions.

The max(A) function returns the algebraically greatest element in **A** if **A** is a vector having all real elements. It returns a row vector containing the greatest elements in each *column* if **A** is a matrix containing all real elements. If *any* of the elements are complex, max(A) returns the element that has the largest

Table 2.1–1 Array functions

Command	Description
cat(n,A,B,C, ...)	Creates a new array by concatenating the arrays A, B, C, and so on along the dimension n.
find(x)	Computes an array containing the indices of the nonzero elements of the array **x**.
[u,v,w] = find(A)	Computes the arrays **u** and **v**, containing the row and column indices of the nonzero elements of the matrix A, and the array **w**, containing the values of the nonzero elements. The array **w** may be omitted.
length(A)	Computes either the number of elements of **A** if **A** is a vector or the largest value of m or n if **A** is an $m \times n$ matrix.
linspace(a,b,n)	Creates a row vector of n regularly spaced values between a and b.
logspace(a,b,n)	Creates a row vector of n logarithmically spaced values between a and b.
max(A)	Returns the algebraically largest element in **A** if **A** is a vector. Returns a row vector containing the largest elements in each column if **A** is a matrix. If any of the elements are complex, max(A) returns the elements that have the largest magnitudes.
[x,k] = max(A)	Similar to max(A) but stores the maximum values in the row vector **x** and their indices in the row vector **k**.
min(A)	Same as max(A) but returns minimum values.
[x,k] = min(A)	Same as [x,k] = max(A) but returns minimum values.
size(A)	Returns a row vector [m n] containing the sizes of the $m \times n$ array **A**.
sort(A)	Sorts each column of the array **A** in ascending order and returns an array the same size as **A**.
sum(A)	Sums the elements in each column of the array **A** and returns a row vector containing the sums.

magnitude. The syntax `[x,k] = max(A)` is similar to `max(A)`, but it stores the maximum values in the row vector **x** and their indices in the row vector **k**.

The functions `min(A)` and `[x,k] = min(A)` are the same as `max(A)` and `[x,k] = max(A)` except that they return minimum values.

The function `size(A)` returns a row vector `[m n]` containing the sizes of the $m \times n$ array **A**. The `length(A)` function computes either the number of elements of **A** if `A` is a vector or the largest value of m or n if **A** is an $m \times n$ matrix.

For example, if

$$\mathbf{A} = \begin{bmatrix} 6 & 2 \\ -10 & -5 \\ 3 & 0 \end{bmatrix}$$

then `max(A)` returns the vector `[6,2]`; `min(A)` returns the vector `[-10,-5]`; `size(A)` returns `[3,2]`; and `length(A)` returns `3`. If `A` has one or more complex elements, `max(A)` returns the element that has the largest magnitude. For example, if

$$\mathbf{A} = \begin{bmatrix} 6 & 2 \\ -10 & -5 \\ 3+4i & 0 \end{bmatrix}$$

then `max(A)` returns the vector `[-10,-5]` and `min(A)` returns the vector `[3+4i,0]`. (The magnitude of $3+4i$ is 5.)

The `sum(A)` function sums the elements in each *column* of the array **A** and returns a row vector containing the sums. The `sort(A)` function sorts each *column* of the array **A** in ascending order and returns an array the same size as **A**.

The `find(x)` command computes an array containing the indices of the *nonzero* elements of the vector **x**. The syntax `[u,v,w] = find(A)` computes the arrays **u** and **v**, containing the row and column indices of the nonzero elements of the matrix **A**, and the array **w**, containing the values of the nonzero elements. The array **w** may be omitted.

For example, if

$$\mathbf{A} = \begin{bmatrix} 6 & 0 & 3 \\ 0 & 4 & 0 \\ 2 & 7 & 0 \end{bmatrix}$$

then the session

```
>>A = [6, 0, 3; 0, 4, 0; 2, 7, 0];
>>[u, v, w] = find(A)
```

returns the vectors

$$\mathbf{u} = \begin{bmatrix} 1 \\ 3 \\ 2 \\ 3 \\ 1 \end{bmatrix} \quad \mathbf{v} = \begin{bmatrix} 1 \\ 1 \\ 2 \\ 2 \\ 3 \end{bmatrix} \quad \mathbf{w} = \begin{bmatrix} 6 \\ 2 \\ 4 \\ 7 \\ 3 \end{bmatrix}$$

The vectors **u** and **v** give the (row, column) indices of the nonzero values, which are listed in **w**. For example, the second entries in **u** and **v** give the indices (3, 1), which specifies the element in row 3, column 1 of **A**, whose value is 2.

These functions are summarized in Table 2.1–1.

Magnitude, Length, and Absolute Value of a Vector

The terms *magnitude, length,* and *absolute value* are often loosely used in everyday language, but you must keep their precise meaning in mind when using MATLAB. The MATLAB `length` command gives the number of elements in the vector. The *magnitude* of a vector **x** having elements $x_1, x_2, \ldots, x_n$ is a scalar, given by $\sqrt{x_1^2 + x_2^2 + \cdots + x_n^2}$, and is the same as the vector's geometric length. The *absolute value* of a vector **x** is a vector whose elements are the absolute values of the elements of **x**. For example, if `x = [2,-4,5]`, its length is 3; its magnitude is $\sqrt{2^2 + (-4)^2 + 5^2} = 6.7082$; and its absolute value is `[2,4,5]`.

Test Your Understanding

T2.1–1 For the matrix **B**, find the array that results from the operation `[B;B']`. Use MATLAB to determine what number is in row 5, column 3 of the result.

$$\mathbf{B} = \begin{bmatrix} 2 & 4 & 10 & 13 \\ 16 & 3 & 7 & 18 \\ 8 & 4 & 9 & 25 \\ 3 & 12 & 15 & 17 \end{bmatrix}$$

T2.1–2 For the same matrix **B**, use MATLAB to (a) find the largest and smallest element in **B** and their indices and (b) sort each column in **B** to create a new matrix **C**.

The Array Editor

In Chapter 1 we saw some commands such as `who`, `whos`, and `exist`, that are useful for managing the workspace. The MATLAB Workspace Browser provides a graphical interface for managing the workspace. You can use the Workspace Browser to view, save, and clear workspace variables. It also includes the *Array Editor,* a graphical interface for working with arrays. To open the Workspace Browser, do one of the following:

- From the **Desktop** menu in the MATLAB Desktop, select **Workspace;**
- Click the Workspace tab below the Current Directory window; or
- Type `workspace` at the Command window prompt.

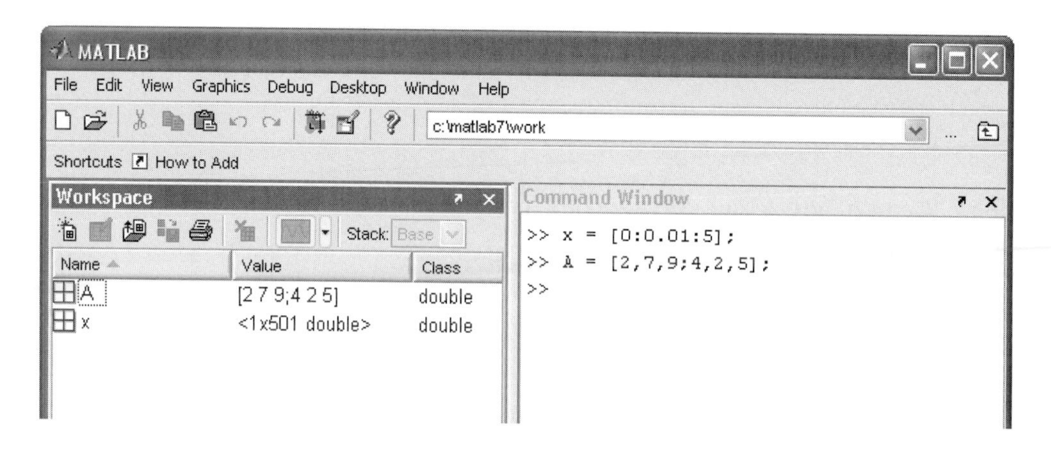

Figure 2.1–2 The Workspace Browser.

The browser appears as shown in Figure 2.1–2. Workspace operations you can perform with the Workspace Browser or with functions are:

1. Viewing and editing the workspace variables.
2. Clearing workspace variables.
3. Plotting workspace variables.
4. Saving the workspace.
5. Loading a saved workspace.
6. Viewing base and function workspaces by using the stack.

Here we discuss the first three operations. The last three operations are discussed in Chapters 3 and 4.

Keep in mind that the Desktop menus are context-sensitive. Thus their contents will change depending on which features of the Browser and Array Editor you are currently using. The Workspace Browser shows the name of each variable, its value, array size, size in bytes, and class. The icon for each variable illustrates its class. To resize the columns of information, drag the column header borders. To show or hide any of the columns, or to specify the sort order, use the **View** menu.

From the Workspace Browser you can open the Array Editor to view and edit a visual representation of two-dimensional numeric arrays, with the rows and columns numbered. To open the Array Editor from the Workspace Browser, double-click on the variable you want to open. The Array Editor opens, displaying the values for the selected variable. The Array Editor appears as shown in Figure 2.1–3.

To open a variable, you can also right-click it and use the **Context** menu. Repeat the steps to open additional variables into the Array Editor. In the Array

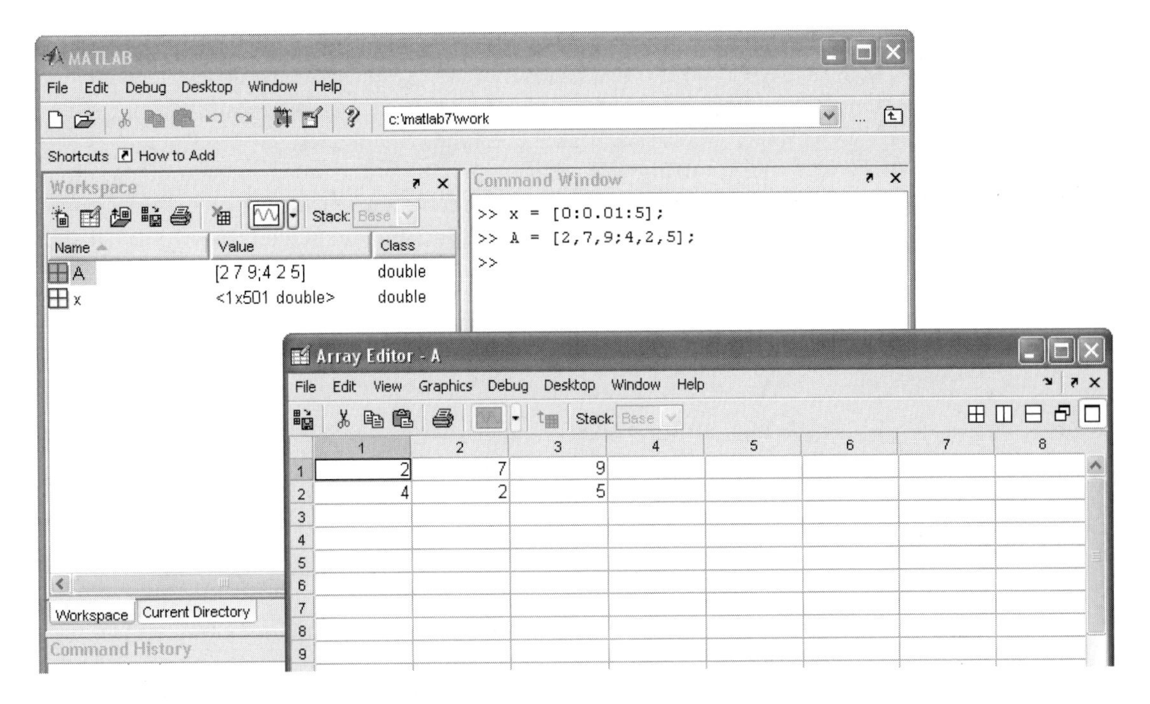

Figure 2.1–3 The Array Editor.

Editor, access each variable via its tab at the bottom of the window, or use the **Window** menu. You can also open the Array Editor directly from the Command window by typing `open('var')`, where `var` is the name of the variable to be edited. Once an array is displayed in the Array Editor, you can change a value in the array by clicking on its location, typing in the new value, and pressing **Enter.**

Right-clicking on a variable brings up the **Context** menu, which can be used to edit, save, or clear the selected variable, or to plot the rows of the variable versus its columns (this type of plot is discussed in Chapter 5).

You can also clear a variable from the Workspace Browser by first highlighting it in the Browser, then clicking on **Delete** in the **Edit** menu.

2.2 Multidimensional Arrays

MATLAB supports multidimensional arrays. Here we present just some of the MATLAB capabilities for such arrays; to obtain more information, type `help datatypes`.

A three-dimensional array has the dimension $m \times n \times q$. A four-dimensional array has the dimension $m \times n \times q \times r$, and so forth. The first two dimensions are the row and column, as with a matrix. The higher dimensions are called *pages*. The elements of a matrix are specified by two indices; three indices are required to specify an element of a three-dimensional array, and so on. You can think of a three-dimensional array as layers of matrices. The first layer is page 1; the second layer is page 2, and so on. If A is a $3 \times 3 \times 2$ array, you can access the element in row 3, column 2 of page 2 by typing `A(3,2,2)`. To access all of page 1, type `A(:,:,1)`. To access all of page 2, type `A(:,:,2)`.

The `ndims` command returns the number of dimensions. For example, for the array A just described, `ndims(A)` returns the value 3.

You can create a multidimensional array by first creating a two-dimensional array and then extending it. For example, suppose you want to create a three-dimensional array whose first page is

$$\begin{bmatrix} 4 & 6 & 1 \\ 5 & 8 & 0 \\ 3 & 9 & 2 \end{bmatrix}$$

and whose second page is

$$\begin{bmatrix} 6 & 2 & 9 \\ 0 & 3 & 1 \\ 4 & 7 & 5 \end{bmatrix}$$

To do so, first create page 1 as a 3×3 matrix and then add page 2, as follows:

```
>>A = [4,6,1;5,8,0;3,9,2];
>>A(:,:,2) = [6,2,9;0,3,1;4,7,5]
```

MATLAB displays the following:

```
A(:,:,1) =
    4    6    1
    5    8    0
    3    9    2
A(:,:,2) =
    6    2    9
    0    3    1
    4    7    5
```

Another way to produce such an array is with the `cat` command. Typing `cat(n,A,B,C, ...)` creates a new array by concatenating the arrays A, B, C, and so on along the dimension n. Note that `cat(1,A,B)` is the same as `[A;B]` and that `cat(2,A,B)` is the same as `[A,B]`. For example, suppose we have the 2×2 arrays **A** and **B**:

$$\mathbf{A} = \begin{bmatrix} 8 & 2 \\ 9 & 5 \end{bmatrix} \qquad \mathbf{B} = \begin{bmatrix} 4 & 6 \\ 7 & 3 \end{bmatrix}$$

Then `C = cat(3,A,B)` produces a three-dimensional array. We can think of this array as composed of two layers; the first layer is the matrix A, and the second layer is the matrix B. The element `C(m,n,p)` is located in row m, column n, and layer p. Thus the element `C(2,1,1)` is 9, and the element `C(2,2,2)` is 3. This function is summarized in Table 2.1–1.

Multidimensional arrays are useful for problems that involve several parameters. For example, if we have data on the temperature distribution in a rectangular object, we could represent the temperatures as an array `T` with three dimensions. Each temperature would correspond to the temperature of a rectangular block within the object, and the array indices would correspond to x, y, z locations within the object. For example, `T(2,4,3)` would be the temperature in the block located in row 2, column 4, page 3 and having the coordinates x_2, y_4, z_3.

2.3 Element-by-Element Operations

To increase the magnitude of a vector, multiply it by a scalar. For example, to double the magnitude of the vector `r = [3,5,2]`, multiply each component by two to obtain `[6,10,4]`. In MATLAB you type `v = 2*r`. See Figure 2.3–1 for the geometric interpretation of scalar multiplication of a vector in three-dimensional space.

Multiplying a matrix **A** by a scalar w produces a matrix whose elements are the elements of **A** multiplied by w. For example:

$$3\begin{bmatrix} 2 & 9 \\ 5 & -7 \end{bmatrix} = \begin{bmatrix} 6 & 27 \\ 15 & -21 \end{bmatrix}$$

This multiplication is performed in MATLAB as follows:

```
>>A = [2,9;5,-7];
>>3*A
ans =
     6    27
    15   -21
```

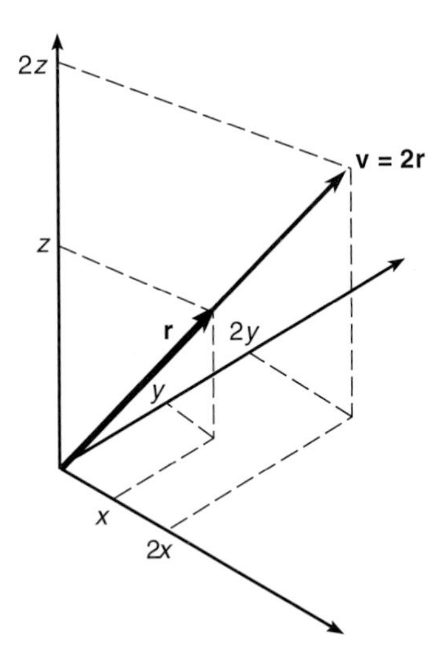

Figure 2.3–1 Geometric interpretation of scalar multiplication of a vector.

ARRAY OPERATIONS

MATRIX OPERATIONS

Thus multiplication of an array by a scalar is easily defined and easily carried out. However, multiplication of two *arrays* is not so straightforward. In fact, MATLAB uses two definitions of multiplication: (1) array multiplication and (2) matrix multiplication. Division and exponentiation must also be carefully defined when you are dealing with operations between two arrays. MATLAB has two forms of arithmetic operations on arrays. In this section we introduce one form, called *array* operations, which are also called *element-by-element* operations. In the next section we introduce *matrix* operations. Each form has its own applications, which we illustrate by examples.

Array Addition and Subtraction

Vector addition can be done either graphically (by using the parallelogram law in two dimensions (see Figure 2.3–2a), or analytically by adding the corresponding

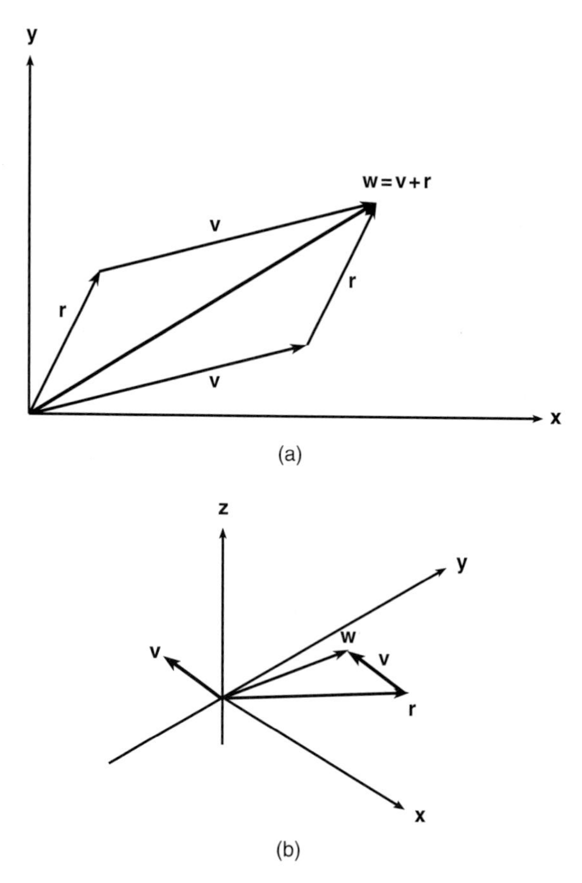

Figure 2.3–2 Vector addition. (a) The parallelogram law. (b) Addition of vectors in three dimensions.

components. To add the vectors `r = [3,5,2]` and `v = [2,-3,1]` to create `w` in MATLAB, you type `w = r + v`. The result is `w = [5,2,3]`. Figure 2.3–2b illustrates vector addition in three dimensions.

When two arrays have identical size, their sum or difference has the same size and is obtained by adding or subtracting their corresponding elements. Thus $\mathbf{C} = \mathbf{A} + \mathbf{B}$ implies that $c_{ij} = a_{ij} + b_{ij}$ if the arrays are matrices. The array **C** has the same size as **A** and **B**. For example:

$$\begin{bmatrix} 6 & -2 \\ 10 & 3 \end{bmatrix} + \begin{bmatrix} 9 & 8 \\ -12 & 14 \end{bmatrix} = \begin{bmatrix} 15 & 6 \\ -2 & 17 \end{bmatrix} \tag{2.3–1}$$

Array subtraction is performed in a similar way.

The addition shown in equation 2.3–1 is performed in MATLAB as follows:

```
>>A = [6,-2;10,3];
>>B = [9,8;-12,14]
>>A+B
ans =
    15    6
    -2   17
```

Array addition and subtraction are associative and commutative. For addition these properties mean that

$$(\mathbf{A} + \mathbf{B}) + \mathbf{C} = \mathbf{A} + (\mathbf{B} + \mathbf{C}) \tag{2.3–2}$$

$$\mathbf{A} + \mathbf{B} + \mathbf{C} = \mathbf{B} + \mathbf{C} + \mathbf{A} = \mathbf{A} + \mathbf{C} + \mathbf{B} \tag{2.3–3}$$

Array addition and subtraction require that both arrays have the same size. The only exception to this rule in MATLAB occurs when we add or subtract a *scalar* to or from an array. In this case the scalar is added or subtracted from each element in the array. Table 2.3–1 gives examples.

Table 2.3–1 Element-by-element operations

Symbol	Operation	Form	Example
`+`	Scalar-array addition	`A + b`	`[6,3]+2=[8,5]`
`-`	Scalar-array subtraction	`A - b`	`[8,3]-5=[3,-2]`
`+`	Array addition	`A + B`	`[6,5]+[4,8]=[10,13]`
`-`	Array subtraction	`A - B`	`[6,5]-[4,8]=[2,-3]`
`.*`	Array multiplication	`A.*B`	`[3,5].*[4,8]=[12,40]`
`./`	Array right division	`A./B`	`[2,5]./[4,8]=[2/4,5/8]`
`.\`	Array left division	`A.\B`	`[2,5].\[4,8]=[2\4,5\8]`

`.^`	Array exponentiation	`A.^B`	`[3,5].^2=[3^2,5^2]`
			`2.^[3,5]=[2^3,2^5]`
			`[3,5].^[2,4]=[3^2,5^4]`

EXAMPLE 2.3–1 Vectors and Relative Velocity

A train is heading east at 60 mi/hr. A car approaches the track crossing heading northeast at 45 mi/hr on a road that makes a 55° angle with the track. See Figure 2.3–3. What is the velocity of the train relative to the car? What is the speed of the train relative to the car?

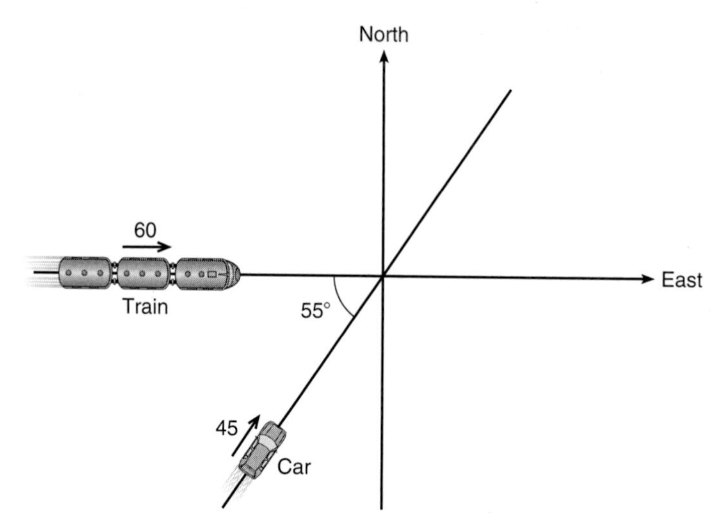

Figure 2.3–3

■ Solution

Velocity is a vector quantity consisting of speed and direction. Speed is the magnitude of the velocity vector. The train's velocity $\mathbf{v_R}$ relative to the car is the difference between the train's velocity $\mathbf{v_T}$ relative to the ground and the car's velocity $\mathbf{v_C}$ relative to the ground. Thus

$$\mathbf{v_R} = \mathbf{v_T} - \mathbf{v_C}$$

Choosing the x direction to be east, and the y direction north, we can write the following velocity vectors:

$$\mathbf{v_T} = 60\mathbf{i} + 0\mathbf{j}$$

$$\mathbf{v_C} = 45\cos(55^\circ)\mathbf{i} + 45\sin(55^\circ)\mathbf{j}$$

In MATLAB we can write these vectors as follows (remembering to convert 55° to radians) and compute $\mathbf{v_R}$:

```
>>v_T = [60, 0];
>>v_C = [45*cos(55*pi/180), 45*sin(55*pi/180)];
>>v_R = v_T-v_C
v_R =
     34.1891   -36.8618
```

Thus $\mathbf{v_R} = 34.1891\mathbf{i} - 36.8618\mathbf{j}$ mi/hr. The velocity of the train relative to the car is approximately 34 mi/hr to the east and 37 mi/hr to the south.

The relative speed s_R is the magnitude of $\mathbf{v_R}$, which can be found as

$$s_R = \sqrt{(34.1891)^2 + (-36.8618)^2} = 50.2761 \text{ mi/hr}$$

In MATLAB the speed can be calculated as follows:

```
>>s_R = sqrt(v_R(1)^2+v_R(2)^2)
```

We will soon see an easier way to compute s_R using array multiplication.

Array Multiplication

The data in Table 2.3–2 illustrates the difference between the two types of multiplication that are defined in MATLAB. The table gives the speed of an aircraft on each leg of a certain trip and the time spent on each leg.

We can define a row vector `s` containing the speeds and a row vector `t` containing the times for each leg. Thus `s = [200, 250, 400, 300]` and `t = [2, 5, 3, 4]`. To find the miles traveled on each leg, we multiply the speed by the time. To do so, we use the MATLAB symbol `.*`, which *specifies* the multiplication `s.*t` to produce the row vector whose elements are the products of the corresponding elements in `s` and `t`:

$$\texttt{s.*t} = [200(2), 250(5), 400(3), 300(4)] = [400, 1250, 1200, 1200]$$

With this notation the symbol `.*` signifies that each element in `s` is multiplied by the corresponding element in `t` and that the resulting products are used to form a row vector having the same number of elements as `s` and `t`. This vector contains the miles traveled by the aircraft on each leg of the trip.

If we had wanted to find only the total miles traveled, we could have used another definition of multiplication, denoted by `s*t'`. In this definition the product is the *sum* of the individual element products; that is

$$\texttt{s*t'} = [200(2) + 250(5) + 400(3) + 300(4)] = 4050$$

These two examples illustrate the difference between *array* multiplication `s.*t`—sometimes called *element-by-element* multiplication—and *matrix* multiplication `s*t'`. We examine matrix multiplication in more detail in Section 2.4.

Table 2.3–2 Aircraft speeds and times per leg

	Leg			
	1	2	3	4
Speed (mi/hr)	200	250	400	300
Time (hr)	2	5	3	4

MATLAB defines element-by-element multiplication only for arrays that have the same size. The definition of the product `x.*y`, where `x` and `y` each have n elements, is

```
x.*y = [x(1)y(1), x(2)y(2), ... , x(n)y(n)]
```

if `x` and `y` are row vectors. For example, if

$$\mathbf{x} = [2,\ 4,\ -5] \qquad \mathbf{y} = [-7,\ 3,\ -8] \tag{2.3–4}$$

then `z = x.*y` gives

$$\mathbf{z} = [2(-7),\ 4(3),\ -5(-8)] = [-14,\ 12,\ 40]$$

If `x` and `y` are column vectors, the result of `x.*y` is a column vector. For example `z = (x').*(y')` gives

$$\mathbf{z} = \begin{bmatrix} 2(-7) \\ 4(3) \\ -5(-8) \end{bmatrix} = \begin{bmatrix} -14 \\ 12 \\ 40 \end{bmatrix}$$

Note that `x'` is a column vector with size 3×1 and thus does not have the same size as `y`, whose size is 1×3. Thus for the vectors `x` and `y` the operations `x'.*y` and `y.*x'` are not defined in MATLAB and will generate an error message.

The generalization of array multiplication to arrays with more than one row or column is straightforward. Both arrays must have the same size. If they are matrices, they must have the same number of rows and the same number of columns. The array operations are performed between the elements in corresponding locations in the arrays. For example, the array multiplication operation `A.*B` results in a matrix `C` that has the same size as `A` and `B` and has the elements $c_{ij} = a_{ij}b_{ij}$. For example, if

$$\mathbf{A} = \begin{bmatrix} 11 & 5 \\ -9 & 4 \end{bmatrix} \qquad \mathbf{B} = \begin{bmatrix} -7 & 8 \\ 6 & 2 \end{bmatrix}$$

then `C = A.*B` gives this result:

$$\mathbf{C} = \begin{bmatrix} 11(-7) & 5(8) \\ -9(6) & 4(2) \end{bmatrix} = \begin{bmatrix} -77 & 40 \\ -54 & 8 \end{bmatrix}$$

With element-by-element multiplication, it is important to remember that the dot (`.`) and the asterisk (`*`) form *one* symbol (`.*`). It might have been better to have defined a single symbol for this operation, but the developers of MATLAB were limited by the selection of symbols on the keyboard.

Vectors and Displacement

EXAMPLE 2.3–2

Suppose two divers start at the surface and establish the following coordinate system: x is to the west, y is to the north, and z is down. Diver 1 swims 55 ft west, 36 ft north, and then dives 25 ft. Diver 2 dives 15 ft, then swims east 20 ft and then north 59 ft. (a) Find the distance between diver 1 and the starting point. (b) How far in each direction must diver 1 swim to reach diver 2? How far in a straight line must diver 1 swim to reach diver 2?

■ Solution

(a) Using the xyz coordinates selected, the position of diver 1 is $\mathbf{r} = 55\mathbf{i} + 36\mathbf{j} + 25\mathbf{k}$, and the position of diver 2 is $\mathbf{r} = -20\mathbf{i} + 59\mathbf{j} + 15\mathbf{k}$. (Note that diver 2 swam east, which is in the negative x direction.) The distance from the origin of a point xyz is given by $\sqrt{x^2 + y^2 + z^2}$, that is, by the magnitude of the vector pointing from the origin to the point xyz. This distance is computed in the following session.

```
>>r = [55,36,25];w = [-20,59,15];
>>dist1 = sqrt(sum(r.*r))
dist1 =
   70.3278
```

The distance is approximately 70 ft.

(b) The location of diver 2 relative to diver 1 is given by the vector $\mathbf{v}$ pointing from diver 1 to diver 2. We can find this vector using vector subtraction: $\mathbf{v} = \mathbf{w} - \mathbf{r}$. Continue the above MATLAB session as follows:

```
>>v = w-r
v =
   -75      23     -10
>>dist2 = sqrt(sum(v.*v))
dist2 =
   79.0822
```

Thus to reach diver 2 by swimming along the coordinate directions, diver 1 must swim 75 ft east, 23 ft north, and 10 ft up. The straight-line distance between them is approximately 79 feet.

The built-in MATLAB functions such as `sqrt(x)` and `exp(x)` automatically operate on array arguments to produce an array result the same size as the array argument `x`. Thus these functions are said to be *vectorized* functions. For example, in the following session the result `y` has the same size as the argument `x`.

```
>>x = [4, 16, 25];
>>y = sqrt(x)
y =
   2    4    5
```

However, when multiplying or dividing these functions, or when raising them to a power, you must use element-by-element operations if the arguments are arrays. For example, to compute $z = (e^y \sin x)\cos^2 x$, you must type `z = exp(y).*sin(x).*(cos(x)).^2`. Obviously, you will get an error message if the size of `x` is not the same as the size of `y`. The result `z` will have the same size as `x` and `y`.

EXAMPLE 2.3–3 Aortic Pressure Model

Biomedical engineers often design instrumentation to measure physiological processes, such as blood pressure. To do this they must develop mathematical models of the process. The following equation is a specific case of one model used to describe the blood pressure in the aorta during systole (the period following the closure of the heart's aortic valve). The variable t represents time in seconds, and the dimensionless variable y represents the pressure difference across the aortic valve, normalized by a constant reference pressure.

$$y(t) = e^{-8t} \sin\left(9.7t + \frac{\pi}{2}\right)$$

Plot this function for $t \geq 0$.

■ Solution

Note that if `t` is a vector, the MATLAB functions `exp(-8*t)` and `sin(9.7*t+pi/2)` will also be vectors the same size as `t`. Thus we must use element-by-element multiplication to compute $y(t)$.

In addition, we must decide on the proper spacing to use for the vector `t` and its upper limit. The sine function $\sin(9.7t + \pi/2)$ oscillates with a frequency of 9.7 rad/sec, which is $9.7/(2\pi) = 1.5$ Hz. Thus its period is $1/1.5 = 2/3$ sec. The spacing of `t` should be a small fraction of the period in order to generate enough points to plot the curve. Thus we select a spacing of 0.003 to give approximately 200 points per period.

The amplitude of the sine wave decays with time because the sine is multiplied by the decaying exponential e^{-8t}. The exponential's initial value is $e^0 = 1$, and it will be 2 percent of its initial value at $t = 0.5$ (because $e^{-8(0.5)} = 0.02$). Thus we select the upper limit of `t` to be 0.5. The session is:

```
>>t = [0:0.003:0.5];
>>y = exp(-8*t).*sin(9.7*t+pi/2);
>>plot(t,y),xlabel('t (sec)'),...
ylabel('Normalized Pressure Difference y(t)')
```

The plot is shown in Figure 2.3–4. Note that we do not see much of an oscillation despite the presence of a sine wave. This is because the period of the sine wave is greater than the time it takes for the exponential e^{-8t} to become essentially zero.

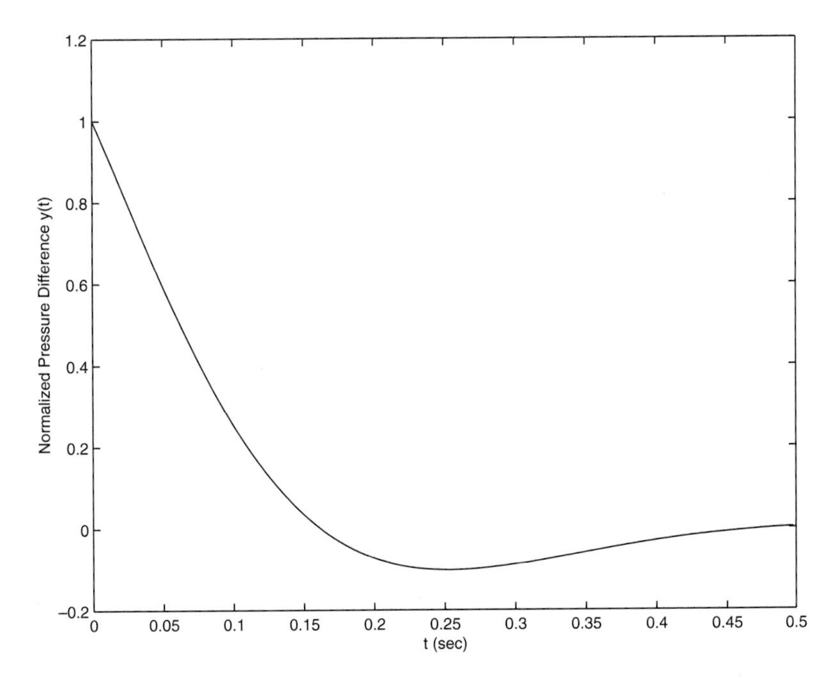

Figure 2.3–4 Aortic pressure response for Example 2.3–3.

Array Division

The definition of array division, also called element-by-element division, is similar to the definition of array multiplication except, of course, that the elements of one array are divided by the elements of the other array. Both arrays must have the same size. The symbol for array right division is `./`. For example, if

$$\mathbf{x} = [8,\ 12,\ 15] \qquad \mathbf{y} = [-2,\ 6,\ 5]$$

then `z = x./y` gives

$$\mathbf{z} = [8/(-2),\ 12/6,\ 15/5] = [-4,\ 2,\ 3]$$

Also, if

$$\mathbf{A} = \begin{bmatrix} 24 & 20 \\ -9 & 4 \end{bmatrix} \qquad \mathbf{B} = \begin{bmatrix} -4 & 5 \\ 3 & 2 \end{bmatrix}$$

then `C = A./B` gives

$$\mathbf{C} = \begin{bmatrix} 24/(-4) & 20/5 \\ -9/3 & 4/2 \end{bmatrix} = \begin{bmatrix} -6 & 4 \\ -3 & 2 \end{bmatrix}$$

The array left division operator (`.\`) is defined to perform element-by-element division using left division. Refer to Table 2.3–1 for examples. Note that `A.\B` is not equivalent to `A./B`.

EXAMPLE 2.3–4 Transportation Route Analysis

The following table gives data for the distance traveled along five truck routes and the corresponding time required to traverse each route. Use the data to compute the average speed required to drive each route. Find the route that has the highest average speed.

	1	2	3	4	5
Distance (mi)	560	440	490	530	370
Time (hr)	10.3	8.2	9.1	10.1	7.5

■ Solution

For example, the average speed on the first route is $560/10.3 = 54.4$ mi/hr. First we define the row vectors `d` and `t` from the distance and time data. Then, to find the average speed on each route using MATLAB, we use array division. The session is

```
>>d = [560, 440, 490, 530, 370]
>>t = [10.3, 8.2, 9.1, 10.1, 7.5]
>>speed = d./t
speed =
   54.3689  53.6585  53.8462  52.4752  49.3333
```

The results are in miles per hour. Note that MATLAB displays more significant figures than is justified by the three-significant-figure accuracy of the given data, so we should round the results to three significant figures before using them. Thus we should report the average speeds to be 54.4, 53.7, 53.8, 52.5, and 49.0 mi/hr, respectively.

To find the highest speed and the corresponding route, continue the session as follows:

```
>>[highest_speed, route] = max(speed)
highest_speed =
   54.3689
route =
   1
```

The first route has the highest speed.

If we did not need the speeds for every route, we could have solved this problem by combining two lines as follows: `[highest_speed, route] = max(d./t)`. As you become more familiar with MATLAB, you will appreciate its power to solve problems with very few lines and keystrokes.

Array Exponentiation

MATLAB enables us not only to raise arrays to powers but also to raise scalars and arrays to *array* powers. To perform exponentiation on an element-by-element basis, we must use the `.^` symbol. For example, if `x = [3, 5, 8]`, then typing `x.^3` produces the array $[3^3, 5^3, 8^3] = [27, 125, 512]$. If `x = [0:2:6]`, then typing `x.^2` returns the array $[0^2, 2^2, 4^2, 6^2] = [0, 4, 16, 36]$. If

$$\mathbf{A} = \begin{bmatrix} 4 & -5 \\ 2 & 3 \end{bmatrix}$$

then `B = A.^3` gives this result:

$$\mathbf{B} = \begin{bmatrix} 4^3 & (-5)^3 \\ 2^3 & 3^3 \end{bmatrix} = \begin{bmatrix} 64 & -125 \\ 8 & 27 \end{bmatrix}$$

We can raise a scalar to an array power. For example, if `p = [2, 4, 5]`, then typing `3.^p` produces the array $[3^2, 3^4, 3^5] = [9, 81, 243]$. This example illustrates a common situation in which it helps to remember that `.^` is a *single* symbol; the dot in `3.^p` is not a decimal point associated with the number 3. The following operations, with the value of `p` given here, are equivalent and give the correct answer:

```
3.^p
3.0.^p
3..^p
(3).^p
3.^[2,4,5]
```

Test Your Understanding

T2.3–1 Given the matrices

$$\mathbf{A} = \begin{bmatrix} 21 & 27 \\ -18 & 8 \end{bmatrix} \qquad \mathbf{B} = \begin{bmatrix} -7 & -3 \\ 9 & 4 \end{bmatrix}$$

find their (a) array product, (b) array right division (**A** divided by **B**), and (c) **B** raised to the third power element by element.
(Answers: *a* `[-147, -81; -162, 32]`, *b* `[-3, -9; -2, 2]`, and *c* `[-343, -27; 729, 64]`.)

Current and Power Dissipation in Resistors

EXAMPLE 2.3–5

The current i passing through an electrical resistor having a voltage v across it is given by Ohm's law: $i = v/R$, where R is the resistance. The power dissipated in the resistor is given by v^2/R. The following table gives data for the resistance and voltage for five resistors. Use the data to compute (a) the current in each resistor and (b) the power dissipated in each resistor.

	1	2	3	4	5
R (Ω)	10^4	2×10^4	3.5×10^4	10^5	2×10^5
v (V)	120	80	110	200	350

■ Solution

(a) First we define two row vectors, one containing the resistance values and one containing the voltage values. To find the current $i = v/R$ using MATLAB, we use array division. The session is

```
>>R = [10000, 20000, 35000, 100000, 200000];
>>v = [120, 80, 110, 200, 350];
>>current = v./R
current =
   0.0120  0.0040  0.0031  0.0020  0.0018
```

The results are in amperes and should be rounded to three significant figures because the voltage data contains only three significant figures.

(b) To find the power $P = v^2/R$, use array exponentiation and array division. The session continues as follows:

```
>>power = v.^2./R
power =
   1.4400  0.3200  0.3457  0.4000  0.6125
```

These numbers are the power dissipation in each resistor in watts. Note that the statement `v.^2./R` is equivalent to `(v.^2)./R`. Although the rules of precedence are unambiguous here, we can always put parentheses around quantities if we are unsure how MATLAB will interpret our commands.

EXAMPLE 2.3–6

A Batch Distillation Process

Chemical and environmental engineers must sometimes design batch processes for producing or purifying liquids and gases. Applications of such processes occur in food and medicine production, and in waste processing and water purification. An example of such a process is a system for heating a liquid benzene/toluene solution to distill a pure benzene vapor. A particular batch distillation unit is charged initially with 100 mol of a 60 percent mol benzene/40 percent mol toluene mixture. Let L (mol) be the amount of liquid remaining in the still, and let x (mol B/mol) be the benzene mole fraction in the remaining liquid. Conservation of mass for benzene and toluene can be applied to derive the following relation [Felder, 1986].

$$L = 100\left(\frac{x}{0.6}\right)^{0.625}\left(\frac{1-x}{0.4}\right)^{-1.625}$$

Determine what mole fraction of benzene remains when $L = 70$. Note that it is difficult to solve this equation directly for x. Use a plot of x versus L to solve the problem.

■ Solution

This equation involves both array multiplication and array exponentiation. Note that MATLAB enables us to use decimal exponents to evaluate L. It is clear that L must be in the range $0 \le L \le 100$; however, we do not know the range of x, except that $x \ge 0$.

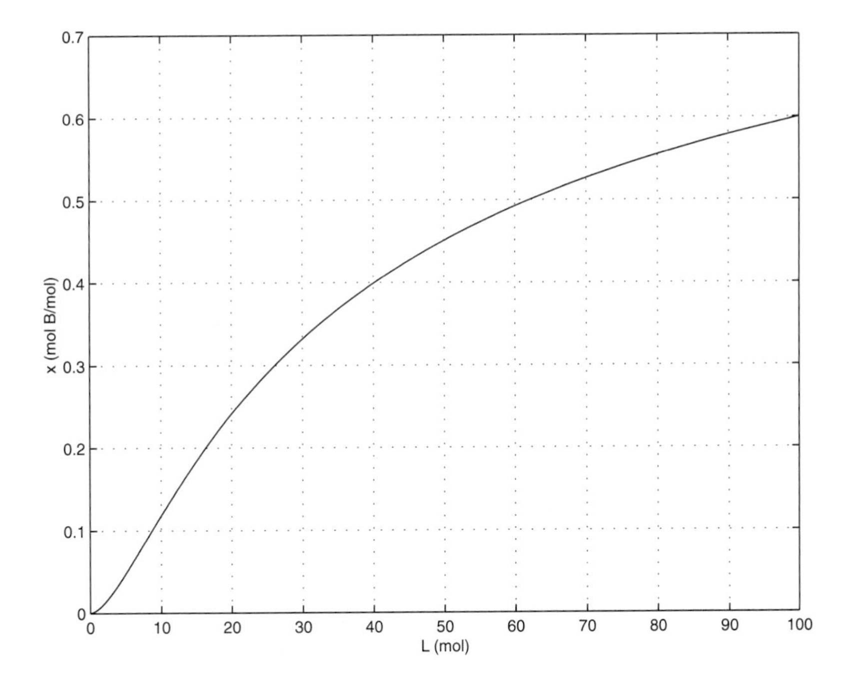

Figure 2.3–5 Plot for Example 2.3–6.

Therefore, we must make a few guesses for the range of x, using a session like the following. We find that $L > 100$ if $x > 0.6$, so we choose `x = [0:0.001:0.6]`. We use the `ginput` function to find the value of x corresponding to $L = 70$.

```
>>x = [0:0.001:0.6];
>>L = 100*(x/0.6).^(0.625).*((1-x)/0.4).^(-1.625);
>>plot(L,x),grid,xlabel('L(mol)'),ylabel('x (mol B/mol)'),...
[L,x] =  ginput(1)
```

The plot is shown in Figure 2.3–5. The answer is $x = 0.52$ if $L = 70$. The plot shows that the remaining liquid becomes leaner in benzene as the liquid amount becomes smaller. Just before the still is empty ($L = 0$), the liquid is pure toluene.

Using Array Operations to Evaluate Multivariable Functions

To evaluate a function of two variables, say, $z = f(x, y)$, using array operations, for the values $x = x_1, x_2, \ldots, x_m$ and $y = y_1, y_2, \ldots, y_m$, define the $m \times n$ matrices:

$$\mathbf{x} = \begin{bmatrix} x_1 & x_1 & \cdots & x_1 \\ x_2 & x_2 & \cdots & x_2 \\ \vdots & \vdots & \vdots & \vdots \\ x_m & x_m & \cdots & x_m \end{bmatrix} \qquad \mathbf{y} = \begin{bmatrix} y_1 & y_2 & \cdots & y_n \\ y_1 & y_2 & \cdots & y_n \\ \vdots & \vdots & \vdots & \vdots \\ y_1 & y_2 & \cdots & y_n \end{bmatrix}$$

When the function $z = f(x, y)$ is evaluated in MATLAB using array operations, the resulting $m \times n$ matrix $\mathbf{z}$ has the elements $z_{ij} = f(x_i, y_j)$. We can extend this technique to functions of more than two variables by using multidimensional arrays.

EXAMPLE 2.3–7 Height versus Velocity

In introductory physics courses Newton's laws of motion are used to derive the following formula for the maximum height h achieved by an object thrown with a speed v at an angle θ to the horizontal.

$$h = \frac{v^2 \sin \theta}{2g}$$

Create a table showing the maximum height for the following values of v and θ:

$$v = 10, 12, 14, 16, 18, 20 \text{ m/s} \qquad \theta = 50^\circ, 60^\circ, 70^\circ, 80^\circ$$

The rows in the table should correspond to the speed values, and the columns should correspond to the angles.

■ Solution

We must first convert the angles to radians before using the `sin` function. In order to use element-by-element operations, we must make sure that the arrays representing speed v and angle θ are the same size (they must have the same number of rows and columns). Because there are six speed values and four angles, and a given speed must correspond to a row, we must create a 6×4 array of speeds, with the *columns* repeated. Similarly, we must create a 6×4 array of angles, with the *rows* repeated. The script is shown below. Note the use of the *empty* array `[]` to provide an initial array to use in the `for` loops.

```
% Input data.
g = 9.8;
v = [10:2:20];th = [50:10:80]
thr = th*(pi/180);
% Create the 6 x 4 array of speeds.
vel=[];
for k=1:length(th)
   vel = [vel,v'];
end
% Create the 6 x 4 array of angles.
theta=[];
for k=1:length(v)
   theta = [theta;thr];
end
% Compute the 6 x 4 array of height values.
h = (vel.^2.*(sin(theta)).^2)/(2*g)
```

```
% Create the 6 x 5 array of speeds and heights.
H = [v',h];
% Create the 7 x 5 array for the table.
table = [0,th;H]
```

A number (in this case, 0) in the last line is required to match the number of columns in H. The following table shows the results, rounded to one decimal place. In Chapter 3 we will see how to control the number of decimal places displayed, and how to format a table with headings. From the table, we can see that the maximum height is 8.8 m if $v = 14$ m/s and $\theta = 70°$.

0	50	60	70	80
10	3.0	3.8	4.5	4.9
12	4.3	5.5	6.5	7.1
14	5.9	7.5	8.8	9.7
16	7.7	9.8	11.5	12.7
18	9.7	12.4	14.6	16.0
20	12.0	15.3	18.0	19.8

2.4 Matrix Operations

Matrix addition and subtraction are identical to element-by-element addition and subtraction. The corresponding matrix elements are summed or subtracted. However, matrix multiplication and division are not the same as element-by-element multiplication and division.

Multiplication of Vectors

Recall that vectors are simply matrices with one row or one column. Thus matrix multiplication and division procedures apply to vectors as well, and we will introduce matrix multiplication by considering the vector case first.

The *vector dot product* $\mathbf{u} \cdot \mathbf{w}$ of the vectors $\mathbf{u}$ and $\mathbf{w}$ is a scalar and can be thought of as the perpendicular projection of $\mathbf{u}$ onto $\mathbf{w}$. It can be computed from $|\mathbf{u}||\mathbf{w}| \cos \theta$, where θ is the angle between the two vectors and $|\mathbf{u}|$, $|\mathbf{w}|$ are the magnitudes of the vectors. Thus if the vectors are parallel and in the same direction, $\theta = 0$ and $\mathbf{u} \cdot \mathbf{w} = |\mathbf{u}||\mathbf{w}|$. If the vectors are perpendicular, $\theta = 90°$ and thus $\mathbf{u} \cdot \mathbf{w} = 0$. Because the unit vectors $\mathbf{i}$, $\mathbf{j}$, and $\mathbf{k}$ have unit length:

$$\mathbf{i} \cdot \mathbf{i} = \mathbf{j} \cdot \mathbf{j} = \mathbf{k} \cdot \mathbf{k} = \mathbf{1} \tag{2.4–1}$$

Because the unit vectors are perpendicular:

$$\mathbf{i} \cdot \mathbf{j} = \mathbf{i} \cdot \mathbf{k} = \mathbf{j} \cdot \mathbf{k} = \mathbf{0} \tag{2.4–2}$$

Thus the vector dot product can be expressed in terms of unit vectors as

$$\mathbf{u} \cdot \mathbf{w} = (u_1\mathbf{i} + u_2\mathbf{j} + u_3\mathbf{k}) \cdot (w_1\mathbf{i} + w_2\mathbf{j} + w_3\mathbf{k})$$

Carrying out the multiplication algebraically and using the properties given by (2.4–1) and (2.4–2), we obtain

$$\mathbf{u} \cdot \mathbf{w} = u_1w_1 + u_2w_2 + u_3w_3$$

The *matrix* product of a *row* vector **u** with a *column* vector **w** is defined in the same way as the vector dot product; the result is a scalar that is the sum of the products of the corresponding vector elements; that is,

$$[u_1 \quad u_2 \quad u_3] \begin{bmatrix} w_1 \\ w_2 \\ w_3 \end{bmatrix} = u_1w_1 + u_2w_2 + u_3w_3$$

if each vector has three elements. Thus the result of multiplying a 1×3 vector times a 3×1 vector is a 1×1 array; that is, a scalar. This definition applies to vectors having any number of elements, as long as both vectors have the same number of elements. Thus

$$[u_1 \quad u_2 \quad u_3 \quad \cdots \quad u_n] \begin{bmatrix} w_1 \\ w_2 \\ w_3 \\ \cdots \\ w_n \end{bmatrix} = u_1w_1 + u_2w_2 + u_3w_3 + \cdots + u_nw_n$$

if each vector has n elements. Thus the result of multiplying a $1 \times n$ vector times an $n \times 1$ vector is a 1×1 array, that is, a scalar.

Vector-Matrix Multiplication

Not all matrix products are scalars. To generalize the preceding multiplication to a column vector multiplied by a matrix, think of the matrix as being composed of row vectors. The scalar result of each row-column multiplication forms an element in the result, which is a column vector:

$$\begin{bmatrix} a_{11} & a_{12} \\ a_{21} & a_{22} \end{bmatrix} \begin{bmatrix} x_1 \\ x_2 \end{bmatrix} = \begin{bmatrix} a_{11}x_1 + a_{12}x_2 \\ a_{21}x_1 + a_{22}x_2 \end{bmatrix}$$

For example:

$$\begin{bmatrix} 2 & 7 \\ 6 & -5 \end{bmatrix} \begin{bmatrix} 3 \\ 9 \end{bmatrix} = \begin{bmatrix} 2(3) + 7(9) \\ 6(3) - 5(9) \end{bmatrix} = \begin{bmatrix} 69 \\ -27 \end{bmatrix} \qquad (2.4\text{–}3)$$

Thus the result of multiplying a 2×2 matrix times a 2×1 vector is a 2×1 array; that is, a column vector. Note that the definition of multiplication requires that the number of columns in the matrix be equal to the number of rows in the vector. In general, the product **Ax**, where **A** has p columns, is defined only if **x** has p rows. If **A** has m rows and **x** is a column vector, the result of **Ax** is a column vector with m rows.

Matrix-Matrix Multiplication

We can expand this definition of multiplication to include the product of two matrices **AB**. The number of columns in **A** must equal the number of rows in **B**. The row-column multiplications form column vectors, and these column vectors form the matrix result. The product **AB** has the same number of rows as **A** and the same number of columns as **B**. For example,

$$\begin{bmatrix} 6 & -2 \\ 10 & 3 \\ 4 & 7 \end{bmatrix} \begin{bmatrix} 9 & 8 \\ -5 & 12 \end{bmatrix} = \begin{bmatrix} (6)(9)+(-2)(-5) & (6)(8)+(-2)(12) \\ (10)(9)+(3)(-5) & (10)(8)+(3)(12) \\ (4)(9)+(7)(-5) & (4)(8)+(7)(12) \end{bmatrix}$$

$$= \begin{bmatrix} 64 & 24 \\ 75 & 116 \\ 1 & 116 \end{bmatrix} \qquad (2.4\text{–}4)$$

Use the operator * to perform matrix multiplication in MATLAB. The following MATLAB session shows how to perform the matrix multiplication shown in (2.4–4).

```
>>A = [6,-2;10,3;4,7];
>>B = [9,8;-5,12];
>>A*B
ans =
     64    24
     75    116
     1     116
```

Element-by-element multiplication is defined for the following product:

$$[3 \quad 1 \quad 7][4 \quad 6 \quad 5] = [12 \quad 6 \quad 35]$$

However, this product is *not* defined for *matrix* multiplication, because the first matrix has three columns, but the second matrix does not have three rows. Thus if we were to type `[3, 1, 7]*[4, 6, 5]` in MATLAB, we would receive an error message.

The following product is defined in matrix multiplication and gives the result shown:

$$\begin{bmatrix} x_1 \\ x_2 \\ x_3 \end{bmatrix} [y_1 \quad y_2 \quad y_3] = \begin{bmatrix} x_1y_1 & x_1y_2 & x_1y_3 \\ x_2y_1 & x_2y_2 & x_2y_3 \\ x_3y_1 & x_3y_2 & x_3y_3 \end{bmatrix}$$

The following product is also defined:

$$[10 \;\; 6] \begin{bmatrix} 7 & 4 \\ 5 & 2 \end{bmatrix} = [10(7)+6(5) \;\; 10(4)+6(2)] = [100 \;\; 52]$$

Test Your Understanding

T2.4–1 Use MATLAB to compute the dot product of the following vectors:

$$\mathbf{u} = 6\mathbf{i} - 8\mathbf{j} + 3\mathbf{k}$$
$$\mathbf{w} = 5\mathbf{i} + 3\mathbf{j} - 4\mathbf{k}$$

Check your answer by hand. (Answer: −6.)

T2.4–2 Use MATLAB to show that

$$\begin{bmatrix} 7 & 4 \\ -3 & 2 \\ 5 & 9 \end{bmatrix} \begin{bmatrix} 1 & 8 \\ 7 & 6 \end{bmatrix} = \begin{bmatrix} 35 & 80 \\ 11 & -12 \\ 68 & 94 \end{bmatrix}$$

EXAMPLE 2.4–1

Manufacturing Cost Analysis

Table 2.4–1 shows the hourly cost of four types of manufacturing processes. It also shows the number of hours required of each process to produce three different products. Use matrices and MATLAB to solve the following. (a) Determine the cost of each process to produce one unit of product 1. (b) Determine the cost to make one unit of each product. (c) Suppose we produce 10 units of product 1, 5 units of product 2, and 7 units of product 3. Compute the total cost.

Table 2.4–1 Cost and time data for manufacturing processes

		Hours required to produce one unit		
Process	**Hourly cost ($)**	**Product 1**	**Product 2**	**Product 3**
Lathe	10	6	5	4
Grinding	12	2	3	1
Milling	14	3	2	5
Welding	9	4	0	3

■ **Solution**

(a) The basic principle we can use here is that cost equals the hourly cost times the number of hours required. For example, the cost of using the lathe for product 1 is ($10/h)(6 h) = $60, and so forth for the other three processes. If we define the row vector of hourly costs to be `hourly_costs` and define the row vector of hours required for product 1 to be `hours_1`, then we can compute the costs of each process for product 1 using *element-by-element* multiplication. In MATLAB the session is

```
>>hourly_cost = [10, 12, 14, 9];
>>hours_1 = [6, 2, 3, 4];
>>process_cost_1 = hourly_cost.*hours_1
process_cost_1 =
   60   24   42   36
```

These are the costs of each of the four processes to produce one unit of product 1.

(b) To compute the total cost of one unit of product 1, we can use the vectors `hourly_costs` and `hours_1` but apply *matrix* multiplication instead of element-by-element multiplication, because matrix multiplication sums the individual products. The matrix multiplication gives

$$[10 \quad 12 \quad 14 \quad 9]\begin{bmatrix}6\\2\\3\\4\end{bmatrix} = 10(6) + 12(2) + 14(3) + 9(4) = 162$$

We can perform similar multiplication for products 2 and 3, using the data in the table. For product 2:

$$[10 \quad 12 \quad 14 \quad 9]\begin{bmatrix}5\\3\\2\\0\end{bmatrix} = 10(5) + 12(2) + 14(3) + 9(0) = 114$$

For product 3:

$$[10 \quad 12 \quad 14 \quad 9]\begin{bmatrix}4\\1\\5\\3\end{bmatrix} = 10(4) + 12(1) + 14(5) + 9(3) = 149$$

These three operations could have been accomplished in one operation by defining a matrix whose columns are formed by the data in the last three columns of the table:

$$[10 \quad 12 \quad 14 \quad 9]\begin{bmatrix}6 & 5 & 4\\2 & 3 & 1\\3 & 2 & 5\\4 & 0 & 3\end{bmatrix} = \begin{bmatrix}60+24+42+36\\50+36+28+\ 0\\40+12+70+27\end{bmatrix} = [162 \quad 114 \quad 149]$$

In MATLAB the session continues as follows. Remember that we must use the transpose operation to convert the row vectors into column vectors.

```
>>hours_2 = [5, 3, 2, 0];
>>hours_3 = [4, 1, 5, 3];
>>unit_cost = hourly_cost*[hours_1', hours_2', hours_3']
unit_cost =
   162     114     149
```

Thus the costs to produce one unit each of products 1, 2, and 3 is \$162, \$114, and \$149, respectively.

(c) To find the total cost to produce 10, 5, and 7 units, respectively, we can use matrix multiplication:

$$[10 \quad 5 \quad 7]\begin{bmatrix}162\\114\\149\end{bmatrix} = 1620 + 570 + 1043 = 3233$$

In MATLAB the session continues as follows. Note the use of the transpose operator on the vector `unit_cost`.

```
>>units = [10, 5, 7];
>>total_cost = units*unit_cost'
total_cost =
   3233
```

The total cost is $3233.

The General Matrix Multiplication Case

We can state the general result for matrix multiplication as follows: Suppose **A** has dimension $m \times p$ and **B** has dimension $p \times q$. If **C** is the product **AB**, then **C** has dimension $m \times q$ and its elements are given by

$$c_{ij} = \sum_{k=1}^{p} a_{ik} b_{kj} \tag{2.4–5}$$

for all $i = 1, 2, \ldots, m$ and $j = 1, 2, \ldots, q$. For the product to be defined, the matrices **A** and **B** must be *conformable;* that is, the number of *rows* in **B** must equal the number of *columns* in **A**. The product has the same number of rows as **A** and the same number of columns as **B**.

The algorithm defined by (2.4–5) is easy to remember. Each element in the ith row of **A** is multiplied by the corresponding element in the jth column of **B**. The sum of the products is the element c_{ij}. If we write the product **AB** in terms of the dimensions, as $(m \times p)(p \times q) = m \times q$, we can easily determine the dimensions of the product by “canceling” the inner dimensions (here p), which must be equal for the product to be defined.

Matrix multiplication does not have the commutative property; that is, in general, $\mathbf{AB} \neq \mathbf{BA}$. A simple example will demonstrate this fact:

$$\mathbf{AB} = \begin{bmatrix} 6 & -2 \\ 10 & 3 \end{bmatrix} \begin{bmatrix} 9 & 8 \\ -12 & 14 \end{bmatrix} = \begin{bmatrix} 78 & 20 \\ 54 & 122 \end{bmatrix} \tag{2.4–6}$$

whereas

$$\mathbf{BA} = \begin{bmatrix} 9 & 8 \\ -12 & 14 \end{bmatrix} \begin{bmatrix} 6 & -2 \\ 10 & 3 \end{bmatrix} = \begin{bmatrix} 134 & 6 \\ 68 & 66 \end{bmatrix} \tag{2.4–7}$$

Reversing the order of matrix multiplication is a common and easily made mistake.

The associative and distributive properties hold for matrix multiplication. The associative property states that

$$\mathbf{A}(\mathbf{B} + \mathbf{C}) = \mathbf{AB} + \mathbf{AC} \tag{2.4–8}$$

The distributive property states that

$$(\mathbf{AB})\mathbf{C} = \mathbf{A}(\mathbf{BC}) \tag{2.4–9}$$

Test Your Understanding

T2.4–3 Use MATLAB to verify the results of equations (2.4–6) and (2.4–7).

Applications to Cost Analysis

Data on costs of engineering projects are often recorded as tables. The data in these tables must often be analyzed in several ways. The elements in MATLAB matrices are similar to the cells in a spreadsheet, and MATLAB can perform many spreadsheet-type calculations for analyzing such tables.

Product Cost Analysis — EXAMPLE 2.4–2

Table 2.4–2 shows the costs associated with a certain product, and Table 2.4–3 shows the production volume for the four quarters of the business year. Use MATLAB to find the quarterly costs for materials, labor, and transportation; the total material, labor, and transportation costs for the year; and the total quarterly costs.

Table 2.4–2 Product costs

	Unit costs ($\$ \times 10^3$)		
Product	**Materials**	**Labor**	**Transportation**
1	6	2	1
2	2	5	4
3	4	3	2
4	9	7	3

Table 2.4–3 Quarterly production volume

Product	Quarter 1	Quarter 2	Quarter 3	Quarter 4
1	10	12	13	15
2	8	7	6	4
3	12	10	13	9
4	6	4	11	5

■ Solution

The costs are the product of the unit cost times the production volume. Thus we define two matrices: U contains the unit costs in Table 2.4–2 in thousands of dollars, and P contains the quarterly production data in Table 2.4–3.

```
>>U = [6, 2, 1;2, 5, 4;4, 3, 2;9, 7, 3];
>>P = [10, 12, 13, 15;8, 7, 6, 4;12, 10, 13, 9;6, 4, 11, 5];
```

Note that if we multiply the first column in U times the first column in P, we obtain the total materials cost for the first quarter. Similarly, multiplying the first column in

U times the *second* column in P gives the total materials cost for the *second* quarter. Also, multiplying the second column in U times the first column in P gives the total *labor* cost for the first quarter, and so on. Extending this pattern, we can see that we must multiply the *transpose* of U times P. This multiplication gives the cost matrix C.

```
>>C = U'*P
```

The result is

$$\mathbf{C} = \begin{bmatrix} 178 & 162 & 241 & 179 \\ 138 & 117 & 172 & 112 \\ 84 & 72 & 96 & 64 \end{bmatrix}$$

Each column in **C** represents one quarter. The total first-quarter cost is the sum of the elements in the first column, the second-quarter cost is the sum of the second column, and so on. Thus because the sum command sums the columns of a matrix, the quarterly costs are obtained by typing:

```
>>Quarterly_Costs = sum(C)
```

The resulting vector, containing the quarterly costs in thousands of dollars, is [400 351 509 355]. Thus the total costs in each quarter are $400,000; $351,000; $509,000; and $355,000.

The elements in the first row of **C** are the material costs for each quarter; the elements in the second row are the labor costs, and those in the third row are the transportation costs. Thus to find the total material costs, we must sum across the first row of **C**. Similarly, the total labor and total transportation costs are the sums across the second and third rows of **C**. Because the sum command sums *columns,* we must use the transpose of **C**. Thus we type the following:

```
>>Category_Costs = sum(C')
```

The resulting vector, containing the category costs in thousands of dollars, is [760 539 316]. Thus the total material costs for the year are $760,000; the labor costs are $539,000; and the transportation costs are $316,000.

We displayed the matrix **C** only to interpret its structure. If we need not display **C**, the entire analysis would consist of only four command lines.

```
>>U = [6, 2, 1;2, 5, 4;4, 3, 2;9, 7, 3];
>>P = [10, 12, 13, 15;8, 7, 6, 4;12, 10, 13, 9;6, 4, 11, 5];
>>Quarterly_Costs = sum(U'*P)
Quarterly_Costs =
   400   351   509   355
>>Category_Costs = sum((U'*P)')
Category_Costs =
   760   539   316
```

This example illustrates the compactness of MATLAB commands.

Special Matrices

NULL MATRIX

IDENTITY MATRIX

Two exceptions to the noncommutative property are the *null* matrix, denoted by **0**, and the *identity,* or *unity,* matrix, denoted by **I**. The null matrix contains all zeros and is not the same as the empty matrix `[]`, which has no elements. The identity matrix is a square matrix whose diagonal elements are all equal to one, with the remaining elements equal to zero. For example, the 2×2 identity matrix is

$$\mathbf{I} = \begin{bmatrix} 1 & 0 \\ 0 & 1 \end{bmatrix}$$

These matrices have the following properties:

$$\mathbf{0A} = \mathbf{A0} = \mathbf{0}$$

$$\mathbf{IA} = \mathbf{AI} = \mathbf{A}$$

MATLAB has specific commands to create several special matrices. Type `help specmat` to see the list of special matrix commands; also check Table 2.4–4. The identity matrix **I** can be created with the `eye(n)` command, where `n` is the desired dimension of the matrix. To create the 2×2 identity matrix, you type `eye(2)`. Typing `eye(size(A))` creates an identity matrix having the same dimension as the matrix **A**.

Sometimes we want to initialize a matrix to have all zero elements. The `zeros` command creates a matrix of all zeros. Typing `zeros(n)` creates an $n \times n$ matrix of zeros, whereas typing `zeros(m,n)` creates an $m \times n$ matrix of zeros. Typing `zeros(size(A))` creates a matrix of all zeros having the same dimension as the matrix **A**. This type of matrix can be useful for applications in which we do not know the required dimension ahead of time. The syntax of the `ones` command is the same, except that it creates arrays filled with ones.

For example, to create and plot the function

$$f(x) = \begin{cases} 10 & 0 \le x \le 2 \\ 0 & 2 < x < 5 \\ -3 & 5 \le x \le 7 \end{cases}$$

Table 2.4–4 Special matrices

Command	Description
`eye(n)`	Creates an $n \times n$ identity matrix.
`eye(size(A))`	Creates an identity matrix the same size as the matrix **A**.
`ones(n)`	Creates an $n \times n$ matrix of ones.
`ones(m,n)`	Creates an $m \times n$ array of ones.
`ones(size(A))`	Creates an array of ones the same size as the array **A**.
`zeros(n)`	Creates an $n \times n$ matrix of zeros.
`zeros(m,n)`	Creates an $m \times n$ array of zeros.
`zeros(size(A))`	Creates an array of zeros the same size as the array **A**.

the script file is

```
x1 = [0:0.01:2];
f1 = 10*ones(size(x1));
x2 = [2.01:0.01:4.99];
f2 = zeros(size(x2));
x3 = [5:0.01:7];
f3 = -3*ones(size(x3));
f = [f1, f2, f3];
x = [x1, x2, x3];
plot(x,f), xlabel('x'),ylabel('y')
```

(Consider what the plot would look like if the command `plot(x,f)` were replaced with the command `plot(x1,f1,x2,f2,x3,f3)`.)

Matrix Division

Matrix division is a more challenging topic than matrix multiplication. Matrix division uses both the right and left division operators, / and \, for various applications, a principal one being the solution of sets of linear algebraic equations. Chapter 6 covers matrix division and a related topic, the matrix inverse.

Matrix Exponentiation

Raising a matrix to a power is equivalent to repeatedly multiplying the matrix by itself, for example, $\mathbf{A}^2 = \mathbf{AA}$. This process requires the matrix to have the same number of rows as columns; that is, it must be a *square* matrix. MATLAB uses the symbol ^ for matrix exponentiation. To find $\mathbf{A}^2$, type `A^2`.

We can raise a scalar n to a matrix power $\mathbf{A}$, if $\mathbf{A}$ is square, by typing `n^A`, but the applications for such a procedure are in advanced courses. However, raising a matrix to a matrix power—that is, $\mathbf{A}^{\mathbf{B}}$—is not defined, even if $\mathbf{A}$ and $\mathbf{B}$ are square.

Note that if n is a scalar and if $\mathbf{B}$ and $\mathbf{C}$ are not square matrices, then the following operations are not defined and will generate an error message in MATLAB:

```
B^n
n^B
B^C
```

Special Products

Many applications in physics and engineering use the cross product and dot product—for example, calculations to compute moments and force components use these special products. If $\mathbf{A}$ and $\mathbf{B}$ are vectors with three elements, the cross-product command `cross(A,B)` computes the three-element vector that is the cross-product $\mathbf{A} \times \mathbf{B}$. If $\mathbf{A}$ and $\mathbf{B}$ are $3 \times n$ matrices, `cross(A,B)` returns a $3 \times n$ array whose columns are the cross products of the corresponding columns in the $3 \times n$ arrays $\mathbf{A}$ and $\mathbf{B}$. For example, the moment $\mathbf{M}$ with respect to a reference point O due to the force $\mathbf{F}$ is given by $\mathbf{M} = \mathbf{r} \times \mathbf{F}$, where $\mathbf{r}$ is the position vector

Table 2.4–5 Special products

Command	Syntax
`cross(A,B)`	Computes a $3 \times n$ array whose columns are the cross products of the corresponding columns in the $3 \times n$ arrays **A** and **B**. Returns a three-element cross-product vector if **A** and **B** are three-element vectors.
`dot(A,B)`	Computes a row vector of length n whose elements are the dot products of the corresponding columns of the $m \times n$ arrays **A** and **B**.

from the point O to the point where the force **F** is applied. To find the moment in MATLAB, you type `M = cross(r,F)`.

The dot-product command `dot(A,B)` computes a row vector of length n whose elements are the dot products of the corresponding columns of the $m \times n$ arrays **A** and **B**. To compute the component of the force **F** along the direction given by the vector **r**, you type `dot(F,r)`. Table 2.4–5 summarizes the dot- and cross-product commands.

2.5 Polynomial Operations Using Arrays

MATLAB has some convenient vector-based tools for working with polynomials, which are used in many advanced courses and applications in engineering. Type `help polyfun` for more information on this category of commands. We will use the following notation to describe a polynomial:

$$f(x) = a_1x^n + a_2x^{n-1} + a_3x^{n-2} + \cdots + a_{n-1}x^2 + a_nx + a_{n+1}$$

This polynomial is a function of x. Its *degree* or *order* is n, the highest power of x that appears in the polynomial. The a_i, $i = 1, 2, \ldots, n+1$ are the polynomial's *coefficients*. We can describe a polynomial in MATLAB with a row vector whose elements are the polynomial's coefficients, *starting with the coefficient of the highest power of* x. This vector is $[a_1, a_2, a_3, \ldots, a_{n-1}, a_n, a_{n+1}]$. For example, the vector `[4,-8,7,-5]` represents the polynomial $4x^3 - 8x^2 + 7x - 5$.

Polynomial roots can be found with the `roots(a)` function, where (a) is the array containing the polynomial coefficients. For example, to obtain the roots of $x^3 + 12x^2 + 45x + 50 = 0$, you type `y = roots([1,12,45,50])`. The answer `(y)` is a column array containing the values $-2, -5, -5$.

The `poly(r)` function computes the coefficients of the polynomial whose roots are specified by the array `r`. The result is a *row* array that contains the polynomial's coefficients. (Note that the `roots` function returns a *column array*.) For example, to find the polynomial whose roots are 1 and $3 \pm 5i$, the session is

```
>>r = [1,3+5i,3-5i];
>>poly(r)
ans =
     1     -7     40          -34
```

Thus the polynomial is $x^3 - 7x^2 + 40x - 34$. The two commands could have been combined into the single command `poly([1,3+5i, 3-5i])`.

Polynomial Addition and Subtraction

To add two polynomials, add the arrays that describe their coefficients. If the polynomials are of different degrees, add zeros to the coefficient array of the lower-degree polynomial. For example, consider

$$f(x) = 9x^3 - 5x^2 + 3x + 7$$

whose coefficient array is `f = [9,-5,3,7]` and

$$g(x) = 6x^2 - x + 2$$

whose coefficient array is `g = [6,-1,2]`. The degree of $g(x)$ is one less that of $f(x)$. Therefore, to add $f(x)$ and $g(x)$, we append one zero to `g` to "fool" MATLAB into thinking $g(x)$ is a third-degree polynomial. That is, we type `g = [0 g]` to obtain `[0,6,-1,2]` for `g`. This vector represents $g(x) = 0x^3 + 6x^2 - x + 2$. To add the polynomials, type `h = f+g`. The result is `h = [9,1,2,9]`, which corresponds to $h(x) = 9x^3 + x^2 + 2x + 9$. Subtraction is done in a similar way.

Polynomial Multiplication and Division

To multiply a polynomial by a scalar, simply multiply the coefficient array by that scalar. For example, $5h(x)$ is represented by `[45,5,10,45]`.

Multiplication of polynomials by hand can be tedious, and polynomial division is even more so, but these operations are easily done with MATLAB. Use the `conv` function (it stands for "convolve") to multiply polynomials and use the `deconv` function (`deconv` stands for "deconvolve") to perform synthetic division. Table 2.5–1 summarizes these functions, as well as the `poly`, `polyval`, and `roots` functions, which we saw in Chapter 1.

Table 2.5–1 Polynomial functions

Command	Description
`conv(a,b)`	Computes the product of the two polynomials described by the coefficient arrays `a` and `b`. The two polynomials need not be the same degree. The result is the coefficient array of the product polynomial.
`[q,r] = deconv(num,den)`	Computes the result of dividing a numerator polynomial, whose coefficient array is `num`, by a denominator polynomial represented by the coefficient array `den`. The quotient polynomial is given by the coefficient array `q`, and the remainder polynomial is given by the coefficient array `r`.
`poly(r)`	Computes the coefficients of the polynomial whose roots are specified by the vector `r`. The result is a *row* vector that contains the polynomial's coefficients arranged in descending order of power.
`polyval(a,x)`	Evaluates a polynomial at specified values of its independent variable `x`, which can be a matrix or a vector. The polynomial's coefficients of descending powers are stored in the array `a`. The result is the same size as `x`.
`roots(a)`	Computes the roots of a polynomial specified by the coefficient array `a`. The result is a *column* vector that contains the polynomial's roots.

The product of the polynomials $f(x)$ and $g(x)$ is

$$\begin{aligned} f(x)g(x) &= (9x^3 - 5x^2 + 3x + 7)(6x^2 - x + 2) \\ &= 54x^5 - 39x^4 + 41x^3 + 29x^2 - x + 14 \end{aligned}$$

Dividing $f(x)$ by $g(x)$ using synthetic division gives a quotient of

$$\frac{f(x)}{g(x)} = \frac{9x^3 - 5x^2 + 3x + 7}{6x^2 - x + 2} = 1.5x - 0.5833$$

with a remainder of $-0.5833x + 8.1667$. Here is the MATLAB session to perform these operations.

```
>>f = [9,-5,3,7];
>>g = [6,-1,2];
>>product = conv(f,g)
product =
   54    -39    41     29     -1     14
>>[quotient, remainder] = deconv(f,g)
quotient =
   1.5     -0.5833
remainder =
   0     0     -0.5833     8.1667
```

The conv and deconv functions do not require that the polynomials have the same degree, so we did not have to fool MATLAB as we did when adding the polynomials. Table 2.5–1 gives the general syntax for the conv and deconv functions.

Plotting Polynomials

The polyval(a,x) function evaluates a polynomial at specified values of its independent variable x, which can be a matrix or a vector. The polynomial's coefficient array is a. The result is the same size as x. For example, to evaluate the polynomial $f(x) = 9x^3 - 5x^2 + 3x + 7$ at the points $x = 0, 2, 4, \ldots, 10$, type

```
>>a = [9,-5,3,7];
>>x = [0:2:10];
>>f = polyval(a,x);
```

The resulting vector f contains six values that correspond to $f(0), f(2), f(4), \ldots, f(10)$. These three commands can be combined into a single command:

```
>>f = polyval([9,-5,3,7],[0:2:10]);
```

Personal preference determines whether to combine terms in this way; some people think that the single, combined command is less readable than three separate commands.

The `polyval` function is very useful for plotting polynomials. To do this you should define an array that contains many values of the independent variable x in order to obtain a smooth plot. For example, to plot the polynomial $f(x) = 9x^3 - 5x^2 + 3x + 7$ for $-2 \le x \le 5$, you type

```
>>a = [9,-5,3,7];
>>x = [-2:0.01:5];
>>f = polyval(a,x);
>>plot(x,f),xlabel('x'),ylabel('f(x)'),grid
```

EXAMPLE 2.5–1 Earthquake-Resistant Building Design

Buildings designed to withstand earthquakes must have natural frequencies of vibration that are not close to the oscillation frequency of the ground motion. A building's natural frequencies are determined primarily by the masses of its floors and by the lateral stiffness of its supporting columns (which act like horizontal springs). We can find these frequencies by solving for the roots of a polynomial called the structure's *characteristic* polynomial (characteristic polynomials are discussed further in Chapter 8). Figure 2.5–1 shows the exaggerated motion of the floors of a three-story building. For such a building, if each floor has a mass m and the columns have stiffness k, the

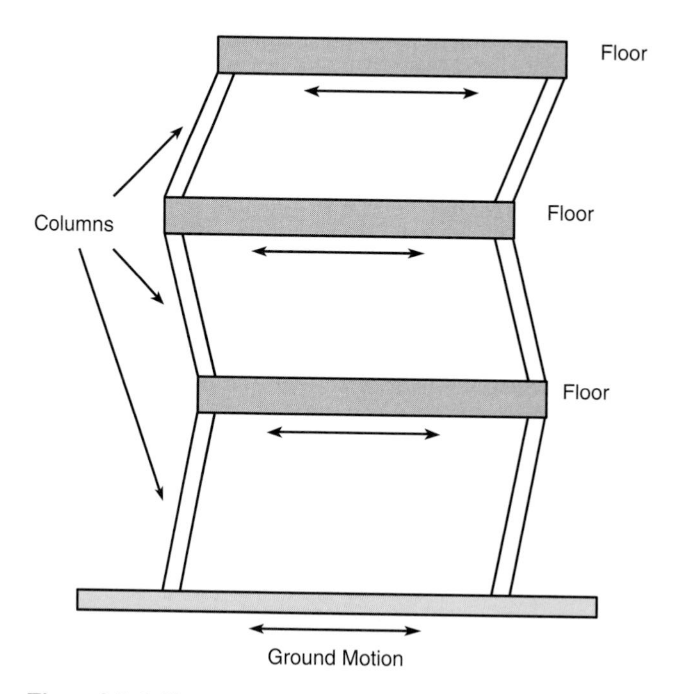

Figure 2.5–1 Simple vibration model of a building subjected to ground motion.

polynomial is

$$(\alpha - f^2)[(2\alpha - f^2)^2 - \alpha^2] + \alpha^2 f^2 - 2\alpha^3$$

where $\alpha = k/4m\pi^2$ (models such as these are discussed in more detail in [Palm, 2005]). The building's natural frequencies in cycles per second are the positive roots of this equation. Find the building's natural frequencies in cycles per second for the case where $m = 1000$ kg and $k = 5 \times 10^6$ N/m.

■ Solution

The characteristic polynomial consists of sums and products of lower-degree polynomials. We can use this fact to have MATLAB do the algebra for us. The characteristic polynomial has the form

$$p_1\left(p_2^2 - \alpha^2\right) + p_3 = 0$$

where

$$p_1 = \alpha - f^2 \qquad p_2 = 2\alpha - f^2 \qquad p_3 = \alpha^2 f^2 - 2\alpha^3$$

The MATLAB script file is

```
k = 5e+6;m = 1000;
alpha = k/(4*m*pi^2);
p1 = [-1,0,alpha];
p2 = [-1,0,2*alpha];
p3 = [alpha^2,0,-2*alpha^3];
p4 = conv(p2,p2)-(0,0,0,0,alpha^2];
p5 = conv(p1,p4);
p6 = p5+[0,0,0,0,p3];
r = roots(p6);
pos = r(r>0)
```

The resulting array is `pos = [20.2789;14.0335;5.0085]`. Thus the frequencies, rounded to the nearest integer, are 20, 14, and 5 Hz.

Test Your Understanding

T2.5–1 Use MATLAB to obtain the roots of

$$x^3 + 13x^2 + 52x + 6 = 0.$$

Use the `poly` function to confirm your answer.

T2.5–2 Use MATLAB to confirm that

$$(20x^3 - 7x^2 + 5x + 10)(4x^2 + 12x - 3)$$
$$= 80x^5 + 212x^4 - 124x^3 + 121x^2 + 105x - 30$$

T2.5–3 Use MATLAB to confirm that

$$\frac{12x^3 + 5x^2 - 2x + 3}{3x^2 - 7x + 4} = 4x + 11$$

with a remainder of $59x - 41$.

T2.5–4 Use MATLAB to confirm that

$$\frac{6x^3 + 4x^2 - 5}{12x^3 - 7x^2 + 3x + 9} = 0.7108$$

when $x = 2$.

T2.5–5 Plot the polynomial

$$y = x^3 + 13x^2 + 52x + 6$$

over the range $-7 \le x \le 1$.

2.6 Cell Arrays

The *cell array* is an array in which each element is a *bin,* or *cell,* which can contain an array. You can store different classes of arrays in a cell array, and you can group data sets that are related but have different dimensions. You access cell arrays using the same indexing operations used with ordinary arrays.

This is the only section in the text that uses cell arrays. Coverage of this section is therefore optional. Some more advanced MATLAB application, such as those found in some of the toolboxes, do use cell arrays.

CELL INDEXING

Creating Cell Arrays

CONTENT INDEXING

You can create a cell array by using assignment statements or by using the `cell` function (see Table 2.6–1). You can assign data to the cells by using either *cell indexing* or *content indexing*. To use cell indexing, enclose in parentheses

Table 2.6–1 Cell array functions

Function	Description
`C = cell(n)`	Creates an $n \times n$ cell array `C` of empty matrices.
`C = cell(n,m)`	Creates an $n \times m$ cell array `C` of empty matrices.
`celldisp(C)`	Displays the contents of cell array `C`.
`cellplot(C)`	Displays a graphical representation of the cell array `C`.
`C = num2cell(A)`	Converts a numeric array `A` into a cell array `C`.
`[X,Y, ...] = deal(A,B, ...)`	Matches up the input and output lists. Equivalent to `X = A, Y = B, ...`.
`[X,Y, ...] = deal(A)`	Matches up the input and output lists. Equivalent to `X = A, Y = A, ...`.
`iscell(C)`	Returns a 1 if `C` is a cell array; otherwise, returns a 0.

the cell subscripts on the left side of the assignment statement and use the standard array notation. Enclose the cell contents on the right side of the assignment statement in braces {}.

An Environmental Database — EXAMPLE 2.6–1

Data collection is important for early detection of changes in our environment. In order to detect such changes, we need to be able to analyze the database efficiently, and this effort requires a database that is set up for easy access. As a simple example, suppose you want to create a 2 × 2 cell array A, whose cells contain the location, the date, the air temperature (measured at 8 A.M., 12 noon, and 5 P.M.), and the water temperatures measured at the same time in three different points in a pond. The cell array looks like the following.

Walden Pond	**June 13, 1997**
[60 72 65]	$\begin{bmatrix} 55 & 57 & 56 \\ 54 & 56 & 55 \\ 52 & 55 & 53 \end{bmatrix}$

■ Solution

You can create this array by typing the following either in interactive mode or in a script file and running it.

```
A(1,1) = {'Walden Pond'};
A(1,2) = {'June 13, 1997'};
A(2,1) = {[60,72,65]};
A(2,2) = {[55,57,56;54,56,55;52,55,53]};
```

If you do not yet have contents for a particular cell, you can type a pair of empty braces { } to denote an empty cell, just as a pair of empty brackets [ ] denotes an empty numeric array. This notation creates the cell but does not store any contents in it.

To use content indexing, enclose in braces the cell subscripts on the left side using the standard array notation. Then specify the cell contents on the right side of the assignment operator. For example:

```
A{1,1} = 'Walden Pond';
A{1,2} = 'June 13, 1997';
A{2,1} = [60,72,65];
A{2,2} = [55,57,56;54,56,55;52,55,53];
```

Type A at the command line. You will see

```
A =
   'Walden Pond'   'June 13, 1997'
    [1x3 double]    [3x3 double]
```

You can use the `celldisp` function to display the full contents. For example, typing `celldisp(A)` displays

```
A{1,1} =
   Walden Pond
A{2,1} =
   60  72  65
   .
   .
   .
   etc.
```

The `cellplot` function produces a graphical display of the cell array's contents in the form of a grid. Type `cellplot(A)` to see this display for the cell array `A`.

Use commas or spaces with braces to indicate columns of cells and use semicolons to indicate rows of cells (just as with numeric arrays). For example, typing

```
B = {[2,4], [6,-9;3,5]; [7;2], 10};
```

creates the following 2 × 2 cell array:

[2 4]	$\begin{bmatrix} 6 & -9 \\ 3 & 5 \end{bmatrix}$
[7 2]	10

You can preallocate empty cell arrays of a specified size by using the `cell` function. For example, type `C = cell(3,5)` to create the 3 × 5 cell array `C` and fill it with empty matrices. Once the array has been defined in this way, you can use assignment statements to enter the contents of the cells. For example, type `C(2,4) = {[6,-3,7]}` to put the 1 × 3 array in cell (2,4) and type `C(1,5) = {1:10}` to put the numbers from 1 to 10 in cell (1,5). Type `C(3,4) = {'30 mph'}` to put the string in cell (3,4).

Do not name a cell array with the same name as a previously used numeric array without first using the `clear` command to clear the name. Otherwise, MATLAB will generate an error. In addition, MATLAB does not clear a cell array when you make a single assignment to it. You can determine if an array is a cell array by using the `iscell` function. You can convert a numeric array to a cell array by using the `num2cell` function.

Accessing Cell Arrays

You can access the contents of a cell array by using either cell indexing or content indexing. For example, to use cell indexing to place the contents of cell (3,4) of the array C in the new variable Speed, type Speed = C(3,4). To place the contents of the cells in rows 1 to 3, columns 2 to 5 in the new cell array D, type D = C(1:3,2:5). The new cell array D will have three rows, four columns, and 12 arrays. To use content indexing to access some or all of the contents in a *single cell,* enclose the cell index expression in braces to indicate that you are assigning the contents, not the cells themselves, to a new variable. For example, typing Speed = C{3,4} assigns the contents '30 mph' in cell (3,4) to the variable Speed. You cannot use content indexing to retrieve the contents of more than one cell at a time. For example, the statements G = C{1,:} and C{1,:} = var, where var is some variable, are both invalid.

You can access subsets of a cell's contents. For example, to obtain the second element in the 1×3-row vector in the (2,4) cell of array C and assign it to the variable r, you type r = C{2,4}(1,2). The result is r = -3.

The deal function accesses elements of a range of cells in a cell array. For example, with the preceding cell array B, x and y can be assigned to the elements in row 2 of B as follows:

```
>>[x,y] = deal(B{2,:})
x =
   7   2
y =
   10
```

Using Cell Arrays

You can use cell arrays in comma-separated lists just as you would use ordinary MATLAB variables. For example, suppose you create the 1×4 cell array H by typing

```
H = {[2,4,8], [6,-8,3], [2:6], [9,2,5]};
```

The expression H{2:4} is equivalent to a comma-separated list of the second through fourth cells in H. To create a numeric array J from the first, second, and fourth cells in the cell array H, you type

```
J = [H{1}; H{2}; H{4}]
```

The result is

$$\mathbf{J} = \begin{bmatrix} 2 & 4 & 8 \\ 6 & -8 & 3 \\ 9 & 2 & 5 \end{bmatrix}$$

Typing H{2:3} displays the arrays in the second and third cells.

```
>>H{2:3}
ans =
      6      -8      3
ans =
      2      3      4      5      6
```

You can also use cell arrays in this manner in function input and output lists, and you can store the results in another cell array, say, K. For example,

```
>>[K{1:2}] = max(J)
K =
   [1x3 double]    [1x3 double]
```

Type K{1} to see the maximum values; type K{2} to see the corresponding indices.

```
>>K{1}
ans =
     9   4   8
>>K{2}
ans =
     3   1   1
```

You can apply functions and operators to cell contents. For example, suppose you create the 3×2 cell array L by typing

```
L = {[2,4,8], [6,-8,3]; [2:6], [9,2,5]; [1,4,5], [7,5,2]};
```

Then, for example:

```
>>max(L{3,2})
ans =
     7
```

Nested cell arrays have cells that contain cell arrays, which may also contain cell arrays, and so on. To create nested arrays, you can use nested braces, the cell function, or assignment statements. For example:

```
N(1,1) = {[2,7,5]};
N(1,2) = {{[5,9,1; 4,8,0], 'Case 1'; {5,8}, [7,3]}};
```

Typing N gives the result

```
N =
   [1x3 double]    {2x2 cell}
```

The following steps create the same array N using the cell function. The method assigns the output of cell to an existing cell.

```
% First create an empty 1x2 cell array.
N = cell(1,2)
% Then create a 2x2 cell array inside N(1,2).
N(1,2) = {cell(2,2)}
% Then fill N using assignment statements.
N(1,1) = {[2,7,5]};
N{1,2}(1,1) = {[5,9,1; 4,8,0]}
N{1,2}(1,2) = {'Case 1'}
```

```
N{1,2}{2,1}(1) = {5}
N{1,2}{2,1}(2) = {8}
N{1,2}(2,2) = {[7,3]}
```

Note that braces are used for subscripts to access cell contents until the lowest "layer" of subscripts is reached. Then parentheses are used because the lowest layer does not contain cell arrays.

As a final example, suppose you create the 3×2 cell array `H` by typing

```
H = {[2,4,8], [6,-8,3]; [2:6], [9,2,5]; [1,4,5], [7,5,2]};
```

You can create a numeric array `J` from the cell array `H` by typing

```
J = [H{1,1}; H{1,2}; H{2,2}]
```

The result is

$$\mathbf{J} = \begin{bmatrix} 2 & 4 & 8 \\ 6 & -8 & 3 \\ 9 & 2 & 5 \end{bmatrix}$$

Typing `H{2:3,:}` displays the arrays in the second and third rows.

```
>>H{2:3,:}
ans =
     2     3     4     5     6
ans =
     9     2     5
ans =
     1     4     5
ans =
     7     5     2
```

Test Your Understanding

T2.6–1 Create the following cell array:

```
A = {[1:4], [0, 9, 2], [2:5], [6:8]}
```

What is `A{1:2}`? What is `[A{2}; A{4}]`? What is `min [A{2}, A{3}`?

2.7 Structure Arrays

Structure arrays are composed of *structures*. This class of arrays enables you to store dissimilar arrays together. The elements in structures are accessed using *named fields*. This feature distinguishes them from cell arrays, which are accessed using the standard array indexing operations.

FIELD

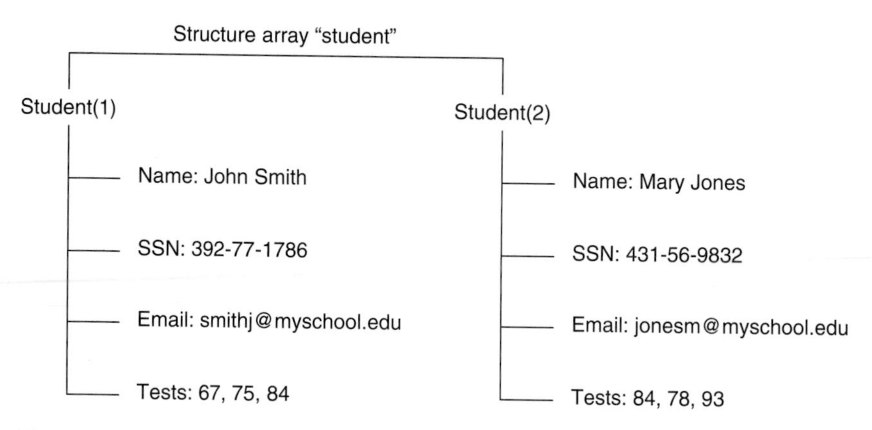

Figure 2.7–1 Arrangement of data in the structure array `student`.

Structure arrays are used in this text only in this section and in Chapter 10. Some MATLAB toolboxes do use structure arrays.

A specific example is the best way to introduce the terminology of structures. Suppose you want to create a database of students in a course, and you want to include each student's name, Social Security number, email address, and test scores. Figure 2.7–1 shows a diagram of this data structure. Each type of data (name, Social Security number, and so on) is a *field,* and its name is the *field name.* Thus our database has four fields. The first three fields each contain a text string, while the last field (the test scores) contains a vector having numerical elements. A *structure* consists of all this information for a single student. A *structure array* is an array of such structures for different students. The array shown in Figure 2.7–1 has two structures arranged in one row and two columns.

Creating Structures

You can create a structure array by using assignment statements or by using the `struct` function. The following example uses assignment statements to build a structure. Structure arrays use the dot notation (.) to specify and to access the fields. You can type the commands either in the interactive mode or in a script file.

EXAMPLE 2.7–1 A Student Database

Create a structure array to contain the following types of student data:

- Student name.
- Social Security number.
- Email address.
- Test scores.

Enter the data shown in Figure 2.7–1 into the database.

■ Solution

You can create the structure array by typing the following either in the interactive mode or in a script file. Start with the data for the first student.

```
student.name = 'John Smith';
student.SSN = '392-77-1786';
student.email = 'smithj@myschool.edu';
student.tests = [67,75,84];
```

If you then type

```
>>student
```

at the command line, you will see the following response:

```
name:   'John Smith'
SSN: = '392-77-1786'
email: = 'smithj@myschool.edu'
tests: = [67 75 84]
```

To determine the size of the array, type `size(student)`. The result is `ans = 1 1`, which indicates that it is a 1×1 structure array.

To add a second student to the database, use a subscript 2 enclosed in parentheses after the structure array's name and enter the new information. For example, type

```
student(2).name = 'Mary Jones';
student(2).SSN = '431-56-9832';
student(2).email = 'jonesm@myschool.edu';
student(2).tests = [84,78,93];
```

This process "expands" the array. Before we entered the data for the second student, the *dimension* of the structure array was 1×1 (it was a single structure). Now it is a 1×2 array consisting of two structures, arranged in one row and two columns. You can confirm this information by typing `size(student)`, which returns `ans = 1 2`. If you now type `length(student)`, you will get the result `ans = 2`, which indicates that the array has two elements (two structures). When a structure array has more than one structure, MATLAB does not display the individual field contents when you type the structure array's name. For example, if you now type `student`, MATLAB displays

```
>>student =

1x2 struct array with fields:
   name
   SSN
   email
   tests
```

Table 2.7–1 Structure functions

Function	Description
`names = fieldnames(S)`	Returns the field names associated with the structure array `S` as `names`, a cell array of strings.
`F = getfield(S,'field')`	Returns the contents of the field `'field'` in the structure array `S`. Equivalent to `F = S.field`.
`isfield(S,'field')`	Returns 1 if `'field'` is the name of a field in the structure array `S`, and 0 otherwise.
`isstruct(S)`	Returns 1 if the array `S` is a structure array, and 0 otherwise.
`S = rmfield(S,'field')`	Removes the field `'field'` from the structure array `S`.
`S = setfield(S,'field',V)`	Sets the contents of the field `'field'` to the value `V` in the structure array `S`.
`S = struct('f1','v1','f2','v2',...)`	Creates a structure array with the fields `'f1'`, `'f2'`,... having the values `'v1'`, `'v2'`,....

You can also obtain information about the fields by using the `fieldnames` function (see Table 2.7–1). For example:

```
>>fieldnames(student)
ans =
    'name'
    'SSN'
    'email'
    'tests'
```

As you fill in more student information, MATLAB assigns the same number of fields and the same field names to each element. If you do not enter some information—for example, suppose you do not know someone's email address—MATLAB assigns an empty matrix to that field for that student.

The fields can have different sizes. For example, each name field can contain a different number of characters, and the arrays containing the test scores can be different sizes, as would be the case if a certain student did not take the second test.

In addition to the assignment statement, you can also build structures using the `struct` function, which lets you "preallocate" a structure array. To build a structure array named `sa_1`, the syntax is

```
sa_1 = struct('field1','values1','field2','values2', ...)
```

where the arguments are the field names and their values. The values arrays `values1`, `values2`, ... must all be arrays of the same size, scalar cells, or single values. The elements of the values arrays are inserted into the corresponding elements of the structure array. The resulting structure array has the same size as the values arrays, or is 1×1 if none of the values arrays is a cell. For example, to preallocate a 1×1 structure array for the student database, you type

```
student = struct('name','John Smith', 'SSN',...
'392-77-1786','email','smithj@myschool.edu',...
'tests',[67,75,84])
```

Accessing Structure Arrays

To access the contents of a particular field, type a period after the structure array name, followed by the field name. For example, typing `student(2).name` displays the value `'Mary Jones'`. Of course, we can assign the result to a variable in the usual way. For example, typing `name2 = student(2).name` assigns the value `'Mary Jones'` to the variable `name2`. To access elements within a field, for example, John Smith's second test score, type `student(1).tests(2)`. This entry returns the value `75`. In general, if a field contains an array, you use the array's subscripts to access its elements. In this example the statement `student(1).tests(2)` is equivalent to `student(1,1).tests(2)` because `student` has one row.

To store all the information for a particular structure—say, all the information about Mary Jones—in another structure array named `M`, you type `M = student(2)`.

You can also assign or change values of field elements. For example, typing `student(2).tests(2) = 81` changes Mary Jones's second test score from 78 to 81. Direct indexing is usually the best way to create or access field values. However, suppose you used the `fieldnames` function in an M-file to obtain a field name. You would know the field name only as a string. In this situation you can use the `setfield` and `getfield` functions for assigning and retrieving field values. For example, typing `setfield(M,'name', 'Mary Lee Jones')` inserts the new name. Typing `getfield(M,'name')` returns the result `ans = Mary Lee Jones`.

The preceding syntax for the `getfield` and `setfield` functions works on 1×1 arrays only. The alternate syntax, which works for an $i \times j$ array `S`, is

```
F = getfield(S, {i,j}, 'field', {k})
```

which is equivalent to `F = S(i,j).field(k)`. For example,

```
getfield(student, {1,1}, 'tests', {2})
```

returns the result `ans = 75`. Similarly,

```
S = setfield(S, {i,j}, 'field', {k})
```

is equivalent to `S(i,j).field(k) = S`.

Modifying Structures

Suppose you want to add phone numbers to the database. You can do this by typing the first student's phone number as follows:

```
student(1).phone = '555-1653'
```

All the other structures in the array will now have a phone field, but these fields will contain the empty array until you give them values.

To delete a field from every structure in the array, use the rmfield function. Its basic syntax is

```
new_struc = rmfield(array,'field');
```

where array is the structure array to be modified, 'field' is the field to be removed, and new_struc is the name of the new structure array so created by the removal of the field. For example, to remove the Social Security field and call the new structure array new_student, type

```
new_student = rmfield(student,'SSN');
```

Using Operators and Functions with Structures

You can apply the MATLAB operators to structures in the usual way. For example, to find the maximum test score of the second student, you type max(student(2).tests). The answer is 93.

The isfield function determines whether or not a structure array contains a particular field. Its syntax is isfield(S, 'field'). It returns a value of 1 (which means "true") if 'field' is the name of a field in the structure array S. For example, typing isfield(student, 'name') returns the result ans = 1.

The isstruct function determines whether or not an array is a structure array. Its syntax is isstruct(S). It returns a value of 1 if S is a structure array, and 0 otherwise. For example, typing isstruct(student) returns the result ans = 1, which is equivalent to "true."

Dynamic Field Names and New Structure Syntax

Prior to MATLAB 6.5, the elements of a structure could be referenced by using *fixed* field names only. As of MATLAB 6.5 you can reference structures using field names that express the field as a *variable* expression that is evaluated at run time.

Structures can now be referenced by using *dynamic* field names that express the fields as variable expressions that are computed at run time. Use the following parentheses syntax to specify which fields are to be dynamic:

```
structure_name.(expression)
```

where expression is the dynamic field name. Use the standard indexing method to access the information in the structure. For example, typing

```
struct_name.(expression)(5,1:10)
```

accesses the information in the field in row 5, columns 1 through 10.

The following table compares the various static and dynamic data types.

Data type	Static example	Dynamic example
Matrix	A(2,5)	A(r,c)
Cell array	C{5}	C{k*2}
Structure	S.name	S.(field)

Because of the new dynamic names feature, setfield and getfield are less useful than before. The setfield and getfield functions execute more slowly because they are not built-in MATLAB functions.

In addition to increased execution speed, dynamic field names offer improved readability over setfield and getfield. For example, consider the following:

1. Using setfield:

   ```
   S = setfield(S,{m,n},fieldname,{k},value)
   ```

2. Using a dynamic field name:

   ```
   S(m,n).(fieldname)(k) = value
   ```

Test Your Understanding

T2.7–1 Create the structure array student shown in Figure 2.7–1 and add the following information about a third student: name: Alfred E. Newman; SSN: 555-12-3456; e-mail: newmana@myschool.edu; tests: 55, 45, 58.

T2.7–2 Edit your structure array to change Mr. Newman's second test score from 45 to 53.

T2.7–3 Edit your structure array to remove the SSN field.

2.8 Summary

You should now be able to perform basic operations and use arrays in MATLAB. For example, you should be able to

- Create, address, and edit arrays.
- Perform array operations including addition, subtraction, multiplication, division, and exponentiation.

Table 2.8–1 Guide to commands introduced in Chapter 2

Special characters	Use	
'	Transposes a matrix, creating complex conjugate elements.	
.'	Transposes a matrix without creating complex conjugate elements.	
;	Suppresses screen printing; also denotes a new row in an array.	
:	Represents an entire row or column of an array.	
	Tables	
	Array functions	Table 2.1–1
	Element-by-element operations	Table 2.3–1
	Special matrices	Table 2.4–4
	Special products	Table 2.4–5
	Polynomial functions	Table 2.5–1
	Cell array functions	Table 2.6–1
	Structure functions	Table 2.7–1

- Perform matrix operations including addition, subtraction, multiplication, division, and exponentiation.
- Perform polynomial algebra.
- Create databases using cell and structure arrays.

You should be careful to distinguish between array (element-by-element) operations and matrix operations. Each has its own applications.

Table 2.8–1 is a reference guide to all the MATLAB commands introduced in this chapter.

Key Terms with Page References

Problems

You can find the answers to problems marked with an asterisk at the end of the text.

Section 2.1

1. *a.* Use two methods to create the vector **x** having 100 regularly spaced values starting at 5 and ending at 28.
   b. Use two methods to create the vector **x** having a regular spacing of 0.2 starting at 2 and ending at 14.
   c. Use two methods to create the vector **x** having 50 regularly spaced values starting at −2 and ending at 5.
2. *a.* Create the vector **x** having 50 logarithmically spaced values starting at 10 and ending at 1000.
   b. Create the vector **x** having 20 logarithmically spaced values starting at 10 and ending at 1000.
3.* Use MATLAB to create a vector **x** having six values between 0 and 10 (including the endpoints 0 and 10). Create an array **A** whose first row contains the values $3x$ and whose second row contains the values $5x - 20$.
4. Repeat Problem 3 but make the first column of **A** contain the values $3x$ and the second column contain the values $5x - 20$.
5. Type this matrix in MATLAB and use MATLAB to answer the following questions:

$$\mathbf{A} = \begin{bmatrix} 3 & 7 & -4 & 12 \\ -5 & 9 & 10 & 2 \\ 6 & 13 & 8 & 11 \\ 15 & 5 & 4 & 1 \end{bmatrix}$$

   a. Create a vector **v** consisting of the elements in the second column of **A**.
   b. Create a vector **w** consisting of the elements in the second row of **A**.
6. Type this matrix in MATLAB and use MATLAB to answer the following questions:

$$\mathbf{A} = \begin{bmatrix} 3 & 7 & -4 & 12 \\ -5 & 9 & 10 & 2 \\ 6 & 13 & 8 & 11 \\ 15 & 5 & 4 & 1 \end{bmatrix}$$

   a. Create a 4×3 array **B** consisting of all elements in the second through fourth columns of **A**.
   b. Create a 3×4 array **C** consisting of all elements in the second through fourth rows of **A**.
   c. Create a 2×3 array **D** consisting of all elements in the first two rows and the last three columns of **A**.

7.* Compute the length and absolute value of the following vectors:

a. $\mathbf{x} = [2, 4, 7]$
b. $\mathbf{y} = [2, -4, 7]$
c. $\mathbf{z} = [5 + 3i, -3 + 4i, 2 - 7i]$

8. Given the matrix

$$\mathbf{A} = \begin{bmatrix} 3 & 7 & -4 & 12 \\ -5 & 9 & 10 & 2 \\ 6 & 13 & 8 & 11 \\ 15 & 5 & 4 & 1 \end{bmatrix}$$

a. Find the maximum and minimum values in each column.
b. Find the maximum and minimum values in each row.

9. Given the matrix

$$\mathbf{A} = \begin{bmatrix} 3 & 7 & -4 & 12 \\ -5 & 9 & 10 & 2 \\ 6 & 13 & 8 & 11 \\ 15 & 5 & 4 & 1 \end{bmatrix}$$

a. Sort each column and store the result in an array **B**.
b. Sort each row and store the result in an array **C**.
c. Add each column and store the result in an array **D**.
d. Add each row and store the result in an array **E**.

10. Consider the following arrays.

$$\mathbf{A} = \begin{bmatrix} 1 & 4 & 2 \\ 2 & 4 & 100 \\ 7 & 9 & 7 \\ 3 & \pi & 42 \end{bmatrix} \qquad \mathbf{B} = \ln(\mathbf{A})$$

Write MATLAB expressions to do the following.

a. Select just the second row of **B**.
b. Evaluate the sum of the second row of **B**.
c. Multiply the second column of **B** and the first column of **A**.
d. Evaluate the maximum value in the vector resulting from element-by-element multiplication of the second column of **B** with the first column of **A**.
e. Evaluate the sum of the first row of **A** divided element-by-element by the first three elements of the third column of **B**.

Section 2.2

11.* *a.* Create a three-dimensional array **D** whose three "layers" are these matrices:

$$\mathbf{A} = \begin{bmatrix} 3 & -2 & 1 \\ 6 & 8 & -5 \\ 7 & 9 & 10 \end{bmatrix} \quad \mathbf{B} = \begin{bmatrix} 6 & 9 & -4 \\ 7 & 5 & 3 \\ -8 & 2 & 1 \end{bmatrix} \quad \mathbf{C} = \begin{bmatrix} -7 & -5 & 2 \\ 10 & 6 & 1 \\ 3 & -9 & 8 \end{bmatrix}$$

b. Use MATLAB to find the largest element in each layer of **D** and the largest element in **D**.

Section 2.3

12.* Given the matrices

$$\mathbf{A} = \begin{bmatrix} -7 & 16 \\ 4 & 9 \end{bmatrix} \quad \mathbf{B} = \begin{bmatrix} 6 & -5 \\ 12 & -2 \end{bmatrix} \quad \mathbf{C} = \begin{bmatrix} -3 & -9 \\ 6 & 8 \end{bmatrix}$$

Use MATLAB to:

a. Find $\mathbf{A} + \mathbf{B} + \mathbf{C}$.
b. Find $\mathbf{A} - \mathbf{B} + \mathbf{C}$.
c. Verify the associative law

$$(\mathbf{A} + \mathbf{B}) + \mathbf{C} = \mathbf{A} + (\mathbf{B} + \mathbf{C})$$

d. Verify the commutative law

$$\mathbf{A} + \mathbf{B} + \mathbf{C} = \mathbf{B} + \mathbf{C} + \mathbf{A} = \mathbf{A} + \mathbf{C} + \mathbf{B}$$

13.* Given the matrices

$$\mathbf{A} = \begin{bmatrix} 64 & 32 \\ 24 & -16 \end{bmatrix} \quad \mathbf{B} = \begin{bmatrix} 16 & -4 \\ 6 & -2 \end{bmatrix}$$

Use MATLAB to:

a. Find the result of **A** times **B** using the array product.
b. Find the result of **A** divided by **B** using array right division.
c. Find **B** raised to the third power element-by-element.

14.* The mechanical work W done in using a force F to push a block through a distance D is $W = FD$. The following table gives data on the amount of force used to push a block through the given distance over five segments of a certain path. The force varies because of the differing friction properties of the surface.

	Path segment				
	1	2	3	4	5
Force (N)	400	550	700	500	600
Distance (m)	2	0.5	0.75	1.5	3

Use MATLAB to find (*a*) the work done on each segment of the path and (*b*) the total work done over the entire path.

15. Plane A is heading southwest at 200 mi/hr, while plane B is heading west at 150 mi/hr. What is the velocity and the speed of plane A relative to plane B?

16. The following table shows the hourly wages, hours worked, and output (number of widgets produced) in one week for five widget makers.

	Worker				
	1	2	3	4	5
Hourly wage ($)	5	5.50	6.50	6	6.25
Hours worked	40	43	37	50	45
Output (widgets)	1000	1100	1000	1200	1100

Use MATLAB to answer these questions:

a. How much did each worker earn in the week?
b. What is the total salary amount paid out?
c. How many widgets were made?
d. What is the average cost to produce one widget?
e. How many hours does it take to produce one widget on average?
f. Assuming that the output of each worker has the same quality, which worker is the most efficient? Which is the least efficient?

17. Two divers start at the surface and establish the following coordinate system: x is to the west, y is to the north, and z is down. Diver 1 swims 60 ft east, then 25 ft south, and then dives 30 ft. At the same time, diver 2 dives 20 ft, swims east 30 ft, and then south 55 ft.

a. Compute the distance between diver 1 and the starting point.
b. How far in each direction must diver 1 swim to reach diver 2?
c. How far in a straight line must diver 1 swim to reach diver 2?

18. The potential energy stored in a spring is $kx^2/2$, where k is the spring constant and x is the compression in the spring. The force required to compress the spring is kx. The following table gives the data for five springs:

	Spring				
	1	2	3	4	5
Force (N)	11	7	8	10	9
Spring constant k (N/m)	1000	800	900	1200	700

Use MATLAB to find (*a*) the compression x in each spring and (*b*) the potential energy stored in each spring.

19. A company must purchase five kinds of material. The following table gives the price the company pays per ton for each material, along with the number of tons purchased in the months of May, June, and July:

		Quantity purchased (tons)		
Material	Price ($/ton)	May	June	July
1	300	5	4	6
2	550	3	2	4
3	400	6	5	3
4	250	3	5	4
5	500	2	4	3

Use MATLAB to answer these questions:

a. Create a 5×3 matrix containing the amounts spent on each item for each month.

b. What is the total spent in May? in June? in July?

c. What is the total spent on each material in the three-month period?

d. What is the total spent on all materials in the three-month period?

20. A fenced enclosure consists of a rectangle of length L and width $2R$, and a semicircle of radius R, as shown in Figure P20. The enclosure is to be built to have an area A of 1600 ft^2. The cost of the fence is \$40/ft for the curved portion, and \$30/ft for the straight sides. Use the `min` function to determine with a resolution of 0.01 foot the values of R and L required to minimize the total cost of the fence. Also compute the minimum cost.

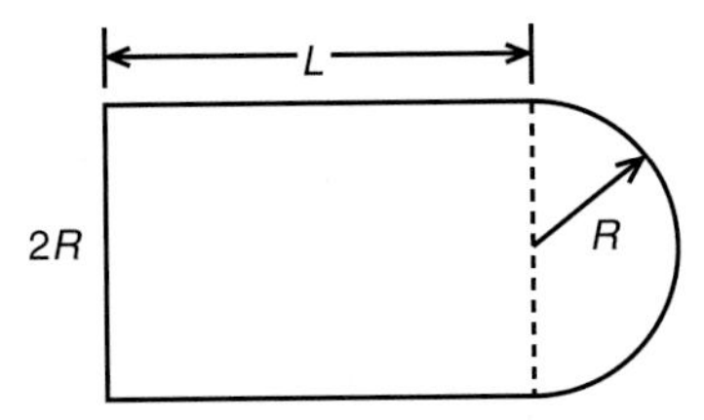

Figure P20

21. A geometric series is defined as the sequence $1, x, x^2, x^3, \ldots,$ in which the powers of x range over the integers from 0 to ∞. The sum of the terms in a geometric series converges to the limiting value of $1/(1 - x)$ if $|x| < 1$; otherwise the terms diverge.

a. For $x = 0.63$, compute the sum of the first 11 terms in the series, and compare the result with the limiting value. Repeat for 51 and 101 terms. Do this by generating a vector of integers to use as the exponent of x; then use the `sum` function.

b. Repeat part (*a*) using $x = -0.63$.

22. A water tank consists of a cylindrical part of radius r and height h, and a hemispherical top. The tank is to be constructed to hold 500 m^3 of fluid when filled. The surface area of the cylindrical part is $2\pi rh$, and its volume is $\pi r^2 h$. The surface area of the hemispherical top is given by $2\pi r^2$, and its volume is given by $2\pi r^3/3$. The cost to construct the cylindrical part of the tank is \$300/m^2 of surface area; the hemispherical part costs \$400/m^2. Plot the cost versus r for $2 \leq r \leq 10$ m, and determine the radius that results in the least cost. Compute the corresponding height h.

23. Write a MATLAB assignment statement for each of the following functions, assuming that w, x, y, and z are vector quantities of equal length, and that c and d are scalars.

$$f = \frac{1}{\sqrt{\frac{2\pi c}{x}}} \qquad E = \frac{x + \frac{w}{y+z}}{x + \frac{w}{y-z}}$$

$$A = \frac{e^{-c/(2x)}}{(\ln y)\sqrt{dz}} \qquad S = \frac{x(2.15 + 0.35y)^{1.8}}{z(1-x)^y}$$

24. In many engineering systems an electrical power source supplies current or voltage to a device called the "load." A common example is an amplifier-speaker system. The load is the speaker, which requires current from the amplifier to produce sound. Figure P24a shows the general representation of a source and load. The resistance R_L is that of the load. Figure P24b shows the circuit representation of the system. The source supplies a voltage v_S and a current i_S and has its own internal resistance R_S. For optimum efficiency, we want to maximize the power supplied to the speaker for given values of v_S and R_S. We can do so by properly selecting the value of the load resistance R_L.

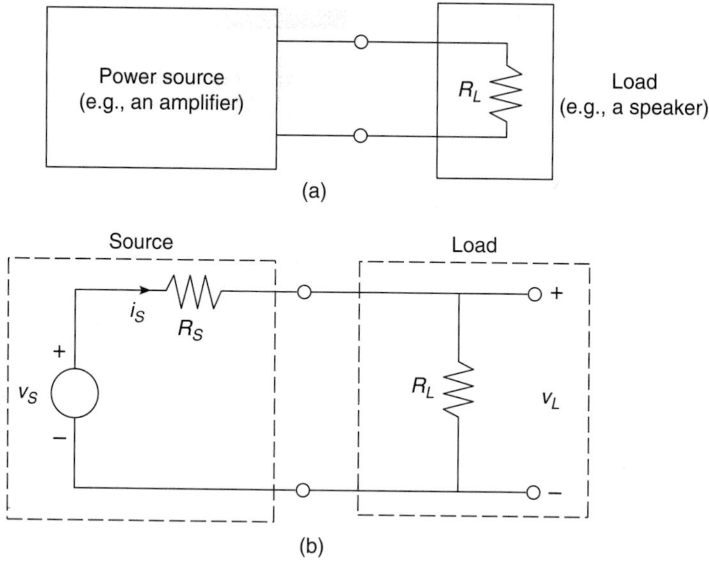

Figure P24

The power consumed by the load is $P_L = i_S^2 R_L = v_L^2/R_L$. Using the relation between v_L and v_S we can express P_L in terms of v_S as

$$P_L = \frac{R_L}{(R_S + R_L)^2} v_S^2$$

To maximize P_L for a fixed value of v_S, we must maximize the ratio

$$r = \frac{R_L}{(R_S + R_L)^2}$$

Consider the specific case where the source resistance can be $R_S = 10, 15, 20,$ or $25\ \Omega$ and where the available load resistances are $R_L = 10, 15, 20, 25,$ and $30\ \Omega$. For a specific value of R_S, determine which value of R_L will maximize the power transfer.

25. Some current research in biomedical engineering deals with devices for measuring the medication levels in the blood and automatically adjusting the intravenous delivery rate to achieve the proper concentration (too high a concentration will cause adverse reactions). To design such devices, engineers must develop a model of the concentration as a function of the dosage and of time.

a. After a dose, the concentration declines due to metabolic processes. The *half-life* of a medication is the time required after an initial dosage for the concentration to be reduced by one-half. A common model for this process is

$$C(t) = C(0)e^{-kt}$$

where $C(0)$ is the initial concentration, t is time (in hours), and k is called the *elimination rate constant,* which varies among individuals. For a particular bronchodilator, k has been estimated to be in the range $0.047 \leq k \leq 0.107$ per hour. Find an expression for the half-life in terms of k, and obtain a plot of the half-life versus k for the indicated range.

b. If the concentration is initially zero, and a constant delivery rate is started and maintained, the concentration as a function of time is described by:

$$C(t) = \frac{a}{k}(1 - e^{-kt})$$

where a is a constant that depends on the delivery rate. Plot the concentration after one hour, $C(1)$, versus k for the case where $a = 1$ and k is in the range $0.047 \leq k \leq 0.107$ per hour.

26. A cable of length L_c supports a beam of length L_b, so that it is horizontal when the weight W is attached at the beam end. The principles of statics can be used to show that the tension force T in the cable is given by

$$T = \frac{L_b L_c W}{D\sqrt{L_b^2 - D^2}}$$

where D is the distance of the cable attachment point to the beam pivot. See Figure P26.

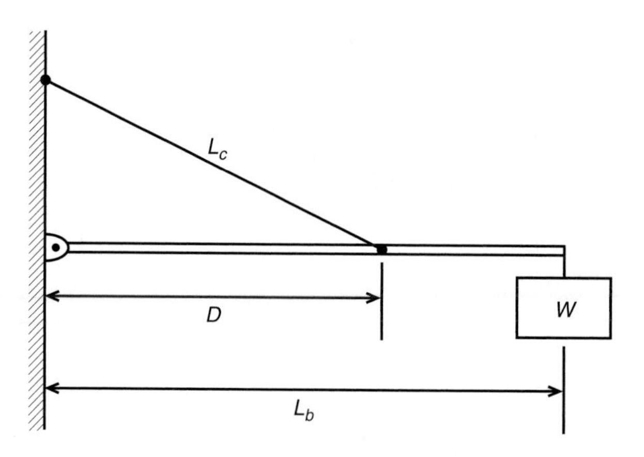

Figure P26

a. For the case where $W = 400$ N, $L_b = 3$ m, and $L_c = 5$ m, use element-by-element operations and the `min` function to compute the value of D that minimizes the tension T. (Do not use a loop.) Compute the minimum tension value.

b. Check the sensitivity of the solution by plotting T versus D. How much can D vary from its optimal value before the tension T increases 10 percent above its minimum value?

Section 2.4

27.* Use MATLAB to find the products **AB** and **BA** for the following matrices:

$$\mathbf{A} = \begin{bmatrix} 11 & 5 \\ -9 & -4 \end{bmatrix} \quad \mathbf{B} = \begin{bmatrix} -7 & -8 \\ 6 & 2 \end{bmatrix}$$

28. Given the matrices

$$\mathbf{A} = \begin{bmatrix} 3 & -2 & 1 \\ 6 & 8 & -5 \\ 7 & 9 & 10 \end{bmatrix} \quad \mathbf{B} = \begin{bmatrix} 6 & 9 & -4 \\ 7 & 5 & 3 \\ -8 & 2 & 1 \end{bmatrix} \quad \mathbf{C} = \begin{bmatrix} -7 & -5 & 2 \\ 10 & 6 & 1 \\ 3 & -9 & 8 \end{bmatrix}$$

Use MATLAB to:

a. Verify the associative property

$$\mathbf{A}(\mathbf{B} + \mathbf{C}) = \mathbf{AB} + \mathbf{AC}$$

b. Verify the distributive property

$$(\mathbf{AB})\mathbf{C} = \mathbf{A}(\mathbf{BC})$$

29. The following tables show the costs associated with a certain product and the production volume for the four quarters of the business year. Use MATLAB to find (*a*) the quarterly costs for materials, labor, and transportation; (*b*) the total material, labor, and transportation costs for the year; and (*c*) the total quarterly costs.

	Unit product costs ($\$ \times 10^3$)		
Product	Materials	Labor	Transportation
1	7	3	2
2	3	1	3
3	9	4	5
4	2	5	4
5	6	2	1

	Quarterly production volume			
Product	Quarter 1	Quarter 2	Quarter 3	Quarter 4
1	16	14	10	12
2	12	15	11	13
3	8	9	7	11
4	14	13	15	17
5	13	16	12	18

30.* Aluminum alloys are made by adding other elements to aluminum to improve its properties, such as hardness or tensile strength. The following table shows the composition of five commonly used alloys, which are known by their alloy numbers (2024, 6061, and so on) [Kutz, 1986]. Obtain a matrix algorithm to compute the amounts of raw materials needed to produce a given amount of each alloy. Use MATLAB to determine how much raw material of each type is needed to produce 1000 tons of each alloy.

	Composition of aluminum alloys				
Alloy	%Cu	%Mg	%Mn	%Si	%Zn
2024	4.4	1.5	0.6	0	0
6061	0	1	0	0.6	0
7005	0	1.4	0	0	4.5
7075	1.6	2.5	0	0	5.6
356.0	0	0.3	0	7	0

31. Redo Example 2.4–2 as a script file to allow the user to examine the effects of labor costs. Allow the user to input the four labor costs in the following table. When you run the file, it should display the quarterly costs and the category costs. Run the file for the case where the unit labor costs are \$3000, \$7000, \$4000, and \$8000, respectively.

Product costs

	Unit costs ($\$ \times 10^3$)		
Product	**Materials**	**Labor**	**Transportation**
1	6	2	1
2	2	5	4
3	4	3	2
4	9	7	3

Quarterly production volume

Product	Quarter 1	Quarter 2	Quarter 3	Quarter 4
1	10	12	13	15
2	8	7	6	4
3	12	10	13	9
4	6	4	11	5

32. Vectors with three elements can represent position, velocity, and acceleration. A mass of 5 kg, which is 3 m away from the x-axis, starts at $x = 2$ m and moves with a speed of 10 m/s parallel to the y-axis. Its velocity is thus described by $\mathbf{v} = [0, 10, 0]$, and its position is described by $\mathbf{r} = [2, 10t + 3, 0]$. Its angular momentum vector $\mathbf{L}$ is found from $\mathbf{L} = m(\mathbf{r} \times \mathbf{v})$, where m is the mass. Use MATLAB to:

a. Compute a matrix $\mathbf{P}$ whose 11 rows are the values of the position vector $\mathbf{r}$ evaluated at the times $t = 0, 0.5, 1, 1.5, \ldots 5$ s.

b. What is the location of the mass when $t = 5$ s?

c. Compute the angular momentum vector $\mathbf{L}$. What is its direction?

33.* The *scalar triple product* computes the magnitude M of the moment of a force vector $\mathbf{F}$ about a specified line. It is $M = (\mathbf{r} \times \mathbf{F}) \cdot \mathbf{n}$, where $\mathbf{r}$ is the position vector from the line to the point of application of the force and $\mathbf{n}$ is a unit vector in the direction of the line.

Use MATLAB to compute the magnitude M for the case where $\mathbf{F} = [10, -5, 4]$ N, $\mathbf{r} = [-3, 7, 2]$ m, and $\mathbf{n} = [6, 8, -7]$.

34. Verify the identity

$$\mathbf{A} \times (\mathbf{B} \times \mathbf{C}) = \mathbf{B}(\mathbf{A} \cdot \mathbf{C}) - \mathbf{C}(\mathbf{A} \cdot \mathbf{B})$$

for the vectors $\mathbf{A} = 5\mathbf{i} - 3\mathbf{j} + 7\mathbf{k}$, $\mathbf{B} = -6\mathbf{i} + 4\mathbf{j} + 3\mathbf{k}$, and $\mathbf{C} = 2\mathbf{i} + 8\mathbf{j} - 9\mathbf{k}$.

35. The area of a parallelogram can be computed from $|\mathbf{A} \times \mathbf{B}|$, where $\mathbf{A}$ and $\mathbf{B}$ define two sides of the parallelogram (see Figure P35). Compute the area of a parallelogram defined by $\mathbf{A} = 7\mathbf{i}$ and $\mathbf{B} = \mathbf{i} + 3\mathbf{j}$.

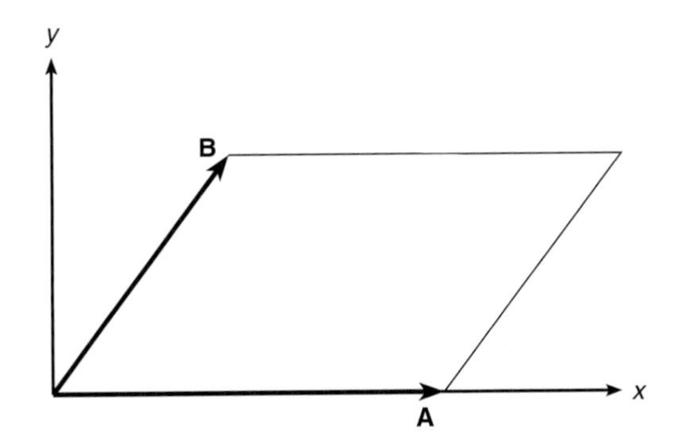

Figure P35

36. The volume of a parallelepiped can be computed from $|\mathbf{A} \cdot (\mathbf{B} \times \mathbf{C})|$, where **A**, **B**, and **C** define three sides of the parallelepiped (see Figure P36). Compute the volume of a parallelepiped defined by $\mathbf{A} = 6\mathbf{i}$, $\mathbf{B} = 2\mathbf{i} + 4\mathbf{j}$, and $\mathbf{C} = 3\mathbf{i} - 2\mathbf{k}$.

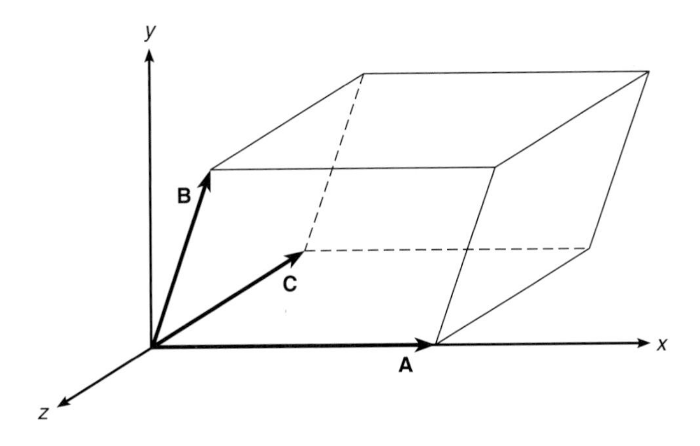

Figure P36

Section 2.5

37. Use MATLAB to plot the polynomials $y = 3x^4 - 6x^3 + 8x^2 + 4x + 90$ and $z = 3x^3 + 5x^2 - 8x + 70$ over the interval $-3 \leq x \leq 3$. Properly label the plot and each curve. The variables y and z represent current in milliamps; the variable x represents voltage in volts.
38. Use MATLAB to plot the polynomial $y = 3x^4 - 5x^3 - 28x^2 - 5x + 200$ on the interval $-1 \leq x \leq 1$. Put a grid on the plot and use the `ginput` function to determine the coordinates of the peak of the curve.

39. Use MATLAB to find the following product:

$$(10x^3 - 9x^2 - 6x + 12)(5x^3 - 4x^2 - 12x + 8)$$

40.* Use MATLAB to find the quotient and remainder of

$$\frac{14x^3 - 6x^2 + 3x + 9}{5x^2 + 7x - 4}$$

41.* Use MATLAB to evaluate

$$\frac{8x^3 - 9x^2 - 7}{10x^3 + 5x^2 - 3x - 7}$$

at $x = 5$.

42. Refer to Example 2.5–1. The polynomial equation that must be solved to find the natural frequencies of a particular building is

$$(\alpha - f^2)[(2\alpha - f^2)^2 - \alpha^2] + \alpha^2 f^2 - 2\alpha^3$$

where $\alpha = k/4m\pi^2$. Suppose $m = 5000$ kg. Consider three cases: (1) $k = 4 \times 10^6$ N/m; (2) $k = 5 \times 10^6$ N/m; and (3) $k = 6 \times 10^6$ N/m. Write a MATLAB script file containing a `for` loop that does the algebra to answer these questions:

a. Which case has the smallest natural frequency?
b. Which case has the largest natural frequency?
c. Which case has the smallest spread between the natural frequencies?

43. Engineers often need to estimate the pressures and volumes of a gas in a container. The *ideal gas law* provides one way of making the estimate. The law is

$$P = \frac{RT}{\hat{V}}$$

More accurate estimates can be made with the *van der Waals* equation:

$$P = \frac{RT}{\hat{V} - b} - \frac{a}{\hat{V}^2}$$

where the term b is a correction for the volume of the molecules, and the term $a/\hat{V}^2$ is a correction for molecular attractions. The values of a and b depend on the type of gas. The gas constant is R, the *absolute* temperature is T, and the gas specific volume is $\hat{V}$. If 1 mol of an ideal gas were confined to a volume of 22.41 L at 0°C (273.2 K), it would exert a pressure of 1 atmosphere. In these units, $R = 0.08206$.

For chlorine (Cl_2), $a = 6.49$ and $b = 0.0562$. Compare the specific volume estimates $\hat{V}$ given by the ideal gas law and the van der Waals equation for 1 mol of Cl_2 at 300 K and a pressure of 0.95 atmosphere.

44. Aircraft A is flying east at 320 mi/hr, while aircraft B is flying south at 160 mi/hr. At 1:00 P.M. the aircraft are located as shown in Figure P44.

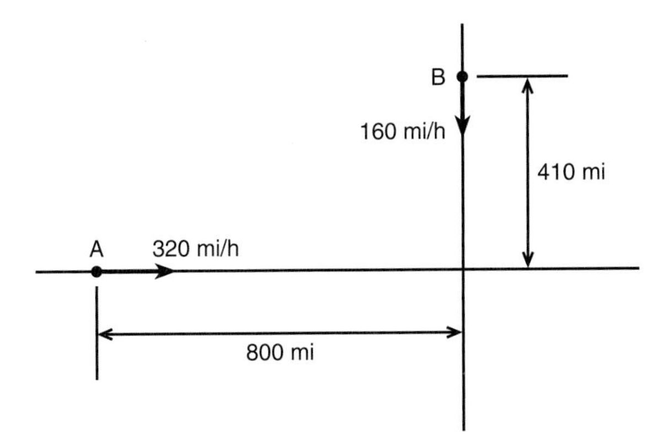

Figure P44

a. Obtain the expression for the distance D between the aircraft as a function of time. Plot D versus time until D reaches its minimum value.

b. Use the `roots` function to compute the time when the aircraft are first within 30 mi of each other.

45. The function

$$y = \frac{3x^2 - 12x + 20}{x^2 - 7x + 10}$$

approaches ∞ as $x \to 2$ and as $x \to 5$. Plot this function over the range $0 \le x \le 7$. Choose an appropriate range for the y-axis.

46. The following formulas are commonly used by engineers to predict the lift and drag of an airfoil.

$$L = \frac{1}{2}\rho C_L S V^2$$

$$D = \frac{1}{2}\rho C_D S V^2$$

where L and D are the lift and drag forces, V is the airspeed, S is the wing span, ρ is the air density, and C_L and C_D are the *lift* and *drag* coefficients. Both C_L and C_D depend on α, the angle of attack, the angle between the relative air velocity and the airfoil's chord line.

Wind tunnel experiments for a particular airfoil have resulted in the following formulas (We will see in Chapter 5 how such formulas can be obtained from data).

$$C_L = 4.47 \times 10^{-5}\alpha^3 + 1.15 \times 10^{-3}\alpha^2 + 6.66 \times 10^{-2}\alpha + 1.02 \times 10^{-1}$$

$$C_D = 5.75 \times 10^{-6}\alpha^3 + 5.09 \times 10^{-4}\alpha^2 + 1.81 \times 10^{-4}\alpha + 1.25 \times 10^{-2}$$

where α is in degrees.

Plot the lift and drag of this airfoil versus V for $0 \leq V \leq 150$ mi/hr (you must convert V to ft/sec; there are 5280 ft/mi). Use the values $\rho = 0.002378$ slug/ft^3 (air density at sea level), $\alpha = 10°$, and $S = 36$ ft. The resulting values of L and D will be in pounds.

47. The lift-to-drag ratio is an indication of the effectiveness of an airfoil. Referring to Problem 46, the equations for lift and drag are

$$L = \frac{1}{2}\rho C_L S V^2$$

$$D = \frac{1}{2}\rho C_D S V^2$$

where, for a particular airfoil, the lift and drag coefficients versus angle of attack α are given by

$$C_L = 4.47 \times 10^{-5}\alpha^3 + 1.15 \times 10^{-3}\alpha^2 + 6.66 \times 10^{-2}\alpha + 1.02 \times 10^{-1}$$

$$C_D = 5.75 \times 10^{-6}\alpha^3 + 5.09 \times 10^{-4}\alpha^2 + 1.81 \times 10^{-4}\alpha + 1.25 \times 10^{-2}$$

Using the first two equations, we see that the lift-to-drag ratio is given simply by the ratio C_L/C_D.

$$\frac{L}{D} = \frac{\frac{1}{2}\rho C_L S V^2}{\frac{1}{2}\rho C_D S V^2} = \frac{C_L}{C_D}$$

Plot L/D versus α for $-2° \leq \alpha \leq 22°$. Determine the angle of attack that maximizes L/D.

Section 2.6

48. *a.* Use both cell indexing and content indexing to create the following 2×2 cell array:

Motor 28C	Test ID 6
$\begin{bmatrix} 3 & 9 \\ 7 & 2 \end{bmatrix}$	[6 5 1]

b. What are the contents of the (1,1) element in the (2,1) cell in this array?

49. The capacitance of two parallel conductors of length L and radius r, separated by a distance d in air, is given by

$$C = \frac{\pi \epsilon L}{\ln\left(\frac{d-r}{r}\right)}$$

where ϵ is the permittitivity of air ($\epsilon = 8.854 \times 10^{-12}$ F/m). Create a cell array of capacitance values versus d, L, and r for $d = 0.003, 0.004, 0.005,$

and 0.01 m; $L = 1, 2, 3$ m; and $r = 0.001, 0.002, 0.003$ m. Use MATLAB to determine the capacitance value for $d = 0.005$, $L = 2$, and $r = 0.001$.

Section 2.7

50. *a.* Create a structure array that contains the conversion factors for converting units of mass, force, and distance between the metric SI system and the British Engineering System.

b. Use your array to compute the following:

- The number of meters in 24 ft.
- The number of feet in 65 m.
- The number of pounds equivalent to 18 N.
- The number of newtons equivalent to 5 lb.
- The number of kilograms in 6 slugs.
- The number of slugs in 15 kg.

51. Create a structure array that contains the following information fields concerning the road bridges in a town: bridge location, maximum load (tons), year built, year due for maintenance. Then enter the following data into the array:

Location	Max load	Year built	Due maintenance
Smith St.	80	1928	1997
Hope Ave.	90	1950	1999
Clark St.	85	1933	1998
North Rd.	100	1960	1998

52. Edit the structure array created in Problem 51 to change the maintenance data for the Clark St. bridge from 1998 to 2000.

53. Add the following bridge to the structure array created in Problem 51.

Location	Max load	Year built	Due maintenance
Shore Rd.	85	1997	2002

Courtesy Henry Guckel (Late)/Dept. of Electrical Engineering, University of Wisconsin.

Engineering in the 21st Century...

Nanotechnology

While large-scale technology is attracting much public attention, many of the engineering challenges and opportunities in the 21st century will involve the development of extremely small devices and even the manipulation of individual atoms. This technology is called *nanotechnology* because it involves processing materials whose size is about 1 nanometer (nm), which is 10^{-9} m, or $1/1{,}000{,}000$ of a ml. The distance between atoms in single-crystal silicon is 0.5 nm.

Nanotechnology is in its infancy, although some working devices have been created. The micromotor with gear train shown above has a dimension of approximately 10^{-4} m. This device converts electrical input power into mechanical motion. It was constructed using the magnetic properties of electroplated metal films.

While we are learning how to make such devices, another challenge is to develop innovative applications for them. Many of the applications proposed thus far are medical; small pumps for drug delivery and surgical tools are two examples. Researchers at the Lawrence Livermore Laboratory have developed a microgripper tool to treat brain aneurysms. It is about the size of a grain of sand and was constructed from silicon cantilever beams powered by a shape-memory alloy actuator. To design and apply these devices, engineers must first model the appropriate mechanical and electrical properties. The features of MATLAB provide excellent support for such analyses. ■

CHAPTER 3

Functions and Files

OUTLINE

MATLAB has many built-in functions, including trigonometric, logarithmic, and hyperbolic functions, as well as functions for processing arrays. These functions are summarized in Section 3.1. In addition, you can define your own functions with a *function* file, and you can use them just as conveniently as the built-in functions. We explain this technique in Section 3.2. Section 3.3 covers advanced topics in function programming, including function handles, anonymous functions, subfunctions, and nested functions. These topics are especially useful for large programming projects. In addition to function files, another type of file that is useful in MATLAB is the data file. Importing and exporting such files is covered in Section 3.4.

3.1 Elementary Mathematical Functions

You can use the `lookfor` command to find functions that are relevant to your application. For example, type `lookfor imaginary` to get a list of the functions that deal with imaginary numbers. You will see listed:

```
imag    Complex imaginary part
i       Imaginary unit
j       Imaginary unit
```

Note that `imaginary` is not a MATLAB function, but the word is found in the help descriptions of the MATLAB function `imag` and the special symbols `i` and `j`. Their names and brief descriptions are displayed when you type `lookfor imaginary`. If you know the correct spelling of a MATLAB function—for example, `disp`—you can type `help disp` to obtain a description of the function.

Some of the functions, like `sqrt` and `sin`, are built-in, and are not M-files. They are part of the MATLAB core so they are very efficient, but the computational details are not readily accessible. Other functions, like `sinh`, are implemented in M-files. You can see the code and even modify it if you want.

Exponential and Logarithmic Functions

Table 3.1–1 summarizes some of the common elementary functions. An example is the square root function `sqrt`. To compute $\sqrt{9}$, you type `sqrt(9)` at the command line. When you press **Enter,** you see the result `ans = 3`. You can use functions with variables. For example, consider the session:

```
>>x = 9;
>>y = sqrt(x)
y =
   3
```

MATLAB automatically handles the square roots of negative numbers and returns a number with an imaginary part as the result. For example, typing `sqrt(-9)` gives the result `ans = 0 + 3.0000i`, which is the *positive* root.

Similarly, we can type `exp(2)` to obtain $e^2 = 7.3891$, where e is the base of the natural logarithms. Typing `exp(1)` gives 2.7183, which is e. Note that in

Table 3.1–1 Some common mathematical functions

Exponential	
`exp(x)`	Exponential; e^x.
`sqrt(x)`	Square root; $\sqrt{x}$.
Logarithmic	
`log(x)`	Natural logarithm; $\ln x$.
`log10(x)`	Common (base 10) logarithm; $\log x = \log_{10} x$.
Complex	
`abs(x)`	Absolute value; x.
`angle(x)`	Angle of a complex number x.
`conj(x)`	Complex conjugate.
`imag(x)`	Imaginary part of a complex number x.
`real(x)`	Real part of a complex number x.
Numeric	
`ceil(x)`	Round to the nearest integer toward ∞.
`fix(x)`	Round to the nearest integer toward zero.
`floor(x)`	Round to the nearest integer toward $-\infty$.
`round(x)`	Round toward the nearest integer.
`sign(x)`	Signum function: $+1$ if $x > 0$; 0 if $x = 0$; -1 if $x < 0$.

mathematics text, ln x denotes the *natural* logarithm, where $x = e^y$ implies that

$$\ln x = \ln(e^y) = y \ln e = y$$

because $\ln e = 1$. However, this notation has not been carried over into MATLAB, which uses `log(x)` to represent ln x.

The *common* (base 10) logarithm is denoted in text by $\log x$ or $\log_{10} x$. It is defined by the relation $x = 10^y$; that is,

$$\log_{10} x = \log_{10} 10^y = y \log_{10} 10 = y$$

because $\log_{10} 10 = 1$. The MATLAB common logarithm function is `log10(x)`. A common mistake is to type `log(x)`, instead of `log10(x)`.

Another common error is to forget to use the array multiplication operator `.*`. Note that in the MATLAB expression `y = exp(x).*log(x)`, we need to use the operator `.*` because both `exp(x)` and `log(x)` are arrays if `x` is an array.

Complex Number Functions

Chapter 1 explained how MATLAB easily handles complex number arithmetic. Several functions facilitate complex number operations. Figure 3.1–1 shows a graphical representation of a complex number in terms of a right triangle. The number $a + ib$ represents a point in the xy plane. In the *rectangular* representation $a + ib$, the number's real part a is the x coordinate of the point, and the imaginary part b is the y coordinate.

The *polar* representation uses the distance M of the point from the origin, which is the length of the hypotenuse, and the angle θ the hypotenuse makes with

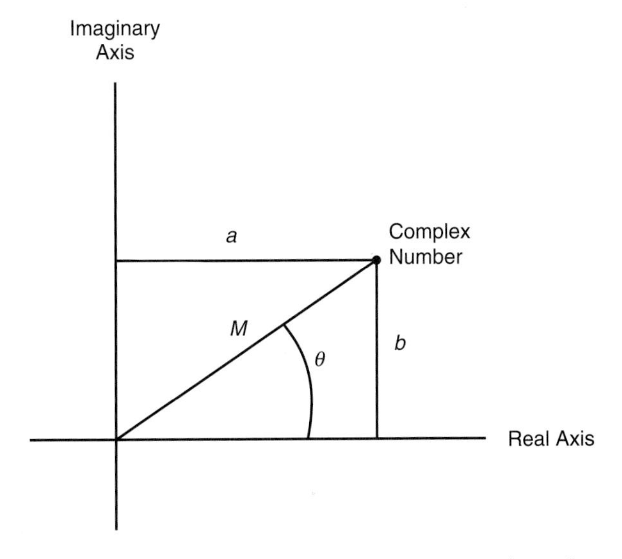

Figure 3.1–1 The rectangular and polar representations of the complex number $a + ib$.

the *positive real* axis. The pair (M, θ) is simply the polar coordinates of the point. In the polar representation the number is written as $M\angle\theta$. From the Pythagorean theorem, the length of the hypotenuse is given by

$$M = \sqrt{a^2 + b^2}$$

which is called the *magnitude* of the number. The angle θ can be found from the trigonometry of the right triangle. It is

$$\theta = \arctan(b/a)$$

Adding and subtracting complex numbers by hand is easy when they are in the rectangular representation. However, the polar representation facilitates multiplication and division of complex numbers by hand. We must enter complex numbers in MATLAB using the rectangular form, and its answers will be given in that form. We can obtain the rectangular representation from the polar representation as follows:

$$a = M\cos\theta \qquad b = M\sin\theta$$

The MATLAB `abs(x)` and `angle(x)` functions calculate the magnitude M and angle θ of the complex number x. The functions `real(x)` and `imag(x)` return the real and imaginary parts of x. The *complex conjugate* of the number $a + ib$ is $a - ib$. It can be shown that the complex conjugate of $M\angle\theta$ is $M\angle(-\theta)$. The function `conj(x)` computes the complex conjugate of `x`.

Note that when **x** is a vector, `abs(x)` gives a *vector* of absolute values. It does not give the *magnitude* $|\mathbf{x}|$ of the vector. The magnitude of **x** is a *scalar* and is given by `sqrt(x'*x)` if **x** is a column vector, and by `sqrt(x*x')` if **x** is a row vector. Thus `abs(x)` does not give $|\mathbf{x}|$.

The magnitude of the product z of two complex numbers x and y is equal to the product of their magnitudes: $|z| = |x||y|$. The angle of the product is equal to the sum of the angles: $\angle z = \angle x + \angle y$. These facts are demonstrated below.

```
>>x = -3 + 4i;
>>y = 6 - 8i;
>>mag_x = abs(x)
mag_x =
   5.0000
>>mag_y = abs(y)
mag_y =
   10.0000
>>mag_product = abs(x*y)
   50.0000
>>angle_x = angle(x)
angle_x =
   2.2143
>>angle_y = angle(y)
angle_y =
   -0.9273
>>sum_angles = angle_x + angle_y
```

```
sum_angles =
   1.2870
>>angle_product = angle(x*y)
angle_product =
   1.2870
```

Similarly, for division, if $z = x/y$, then $|z| = |x|/|y|$ and $\angle z = \angle x - \angle y$.

Numeric Functions

Recall that one of the strengths of MATLAB is that it has been optimized to deal with arrays, and it will treat a variable as an array automatically. For example, to compute the square roots of 5, 7, and 15, type

```
>>x = [5,7,15];
>>y = sqrt(x)
y =
   2.2361   2.6358   3.8730
```

The square root function operates on every element in the array `x`.

The `round` function rounds to the nearest integer. Typing `round(y)` following the preceding session gives the results 2, 3, 4. The `fix` function truncates to the nearest integer toward zero. Typing `fix(y)` following the above session gives the results 2, 2, 3. The `ceil` function (which stands for "ceiling") rounds to the nearest integer toward ∞. Typing `ceil(y)` produces the answers 3, 3, 4.

Suppose `z = [-2.6,-2.3,5.7]`. The `floor` function rounds to the nearest integer toward $-\infty$. Typing `floor(z)` produces the result -3, -3, 5. Typing `fix(z)` produces the answer -2, -2, 5. The `abs` function computes the absolute value. Thus `abs(z)` produces 2.6, 2.3, 5.7.

Test Your Understanding

T3.1–1 For several values of x and y, confirm that $\ln(xy) = \ln x + \ln y$.

T3.1–2 Find the magnitude, angle, real part, and imaginary part of the number $\sqrt{2+6i}$.

Trigonometric Functions

When writing mathematics in text, we use parentheses ( ), brackets [ ], and braces { } to improve the readability of expressions, and we have much latitude over their use. For example, we can write sin 2 in text, but MATLAB requires parentheses surrounding the 2 (which is called the *function argument* or *parameter*). Thus to evaluate sin 2 in MATLAB, we type `sin(2)`. The MATLAB function name must be followed by a pair of parentheses that surround the argument. To express in text the sine of the second element of the array x, we would type `sin[x(2)]`.

FUNCTION ARGUMENT

However, in MATLAB you cannot use brackets or braces in this way, and you must type `sin(x(2))`.

You can include expressions and other functions as arguments. For example, to evaluate $\sin(x^2+5)$, you type `sin(x.^2 + 5)`. To evaluate $\sin(\sqrt{x}+1)$, you type `sin(sqrt(x)+1)`. Using a function as an argument of another function is called *function composition.* Be sure to check the order of precedence and the number and placement of parentheses when typing such expressions. Every left-facing parenthesis requires a right-facing mate. However, this condition does not guarantee that the expression is correct!

Another common mistake involves expressions like $\sin^2 x$, which means $(\sin x)^2$. In MATLAB we write this expression as `(sin(x))^2`, *not* as `sin^2(x)`, `sin^2x`, `sin(x^2)`, or `sin(x)^2`!

Other commonly used functions are `cos(x)`, `tan(x)`, `sec(x)`, and `csc(x)`, which return $\cos x$, $\tan x$, $\sec x$, and $\csc x$, respectively. Table 3.1–2 lists the MATLAB trigonometric functions.

The MATLAB trigonometric functions operate in radian mode. Thus `sin(5)` computes the sine of 5 rad, not the sine of 5°. To convert between degrees and radians, use the relation $\theta_{\text{radians}} = (\pi/180)\theta_{\text{degrees}}$. Similarly, the inverse trigonometric functions return an answer in radians. The inverse sine, arcsin $x = \sin^{-1} x$, is the value y that satisfies $\sin y = x$. To compute the inverse sine, type `asin(x)`. For example, `asin(0.5)` returns the answer: 0.5236 rad. Thus $\sin(0.5236) = 0.5$.

MATLAB has two inverse tangent functions. The function `atan(x)` computes $\arctan x$—the arctangent or inverse tangent—and returns an angle between $-\pi/2$ and $\pi/2$. Another correct answer is the angle that lies in the opposite quadrant. The user must be able to choose the correct answer. For example, `atan(1)` returns the answer 0.7854 rad, which corresponds to 45°. Thus $\tan 45° = 1$. However, $\tan(45° + 180°) = \tan 225° = 1$ also. Thus $\arctan(1) = 225°$ is also correct.

Table 3.1–2 Trigonometric functions

Trigonometric	
`cos(x)`	Cosine; $\cos x$.
`cot(x)`	Cotangent; $\cot x$.
`csc(x)`	Cosecant; $\csc x$.
`sec(x)`	Secant; $\sec x$.
`sin(x)`	Sine; $\sin x$.
`tan(x)`	Tangent; $\tan x$.
Inverse trigonometric	
`acos(x)`	Inverse cosine; $\arccos x = \cos^{-1} x$.
`acot(x)`	Inverse cotangent; $\text{arccot}\, x = \cot^{-1} x$.
`acsc(x)`	Inverse cosecant; $\text{arccsc}\, x = \csc^{-1} x$.
`asec(x)`	Inverse secant; $\text{arcsec}\, x = \sec^{-1} x$.
`asin(x)`	Inverse sine; $\arcsin x = \sin^{-1} x$.
`atan(x)`	Inverse tangent; $\arctan x = \tan^{-1} x$.
`atan2(y,x)`	Four-quadrant inverse tangent.

MATLAB provides the `atan2(y,x)` function to determine the arctangent unambiguously, where `x` and `y` are the coordinates of a point. The angle computed by `atan2(y,x)` is the angle between the positive real axis and line from the origin (0,0) to the point (x, y). For example, the point $x = 1$, $y = -1$ corresponds to $-45°$ or -0.7854 rad, and the point $x = -1$, $y = 1$ corresponds to $135°$ or 2.3562 rad. Typing `atan2(-1,1)` returns -0.7854, while typing `atan2(1,-1)` returns 2.3562. The `atan2(y,x)` function is an example of a function that has two arguments. The order of the arguments is important for such functions. For example, we have seen that `atan2(-1,1)` is not the same as `atan2(1,-1)`.

Test Your Understanding

T3.1–3 For several values of x, confirm that $e^{ix} = \cos x + i \sin x$.

T3.1–4 For several values of x in the range $0 \le x \le 2\pi$, confirm that $\sin^{-1} x + \cos^{-1} x = \pi/2$.

T3.1–5 For several values of x in the range $0 \le x \le 2\pi$, confirm that $\tan(2x) = 2 \tan x/(1 - \tan^2 x)$.

Hyperbolic Functions

The *hyperbolic functions* are the solutions of some common problems in engineering analysis. For example, the *catenary* curve, which describes the shape of a hanging cable supported at both ends, can be expressed in terms of the hyperbolic cosine, $\cosh x$, which is defined as

$$\cosh x = \frac{e^x + e^{-x}}{2}$$

The hyperbolic sine, $\sinh x$, is defined as

$$\sinh x = \frac{e^x - e^{-x}}{2}$$

The inverse hyperbolic sine, $\sinh^{-1} x$, is the value y that satisfies $\sinh y = x$. It can be expressed in terms of the natural logarithm as follows:

$$\sinh^{-1} x = \ln\left(x + \sqrt{x^2 + 1}\right) \qquad -\infty < x < \infty$$

Several other hyperbolic functions have been defined. Table 3.1–3 lists these hyperbolic functions and the MATLAB commands to obtain them.

Test Your Understanding

T3.1–6 For several values of x in the range $0 \le x \le 5$, confirm that $\sin(ix) = i \sinh x$.

Table 3.1–3 Hyperbolic functions

Hyperbolic	
`cosh(x)`	Hyperbolic cosine; $\cosh x = (e^x + e^{-x})/2$.
`coth(x)`	Hyperbolic cotangent; $\cosh x/\sinh x$.
`csch(x)`	Hyperbolic cosecant; $1/\sinh x$.
`sech(x)`	Hyperbolic secant; $1/\cosh x$.
`sinh(x)`	Hyperbolic sine; $\sinh x = (e^x - e^{-x})/2$.
`tanh(x)`	Hyperbolic tangent; $\sinh x/\cosh x$.
Inverse hyperbolic	
`acosh(x)`	Inverse hyperbolic cosine; $\cosh^{-1} x = \ln\left(x + \sqrt{x^2 - 1}\right), x \geq 1$.
`acoth(x)`	Inverse hyperbolic cotangent; $\coth^{-1} x = \frac{1}{2}\ln\left(\frac{x+1}{x-1}\right), x > 1$ or $x < -1$.
`acsch(x)`	Inverse hyperbolic cosecant; $\operatorname{csch}^{-1} x = \ln\left(\frac{1}{x} + \sqrt{\frac{1}{x^2} + 1}\right), x \neq 0$.
`asech(x)`	Inverse hyperbolic secant; $\operatorname{sech}^{-1} x = \ln\left(\frac{1}{x} + \sqrt{\frac{1}{x^2} - 1}\right), 0 < x \leq 1$.
`asinh(x)`	Inverse hyperbolic sine; $\sinh^{-1} x = \ln\left(x + \sqrt{x^2 + 1}\right), -\infty < x < \infty$.
`atanh(x)`	Inverse hyperbolic tangent; $\tanh^{-1} x = \frac{1}{2}\ln\left(\frac{1+x}{1-x}\right), -1 < x < 1$.

3.2 User-Defined Functions

FUNCTION FILE

Another type of M-file is a *function file.* Unlike a script file, all the variables in a function file are *local,* which means their values are available only within the function. Function files are useful when you need to repeat a set of commands several times. Function files are like functions in C, subroutines in FORTRAN and BASIC, and procedures in Pascal. They are the building blocks of larger programs.

FUNCTION DEFINITION LINE

The first line in a function file must begin with a *function definition line* that has a list of inputs and outputs. This line distinguishes a function M-file from a script M-file. Its syntax is as follows:

```
function [output variables] = function_name(input variables)
```

Note that the output variables are enclosed in *square brackets,* while the input variables must be enclosed with *parentheses.* The `function_name` should be the same as the file name in which it is saved (with the `.m` extension). That is, if we name a function `drop`, it should be saved in the file `drop.m`. The function is "called" by typing its name (for example, `drop`) at the command line. The word `function` in the function definition line must be *lowercase.* Note also that even though MATLAB is case-sensitive by default, your computer's operating system might not be case-sensitive with regard to file names. For example, while

MATLAB would recognize `Drop` and `drop` as two different variables, your operating system might treat `Drop.m` and `drop.m` as the same file.

Some Simple Function Examples

Now let us look at some simple user-defined functions. Remember that the name of the function M-file should be the same as the name of the function. Before naming a function, you can use the `exist` function to see if another function has the same name. Functions operate on variables within their own workspace (called *local variables*), which is separate from the workspace you access at the MATLAB command prompt.

Consider the following user-defined function `fun`.

```
function z = fun(x,y)
u = 3*x;
z = u + 6*y.^2;
```

Note the use of a semicolon at the end of the lines. This prevents the values of `u` and `z` from being displayed. Note also the use of the array exponentiation operator (`.^`). This enables the function to accept `y` as an array.

Now consider what happens when you call this function in various ways in the Command window. Call the function with its output argument:

```
>>z = fun(3,7)
z =
   303
```

The function uses `x` = 3 and `y` = 7 to compute `z`.

Call the function without its output argument and try to access its value. You see an error message.

```
>>fun(3,7)
ans =
     303
>>z
???  Undefined function or variable 'z'.
```

Assign the output argument to another variable:

```
>>q = fun(3,7)
q =
   303
```

You can suppress the output by putting a semicolon after the function call. For example, if you type `q = fun(3,7);` the value of `q` will be computed but not displayed.

The variables `x` and `y` are *local* to the function `fun`, so unless you pass their values by naming them `x` and `y`, their values will not be available in the workspace outside the function. The variable `u` is also local to the function. For example,

```
>>x = 3;y = 7;
>>q = fun(x,y);
>>x
x =
   3
>>y
y =
   7
>>u
???  Undefined function or variable 'u'.
```

Compare this to

```
>>q = fun(3,7);
>>x
???  Undefined function or variable 'x'.
>>y
???  Undefined function or variable 'y'.
```

Only the order of the arguments is important, not the names of the arguments:

```
>>x = 7;y = 3;
>>z = fun(y,x)     % This is equivalent to z = fun(3,7)
z =
   303
```

You can use arrays as input arguments:

```
>>r = fun([2:4],[7:9])
r =
   300      393      498
```

A function may have more than one output. These are enclosed in square brackets. For example, the function `circle` computes the area A and circumference C of a circle, given its radius as an input argument.

```
function [A, C] = circle(r)
A = pi*r.^2;
C = 2*pi*r;
```

The function is called as follows, if $r = 4$.

```
>>[A, C] = circle(4)
A =
   50.2655
C =
   25.1327
```

If you omit an output argument, for example, and type `[C]=circle(4)`, MATLAB displays the value (50.2655) of the *first* output variable (which is `A`), but names it `C`.

A function may have no input arguments and no output list. For example, the function show_date computes and stores the date in the variable today, and displays the value of today.

```
function show_date
today = date
```

Variations in the Function Line

The following examples show permissible variations in the format of the function line. The differences depend on whether there is no output, a single output, or multiple outputs.

Function definition line	File name
1. function [area_square] = square(side);	square.m
2. function area_square = square(side);	square.m
3. function [volume_box] = box(height,width,length);	box.m
4. function [area_circle,circumf] = circle(radius);	circle.m
5. function sqplot(side);	sqplot.m

Example 1 is a function with one input and one output. The square brackets are optional when there is only one output (see example 2). Example 3 has one output and three inputs. Example 4 has two outputs and one input. Example 5 has no output variable (for example, a function that generates a plot). In such cases the equal sign may be omitted.

Comment lines starting with the % sign can be placed anywhere in the function file. However, if you use help to obtain information about the function, MATLAB displays all comment lines immediately following the function definition line up to the first blank line or first executable line. The first comment line can be accessed by the lookfor command.

We can call both built-in and user-defined functions either with the output variables explicitly specified, as in examples 1 through 4, or without any output variables specified. For example, we can call the function square as square(side) if we are not interested in its output variable area_square. (The function might perform some other operation that we want to occur, such as toggling diary on or off.) Note that if we omit the semicolon at the end of the function call statement, the first variable in the output variable list will be displayed using the default variable name ans.

Variations in Function Calls

The following function, called drop, computes a falling object's velocity and distance dropped. The input variables are the acceleration g, the initial velocity v_0, and the elapsed time t. Note that we must use the element-by-element operations for any operations involving function inputs that are arrays. Here we anticipate that t will be an array, so we use the element-by-element operator (.^).

```
function [dist,vel] = drop(g,v0,t);
% Computes the distance travelled and the
% velocity of a dropped object, as functions
% of g, the initial velocity v0, and the time t.
vel = g*t + v0;
dist = 0.5*g*t.^2 + v0*t;
```

The following examples show various ways to call the function `drop`:

1. The variable names used in the function definition may, but need not, be used when the function is called:

```
a = 32.2;
initial_speed = 10;
time = 5;
[feet_dropped,speed] = drop(a,initial_speed,time)
```

2. The input variables need not be assigned values outside the function prior to the function call:

```
[feet_dropped,speed] = drop(32.2,10,5)
```

3. The inputs and outputs may be arrays:

```
[feet_dropped,speed]=drop(32.2,10,[0:1:5])
```

This function call produces the arrays `feet_dropped` and `speed`, each with six values corresponding to the six values of time in the array `time`.

Function Arguments

When invoking, or "calling" a function, the caller (either an entry from the keyboard or a running file) must provide the function with any input data it needs by passing the data in the argument list. Data that needs to be returned to the caller is passed back in a list of output or return values. MATLAB uses two ways to pass argument data. It passes it *by value* (for example, as an array) or by an *internal pointer* to the data. If the function does not modify the value of the input data, it uses a more efficient pointer mechanism. If, however, the function modifies the value of the input data, MATLAB passes the argument by value.

If the function changes the value of an input variable, its value in the calling workspace might need be updated. This can be done by having the function return the updated value as an output argument. For example, consider the following function, which computes the radius of a circle whose area is a specified factor `f` of the original area.

```
function [r, A] = area_change(A, f)
A = f*A;
r = sqrt(A/pi);
```

If, on the other hand, we want to use the value of the new area but also preserve the value of the previous area, we could call the function as follows.

```
>>[rnew, Anew] = area_change(A, f)
```

After the function is called, the original value of A remains in the workspace.

Local Variables

The names of the input variables given in the function definition line are local to that function. This means that other variable names can be used when you call the function. All variables inside a function are erased after the function finishes executing, except when the same variable names appear in the output variable list used in the function call.

For example, when using the drop function in a program, we can assign a value to the variable dist before the function call, and its value will be unchanged after the call because its name was not used in the output list of the call statement (the variable feet_dropped was used in the place of dist). This is what is meant by the function's variables being "local" to the function. This feature allows us to write generally useful functions using variables of our choice, without being concerned that the calling program uses the same variable names for other calculations. This means that our function files are "portable," and need not be rewritten every time they are used in a different program.

You might find the M-file Debugger to be useful for locating errors in function files. Runtime errors in functions are more difficult to locate because the function's local workspace is lost when the error forces a return to the MATLAB base workspace. The Debugger provides access to the function workspace, and allows you to change values. It also enables you to execute lines one at a time and to set *breakpoints,* which are specific locations in the file where execution is temporarily halted. The applications in this text will probably not require use of the Debugger, which is useful mainly for very large programs, or programs containing *nested* functions (that is, functions that call other functions). For more information, see Section 4.7 in Chapter 4.

Global Variables

The global command declares certain variables global, and therefore their values are available to the basic workspace and to other functions that declare these variables global. The syntax to declare the variables a, x, and q is global a x q. Any assignment to those variables, in any function or in the base workspace, is available to all the other functions declaring them global. If the global variable doesn't exist the first time you issue the global statement, it will be initialized to the empty matrix. If a variable with the same name as the global variable already exists in the current workspace, MATLAB issues a warning and changes the value of that variable to match the global. In a user-defined function, make the global command the first executable line.

The decision whether or not to declare a variable global is not always clear cut. Often there is more than one effective way to solve a given problem. The following example illustrates this.

EXAMPLE 3.2–1 Using Global Variables

Engineers often need to estimate the pressures and volumes of a gas in a container. The *ideal gas law* provides one way of making the estimate. The law is

$$P = \frac{RT}{\hat{V}}$$

More accurate estimates can be made with the *van der Waals* equation:

$$P = \frac{RT}{\hat{V} - b} - \frac{a}{\hat{V}^2}$$

where the term b is a correction for the volume of the molecules, and the term $a/\hat{V}^2$ is a correction for molecular attractions. The gas constant is R, the *absolute* temperature is T, and the gas specific volume is $\hat{V}$. The value of R is the same in both equations and for all gases; it is $R = 0.08206$ L atmosphere/mol K. The values of a and b depend on the type of gas. For example, for chlorine (Cl_2), $a = 6.49$ and $b = 0.0562$ in these units.

Write two user-defined functions, one to compute the pressure using the ideal gas law, and one using the van der Waals equation. Develop a solution without using global variables, and one using global variables.

■ Solution

We can write the following function for the ideal gas law:

```
function P = ideal_1(T,Vhat,R)
P = R*T./Vhat;
```

Note that we should use array division because the user might call the function using `Vhat` as a vector. However, we need not use array multiplication between `R` and `T` because `R` will always be a scalar. This function is independent of the units used for T, $\hat{V}$, and R, as long as they are consistent with each other.

The difficulty with this function is that we must always enter the value of `R` when we call the function. To avoid this, we can rewrite the function as follows:

```
function P = ideal_2(T,Vhat)
% This requires liter, atmosphere, and mole units!
R = 0.08206;
P = R*T./Vhat;
```

Now the value of R is "hard-wired" into the function. However, the danger with this approach is that the value of R depends on the units being used for $\hat{V}$ and T. For example, in SI metric units, $R = 8.314$ J/mol-K. So if the user enters values for `Vhat` and `T` in SI units, the computed value of `P` will be incorrect. Thus it is a good idea to put a comment to that effect within the function. However, it is not necessary to read the file in order to call the function, and users will not see the comment unless they read the file!

A solution to this problem is to declare `R` a global variable. Then the function file would be:

```
function P = ideal_3(T,Vhat)
global R
P = R*T./Vhat;
```

With this approach, you must then declare R global wherever you call the function, either from the base workspace or from another function. For example, the following session computes the gas pressures in gases having $\hat{V} = 20$, at two temperatures: $T = 300$ K and 330 K, respectively.

```
>>global R
>>R = 0.08206;
>>ideal_3([300,330],20)
ans =
     1.2309     1.3540
```

The pressures are approximately 1.23 and 1.35 atmospheres. With this approach, you must always enter the value of R, but this might force you to think about what units are being used for Vhat, T, and P. If you are sure that you will always use the same units, then the simplest approach is to hard-wire the value of R into the function, as was done with ideal_2.

In the van der Waals equation, the constants a and b depend on the particular gas. A function that is independent of the units used and the particular gas is

```
function P = vdwaals_1(T,Vhat,R,a,b)
P = R*T./(Vhat-b)-a./Vhat.^2;
```

The difficulty with this function is that you must always enter the values of R, a, and b. If you use the equation to analyze chlorine only, then you can hard-wire the constants into the function along with the value of R, as follows, but if you do so be sure you always use the proper units in the function's arguments!

```
function P = vdwaals_2(T,Vhat)
% For chlorine only, and liter, atmosphere, and mole units!
R = 0.08206;
a = 6.49;b = 0.0562;
P = R*T./(Vhat-b)-a./Vhat.^2;
```

Note that the variable P is local to all these functions, and is available only if you assign it a value when calling the function. For example, consider the following session.

```
>>vdwaals_3(300,20)
ans =
     1.3416
>>P
?? Undefined function or variable 'P'.
>>P = vdwaals_3(300,20)
P =
   1.3416
```

In addition to using the global command to make values accessible within functions, we can also use it to make global variables that would otherwise be local. For example, suppose we want to make the chlorine values of a and b available to other functions. Then we could modify the vdwaals_2 function by adding the statement global a b. Then any other function declaring a and b global would have access to those values.

Applications

Some MATLAB commands act on functions. If the function of interest is not a simple function, it is more convenient to define the function in an M-file when using one of these commands.

Finding the Zeros of a Function The `roots` function finds the zeros of polynomial functions only. Otherwise, you can use the `fzero` function to find the zero of a function of a single variable, which is denoted by `x`. One form of its syntax is

```
fzero('function', x0)
```

where `function` is a string containing the name of the function, and `x0` is a user-supplied guess for the zero. The `fzero` function returns a value of `x` that is near `x0`. It identifies only points where the function crosses the x-axis, not points where the function just touches the axis. For example, `fzero('cos',2)` returns the value $x = 1.5708$.

The function `fzero('function',x0)` tries to find a zero of `function` near `x0`, if `x0` is a scalar. The value returned by `fzero` is near a point where `function` changes sign, or `NaN` if the search fails. In this case, the search terminates when the search interval is expanded until an `Inf`, `NaN`, or a complex value is found (`fzero` cannot find complex zeros). If `x0` is a *vector* of length two, `fzero` assumes that `x0` is an interval where the sign of `function(x0(1))` differs from the sign of `function(x0(2))`. An error occurs if this is not true. Calling `fzero` with such an interval guarantees that `fzero` will return a value near a point where `function` changes sign. Plotting the function first is a good way to get a value for the vector `x0`.

An alternate syntax is

```
[x,fval] = fzero('function',x0)
```

This returns the value of `function` at the solution `x`. The syntax

```
[x,fval, exitflag] = fzero('function',x0)
```

returns a value of `exitflag` that describes the exit condition of `fzero`. A positive value of `exitflag` indicates that the function found a zero, which is the value `x`. A negative value of `exitflag` indicates that no interval was found with a sign change, or that a `NaN` or an `Inf` function value was encountered during the search for an interval containing a sign change, or that a complex function value was encountered during the search for an interval containing a sign change.

The `fzero` function finds a point where the function changes sign. If the function is continuous, this is also a point where the function has a value near zero. If the function is not continuous, `fzero` might return values that are discontinuous points instead of zeros. For example, `x = fzero('tan',1)` returns `x = 1.5708`, a discontinuous point in $\tan(x)$.

Furthermore, the `fzero` command defines a zero as a point where the function crosses the x-axis. Points where the function touches, but does not cross, the

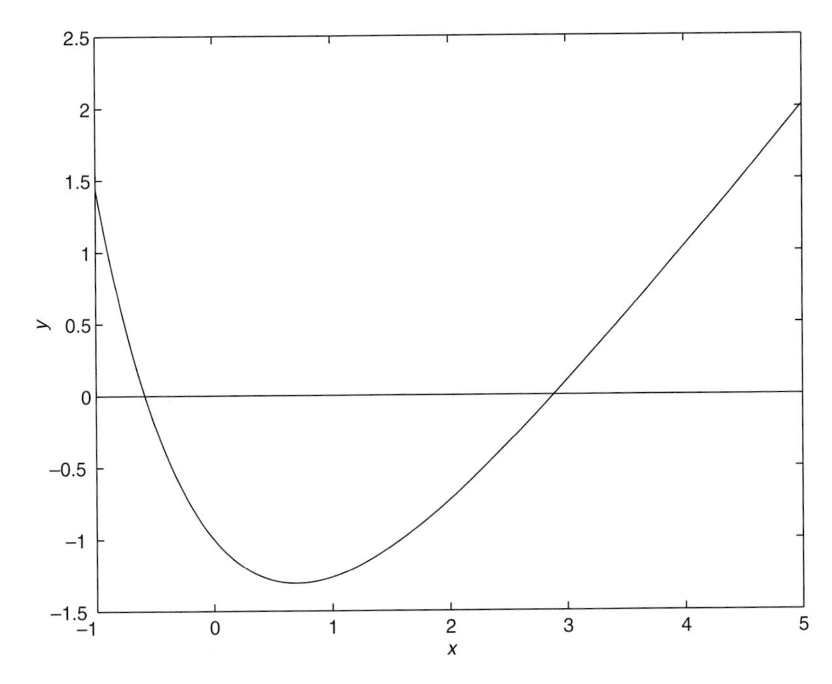

Figure 3.2–1 Plot of the function $y = x + 2e^{-x} - 3$.

x-axis are not valid zeros. For example, $y = x^2$ is a parabola that touches the x-axis at $x = 0$. Because the function never crosses the x-axis, however, no zero is found. For functions with no valid zeros, `fzero` executes until `Inf`, `NaN`, or a complex value is detected.

To use this function to find the zeros of more complicated functions, it is more convenient to define the function in a function file. For example, if $y = x + 2e^{-x} - 3$, define the following function file:

```
function y = f1(x)
y = x + 2*exp(-x) - 3;
```

Functions can have more than one zero, so it helps to plot the function first and then use `fzero` to obtain an answer that is more accurate than the answer read off the plot. Figure 3.2–1 shows the plot of this function, which has two zeros, one near $x = -0.5$ and one near $x = 3$. To find the zero near $x = -0.5$, type `x = fzero('f1',-0.5)`. The answer is $x = -0.5831$. To find the zero near $x = 3$, type `x = fzero('f1',3)`. The answer is $x = 2.8887$.

Minimizing a Function of One Variable The `fminbnd` function finds the minimum of a function of a single variable, which is denoted by `x`. One form of its syntax is

```
fminbnd('function', x1, x2)
```

where `function` is a string containing the name of the function. The `fminbnd` function returns a value of `x` that minimizes the function in the interval `x1` $\leq$ `x` $\leq$ `x2`. For example, `fminbnd('cos',0,4)` returns the value $x = 3.1416$.

However, to use this function to find the minimum of more-complicated functions, it is more convenient to define the function in a function file. For example, if $y = 1 - xe^{-x}$, define the following function file:

```
function y = f2(x)
y = 1-x.*exp(-x);
```

To find the value of x that gives a minimum of y for $0 \leq x \leq 5$, type `x = fminbnd('f2',0,5)`. The answer is $x = 1$. To find the minimum value of y, type `y = f2(x)`. The result is `y = 0.6321`.

The alternate syntax

```
[x,fval] = fminbnd('function',x1,x2)
```

returns the value of `function` at the solution `x`. The syntax

```
[x,fval, exitflag] = fminbnd('function',x1,x2)
```

returns a value of `exitflag`, which describes the exit condition of `fminbnd`. The value `exitflag = 1` indicates that the function converged to a solution `x`; a value other than `exitflag = 1` indicates that the function did not converge to a solution.

To find the maximum of a function, use the `fminbnd` function with the negative of the function of interest. For example, to find the maximum of $y = xe^{-x}$ over the interval $0 \leq x \leq 5$, you must define the function file as follows:

```
function y = f3(x)
y = -x.*exp(-x);
```

Typing `fminbnd('f3',0,5)` gives the result $x = 1$. The function $y = xe^{-x}$ has a maximum at $x = 1$.

Whenever we use a minimization technique, we should check to make sure that the solution is a true minimum. For example, consider the following polynomial:

$$y = 0.025x^5 - 0.0625x^4 - 0.333x^3 + x^2$$

Its plot is shown in Figure 3.2–2. The function has two minimum points in the interval $-1 < x < 4$. The minimum near $x = 3$ is called a *relative* or *local* minimum because it forms a valley whose lowest point is higher than the minimum at $x = 0$. The minimum at $x = 0$ is the true minimum and is also called the *global* minimum. If we specify the interval $-1 \leq x \leq 4$ by typing

```
>>x = fminbnd('0.025*x.^5-0.0625*x.^4-0.333*x.^3+x.^2',-1,4)
```

MATLAB gives the answer `x = 2.0438e-006`, which is essentially 0, the true minimum point. If we specify the interval $0.1 \leq x \leq 2.5$, MATLAB gives the answer `x = 0.1001`, which corresponds to the minimum value of y on the

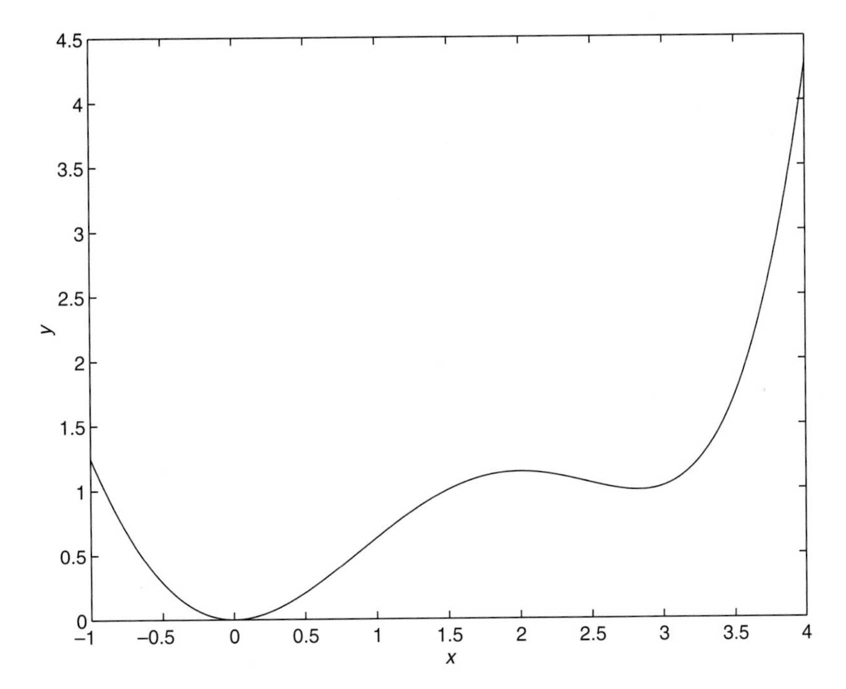

Figure 3.2–2 Plot of the function $y = 0.025x^5 - 0.0625x^4 - 0.333x^3 + x^2$.

interval $0.1 \leq x \leq 2.5$. Thus we will miss the true minimum point if our specified interval does not include it.

Also, `fminbnd` can give incorrect answers. In earlier versions of MATLAB, if we specified the interval $1 \leq x \leq 4$, MATLAB gave the answer `x = 2.8236`, which corresponds to the "valley" shown in the plot, but which is not the minimum point on the interval $1 \leq x \leq 4$. On this interval the minimum point is at the boundary $x = 1$, and MATLAB 7 correctly computes this answer. The `fminbnd` procedure now looks for a minimum point corresponding to a zero slope, and it looks at the function values at the boundaries of the specified interval for x. In this example, MATLAB 7 identified the true minimum at the boundary, but in other problems it might incorrectly identify a local minimum as the true minimum.

No one has yet developed a numerical minimization method that works for every possible function. Therefore, we must use any such method with care. In practice, the best use of the `fminbnd` function is to determine precisely the location of a minimum point whose approximate location was found by other means, such as by plotting the function.

Minimizing a Function of Several Variables To find the minimum of a function of more than one variable, use the `fminsearch` function. One form of its syntax is

```
fminsearch('function', x0)
```

where function is a string containing the name of the function. The vector x0 is a guess that must be supplied by the user. For example, to minimize the function $f = xe^{-x^2-y^2}$, we first define it in an M-file, using the vector x whose elements are x(1) = x and x(2) = y.

```
function f = f4(x)
f = x(1).*exp(-x(1).^2-x(2).^2);
```

Suppose we guess that the minimum is near $x = y = 0$. The session is

```
>>fminsearch('f4',[0,0])
ans =
   -0.7071     0.000
```

Thus the minimum occurs at $x = -0.7071$, $y = 0$.

The alternate syntax

```
[x,fval] = fminsearch('function',x0)
```

returns the value of function at the solution x. The syntax

```
[x,fval, exitflag] = fminsearch('function',x1,x2)
```

returns a value of exitflag, which describes the exit condition of fminsearch. The value exitflag = 1 indicates that the function converged to a solution x; a value other than exitflag = 1 indicates that the function did not converge to a solution.

The fminsearch function can often handle discontinuities, particularly if they do not occur near the solution. The fminsearch function might give local solutions only, and it minimizes over the real numbers only; that is, x must consist of real variables only and the function must return real numbers only. When x has complex variables, they must be split into real and imaginary parts.

Table 3.2–1 summarizes the fminbnd, fminsearch, and fzero commands.

Table 3.2–1 Minimization and root-finding functions

Function	Description
fminbnd('function',x1,x2)	Returns a value of x in the interval x1 ≤ x ≤ x2 that corresponds to a minimum of the single-variable function described by the string 'function'.
fminsearch('function',x0)	Uses the starting vector x0 to find a minimum of the multivariable function described by the string 'function'.
fzero('function',x0)	Uses the starting value x0 to find a zero of the single-variable function described by the string 'function'.

Design Optimization

One way to improve engineering designs is by formulating the equations describing the design in the form of a minimization or maximization problem. This approach is called *design optimization.* Examples of quantities we would like to minimize are energy consumption and construction materials. Items we would like to maximize are useful life and capacity (such as the vehicle weight that can be supported by a bridge). The following example illustrates the concept of design optimization.

Optimization of an Irrigation Channel — EXAMPLE 3.2–2

Figure 3.2–3 shows the cross section of an irrigation channel. A preliminary analysis has shown that the cross-sectional area of the channel should be 100 ft^2 to carry the desired water-flow rate. To minimize the cost of concrete used to line the channel, we want to minimize the length of the channel's perimeter. Find the values of d, b, and θ that minimize this length.

■ Solution

The perimeter length L can be written in terms of the base b, depth d, and angle θ as follows:

$$L = b + \frac{2d}{\sin\theta}$$

The area of the trapezoidal cross section is

$$100 = db + \frac{d^2}{\tan\theta}$$

The variables to be selected are b, d, and θ. We can reduce the number of variables by solving the latter equation for b to obtain

$$b = \frac{1}{d}\left(100 - \frac{d^2}{\tan\theta}\right)$$

Substitute this expression into the equation for L. The result is

$$L = \frac{100}{d} - \frac{d}{\tan\theta} + \frac{2d}{\sin\theta}$$

We must now find the values of d and θ to minimize L.

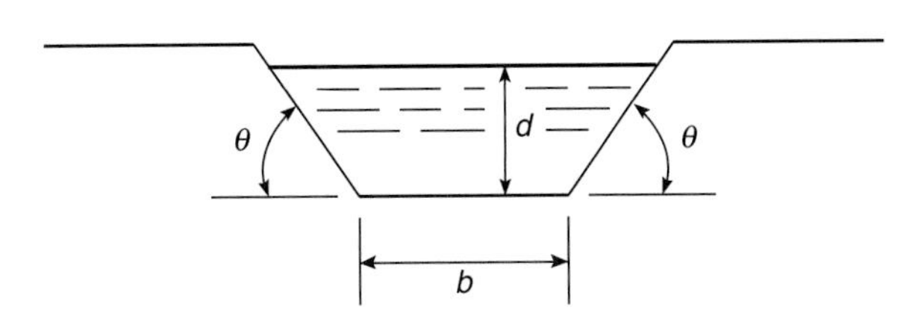

Figure 3.2–3 Cross section of an irrigation channel.

First define the function file for the perimeter length. Let the vector **x** be $[d\ \theta]$.

```
function L = channel(x)
L = 100./x(1) - x(1)./tan(x(2)) + 2*x(1)./sin(x(2));
```

Then use the `fminsearch` function. Using a guess of $d = 20$ and $\theta = 1$ rad, the session is

```
>>x = fmins('channel',[20,1])
x =
   7.5984    1.0472
```

Thus the minimum perimeter length is obtained with $d = 7.5984$ ft and $\theta = 1.0472$ rad, or $\theta = 60^\circ$. Using a different guess, $d = 1, \theta = 0.1$, produces the same answer. The value of the base b corresponding to these values is $b = 8.7738$.

However, using the guess $d = 20, \theta = 0.1$ produces the physically meaningless result $d = -781, \theta = 3.1416$. The guess $d = 1, \theta = 1.5$ produces the physically meaningless result $d = 3.6058, \theta = -3.1416$.

The equation for L is a function of the two variables d and θ, and it forms a surface when L is plotted versus d and θ on a three-dimensional coordinate system. This surface might have multiple peaks, multiple valleys, and "mountain passes" called saddle points that can fool a minimization technique. Different initial guesses for the solution vector can cause the minimization technique to find different valleys and thus report different results. We can use the surface-plotting functions covered in Chapter 5 to look for multiple valleys, or we can use a large number of initial values for d and θ, say, over the physically realistic ranges $0 < d < 30$ and $0 < \theta < \pi/2$. If all the physically meaningful answers are identical, then we can be reasonably sure that we have found the minimum.

The `fzero`, `fminbnd`, and `fminsearch` functions have alternative forms not described here. With these forms you can specify the accuracy required for the solution, as well as the number of steps to use before stopping. Use the `help` facility to find out more about these functions.

Test Your Understanding

T3.2–1 The equation $e^{-0.2x}\sin(x + 2) = 0.1$ has three solutions in the interval $0 < x < 10$. Find these three solutions.

T3.2–2 The function $y = 1 + e^{-0.2x}\sin(x + 2)$ has two minimum points in the interval $0 < x < 10$. Find the values of x and y at each minimum.

T3.2–3 Find the depth d and angle θ to minimize the perimeter length of the channel shown in Figure 3.2–3 to provide an area of 200 ft^2. (Answer: $d = 10.7457$ ft, $\theta = 60^\circ$.)

3.3 Advanced Function Programming

Since the introduction of *function handles* in MATLAB 6.0, their use has become quite widespread in the MATLAB documentation and elsewhere, and so it is a good idea to become familiar with the concept. In addition, *anonymous functions, subfunctions,* and *nested functions* are also now seen more often in documentation and examples. This section covers the basic features of function handles and these new types of functions. The topics covered in this section are most useful for large programming projects. An understanding of these topics is not necessary to master the remaining topics in this text. Therefore, this section may be omitted, if desired.

Function Handles

You can create a function handle to any function by using the *at* sign, `@`, before the function name. You can then name the handle if you wish, and use the handle to reference the function. For example, to create a handle to the sine function, you type

```
>>sine_handle = @sin;
```

where `sine_handle` is a user-selected name for the handle.

A common use of a function handle is to pass the function as an argument to another function. For example, we can plot $\sin x$ over $0 \leq x \leq 6$ as follows:

```
>>plot([0:0.01:6],sine_handle,[0:0.01:6])
```

where we have used the `feval` function to evaluate the function at the given argument `[0:0.01:6]`. This is a rather cumbersome way to plot the sine function, but the concept can be extended to create a general purpose plotting function that accepts a function as an input. For example,

```
function x = gen_plot(fun_handle, interval)
plot(interval, fun_handle, interval)
```

You may call this function to plot the $\sin x$ over $0 \leq x \leq 6$ as follows:

```
>>gen_plot(sine_handle,[0:0.01:6])
```

or

```
>>gen_plot(@sin,[0:0.01:6])
```

There are several advantages to using function handles; these are described in the MATLAB documentation. Two advantages that we will discuss later are speed of execution and providing access to subfunctions, which are normally not visible outside of their defining M-file. Another advantage is that function handles are a standard MATLAB data type, and thus they can be used in the same manner as other data types. For example, you can create arrays, cell arrays, or structures of function handles. You can access individual function handles just as you access elements of a numeric array or structure.

For example, the function gen_plot may be used as follows with a function handle array, to create two subplots, one for the sine and one for the cosine.

```
fh(1) = @sin;
fh(2) = @cos;
for k = 1:2
  subplot(2,1,k)
  gen_plot(fh(k),[0:0.01:8])
end
```

Methods for Calling Functions

There are four ways to invoke, or "call," a function into action. These are:

1. As a character string identifying the appropriate function M-file,
2. As a function handle,
3. As an "inline" function object, or
4. As a string expression.

Examples of these ways follow for the fzero function used with the user-defined function fun1, which computes $y = x^2 - 4$.

1. As a character string identifying the appropriate function M-file, which is

   ```
   function y = fun1(x)
   y = x.^2-4;
   ```

   The function may be called as follows, to compute the zero over the range $0 \leq x \leq 3$:

   ```
   >>[x, value] = fzero('fun1',[0, 3])
   ```

2. As a function handle to an existing function M-file:

   ```
   >>[x, value] = fzero(@fun1,[0, 3])
   ```

3. As an "inline" function object:

   ```
   >>fun1 = 'x.^2-4';
   >>fun_inline = inline(fun1);
   >>[x, value] = fzero(fun_inline,[0, 3])
   ```

4. As a string expression:

   ```
   >>fun1 = 'x.^2-4';
   >>[x, value] = fzero(fun1,[0, 3])
   ```

   or as

   ```
   >>[x, value] = fzero('x.^2-4',[0, 3])
   ```

The third method, which uses an inline object, is not discussed in the text and will not be covered here because it is a slower method than the first two. The third and fourth methods are equivalent because they both utilize the `inline` function; the only difference is that with the fourth method MATLAB determines that the first argument of `fzero` is a string variable and calls `inline` to convert the string variable to an inline function object. The function handle method (method 2) is the fastest method, followed by method 1.

In addition to speed improvement, another advantage of using a function handle is that it provides access to subfunctions, which are normally not visible outside of their defining M-file. This is discussed later in this section.

Types of Functions

At this point it is helpful to review the types of functions provided for in MATLAB. MATLAB provides built-in functions, such as `clear`, `sin`, and `plot`, which are not M-files, and also some functions that are M-files, such as the function `mean`. In addition, the following types of *user-defined* functions can be created in MATLAB.

PRIMARY FUNCTION

- The *primary* function is the first function in an M-file and typically contains the main program. Following the primary function in the same file can be any number of subfunctions, which can serve as subroutines to the primary function. Usually the primary function is the only function in an M-file that you can call from the MATLAB command line or from another M-file function. You invoke this function using the name of the M-file in which it is defined. We normally use the same name for the function and its file, but if the function name differs from the file name, you must use the file name to invoke the function.
- *Anonymous* functions enable you to create a simple function without needing to create an M-file for it. You can construct an anonymous function either at the MATLAB command line or from within another function or script. Thus, anonymous functions provide a quick way of making a function from any MATLAB expression without the need to create, name, and save a file.
- *Subfunctions* are placed in the primary function and are called by the primary function. You can use multiple functions within a single primary function M-file.
- *Nested* functions are functions defined within another function. They can help to improve the readability of your program and also give you more flexible access to variables in the M-file. The difference between nested functions and subfunctions is that subfunctions normally cannot be accessed outside of their primary function file.
- *Overloaded* functions are functions that respond differently to different types of input arguments. They are similar to overloaded functions in any object-oriented language. For example, an overloaded function can be created to treat integer inputs differently than inputs of class double.

PRIVATE FUNCTION

- *Private* functions enable you to restrict access to a function. They can be called only from an M-file function in the parent directory.

The term *function function* is not a separate function type but refers to any function that accepts another function as an input argument, such as the function `fzero`. You can pass a function to another function using a function handle.

Anonymous Functions

Anonymous functions enable you to create a simple function without needing to create an M-file for it. You can construct an anonymous function either at the MATLAB command line or from within another function or script. The syntax for creating an anonymous function from an expression is

```
fhandle = @(arglist) expr
```

where `arglist` is a comma-separated list of input arguments to be passed to the function, and `expr` is any single, valid MATLAB expression. This syntax creates the function handle `fhandle`, which enables you to invoke the function. Note that this syntax is different from that used to create other function handles, `fhandle = @functionname`. The handle is also useful for passing the anonymous function in a call to some other function in the same way as any other function handle.

For example, to create a simple function called `sq` to calculate the square of a number, type

```
sq = @(x) x.^2;
```

To improve readability, you may enclose the expression in parentheses, as `sq = @(x) (x.^2);`. To execute the function, type the name of the function handle, followed by any input arguments enclosed in parentheses. For example,

```
>>sq(5)
ans =
     25
>>sq([5,7])
ans =
     25     49
```

You might think that this particular anonymous function will not save you any work because typing `sq([5,7])` requires nine keystrokes, one more than is required to type `[5,7].^2`. Here, however, the anonymous function protects you from forgetting to type the period (`.`) required for array exponentiation. Anonymous functions are useful, however, for more complicated functions involving numerous keystrokes.

You can pass the handle of an anonymous function to other functions. For example, to find the minimum of the polynomial $4x^2 - 50x + 5$ over the interval $[-10, 10]$, you type

```
>>poly1 = @(x)  4*x.^2 - 50*x + 5;
>>fminbnd(poly1, -10, 10)
ans =
     6.2500
```

If you are not going to use that polynomial again, you can omit the handle definition line and type instead

```
>>fminbnd(@(x)  4*x.^2 - 50*x + 5, -10, 10)
```

Multiple Input Arguments You can create anonymous functions having more than one input. For example, to define the function $\sqrt{x^2 + y^2}$, type

```
>>sqrtsum = @(x,y) sqrt(x.^2 + y.^2);
```

Then

```
>>sqrtsum(3, 4)
ans =
     5
```

As another example, consider the function defining a plane, $z = Ax + By$. The scalar variables A and B must be assigned values before you create the function handle. For example,

```
>>A = 6; B = 4:
>>plane = @(x,y) A*x + B*y;
>>z = plane(2,8)
z =
   44
```

No Input Arguments To construct a handle for an anonymous function that has no input arguments, use empty parentheses for the input argument list, as shown by the following: `d = @() date;`.

Use empty parentheses when invoking the function, as follows:

```
>>d()
ans =
     01-Mar-2004
```

You must include the parentheses. If you do not, MATLAB just identifies the handle; it does not execute the function.

Calling One Function within Another One anonymous function can call another to implement function composition. Consider the function 5 $\sin(x^3)$. It is composed of the functions $g(y) = 5\sin(y)$ and $f(x) = x^3$. In the following session the function whose handle is `h` calls the functions whose handles are `f` and `g`.

```
>>f = @(x) x.^3;
>>g = @(x) 5*sin(x);
>>h = @(x) g(f(x));
>>h(2)
ans =
     4.9468
```

To preserve an anonymous function from one MATLAB session to the next, save the function handle to a MAT-file. For example, to save the function associated with the handle `h`, type `save anon.mat h`. To recover it in a later session, type `load anon.mat h`.

Variables and Anonymous Functions Variables can appear in anonymous functions in two ways:

- As variables specified in the argument list, as for example `f = @(x) x.^3;`, and
- As variables specified in the body of the expression, as for example with the variables `A` and `B` in `plane = @(x,y) A*x + B*y`. In this case, when the function is created MATLAB captures the values of these variables and retains those values for the lifetime of the function handle. In this example, if the values of `A` or `B` are changed after the handle is created, their values associated with the handle do not change. This feature has both advantages and disadvantages, so you must keep it in mind.

Subfunctions

A function M-file may contain more than one user-defined function. The first defined function in the file is called the *primary function,* whose name is the same as the M-file name. All other functions in the file are called *subfunctions*. Subfunctions are normally "visible" only to the primary function and other subfunctions in the same file; that is, they normally cannot be called by programs or functions outside the file. However, this limitation can be removed with the use of function handles, as we will see later in this section.

Create the primary function first with a function definition line and its defining code, and name the file with this function name as usual. Then create each subfunction with its own function definition line and defining code. The order of the subfunctions does not matter, but function names must be unique within the M-file.

The order in which MATLAB checks for functions is very important. When a function is called from within an M-file, MATLAB first checks to see if the function is a built-in function such as `sin`. If not, it checks to see if it is a *subfunction* in the file, then checks to see if it is a *private* function (which is a function M-file residing in the `private` subdirectory of the calling function). Then MATLAB checks for a standard M-file on your search path. Thus, because MATLAB checks for a subfunction before checking for private and standard M-file functions, you may use subfunctions with the same name as another existing M-file. This feature allows you to name subfunctions without being concerned about whether another function exists with the same name, so you need not choose long function names to avoid conflict. This feature also protects you from using another function unintentionally.

Note that you may even supercede a MATLAB M-function in this way. The following example shows how the MATLAB M-function `mean` can be superceded by our own definition of the mean, one which gives the root-mean-square value. The function `mean` is a subfunction. The function `subfun_demo` is the primary function.

```
function y = subfun_demo(a)
y = a - mean(a);
%
function w = mean(x)
w = sqrt(sum(x.^2))/length(x);
```

A sample session follows.

```
>>y = subfn_demo([4, -4])
y =
   1.1716    -6.8284
```

If we had used the MATLAB M-function `mean`, we would have obtained a different answer; that is,

```
>>a=[4,-4];
>>b = a - mean(a)
b =
   4   -4
```

Thus the use of subfunctions enables you to reduce the number of files that define your functions. For example, if it were not for the subfunction `mean` in the previous example, we would have had to define a separate M-file for our `mean` function and give it a different name so as not to confuse it with the MATLAB function of the same name.

Subfunctions are normally visible only to the primary function and other subfunctions in the same file. However, we can use a function handle to allow access to the subfunction from outside the M-file, as the following example shows. Create the following M-file with the primary function `fn_demo1(range)` and the subfunction `testfun(x)` to compute the zeros of the function $(x^2 - 4)\cos x$ over the range specified in the input variable `range`. Note the use of a function handle in the second line.

```
function yzero = fn_demo1(range)
fun = @testfun;
[yzero,value] = fzero(fun,range);
%
function y = testfun(x)
y = (x.^2-4).*cos(x);
```

A test session gives the following results.

```
>>yzero = fn_demo1([3, 6])
yzero =
   4.7124
```

So the zero of $(x^2 - 4)\cos x$ over $3 \leq x \leq 6$ occurs at $x = 4.7124$.

Suppose we had not used the function handle, as in

```
function yzero = fn_demo2(range)
[yzero,value] = fzero(testfun,range);
%
function y = testfun(x)
y = (x.^2-4).*cos(x);
```

Then the following session produces an error message.

```
>>yzero = fn_demo2([3, 6])
??? Input argument 'x' is undefined.
```

We get an error because the MATLAB function `fzero` must call the function `testfun` repeatedly, but without the function handle, `testfun` is not visible to `fzero`.

Nested Functions

With MATLAB 7 you can now place the definitions of one or more functions within another function. Functions so defined are said to be *nested* within the main function. You can also nest functions within other nested functions. Like any M-file function, a nested function contains the usual components of an M-file function. You must, however, always terminate a nested function with an `end` statement. In fact, if an M-file contains at least one nested function, you must terminate *all* functions, including subfunctions, in the file with an `end` statement, whether or not they contain nested functions.

The following example constructs a function handle for a nested function and then passes the handle to the MATLAB function `fminbnd` to find the minimum point on the parabola. The `parabola` function constructs and returns a function handle `f` for the nested function `p`. This handle gets passed to `fminbnd`.

```
function f = parabola(a, b, c)
f = @p;
   function y = p(x)
     y = a*x^2 + b*x + c;
   end
end
```

In the Command window type

```
>>f = parabola(4, -50, 5);
>>fminbnd(f, -10, 10)
ans =
      6.2500
```

Note than the function `p(x)` can see the variables `a`, `b`, and `c` in the calling function's workspace.

Nested functions might seem to be the same as subfunctions, but they are not. Nested functions have two unique properties:

1. A nested function can access the workspaces of all functions inside of which it is nested. So for example, a variable that has a value assigned to it by the primary function can be read or overwritten by a function nested at any level within the main function. In addition, a variable assigned in a nested function can be read or overwritten by any of the functions containing that function.
2. If you construct a function handle for a nested function, the handle not only stores the information needed to access the nested function; it also stores the values of all variables shared between the nested function and those functions that contain it. This means that these variables persist in memory between calls made by means of the function handle.

Consider the following representation of some functions named A, B, . . . , E.

```
function A(x, y)          % The primary function
B(x, y);
D(y);

    function B(x, y)      % Nested in A
    C(x);
    D(y);

       function C(x)      % Nested in B
       D(x);
       end    % This terminates C
    end    % This terminates B

    function D(x)      % Nested in A
    E(x);

       function E     % Nested in D
       ...
       end    % This terminates E
    end    % This terminates D
end    % This terminates A
```

You call a nested function in several ways.

1. You can call it from the level immediately above it. (In the previous code, function A can call B or D, but not C or E.)
2. You can call it from a function nested at the same level within the same parent function. (Function B can call D, and D can call B.)
3. You can call it from a function at any lower level. (Function C can call B or D, but not E.)

4. If you construct a function handle for a nested function, you can call the nested function from any MATLAB function that has access to the handle.

You can call a subfunction from any nested function in the same M-file.

Private Functions

Private functions reside in subdirectories with the special name `private`, and they are visible only to functions in the parent directory. Assume the directory `rsmith` is on the MATLAB search path. A subdirectory of `rsmith` called `private` may contain functions that only the functions in `rsmith` can call. Because private functions are invisible outside the parent directory `rsmith`, they can use the same names as functions in other directories. This is useful if the main directory used by several individuals including R. Smith, but R. Smith wants to create a personal version of a particular function while retaining the original in the main directory. Because MATLAB looks for private functions before standard M-file functions, it will find a private function named, say `cylinder.m`, before a nonprivate M-file named `cylinder.m`.

Primary functions and subfunctions can be implemented as private functions. Create a private directory by creating a subdirectory called `private` using the standard procedure for creating a directory or a folder on your computer, but do not place the private directory on your path.

3.4 Working with Data Files

A typical ASCII data file has one or more lines of text at the beginning. These might be comments that describe what the data represents, the date it was created, and who created the data, for example. These lines are called the *header.* One or more lines of data, arranged in rows and columns, follow the header. The numbers in each row might be separated by spaces or by commas.

If it is inconvenient to edit the data file, the MATLAB environment provides many ways to bring data created by other applications into the MATLAB workspace, a process called *importing data,* and to package workspace variables so that they can be used by other applications, a process called *exporting data.* Your choice of which mechanism to use depends on which operation you are performing, importing or exporting, and whether you are working with ASCII data or binary data. To make importing data easier, both ASCII and binary, MATLAB includes a graphical user interface, called the *Import Wizard,* that leads you through the import process.

Importing Data from Externally Generated Files

You can enter data into MATLAB by typing it into an array or you can enter and store the data in an M-file. However, either method is inconvenient when you have a lot of data. Such data is often generated by some other application. For example, you might be given a data file generated by a spreadsheet program, or you might have to analyze data collected by a laboratory instrumentation system. Many applications support the ASCII file format, so you are likely to receive a data file stored in this format. If the file has a header or the data is separated by commas, MATLAB will produce an error message. To correct this situation, first load the data file into a text editor, remove the header, and replace the commas with spaces (the number of spaces does not matter as long as there is at least one). To retrieve this data into MATLAB, type `load filename`. If the file has m lines with n values in each line, the data will be assigned to an $m \times n$ matrix having the same name as the file with the extension stripped off. For example, if your data file contains 10 lines and 3 columns of data and is named `force.dat`, typing `load force.dat` creates the 10×3 matrix `force`, which you can use in your MATLAB session just as you would use any other variable. Your data file can have any extension except `.mat`, so that MATLAB will not try to load the file as a workspace file.

Importing Spreadsheet Files

Some spreadsheet programs store data in the `.wk1` format. You can use the command `M = wk1read('filename')` to import this data into MATLAB and store it in the matrix `M`.

The command `A = xlsread('filename')` imports the Microsoft Excel workbook file `filename.xls` into the array `A`. The command `[A, B] = xlsread('filename')` imports all numeric data into the array `A` and all text data into the cell array `B`.

The Import Wizard

To import ASCII data, you must know how the data in the file is formatted. For example, many ASCII data files use a fixed (or uniform) format of rows and columns. For these files, you should know the following.

- How many data items are in each row?
- Are the data items numeric, text strings, or a mixture of both types?
- Does each row or column have a descriptive text header?
- What character is used as the *delimiter*, that is, the character used to separate the data items in each row? The delimiter is also called the *column separator*.

To find out how your ASCII data file is formatted, view it in a text editor. The data format will usually fall into one of the following categories:

1. Space-delimited ASCII data files,
2. Mixed text and numeric ASCII data files,
3. ASCII data files with text headers (the only text is at the head of each data column), or
4. ASCII data files with nonspace delimiters (usually semicolons).

You can use the Import Wizard to import many types of ASCII data formats, including data on the clipboard. Note that when you use the Import Wizard to create a variable in the MATLAB workspace, it overwrites any existing variable in the workspace with the same name without issuing a warning. The Import Wizard presents a series of dialog boxes in which you:

1. Specify the name of the file you want to import,
2. Specify the delimiter used in the file, and
3. Select the variables that you want to import.

The following provides a step-by-step procedure for using the Import Wizard to import this sample tab-delimited, ASCII data file, `testdata.txt`.

```
1     2     3     4     5;
17    12    8     15    25;
```

1. Activate the Import Wizard by selecting the **Import Data** option on the MATLAB Desktop **File** menu. The Import Wizard displays a dialog box that asks you to specify the name of the file you want to import. You can enter the file name in the text entry field or click the browse button to find the file you want to import. When the Import Wizard opens the file, it displays a preview of the data in the file. You can use the preview to verify that you have specified the correct file. To continue with the import process, click **Next.**
2. The Import Wizard processes the contents of the file and displays tabs identifying the variables it recognizes in the file, and displays a portion of the data in a grid, similar to a spreadsheet (see Figure 3.4–1). The Import Wizard uses the space character as the default delimiter. If your file uses another character as a delimiter, the Import Wizard attempts to identify the delimiter. Make sure the correct delimiter button is highlighted; if not, then

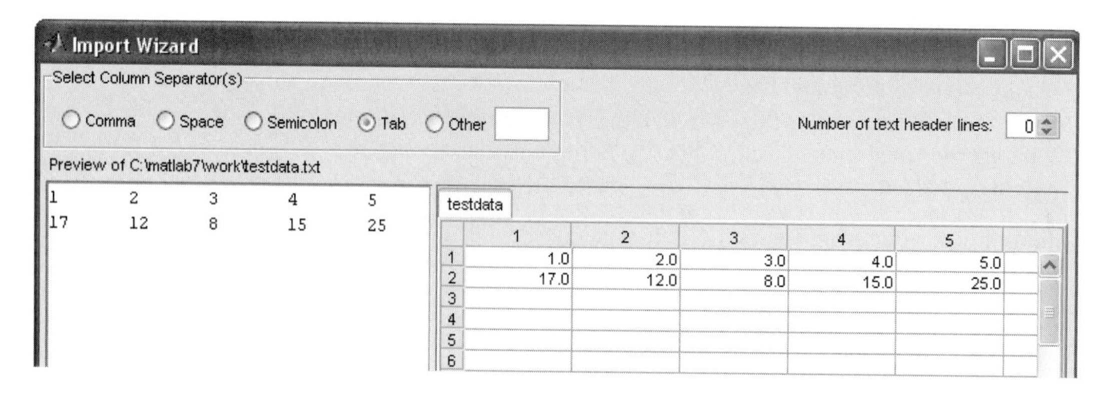

Figure 3.4–1 The first screen in the Import Wizard.

click on the correct one. In this example, the Import Wizard correctly interprets the contents of the tab-delimited sample file. If the delimiter is not listed, check the **Other** button and type the character in the text field. Click **Next** to continue the import operation.

3. In the next dialog box, the Import Wizard displays a list of the variables it found in the file. It also displays the contents of the first variable in the list. To view the contents of a variable, click on its name. (The variable displayed in the dialog box is highlighted in the list.) In this example there is only one variable, named `testdata`.
4. Choose the variables you want to import by clicking the check box next to their names. By default, all variables are checked for import. After selecting the variables you want to import, click the **Finish** button to import the data into the MATLAB workspace. This button dismisses the Import Wizard.

To import data from the clipboard, select **Paste Special** from the **Edit** menu. The proceed with step 2. The default variable name is `A_pastespecial`.

Importing ASCII Data Files with Text Headers

Follow the same procedure to use the Import Wizard to import an ASCII data file that contains text headers. You will see a tab for the numeric data, a tab for all the text in the file, and a tab for the text headers. For example, the following file, `temperature.dat`, contains space-delimited numeric data with a one-line text header.

```
Temp1    Temp2  Temp3
78.8      55.9   45.9
99.5      66.8   78.0
89.5      77.0   56.7
```

The default name given to the text headers by the Import Wizard is `colheaders`, and that given to the numeric data is `data`. In the Import Wizard screen that asks you to select variables to load, if you select "Create variables matching preview," when you click **Finish** you will have the variables `colheaders` and `data` in the workspace. The variable `data` is a numeric array containing the temperatures, and the variable `colheaders` is a cell array containing the headers. On the other hand, if you select "Create vectors from each column using column names," when you click **Finish** you will have the variables `Temp1`, `Temp2`, and `Temp3` in the workspace. These are numeric arrays. The appropriate choice depends on what you want to do with the data.

Importing Binary Data Files

You can use the Import Wizard to import many types of binary data formats, including MAT-files in which you saved previous sessions. The procedure is the same as for loading ASCII data files. You will see a text tab if your file contains text data, and a data tab if it contains numeric data.

Exporting Delimited ASCII Data Files

You might want to export a MATLAB matrix as an ASCII data file where the rows and columns are represented as space-delimited, numeric values. To export an array as a delimited ASCII data file, you can use either the `save` command, specifying the `-ASCII` qualifier, or you can use the `dlmwrite` function. The `save` command is easy to use; however, the `dlmwrite` function provides more flexibility, allowing you to specify any character as a delimiter and to export subsets of an array by specifying a range of values.

Suppose you have created the array `A = [1 2 3 4; 5 6 7 8]` in MATLAB. To export the array using the `save` command, type the following in the Command window.

```
>>save my_data.out A -ASCII
```

If you view the created file in a text editor, it looks like this

```
1.0000000e+000  2.0000000e+000  3.0000000e+000  4.0000000e+000
5.0000000e+000  6.0000000e+000  7.0000000e+000  8.0000000e+000
```

By default, `save` uses spaces as delimiters, but you can use tabs instead of spaces by specifying the `-tab` qualifier.

To export an array in ASCII format and specify the delimiter used in the file, use the `dlmwrite` function. For example, to export the array `A` as an ASCII data file that uses semicolons as a delimiter, type the following.

```
>>dlmwrite('my_data.out',A, ';')
```

If you view the created file in a text editor, it looks like this:

```
1;2;3;4 5;6;7;8
```

Note that `dlmwrite` does not insert delimiters at the end of rows. By default, if you do not specify a delimiter, `dlmwrite` uses commas as a delimiter. You can specify a space (`' '`) as a delimiter or you can specify no delimiter by using a pair of single quotes (`''`).

3.5 Summary

MATLAB supplies very many functions. In Section 3.1 we introduced just some of the most commonly used mathematical functions. You should now be able to use the MATLAB help to find other functions you need. If necessary, you can create your own functions, using the methods of Section 3.2.

Function handles, anonymous functions, subfunctions, and nested functions extend the capabilities of MATLAB and are especially useful for large programming projects. These topics were treated in Section 3.3.

Table 3.5–1 Guide to MATLAB commands introduced in Chapter 3

Some common mathematical functions	Table 3.1–1
Trigonometric functions	Table 3.1–2
Hyperbolic functions	Table 3.1–3
Minimization and root-finding functions	Table 3.2–1

In addition to function files, data files are also useful for many applications. Section 3.4 shows how to import and export such files in MATLAB.

Table 3.5–1 is a guide to all the commands introduced in this chapter.

Key Terms with Page References

Problems

You can find the answers to problems marked with an asterisk at the end of the text.

Section 3.1

1.* Suppose that $y = -3 + ix$. For $x = 0$, 1, and 2, use MATLAB to compute the following expressions. Hand check the answers.

a. $|y|$
b. $\sqrt{y}$
c. $(-5 - 7i)y$
d. $\frac{y}{6-3i}$

2.* Let $x = -5 - 8i$ and $y = 10 - 5i$. Use MATLAB to compute the following expressions. Hand check the answers.

a. The magnitude and angle of xy.
b. The magnitude and angle of $\frac{x}{y}$.

3.* Use MATLAB to find the angles corresponding to the following coordinates. Hand check the answers.

a. $(x, y) = (5, 8)$
b. $(x, y) = (-5, 8)$
c. $(x, y) = (5, -8)$
d. $(x, y) = (-5, -8)$

4. For several values of x, use MATLAB to confirm that $\sinh x = (e^x - e^{-x})/2$.

5. For several values of x, use MATLAB to confirm that $\sinh^{-1} x = \ln(x + \sqrt{x^2 + 1})$, $-\infty < x < \infty$.

6. The capacitance of two parallel conductors of length L and radius r, separated by a distance d in air, is given by

$$C = \frac{\pi \epsilon L}{\ln\left(\frac{d-r}{r}\right)}$$

where ϵ is the permittivity of air ($\epsilon = 8.854 \times 10^{-12}$ F/m).

Write a script file that accepts user input for d, L, and r, and computes and displays C. Test the file with the values: $L = 1$ m, $r = 0.001$ m, and $d = 0.004$ m.

7.* When a belt is wrapped around a cylinder, the relation between the belt forces on each side of the cylinder is

$$F_1 = F_2 e^{\mu\beta}$$

where β is the angle of wrap of the belt and μ is the friction coefficient. Write a script file that first prompts a user to specify β, μ, and F_2 and then computes the force F_1. Test your program with the values $\beta = 130°$, $\mu = 0.3$, and $F_2 = 100$ N. (Hint: Be careful with β!)

Section 3.2

8. The MATLAB trigonometric functions expect their argument to be in radians. Write a function called `sind` that accepts an angle x in degrees and computes $\sin x$. Test your function.

9. Write a function that accepts temperature in degrees F and computes the corresponding value in degrees C. The relation between the two is

$$T\ °\text{C} = \frac{5}{9}(T\ °\text{F} - 32)$$

Be sure to test your function.

10.* An object thrown vertically with a speed v_0 reaches a height h at time t, where

$$h = v_0 t - \frac{1}{2} g t^2$$

Write and test a function that computes the time t required to reach a specified height h, for a given value of v_0. The function's inputs should be h, v_0, and g. Test your function for the case where $h = 100$ m, $v_0 = 50$ m/s, and $g = 9.81$ m/s^2. Interpret both answers.

11. A water tank consists of a cylindrical part of radius r and height h, and a hemispherical top. The tank is to be constructed to hold 500 m^3 when filled. The surface area of the cylindrical part is $2\pi rh$, and its volume is $\pi r^2 h$. The surface area of the hemispherical top is given by $2\pi r^2$, and its

volume is given by $2\pi r^3/3$. The cost to construct the cylindrical part of the tank is \$300 per square meter of surface area; the hemispherical part costs \$400 per square meter. Use the `fminbnd` function to compute the radius that results in the least cost. Compute the corresponding height h.

12. A fence around a field is shaped as shown in Figure P12. It consists of a rectangle of length L and width W, and a right triangle that is symmetrical about the central horizontal axis of the rectangle. Suppose the width W is known (in meters), and the enclosed area A is known (in square meters). Write a user-defined function file with W and A as inputs. The outputs are the length L required so that the enclosed area is A, and the total length of fence required. Test your function for the values $W = 6$ m and $A = 80$ m^2.

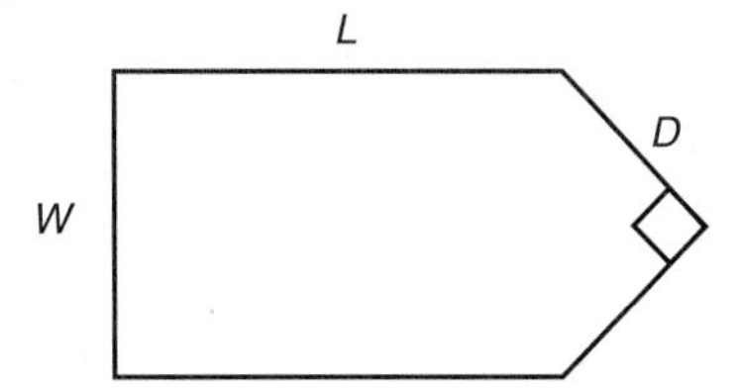

Figure P12

13. A fenced enclosure consists of a rectangle of length L and width $2R$, and a semicircle of radius R, as shown in Figure P13. The enclosure is to be built to have an area A of 1600 ft^2. The cost of the fence is \$40 per foot for the curved portion, and \$30 per foot for the straight sides. Use the `fminbnd` function to determine with a resolution of 0.01 ft the values of R and L required to minimize the total cost of the fence. Also compute the minimum cost.

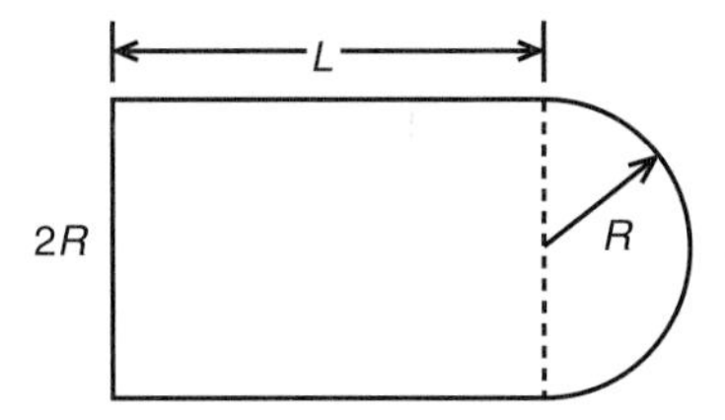

Figure P13

14. Using estimates of rainfall, evaporation, and water consumption, the town engineer developed the following model of the water volume in the reservoir as a function of time.

$$V(t) = 10^9 + 10^8(1 - e^{-t/100}) - rt$$

where V is the water volume in liters, t is time in days, and r is the town's consumption rate in liters/day. Write two user-defined functions. The first function should define the function $V(t)$ for use with the `fzero` function.

The second function should use `fzero` to compute how long it will take for the water volume to decrease to x percent of its initial value of 10^9 L. The inputs to the second function should be x and r. Test your functions for the case where $x = 50$ percent and $r = 10^7$ L/day.

15. The volume V and paper surface area A of a conical paper cup are given by

$$V = \frac{1}{3}\pi r^2 h \qquad A = \pi r \sqrt{r^2 + h^2}$$

where r is the radius of the base of the cone and h is the height of the cone.

a. By eliminating h, obtain the expression for A as a function of r and V.

b. Create a user-defined function that accepts R as the only argument and computes A for a given value of V. Declare V to be global within the function.

c. For $V = 10$ in.^{3}, use the function with the `fminbnd` function to compute the value of r that minimizes the area A. What is the corresponding value of the height h? Investigate the sensitivity of the solution by plotting V versus r. How much can R vary about its optimal value before the area increases 10 percent above its minimum value?

16. A torus is a shaped like a doughnut. If its inner radius is a and its outer radius is b, its volume and surface area are given by

$$V = \frac{1}{4}\pi^2(a + b)(b - a)^2 \qquad A = \pi^2(b^2 - a^2)$$

a. Create a user-defined function that computes V and A from the arguments a and b.

b. Suppose that the outer radius is constrained to be 2 in. greater than the inner radius. Write a script file that uses your function to plot A and V versus a for $0.25 \le a \le 4$ in.

17. Suppose it is known that the graph of the function $y = ax^3 + bx^2 + cx + d$ passes through four given points (x_i, y_i), $i = 1, 2, 3, 4$. Write a user-defined function that accepts these four points as input and computes the coefficients a, b, c, and d. The function should solve four linear equations in terms of the four unknowns a, b, c, and d. Test your function for the case where $(x_i, y_i) = (-2, -20)$, $(0, 4)$, $(2, 68)$, and $(4, 508)$, whose answer is $a = 7$, $b = 5$, $c = -6$, and $d = 4$.

Section 3.3

18. Use the `gen_plot` function described in Section 3.3 to obtain two subplots, one plot of the function $10e^{-2x}$ over the range $0 \le x \le 2$, and the other a plot of $5\sin(2\pi x/3)$ over the range $0 \le x \le 6$.

19. Create an anonymous function for $10e^{-2x}$ and use it to plot the function over the range $0 \le x \le 2$.

20. Create an anonymous function for $20x^2 - 200x + 3$ and use it

a. to plot the function to determine the approximate location of its minimum, and

b. with the `fminbnd` function to precisely determine the location of the minimum.

21. Create four anonymous functions to represent the function $6e^{3\cos x^2}$, which is composed of the functions $h(z) = 6e^z$, $g(y) = 3\cos y$, and $f(x) = x^2$. Use the anonymous functions to plot $6e^{3\cos x^2}$ over the range $0 \le x \le 4$.

22. Use a primary function with a subfunction to compute the zeros of the function $3x^3 - 12x^2 - 33x + 90$ over the range $-10 \le x \le 10$.

23. Create a primary function that uses a function handle with a nested function to compute the minimum of the function $20x^2 - 200x + 3$ over the range $0 \le x \le 10$.

Section 3.4

24. Use a text editor to create a file containing the following data. Then use the `load` function to load the file into MATLAB, and use the `mean` function to compute the mean value of each column.

```
55  42  98
51  39  95
63  43  94
58  45  90
```

25. Enter and save the data given in Problem 24 in a spreadsheet. Then import the spreadsheet file into the MATLAB variable `A`. Use MATLAB to compute the sum of each column.

26. Use a text editor to create a file from the data given in Problem 24, but separate each number with a semicolon. Then use the Import Wizard to load and save the data in the MATLAB variable `A`.

27. Use a text editor to create a file `temperature.dat` containing the temperature data given on page 175. Then use the Import Wizard to load and save the data in the MATLAB variable `temperature`. Compute the mean value of each column.

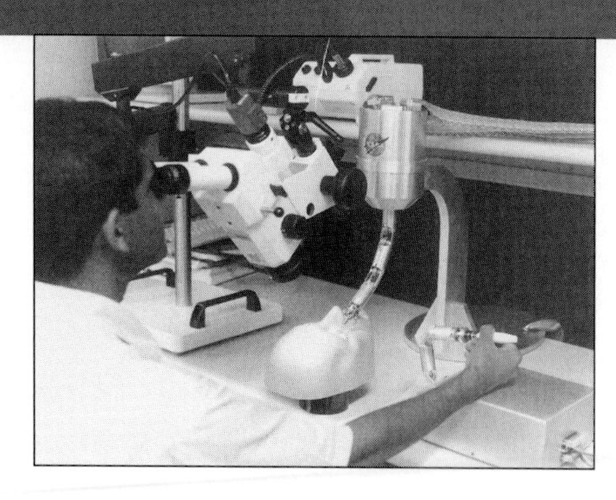

Courtesy Jet Propulsion Lab/NASA.

Engineering in the 21st Century...

Robot-Assisted Microsurgery

You need not be a medical doctor to participate in the exciting developments now taking place in the health field. Many advances in medicine and surgery are really engineering achievements, and many engineers are contributing their talents in this area. Recent achievements include

- Laparoscopic surgery in which a fiber-optic scope guides a small surgical device. This technology eliminates the need for large incisions and the resulting long recuperation.
- Computerized axial tomography (CAT) scans and magnetic resonance imaging (MRI), which provide noninvasive tools for diagnosing medical problems.
- Medical instrumentation, such as a fingertip sensor for continuously measuring oxygen in the blood and automatic blood pressure sensors.

As we move into the 21st century, an exciting challenge will be the development of robot-assisted surgery in which a robot, supervised by a human, performs operations requiring precise, steady motions. Robots have already assisted in hip surgery on animals, but much more development is needed. Another developing technology is *telesurgery* in which a surgeon uses a television interface to remotely guide a surgical robot. This technology would allow delivery of medical services to remote areas.

Robot-assisted microsurgery, which uses a robot capable of very small, precise motions, shows great promise. One application is in eye surgery, and the photo above shows a test of such a device on a dummy head. Designing such devices requires geometric analysis, control system design, and image processing. The MATLAB Image Processing toolbox and the several MATLAB toolboxes dealing with control system design are useful for such applications. ■

CHAPTER 4

Programming with MATLAB

OUTLINE

The MATLAB interactive mode is very useful for simple problems, but more complex problems require a script file. Such a file can be called a *computer program,* and writing such a file is called *programming.* Section 4.1 presents a general and efficient approach to the design and development of programs.

The usefulness of MATLAB is greatly increased by the use of decision-making functions in its programs. These functions enable you to write programs whose operations depend on the results of calculations made by the program. Sections 4.2, 4.3, and 4.4 deal with these decision-making functions.

MATLAB can also repeat calculations a specified number of times or until some condition is satisfied. This feature enables engineers to solve problems of great complexity or requiring numerous calculations. These "loop" structures are covered in Section 4.5.

The `switch` structure enhances the MATLAB decision-making capabilities. This topic is covered in Section 4.6. Use of the MATLAB Editor/Debugger for debugging programs is covered in Section 4.7.

Section 4.8 discusses "simulation," a major application of MATLAB programs that enables us to study the operation of complicated systems, processes, and organizations. Tables summarizing the MATLAB commands introduced in this chapter appear throughout the chapter, and Table 4.9–1 will help you locate the information you need.

4.1 Program Design and Development

In Chapter 1 we introduced *relational operators,* such as > and ==, and the two types of loops used in MATLAB, the `for` loop and the `while` loop. These features, plus MATLAB functions and the *logical operators* to be introduced in Section 4.3, form the basis for constructing MATLAB programs to solve complex problems. Design of computer programs to solve complex problems needs to be done in a systematic manner from the start to avoid time-consuming and frustrating difficulties later in the process. In this section we show how to structure and manage the design process.

Algorithms and Control Structures

An *algorithm* is an ordered sequence of precisely defined instructions that performs some task in a finite amount of time. An ordered sequence means that the instructions can be numbered, but an algorithm must have the ability to alter the order of its instructions using what is called a *control structure.* There are three categories of algorithmic operations:

Sequential operations. These are instructions that are executed in order.

Conditional operations. These are control structures that first ask a question to be answered with a true/false answer and then select the next instruction based on the answer.

Iterative operations (loops). These are control structures that repeat the execution of a block of instructions.

Not every problem can be solved with an algorithm and some potential algorithmic solutions can fail because they take too long to find a solution.

Structured Programming

Structured programming is a technique for designing programs in which a hierarchy of *modules* is used, each having a single entry and a single exit point, and in which control is passed downward through the structure without unconditional branches to higher levels of the structure. In MATLAB these modules can be built-in or user-defined functions.

Control of the program flow uses the same three types of control structures used in algorithms: sequential, conditional, and iterative. In general, any computer program can be written with these three structures. This realization led to the development of structured programming. Languages suitable for structured programming, such as MATLAB, thus do not have an equivalent to the `goto` statement that you might have seen in the BASIC and FORTRAN languages. An unfortunate result of the `goto` statement was confusing code, called *spaghetti code,* composed of a complex tangle of branches.

Structured programming, if used properly, results in programs that are easy to write, understand, and modify. The advantages of structured programming are as follows.

1. Structured programs are easier to write because the programmer can study the overall problem first and then deal with the details later.
2. Modules (functions) written for one application can be used for other applications (this is called *reusable code*).
3. Structured programs are easier to debug because each module is designed to perform just one task and thus it can be tested separately from the other modules.
4. Structured programming is effective in a teamwork environment because several people can work on a common program, each person developing one or more modules.
5. Structured programs are easier to understand and modify, especially if meaningful names are chosen for the modules and if the documentation clearly identifies the module's task.

Top-down Design and Program Documentation

A method for creating structured programs is *top-down design,* which aims to describe a program's intended purpose at a very high level initially, and then partition the problem repeatedly into more detailed levels, one level at a time, until enough is understood about the program structure to enable it to be coded. Table 4.1–1, which is repeated from Chapter 1, summarizes the process of top-down design. In step 4 you create the algorithms used to obtain the solution. Note that step 5, Write and Run the Program, is only part of the top-down design process. In this step you create the necessary modules and test them separately.

Table 4.1–1 Steps for developing a computer solution

1. State the problem concisely.
2. Specify the data to be used by the program. This is the "input."
3. Specify the information to be generated by the program. This is the "output."
4. Work through the solution steps by hand or with a calculator; use a simpler set of data if necessary.
5. Write and run the program.
6. Check the output of the program with your hand solution.
7. Run the program with your input data and perform a reality check on the output.
8. If you will use the program as a general tool in the future, test it by running it for a range of reasonable data values; perform a reality check on the results.

STRUCTURE CHART

FLOWCHART

Two types of charts aid in developing structured programs and in documenting them. These are *structure charts* and *flowcharts*. A structure chart is a graphical description showing how the different parts of the program are connected together. This type of diagram is particularly useful in the initial stages of top-down design.

A structure chart displays the organization of a program without showing the details of the calculations and decision processes. For example, we can create program modules using function files that do specific, readily identifiable tasks. Larger programs are usually composed of a main program that calls on the modules to do their specialized tasks as needed. A structure chart shows the connection between the main program and the modules.

For example, suppose you want to write a program that plays a game, say Tic-Tac-Toe. You would need a module to allow the human player to input a move, a module to update and display the game grid, and a module that contains the computer's strategy for selecting its moves. Figure 4.1–1 shows the structure chart of such a program.

Flowcharts are useful for developing and documenting programs that contain conditional statements, because they can display the various paths (called "branches") that a program can take, depending on how the conditional statements are executed. The flowchart representation of the `if` statement is shown in Figure 4.1–2. Flowcharts use the diamond symbol to indicate decision points.

The usefulness of structure charts and flowcharts is limited by their size. For large, more complicated programs, it might be impractical to draw such charts. Nevertheless, for smaller projects, sketching a flowchart and/or a structure chart might help you organize your thoughts before beginning to write the specific MATLAB code. Because of the space required for such charts we do not use them in this text. You are encouraged, however, to use them when solving problems.

Documenting programs properly is very important, even if you never give your programs to other people. If you need to modify one of your programs, you

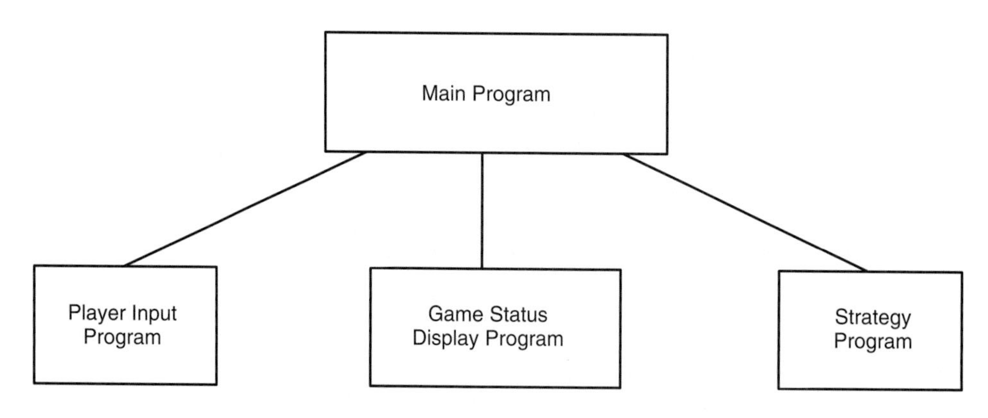

Figure 4.1–1 Structure chart of a game program.

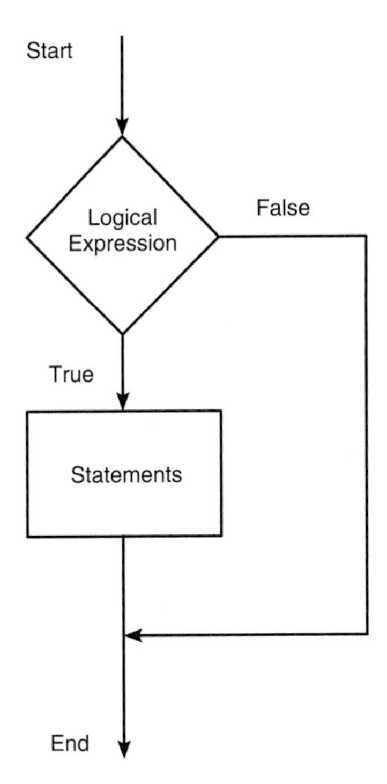

Figure 4.1–2 Flowchart representation of the `if` statement.

will find that it is often very difficult to recall how it operates if you have not used it for some time. Effective documentation can be accomplished with the use of

1. Proper selection of variable names to reflect the quantities they represent.
2. Use of comments within the program.
3. Use of structure charts.
4. Use of flowcharts.
5. A verbal description of the program, often in *pseudocode*.

The advantage of using suitable variable names and comments is that they reside with the program; anyone who gets a copy of the program will see such documentation. However, they often do not provide enough of an overview of the program. The latter three elements can provide such an overview.

Pseudocode

Use of natural language, such as English, to describe algorithms often results in a description that is too verbose and is subject to misinterpretation. To avoid dealing immediately with the possibly complicated syntax of the programming language, we can instead use *pseudocode,* in which natural language and mathematical expressions are used to construct statements that look like computer statements but without detailed syntax. Pseudocode may also use some simple MATLAB syntax to explain the operation of the program.

As its name implies, pseudocode is an imitation of the actual computer code. The pseudocode can provide the basis for comments within the program. In addition to providing documentation, pseudocode is useful for outlining a program before writing the detailed code, which takes longer to write because it must conform to the strict rules of MATLAB.

Each pseudocode instruction may be numbered, but should be unambiguous and computable. Note that MATLAB does not use line numbers except in the Debugger. Each of the following examples illustrates how pseudocode can document each of the control structures used in algorithms: sequential, conditional, and iterative operations.

Example 1. Sequential Operations Compute the perimeter p and the area A of a triangle whose sides are a, b, and c. The formulas are:

$$p = a + b + c \qquad s = \frac{p}{2} \qquad A = \sqrt{s(s-a)(s-b)(s-c)}$$

1. Enter the side lengths a, b, and c.
2. Compute the perimeter p.

$$p = a + b + c$$

3. Compute the semiperimeter s.

$$s = p/2$$

4. Compute the area A.

$$A = \sqrt{s(s-a)(s-b)(s-c)}$$

5. Display the results p and A.
6. Stop

The program is

```
a = input('Enter the value of side a: ');
b = input('Enter the value of side b: ');
c = input('Enter the value of side c: ');
p = a + b + c;
s = p/2;
A = sqrt(s*(s-a)*(s-b)*(s-c));
disp('The perimeter is:')
p
disp('The area is:')
A
```

Example 2. Conditional Operations Given the (x, y) coordinates of a point, compute its polar coordinates (r, θ), where

$$r = \sqrt{x^2 + y^2} \qquad \theta = \tan^{-1}\left(\frac{y}{x}\right)$$

1. Enter the coordinates x and y.
2. Compute the hypoteneuse r.

```
r = sqrt(x^2+y^2)
```

3. Compute the angle θ.
   3.1 If $x \geq 0$

   ```
   theta = atan(y/x)
   ```

   3.2 Else

   ```
   theta = atan(y/x) + pi
   ```

4. Convert the angle to degrees.

```
theta = theta*(180/pi)
```

5. Display the results `r` and `theta`.
6. Stop

Note the use of the numbering scheme 3.1 and 3.2 to indicate subordinate clauses. Note also that MATLAB syntax may be used for clarity where needed. The program is

```
x = input('Enter the value of x: ');
y = input('Enter the value of y: ');
r = sqrt(x^2+y^2);
if x >= 0
   theta = atan(y/x);
else
   theta = atan(y/x) + pi;
end
disp('The hypoteneuse is:')
disp(r)
theta = theta*(180/pi);
disp('The angle is degrees is:')
disp(theta)
```

Example 3. Iterative Operations Determine how many terms are required for the sum of the series $10k^2 - 4k + 2$, $k = 1, 2, 3, \ldots$ to exceed 20,000. What is the sum for this many terms?

Because we do not know how many times we must evaluate the expression $10k^2 - 4k + 2$, we use a `while` loop.

1. Initialize the total to zero.
2. Initialize the counter to zero.
3. While the total is less than 20,000 compute the total.
   3.1 Increment the counter by 1.

   ```
   k = k + 1
   ```

   3.2 Update the total.

   ```
   total = 10*k^2 - 4*k + 2 + total
   ```

4. Display the current value of the counter.
5. Display the value of the total.
6. Stop

The program is

```
total = 0;
k = 0;
while total < 2e+4
   k = k+1;
   total = 10*k^2 - 4*k + 2 + total;
end
disp('The number of terms is:')
disp(k)
disp('The sum is:')
disp(total)
```

Finding Bugs

Debugging a program is the process of finding and removing the "bugs," or errors, in a program. Such errors usually fall into one of the following categories.

1. Syntax errors such as omitting a parenthesis or comma, or spelling a command name incorrectly. MATLAB usually detects the more obvious errors and displays a message describing the error and its location.
2. Errors due to an incorrect mathematical procedure. These are called *runtime errors*. They do not necessarily occur every time the program is executed; their occurrence often depends on the particular input data. A common example is division by zero.

The MATLAB error messages usually enable you to find syntax errors. However, runtime errors are more difficult to locate. To locate such an error, try the following:

1. Always test your program with a simple version of the problem, whose answers can be checked by hand calculations.
2. Display any intermediate calculations by removing semicolons at the end of statements.

3. To test user-defined functions, try commenting out the `function` line and running the file as a script.
4. Use the debugging features of the Editor/Debugger, which is discussed in Section 4.7.

Development of Large Programs

Large programs and software, including commercial software such as MATLAB, undergo a rigorous process of development and testing before finally being released and approved for general use. The stages in this process are typically the following.

1. Writing and testing of individual modules (the unit-testing phase).
2. Writing of the top-level program that uses the modules (the build phase). Not all modules are included in the initial testing. As the build proceeds, more modules are included.
3. Testing of the first complete program (the alpha release phase). This is usually done only in-house by technical people closely involved with the program development. There might be several alpha releases as bugs are discovered and removed.
4. Testing of the final alpha release by in-house personnel and by familiar and trusted outside users, who often must sign a confidentiality agreement. This is the beta release phase, and there might be several beta releases.

This process can take quite a while, depending on the complexity of the program and the company's dedication to producing quality software. In the case of MATLAB 7, there were approximately nine months between the first beta release and the final release of the completed software.

4.2 Relational Operators and Logical Variables

MATLAB has six *relational* operators to make comparisons between arrays. These operators are shown in Table 4.2–1 and were introduced in Section 1.3. Recall that the *equal to* operator consists of two = signs, not a single = sign as you might expect. The single = sign is the *assignment,* or *replacement,* operator in MATLAB.

Table 4.2–1 Relational operators

Relational operator	Meaning
<	Less than.
<=	Less than or equal to.
>	Greater than.
>=	Greater than or equal to.
==	Equal to.
~=	Not equal to.

The result of a comparison using the relational operators is either 0 (if the comparison is *false*), or 1 (if the comparison is *true*), and the result can be used as a variable. For example, if `x = 2` and `y = 5`, typing `z = x < y` returns the value `z = 1` and typing `u = x==y` returns the value `u = 0`. To make the statements more readable, we can group the logical operations using parentheses. For example, `z = (x < y)` and `u = (x==y)`.

When used to compare arrays, the relational operators compare the arrays on an element-by-element basis. The arrays being compared must have the same dimension. The only exception occurs when we compare an array to a scalar. In that case all the elements of the array are compared to the scalar. For example, suppose that `x = [6,3,9]` and `y = [14,2,9]`. The following MATLAB session shows some examples.

```
>>z = (x < y)
z =
   1   0   0
>>z = (x ~= y)
z =
   1   1   0
>>z = (x > 8)
z =
   0   0   1
```

The relational operators can be used for array addressing. For example, with `x = [6,3,9]` and `y = [14,2,9]`, typing `z = x(x<y)` finds all the elements in `x` that are less than the corresponding elements in `y`. The result is `z = 6`.

The arithmetic operators +, -, *, /, and \ have precedence over the relational operators. Thus the statement `z = 5 > 2 + 7` is equivalent to `z = 5 >(2+7)` and returns the result `z = 0`. We can use parentheses to change the order of precedence; for example, `z = (5 > 2) + 7` evaluates to `z = 8`.

The relational operators have equal precedence among themselves, and MATLAB evaluates them in order from left to right. Thus the statement

```
z = 5 > 3 ~= 1
```

is equivalent to

```
z = (5>3) ~= 1
```

Both statements return the result `z = 0`.

With relational operators that consist of more than one character, such as == or >=, be careful not to put a space between the characters.

The `logical` Class

When the relational operators are used, such as `x = (5 > 2)`, they create a logical variable, in this case, `x`. Prior to MATLAB 6.5 `logical` was an attribute of any numeric data type. Now `logical` is a first-class data type and a MATLAB class, and so `logical` is now equivalent to other first-class types such as character and cell arrays. Logical variables may have only the values 1 (true) and 0 (false).

Just because an array contains only 0s and 1s, however, it is not necessarily a logical array. For example, in the following session `k` and `w` appear the same, but `k` is a logical array and `w` is a numeric array, and thus an error message is issued.

```
>>x = [-2:2]
x =
   -2    -1    0    1    2
>>k = (abs(x)>1)
k =
    1    0    0     0    1
>>z = x(k)
z =
   -2    2
>>w = [1,0,0,0,1];
>>v = x(w)
??? Subscript indices must either be real positive...
   integers or logicals.
```

The `logical` Function

Logical arrays can be created with the relational and logical operators and with the `logical` function. The `logical` function returns an array that can be used for logical indexing and logical tests. Typing `B = logical(A)`, where `A` is a numeric array, returns the logical array `B`. So to correct the error in the previous session, you may type instead `w = logical([1,0,0,0,1])` before typing `v = x(w)`.

When a finite, real value other than 1 or 0 is assigned to a logical variable, the value is converted to logical 1 and a warning message is issued. For example, when you type `y = logical(9)`, `y` will be assigned the value logical 1 and a warning will be issued. You may use the `double` function to convert a logical array to an array of class `double`. For example, `x = (5>3); y = double(x);`. Some arithmetic operations convert a logical array to a double array. For example, if we add zero to each element of `B` by typing `B = B + 0`, `B` will be converted to a numeric (double) array. However, not all mathematical operations are defined for logical variables. For example, typing

```
>>x = ([2, 3] > [1, 6]);
>>y = sin(x)
```

will generate an error message.

Accessing Arrays Using Logical Arrays

When a logical array is used to address another array, it extracts from that array the elements in the locations where the logical array has 1s. So typing `A(B)`, where `B` is a logical array of the same size as `A`, returns the values of `A` at the indices where `B` is 1.

Given `A = [5,6,7;8,9,10;11,12,13]` and `B = logical(eye(3))`, we can extract the diagonal elements of `A` by typing `C = A(B)` to obtain `C = [5;9;13]`. Specifying array subscripts with logical arrays extracts the elements that correspond to the true (1) elements in the logical array.

Note, however, that using the *numeric* array `eye(3)`, as `C = A(eye(3))`, results in an error message because the elements of `eye(3)` do not correspond to locations in `A`. If the numeric array values correspond to valid locations, you may use a numeric array to extract the elements. For example, to extract the diagonal elements of `A` with a numeric array, type `C = A([1,5,9])`.

MATLAB data types are preserved when indexed assignment is used. So now that `logical` is a MATLAB data type, if `A` is a logical array, for example `A = logical(eye(4))`, then typing `A(3,4) = 1` does not change `A` to a double array. However, typing `A(3,4) = 5` will set `A(3,4)` to logical 1 and cause a warning to be issued.

4.3 Logical Operators and Functions

MATLAB has five *logical operators,* which are sometimes called *Boolean* operators (see Table 4.3–1). These operators perform element-by-element operations. With the exception of the NOT operator (~), they have a lower precedence than the arithmetic and relational operators (see Table 4.3–2). The NOT symbol is called the *tilde.*

The NOT operation `~A` returns an array of the same dimension as `A`; the new array has ones where `A` is zero and zeros where `A` is nonzero. If `A` is logical, then `~A` replaces ones with zeros and zeros with ones. For example,

Table 4.3–1 Logical operators

Operator	Name	Definition
`~`	NOT	`~A` returns an array the same dimension as `A`; the new array has ones where `A` is zero and zeros where `A` is nonzero.
`&`	AND	`A & B` returns an array the same dimension as `A` and `B`; the new array has ones where both `A` and `B` have nonzero elements and zeros where either `A` or `B` is zero.
`|`	OR	`A | B` returns an array the same dimension as `A` and `B`; the new array has ones where at least one element in `A` or `B` is nonzero and zeros where `A` and `B` are both zero.
`&&`	Short-Circuit AND	Operator for scalar logical expressions. `A && B` returns true if both `A` and `B` evaluate to true, and false if they do not.
`||`	Short-Circuit OR	Operator for scalar logical expressions. `A || B` returns true if either `A` or `B` or both evaluate to true, and false if they do not.

Table 4.3–2 Order of precedence for operator types

Precedence	Operator type
First	Parentheses; evaluated starting with the innermost pair.
Second	Arithmetic operators and logical NOT (~); evaluated from left to right.
Third	Relational operators; evaluated from left to right.
Fourth	Logical AND.
Fifth	Logical OR.

if `x = [0,3,9]` and `y = [14,-2,9]`, then `z = ~x` returns the array `z = [1,0,0]` and the statement `u = ~x > y` returns the result `u = [0,1,0]`. This expression is equivalent to `u = (~x) > y`, whereas `v = ~(x > y)` gives the result `v = [1,0,1]`. This expression is equivalent to `v = (x <= y)`.

The `&` and `|` operators compare two arrays of the same dimension. The only exception, as with the relational operators, is that an array can be compared to a scalar. The AND operation `A&B` returns ones where both `A` and `B` have nonzero elements and zeros where any element of `A` or `B` is zero. The expression `z = 0&3` returns `z = 0`; `z = 2&3` returns `z = 1`; `z = 0&0` returns `z = 0`, and `z = [5,-3,0,0]&[2,4,0,5]` returns `z = [1,1,0,0]`. Because of operator precedence, `z = 1&2+3` is equivalent to `z = 1&(2+3)`, which returns `z = 1`. Similarly, `z = 5<6&1` is equivalent to `z = (5<6)&1`, which returns `z = 1`.

Let `x = [6,3,9]` and `y = [14,2,9]` and let `a = [4,3,12]`. The expression

```
z = (x>y) & a
```

gives `z = [0,1,0]`, and

```
z = (x>y)&(x>a)
```

returns the result `z = [0,0,0]`. This is equivalent to

```
z = x>y&x>a
```

which is much less readable.

Be careful when using the logical operators with inequalities. For example, note that `~(x > y)` is equivalent to `x <= y`. It is *not* equivalent to `x < y`. As another example, the relation $5 < x < 10$ must be written as

```
(5 < x) & (x < 10)
```

in MATLAB.

The OR operation `A|B` returns ones where at least one of `A` and `B` has nonzero elements and zeros where both `A` and `B` are zero. The expression `z = 0|3` returns `z = 1`; the expression `z = 0|0` returns `z = 0`; and

```
z = [5,-3,0,0]|[2,4,0,5]
```

returns `z = [1,1,0,1]`. Because of operator precedence,

```
z = 3<5|4==7
```

is equivalent to

```
z =(3<5)|(4==7)
```

which returns `z = 1`. Similarly, `z = 1|0&1` is equivalent to `z = (1|0)&1`, which returns `z = 1`, while `z = 1|0&0` returns `z = 0`, and `z = 0&0|1` returns `z = 1`.

Because of the precedence of the NOT operator, the statement

```
z = ~3==7|4==6
```

returns the result `z = 0`, which is equivalent to

```
z = ((~3)==7)|(4==6)
```

The exclusive OR function `xor(A,B)` returns zeros where `A` and `B` are either both nonzero or both zero, and ones where either `A` or `B` is nonzero, *but not both.* The function is defined in terms of the AND, OR, and NOT operators as follows.

```
function z = xor(A,B)
z = (A|B) & ~(A&B);
```

The expression

```
z = xor([3,0,6],[5,0,0])
```

returns `z = [0,0,1]`, whereas

```
z = [3,0,6]|[5,0,0]
```

returns `z = [1,0,1]`.

TRUTH TABLE

Table 4.3–3 is a so-called *truth table* that defines the operations of the logical operators and the function `xor`. Until you acquire more experience with the logical operators, you should use this table to check your statements. Remember that *true* is equivalent to logical 1, and *false* is equivalent to logical 0. We can test the truth table by building its numerical equivalent as follows. Let `x` and `y` represent the first two columns of the truth table in terms of ones and zeros.

Table 4.3–3 Truth table

x	y	~x	x|y	x&y	xor(x,y)
true	true	false	true	true	false
true	false	false	true	false	true
false	true	true	true	false	true
false	false	true	false	false	false

The following MATLAB session generates the truth table in terms of ones and zeros.

```
>>x = [1,1,0,0]';
>>y = [1,0,1,0]';
>>Truth_Table = [x,y,~x,x|y,x&y,xor(x,y)]
Truth_Table =
   1 1 0 1 1 0
   1 0 0 1 0 1
   0 1 1 1 0 1
   0 0 1 0 0 0
```

Starting with MATLAB 6, the AND operator (`&`) was given a higher precedence than the OR operator (`|`). This was not true in earlier versions of MATLAB, so if you are using code created in an earlier version, you should make the necessary changes before using it in MATLAB 6 or higher. For example, now the statement `y = 1|5&0` is evaluated as `y = 1|(5&0)`, yielding the result `y = 1`, whereas in MATLAB 5.3 and earlier, the statement would have been evaluated as `y = (1|5)&0`, yielding the result `y = 0`. To avoid potential problems due to precedence, it is important to use parentheses in statements containing arithmetic, relational, or logical operators, even where parentheses are optional. MATLAB now provides a feature to enable the system to produce either an error message or a warning for any expression containing `&` and `|` that would be evaluated differently than in earlier versions. If you do not use this feature, MATLAB will issue a warning as the default. To activate the error feature, type `feature('OrAndError',1)`. To reinstate the default, type `feature('OrAndError',0)`.

Short-Circuit Operators

The following operators perform AND and OR operations on logical expressions containing *scalar* values only. They are called short-circuit operators because they evaluate their second operand only when the result is not fully determined by the first operand. They are defined as follows in terms of the two logical variables `A` and `B`.

`A&&B` Returns true (logical 1) if both `A` and `B` evaluate to true, and false (logical 0) if they do not.

`A||B` Returns true (logical 1) if either `A` or `B`, or both, evaluate to true, and false (logical 0) if they do not.

Thus in the statement `A&&B`, if `A` equals logical zero, then the entire expression will evaluate to false, regardless of the value of `B`, and therefore there is no need to evaluate `B`.

For `A||B`, if `A` is true, regardless of the value of `B`, the statement will evaluate to true.

Table 4.3–4 Logical functions

Logical function	Definition
`all(x)`	Returns a scalar, which is 1 if all the elements in the vector `x` are nonzero and 0 otherwise.
`all(A)`	Returns a row vector having the same number of columns as the matrix `A` and containing ones and zeros, depending on whether or not the corresponding column of `A` has all nonzero elements.
`any(x)`	Returns a scalar, which is 1 if any of the elements in the vector `x` is nonzero and 0 otherwise.
`any(A)`	Returns a row vector having the same number of columns as `A` and containing ones and zeros, depending on whether or not the corresponding column of the matrix `A` contains any nonzero elements.
`find(A)`	Computes an array containing the indices of the nonzero elements of the array `A`.
`[u,v,w] = find(A)`	Computes the arrays `u` and `v` containing the row and column indices of the nonzero elements of the array `A` and computes the array `w` containing the values of the nonzero elements. The array `w` may be omitted.
`finite(A)`	Returns an array of the same dimension as `A` with ones where the elements of `A` are finite and zeros elsewhere.
`ischar(A)`	Returns a 1 if `A` is a character array and 0 otherwise.
`isempty(A)`	Returns a 1 if `A` is an empty matrix and 0 otherwise.
`isinf(A)`	Returns an array of the same dimension as `A`, with ones where `A` has 'inf' and zeros elsewhere.
`isnan(A)`	Returns an array of the same dimension as `A` with ones where `A` has 'NaN' and zeros elsewhere. ('NaN' stands for "not a number," which means an undefined result.)
`isnumeric(A)`	Returns a 1 if `A` is a numeric array and 0 otherwise.
`isreal(A)`	Returns a 1 if `A` has no elements with imaginary parts and 0 otherwise.
`logical(A)`	Converts the elements of the array `A` into logical values.
`xor(A,B)`	Returns an array the same dimension as `A` and `B`; the new array has ones where either `A` or `B` is nonzero, but not both, and zeros where `A` and `B` are either both nonzero or both zero.

Table 4.3–4 lists several useful logical functions. You learned about the `find` function in Chapter 1.

Logical Operators and the `find` Function

The `find` function is very useful for creating decision-making programs, especially when combined with the relational or logical operators. The function `find(x)` computes an array containing the indices of the nonzero elements of the array `x`. We saw examples of its use with relational operators in Chapter 1. It is also useful when combined with the logical operators. For example, consider the session

```
>>x = [5, -3, 0, 0, 8]; y = [2, 4, 0, 5, 7];
>>z = find(x&y)
z =
   1    2    5
```

The resulting array `z = [1, 2, 5]` indicates that the first, second, and fifth elements of `x` and `y` are both nonzero. Note that the `find` function returns the *indices,* and not the *values.* In the following session, note the difference between the result obtained by `y(x&y)` and the result obtained by `find(x&y)` above.

```
>>x = [5, -3, 0, 0, 8];y = [2, 4, 0, 5, 7];
>>values = y(x&y)
values =
   2     4     7
>>how_many = length(values)
how_many =
   3
```

Thus there are three nonzero values in the array `y` that correspond to nonzero values in the array `x`. They are the first, second, and fifth values, which are 2, 4, and 7.

In the above example, there were only a few numbers in the arrays `x` and `y`, and thus we could have obtained the answers by visual inspection. However, these MATLAB methods are very useful either where there is so much data that visual inspection would be very time-consuming, or where the values are generated internally in a program.

Test Your Understanding

T4.3–1 If `x = [5,-3,18,4]` and `y = [-9,13,7,4]`, what will be the result of the following operations? Use MATLAB to check your answer.

a. `z = ~y > x`
b. `z = x&y`
c. `z = x|y`
d. `z = xor(x,y)`

Height and Speed of a Projectile — EXAMPLE 4.3–1

The height and speed of a projectile (such as a thrown ball) launched with a speed of v_0 at an angle A to the horizontal are given by

$$h(t) = v_0 t \sin A - 0.5 g t^2$$

$$v(t) = \sqrt{v_0^2 - 2 v_0 g t \sin A + g^2 t^2}$$

where g is the acceleration due to gravity. The projectile will strike the ground when $h(t) = 0$, which gives the time to hit, $t_{hit} = 2(v_0/g)\sin A$. Suppose that $A = 40°$, $v_0 = 20$ m/s, and $g = 9.81$ m/s^2. Use the MATLAB relational and logical operators to find the times when the height is no less than 6 m and the speed is simultaneously no greater than 16 m/s. In addition, discuss another approach to obtaining a solution.

■ Solution

The key to solving this problem with relational and logical operators is to use the `find` command to determine the times at which the logical expression `(h >= 6)&(v <= 16)` is true. First we must generate the vectors `h` and `v` corresponding to times t_1 and t_2 between $0 \le t \le t_{hit}$, using a spacing for time t that is small enough to achieve sufficient accuracy for our purposes. We will choose a spacing of $t_{hit}/100$, which provides 101 values of time. The program follows. When computing the times t_1 and t_2, we must subtract 1 from `u(1)` and from `length(u)` because the first element in the array `t` corresponds to $t = 0$ (that is, `t(1)` is 0).

```
% Set the values for initial speed, gravity, and angle.
v0 = 20; g = 9.81; A = 40*pi/180;
% Compute the time to hit.
t_hit = 2*v0*sin(A)/g;
% Compute the arrays containing time, height, and speed.
t = [0:t_hit/100:t_hit];
h = v0*t*sin(A) - 0.5*g*t.^2;
v = sqrt(v0^2 - 2*v0*g*sin(A)*t + g^2*t.^2);
% Determine when the height is no less than 6,
% and the speed is no greater than 16.
u = find(h>=6&v<=16);
% Compute the corresponding times.
t_1 = (u(1)-1)*(t_hit/100)
t_2 = u(length(u)-1)*(t_hit/100)
```

The results are $t_1 = 0.8649$ and $t_2 = 1.7560$. Between these two times $h \ge 6$ m and $v \le 16$ m/s.

We could have solved this problem by plotting $h(t)$ and $v(t)$, but the accuracy of the results would be limited by our ability to pick points off the graph; in addition, if we had to solve many such problems, the graphical method would be more time-consuming.

Test Your Understanding

T4.3–2 Consider the problem given in Example 4.3–1. Use relational and logical operators to find the times for which either the projectile's height is less than 4 m or the speed is greater than 17 m/s. Plot $h(t)$ and $v(t)$ to confirm your answer.

4.4 Conditional Statements

In everyday language we describe our decision making by using conditional phrases such as, If I get a raise, I will buy a new car. If the statement, I get a raise, is true, the action indicated (buy a new car) will be executed. Here is another example: If I get at least a $100 per week raise, I will buy a new car; else, I will put the raise into savings. A slightly more involved example is: If I get at least a $100 per week raise, I will buy a new car; else, if the raise is greater than $50, I will buy a new stereo; otherwise, I will put the raise into savings.

We can illustrate the logic of the first example as follows:

```
If I get a raise,
    I will buy a new car
. (period)
```

Note how the period marks the end of the statement.

The second example can be illustrated as follows:

```
If I get at least a $100 per week raise,
    I will buy a new car;
else,
    I will put the raise into savings
. (period)
```

The third example follows.

```
If I get at least a $100 per week raise,
    I will buy a new car;
else, if the raise is greater than $50,
    I will buy a new stereo;
otherwise,
    I will put the raise into savings
. (period)
```

The MATLAB *conditional statements* enable us to write programs that make decisions. Conditional statements contain one or more of the `if`, `else`, and `elseif` statements. The `end` statement denotes the end of a conditional statement, just as the period was used in the preceding examples. These conditional statements have a form similar to the examples, and they read somewhat like their English-language equivalents.

The `if` Statement

The `if` statement's basic form is

`if` *logical expression*
    statements
`end`

Every `if` statement must have an accompanying `end` statement. The `end` statement marks the end of the *statements* that are to be executed if the *logical expression* is true. A space is required between the `if` and the *logical expression,* which may be a scalar, a vector, or a matrix.

For example, suppose that x is a scalar and that we want to compute $y = \sqrt{x}$ only if $x \geq 0$. In English, we could specify this procedure as follows: If x is greater than or equal to zero, compute y from $y = \sqrt{x}$. The following `if` statement implements this procedure in MATLAB assuming x already has a scalar value.

```
if x >= 0
   y = sqrt(x)
end
```

If `x` is negative, the program takes no action. The *logical expression* here is `x >= 0`, and the *statement* is the single line `y = sqrt(x)`.

The `if` structure may be written on a single line; for example:

```
if x >= 0, y = sqrt(x), end
```

However, this form is less readable than the previous form. The usual practice is to indent the *statements* to clarify which statements belong to the `if` and its corresponding `end` and thereby improve readability.

The *logical expression* may be a compound expression; the *statements* may be a single command or a series of commands separated by commas or semicolons or on separate lines. For example if x and y have scalar values:

```
z = 0;w = 0;
if (x >= 0)&(y >= 0)
   z = sqrt(x) + sqrt(y)
   w = log(x) - 3*log(y)
end
```

The values of `z` and `w` are computed only if both `x` and `y` are nonnegative. Otherwise, z and w retain their values of zero. The flowchart is shown in Figure 4.4–1.

We may "nest" `if` statements, as shown by the following example.

```
if logical expression 1
   statement group 1
   if logical expression 2
      statement group 2
   end
end
```

Note that each `if` statement has an accompanying `end` statement.

The `else` Statement

When more than one action can occur as a result of a decision, we can use the `else` and `elseif` statements along with the `if` statement. The basic structure

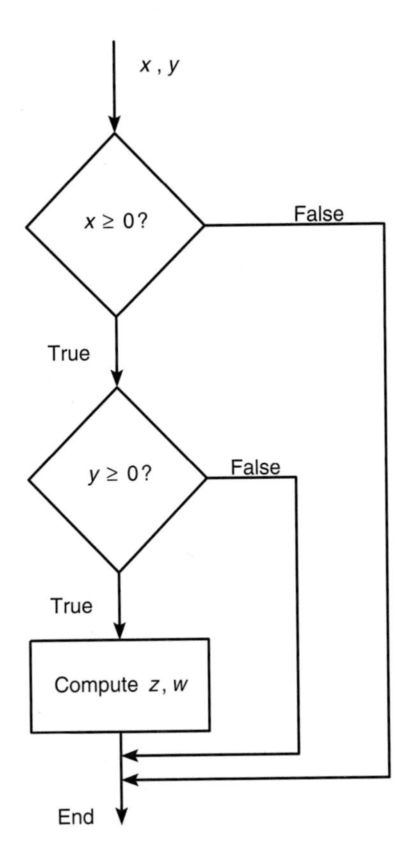

Figure 4.4–1 Flowchart corresponding to the pseudocode example.

for the use of the `else` statement is

`if` *logical expression*
    statement group 1
`else`
    statement group 2
`end`

Figure 4.4–2 shows the flowchart of this structure.

For example, suppose that $y = \sqrt{x}$ for $x \geq 0$ and that $y = e^x - 1$ for $x < 0$. The following statements will calculate y, assuming that x already has a scalar value.

```
if x >= 0
   y = sqrt(x)
else
   y = exp(x) - 1
end
```

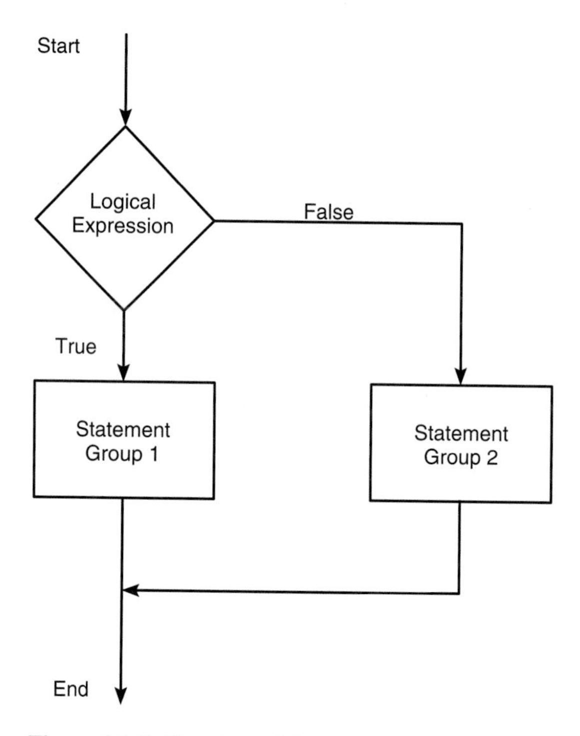

Figure 4.4–2 Flowchart of the `else` structure.

When the test, `if` *logical expression,* is performed, where the logical expression may be an *array,* the test returns a value of true only if *all* the elements of the logical expression are true! For example, if we fail to recognize how the test works, the following statements do not perform the way we might expect.

```
x = [4,-9,25];
if x < 0
   disp('Some of the elements of x are negative.')
else
   y = sqrt(x)
end
```

When this program is run it gives the result

```
y =
   2      0 + 3.000i      5
```

The program does not test each element in `x` in sequence. Instead it tests the truth of the vector relation `x < 0`. The test `if x < 0` returns a false value because it generates the vector `[0,1,0]`. Compare the preceding program with the following program.

```
x = [4,-9,25];
if x >= 0
   y = sqrt(x)
else
   disp('Some of the elements of x are negative.')
end
```

When executed, it produces the following result: `Some of the elements of x are negative.` The test `if x < 0` is false, and the test `if x >= 0` also returns a false value because `x >= 0` returns the vector `[1,0,1]`.

We sometimes must choose between a program that is concise, but perhaps more difficult to understand, and one that uses more statements than is necessary. For example, the statements

```
if logical expression 1
   if logical expression 2
      statements
   end
end
```

can be replaced with the more concise program

```
if logical expression 1 & logical expression 2
   statements
end
```

The `elseif` Statement

The general form of the `if` statement is

```
if logical expression 1
   statement group 1
elseif logical expression 2
   statement group 2
else
   statement group 3
end
```

The `else` and `elseif` statements may be omitted if not required. However, if both are used, the `else` statement must come after the `elseif` statement to take care of all conditions that might be unaccounted for. Figure 4.4–3 is the flowchart for the general `if` structure.

For example, suppose that $y = \ln x$ if $x \geq 5$ and that $y = \sqrt{x}$ if $0 \leq x < 5$. The following statements will compute y if x has a scalar value.

```
if x >= 5
   y = log(x)
```

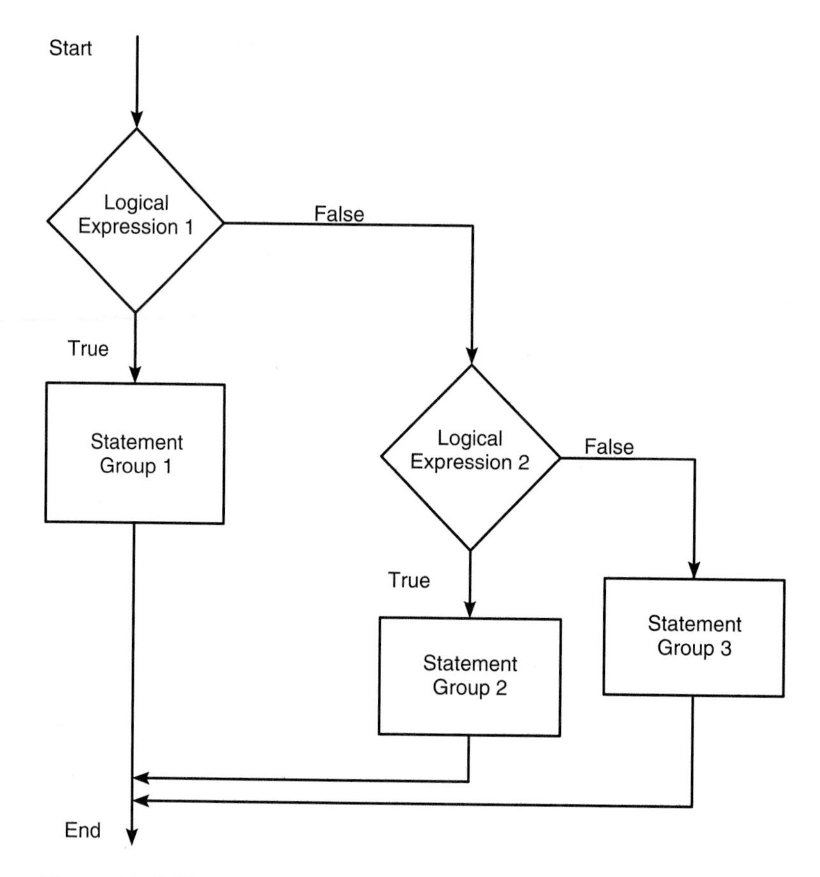

Figure 4.4–3 Flowchart for the general `if` structure.

```
else
   if x >= 0
      y = sqrt(x)
   end
end
```

If $x = -2$, for example, no action will be taken. If we use an `elseif`, we need fewer statements. For example:

```
if x >= 5
   y = log(x)
elseif x >= 0
   y = sqrt(x)
end
```

Note that the `elseif` statement does not require a separate `end` statement.

The `else` statement can be used with `elseif` to create detailed decision-making programs. For example, suppose that $y = \ln x$ for $x > 10$, $y = \sqrt{x}$ for

$0 \le x \le 10$, and $y = e^x - 1$ for $x < 0$. The following statements will compute y if x already has a scalar value.

```
if x > 10
   y = log(x)
elseif x >= 0
   y = sqrt(x)
else
   y = exp(x) - 1
end
```

Decision structures may be *nested;* that is, one structure can contain another structure, which in turn can contain another, and so on. The flowchart in Figure 4.4–4 describes the following code, which contains an example of nested `if` statements and assumes that x already has a scalar value.

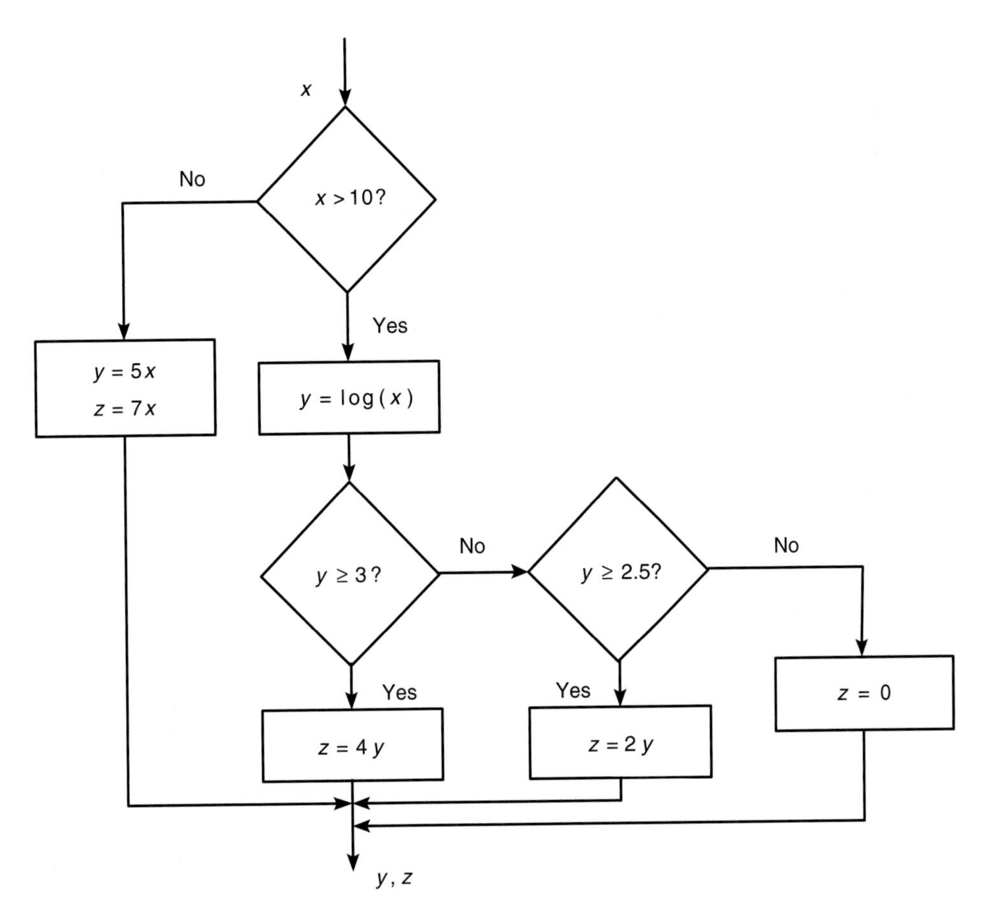

Figure 4.4–4 Flowchart illustrating nested `if` statements.

```
if x > 10
   y = log(x)
   if y >= 3
      z = 4*y
   elseif y >= 2.5
      z = 2*y
   else
      z = 0
   end
else
   y = 5*x
   z = 7*x
end
```

Note how the indentations emphasize the statement groups associated with each `end` statement. The flowchart required to represent this code is quite large. In practice, flowcharts often must be condensed by omitting some details to effectively describe the overall program.

Test Your Understanding

T4.4–1 Enter the script file whose flowchart is shown in Figure 4.4–4. Run the file for the following values of x. Check the program results by hand: $x = 11, 25, 2, 13$.

T4.4–2 Given a number x and the quadrant q ($q = 1, 2, 3, 4$), write a program to compute $\sin^{-1}(x)$ in degrees, taking into account the quadrant. The program should display an error message if $|x| > 1$.

Checking the Number of Input and Output Arguments

Sometimes you will want to have a function act differently depending on how many inputs it has. You can use the function `nargin`, which stands for "number of input arguments." Within the function you can use conditional statements to direct the flow of the computation depending on how many input arguments there are. For example, suppose you want to compute the square root of the the input if there is only one, but compute the square root of the average if there are two inputs. The following function does this.

```
function z = sqrtfun(x, y)
if (nargin == 1)
  z = sqrt(x);
elseif (nargin == 2)
  z = sqrt((x + y)/2);
end
```

The `nargout` function can be used to determine the number of output arguments.

Strings and Conditional Statements

A string is a variable that contains characters. Strings are useful for creating input prompts and messages and for storing and operating on data such as names and addresses. To create a string variable, enclose the characters in single quotes. For example, the string variable `name` is created as follows:

```
>>name = 'Leslie Student'
name =
     Leslie Student
```

The following string, `number`

```
>>number = '123'
number =
     123
```

is *not* the same as the variable `number` created by typing `number = 123`.

Strings are stored as row vectors in which each column represents a character. For example, the variable `name` has 1 row and 14 columns (each blank space occupies one column). Thus

```
>>size(name)
ans =
     1 14
```

We can access any column the way we access any other vector. For example, the letter S in the name Leslie Student occupies the eighth column in the vector `name`. It can be accessed as follows:

```
>>name(8)
ans =
     S
```

The colon operator can be used with string variables as well. For example:

```
>>first_name = name(1:6)
first_name =
     Leslie
```

We can manipulate the columns of string variables just as we do vectors. For example, to insert a middle initial, we type

```
>>full_name = [name(1:6),' C.',name(7:14)]
full_name =
     Leslie C. Student
>>full_name(8) = 'F'
full_name =
     Leslie F. Student
```

The `findstr` function (which stands for *find string*) is useful for finding the locations of certain characters. For example:

```
>>findstr(full_name,'e')
ans =
    2 6 15
```

This session tells us that the letter `e` occurs in the 2nd, 6th, and 15th columns.

Two string variables are equal if and only if every character is the same, including blank spaces. Note that uppercase and lowercase letters are *not* the same. Thus the strings `'Hello'` and `'hello'` are not equal, and the strings `'can not'` and `'cannot'` are not equal. The function `strcmp` (for *string compare*) determines whether two strings are equal. Typing `strcmp('string1', 'string2')` returns a 1 if the strings `'string1'` and `'string2'` are equal and 0 otherwise. The functions `lower('string')` and `upper('string')` convert `'string'` to all lowercase or all uppercase letters. These functions are useful for accepting keyboard input without forcing the user to distinguish between lowercase and uppercase.

One of the most important applications for strings is to create input prompts and output messages. The following prompt program uses the `isempty(x)` function, which returns a 1 if the array `x` is empty and 0 otherwise. It also uses the `input` function, whose syntax is

`x = input('`*prompt*`', '`*string*`')`

This function displays the string *prompt* on the screen, waits for input from the keyboard, and returns the entered value in the string variable `x`. The function returns an empty matrix if you press the **Enter** key without typing anything.

The following prompt program is a script file that allows the user to answer Yes by typing either Y or y or by pressing the **Enter** key. Any other response is treated as a No answer.

```
response = input('Do you want to continue? Y/N [Y]: ','s');
if (isempty(response))|(response == 'Y')|(response == 'y')
   response = 'Y'
else
   response = 'N'
end
```

Many more string functions are available in MATLAB. Type `help strfun` to obtain information on these.

4.5 Loops

A *loop* is a structure for repeating a calculation a number of times. Each repetition of the loop is a *pass*. MATLAB uses two types of explicit loops: the `for` loop, when the number of passes is known ahead of time, and the `while` loop, when the looping process must terminate when a specified condition is satisfied, and thus the number of passes is not known in advance.

`for` Loops

A simple example of a `for` loop is

```
for k = 5:10:35
   x = k^2
end
```

The *loop variable* `k` is initially assigned the value 5, and `x` is calculated from `x = k^2`. Each successive pass through the loop increments `k` by 10 and calculates `x` until `k` exceeds 35. Thus `k` takes on the values 5, 15, 25, and 35, and `x` takes on the values 25, 225, 625, and 1225. The program then continues to execute any statements following the `end` statement.

The typical structure of a `for` loop is

`for` *loop variable* = *m:s:n*
   statements
`end`

The expression `m:s:n` assigns an initial value of `m` to the loop variable, which is incremented by the value `s`—called the *step value* or *incremental value.* The *statements* are executed once during each pass, using the current value of the loop variable. The looping continues until the loop variable exceeds the *terminating value* `n`. For example, in the expression `for k = 5:10:36`, the final value of `k` is 35. Note that we need not place a semicolon after the `for m:s:n` statement to suppress printing `k`. Figure 4.5–1 shows the flowchart of a `for` loop.

Note that a `for` statement needs an accompanying `end` statement. The `end` statement marks the end of the *statements* that are to be executed. A space is required between the `for` and the *loop variable,* which may be a scalar, a vector, or a matrix, although the scalar case is by far the most common.

The `for` loop may be written on a single line; for example:

```
for x = 0:2:10, y = sqrt(x), end
```

However, this form is less readable than the previous form. The usual practice is to indent the *statements* to clarify which statements belong to the `for` and its corresponding `end` and thereby improve readability.

NESTED LOOPS

We may nest loops and conditional statements, as shown by the following example. (Note that each `for` and `if` statement needs an accompanying `end` statement.)

Suppose we want to create a special square matrix that has ones in the first row and first column, and whose remaining elements are the sum of two elements, the element above and the element to the left, if the sum is less than 20. Otherwise, the element is the maximum of those two element values. The following function

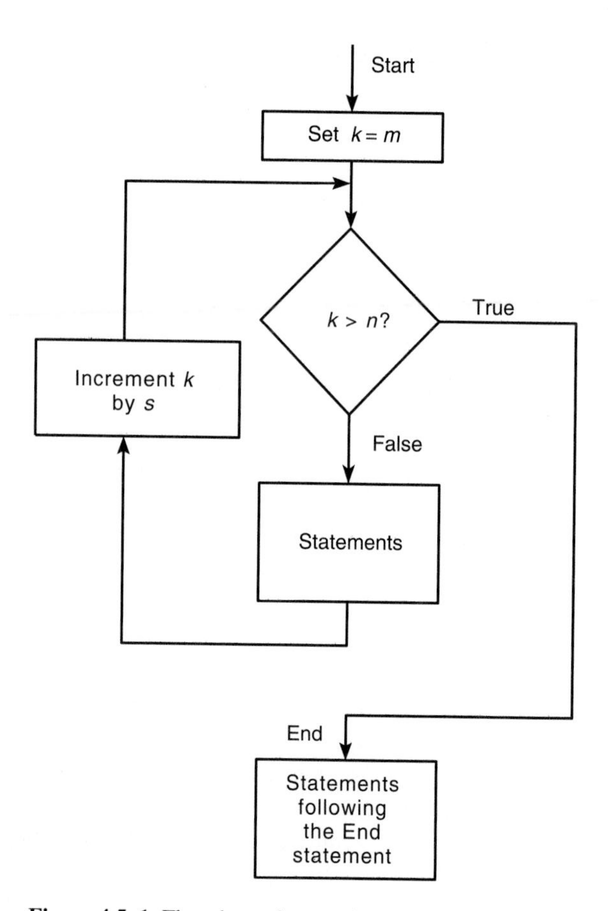

Figure 4.5–1 Flowchart of a `for` loop.

creates this matrix. The row index is `r`; the column index is `c`. Note how indenting improves the readability.

```
function A = specmat(n)
A = ones(n);
for r = 1:n
   for c = 1:n
      if (r>1)&(c>1)
        s = A(r-1,c) + A(r,c-1);
        if s<20
           A(r,c) = s;
        else
           A(r,c) = max(A(r-1,c),A(r,c-1));
        end
      end
   end
end
```

Typing `specmat(5)` produces the following matrix

$$\begin{bmatrix} 1 & 1 & 1 & 1 & 1 \\ 1 & 2 & 3 & 4 & 5 \\ 1 & 3 & 6 & 10 & 15 \\ 1 & 4 & 10 & 10 & 15 \\ 1 & 5 & 15 & 15 & 15 \end{bmatrix}$$

Test Your Understanding

T4.5–1 Write a program to produce the following matrix:

$$\mathbf{A} = \begin{bmatrix} 4 & 8 & 12 \\ 10 & 14 & 18 \\ 16 & 20 & 24 \\ 22 & 26 & 30 \end{bmatrix}$$

Note the following rules when using `for` loops with the loop variable expression `k = m:s:n`:

- The step value `s` may be negative. For example, `k = 10:-2:4` produces `k` = 10, 8, 6, 4.
- If `s` is omitted, the step value defaults to one.
- If `s` is positive, the loop will not be executed if `m` is greater than `n`.
- If `s` is negative, the loop will not be executed if `m` is less than `n`.
- If `m` equals `n`, the loop will be executed only once.
- If the step value `s` is not an integer, round-off errors can cause the loop to execute a different number of passes than intended.

When the loop is completed, `k` retains its last value. You should not alter the value of the loop variable `k` within the *statements*. Doing so can cause unpredictable results.

A common practice in traditional programming languages like BASIC and FORTRAN is to use the symbols `i` and `j` as loop variables. However, this convention is not good practice in MATLAB, which uses these symbols for the imaginary unit $\sqrt{-1}$. For example, what do you think is the result of the following program? Try it and see!

```
x = 1;
y = 1;
for i = 1:5
    x = x + 6i
    y = y + 5/i
end
```

The break and continue Statements

It is permissible to use an `if` statement to "jump" out of the loop before the loop variable reaches its terminating value. The `break` command, which terminates the loop but does not stop the entire program, can be used for this purpose. For example:

```
for k = 1:10
   x = 50 - k^2;
   if x < 0
      break
   end
   y = sqrt(x)
end
% The program execution jumps to here
% if the break command is executed.
```

However, it is usually possible to write the code to avoid using the `break` command. This can often be done with a `while` loop as explained in the next section.

The `break` statement stops the execution of the loop. There can be applications where we want to not execute the case producing an error but continue executing the loop for the remaining passes. We can use the `continue` statement to do this. The `continue` statement passes control to the next iteration of the `for` or `while` loop in which it appears, skipping any remaining statements in the body of the loop. In nested loops, `continue` passes control to the next iteration of the `for` or `while` loop enclosing it.

For example, the following code uses a `continue` statement to avoid computing the logarithm of a negative number.

```
x = [10,1000,-10,100];
y = NaN*x;
for k = 1:length(x)
   if x(k) < 0
      continue
   end
   y(k) = log10(x(k));
end
y
```

The result is `y = 1, 3, NaN, 2.`

Using an Array as a Loop Index

It is permissible to use a matrix expression to specify the number of passes. In this case the loop variable is a vector that is set equal to the successive columns of the matrix expression during each pass. For example,

```
A = [1,2,3;4,5,6];
for v = A
   disp(v)
end
```

is equivalent to

```
A = [1,2,3;4,5,6];
n = 3;
for k = 1:n
   v = A(:,k)
end
```

The common expression `k = m:s:n` is a special case of a matrix expression in which the columns of the expression are scalars, not vectors.

For example, suppose we want to compute the distance from the origin to a set of three points specified by their xy coordinates (3,7), (6,6), and (2,8). We can arrange the coordinates in the array `coord` as follows.

$$\begin{bmatrix} 3 & 6 & 2 \\ 7 & 6 & 8 \end{bmatrix}$$

Then `coord = [3,6,2;7,6,8]`. The following program computes the distance and determines which point is farthest from the origin. The first time through the loop the index `coord` is `[3, 7]'`. The second time the index is `[6, 6]'`, and during the final pass it is `[2, 8]'`.

```
k = 0;
for coord = [3,6,2;7,6,8]
   k = k + 1;
   distance(k) = sqrt(coord'*coord)
end
[max_distance,farthest] = max(distance)
```

The previous program illustrates the use of an array index but the problem can be solved more concisely with the following program, which uses the `diag` function to extract the diagonal elements of an array.

```
coord = [3,6,2;7,6,8];
distance = sqrt(diag(coord'*coord))
[max_distance,farthest] = max(distance)
```

Implied Loops

Many MATLAB commands contain *implied loops*. For example, consider these statements.

```
x = [0:5:100];
y = cos(x);
```

To achieve the same result using a `for` loop, we must type

```
for k = 1:21
   x = (k-1)*5;
   y(k) = cos(x);
end
```

The `find` command is another example of an implied loop. The statement `y = find(x>0)` is equivalent to

```
m=0;
for k=1:length(x)
   if x(k)>0
      m = m + 1;
      y(m)=k;
   end
end
```

If you are familiar with a traditional programming language such as FORTRAN or BASIC, you might be inclined to solve problems in MATLAB using loops, instead of using the powerful MATLAB commands such as `find`. To use these commands and to maximize the power of MATLAB, you might need to adopt a new approach to problem solving. As the preceding example shows, you often can save many lines of code by using MATLAB commands, instead of using loops. Your programs will also run faster because MATLAB was designed for high-speed vector computations.

Test Your Understanding

T4.5–2 Write a `for` loop that is equivalent to the command `sum(A)`, where `A` is a matrix.

EXAMPLE 4.5–1 Data Sorting

A vector `x` has been obtained from measurements. Suppose we want to consider any data value in the range $-0.1 < x < 0.1$ as being erroneous. We want to remove all such elements and replace them with zeros at the end of the array. Develop two ways of doing this. An example is given in the following table.

	Before	After
x(1)	1.92	1.92
x(2)	0.05	−2.43
x(3)	−2.43	0.85
x(4)	−0.02	0
x(5)	0.09	0
x(6)	0.85	0
x(7)	−0.06	0

■ **Solution**

The following script file uses a `for` loop with conditional statements. Note how the null array `[]` is used.

```
x = [1.92,0.05,-2.43,-0.02,0.09,0.85,-0.06];
y = [];z = [];
for k = 1:length(x)
   if abs(x(k)) >= 0.1
      y = [y,x(k)];
   else
      z = [z,x(k)];
   end
end
xnew = [y,zeros(size(z))]
```

The next script file uses the `find` function.

```
x = [1.92,0.05,-2.43,-0.02,0.09,0.85,-0.06];
y = x(find(abs(x) >= 0.1));
z = zeros(size(find(abs(x)<0.1)));
xnew = [y,z]
```

Use of Logical Arrays as Masks

Consider the array **A**.

$$\mathbf{A} = \begin{bmatrix} 0 & -1 & 4 \\ 9 & -14 & 25 \\ -34 & 49 & 64 \end{bmatrix}$$

The following program computes the array **B** by computing the square roots of all the elements of **A** whose value is no less than 0, and adding 50 to each element that is negative.

```
A = [0, -1, 4; 9, -14, 25; -34, 49, 64];
for m = 1:size(A,1)
   for n = 1:size(A,2)
      if A(m,n) >= 0
```

```
            B(m,n) = sqrt(A(m,n));
        else
            B(m,n) = A(m,n) + 50;
        end
    end
end
B
```

The result is

$$\mathbf{B} = \begin{bmatrix} 0 & 49 & 2 \\ 3 & 36 & 5 \\ 16 & 7 & 8 \end{bmatrix}$$

When a logical array is used to address another array, it extracts from that array the elements in the locations where the logical array has 1s. We can often avoid the use of loops and branching and thus create simpler and faster programs by using a logical array as a *mask* that selects elements of another array. Any elements not selected will remain unchanged.

The following session creates the logical array C from the numeric array A given previously.

```
>>A = [0, -1, 4; 9, -14, 25; -34, 49, 64];
>>C = (A >= 0);
```

The result is

$$\mathbf{C} = \begin{bmatrix} 1 & 0 & 1 \\ 1 & 0 & 1 \\ 0 & 1 & 1 \end{bmatrix}$$

We can use this technique to compute the square root of only those elements of A given in the previous program that are no less than 0 and add 50 to those elements that are negative. The program is

```
A = [0, -1, 4; 9, -14, 25; -34, 49, 64];
C = (A >= 0);
A(C) = sqrt(A(C))
A(~C) = A(~C) + 50
```

The result after the third line is executed is

$$\mathbf{A} = \begin{bmatrix} 0 & -1 & 2 \\ 3 & -14 & 25 \\ -34 & 49 & 64 \end{bmatrix}$$

The result after the last line is executed is

$$\mathbf{A} = \begin{bmatrix} 0 & 49 & 2 \\ 3 & 36 & 5 \\ 16 & 7 & 8 \end{bmatrix}$$

Flight of an Instrumented Rocket

EXAMPLE 4.5–2

All rockets lose weight as they burn fuel; thus the mass of the system is variable. The following equations describe the speed v and height h of a rocket launched vertically, neglecting air resistance. They can be derived from Newton's law.

$$v(t) = u \ln \frac{m_0}{m_0 - qt} - gt \qquad (4.5\text{–}1)$$

$$h(t) = \frac{u}{q}(m_0 - qt)\ln(m_0 - qt) + u(\ln m_0 + 1)t - \frac{gt^2}{2} - \frac{m_0 u}{q}\ln m_0 \qquad (4.5\text{–}2)$$

where m_0 is the rocket's initial mass, q is the rate at which the rocket burns fuel mass, u is the exhaust velocity of the burned fuel relative to the rocket, and g is the acceleration due to gravity. Let b be the *burn time,* after which all the fuel is consumed. Thus the rocket's mass without fuel is $m_e = m_0 - qb$.

For $t > b$ the rocket engine no longer produces thrust, and the speed and height are given by

$$v(t) = v(b) - g(t - b) \qquad (4.5\text{–}3)$$

$$h(t) = h(b) + v(b)(t - b) - \frac{g(t - b)^2}{2} \qquad (4.5\text{–}4)$$

The time t_p to reach the peak height is found by setting $v(t) = 0$. The result is $t_p = b + v(b)/g$. Substituting this expression into the expression (4.5–4) for $h(t)$ gives the following expression for the peak height: $h_p = h(b) + v^2(b)/(2g)$. The time at which the rocket hits the ground is $t_{\text{hit}} = t_p + \sqrt{2h_p/g}$.

Suppose the rocket is carrying instruments to study the upper atmosphere, and we need to determine the amount of time spent above 50,000 feet as a function of the burn

Table 4.5–1 Pseudocode for Example 4.5–2

Enter data.
Increment burn time from 0 to 100. For each burn-time value:
    Compute m_0, v_b, h_b, h_p.
    If $h_p \geq h_{\text{desired}}$,
        Compute t_p, t_{hit}.
        Increment time from 0 to t_{hit}.
            Compute height as a function of time, using the appropriate equation, depending on whether burnout has occurred.
        Compute the duration above desired height.
        End of the time loop.
    If $h_p < h_{\text{desired}}$, set duration equal to zero.
End of the burn-time loop.
Plot the results.

time b (and thus as a function of the fuel mass qb). Assume that we are given the following values: $m_e = 100$ slugs, $q = 1$ slug/sec, $u = 8000$ ft/sec, and $g = 32.2$ ft/sec^2. If the rocket's maximum fuel load is 100 slugs, the maximum value of b is $100/q = 100$. Write a MATLAB program to solve this problem.

■ Solution

Pseudocode for developing the program appears in Table 4.5–1. A `for` loop is a logical choice to solve this problem because we know the burn time b and t_{hit}, the time it takes to hit the ground. A MATLAB program to solve this problem appears in Table 4.5–2. It has two nested `for` loops. The inner loop is over time and evaluates the equations of

Table 4.5–2 MATLAB program for Example 4.5–2

```
% Script file rocket1.m
% Computes flight duration as a function of burn time.
% Basic data values.
m_e = 100; q = 1; u = 8000; g = 32.2;
dt = 0.1; h_desired = 50000;
for b = 1:100 % Loop over burn time.
   burn_time(b) = b;
   % The following lines implement the formulas in the text.
   m_0 = m_e + q*b; v_b = u*log(m_0/m_e) - g*b;
   h_b = ((u*m_e)/q)*log(m_e/(m_e+q*b))+u*b - 0.5*g*b^2;
   h_p = h_b + v_b^2/(2*g);
   if h_p >= h_desired
   % Calculate only if peak height > desired height.
      t_p = b + v_b/g; % Compute peak time.
      t_hit = t_p + sqrt(2*h_p/g); % Compute time to hit.
      for p = 0:t_hit/dt
         % Use a loop to compute the height vector.
         k = p + 1; t = p*dt; time(k) = t;
         if t <= b
            % Burnout has not yet occurred.
            h(k) = (u/q)*(m_0 - q*t)*log(m_0 - q*t)...
               + u*(log(m_0) + 1)*t - 0.5*g*t^2 ...
               - (m_0*u/q)*log(m_0);
         else
            % Burnout has occurred.
            h(k) = h_b - 0.5*g*(t - b)^2 + v_b*(t - b);
         end
      end
      % Compute the duration.
      duration(b) = length(find(h>=h_desired))*dt;
   else
      % Rocket did not reach the desired height.
      duration(b) = 0;
   end
end % Plot the results.
plot(burn_time,duration),xlabel('Burn Time (sec)'),...
ylabel('Duration (sec)'),title('Duration Above 50,000 Feet')
```

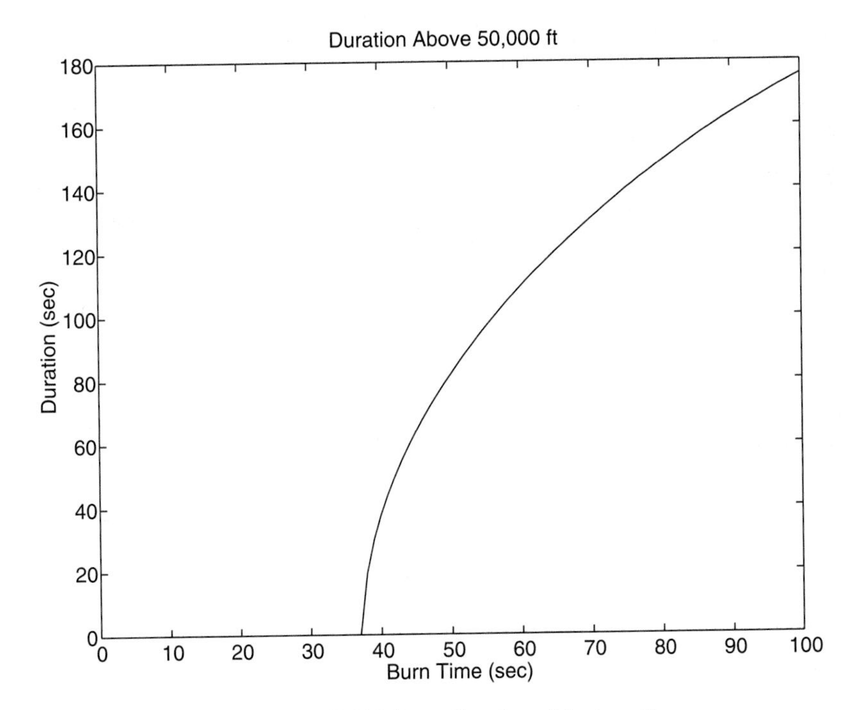

Figure 4.5–2 Duration above 50,000 ft as a function of the burn time.

motion at times spaced 1/10 of a second apart. This loop calculates the duration above 50,000 feet for a specific value of the burn time b. We can obtain more accuracy by using a smaller value of the time increment dt. The outer loop varies the burn time in integer values from $b = 1$ to $b = 100$. The final result is the vector of durations for the various burn times. Figure 4.5–2 gives the resulting plot.

`while` Loops

The `while` loop is used when the looping process terminates because a specified condition is satisfied, and thus the number of passes is not known in advance. A simple example of a `while` loop is

```
x = 5;
while x < 25
   disp(x)
   x = 2*x - 1;
end
```

The results displayed by the `disp` statement are 5, 9, and 17. The *loop variable* `x` is initially assigned the value 5, and it has this value until the statement `x = 2*x - 1` is encountered the first time. The value then changes to 9. Before each pass through the loop, `x` is checked to see whether its value is less than 25. If so, the pass is made. If not, the loop is skipped and the program continues to execute any statements following the `end` statement.

A principal application of `while` loops is when we want the loop to continue as long as a certain statement is true. Such a task is often more difficult to do with a `for` loop. For example:

```
x = 1;
while x ~= 5
   disp(x)
   x = x + 1;
end
```

The statements between the `while` and the `end` are executed once during each pass, using the current value of the loop variable `x`. The looping continues until the condition `x~=5` is false. The results displayed by the `disp` statement are 1, 2, 3, and 4.

The typical structure of a `while` loop follows.

`while` *logical expression*
   statements
`end`

MATLAB first tests the truth of the *logical expression*. A loop variable must be included in the *logical expression*. For example, `x` is the loop variable in the statement `while x ~= 5`. If the *logical expression* is true, the *statements* are executed. For the `while` loop to function properly, the following two conditions must occur:

1. The loop variable must have a value before the `while` statement is executed.
2. The loop variable must be changed somehow by the *statements*.

The *statements* are executed once during each pass, using the current value of the loop variable. The looping continues until the *logical expression* is false. Figure 4.5–3 shows the flowchart of the `while` loop.

Each `while` statement must be matched by an accompanying `end`. As with `for` loops, the *statements* should be indented to improve readability. You may nest `while` loops, and you may nest them with `for` loops and `if` statements.

Always make sure that the loop variable has a value assigned to it before the start of the loop. For example, the following loop can give unintended results if `x` has an overlooked previous value.

```
while x < 10
   x = x + 1;
   y = 2*x;
end
```

If `x` has not been assigned a value prior to the loop, an error message will occur. If we intend `x` to start at zero, then we should place the statement `x = 0;` before the `while` statement.

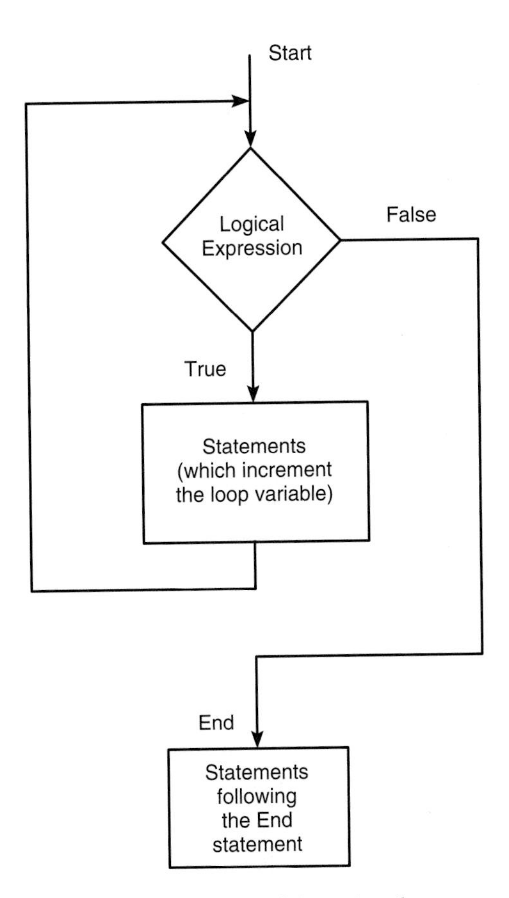

Figure 4.5–3 Flowchart of the `while` loop.

It is possible to create an *infinite loop,* which is a loop that never ends. For example:

```
x = 8;
while x ~= 0
   x = x - 3;
end
```

Within the loop the variable `x` takes on the values 5, 2, -1, -4, . . . , and the condition `x ~= 0` is always satisfied, so the loop never stops.

Time to Reach a Specified Height — EXAMPLE 4.5–3

Consider the variable-mass rocket treated in Example 4.5–2. Write a program to determine how long it takes for the rocket to reach 40,000 ft if the burn time is 50 sec.

■ Solution

The pseudocode appears in Table 4.5–3. Because we do not know the time required, a `while` loop is convenient to use. The program in Table 4.5–4 performs the task and

Table 4.5–3 Pseudocode for Example 4.5–3

Enter data.
Compute m_0, v_b, h_b, h_p.
If $h_p \geq h_{\text{desired}}$,
    Use a `while` loop to increment time and compute height until desired height is reached.
        Compute height as a function of time, using the appropriate equation, depending on whether burnout has occurred.
    End of the time loop.
    Display the results.
If $h_p < h_{\text{desired}}$, rocket cannot reach desired height.

Table 4.5–4 MATLAB program for Example 4.5–3

```
% Script file rocket2.m
% Computes time to reach desired height.
% Set the data values.
h_desired = 40000; m_e = 100; q = 1;
u = 8000; g = 32.2; dt = 0.1; b = 50;
% Compute values at burnout, peak time, and height.
m_0 = m_e + q*b; v_b = u*log(m_0/m_e) - g*b;
h_b = ((u*m_e)/q)*log(m_e/(m_e+q*b))+u*b - 0.5*g*b^2;
t_p = b + v_b/g;
h_p = h_b + v_b^2/(2*g);
% If h_p > h_desired, compute time to reached h_desired.
if h_p > h_desired
   h = 0; k = 0;
   while h < h_desired % Compute h until h = h_desired.
      t = k*dt; k = k + 1;
      if t <= b
         % Burnout has not yet occurred.
         h = (u/q)*(m_0 - q*t)*log(m_0 - q*t)...
             + u*(log(m_0) + 1)*t - 0.5*g*t^2 ...
             - (m_0*u/q)*log(m_0);
      else
         % Burnout has occurred.
         h = h_b - 0.5*g*(t - b)^2 + v_b*(t - b);
      end
   end
   % Display the results.
   disp('The time to reach the desired height is:')
   disp(t)
else
   disp('Rocket cannot achieve the desired height.')
end
```

is a modification of the program in Table 4.5–2. Note that the new program allows for the possibility that the rocket might not reach 40,000 ft. It is important to write your programs to handle all such foreseeable circumstances. The answer given by the program is 53 sec.

Test Your Understanding

T4.5–3 Rewrite the following code using a `while` loop to avoid using the `break` command.

```
for k = 1:10
   x = 50 - k^2;
   if x < 0
      break
   end
   y = sqrt(x)
end
```

T4.5–4 Find to two decimal places the largest value of x before the error in the series approximation $e^x \approx 1 + x + x^2/2 + x^3/6$ exceeds 1 percent. (Answer: $x = 0.83$.)

4.6 The switch Structure

The `switch` structure provides an alternative to using the `if`, `elseif`, and `else` commands. Anything programmed using `switch` can also be programmed using `if` structures. However, for some applications the `switch` structure is more readable than code using the `if` structure. The syntax is

```
switch input expression (scalar or string)
   case value1
      statement group 1
   case value2
      statement group 2
   .
   .
   .
   otherwise
      statement group n
end
```

The *input expression* is compared to each `case` value. If they are the same, then the statements following that `case` statement are executed and processing continues with any statements after the `end` statement. If the *input expression* is a string, then it is equal to the `case` *value* if `strcmp` returns a value of 1 (true). Only the *first* matching `case` is executed. If no match occurs, the statements following the otherwise statement are executed. However, the otherwise statement is optional. If it is absent, execution continues with the statements following the end statement if no match exists. Each case *value* statement must be on a single line.

For example, suppose the variable angle has an integer value that represents an angle measured in degrees from North. The following switch block displays the point on the compass that corresponds to that angle.

```
switch angle
   case 45
      disp('Northeast')
   case 135
      disp('Southeast')
   case 225
      disp('Southwest')
   case 315
      disp('Northwest')
   otherwise
      disp('Direction Unknown')
end
```

The use of a string variable for the *input expression* can result in very readable programs. For example, in the following code the numeric vector x has values, and the user enters the value of the string variable response; its intended values are min, max, or sum. The code then either finds the minimum or maximum value of x or sums the elements of x, as directed by the user.

```
t = [0:100]; x = exp(-t).*sin(t);
response = input('Type min, max, or sum.','s')
response = lower('response');
switch response
   case min
      minimum = min(x)
   case max
      maximum = max(x)
   case sum
      total = sum(x)
   otherwise
      disp('You have not entered a proper choice.')
end
```

The switch statement can handle multiple conditions in a single case statement by enclosing the case *value* in a cell array. For example, the following switch block displays the corresponding point on the compass, given the integer angle measured from North.

```
switch angle
   case {0,360}
      disp('North')
```

```
   case {-180,180}
      disp('South')
   case {-270,90}
      disp('East')
   case {-90,270}
      disp('West')
   otherwise
      disp('Direction Unknown')
end
```

Test Your Understanding

T4.6–1 Write a program using the `switch` structure to input one angle, whose value may be 45, −45, 135, or −135°, and display the quadrant (1, 2, 3, or 4) containing the angle.

Using the `switch` Structure for Calendar Calculations

EXAMPLE 4.6–1

Use the `switch` structure to compute the total elapsed days in a year, given the number (1–12) of the month, the day, and an indication of whether or not the year is a leap year.

■ Solution

Note that February has an extra day if the year is a leap year. The following function computes the total elapsed number of days in a year, given the month, the day of the month, and the value of `extra_day`, which is 1 for a leap year, and 0 otherwise.

```
function total_days = total(month,day,extra_day)
total_days = day;
for k = 1:month - 1
   switch k
      case {1,3,5,7,8,10,12}
         total_days = total_days + 31;
      case {4,6,9,11}
         total_days = total_days + 30;
      case 2
         total_days = total_days + 28 + extra_day;
   end
end
```

The function can be used as shown in the following program.

```
month = input('Enter month (1 - 12): ');
day = input('Enter day (1 - 31): ');
extra_day = input('Enter 1 for leap year; 0 otherwise: ');
total_days = total(month,day,extra_day)
```

One of the chapter problems for Section 4.4 (Problem 18) asks you to write a program to determine whether or not a given year is a leap year.

4.7 Debugging MATLAB Programs

Use of the MATLAB Editor/Debugger as an M-file *editor* was discussed in Section 1.4 of Chapter 1. Figure 1.4–1 (in Chapter 1) shows the Editor/Debugger screen. Figure 4.7–1 shows the Debugger containing two programs to be analyzed. Here we discuss its use as a *debugger*. The Editor/Debugger menu bar contains the following items: **File, Edit, Text, Cell, Tools, Debug, Desktop, Window,** and **Help.**

The **File, Edit, Desktop, Window,** and **Help** menus are similar to those in the Desktop, with a few exceptions. For example, the **File** menu in the Editor/Debugger contains the item **Source Control,** a file management system for integrating files from different sources. This system is used by developers of very large programs. Another example is the item **Go to Line** on the **Edit** menu. Click on it and enter a line number in the dialog box, then click **OK.** This feature is useful for navigating through files with many lines. An additional example is the **Help** menu in the Editor/Debugger, which contains the specific help item **Using the M-file Editor.**

The **Cell** and **Tools** menus involve advanced topics that will not be treated in this text. The **Desktop** menu is similar to that in the Command window. It enables you to dock and undock windows, arrange the Editor window, and turn the Editor toolbar on and off.

Below the menu bar is the Editor/Debugger toolbar. It enables you to access several of the items in the menus with one click of the mouse. Hold the mouse cursor over a button on the toolbar to see its function. For example, clicking the button with the binoculars icon is equivalent to selecting **Find and Replace** from the **Edit** menu. One item on the toolbar that is not in the menus is the function button with the script f icon (f). Use this button to go to a particular function in the M-file. The list of functions that you will see includes only those functions whose function statements are in the program. The list does not include functions that are called from the M-file.

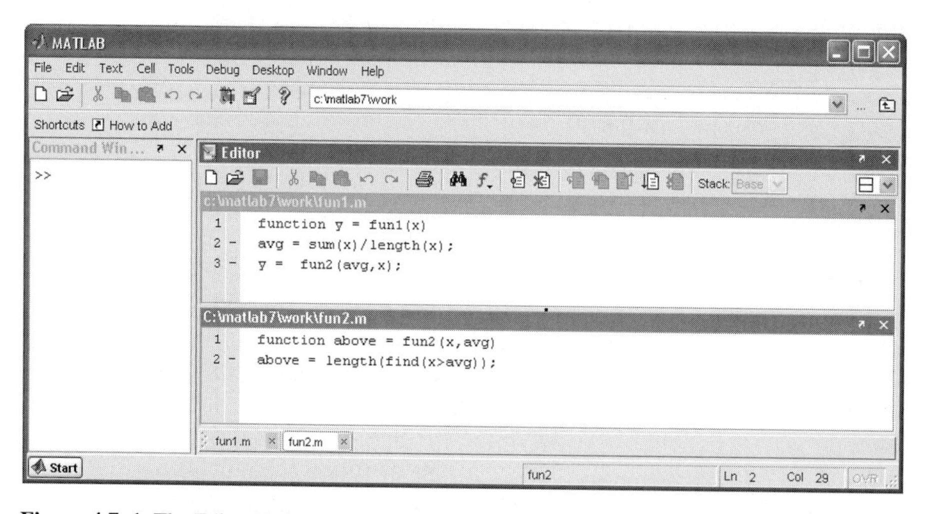

Figure 4.7–1 The Editor/Debugger containing two programs to be analyzed.

At the far right of the toolbar is the **Stack** menu. Here we discuss the **Text, Debug,** and **Stack** menus of the Editor/Debugger.

To create a new M-file in the MS Windows environment, in the Command window select **New** from the **File** menu, then select **M-file.** You will then see the Editor/Debugger window. You can use the keyboard and the **Edit** menu in the Editor/Debugger as you would in most word processors to create and edit the file. Note that each line in the file is numbered on the left. When finished, select **Save** from the **File** menu in the Editor/Debugger. In the dialog box that appears, replace the default name provided (usually named `Untitled`) with the name `example1`, and click on **Save.** The Editor/Debugger will automatically provide the extension `.m` and save the file in the MATLAB current directory, which for now we will assume is on the hard drive.

To open an existing file, in the Command window select **Open** from the **File** menu. Type in the name of the file or use the browser to select it. The Editor/Debugger window then opens. Once in the Editor/Debugger you can open more than one file. If you do, each file has a tab at the bottom of the window. Click on a tab to make that file the active one for editing and debugging.

The Text Menu

The **Text** menu supplements the **Edit** menu for creating M-files. With the **Text** menu you can insert or remove comments, increase or decrease the amount of indenting, turn on smart indenting, and evaluate and display the values of selected variables in the Command window. Click *anywhere* in a previously typed line, and then click **Comment** in the **Text** menu. This makes the entire line a comment. To turn a commented line into an executable line, click *anywhere* in the line, and then click **Uncomment** in the **Text** menu.

The **Increase Indent** and **Decrease Indent** items on the **Text** menu work in a similar way. Just click *anywhere* in a previously typed line, and then click **Increase Indent** or **Decrease Indent** to change the indentation of the line.

The Editor/Debugger automatically indents any lines typed after a conditional statement, a `for` statement, or a `while` statement, up to where you type the corresponding `end` statement. Use the **Smart Indent** item on the **Text** menu to start automatic indenting in a previously typed line and in any lines typed thereafter. Click *anywhere* in the line, and then click **Smart Indent** in the **Text** menu.

Use the **Evaluate Selection** item in the **Text** menu to display the values of selected variables in the Command window. Highlight the variable in the file, and click on **Evaluate Selection.** The variable name and its value appear in the Command window. After highlighting the variable, you can also right-click and the **Context** menu will appear. Then select **Evaluate Selection** on this menu. A third way to evaluate the value of a variable is to go to the Command window and type its name; however, this process requires you to leave the Editor/Debugger window.

After the file has been executed you can view a variable's value in Datatips, a window that appears when you position the cursor to the left of a variable. The variable's value stays in view until you move the cursor. Datatips are always on in debug mode, but are off by default in edit mode. You can turn them on by using the **Preferences** choice under the **File** menu. You can also view values in the Array Editor. In the following discussion, when we say you should "evaluate the variable," you can use any of these methods.

The Debug Menu

BREAKPOINT

Breakpoints are points in the file where execution stops temporarily so that you can examine the values of the variables up to that point. You set breakpoints with the **Set/Clear Breakpoint** item on the **Debug** menu. Use the **Step, Step In,** and **Step Out** items on the **Debug** menu to step through your file after you have set breakpoints and run the file. Click **Step** to watch the script execute one step at a time. Click **Step In** to step into the first executable line in a function being called. Click **Step Out** in a called function to run the rest of the function and then return to the calling program.

The solid green arrow to the left of the line text indicates the next line to be executed. When this arrow changes to a hollow green arrow, MATLAB control is now in a function being called. Execution returns to the line with the solid green arrow after the function completes its operation. The arrow turns yellow at a line where execution pauses or where a function completes its operation. When the program pauses you can assign new values to a variable, using either the Command window or the Array Editor.

Click on the **Go Until Cursor** item to run the file until it reaches the line where the cursor is; this process sets a temporary breakpoint at the cursor. You can save and execute your program directly from the **Debug** menu if you want, by clicking on **Run** (or **Save and Run** if you have made changes). You need not set any breakpoints beforehand. Click **Exit Debug Mode** to return to normal editing. To save any changes you have made to the program, first exit the debug mode, and then save the file.

Using Breakpoints

Most debugging sessions start by setting a breakpoint. A breakpoint stops M-file execution at a specified line and allows you to view or change values in the function's workspace before resuming execution. To set a breakpoint, position the cursor in the line of text and click on the breakpoint icon in the toolbar or select **Set/Clear Breakpoints** from the **Debug** menu. You can also set a breakpoint by right-clicking on the line of text to bring up the **Context** menu and choose **Set/Clear Breakpoint.** A red circle next to a line indicates that a breakpoint is set at that line. If the line selected for a breakpoint is not an executable line, then the breakpoint is set at the next executable line. The **Debug** menu enables you to clear all the breakpoints (select **Clear Breakpoints in All Files**). The **Debug** menu also lets you halt M-file execution if your code generates a warning, an error, or a `NaN` or `Inf` value (select **Stop if Errors/Warnings**).

The Stack Menu

MATLAB assigns each M-file function its own workspace, called the function workspace, which is separate from the MATLAB base workspace. You can access the base and function workspaces when debugging M-files by using the **Stack** menu in the Editor/Debugger. The **Stack** menu is only available in debug mode; otherwise it is grayed out. The base workspace is the workspace used by the Command window.

Unless explicitly declared to be a global variable with the `global` command, all variables created in a function are local to that function. We think of the base workspace and the function workspaces as a "stack" of objects. Going up and down the stack means going into and out of the various workspaces.

STACK

Setting Preferences

To set preferences for the Editor/Debugger, select **Preferences** from the **File** menu. This opens a dialog box with several items. Here we mention two useful items to keep in mind. Under the **Display** preferences item, you can choose to display or not display line numbers and Datatips.

Under **Keyboard** you can choose to have the Editor match parentheses while editing.

Finding Bugs

The Editor/Debugger is useful for correcting runtime errors because it enables you to access function workspaces and examine or change the values they contain. We will now step you through an example debugging session. Although the example M-files are simpler than most MATLAB code, the debugging concepts demonstrated here apply in general. First create an M-file called `fun1.m` that accepts an input vector and returns the number of values in the vector that are above its average (mean) value. This file calls another M-file, `fun2.m`, that computes the number of values above the average, given the vector and its average.

```
function y = fun1(x)
avg = sum(x)/length(x);
y = fun2(avg,x);
```

Create the `fun1.m` file exactly as it is shown above, complete with a planted bug. Then create the file `fun2.m`, which is shown below.

```
function above = fun2(x,avg)
above = length(find(x>avg));
```

Use a simple test case that can be calculated by hand. For example, use the vector `v = [1, 2, 3, 4, 10]`. Its average is 4, and it contains one value (10) above the average. Now call the function `fun1` to test it.

```
>>above = fun1([1,2,3,4,10])
above =
  3
```

The answer should be 1, and therefore at least one of the functions, `fun1.m` or `fun2.m`, is working incorrectly. We will use the Editor/Debugger graphical interface to isolate the error. You could also use the debugging functions from the

Command window prompt. For information about these functions, use the Search tab in the Help Navigator to search for the phrase "functions for debugging."

If you have just created the M-files using the Editor/Debugger window, you can continue from this point. If you've created the M-files using an external text editor, start the Editor/Debugger and then open both M-files. You will see two tabs at the bottom of the screen, named `fun1.m` and `fun2.m`. Use these to switch between the two files. See Figure 4.7–1.

At this point you might find it convenient to close all windows on the Desktop except for the Command window, and then dock the Debugger windows for both `fun1.m` and `fun2.m` in the Desktop (to do this, select **Dock** from the Debugger's **Desktop** menu, once for each file). Then click on one of the tiling icons on the far right of the menu bar. Select the tiling pattern you want. Figure 4.7–1 shows the two functions split top to bottom. This displays the Command window and the two Editor/Debugger windows, one for `fun1.m` and one for `fun2.m`. This enables you to easily see the computation results in the Command window. Be sure to reduce the width of the Command window so that you can see the **Stack** menu in the Editor/Debugger. If the Editor/Debugger is not docked, then you will have to switch back and forth between the three windows (by clicking on the desired window, or by pressing **Alt** and **Tab** simultaneously, for example).

Setting Breakpoints

At the beginning of the debugging session, you are not sure where the error is. A logical place to insert a breakpoint is after the computation of the average in `fun1.m`. Go to the Editor/Debugger window for `fun1.m` and set a breakpoint at line 3 (`y = fun2(avg,x)`) by using the **Set Breakpoint** button on the toolbar. The line number is indicated at the left. Note that to evaluate the value of the variable `avg`, we must set a breakpoint in any line *following* the line in which `avg` is computed.

Examining Variables

To get to the breakpoint and check the values of interest, first execute the function from the Command window by typing `fun1([1, 2, 3, 4, 10])`. When execution of an M-file pauses at a breakpoint, the green arrow to the left of the text indicates the next line to be executed. Check the value of `avg` by highlighting the name of the variable, then right-clicking to bring up the **Context** menu, and choosing **Evaluate Selection.** You should now see `avg = 4` displayed in the Command window. Because the value of `avg` is correct, the error must lie either in the function call to `fun2` in line 3 in `fun1.m`, or in the `fun2.m` file.

Note that the prompt has changed to `K>>`, which stands for "keyboard." With this prompt you can enter commands in the Command window without disturbing the execution of the program. Suppose you find that the output of a function is incorrect. To continue debugging, you can enter the correct value for the variable at the `K>>` prompt.

Changing Workspaces

Use the **Stack** pull-down menu in the upper-right corner of the Debugger window to change workspaces. To see the base workspace contents, select **Base Workspace** from the **Stack** menu. Check the workspace contents using `whos` or the graphical Workspace Browser. Any variables you may have created in the current session will show up in the listing. Note that the variables `avg` and `x` do not show up because they are local to the function `fun1`. Similarly, to see the contents of the workspace of `fun1.m`, select **fun1** from the **Stack** menu and type `whos` in the Command window. You will see the local variables `avg` and `x` displayed.

Stepping through Code and Continuing Execution

Clear the breakpoint at line 3 in `fun1.m` by placing the cursor on the line and clicking on the **Clear Breakpoints** button. (Or right-click on the line to bring up the **Context** menu and choose **Set/Clear Breakpoint**). Continue executing the M-file by clicking the **Continue** button on the Debugger's toolbar. Open the `fun2.m` file and set a breakpoint at line 2 to see if the correct values of `x` and `avg` are being passed to the function. In the Command window, type `above = fun1([1,2,3,4,10])`. Highlight the variable `x` in the expression `above = length(find(x>avg));` in line 2, right-click on it, and select **Evaluate Selection** from the **Context** menu. You should see `x = 4` in the Command window. This value is incorrect because `x` should be `[1,2,3,4,10]`. Now evaluate the variable `avg` in line 2 the same way. You should see `avg = [1,2,3,4,10]` in the Command window. This is incorrect because `avg` should be `4`.

So the values of `x` and `avg` have been reversed in the function call in line 3 of `fun1.m`. This line should be `y = fun2(x,avg)`. Clear all breakpoints, exit the debug mode by selecting **Exit Debug Mode** on the **Debug** menu. Edit the line to correct the error, save the file, and run the test case again. You should get the correct answer.

Debugging a Loop

Loops such as `for` and `while` loops that do not execute the proper number of times are a common source of errors. The following function file `invest.m`, which has a planted bug, is intended to calculate how much money will be accumulated in a savings account that draws interest at the rate r percent compounded annually, if an amount $x(k)$, $k = 1, 2, 3, \ldots$ is deposited at the end of year k (this amount is not included in the interest calculation for that year).

```
function z = invest(x,r)
z = 0;
y = 1 + 0.01*r;
for k = 1:length(y)
  z = z*y + x(k);
end
```

To check the function, use the following test case, which is easily computed by hand. Suppose we deposit \$1000, \$1500, and \$2000 over three years, in a bank paying 10 percent interest. At the end of the first year the amount will be \$1000; at the end of the second year it will be \$1000(1.1) + \$1500 = \$2600, and at the end of the third year it will be \$2600(1.1) + \$2000 = \$4860. After creating and saving the function `invest.m`, call it in the Command window as follows:

```
>>total = invest([1000,1500,2000],10)
total =
  1000
```

which is incorrect (the answer should be 4860). To find the error, set a breakpoint at line 5 (the line containing the text `z = z*y + x(k);`). Run the function from the Command window by typing `total = invest([1000,1500,2000], 10))`. Execution stops at the breakpoint. Check the values of `z`, `y`, and `k`. These are `z = 0`, `y = 1.1`, and `k = 1`, which are correct. Next, select **Step** on the **Debug** menu. The green arrow moves to the line containing the `end` statement. Check the values. They are `z = 1000` and `k = 1`, which are correct. Select **Step** one more time, and again check the values of `z` and `k`. They are still `z = 1000` and `k = 1`, which are correct. Finally, select **Step** again, and check the values. You should see the following in the Command window.

```
K>> z??? Undefined function or variable 'z'.
K>> k??? Undefined function or variable 'k'.
```

Therefore, the program has gone through the loop only once, instead of three times. The error is in the upper limit of `k`, which should be `length(x)`, not `length(y)`.

4.8 Applications to Simulation

OPERATIONS RESEARCH

Simulation is the process of building and analyzing the output of computer programs that describe the operations of an organization, process, or physical system. Such a program is called a *computer model*. Simulation is often used in *operations research,* which is the quantitative study of an organization in action, to find ways to improve the functioning of the organization. Simulation enables engineers to study the past, present, and future actions of the organization for this purpose. Operations research techniques are useful in all engineering fields. Common examples include airline scheduling, traffic-flow studies, and production lines. The MATLAB logical operators and loops are excellent tools for building simulation programs.

EXAMPLE 4.8–1 A College Enrollment Model: Part I

As an example of how simulation can be used for operations research, consider the following college enrollment model. A certain college wants to analyze the effect of admissions and freshman retention rate on the college's enrollment so that it can predict the future need for instructors and other resources. Assume that the college has estimates of the percentages of students repeating a grade or leaving school before graduating. Develop a matrix equation on which to base a simulation model that can help in this analysis.

■ Solution

Suppose that the current freshman enrollment is 500 students and the college decides to admit 1000 freshmen per year from now on. The college estimates that 10 percent of the freshman class will repeat the year. The number of freshmen in the following year will be $0.1(500) + 1000 = 1050$, then it will be $0.1(1050) + 1000 = 1105$, and so on. Let $x_1(k)$ be the number of freshmen in year k, where $k = 1, 2, 3, 4, 5, 6, \ldots$. Then in year $k + 1$, the number of freshmen is given by

$$\begin{aligned} x_1(k+1) &= \text{10 percent of previous freshman class repeating freshman year} \\ &\quad + \text{1000 new freshmen} \\ &= 0.1x_1(k) + 1000 \qquad (4.8\text{–}1) \end{aligned}$$

Because we know the number of freshmen in the first year of our analysis (which is 500), we can solve this equation step by step to predict the number of freshmen in the future.

Let $x_2(k)$ be the number of sophomores in year k. Suppose that 15 percent of the freshmen do not return and that 10 percent repeat freshman year. Thus 75 percent of the freshman class returns as sophomores. Suppose also 5 percent of the sophomores repeat the sophomore year and that 200 sophomores each year transfer from other schools. Then in year $k + 1$, the number of sophomores is given by

$$x_2(k+1) = 0.75x_1(k) + 0.05x_2(k) + 200$$

To solve this equation we need to solve the "freshman" equation (4.8–1) at the same time, which is easy to do with MATLAB. Before we solve these equations, let us develop the rest of the model.

Let $x_3(k)$ and $x_4(k)$ be the number of juniors and seniors in year k. Suppose that 5 percent of the sophomores and juniors leave school and that 5 percent of the sophomores, juniors, and seniors repeat the grade. Thus 90 percent of the sophomores and juniors return and advance in grade. The models for the juniors and seniors are

$$x_3(k+1) = 0.9x_2(k) + 0.05x_3(k)$$
$$x_4(k+1) = 0.9x_3(k) + 0.05x_4(k)$$

These four equations can be written in the following matrix form:

$$\begin{bmatrix} x_1(k+1) \\ x_2(k+1) \\ x_3(k+1) \\ x_4(k+1) \end{bmatrix} = \begin{bmatrix} 0.1 & 0 & 0 & 0 \\ 0.75 & 0.05 & 0 & 0 \\ 0 & 0.9 & 0.05 & 0 \\ 0 & 0 & 0.9 & 0.05 \end{bmatrix} \begin{bmatrix} x_1(k) \\ x_2(k) \\ x_3(k) \\ x_4(k) \end{bmatrix} + \begin{bmatrix} 1000 \\ 200 \\ 0 \\ 0 \end{bmatrix}$$

In Example 4.8–2 we will see how to use MATLAB to solve such equations.

Test Your Understanding

T4.8–1 Suppose that 70 percent of the freshmen, instead of 75 percent, return for the sophomore year. How does the previous equation change?

EXAMPLE 4.8–2 A College Enrollment Model: Part II

To study the effects of admissions and transfer policies, generalize the enrollment model in Example 4.8–1 to allow for varying admissions and transfers.

■ Solution

Let $a(k)$ be the number of new freshmen admitted in the spring of year k for the following year $k + 1$ and let $d(k)$ be the number of transfers into the following year's sophomore class. Then the model becomes

$$x_1(k+1) = c_{11}x_1(k) + a(k)$$

$$x_2(k+1) = c_{21}x_1(k) + c_{22}x_2(k) + d(k)$$

$$x_3(k+1) = c_{32}x_2(k) + c_{33}x_3(k)$$

$$x_4(k+1) = c_{43}x_3(k) + c_{44}x_4(k)$$

where we have written the coefficients c_{21}, c_{22}, and so on in symbolic, rather than numerical, form so that we can change their values if desired.

STATE TRANSITION DIAGRAM

This model can be represented graphically by a *state transition diagram,* like the one shown in Figure 4.8–1. Such diagrams are widely used to represent time-dependent and probabilistic processes. The arrows indicate how the model's calculations are updated for each new year. The enrollment at year k is described completely by the values of $x_1(k)$, $x_2(k)$, $x_3(k)$, and $x_4(k)$; that is, by the vector $\mathbf{x}(k)$, which is called the *state vector.* The elements of the state vector are the *state variables.* The state transition diagram shows how the new values of the state variables depend on both the previous values and the inputs $a(k)$ and $d(k)$.

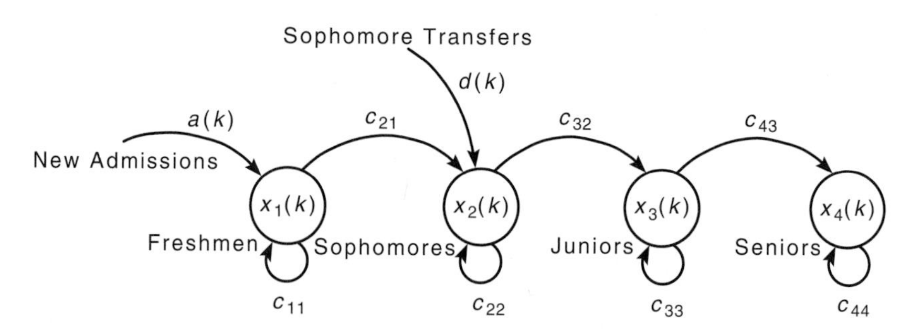

Figure 4.8–1 The state transition diagram for the college enrollment model.

The four equations can be written in the following matrix form:

$$\begin{bmatrix} x_1(k+1) \\ x_2(k+1) \\ x_3(k+1) \\ x_4(k+1) \end{bmatrix} = \begin{bmatrix} c_{11} & 0 & 0 & 0 \\ c_{21} & c_{22} & 0 & 0 \\ 0 & c_{32} & c_{33} & 0 \\ 0 & 0 & c_{43} & c_{44} \end{bmatrix} \begin{bmatrix} x_1(k) \\ x_2(k) \\ x_3(k) \\ x_4(k) \end{bmatrix} + \begin{bmatrix} a(k) \\ d(k) \\ 0 \\ 0 \end{bmatrix}$$

or more compactly as

$$\mathbf{x}(k+1) = \mathbf{C}\mathbf{x}(k) + \mathbf{b}(k)$$

where

$$\mathbf{x}(k) = \begin{bmatrix} x_1(k) \\ x_2(k) \\ x_3(k) \\ x_4(k) \end{bmatrix} \qquad \mathbf{b}(k) = \begin{bmatrix} a(k) \\ d(k) \\ 0 \\ 0 \end{bmatrix}$$

and

$$\mathbf{C} = \begin{bmatrix} c_{11} & 0 & 0 & 0 \\ c_{21} & c_{22} & 0 & 0 \\ 0 & c_{32} & c_{33} & 0 \\ 0 & 0 & c_{43} & c_{44} \end{bmatrix}$$

Suppose that the initial total enrollment of 1480 consists of 500 freshmen, 400 sophomores, 300 juniors, and 280 seniors. The college wants to study, over a 10-year period, the effects of increasing admissions by 100 each year and transfers by 50 each year until the total enrollment reaches 4000; then admissions and transfers will be held constant. Thus the admissions and transfers for the next 10 years are given by

$$a(k) = 900 + 100k$$

$$d(k) = 150 + 50k$$

for $k = 1, 2, 3, \ldots$ until the college's total enrollment reaches 4000; then admissions and transfers are held constant at the previous year's levels. We cannot determine when this event will occur without doing a simulation. Table 4.8–1 gives the pseudocode for solving this problem. The enrollment matrix **E** is a 4×10 matrix whose columns represent the enrollment in each year.

Table 4.8–1 Pseudocode for Example 4.8–2

Enter the coefficient matrix **C** and the initial enrollment vector **x**.
Enter the initial admissions and transfers, $a(1)$ and $d(1)$.
Set the first column of the enrollment matrix **E** equal to **x**.
Loop over years 2 to 10.
    If the total enrollment is ≤ 4000, increase admissions by 100 and transfers by 50 each year.
    If the total enrollment is > 4000, hold admissions and transfers constant.
    Update the vector **x**, using $\mathbf{x} = \mathbf{C}\mathbf{x} + \mathbf{b}$.
    Update the enrollment matrix **E** by adding another column composed of **x**.
End of the loop over years 2 to 10.
Plot the results.

Because we know the length of the study (10 years), a `for` loop is a natural choice. We use an `if` statement to determine when to switch from the increasing admissions and transfer schedule to the constant schedule. A MATLAB script file to predict the enrollment for the next 10 years appears in Table 4.8–2. Figure 4.8–2 shows the resulting plot. Note that after year 4 there are more sophomores than freshmen. The reason is that the increasing transfer rate eventually overcomes the effect of the increasing admission rate.

In actual practice this program would be run many times to analyze the effects of different admissions and transfer policies and to examine what happens if different values are used for the coefficients in the matrix **C** (indicating different dropout and repeat rates).

Table 4.8–2 College enrollment model

```
% Script file enroll1.m. Computes college enrollment.
% Model's coefficients.
C = [0.1,0,0,0;0.75,0.05,0,0;0,0.9,0.05,0;0,0,0.9,0.05];
% Initial enrollment vector.
x = [500;400;300;280];
% Initial admissions and transfers.
a(1) = 1000; d(1) = 200;
% E is the 4 x 10 enrollment matrix.
E(:,1) = x;
% Loop over years 2 to 10.
for k = 2:10
   % The following describes the admissions
   % and transfer policies.
   if sum(x) <= 4000
      % Increase admissions and transfers.
      a(k) = 900+100*k;
      d(k) = 150+50*k;
   else
      % Hold admissions and transfers constant.
      a(k) = a(k-1);
      d(k) = d(k-1);
   end
   % Update enrollment matrix.
   b = [a(k);d(k);0;0];
   x = C*x+b;
   E(:,k) = x;
end
% Plot the results.
plot(E'),hold,plot(E(1,:),'o'),plot(E(2,:),'+'),plot(E(3,:),'*'),...
plot(E(4,:),'x'),xlabel('Year'),ylabel('Number of Students'),...
gtext('Frosh'),gtext('Soph'),gtext('Jr'),gtext('Sr'),...
title('Enrollment as a Function of Time')
```

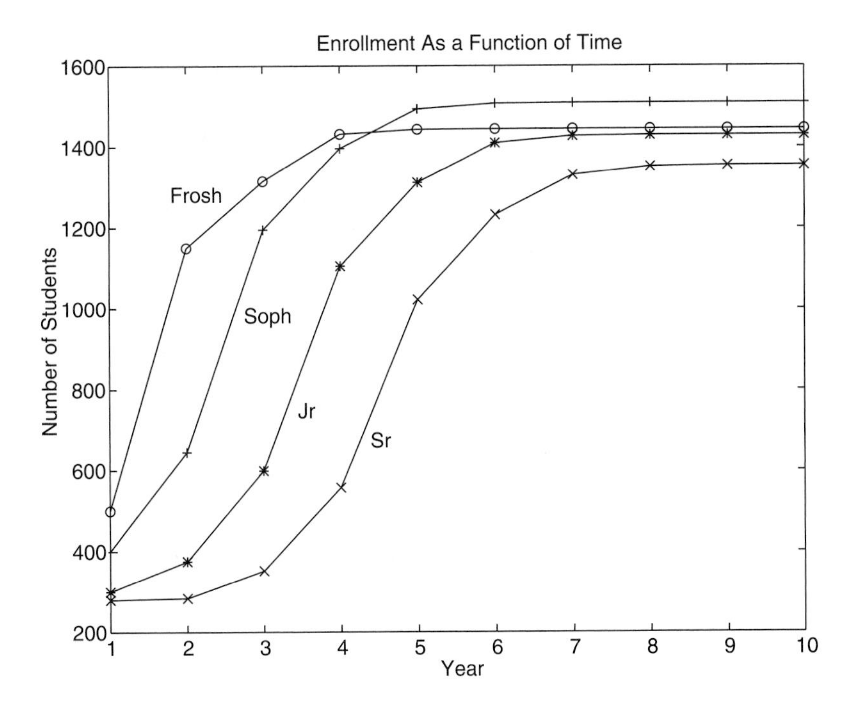

Figure 4.8–2 Class enrollments versus time.

Test Your Understanding

T4.8–2 In the program in Table 4.8–2, lines 16 and 17 compute the values of `a(k)` and `d(k)`. These lines are repeated here:

```
a(k) = 900+100*k
d(k) = 150+50*k;
```

Why does the program contain the line `a(1)=1000; d(1)=200;`?

4.9 Summary

Now that you have finished this chapter, you should be able to write programs that can perform decision-making procedures; that is, the program's operations depend on results of the program's calculations or on input from the user. Sections 4.2, 4.3, and 4.4 covered the necessary functions: the relational operators, the logical operators and functions, and the conditional statements.

You should also be able to use MATLAB loop structures to write programs that repeat calculations a specified number of times or until some condition is satisfied. This feature enables engineers to solve problems of great complexity or requiring numerous calculations. The `for` loop and `while` loop structures were covered in Section 4.5. Section 4.6 covered the `switch` structure.

Table 4.9–1 Guide to MATLAB commands introduced in Chapter 4

Relational operators	Table 4.2–1
Logical operators	Table 4.3–1
Order of precedence for operator types	Table 4.3–2
Truth table	Table 4.3–3
Logical functions	Table 4.3–4

Miscellaneous commands

Command	Description	Section
`break`	Terminates the execution of a `for` or a `while` loop.	4.5
`case`	Used with `switch` to direct program execution.	4.6
`continue`	Passes control to the next iteration of a `for` or `while` loop.	4.5
`double`	Converts a logical array to class double.	4.2
`else`	Delineates an alternate block of statements.	4.4
`elseif`	Conditionally executes statements.	4.4
`end`	Terminates `for`, `while`, and `if` statements.	4.4, 4.5
`findstr('s1','s2')`	For character strings `s1` and `s2`, finds the starting indices of any occurrences of the shorter string within the longer string of the pair.	4.4
`for`	Repeats statements a specific number of times.	4.5
`if`	Executes statements conditionally.	4.4
`input('s1', 's')`	Display the prompt string `s1` and stores user input as a string.	4.4
`logical`	Converts numeric values to logical values.	4.2
`lower('s')`	Converts the string `s` to all lowercase.	4.4
`nargin`	Determines the number of input arguments of a function.	4.4
`nargout`	Determines the number of output arguments of a function.	4.4
`strcmp('s1', 's2')`	Compares strings `s1` and `s2`.	4.4
`switch`	Directs program execution by comparing the input expression with the associated `case` expressions.	4.6
`upper('s')`	Converts the string `s` to all uppercase.	4.4
`while`	Repeats statements an indefinite number of times.	4.5
`xor`	Exclusive OR function.	4.3

Section 4.7 gave an overview and an example of how to debug programs using the Editor/Debugger. Section 4.8 presented an application of these methods to simulation, which enables engineers to study the operation of complicated systems, processes, and organizations.

Tables summarizing the MATLAB commands introduced in this chapter are located throughout the chapter. Table 4.9–1 will help you locate these tables. It also summarizes those commands not found in the other tables.

Key Terms with Page References

Breakpoint, 230
Conditional statement, 201
Flowchart, 186
`for` loop, 211
Implied loop, 216
Logical operator, 194
Masks, 217
Nested loops, 211
Operations research, 234
Pseudocode, 188
Relational operator, 191
Simulation, 234

Problems

You can find answers to problems marked with an asterisk at the end of the text.

Section 4.1

1. The volume V and surface area A of a sphere of radius r are given by

$$V = \frac{4}{3}\pi r^3 \qquad A = 4\pi r^2$$

a. Develop a pseudocode description of a program to compute V and A for $0 \le r \le 3$ m, and to plot V versus A.
b. Write and run the program described in part *a*.

2. The roots of the quadratic equation $ax^2 + bx + c = 0$ are given by

$$x = \frac{-b \pm \sqrt{b^2 - 4ac}}{2a}$$

a. Develop a pseudocode description of a program to compute both roots given the values of a, b, and c. Be sure to identify the real and imaginary parts.
b. Write the program described in part *a* and test it for the following cases:
1. $a = 2, b = 10, c = 12$
2. $a = 3, b = 24, c = 48$
3. $a = 4, b = 24, c = 100$

3. It is desired to compute the sum of the first ten terms of the series $14k^3 - 20k^2 + 5k$, $k = 1, 2, 3, \ldots$.
a. Develop a pseudocode description of the required program.
b. Write and run the program described in part *a*.

Section 4.2

4.* Suppose that `x = 6`. Find the results of the following operations by hand and use MATLAB to check your results.
a. `z = (x<10)`
b. `z = (x==10)`
c. `z = (x>=4)`
d. `z = (x~=7)`

5.* Find the results of the following operations by hand and use MATLAB to check your results.

a. `z = 6>3+8`
b. `z = 6+3>8`
c. `z = 4>(2+9)`
d. `z = (4<7)+3`
e. `z = 4<7+3`
f. `z = (4<7)*5`
g. `z = 4<(7*5)`
h. `z = 2/5>=5`

6.* Suppose that `x = [10, -2, 6, 5, -3]` and `y = [9, -3, 2, 5, -1]`. Find the results of the following operations by hand and use MATLAB to check your results.

a. `z = (x<6)`
b. `z = (x<=y)`
c. `z = (x==y)`
d. `z = (x~=y)`

7. For the arrays `x` and `y` given below, use MATLAB to find all the elements in `x` that are greater than the corresponding elements in `y`.

`x = [-3, 0, 0, 2, 6, 8]` `y = [-5, -2, 0, 3, 4, 10]`

8. The array `price` given below contains the price in dollars of a certain stock over 10 days. Use MATLAB to determine how many days the price was above $20.

`price = [19, 18, 22, 21, 25, 19, 17, 21, 27, 29]`

9. The arrays `price_A` and `price_B` given below contain the price in dollars of two stocks over 10 days. Use MATLAB to determine how many days the price of stock A was above the price of stock B.

`price_A = [19, 18, 22, 21, 25, 19, 17, 21, 27, 29]`

`price_B = [22, 17, 20, 19, 24, 18, 16, 25, 28, 27]`

10. The arrays `price_A`, `price_B`, and `price_C` given below contain the price in dollars of three stocks over 10 days.

a. Use MATLAB to determine how many days the price of stock A was above both the price of stock B and the price of stock C.
b. Use MATLAB to determine how many days the price of stock A was above either the price of stock B or the price of stock C.
c. Use MATLAB to determine how many days the price of stock A was above either the price of stock B or the price of stock C, but not both.

`price_A = [19, 18, 22, 21, 25, 19, 17, 21, 27, 29]`

`price_B = [22, 17, 20, 19, 24, 18, 16, 25, 28, 27]`

`price_C = [17, 13, 22, 23, 19, 17, 20, 21, 24, 28]`

Section 4.3

11.* Suppose that `x = [-3, 0, 0, 2, 5, 8]` and `y = [-5, -2, 0, 3, 4, 10]`. Find the results of the following operations by hand and use MATLAB to check your results.

a. `z = y<~x`
b. `z = x&y`
c. `z = x|y`
d. `z = xor(x,y)`

12. The height and speed of a projectile (such as a thrown ball) launched with a speed of v_0 at an angle A to the horizontal are given by

$$h(t) = v_0 t \sin A - 0.5gt^2$$
$$v(t) = \sqrt{v_0^2 - 2v_0 gt \sin A + g^2 t^2}$$

where g is the acceleration due to gravity. The projectile will strike the ground when $h(t) = 0$, which gives the time to hit $t_{\text{hit}} = 2(v_0/g)\sin A$.

Suppose that $A = 30°$, $v_0 = 40$ m/s, and $g = 9.81\text{m/s}^2$. Use the MATLAB relational and logical operators to find the times when

a. The height is no less than 15 m.
b. The height is no less than 15 m and the speed is simultaneously no greater than 36 m/s.
c. The height is less than 5 m or the speed is greater than 35 m/s.

13.* The price, in dollars, of a certain stock over a 10-day period is given in the following array.

```
price = [19, 18, 22, 21, 25, 19, 17, 21, 27, 29]
```

Suppose you owned 1000 shares at the start of the 10-day period, and you bought 100 shares every day the price was below $20 and sold 100 shares every day the price was above $25. Use MATLAB to compute (*a*) the amount you spent in buying shares, (*b*) the amount you received from the sale of shares, (*c*) the total number of shares you own after the 10th day, and (*d*) the net increase in the worth of your portfolio.

14. Let `e1` and `e2` be logical expressions. DeMorgan's laws for logical expressions state that

NOT(e1 AND e2) implies that (NOT e1) OR (NOT e2)

and

NOT(e1 OR e2) implies that (NOT e1) AND (NOT e2)

Use these laws to find an equivalent expression for each of the following expressions and use MATLAB to verify the equivalence.

a. `~((x < 10)&(x>=6))`
b. `~((x == 2) | (x > 5))`

15. Are these following expressions equivalent? Use MATLAB to check your answer for specific values of a, b, c, and d.

a. 1. `(a==b)&((b==c)|(a==c))`
   2. `(a==b)|((b==c)&(a==c))`

b. 1. `(a<b)&((a>c)|(a>d))`
   2. `(a<b)&(a>c)|((a<b)&(a>d))`

Section 4.4

16. Rewrite the following statements to use only one `if` statement.

```
if x < y
   if z < 10
      w = x*y*z
   end
end
```

17. Write a program that accepts a numerical value x from 0 to 100 as input and computes and displays the corresponding letter grade given by the following table.

A $x \geq 90$
B $80 \leq x \leq 89$
C $70 \leq x \leq 79$
D $60 \leq x \leq 69$
F $x < 60$

a. Use nested `if` statements in your program (do not use `elseif`).
b. Use only `elseif` clauses in your program.

18. Write a program that accepts a year and determines whether or not the year is a leap year. Use the `mod` function. The output should be the variable `extra_day`, which should be 1 if the year is a leap year and 0 otherwise. The rules for determining leap years in the Gregorian calendar are:

1. All years evenly divisible by 400 are leap years.
2. Years evenly divisible by 100 but not by 400 are not leap years.
3. Years divisible by 4 but not by 100 are leap years.
4. All other years are not leap years.

For example, the years 1800, 1900, 2100, 2300, and 2500 are not leap years, but 2400 is a leap year.

19. Figure P19a shows a mass-spring model of the type used to design packaging systems and vehicle suspensions, for example. The springs exert a force that is proportional to their compression, and the proportionality constant is the spring constant k. The two side springs provide additional resistance if the weight W is too heavy for the center spring. When the weight W is gently placed, it moves through a distance x

before coming to rest. From statics, the weight force must balance the spring forces at this new position. Thus

$$W = k_1 x \qquad \text{if } x < d$$
$$W = k_1 x + 2k_2(x - d) \qquad \text{if } x \geq d$$

These relations can be used to generate the plot of W versus x, shown in Figure P19b.

a. Create a function file that computes the distance x, using the input parameters W, k_1, k_2, and d. Test your function for the following two cases, using the values $k_1 = 10^4$ N/m; $k_2 = 1.5 \times 10^4$ N/m; $d = 0.1$ m.

$$W = 500 \text{ N}$$
$$W = 2000 \text{ N}$$

b. Use your function to plot x versus W for $0 \leq W \leq 3000$ N for the values of k_1, k_2, and d given in part *a*.

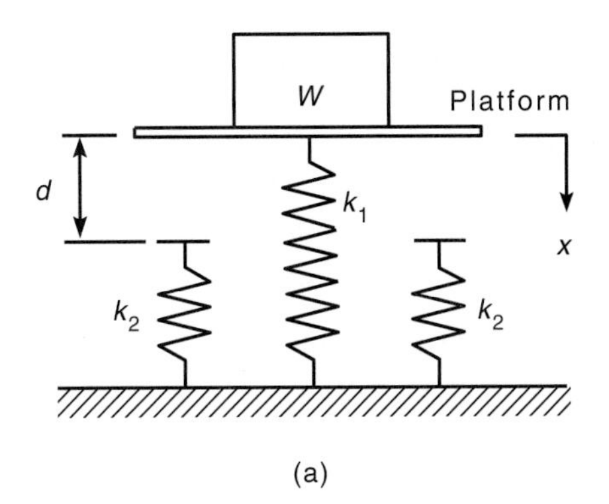

(a)

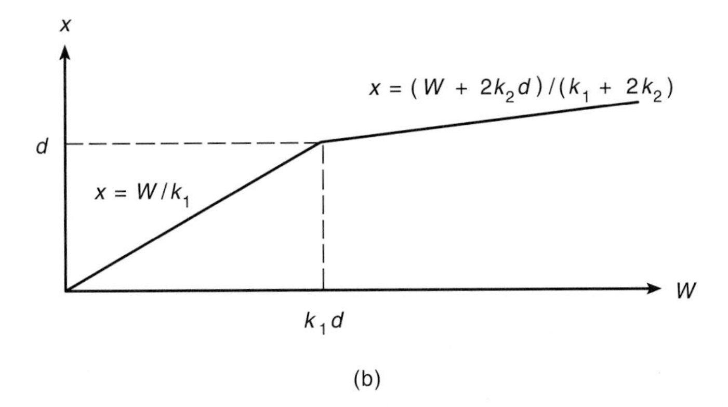

(b)

Figure P19

Section 4.5

20. The (x, y) coordinates of a certain object as a function of time t are given by

$$x(t) = 5t - 10 \qquad y(t) = 25t^2 - 120t + 144$$

for $0 \le t \le 4$. Write a program to determine the time at which the object is the closest to the origin at (0, 0). Determine also the minimum distance. Do this two ways:

a. By using a `for` loop.
b. By not using a `for` loop.

21. Consider the array **A**.

$$\mathbf{A} = \begin{bmatrix} 3 & 5 & -4 \\ -8 & -1 & 33 \\ -17 & 6 & -9 \end{bmatrix}$$

Write a program that computes the array **B** by computing the natural logarithm of all the elements of **A** whose value is no less than 1, and adding 20 to each element that is equal to or greater than 1. Do this two ways:

a. By using a `for` loop with conditional statements.
b. By using a logical array as a mask.

22. We want to analyze the mass-spring system discussed in Problem 19 for the case in which the weight W is dropped onto the platform attached to the center spring. If the weight is dropped from a height h above the platform, we can find the maximum spring compression x by equating the weight's gravitational potential energy $W(h + x)$ with the potential energy stored in the springs. Thus

$$W(h + x) = \tfrac{1}{2}k_1 x^2 \qquad \text{if } x < d$$

which can be solved for x as

$$x = \frac{W \pm \sqrt{W^2 + 2k_1 W h}}{k_1} \qquad \text{if } x < d$$

and

$$W(h + x) = \tfrac{1}{2}k_1 x^2 + \tfrac{1}{2}(2k_2)(x - d)^2 \qquad \text{if } x \ge d$$

which gives the following quadratic equation to solve for x:

$$(k_1 + 2k_2)x^2 - (4k_2 d + 2W)x + 2k_2 d^2 - 2Wh = 0 \qquad \text{if } x \ge d$$

a. Create a function file that computes the maximum compression x due to the falling weight. The function's input parameters are k_1, k_2, d, W, and h. Test your function for the following two cases, using the values $k_1 = 10^4$ N/m; $k_2 = 1.5 \times 10^4$ N/m; and $d = 0.1$ m.

$$W = 100 \text{ N}, h = 0.5 \text{ m}$$

$$W = 2000 \text{ N}, h = 0.5 \text{ m}$$

b. Use your function file to generate a plot of x versus h for $0 \le h \le 2$ m. Use $W = 100$ N and the preceding values for k_1, k_2, and d.

23. Electrical resistors are said to be connected "in series" if the same current passes through each and "in parallel" if the same voltage is applied across each. If in series, they are equivalent to a single resistor whose resistance is given by

$$R = R_1 + R_2 + R_3 + \cdots + R_n$$

If in parallel, their equivalent resistance is given by

$$\frac{1}{R} = \frac{1}{R_1} + \frac{1}{R_2} + \frac{1}{R_3} + \cdots + \frac{1}{R_n}$$

Write an M-file that prompts the user for the type of connection (series or parallel) and the number of resistors n and then computes the equivalent resistance.

24. *a.* An *ideal* diode blocks the flow of current in the direction opposite that of the diode's arrow symbol. It can be used to make a *half-wave rectifier* as shown in Figure P24a. For the ideal diode, the voltage v_L across the load R_L is given by

$$v_L = \begin{cases} v_S & \text{if } v_S > 0 \\ 0 & \text{if } v_S \le 0 \end{cases}$$

Suppose the supply voltage is

$$v_S(t) = 3e^{-t/3}\sin(\pi t) \quad \text{volts}$$

where time t is in seconds. Write a MATLAB program to plot the voltage v_L versus t for $0 \le t \le 10$.

b. A more accurate model of the diode's behavior is given by the *offset diode* model, which accounts for the offset voltage inherent in semiconductor diodes. The offset model contains an ideal diode and a battery whose voltage equals the offset voltage (which is approximately 0.6 V for silicon diodes) [Rizzoni, 1996]. The half-wave rectifier using this model is shown in Figure P24b. For this circuit,

$$v_L = \begin{cases} v_S - 0.6 & \text{if } v_S > 0.6 \\ 0 & \text{if } v_S \le 0.6 \end{cases}$$

Using the same supply voltage given in part *a*, plot the voltage v_L versus t for $0 \le t \le 10$; then compare the results with the plot obtained in part *a*.

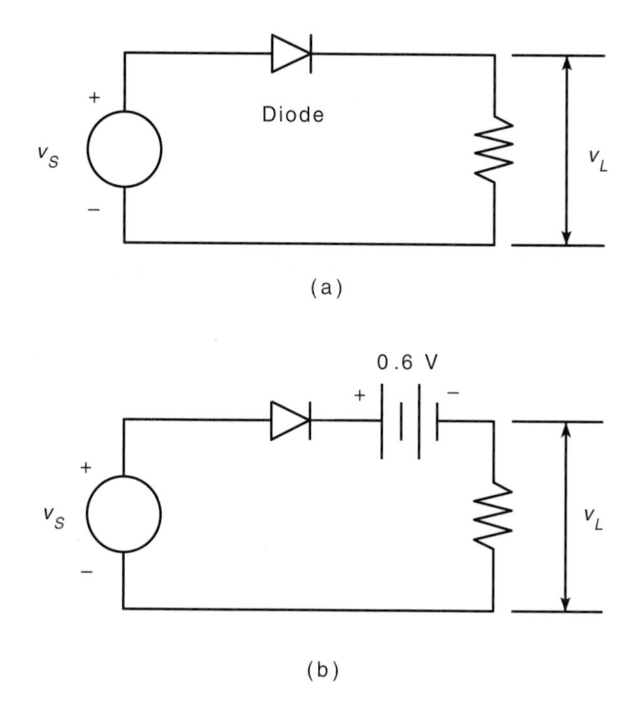

Figure P24

25.* Engineers in industry must continually look for ways to make their designs and operations more efficient. One tool for doing so is *optimization,* which uses a mathematical description of the design or operation to select the best values of certain variables. Many sophisticated mathematical tools have been developed for this purpose, and some are in the MATLAB Optimization toolbox. However, problems that have a limited number of possible variable values can use MATLAB loop structures to search for the optimum solution. This problem and the next two are examples of multivariable optimization that can be done with the basic MATLAB program.

A company wants to locate a distribution center that will serve six of its major customers in a 30 × 30 mi area. The locations of the customers relative to the southwest corner of the area are given in the following table in terms of (x, y) coordinates (the x direction is east; the y direction is north) (see Figure P25). Also given is the volume in tons per week that must be delivered from the distribution center to each customer. The weekly delivery cost c_i for customer i depends on the volume V_i and the distance d_i from the distribution center. For simplicity we will assume that this distance is the straight-line distance. (This assumes that

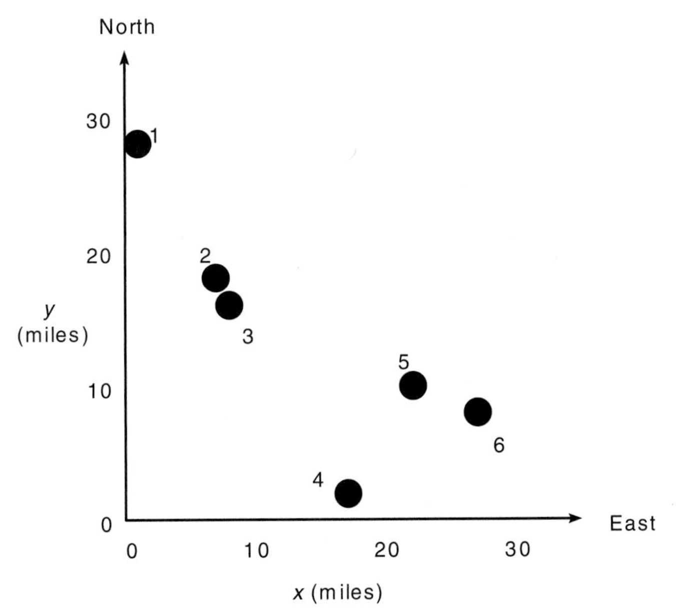

Figure P25

the road network is dense.) The weekly cost is given by $c_i = 0.5d_i V_i$; $i = 1, \ldots, 6$. Find the location of the distribution center (to the nearest mile) that minimizes the total weekly cost to service all six customers.

Customer	*x* location (miles)	*y* location (miles)	Volume (tons/week)
1	1	28	3
2	7	18	7
3	8	16	4
4	17	2	5
5	22	10	2
6	27	8	6

26. A company has the choice of producing up to four different products with its machinery, which consists of lathes, grinders, and milling machines. The number of hours on each machine required to produce a product is given in the following table, along with the number of hours available per week on each type of machine. Assume that the company can sell everything it produces. The profit per item for each product appears in the last line of the table.

	Product				Hours available
	1	2	3	4	
Hours required					
Lathe	1	2	0.5	3	40
Grinder	0	2	4	1	30
Milling	3	1	5	2	45
Unit profit ($)	100	150	90	120	

a. Determine how many units of each product the company should make to maximize its total profit and then compute this profit. Remember, the company cannot make fractional units, so your answer must be in integers. (Hint: First estimate the upper limits on the number of products that can be produced without exceeding the available capacity.)

b. How sensitive is your answer? How much does the profit decrease if you make one more or one less item than the optimum?

27. A certain company makes televisions, stereo units, and speakers. Its parts inventory includes chassis, picture tubes, speaker cones, power supplies, and electronics. The inventory, required components, and profit for each product appear in the following table. Determine how many of each product to make in order to maximize the profit.

	Product			
	Television	Stereo unit	Speaker unit	Inventory
Requirements				
Chassis	1	1	0	450
Picture Tube	1	0	0	250
Speaker Cone	2	2	1	800
Power Supply	1	1	0	450
Electronics	2	2	1	600
Unit profit ($)	80	50	40	

28.* Use a loop in MATLAB to determine how long it will take to accumulate $1,000,000 in a bank account if you deposit $10,000 initially and $10,000 at the end of each year; the account pays 6 percent annual interest.

29. A weight W is supported by two cables anchored a distance D apart (see Figure P29). The cable length L_{AB} is given, but the length L_{AC} is to be selected. Each cable can support a maximum tension force equal to W. For the weight to remain stationary, the total horizontal force and total vertical force must each be zero. This principle gives the equations

$$-T_{AB}\cos\theta + T_{AC}\cos\phi = 0$$
$$T_{AB}\sin\theta + T_{AC}\sin\phi = W$$

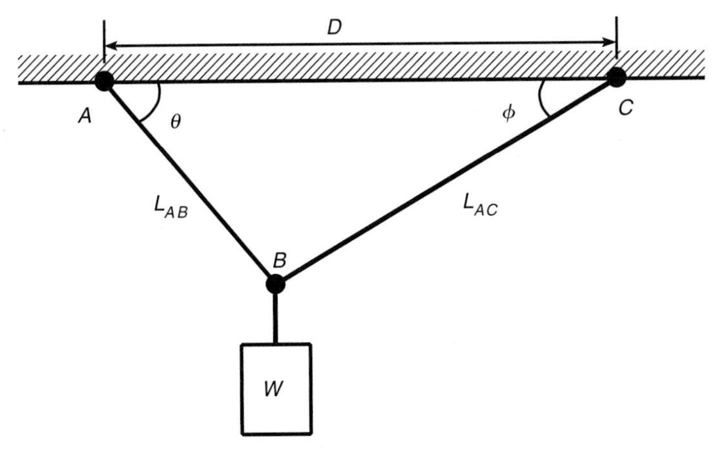

Figure P29

We can solve these equations for the tension forces T_{AB} and T_{AC} if we know the angles θ and ϕ. From the law of cosines

$$\theta = \cos^{-1}\left(\frac{D^2 + L_{AB}^2 - L_{AC}^2}{2DL_{AB}}\right)$$

From the law of sines

$$\phi = \sin^{-1}\left(\frac{L_{AB}\sin\theta}{L_{AC}}\right)$$

For the given values $D = 6$ ft, $L_{AB} = 3$ ft, and $W = 2000$ lb, use a loop in MATLAB to find $L_{AC\,\min}$, the shortest length L_{AC} we can use without T_{AB} or T_{AC} exceeding 2000 lb. Note that the largest L_{AC} can be is 6.7 ft (which corresponds to $\theta = 90°$). Plot the tension forces T_{AB} and T_{AC} on the same graph versus L_{AC} for $L_{AC\,\min} \leq L_{AC} \leq 6.7$.

30.* In the structure in Figure P30a, six wires support three beams. Wires 1 and 2 can support no more than 1200 N each, wires 3 and 4 can support no more than 400 N each, and wires 5 and 6 no more than 200 N each. Three equal weights W are attached at the points shown. Assuming that the structure is stationary and that the weights of the wires and the beams are very small compared to W, the principles of statics applied to a particular beam state that the sum of vertical forces is zero and that the sum of moments about any point is also zero. Applying these principles to each beam using the free-body diagrams shown in Figure P30b, we obtain the following equations. Let the tension force in wire i be T_i. For beam 1

$$T_1 + T_2 = T_3 + T_4 + W + T_6$$

$$-T_3 - 4T_4 - 5W - 6T_6 + 7T_2 = 0$$

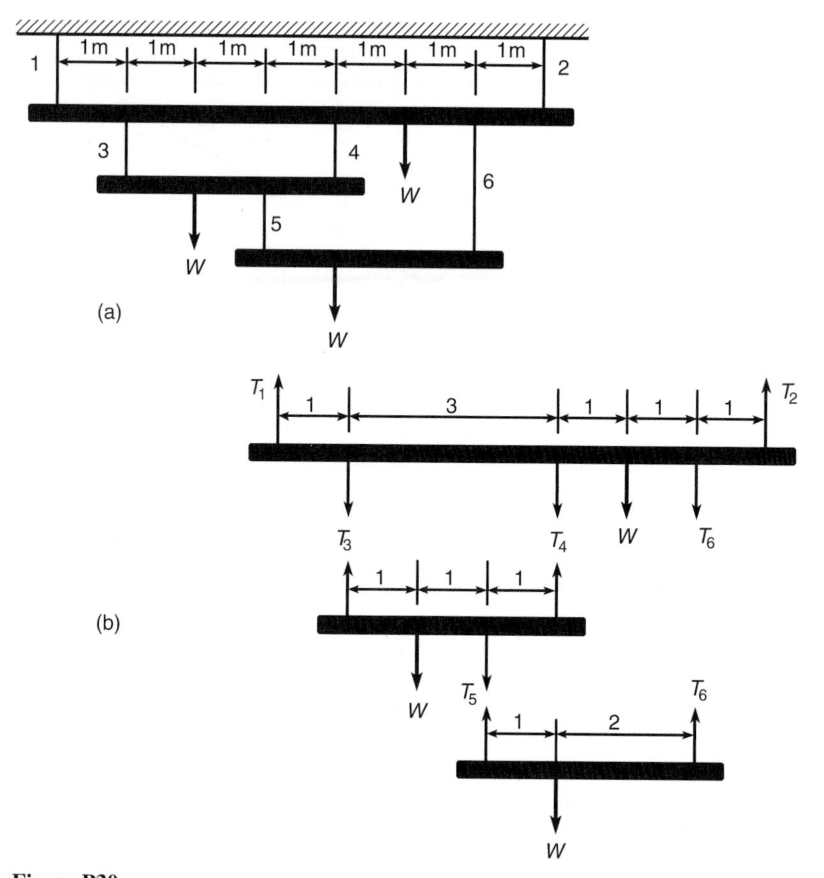

Figure P30

For beam 2

$$T_3 + T_4 = W + T_5$$
$$-W - 2T_5 + 3T_4 = 0$$

For beam 3

$$T_5 + T_6 = W$$
$$-W + 3T_6 = 0$$

Find the maximum value of the weight W the structure can support. Remember that the wires cannot support compression, so T_i must be nonnegative.

31. The equations describing the circuit shown in Figure P31 are:

$$-v_1 + R_1 i_1 + R_4 i_4 = 0$$

$$-R_4 i_4 + R_2 i_2 + R_5 i_5 = 0$$

$$-R_5 i_5 + R_3 i_3 + v_2 = 0$$

$$i_1 = i_2 + i_4$$

$$i_2 = i_3 + i_5$$

a. The given values of the resistances and the voltage v_1 are $R_1 = 5$, $R_2 = 100$, $R_3 = 200$, $R_4 = 150$, $R_5 = 250\,\text{k}\Omega$, and $v_1 = 100$ V. (Note that 1 kΩ = 1000 Ω.) Suppose that each resistance is rated to carry a current of no more than 1 mA (= 0.001 A). Determine the allowable range of positive values for the voltage v_2.

b. Suppose we want to investigate how the resistance R_3 limits the allowable range for v_2. Obtain a plot of the allowable limit on v_2 as a function of R_3 for $150 \le R_3 \le 250\,\text{k}\Omega$.

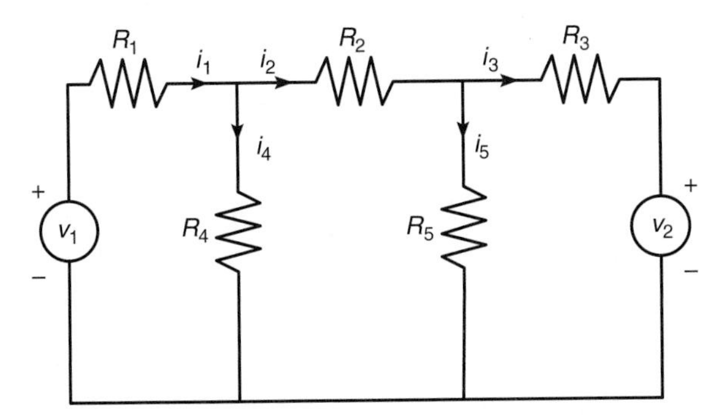

Figure P31

32. Many applications require us to know the temperature distribution in an object. For example, this information is important for controlling the material properties, such as hardness, when cooling an object formed from molten metal. In a heat transfer course, the following description of the temperature distribution in a flat, rectangular metal plate is often derived. The temperature is held constant at T_1 on three sides, and at T_2 on the fourth side (see Figure P32). The temperature $T(x, y)$ as a function of the xy coordinates shown is given by

$$T(x, y) = (T_2 - T_1)w(x, y) + T_1$$

where

$$w(x, y) = \frac{2}{\pi} \sum_{n \text{ odd}}^{\infty} \frac{2}{n} \sin\left(\frac{n\pi x}{L}\right) \frac{\sinh(n\pi y/L)}{\sinh(n\pi W/L)}$$

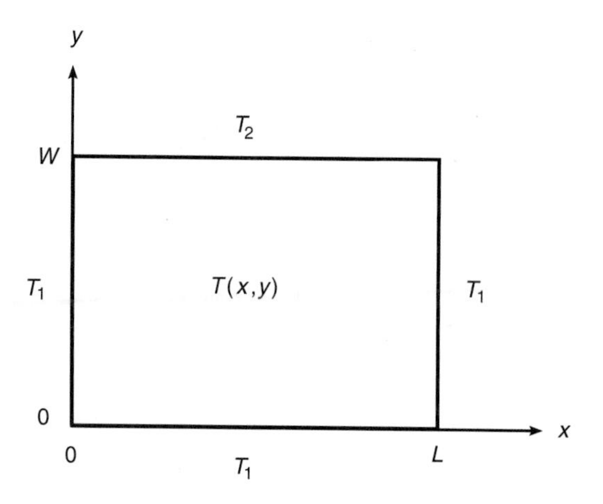

Figure P32

Use the following data: $T_1 = 70°\text{F}$, $T_2 = 200°\text{F}$, and $W = L = 2$ ft.

a. The terms in the preceding series become smaller in magnitude as n increases. Write a MATLAB program to verify this fact for $n = 1, \ldots, 19$ for the center of the plate ($x = y = 1$).

b. Using $x = y = 1$, write a MATLAB program to determine how many terms are required in the series to produce a temperature calculation that is accurate to within 1 percent. (That is, for what value of n will the addition of the next term in the series produce a change in T of less than 1 percent.) Use your physical insight to determine whether this answer gives the correct temperature at the center of the plate.

c. Modify the program from part b to compute the temperatures in the plate; use a spacing of 0.2 for both x and y.

33. Consider the following script file. Fill in the lines of the following table with the values that would be displayed immediately after the `while` statement if you ran the script file. Write in the values the variables have each time the `while` statement is executed. You might need more or fewer lines in the table. Then type in the file, and run it to check your answers.

```
k = 1;b = -2;x = -1;y = -2;
while k <= 3
   k, b, x, y
   y = x^2 - 3;
   if y < b
      b = y;
   end
   x = x + 1;
   k = k + 1;
end
```

Pass	k	b	x	y
First				
Second				
Third				
Fourth				
Fifth				

34. Assume that the human player makes the first move against the computer in a game of Tic-Tac-Toe, which has a 3×3 grid. Write a MATLAB function that lets the computer respond to that move. The function's input argument should be the cell location of the human player's move. The function's output should be the cell location of the computer's first move. Label the cells as 1, 2, 3 across the top row; 4, 5, 6 across the middle row, and 7, 8, 9 across the bottom row.

Section 4.6

35. The following table gives the approximate values of the static coefficient of friction μ for various materials.

Materials	μ
Metal on metal	0.20
Wood on wood	0.35
Metal on wood	0.40
Rubber on concrete	0.70

To start a weight W moving on a horizontal surface, you must push with a force F, where $F = \mu W$. Write a MATLAB program that uses the `switch` structure to compute the force F. The program should accept as input the value of W and the type of materials.

36. The height and speed of a projectile (such as a thrown ball) launched with a speed of v_0 at an angle A to the horizontal are given by

$$h(t) = v_0 t \sin A - 0.5gt^2$$

$$v(t) = \sqrt{v_0^2 - 2v_0 gt \sin A + g^2 t^2}$$

where g is the acceleration due to gravity. The projectile will strike the ground when $h(t) = 0$, which gives the time to hit $t_{hit} = 2(v_0/g) \sin A$.

Use the `switch` structure to write a MATLAB program to compute either the maximum height reached by the projectile, the total horizontal distance traveled, or the time to hit. The program should accept as input the user's choice of which quantity to compute and the values of v_0, A, and g. Test the program for the case where $v_0 = 40$ m/s, $A = 30°$, and $g = 9.81$ m/s^2.

37. Use the `switch` structure to write a MATLAB program to compute how much money accumulates in a savings account in one year. The program should accept the following input: the initial amount of money deposited in the account; the frequency of interest compounding (monthly, quarterly, semiannually, or annually); and the interest rate. Run your program for a $1000 initial deposit for each case; use a 5 percent interest rate. Compare the amounts of money that accumulate for each case.

38. Engineers often need to estimate the pressures and volumes of a gas in a container. The *van der Waals* equation is often used for this purpose. It is

$$P = \frac{RT}{\hat{V} - b} - \frac{a}{\hat{V}^2}$$

where the term b is a correction for the volume of the molecules, and the term $a/\hat{V}^2$ is a correction for molecular attractions. The gas constant is R, the *absolute* temperature is T, and the gas specific volume is $\hat{V}$. The value of R is the same for all gases; it is $R = 0.08206$ L-atm/mol-K. The values of a and b depend on the type of gas. Some values are given in the following table. Write a user-defined function using the `switch` structure that computes the pressure P on the basis of the van der Waals equation. The function's input arguments should be T, $\hat{V}$, and a string variable containing the name of a gas listed in the table. Test your function for chlorine (Cl_2) for $T = 300$ K and $\hat{V} = 20$ L/mol.

Gas	a (L^2-atm/mol^2)	b (L/mol)
Helium, He	0.0341	0.0237
Hydrogen, H_2	0.244	0.0266
Oxygen, O_2	1.36	0.0318
Chlorine, Cl_2	6.49	0.0562
Carbon dioxide, CO_2	3.59	0.0427

39. Using the program developed in Problem 18, write a program that uses the `switch` structure to compute the number of days in a year up to a given date, given the year, the month, and the day of the month.

Section 4.8

40. Consider the college enrollment model discussed in Example 4.8–2. Suppose the college wants to limit freshmen admissions to 120 percent of the current sophomore class and limit sophomore transfers to 10 percent of the current freshman class. Rewrite and run the program given in the example to examine the effects of these policies over a 10-year period. Plot the results.

41. Suppose you project that you will be able to deposit the following monthly amounts into a savings account for a period of five years. The account initially has no money in it.

Year	1	2	3	4	5
Monthly deposit ($)	300	350	350	350	400

At the end of each year in which the account balance is at least $3000, you withdraw $2000 to buy a certificate of deposit (CD), which pays 6 percent interest compounded annually.

Write a MATLAB program to compute how much money will accumulate in five years in the account and in any CDs you buy. Run the program for two different savings interest rates: 4 percent and 5 percent.

42.* A certain company manufactures and sells golf carts. At the end of each week, the company transfers the carts produced that week into storage (inventory). All carts that are sold are taken from the inventory. A simple model of this process is

$$I(k+1) = P(k) + I(k) - S(k)$$

where

$P(k)$ = the number of carts produced in week k
$I(k)$ = the number of carts in inventory in week k
$S(k)$ = the number of carts sold in week k

The projected weekly sales for 10 weeks are

Week	1	2	3	4	5	6	7	8	9	10
Sales	50	55	60	70	70	75	80	80	90	55

Suppose the weekly production is based on the previous week's sales so that $P(k) = S(k-1)$. Assume that the first week's production is 50 carts; that is, $P(1) = 50$. Write a MATLAB program to compute and plot the number of carts in inventory for each of the 10 weeks or until the inventory drops below zero. Run the program for two cases: *a.* an initial inventory of 50 carts so that $I(1) = 50$, and *b.* an initial inventory of 30 carts so that $I(1) = 30$.

43. Redo Problem 42 with the restriction that the next week's production is set to zero if the inventory exceeds 40 carts.

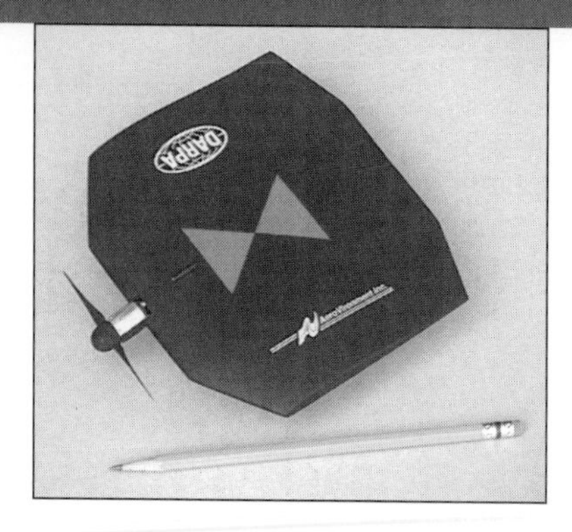

Courtesy of Aero Vironment, Inc.

Engineering in the 21st Century...

Low-Speed Aeronautics

Sometimes just when we think a certain technical area is mature and the possibility of further development is unlikely, we are surprised by a novel design. Recent developments in low-speed aeronautics are examples of this phenomenon. Even though engineers have known for years that a human could generate enough power to propel an aircraft, the feat remained impossible until the availability of lightweight materials that enabled the Gossamer Challenger to fly across the English Channel. Solar-powered aircraft that can stay aloft for over a day are other examples.

Another example is the recent appearance of wing-in-ground effect (WIG) vehicles. WIG vehicles make use of an air cushion to create lift. They are a hybrid between an aircraft and a hovercraft and most are intended for over-water flight only. A hovercraft rides on an air cushion created by fans, but the air cushion of a WIG vehicle is due to the air that is captured under its stubby wings.

Small aircraft with cameras will be useful for search and reconnaissance. An example of such a "micro air vehicle" (MAV) is the 6-inch long Black Widow shown in the photo. It carries a 2-gram video camera the size of a sugar cube, and flies at about 65 km/h with a range of 10 km. Proper design of such vehicles requires a systematic methodology to find the optimum combination of airfoil shape, motor type, battery type, and most importantly, the propeller shape.

The MATLAB advanced graphics capabilities make it useful for visualizing flow patterns, and the Optimization toolbox is useful for designing such vehicles. ■

CHAPTER 5

Advanced Plotting and Model Building

OUTLINE

The popular phrase "A picture is worth a thousand words" emphasizes the importance of graphical representation in communicating information. It is easier to identify patterns in a plot than in a table of numbers. Engineers frequently use plots both to gain insight and to communicate their findings and ideas to others. Plotting, like any language, has a set of rules, standards, and practices that the engineer should follow to produce effective plots. Failure to do so will diminish one's reputation with colleagues, at least, and at worst, could lead others to draw incorrect conclusions about the data presented.

MATLAB has many functions that are useful for creating plots. In this chapter you will learn how to use them to create two-dimensional plots, which are also called *xy plots,* and three-dimensional plots called *xyz plots,* or *surface* plots. These plotting functions are described in the `graph2d` and `graph3d` help categories, so typing `help graph2d` or `help graph3d` will display a list of the relevant plotting functions.

This chapter also discusses the elements of a correct graph and how to use MATLAB to create effective graphs that convey the desired information. An important application of plotting is *function discovery,* the technique for using data plots to obtain a mathematical function or "mathematical model" that describes the process that generated the data. This feature is very useful for engineering applications because engineers frequently need to use mathematical models to predict how their proposed designs will work. A systematic method for obtaining models is *regression,* which is also covered in this chapter.

5.1 xy Plotting Functions

The most common plot is the xy plot. Its name assumes that we are plotting a function $y = f(x)$, although, of course, other symbols may be used. We plot the x values on the horizontal axis (the *abscissa*), and the y values on the vertical axis (the *ordinate*). Usually we plot the independent variable, which is the one more easily varied, on the abscissa, and the dependent variable on the ordinate.

ABSCISSA

ORDINATE

MATLAB has many functions and commands to produce various plots with special features. In this section we introduce the commands that are useful for making xy plots. In Section 5.8 we treat three-dimensional plots.

The Anatomy of a Plot

The "anatomy" and nomenclature of a typical xy plot is shown in Figure 5.1–1, in which the plot of a data set and a curve generated from an equation appear. The *scale* on each axis refers to the range and spacing of the numbers. Both axes in this plot are said to be "rectilinear"—often shortened to *linear*—because the spacing of the numbers is regular; for example, the distance between the numbers 2 and 3 is the same as the distance between the numbers 4 and 5. Another type of scale is the *logarithmic,* which we explain later in this chapter. *Tick marks* are placed on the axis to help visualize the numbers being plotted. The *tick-mark labels* are the numbers that correspond to the tick-mark locations. (Some plots will have tick-mark labels that are not numbers; for example, if we plot temperature versus time of year, the tick-mark labels on the horizontal axis could be the names of months.) The spacing of the tick marks and their labels is important. We cover this topic later in the chapter.

SCALE

TICK MARK

Each axis must have an *axis label*—also called an *axis title.* This label gives the name and units of the quantity plotted on that axis. An exception occurs when plotting a mathematical expression that has no physical interpretation; in that case the variables have no units. In addition, the plot often must have a plot title as well. The plot title is placed above the plot.

AXIS LABEL

A plot can be made from measured data or from an equation. When data is plotted, each data point is plotted with a *data symbol,* or *point marker,* such as the

DATA SYMBOL

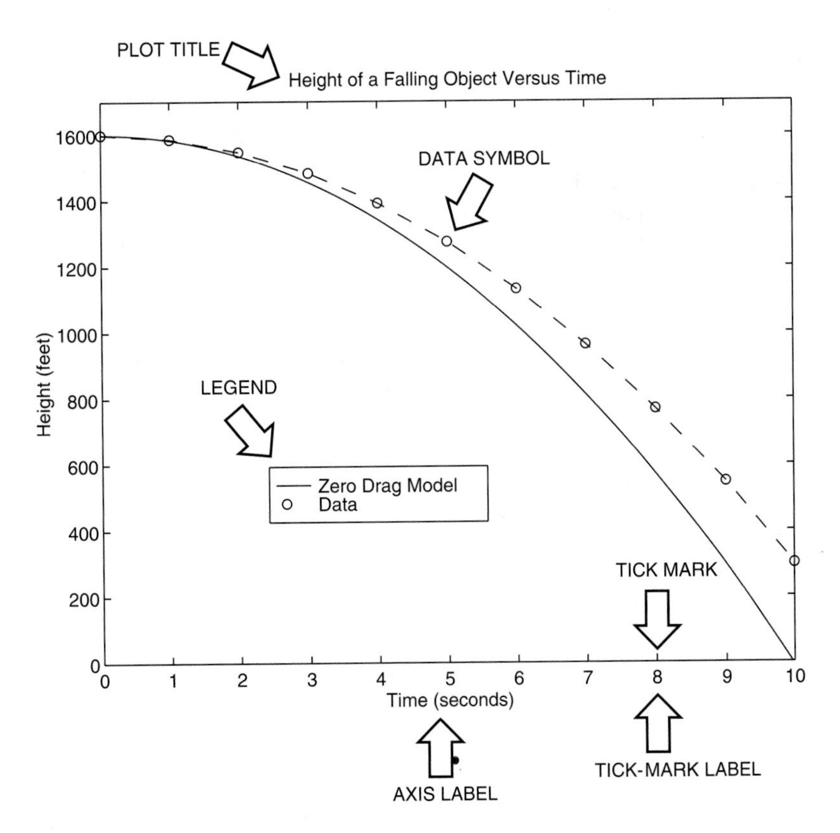

Figure 5.1–1 Nomenclature for a typical xy plot.

small circle shown in Figure 5.1–1. A rare exception to this rule would be when there are so many data points that the symbols would be too densely packed. In that case, the data points should be plotted with a dot. However, when the plot is generated from a function, data symbols must *never* be used! Lines are always used to plot a function.

Sometimes data symbols are connected by lines to help the viewer visualize the data, especially if there are few data points. However, connecting the data points—especially with a solid line—might imply knowledge of what occurs between the data points, and thus you should be careful to prevent such misinterpretation.

LEGEND

When multiple curves or data sets are plotted, they must be distinguished from each other. One way of doing so is with a *legend,* which relates the data set symbol or the curve's line type to the quantity being plotted. Another method is to place a description (either text or an equation) near the curve or data symbols. We show examples of both methods later in the chapter.

Requirements for a Correct Plot

The following list describes the essential features of any plot:

1. Each axis must be labeled with the name of the quantity being plotted *and its units!* If two or more quantities having different units are plotted (such as when plotting both speed and distance versus time), indicate the units in the axis label if there is room, or in the legend or labels for each curve.
2. Each axis should have regularly spaced tick marks at convenient intervals—not too sparse, but not too dense—with a spacing that is easy to interpret and interpolate. For example, use 0.1, 0.2, and so on, rather than 0.13, 0.26, and so on.
3. If you are plotting more than one curve or data set, label each on its plot or use a legend to distinguish them.
4. If you are preparing multiple plots of a similar type or if the axes' labels cannot convey enough information, use a title.
5. If you are plotting measured data, plot each data point with a symbol such as a circle, square, or cross (use the same symbol for every point in the same data set). If there are many data points, plot them using the dot symbol.
6. Sometimes data symbols are connected by lines to help the viewer visualize the data, especially if there are few data points. However, connecting the data points, especially with a solid line, might be interpreted to imply knowledge of what occurs between the data points. Thus you should be careful to prevent such misinterpretation.
7. If you are plotting points generated by evaluating a function (as opposed to measured data), do *not* use a symbol to plot the points. Instead, be sure to generate many points, and connect the points with solid lines.

Plot, Label, and Title Commands

The MATLAB basic xy plotting function is `plot(x,y)`. If `x` and `y` are vectors, a single curve is plotted with the `x` values on the abscissa and the `y` values on the ordinate. The `xlabel` and `ylabel` commands put labels on the abscissa and the ordinate, respectively. The syntax is `xlabel('text')`, where `text` is the text of the label. Note that you must enclose the label's text in single quotes. The syntax for `ylabel` is the same. The `title` command puts a title at the top of the plot. Its syntax is `title('text')`, where `text` is the title's text.

The following MATLAB session plots $y = 0.4\sqrt{1.8x}$ for $0 \leq x \leq 52$, where y represents the height of a rocket after launch, in miles, and x is the horizontal (downrange) distance in miles.

```
>>x = [0:0.1:52];
>>y = 0.4*sqrt(1.8*x);
>>plot(x,y)
>>xlabel('Distance (miles)')
```

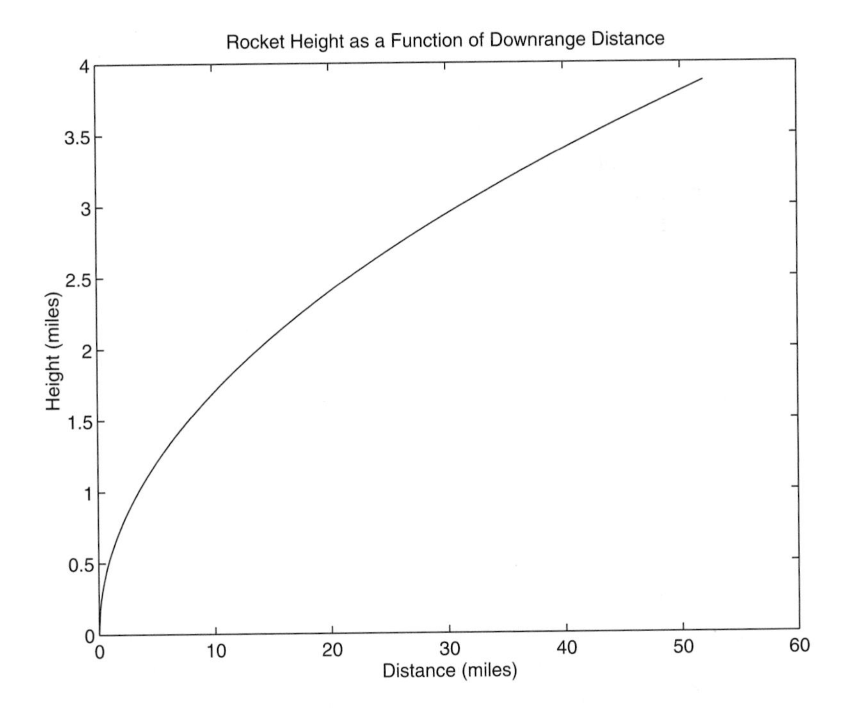

Figure 5.1–2 The autoscaling feature in MATLAB selects tick-mark spacing.

```
>>ylabel('Height (miles)')
>>title('Rocket Height as a Function of Downrange Distance')
```

Figure 5.1–2 shows the plot. A spacing of 0.1 was selected for the x values to generate several hundred plotting points to produce a smooth curve. The `plot(x,y)` function in MATLAB automatically selects a tick-mark spacing for each axis and places appropriate tick labels. This feature is called autoscaling.

MATLAB also chose an upper limit for the x-axis, which is beyond the maximum value of 52 in the x values, to obtain a convenient spacing of 10 for the tick labels. A tick-label spacing of two would generate 27 labels, which gives a spacing so dense that the labels would overlap one another. A spacing of 13 would work, but is not as convenient as a spacing of 10. Later you will learn to override the values selected by MATLAB.

The axis labels and plot title are produced by the `xlabel`, `ylabel`, and `title` commands. The order of the `xlabel`, `ylabel`, and `title` commands does not matter, but we must place them *after* the `plot` command, either on separate lines using ellipses or on the same line separated by commas, as

```
>>x = [0:0.1:52];
>>y = 0.4*sqrt(1.8*x);
>>plot(x,y),xlabel('Distance (miles)'),ylabel('Height (miles)'),...
title('Rocket Height as a Function of Downrange Distance')
```

The plot will appear in the Figure window. You can obtain a hard copy of the plot in one of several ways:

1. Use the menu system. Select **Print** on the **File** menu in the Figure window. Answer **OK** when you are prompted to continue the printing process.
2. Type `print` at the command line. This command sends the current plot directly to the printer.
3. Save the plot to a file to be printed later or imported into another application such as a word processor. You need to know something about graphics file formats to use this file properly. See the subsection **Exporting Figures** later in this section.

Type `help print` to obtain more information.

MATLAB assigns the output of the `plot` command to figure window number 1. When another `plot` command is executed, MATLAB overwrites the contents of the existing figure window with the new plot. Although you can keep more than one figure window active, we do not use this feature in this text.

When you have finished with the plot, close the figure window by selecting **Close** from the **File** menu in the figure window. Note that using the **Alt-Tab** key combination in Windows-based systems will return you to the Command window without closing the figure window. If you do not close the window, it will not reappear when a new `plot` command is executed. However, the figure will still be updated.

`grid` and `axis` Commands

The `grid` command displays gridlines at the tick marks corresponding to the tick labels. Type `grid on` to add gridlines; type `grid off` to stop plotting gridlines. When used by itself, `grid` toggles this feature on or off, but you might want to use `grid on` and `grid off` to be sure.

AXIS LIMITS

You can use the `axis` command to override the MATLAB selections for the axis limits. The basic syntax is `axis([xmin xmax ymin ymax])`. This command sets the scaling for the x- and y-axes to the minimum and maximum values indicated. Note that, unlike an array, this command does not use commas to separate the values.

The `axis` command has the following variants:

- `axis square`, which selects the axes' limits so that the plot will be square.
- `axis equal`, which selects the scale factors and tick spacing to be the same on each axis. This variation makes `plot(sin(x),cos(x))` look like a circle, instead of an oval.
- `axis auto`, which returns the axis scaling to its default autoscaling mode in which the best axes limits are computed automatically.

For example, to add a grid and to change the axes' limits on the previous plot to $0 \leq x \leq 52$ and $0 \leq y \leq 5$, the session would look like

```
>>x = [0:0.1:52];
>>y = 0.4*sqrt(1.8*x);
```

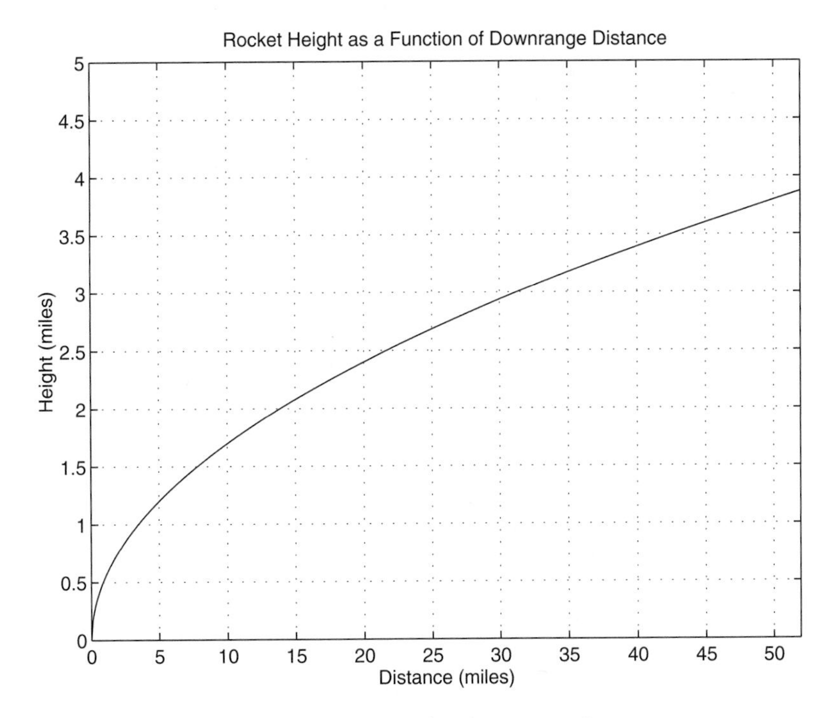

Figure 5.1–3 The effects of the `axis` and `grid` commands.

```
>>plot(x,y),xlabel('Distance (miles)'),ylabel('Height (miles)'),...
title('Rocket Height as a Function of Downrange Distance'),...
grid on, axis([0 52 0 5])
```

Figure 5.1–3 shows this plot. Notice how MATLAB chose a tick-label spacing of 5, not 13, for the x-axis.

This example illustrates how the printed plot can look different from the plot on the computer screen. MATLAB determines the number of tick-mark labels that can reasonably fit on the axis without being too densely spaced. A reasonable number for the computer screen is often different from the number for the printed output. In the preceding example, the screen plot showed labels on the x-axis at 0, 10, 20, . . . , whereas the printed plot had labels at the intervals 0, 5, 10, 15, 20, ... You can eliminate this effect by using the tick-mark commands discussed later in the chapter.

Plots of Complex Numbers

With only one argument, say, `plot(y)`, the `plot` function will plot the values in the vector `y` versus their indices 1, 2, 3, . . . , and so on. If `y` is complex, `plot(y)` plots the imaginary parts versus the real parts. Thus `plot(y)` in this case is equivalent to `plot(real(y),imag(y))`. This situation is the only time when the `plot` function handles the imaginary parts; in all other variants of the `plot`

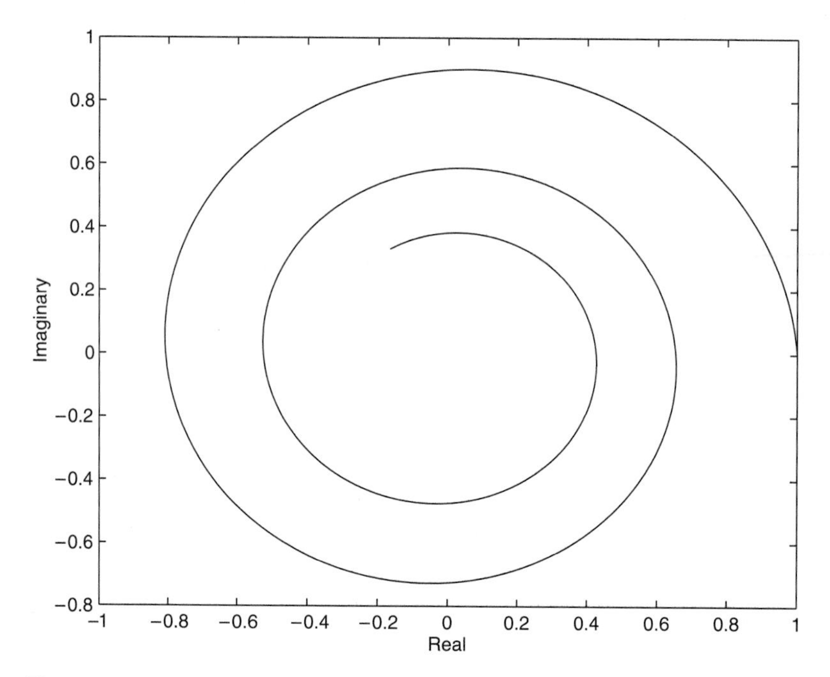

Figure 5.1–4 Application of the `plot(y)` function.

function, it ignores the imaginary parts. For example, the script file

```
z = 0.1 + 0.9i;
n = [0:0.01:10];
plot(z.^n),xlabel('Real'),ylabel('Imaginary')
```

generates the spiral shown in Figure 5.1–4. As you become more familiar with MATLAB, you will feel comfortable combining these commands as follows:

```
plot((0.1+0.9i).^[0:0.01:10]),xlabel('Real'),ylabel('Imaginary')
```

The Function Plot Command `fplot`

MATLAB has a "smart" command for plotting functions. The `fplot` command automatically analyzes the function to be plotted and decides how many plotting points to use so that the plot will show all the features of the function. Its syntax is `fplot ('string', [xmin xmax])`, where `'string'` is a text string that describes the function to be plotted and `[xmin xmax]` specifies the minimum and maximum values of the independent variable. The range of the dependent variable can also be specified. In this case the syntax is `fplot('string', [xmin xmax ymin ymax])`.

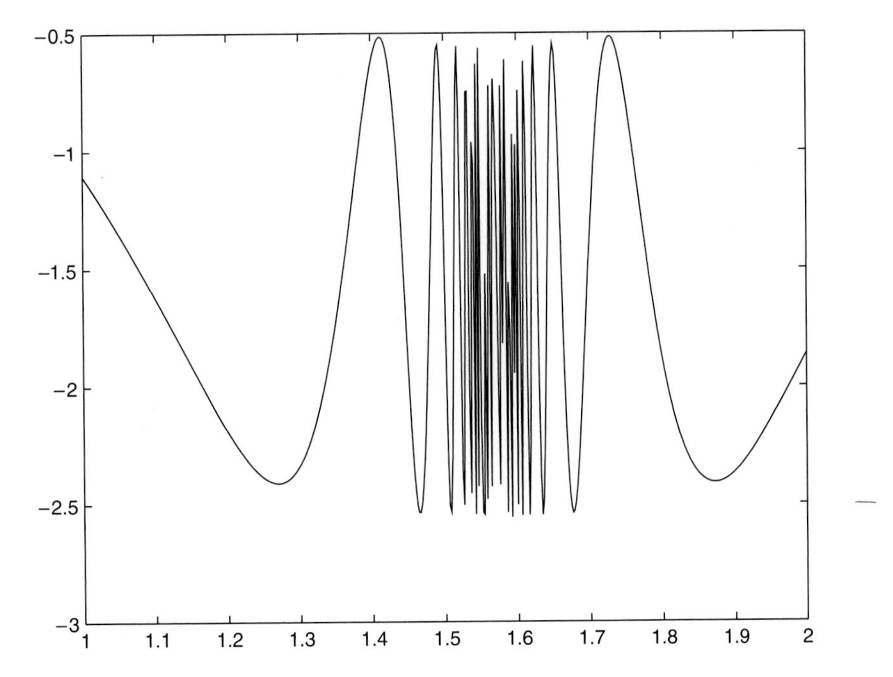

Figure 5.1–5 A plot generated with the `fplot` command.

For example, the session

```
>>f = 'cos(tan(x)) - tan(sin(x))';
>>fplot(f,[1 2])
```

produces the plot shown in Figure 5.1–5. You may combine the two commands into a single command as follows: `fplot('cos(tan(x)) - tan(sin(x))',[1 2])`. Always remember to enclose the function in single quotes.

Contrast this plot with the one shown in Figure 5.1–6, which is produced by the `plot` command using 101 plotting points.

```
>>x = [1:0.01:2];
>>y = cos(tan(x)) - tan(sin(x));
>>plot(x,y)
```

We can see that the `fplot` command automatically chose enough plotting points to display all the variations in the function. We can achieve the same plot using the `plot` command, but we need to know how many values to use in specifying the `x` vector.

Another form is `[x,y] = fplot('string', limits)`, where `limits` may be either `[xmin xmax]` or `[xmin xmax ymin ymax]`. With this form the command returns the abscissa and ordinate values in the column

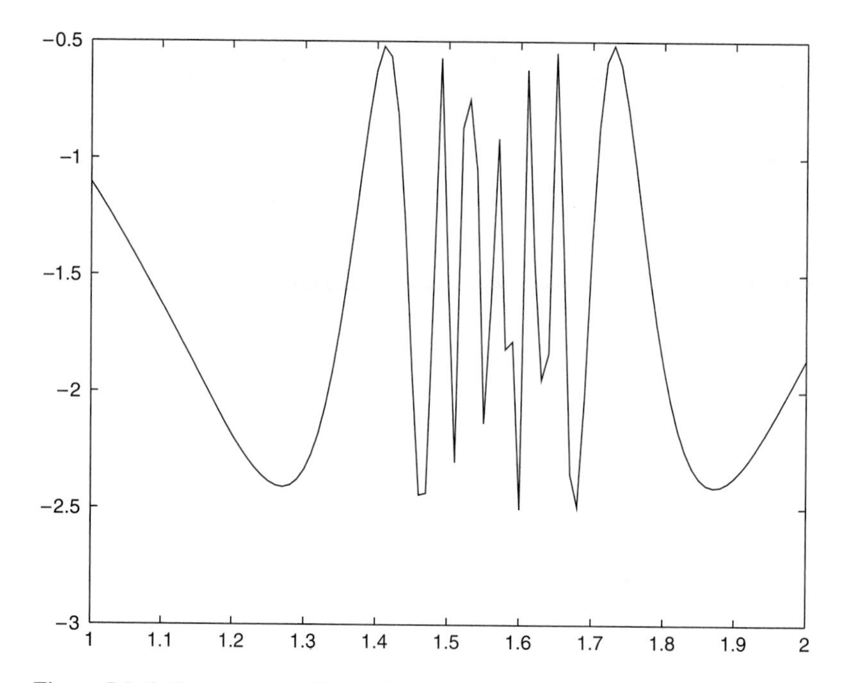

Figure 5.1–6 The function in Figure 5.1–5 generated with the `plot` command.

vectors `x` and `y`, but no plot is produced. The returned values can then be used for other purposes, such as plotting multiple curves, which is the topic of the next section.

Other commands can be used with the `fplot` command to enhance a plot's appearance, for example, the `title`, `xlabel`, and `ylabel` commands and the line type commands to be introduced in the next section.

Plotting Polynomials

We can plot polynomials more easily by using the `polyval` function, introduced in Chapter 2. This function evaluates the polynomial at specified values of the independent variable. It requires only the polynomial's coefficients and thus eliminates the need to type in the polynomial's expression. For example, to plot the polynomial $3x^5 + 2x^4 - 100x^3 + 2x^2 - 7x + 90$ over the range $-6 \le x \le 6$ with a spacing of 0.01, you type

```
>>x = [-6:0.01:6];
>>p = [3,2,-100,2,-7,90];
>>plot(x,polyval(p,x)),xlabel('x'),ylabel('p')
```

Table 5.1–1 summarizes the xy plotting commands introduced in this section.

Table 5.1–1 Basic xy plotting commands

Command	Description
`axis([xmin xmax ymin ymax])`	Sets the minimum and maximum limits of the x- and y-axes.
`fplot('string', [xmin xmax])`	Performs intelligent plotting of functions, where `'string'` is a text string that describes the function to be plotted and `[xmin xmax]` specifies the minimum and maximum values of the independent variable. The range of the dependent variable can also be specified. In this case the syntax is `fplot('string', [xmin xmax ymin ymax])`.
`grid`	Displays gridlines at the tick marks corresponding to the tick labels.
`plot(x,y)`	Generates a plot of the array `y` versus the array `x` on rectilinear axes.
`plot(y)`	Plots the values of `y` versus their indices if `y` is a vector. Plots the imaginary parts of `y` versus the real parts if `y` is a vector having complex values.
`print`	Prints the plot in the Figure window.
`title('text')`	Puts text in a title at the top of a plot.
`xlabel('text')`	Adds a text label to the x-axis (the abscissa).
`ylabel('text')`	Adds a text label to the y-axis (the ordinate).

Test Your Understanding

T5.1–1 Redo the plot of the equation $y = 0.4\sqrt{1.8x}$ shown in Figure 5.1–2 for $0 \le x \le 35$ and $0 \le y \le 3.5$.

T5.1–2 Use the `fplot` command to investigate the function $\tan(\cos x) - \sin(\tan x)$ for $0 \le x \le 2\pi$. How many values of x are needed to obtain the same plot using the `plot` command? (Answer: 292 values.)

T5.1–3 Plot the imaginary part versus the real part of the function $(0.2 + 0.8i)^n$ for $0 \le n \le 20$. Choose enough points to obtain a smooth curve. Label each `axis` and put a title on the plot. Use the `axis` command to change the tick-label spacing.

Saving Figures

When you create a plot, the Figure window appears (see Figure 5.1–7). This window has eight menus, which are discussed in detail in Section 5.4. The **File** menu is used for saving and printing the figure. You can save your figure in a format that can be opened during another MATLAB session or in a format that can be used by other applications.

To save a figure that can be opened in subsequent MATLAB sessions, save it in a figure file with the .fig file name extension. To do this, select **Save** from the Figure window **File** menu or click the **Save** button (the disk icon) on the toolbar. If this is the first time you are saving the file, the **Save As** dialog box appears. Make sure that the type is MATLAB Figure (*.fig). Specify the name you want assigned to the figure file. Click OK. You can also use the `saveas` command.

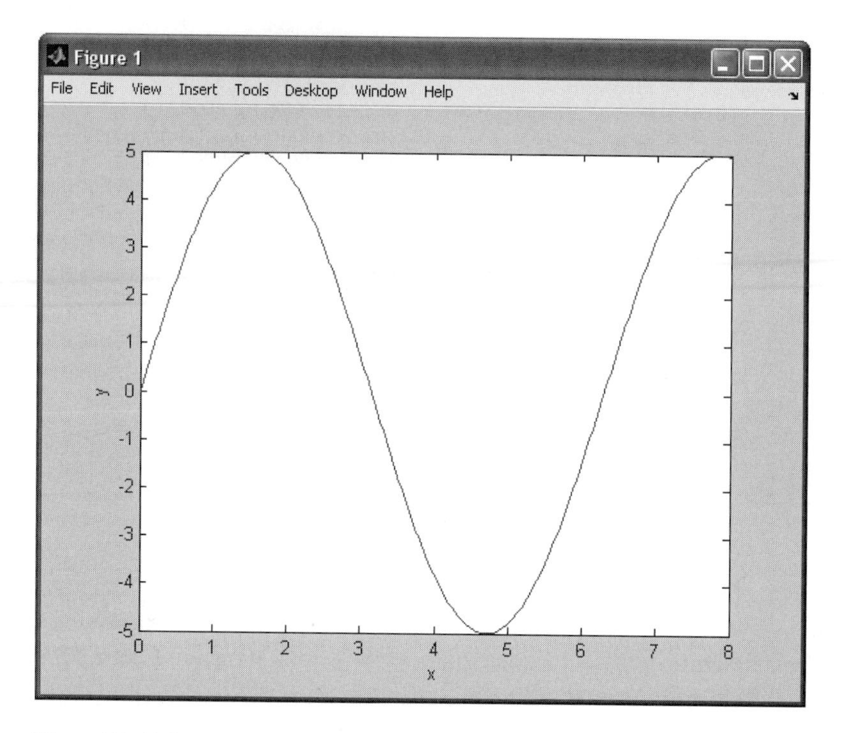

Figure 5.1–7 An example of a Figure window.

To open a figure file, select **Open** from the **File** menu or click the **Open** button (the opened folder icon) on the toolbar. Select the figure file you want to open and click OK. The figure file appears in a new figure window. You can also use the `open` command.

Exporting Figures

If you want to save the figure in a format that can be used by another application, such as the standard graphics file formats TIFF or EPS, perform these steps.

1. Select **Export Setup** from the **File** menu. This dialog provides options you can specify for the output file, such as the figure size, fonts, line size and style, and output format.
2. Select **Export** from the **Export Setup** dialog. A standard **Save As** dialog appears.
3. Select the format from the list of formats in the **Save As** type menu. This selects the format of the exported file and adds the standard file name extension given to files of that type.
4. Enter the name you want to give the file, less the extension.
5. Click **Save.**

You can also export the figure from the command line, by using the `print` command. See MATLAB help for more information about exporting figures in different formats.

On Windows systems, you can also copy a figure to the clipboard and then paste it into another application:

1. Select **Copy Options** from the **Edit** menu. The **Copying Options** page of the **Preferences** dialog box appears.
2. Complete the fields on the **Copying Options** page and click **OK.**
3. Select **Copy Figure** from the **Edit** menu.

The figure is copied to the Windows clipboard and can be pasted into another application.

MATLAB also enables you to save figures in formats compatible with PowerPoint and MSWord. See the MATLAB help for more information.

The graphics functions covered in this section and in Sections 5.2 and 5.3 are sufficient to create detailed, professional-looking plots in MATLAB. These functions can be placed in script files that can reused to create similar plots. This feature gives them an advantage over the interactive plotting tools that are discussed in Section 5.4

5.2 Subplots and Overlay Plots

MATLAB can create figures that contain an array of plots, called *subplots.* These are useful when you want to compare the same data plotted with different axis types, for example. The MATLAB `subplot` command creates such figures.

We frequently need to plot more than one curve or data set on a single plot. Such a plot is called an *overlay plot.* This section describes several MATLAB commands for creating overlay plots.

Subplots

You can use the `subplot` command to obtain several smaller "subplots" in the same figure. The syntax is `subplot(m,n,p)`. This command divides the Figure window into an array of rectangular panes with m rows and n columns. The variable `p` tells MATLAB to place the output of the `plot` command following the `subplot` command into the pth pane. For example, `subplot(3,2,5)` creates an array of six panes, three panes deep and two panes across, and directs the next plot to appear in the fifth pane (in the bottom-left corner). The following script file created Figure 5.2–1, which shows the plots of the functions $y = e^{-1.2x}\sin(10x+5)$ for $0 \le x \le 5$ and $y = |x^3 - 100|$ for $-6 \le x \le 6$.

```
x = [0:0.01:5];
y = exp(-1.2*x).*sin(10*x+5);
subplot(1,2,1)
plot(x,y),xlabel('x'),ylabel('y'),axis([0 5 -1 1])
x = [-6:0.01:6];
```

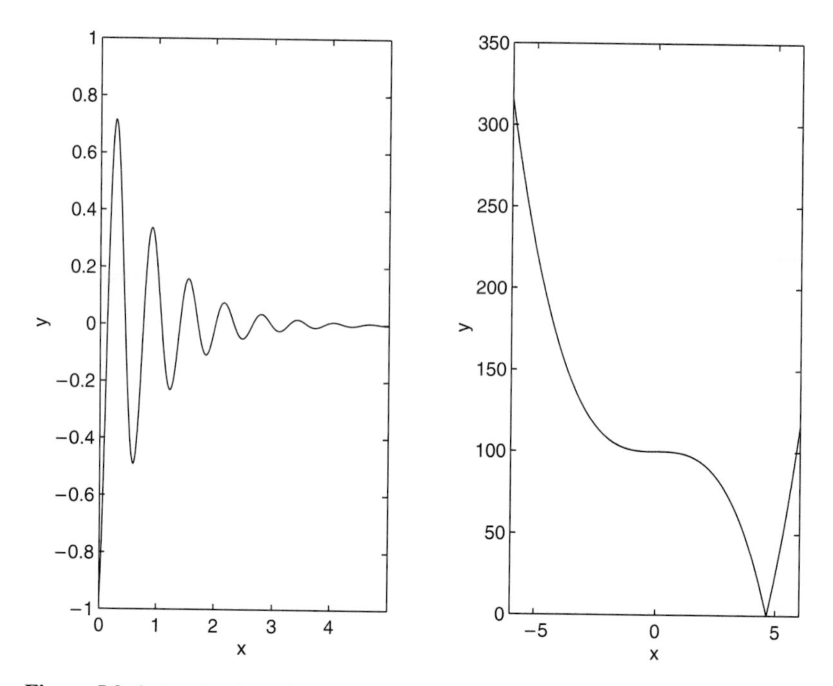

Figure 5.2–1 Application of the `subplot` command.

```
y = abs(x.^3-100);
subplot(1,2,2)
plot(x,y),xlabel('x'),ylabel('y'),axis([-6 6 0 350])
```

Test Your Understanding

T5.2–1 Pick a suitable spacing for t and v, and use the `subplot` command to plot the function $z = e^{-0.5t}\cos(20t - 6)$ for $0 \le t \le 8$ and the function $u = 6\log_{10}(v^2 + 20)$ for $-8 \le v \le 8$. Label each axis.

Overlay Plots

You can use the following variants of the MATLAB basic plotting functions `plot(x,y)` and `plot(y)` to create overlay plots:

- `plot(A)` plots the columns of `A` versus their indices and generates n curves where `A` is a matrix with m rows and n columns.
- `plot(x,A)` plots the matrix `A` versus the vector `x`, where `x` is either a row vector or column vector and `A` is a matrix with m rows and n columns. If the length of `x` is m, then each *column* of `A` is plotted versus the vector `x`. There will be as many curves as there are columns of `A`. If `x` has length n, then

each *row* of A is plotted versus the vector x. There will be as many curves as there are rows of A.

- plot(A,x) plots the vector x versus the matrix A. If the length of x is m, then x is plotted versus the *columns* of A. There will be as many curves as there are columns of A. If the length of x is n, then x is plotted versus the *rows* of A. There will be as many curves as there are rows of A.
- plot(A,B) plots the columns of the matrix B versus the columns of the matrix A.

Data Markers and Line Types

To plot the vector y versus the vector x and mark each point with a data marker, enclose the symbol for the marker in single quotes in the plot function. Table 5.2–1 shows the symbols for some of the available data markers. For example, to use a small circle, which is represented by the lowercase letter o, type plot(x,y,'o'). This notation results in a plot like the one on the left in Figure 5.2–2. To connect each data marker with a straight line, we must plot the data twice, by typing plot(x,y,x,y,'o'). See the plot on the right in Figure 5.2–2.

Suppose we have two curves or data sets stored in the vectors x, y, u, and v. To plot y versus x and v versus u on the same plot, type plot(x,y,u,v). Both sets will be plotted with a solid line, which is the default line style. To distinguish the sets, we can plot them with different line types. To plot y versus x with a solid line and u versus v with a dashed line, type plot(x,y,u,v,'--'), where the symbols '--' represent a dashed line. Table 5.2–1 gives the symbols for other line types. To plot y versus x with asterisks (*) connected with a dotted line, you must plot the data twice by typing plot(x,y,'*',x,y,':'). See Figure 5.2–3.

You can obtain symbols and lines of different colors by using the color symbols shown in Table 5.2–1. The color symbol can be combined with the data-marker symbol and the line-type symbol. For example, to plot y versus x with green asterisks (∗) connected with a red dashed line, you must plot the data twice by typing plot(x,y,'g*',x,y,'r--'). (Do not use colors if you are going to print the plot on a black-and-white printer.)

Table 5.2–1 Specifiers for data markers, line types, and colors

Data markers†		Line types		Colors	
Dot (·)	.	Solid line	-	Black	k
Asterisk (*)	*	Dashed line	- -	Blue	b
Cross (×)	×	Dash-dotted line	-.	Cyan	c
Circle (○)	o	Dotted line	:	Green	g
Plus sign (+)	+			Magenta	m
Square (□)	s			Red	r
Diamond (◇)	d			White	w
Five-pointed star (★)	p			Yellow	y

†Other data markers are available. Search for "markers" in MATLAB help.

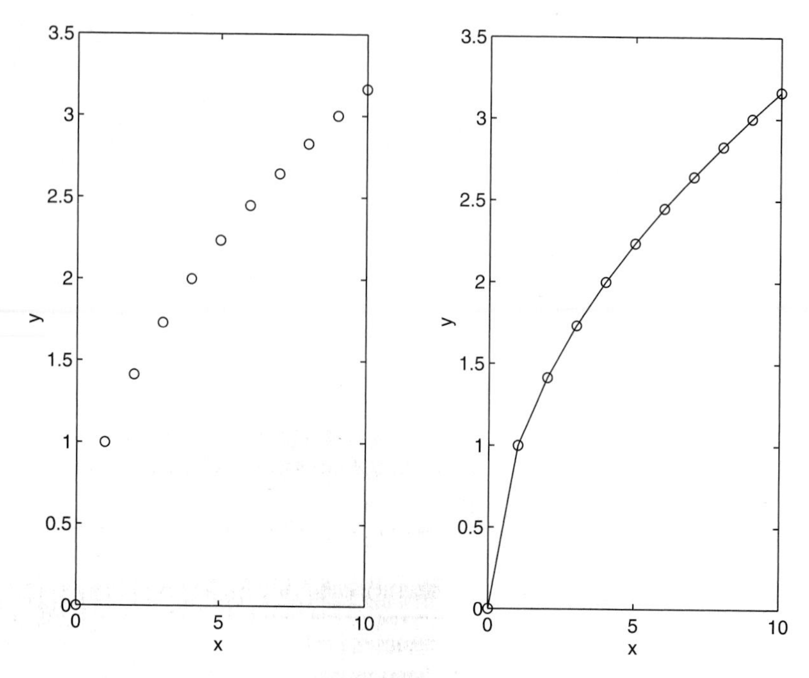

Figure 5.2–2 Use of data markers.

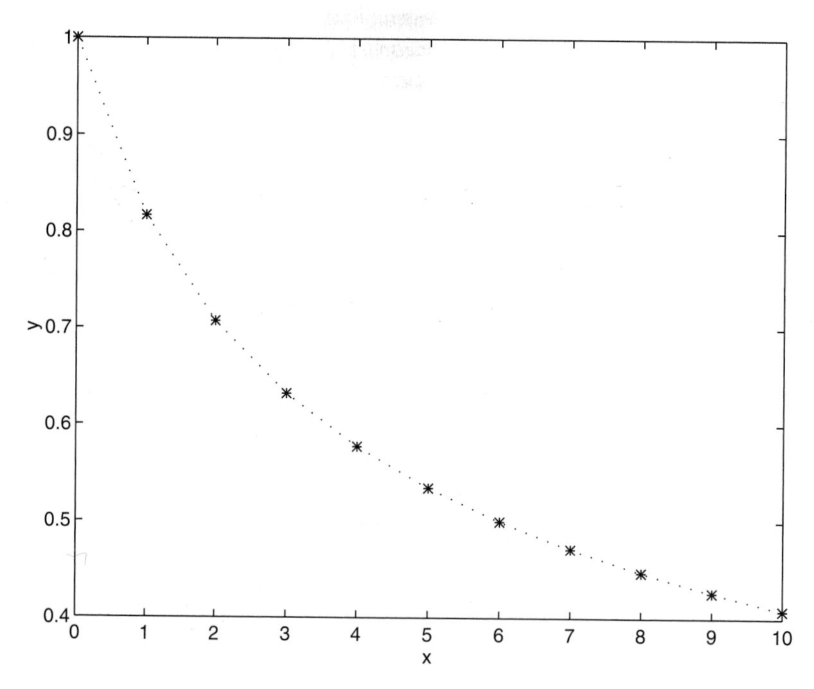

Figure 5.2–3 Data plotted using asterisks connected with a dotted line.

Labeling Curves and Data

When more than one curve or data set is plotted on a graph, we must distinguish between them. If we use different data symbols or different line types, then we must either provide a legend or place a label next to each curve. To create a legend, use the `legend` command. The basic form of this command is `legend('string1','string2')`, where `string1` and `string2` are text strings of your choice. The `legend` command automatically obtains from the plot the line type used for each data set and displays a sample of this line type in the legend box next to the string you selected. The following script file produced the plot in Figure 5.2–4.

```
x = [0:0.01:2];
y = sinh(x);
z = tanh(x);
plot(x,y,x,z,'--'),xlabel('x'), ...
ylabel('Hyperbolic Sine and Tangent'), ...
legend('sinh(x)','tanh(x)')
```

The `legend` command must be placed somewhere after the `plot` command. When the plot appears in the Figure window, use the mouse to position the legend box. (Hold down the left button on a two-button mouse to move the box.)

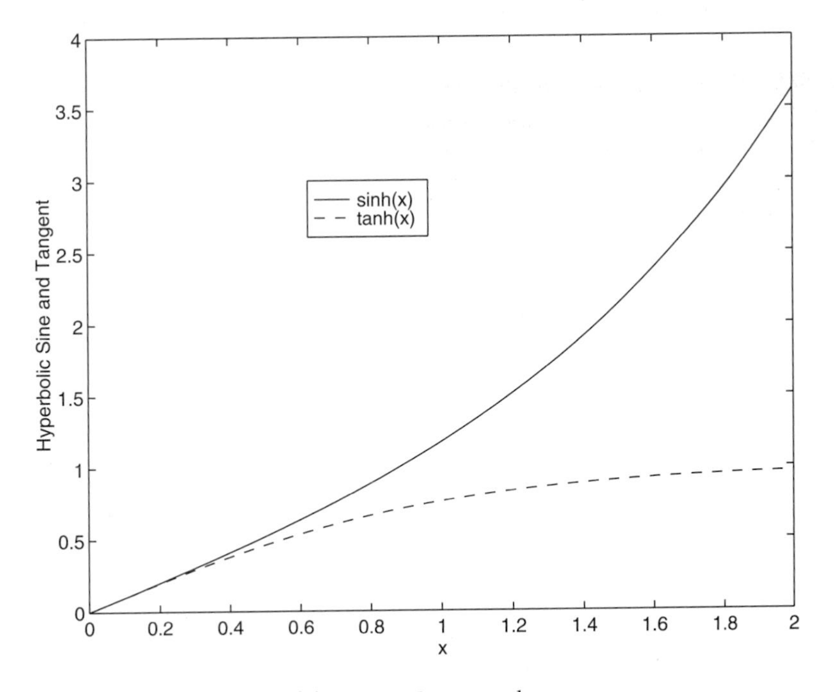

Figure 5.2–4 Application of the `legend` command.

Gridlines can obscure the legend box. To prevent this situation, instead of placing the `legend` command as shown in the preceding session, type the following lines in the Command window, after the plot appears in the Figure window but before printing the plot:

```
>>axes(legend('string1','string2'))
>>refresh
```

The first line makes the legend box act as the current set of drawing axes. The `refresh` command forces the plot to be redrawn in the Figure window. You can then print the plot. The `axes` command, not to be confused with the `axis` command, is a powerful command with many features for manipulating figures in MATLAB. However, this advanced topic is not covered in this text.

Another way to distinguish curves is to place a label next to each. The label can be generated with either the `gtext` command, which lets you place the label using the mouse, or with the `text` command, which requires you to specify the coordinates of the label. The syntax of the `gtext` command is `gtext('string')`, where `string` is a text string that specifies the label of your choice. When this command is executed, MATLAB waits for a mouse button or a key to be pressed while the mouse pointer is within the Figure window; the label is placed at that position of the mouse pointer. You may use more than one `gtext` command for a given plot.

The `text` command, `text(x,y,'string')`, adds a text string to the plot at the location specified by the coordinates `x,y`. These coordinates are in the same units as the plot's data. The following script file illustrates the uses of the `gtext` and `text` commands and was used to create the plot shown in Figure 5.2–5.

```
x = [0:0.01:1];
y = tan(x);
z = sec(x);
plot(x,y,x,z),xlabel('x'), ...
ylabel('Tangent and Secant'),gtext('tan(x)'), ...
text(0.3,1.2,'sec(x)')
```

Of course, finding the proper coordinates to use with the `text` command usually requires some trial and error.

Graphical Solution of Equations

When we need to solve two equations in two unknown variables, we can plot the equations. The solution corresponds to the intersection of the two lines. If they do not intersect, there is no solution. If they intersect more than once, there are multiple solutions. A limitation of this approach is that we must know the approximate ranges of the two variables so that we can generate the plot. Another limitation is that the accuracy of the solution is limited by the accuracy with which we can read the plot. Of course, we can always expand the plot to increase the accuracy.

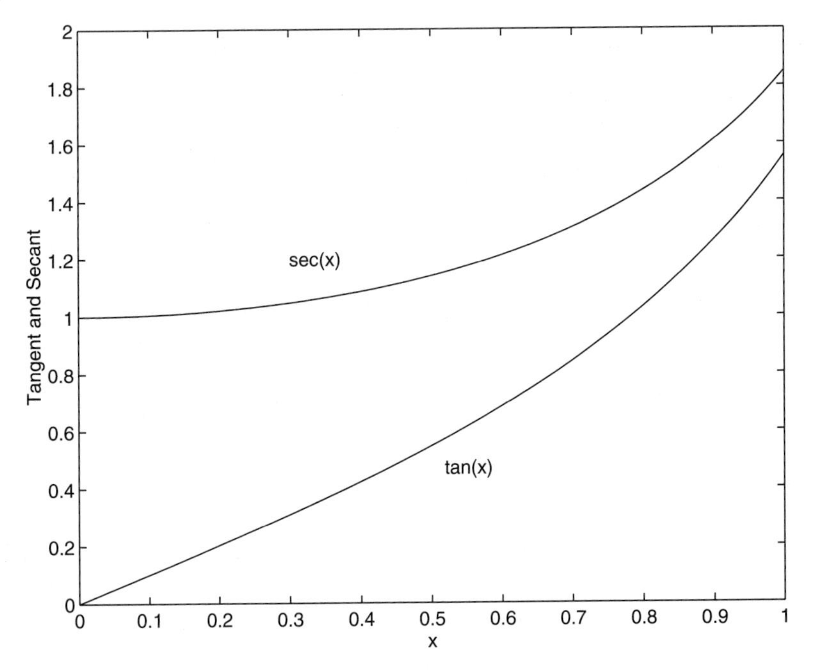

Figure 5.2–5 Application of the `gtext` and `text` commands.

Load-Line Analysis of Electrical Circuits

EXAMPLE 5.2–1

Figure 5.2–6 is a representation of an electrical system with a power supply and a load. The power supply produces the fixed voltage v_1 and supplies the current i_1 required by the load, whose voltage drop is v_2. The current-voltage relationship for a specific load is found from experiments to be

$$i_1 = 0.16(e^{0.12v_2} - 1) \tag{5.2–1}$$

Suppose that the supply resistance is $R_1 = 30\ \Omega$ and the supply voltage is $v_1 = 15$ V. To select or design an adequate power supply, we need to determine how much current will be drawn from the power supply when this load is attached. Find the voltage drop v_2 as well.

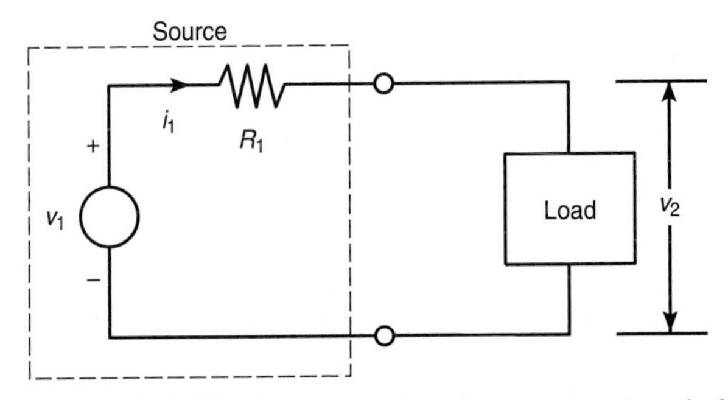

Figure 5.2–6 Circuit representation of a power supply and a load.

■ Solution

Using Kirchhoff's voltage law, we obtain

$$v_1 - i_1 R_1 - v_2 = 0$$

Solve this for i_1.

$$i_1 = -\frac{1}{R_1} v_2 + \frac{v_1}{R_1} = -\frac{1}{30} v_2 + \frac{15}{30} \tag{5.2–2}$$

The plot of this equation is a straight line called the *load line*. The load line is so named because it shows how the current drawn by the load changes as the load's voltage changes. To find i_1 and v_2, we need to solve equations (5.2–1) and (5.2–2). Because of the term $e^{0.12 v_2}$, it is not possible to obtain a solution using algebra. However, we can plot the curves corresponding to these equations and find their intersection. The MATLAB script file to do so follows, and Figure 5.2–7 shows the resulting plot.

```
v_2=[0:0.01:20];
i_11=.16*(exp(0.12*v_2)-1);
i_12=-(1/30)*v_2+0.5;
plot(v_2,i_11,v_2,i_12),grid,xlabel('v_2 (volts)'),...
ylabel('i_1 (amperes)'),axis([0 20 0 1]),...
gtext('Load Line'),gtext('Device Curve')
```

From the figure we can see that the curves intersect at approximately $i_1 = 0.25$ A, $v_2 = 7.5$ V. For a more accurate answer, change the `axis` statement to `axis ([7 8 0.2 0.3])` and obtain a new plot.

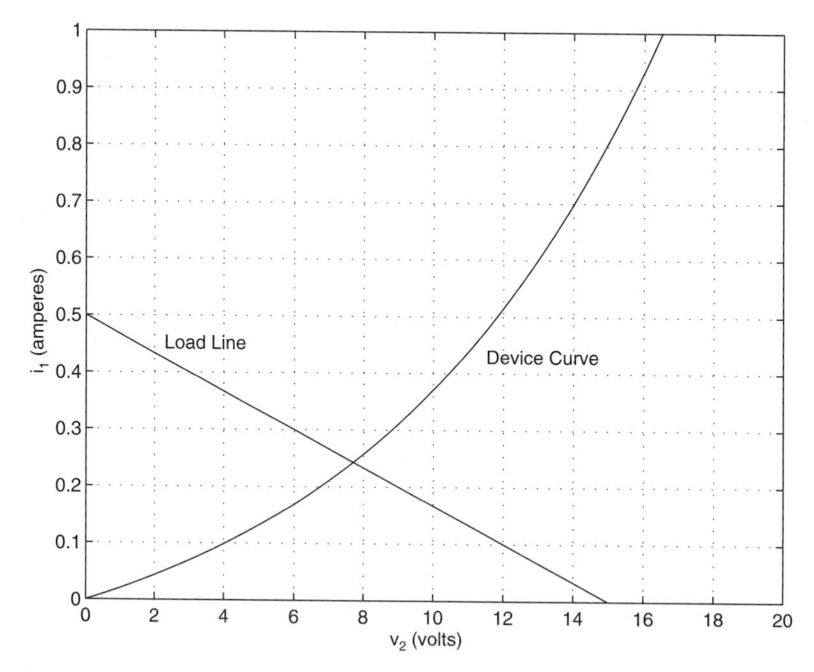

Figure 5.2–7 Plot of the load line and the device curve for Example 5.2–1.

The `hold` Command

The `hold` command creates a plot that needs two or more `plot` commands. Suppose we wanted to plot $y_2 = 4 + e^{-x} \cos 6x$ versus $y_1 = 3 + e^{-x} \sin 6x$, $-1 \le x \le 1$ on the same plot with $z = (0.1 + 0.9i)^n$, where $0 \le n \le 10$. This plot requires two plot commands. The script file to create this plot using the `hold` command follows.

```
x = [-1:0.01:1];
y1 = 3+exp(-x).*sin(6*x);
y2 = 4+exp(-x).*cos(6*x);
plot((0.1+0.9i).^[0:0.01:10]),hold,plot(y1,y2), ...
gtext('y2 versus y1'),gtext('Imag(z) versus Real(z)')
```

Figure 5.2–8 shows the result.

Although it is not needed to generate multiple plots with the `plot(x,y,u,v)` type command, the `hold` command is especially useful with some of the advanced MATLAB toolbox commands that generate specialized plots. Some of these commands do not allow for more than one curve to be generated at a time, and so they must be executed more than once to generate multiple curves. The `hold` command is used to do this.

When more than one `plot` command is used, do not place any of the `gtext` commands before any `plot` command. Because the scaling changes as each `plot` command is executed, the label placed by the `gtext` command might

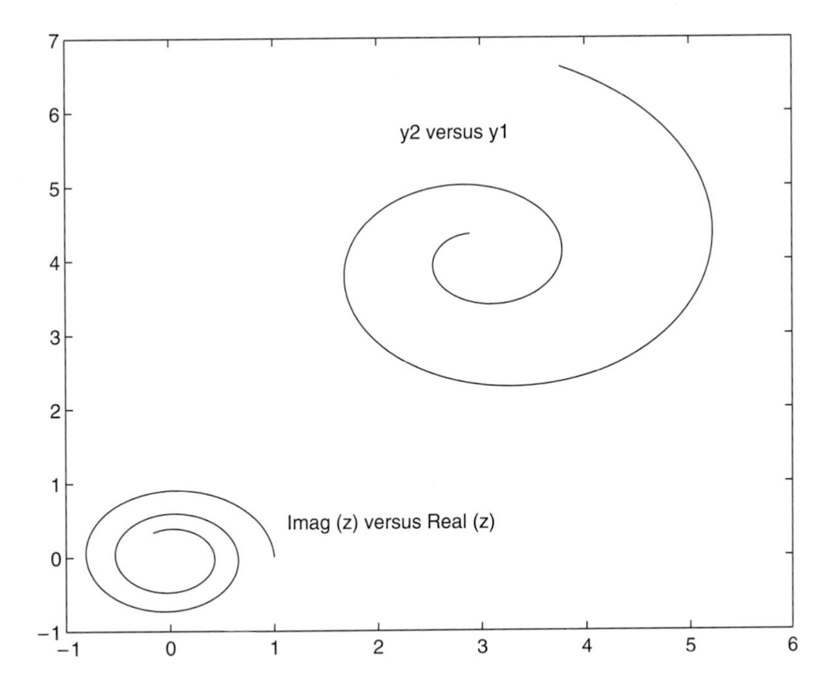

Figure 5.2–8 Application of the `hold` command.

Table 5.2–2 Plot enhancement commands

Command	Description
`axes`	Creates axes objects.
`gtext('text')`	Places the string `text` in the Figure window at a point specified by the mouse.
`hold`	Freezes the current plot for subsequent graphics commands.
`legend('leg1','leg2', ...)`	Creates a legend using the strings `leg1, leg2`, and so on and specifies its placement with the mouse.
`plot(x,y,u,v)`	Plots, on rectilinear axes, four arrays: `y` versus `x` and `v` versus `u`.
`plot(x,y,'type')`	Plots the array `y` versus the array `x` on rectilinear axes, using the line type, data marker, and colors specified in the string `type`. See Table 5.2–1.
`plot(A)`	Plots the columns of the $m \times n$ array `A` versus their indices and generates n curves.
`plot(P,Q)`	Plots array `Q` versus array `P`. See the text for a description of the possible variants involving vectors and/or matrices: `plot(x,A), plot(A,x),` and `plot(A,B).`
`refresh`	Redraws the current Figure window.
`subplot(m,n,p)`	Splits the Figure window into an array of subwindows with m rows and n columns and directs the subsequent plotting commands to the pth subwindow.
`text(x,y,'text')`	Places the string `text` in the Figure window at a point specified by coordinates `x, y`.

end up in the wrong position. Table 5.2–2 summarizes the plot enhancement introduced in this section.

Test Your Understanding

T5.2–2 Plot the following two data sets on the same plot. For each set, $x = 0, 1, 2, 3, 4, 5$. Use a different data marker for each set. Connect the markers for the first set with solid lines. Connect the markers for the second set with dashed lines. Use a legend, and label the plot axes appropriately. The first set is $y = 11, 13, 8, 7, 5, 9$. The second set is $y = 2, 4, 5, 3, 2, 4$.

T5.2–3 Plot $y = \cosh x$ and $y = 0.5e^x$ on the same plot for $0 \le x \le 2$. Use different line types and a legend to distinguish the curves. Label the plot axes appropriately.

T5.2–4 Plot $y = \sinh x$ and $y = 0.5e^x$ on the same plot for $0 \le x \le 2$. Use a solid line type for each, the `gtext` command to label the $\sinh x$ curve, and the `text` command to label the $0.5e^x$ curve. Label the plot axes appropriately.

T5.2–5 Use the `hold` command and the `plot` command twice to plot $y = \sin x$ and $y = x - x^3/3$ on the same plot for $0 \le x \le 1$. Use a solid line type for each and use the `gtext` command to label each curve. Label the plot axes appropriately.

Annotating Plots

You can create text, titles, and labels that contain mathematical symbols, Greek letters, and other effects such as italics. The features are based on the TEX typesetting language. Here we give a summary of these features. For more information, including a list of the available characters, search the online help for "text properties."

The text, gtext, title, xlabel, and ylabel commands all require a string as their argument. For example, typing

```
>>title('A*exp(-t/tau)sin(omega t)')
```

produces a title that looks like A*exp(-t/tau)sin(omega t) but is supposed to represent the function $Ae^{-t/\tau}\sin(\omega t)$. You can create a title that looks like the mathematical function by typing

```
>>title('Ae^{- t/\tau}sin(\omega t)')
```

The backslash character \ precedes all TEX character sequences. Thus the strings \tau and \omega represent the Greek letters τ and ω. Superscripts are created by typing ^; subscripts are created by typing _. To set multiple characters as superscripts or subscripts, enclose them in braces. For example, type x_{13} to produce x_{13}.

In mathematical text variables are usually set in italic, and functions, like sin, are set in roman type. To set a character, say, x, in italic using the TEX commands, you type {\it x}. To set the title function using these conventions, you would type

```
>>title('{\it Ae}^{-{\it t/\tau}}\sin({\it \omega t})')
```

Hints for Improving Plots

The following actions, while not required, can nevertheless improve the appearance of your plots:

1. Start scales from zero whenever possible. This technique prevents a false impression of the magnitudes of any variations shown on the plot.
2. Use sensible tick-mark spacing. For example, if the quantities are months, choose a spacing of 12 because 1/10 of a year is not a convenient division. Space tick marks as close as is useful, but no closer. For example, if the data is given monthly over a range of 24 months, 48 tick marks would be too dense, and also unnecessary.
3. Minimize the number of zeros in the data being plotted. For example, use a scale in millions of dollars when appropriate, instead of a scale in dollars with six zeros after every number.
4. Determine the minimum and maximum data values for each axis before plotting the data. Then set the axis limits to cover the entire data range plus an additional amount to allow convenient tick-mark spacing to be selected.

For example, if the data on the x-axis ranges from 1.2 to 9.6, a good choice for axis limits is 0 to 10. This choice allows you to use a tick spacing of 1 or 2.

5. Use a different line type for each curve when several are plotted on a single plot and they cross each other; for example, use a solid line, a dashed line, and combinations of lines and symbols. Beware of using colors to distinguish plots if you are going to make black-and-white printouts and photocopies.
6. Do not put many curves on one plot, particularly if they will be close to each other or cross one another at several points.
7. Use the same scale limits and tick spacing on each plot if you need to compare information on more than one plot.

5.3 Special Plot Types

In this section we show how to obtain logarithmic axes; how to change the default tick-mark spacing and labels; and how to produce other specialized plots.

Logarithmic Plots

Thus far we have used only rectilinear scales. However, *logarithmic* scales are also widely used. (We often refer to them with the shorter term, *log* scale.) Two common reasons for choosing a log scale are (1) to represent a data set that covers a wide range of values and (2) to identify certain trends in data. As you will see, certain types of functional relationships appear as straight lines when plotted using a log scale. This method makes it easier to identify the function. A *log-log* plot has log scales on both axes. A *semilog* plot has a log scale on only one axis.

For example, Figures 5.3–1 and 5.3–2 show plots of the function:

$$y = \sqrt{\frac{100(1 - 0.01x^2)^2 + 0.02x^2}{(1 - x^2)^2 + 0.1x^2}}$$

The first plot uses rectilinear scales, and the second is a log-log plot. Because of the wide range in values on both the abscissa and ordinate, rectilinear scales do not reveal the important features.

It is important to remember the following points when using log scales:

1. You cannot plot negative numbers on a log scale, because the logarithm of a negative number is not defined as a real number.
2. You cannot plot the number 0 on a log scale, because $\log_{10} 0 = \ln 0 = -\infty$. You must choose an appropriately small number as the lower limit on the plot.

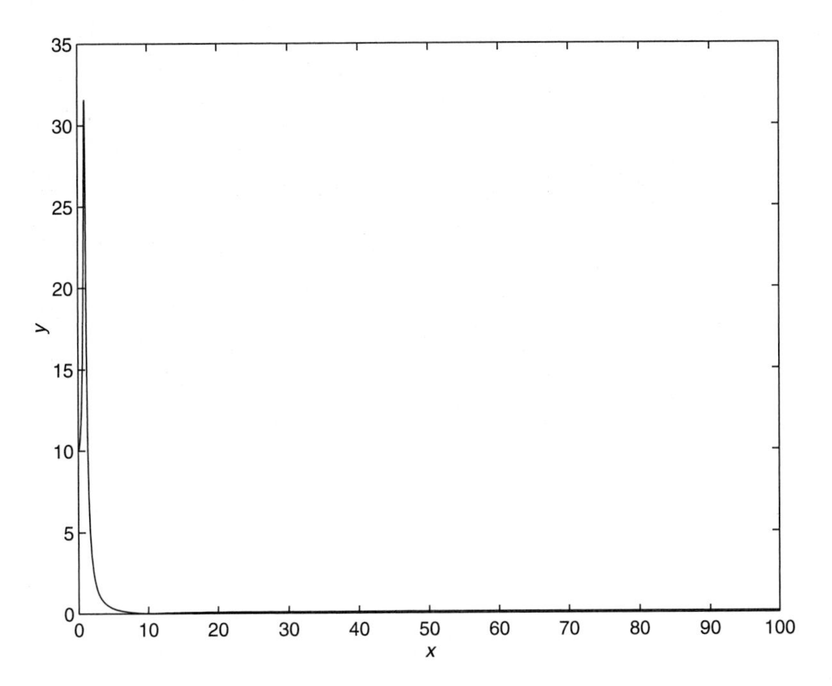

Figure 5.3–1 Rectilinear scales cannot properly display variations over wide ranges.

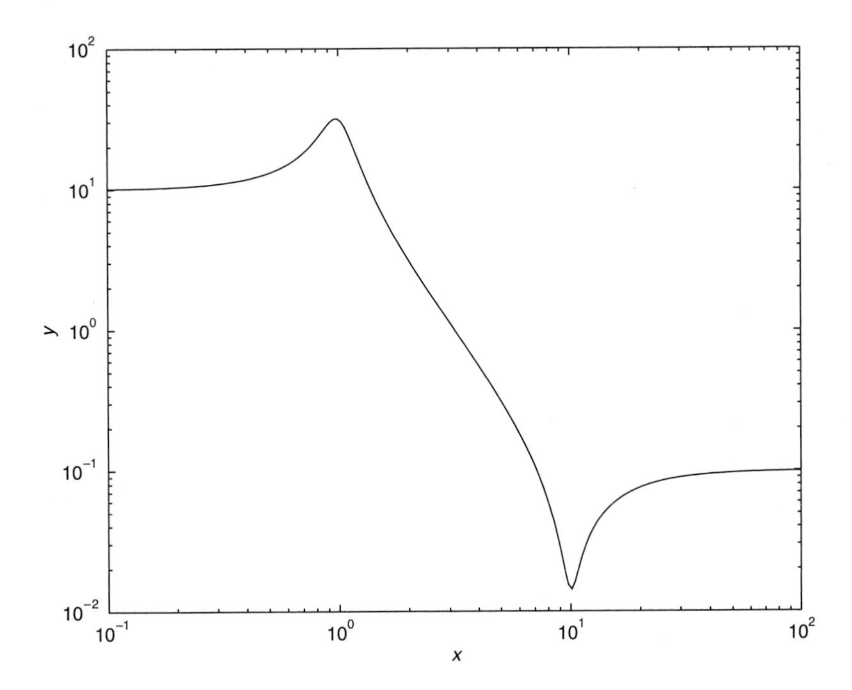

Figure 5.3–2 A log-log plot can display wide variations in data values.

3. The tick-mark labels on a log scale are the actual values being plotted; they are not the logarithms of the numbers. For example, the range of x values in the plot in Figure 5.3–2 is from $10^{-1} = 0.1$ to $10^2 = 100$.
4. Equal distances on a log scale correspond to multiplication by the same constant (as opposed to addition of the same constant on a rectilinear scale). For example, all numbers that differ by a factor of 10 are separated by the same distance on a log scale. That is, the distance between 0.3 and 3 is the same as the distance between 30 and 300. This separation is referred to as a *decade* or *cycle*. The plot shown in Figure 5.3–2 covers three decades in x (from 0.1 to 100) and four decades in y and is thus called a *four-by-three-cycle plot.*
5. Gridlines and tick marks within a decade are unevenly spaced. If 8 gridlines or tick marks occur within the decade, they correspond to values equal to 2, 3, 4, . . . , 8, 9 times the value represented by the first gridline or tick mark of the decade.

MATLAB has three commands for generating plots having log scales. The appropriate command depends on which axis must have a log scale. Follow these rules:

1. Use the `loglog(x,y)` command to have both scales logarithmic.
2. Use the `semilogx(x,y)` command to have the x scale logarithmic and the y scale rectilinear.
3. Use the `semilogy(x,y)` command to have the y scale logarithmic and the x scale rectilinear.

Table 5.3–1 summarizes these functions. For other 2D plot types, type `help specgraph`.

We can plot multiple curves with these commands just as with the `plot` command. In addition, we can use the other commands, such as `grid`, `xlabel`, and `axis`, in the same manner.

Table 5.3–1 Specialized plot commands

Command	Description
`bar(x,y)`	Creates a bar chart of `y` versus `x`.
`loglog(x,y)`	Produces a log-log plot of `y` versus `x`.
`plotyy(x1,y1,x2,y2)`	Produces a plot with two y-axes, `y1` on the left and `y2` on the right.
`polar(theta,r,'type')`	Produces a polar plot from the polar coordinates `theta` and `r`, using the line type, data marker, and colors specified in the string `type`.
`semilogx(x,y)`	Produces a semilog plot of `y` versus `x` with logarithmic abscissa scale.
`semilogy(x,y)`	Produces a semilog plot of `y` versus `x` with logarithmic ordinate scale.
`stairs(x,y)`	Produces a stairs plot of `y` versus `x`.
`stem(x,y)`	Produces a stem plot of `y` versus `x`.

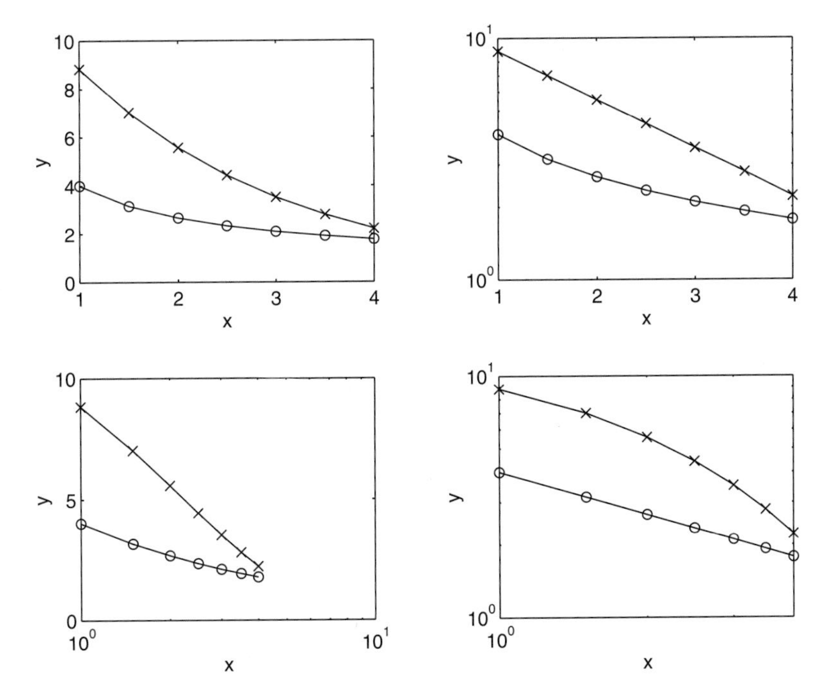

Figure 5.3–3 Two data sets plotted on four types of plots.

Figure 5.3–3 shows plots made with the `plot` command and the three logarithmic plot commands. The same two data sets were used for each plot. The session follows.

```
>>x = [1,1.5,2,2.5,3,3.5,4];
>>y1 = [4,3.16,2.67,2.34,2.1,1.92,1.78];
>>y2 = [8.83,7.02,5.57,4.43,3.52,2.8,2.22];
>>subplot(2,2,1)
>>plot(x,y1,x,y1,'o',x,y2,x,y2,'x'),xlabel('x'),ylabel('y') ...
axis([1 4 0 10])
>>subplot(2,2,2)
>>semilogy(x,y1,x,y1,'o',x,y2,x,y2,'x'),xlabel('x'),ylabel('y')
>>subplot(2,2,3)
>>semilogx(x,y1,x,y1,'o',x,y2,x,y2,'x'),xlabel('x'),ylabel('y')
>>subplot(2,2,4)
>>loglog(x,y1,x,y1,'o',x,y2,x,y2,'x'),xlabel('x'),ylabel('y'), ...
axis([1 4 1 10])
```

Note that the first data set lies close to a straight line only when plotted with both scales logarithmic, and the second data set nearly forms a straight line only on the semilog plot where the vertical axis is logarithmic. In Section 5.5 we explain how to use these observations to derive a mathematical model for the data.

Frequency-Response Plots and Filter Circuits

Many electrical applications use specialized circuits called *filters* to remove signals having certain frequencies. Filters work by responding only to signals that have the desired frequencies. These signals are said to "pass through" the circuit. The signals that do not pass through are said to be "filtered out." For example, a particular circuit in a radio is designed to respond only to signals having the broadcast frequency of the desired radio station. Other circuits, such as those constituting the graphic equalizer, enable the user to select certain musical frequencies such as bass or treble to be passed through to the speakers.

The mathematics required to design filter circuits is covered in upper-level engineering courses. However, a simple plot often describes the characteristics of filter circuits. Such a plot, called a *frequency-response plot,* is often provided when you buy a speaker-amplifier system.

EXAMPLE 5.3–1 Frequency-Response Plot of a Low-Pass Filter

The circuit shown in Figure 5.3–4 consists of a resistor and a capacitor and is thus called an RC circuit. If we apply a sinusoidal voltage v_i, called the input voltage, to the circuit as shown, then eventually the output voltage v_o will be sinusoidal also, with the same frequency but with a different amplitude and shifted in time relative to the input voltage. Specifically, if $v_i = A_i \sin \omega t$, then $v_o = A_o \sin(\omega t + \phi)$. The frequency-response plot is a plot of A_o/A_i versus frequency ω. It is usually plotted on logarithmic axes. Upper-level engineering courses explain that for the RC circuit shown, this ratio depends on ω and RC as follows:

$$\frac{A_o}{A_i} = \left| \frac{1}{RCs + 1} \right| \tag{5.3–1}$$

where $s = \omega i$. For $RC = 0.1$ second, obtain the log-log plot of $|A_o/A_i|$ versus ω and use it to find the range of frequencies for which the output amplitude A_o is less than 70 percent of the input amplitude A_i.

■ Solution

As with many graphical procedures, you must guess a range for the parameters in question. Here we must guess a range to use for the frequency ω. If we use $1 \le \omega \le 100$ rad/s, we

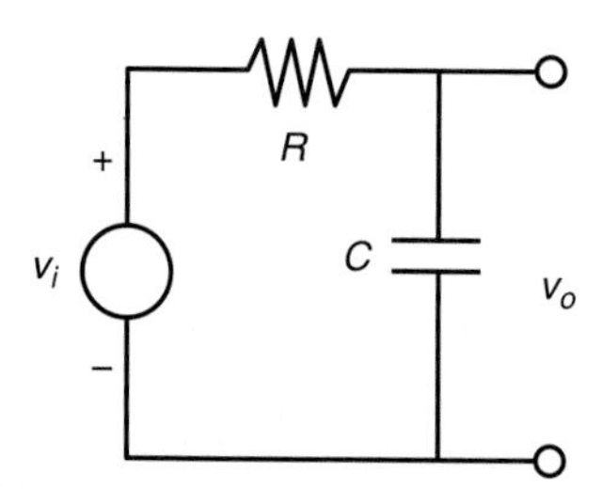

Figure 5.3–4 An RC circuit.

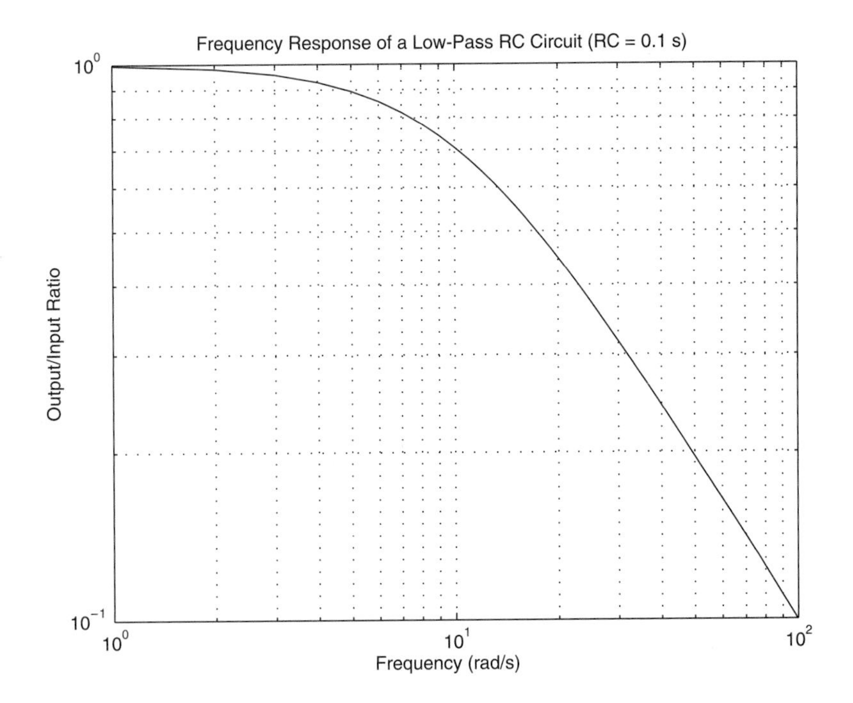

Figure 5.3–5 Frequency-response plot of a low-pass RC circuit.

will see the part of the curve that is of interest. The MATLAB script file is as follows:

```
RC = 0.1;
s = [1:100]*i;
M = abs(1./(RC*s+1));
loglog(imag(s),M),grid,xlabel('Frequency(rad/s)'),...
ylabel('Output/Input Ratio'),...
title('Frequency Response of a Low-Pass RC Circuit (RC = 0.1 s)')
```

Figure 5.3–5 shows the plot. We can see that the output/input ratio A_o/A_i decreases as the frequency ω increases. The ratio is approximately 0.7 at $\omega = 10$ rad/s. The amplitude of any input signal having a frequency greater than this frequency will decrease by at least 30 percent. Thus this circuit is called a *low-pass* filter because it passes low-frequency signals better than it passes high-frequency signals. Such a circuit is often used to filter out noise from nearby electrical machinery.

Controlling Tick-Mark Spacing and Labels

The MATLAB `set` command is a powerful command for changing the properties of MATLAB "objects," such as plots. We will not cover this command in depth, but will show how to use it to specify the spacing and labels of the tick marks. To explore this command further, type `help set` and `help axes`. Many of the properties that affect the appearance of plot axes are described under the `axes` command, which should not be confused with the `axis` command.

Up to now we changed the tick-mark spacing by using the `axis` command and hoped that the MATLAB autoscaling feature chose a proper tick-mark spacing. We can also use the following command to specify this spacing.

```
set(gca,'XTick',[xmin:dx:xmax],'YTick',[ymin:dy:ymax])
```

Here `xmin` and `xmax` are the x values that specify the placement of the first and the last tick marks on the x-axis, and `dx` specifies the spacing between tick marks. You would normally use the same values for `xmin` and `xmax` in both the `set` and `axis` commands. Similar definitions apply to the y-axis values `ymin`, `ymax`, and `dy`. The term `gca` stands for "get current axes." It tells MATLAB to apply the new values to the axes currently used for plotting. For example, to plot $y = 0.25x^2$ for $0 \le x \le 2$, with tick marks spaced at intervals of 0.2 on the x-axis and 0.1 on the y-axis, you would type:

```
>>x = [0:0.01:2];
>>y =0.25*x.^2;
>>plot(x,y),set(gca,'XTick',[0:0.2:2],'YTick',[0:0.1:1]), ...
xlabel('x'),ylabel('y')
```

You can also use the Plot Editor to change the tick spacing. This is discussed in Section 5.4.

The `set` command can also be used to change the tick-mark labels, for example, from numbers to text. Suppose we sell printers, and we want to plot the monthly sales in thousands of dollars from January to June. We can use the `set` command to label the x-axis with the names of the months, as shown in the following session. The vector `x` contains the number of the month, and the vector `y` contains the monthly sales in thousands of dollars.

```
>>x = [1:6];
>>y = [13,5,7,14,10,12];
>>plot(x,y,'o',x,y), ...
set(gca,'XTicklabel',['Jan';'Feb';'Mar';'Apr';'May';'Jun']),...
set(gca,'XTick',[1:6]),axis([1 6 0 15]),xlabel('Month'), ...
ylabel('Monthly Sales ($1000)'), ...
title('Printer Sales for January to June, 1997')
```

The plot appears in Figure 5.3–6. You can also use the Plot Editor to change labels.

Note the labels in the `set` command must be enclosed in single quotes and are specified as a column vector; thus they are separated by semicolons. Another requirement is that all the labels must have the same number of characters (here, three characters). Table 5.3–2 summarizes the `set` command.

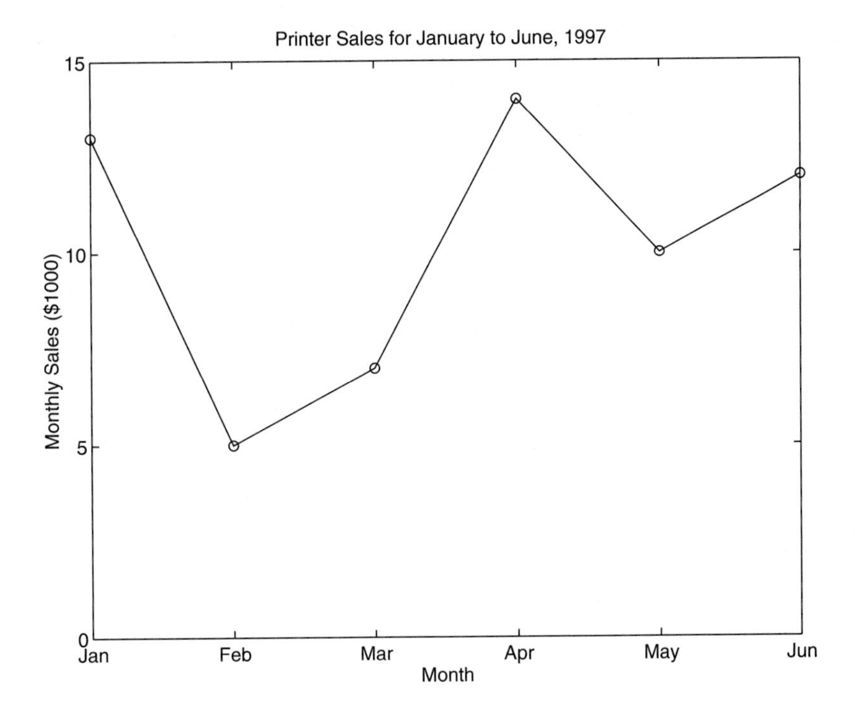

Figure 5.3–6 An example of controlling the tick-mark labels with the `set` command.

Table 5.3–2 The `set` command

The `set` command specifies properties of objects such as axes. For example,

```
set(gca,'XTick',[xmin:dx:xmax],'YTick',[ymin:dy:ymax])
```

specifies the axis limits `xmin`, `xmax`, `ymin`, and `ymax` and the tick spacing `dx` and `dy`. The command

```
set(gca,'XTicklabel',['text'])
```

specifies the tick labels on the x-axis, where the string `text` is a column vector that specifies the tick labels. Each label must be enclosed in single quotes, and all labels must have the same number of characters. For more information, type `help axes`.

Stem, Stairs, and Bar Plots

MATLAB has several other plot types that are related to xy plots. These include the stem, stairs, and bar plots. Their syntax is very simple; namely, `stem(x,y)`, `stairs(x,y)`, and `bar(x,y)`. See Table 5.3–1.

Separate *y*-Axes

The `plotyy` function generates a graph with two y-axes. The syntax `plotyy(x1,y1,x2,y2)` plots `y1` versus `x1` with y-axis labeling on the left, and plots `y2` versus `x2` with y-axis labeling on the right. The syntax `plotyy(x1,y1,x2,y2,'type1','type2')` generates a `'type1'` plot of `y1` versus `x1` with y-axis labeling on the left, and generates a `'type2'` plot of `y2` versus `x2` with y-axis labeling on the right. For example, `plotyy(x1,y1,x2,y2, 'plot','stem')` uses `plot(x1,y1)` to generate a plot for the left axis, and `stem(x2,y2)` to generate a plot for the right axis.

Polar Plots

Polar plots are two-dimensional plots made using polar coordinates. If the polar coordinates are (θ, r), where θ is the angular coordinate and r is the radial coordinate of a point, then the command `polar(theta,r)` will produce the polar plot. A grid is automatically overlaid on a polar plot. This grid consists of concentric circles and radial lines every 30°. The `title` and `gtext` commands can be used to place a title and text. The variant command `polar(theta,r,'type')` can be used to specify the line type or data marker, just as with the `plot` command.

EXAMPLE 5.3–2

Plotting Orbits

The equation

$$r = \frac{p}{1 - \epsilon \cos \theta}$$

describes the polar coordinates of an orbit measured from one of the orbit's two focal points. For objects in orbit around the sun, the sun is at one of the focal points. Thus r is the distance of the object from the sun. The parameters p and ϵ determine the size of the orbit and its eccentricity, respectively. A circular orbit has an eccentricity of 0; if $0 < \epsilon < 1$, the orbit is elliptical; and if $\epsilon > 1$, the orbit is hyperbolic. Obtain the polar plot that represents an orbit having $\epsilon = 0.5$ and $p = 2$ AU (AU stands for "astronomical unit"; 1 AU is the mean distance from the sun to Earth). How far away does the orbiting object get from the sun? How close does it approach Earth's orbit?

■ Solution

Figure 5.3–7 shows the polar plot of the orbit. The plot was generated by the following session.

```
>>theta = [0:pi/90:2*pi];
>>r = 2./(1-0.5*cos(theta));
>>polar(theta,r),title('Orbital Eccentricity = 0.5')
```

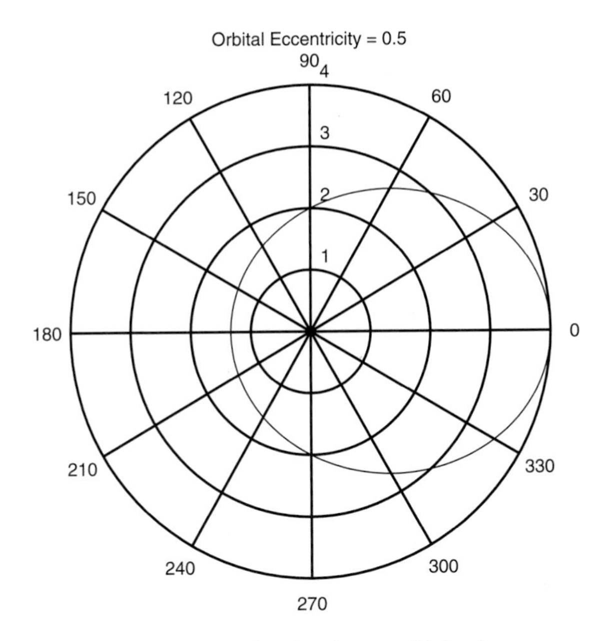

Figure 5.3–7 A polar plot showing an orbit having an eccentricity of 0.5.

The sun is at the origin, and the plot's concentric circular grid enables us to determine that the closest and farthest distances the object is from the sun are approximately 1.3 and 4 AU. Earth's orbit, which is nearly circular, is represented by the innermost circle. Thus the closest the object gets to Earth's orbit is approximately 0.3 AU. The radial gridlines allow us to determine that when $\theta = 90^\circ$ and 270°, the object is 2 AU from the sun.

Test Your Understanding

T5.3–1 Obtain the plots shown in Figure 5.3–8. The power function is $y = 2x^{-0.5}$, and the exponential function is $y = 10^{1-x}$.

T5.3–2 Plot the function $y = 8x^3$ for $-1 \leq x \leq 1$ with a tick spacing of 0.25 on the x-axis and 2 on the y-axis.

T5.3–3 The *spiral of Archimedes* is described by the polar coordinates (θ, r), where $r = a\theta$. Obtain a polar plot of this spiral for $0 \leq \theta \leq 4\pi$, with the parameter $a = 2$.

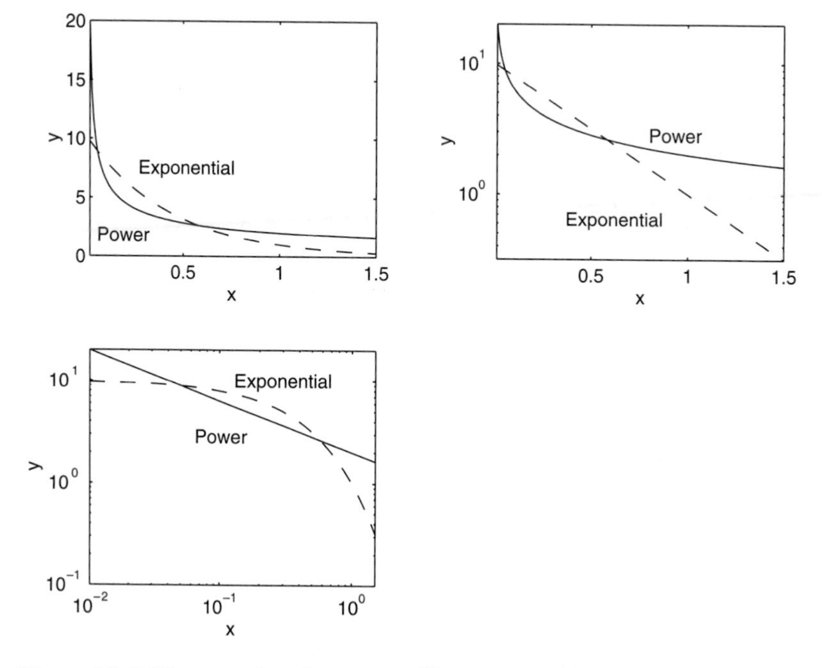

Figure 5.3–8 The power function $y = 2x^{-0.5}$ and the exponential function $y = 10^{1-x}$.

5.4 Interactive Plotting in MATLAB

This is an optional section that may be omitted without affecting your understanding of the material in subsequent sections and chapters. The graphics functions covered in Sections 5.1 through 5.3 are powerful enough to create detailed, professional-looking plots in MATLAB, and they can be placed in reusable script files to create similar plots. This feature gives them an advantage over the interactive plotting interface discussed in this section. This interface can, however, can be advantageous in situations where:

- You need to create a large number of different types of plots,
- You must construct plots involving many data sets,
- You want to add annotations such as rectangles and ellipses, or
- You want to change plot characteristics such as tick spacing, fonts, bolding, italics, and colors.

The interactive plotting environment in MATLAB is a set of tools for:

- Creating different types of graphs,
- Selecting variables to plot directly from the Workspace Browser,
- Creating and editing subplots,

- Adding annotations such as lines, arrows, text, rectangles, and ellipses, and
- Editing properties of graphics objects, such as their color, line weight, and font.

The Plot Tools interface includes the following three panels associated with a given figure.

- **The Figure Palette:** Use this to create and arrange subplots, to view and plot workspace variables, and to add annotations.
- **The Plot Browser:** Use this to select and control the visibility of the axes or graphics objects plotted in the figure, and to add data for plotting.
- **The Property Editor:** Use this to set basic properties of the selected object and to obtain access to all properties through the Property Inspector.

Space limitations prevent us from discussing in detail all the features of the MATLAB interactive plotting environment. The following overview, however, should be sufficient to get you started. It is recommended that as you read this section you follow along and perform the steps in MATLAB. Note that selecting **Help** from the Figure window enables you to go directly to graphics-specific sections of the MATLAB help.

The Figure Window

When you create a plot, the Figure window appears with the Figure toolbar visible (see Figure 5.4–1). This window has eight menus.

The File Menu The **File** menu is used for saving and printing the figure. This menu was discussed in Section 5.1 under **Saving Figures** and **Exporting Figures.**

The Edit Menu You can use the **Edit** menu to cut, copy, and paste items, such as legend or title text, that appear in the figure. Click on **Figure Properties** to open the Property Editor—Figure dialog box to change certain properties of the figure.

Three items on the **Edit** menu are very useful for editing the figure. Clicking the **Axes Properties** item brings up the Property Editor—Axes dialog box. Double-clicking on any axis also brings up this box. You can change the scale type (linear, log, etc.), the labels, and the tick marks by selecting the tab for the desired axis or the font to be edited.

The **Current Object Properties** item enables you to change the properties of an object in the figure. To do this, first click on the object, such as a plotted line, then click on **Current Object Properties** in the **Edit** menu. You will see the Property Editor—Lineseries dialog box that lets you change properties such as line weight and color, data-marker type, and plot type.

Clicking on any text, such as that placed with the `title`, `xlabel`, `ylabel`, `legend`, or `gtext` commands, then selecting **Current Object Properties** in the **Edit** menu brings up the Property Editor—Text dialog box, which enables you to edit the text.

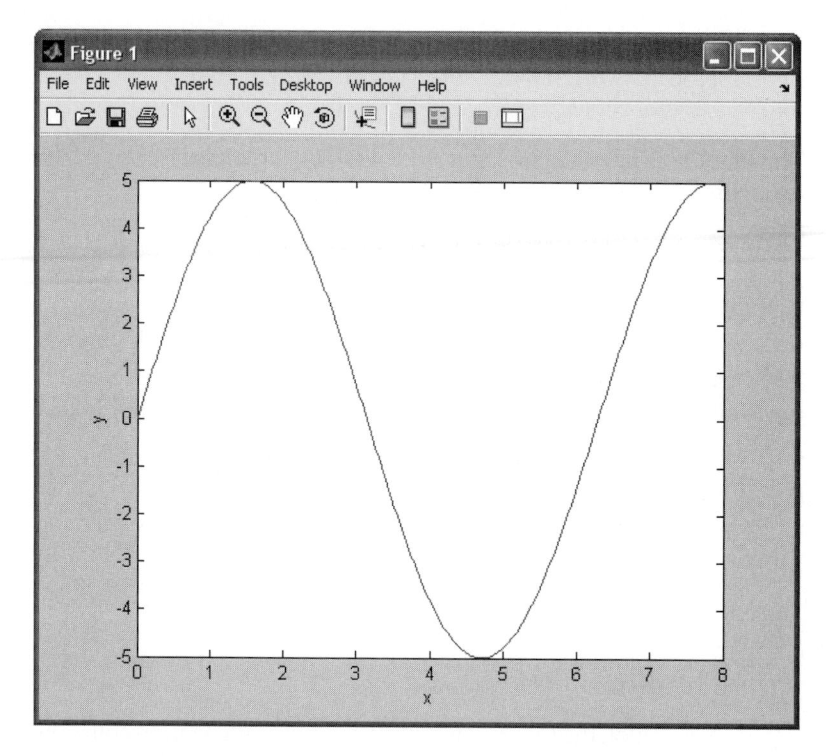

Figure 5.4–1 The Figure window with the Figure toolbar displayed.

The View Menu The items on the **View** menu are the three toolbars (**Figure toolbar, Plot Edit toolbar,** and **Camera toolbar**), the **Figure Palette,** the **Plot Browser,** and the **Property Editor.** These will be discussed later in this section.

The Insert Menu The **Insert** menu enables you to insert labels, legends, titles, text, and drawing objects, rather than using the relevant commands from the Command window. To insert a label on the *y*-axis, for example, click on the **Y Label** item on the menu; a box will appear on the *y*-axis. Type the label in this box, and then click outside the box to finish.

The **Insert** menu also enables you to insert arrows, lines, text, rectangles, and ellipses in the figure. To insert an arrow, click on the **Arrow** item; the mouse cursor changes to a crosshair style. Then click the mouse button, and move the cursor to create the arrow. The arrowhead will appear at the point where you release the mouse button. Be sure to add arrows, lines, and other annotations only after you are finished moving or resizing your axes, because these objects are not anchored to the axes. (They can be anchored to the plot by *pinning;* see the MATLAB help.)

To delete or move a line or arrow, click on it, then press the **Delete** key to delete it, or press the mouse button and move it to the desired location. The **Axes** item lets you use the mouse to place a new set of axes within the existing plot.

Click on the new axes, and a box will surround them. Any further plot commands issued from the Command window will direct the output to these axes.

The **Light** item applies to 3D plots.

The Tools Menu The **Tools** menu includes items for adjusting the view (by zooming and panning) and the alignment of objects on the plot. The **Edit Plot** item starts the plot editing mode, which can also be started by clicking on the northwest-facing arrow on the Figure toolbar. The **Tools** menu also gives access to the **Data Cursor,** which is discussed later in this section. The last two items, **Basic Fitting** and **Data Statistics,** will be discussed in Sections 5.7 and 7.1 respectively.

Other Menus The **Desktop** menu enables you to dock the Figure window within the desktop. The **Window** menu lets you switch between the Command window and any other Figure windows. The **Help** menu accesses the general MATLAB Help System, as well as help features specific to plotting.

There are three toolbars available in the Figure window: the Figure toolbar, the Plot Edit toolbar, and the Camera toolbar. The **View** menu lets you select which ones you want to appear. We will discuss the Figure toolbar and the Plot Edit toolbar in this section. The Camera toolbar is useful for 3D plots, which are discussed at the end of this chapter.

The Figure Toolbar

To activate the Figure toolbar, select it from the **View** menu (see Figure 5.4–1). The four left-most buttons are for opening, saving, and printing the figure. Clicking on the northwest-facing arrow button toggles the plot edit mode on and off.

The **Zoom-in** and **Zoom-out** buttons let you obtain a close-up or faraway view of the figure. The **Pan** and **Rotate 3D** buttons are used for 3D plots.

The **Data Cursor** button enables you to read data directly from a graph by displaying the values of points you select on plotted lines, surfaces, images, and so on.

The **Insert Colorbar** button inserts a color map strip in the graph and is useful for 3D surface plots. The **Insert Legend** button enables you to insert a legend in the plot.

The Plot Edit Toolbar

Once a plot is in the window you can enable plot editing by clicking on the northwest-facing arrow on the Figure toolbar. Then double-click on an axis, a plotted line, or a label to activate the appropriate property editor. Select **Plot Edit toolbar** from the **View** menu (see Figure 5.4–2). To add text that is not a label, title, or legend, click the button labeled **T,** move the cursor to the desired location for the text, click the mouse button, and type the text. When finished, click outside the text box and note that the nine left-most buttons become highlighted and available. These enable you to modify the color, font, and other attributes of the text.

To insert arrows, lines, rectangles, and ellipses, click on the appropriate button and follow the instructions given previously for the **Insert** menu.

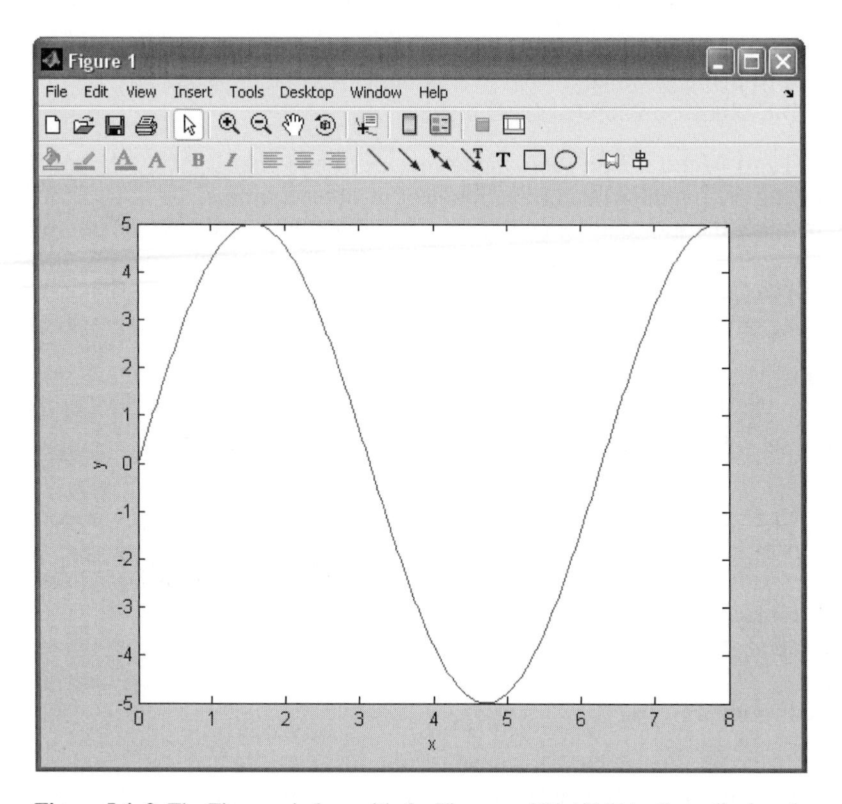

Figure 5.4–2 The Figure window with the Figure and Plot Edit toolbars displayed.

The Plot Tools

Once a figure has been created you can display any or all of the three Plot Tools (Figure Palette, Plot Browser, and Property Editor) by selecting them from the **View** menu. You can also start the environment by first creating a plot and then clicking on the **Show Plot Tools** icon in the Figure toolbar (see Figure 5.4–3), or by creating a figure with the plotting tools attached by using the `plottools` command. Remove the tools by clicking on the **Hide Tools** icon.

Figure 5.4–3 shows the result of clicking on the plotted line before clicking the **Show Plot Tools** icon. The plotting interface then displays the Property Editor—Lineseries.

The Figure Palette

The Figure Palette contains three panels, which are selected and expanded by clicking the appropriate button. Click on the grid icon in the **New Subplots** panel to display the selector grid that enables you to specify the layout of the subplots. In the Variables panel you can select a graphics function to plot the variable by selecting the variable and right-clicking to display the context menu. This menu contains a list of possible plot types based on the type of variable you select. You can also drag the variable into an axes set and MATLAB will select an appropriate plot type.

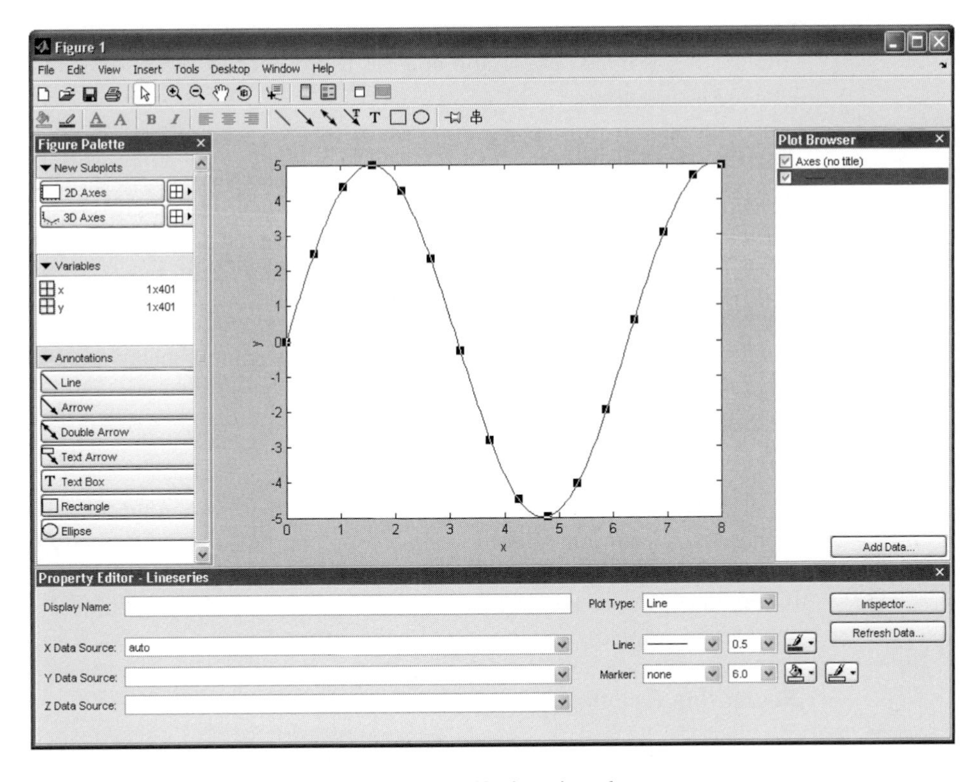

Figure 5.4–3 The Figure window with the Plot Tools activated.

Selecting **More Plots** from the context menu activates the Plot Catalog tool, which provides access to most of the plotting functions. After selecting a plot category, and a plot type from that category, you will see its description in the right-most display. Type the name of a one or more variables in the Plotted Variables field, separated by commas, and they will be passed to the selected plotting function as arguments. You can also type a MATLAB expression that uses any workspace variables shown in the Figure Palette.

Click on the **Annotations** panel to display a menu of objects such as lines, arrows, etc. Click on the desired object and use the mouse to position and size it.

The Plot Browser

The Plot Browser provides a legend of all the graphs in the figure. For example, if you plot an array with multiple rows and columns, the Browser lists each axis and the objects (lines, surfaces, etc.) used to create the graph. To set the properties of an individual line, double-click on the line. Its properties are displayed in the Property Editor—Lineseries box, which opens on the bottom of the figure.

If you select a line in the graph, the corresponding entry in the Plot Browser is highlighted, indicating which column in the variable produced the line. The check box next to each item in the Browser controls the object's visibility. For example, if you want to plot only certain columns of data, you can uncheck the columns not wanted. The graph updates as you uncheck each box and rescales the axes as required.

The Property Editor

The Property Editor enables you to access a subset of the selected object's properties. When no object is selected, the Property Editor displays the figure's properties. There are several ways to display the Property Editor.

1. Double-click an object when plot edit mode is enabled.
2. Select an object and right-click to display its context menu, then select **Properties.**
3. Select **Property Editor** from the **View** menu.
4. Use the `propertyeditor` command.

The Property Editor enables you to change the most commonly used object properties. If you want to access all object properties, use the Property Inspector. To display the Property Inspector, click the **Inspector** button on any Property Editor panel. Use of this feature requires detailed knowledge of object properties and handle graphics, and thus will not be covered here.

Recreating Graphs from M-Files

Once your graph is finished, you can generate MATLAB code to reproduce the graph by selecting **Generate M-File** from the **File** menu. MATLAB creates a function that recreates the graph and opens the generated M-File in the editor. This feature is particularly useful for capturing property settings and other modifications made in the plot editor. You can also use the `makemcode` function.

Adding Data to Axes

The Plot Browser provides the mechanism by which you add data to axes. The procedure is as follows:

1. Select a 2D or 3D axis from the New Subplots subpanel.
2. After creating the axis, select it in the Plot Browser panel to enable the **Add Data** button at the bottom of the panel.
3. Click the **Add Data** button to display the Add Data to Axes dialog box. The Add Data to Axes dialog enables you to select a plot type and specify the workspace variables to pass to the plotting function. You can also specify a MATLAB expression, which is evaluated to produce the data to plot.

5.5 Function Discovery

Function discovery is the process of finding, or "discovering," a function that can describe a particular set of data. The following three function types can often describe physical phenomena. For example, the *linear* function describes the voltage-current relation for a resistor ($v = iR$). The linear relation also describes the velocity versus time relation for an object with constant acceleration a ($v = at + v_0$). A *power* function describes the distance d traveled by a falling object versus time ($d = 0.5gt^2$). An *exponential* function can describe the relative temperature ΔT of a cooling object ($\Delta T = \Delta T_0 e^{-ct}$). The general forms of these functions are:

1. The *linear* function:

$$y(x) = mx + b \tag{5.5–1}$$

Note that $y(0) = b$.

2. The *power* function:

$$y(x) = bx^m \tag{5.5–2}$$

Note that $y(0) = 0$ if $m \geq 0$, and $y(0) = \infty$ if $m < 0$.

3. The *exponential* function:

$$y(x) = b(10)^{mx} \tag{5.5–3}$$

or its equivalent form:

$$y = be^{mx} \tag{5.5–4}$$

where e is the base of the natural logarithm ($\ln e = 1$). Note that $y(0) = b$ for both forms.

Each function gives a straight line when plotted using a specific set of axes:

1. The linear function $y = mx + b$ gives a straight line when plotted on rectilinear axes. Its slope is m and its intercept is b.

2. The power function $y = bx^m$ gives a straight line when plotted on log-log axes.

3. The exponential function $y = b(10)^{mx}$ and its equivalent form $y = be^{mx}$ give a straight line when plotted on a semilog plot whose y-axis is logarithmic.

These properties were illustrated in Figure 5.3–8, which shows the power function $y = 2x^{-0.5}$ and the exponential function $y = 10^{1-x}$.

We look for a straight line on the plot because it is relatively easy to recognize, and therefore we can easily tell whether the function will fit the data well. Using the following properties of base 10 logarithms, which are shared with natural logarithms, we have

$$\log_{10}(ab) = \log_{10} a + \log_{10} b$$
$$\log_{10}(a^m) = m \log_{10} a$$

Take the logarithm of both sides of the power equation $y = bx^m$ to obtain

$$\log_{10} y = \log_{10}(bx^m) = \log_{10} b + m \log_{10} x$$

This has the form

$$Y = B + mX$$

if we let $Y = \log_{10} y$, $X = \log_{10} x$, and $B = \log_{10} b$. Thus if we plot Y versus X on rectilinear scales, we will obtain a straight line whose slope is m and whose intercept is B. This is the same as plotting $\log_{10} y$ versus $\log_{10} x$ on rectilinear scales, so we will obtain a straight line whose slope is m and whose intercept is $\log_{10} b$. This is equivalent to plotting y versus x on *log-log* axes. Thus if the data can be described by the power function, it will form a straight line when plotted on log-log axes.

Taking the logarithm of both sides of the exponential equation $y = b(10)^{mx}$ we obtain

$$\log_{10} y = \log_{10}[b(10)^{mx}] = \log_{10} b + mx \log_{10} 10 = \log_{10} b + mx$$

because $\log_{10} 10 = 1$. This has the form

$$Y = B + mx$$

if we let $Y = \log_{10} y$ and $B = \log_{10} b$. Thus if we plot Y versus x on rectilinear scales, we will obtain a straight line whose slope is m and whose intercept is B. This is the same as plotting $\log_{10} y$ versus x on rectilinear scales, so we will obtain a straight line whose slope is m and whose intercept is $\log_{10} b$. This is equivalent to plotting y on a log axis and x on a rectilinear axis (that is, *semilog* axes). Thus if the data can be described by the exponential function, it will form a straight line when plotted on semilog axes (with the log axis used for the ordinate).

Taking the logarithm of both sides of the equivalent exponential form $y = be^{mx}$ gives

$$\log_{10} y = \log_{10}(be^{mx}) = \log_{10} b + mx \log_{10} e$$

This has the form

$$Y = B + Mx$$

if we let $Y = \log_{10} y$, $B = \log_{10} b$, and $M = m \log_{10} e$. Thus if we plot Y versus x on rectilinear scales, we will obtain a straight line whose slope is M and whose intercept is B. This is the same as plotting $\log_{10} y$ versus x on rectilinear scales, so we will obtain a straight line whose slope is $m \log_{10} e$ and whose intercept is $\log_{10} b$. This is equivalent to plotting y on a log axis and x on a rectilinear axis. Thus both equivalent exponential forms (5.5–3) and (5.5–4) will plot as a straight line on semilog axes.

Steps for Function Discovery

Here is a summary of the procedure to find a function that describes a given set of data. We assume that one of the function types (linear, exponential, or power) can describe the data. Fortunately, many applications generate data that these functions can describe. We assume that there is enough data and that it is accurate enough to identify the function.

1. Examine the data near the origin. The exponential function can never pass through the origin (unless of course $b = 0$, which is a trivial case). (See Figure 5.5–1 for examples with $b = 1$.) The linear function can pass through the origin only if $b = 0$. The power function can pass through the origin but only if $m > 0$. (See Figure 5.5–2 for examples with $b = 1$.)
2. Plot the data using rectilinear scales. If it forms a straight line, then it can be represented by the linear function and you are finished. Otherwise, if you have data at $x = 0$, then
   a. If $y(0) = 0$, try the power function.
   b. If $y(0) \neq 0$, try the exponential function.
   If data is not given for $x = 0$, proceed to step 3.
3. If you suspect a power function, plot the data using log-log scales. Only a power function will form a straight line on a log-log plot. If you suspect an exponential function, plot the data using the semilog scales. Only an exponential function will form a straight line on a semilog plot.

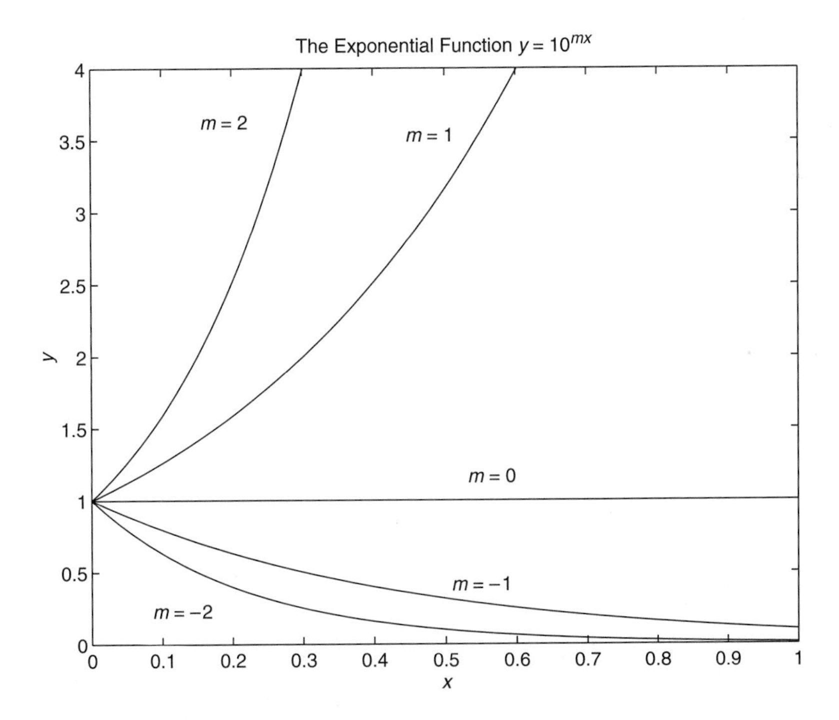

Figure 5.5–1 Examples of exponential functions.

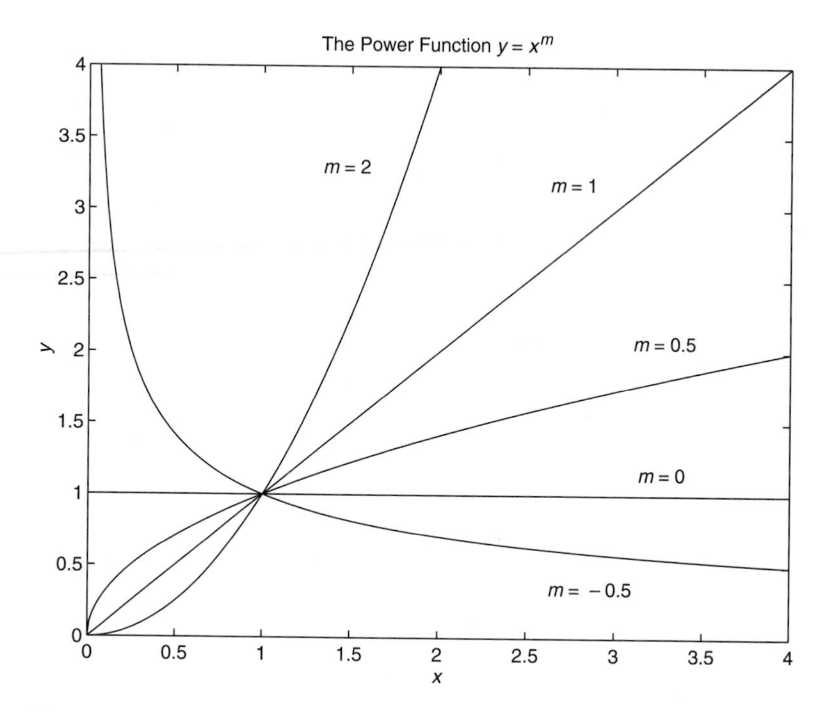

Figure 5.5–2 Examples of power functions.

4. In function discovery applications, we use the log-log and semilog plots *only* to identify the function type, but not to find the coefficients b and m. The reason is that it is difficult to interpolate on log scales.

We can find the values of b and m with the MATLAB `polyfit` function. This function finds the coefficients of a polynomial of specified degree n that best fits the data, in the so-called least squares sense. You will see what this means in Section 5.6. The syntax appears in Table 5.5–1.

Because we are assuming that our data will form a straight line on either a rectilinear, semilog, or log-log plot, we are interested only in a polynomial that corresponds to a straight line; that is, a first-degree polynomial, which we will denote as $w = p_1 z + p_2$. Thus, referring to Table 5.5–1, we see that the vector `p` will be $[p_1, p_2]$ if `n` is 1. This polynomial has a different interpretation in each of the three cases:

- **The linear function:** $y = mx + b$. In this case the variables w and z in the polynomial $w = p_1 z + p_2$ are the original data variables x and y, and we can find the linear function that fits the data by typing `p = polyfit(x,y,1)`. The first element p_1 of the vector `p` will be m, and the second element p_2 will be b.

Table 5.5–1 The `polyfit` function

Command	Description
`p = polyfit(x,y,n)`	Fits a polynomial of degree n to data described by the vectors x and y, where x is the independent variable. Returns a row vector `p` of length $n + 1$ that contains the polynomial coefficients in order of descending powers.

- **The power function:** $y = bx^m$. In this case

$$\log_{10}y = m \log_{10}x + \log_{10}b \tag{5.5–5}$$

which has the form

$$w = p_1 z + p_2$$

where the polynomial variables w and z are related to the original data variables x and y by $w = \log_{10}y$ and $z = \log_{10}x$. Thus we can find the power function that fits the data by typing `p = polyfit(log10(x), log10(y),1)`. The first element p_1 of the vector `p` will be m, and the second element p_2 will be $\log_{10}b$. We can find b from $b = 10^{p_2}$.

- **The exponential function:** $y = b(10)^{mx}$. In this case

$$\log_{10}y = mx + \log_{10}b \tag{5.5–6}$$

which has the form

$$w = p_1 z + p_2$$

where the polynomial variables w and z are related to the original data variables x and y by $w = \log_{10}y$ and $z = x$. Thus we can find the exponential function that fits the data by typing `p = polyfit(x, log10(y),1)`. The first element p_1 of the vector `p` will be m, and the second element p_2 will be $\log_{10}b$. We can find b from $b = 10^{p_2}$.

Applications

Function discovery is useful in all branches of engineering. Here we give three examples of applications in structural vibration, heat transfer, and fluid mechanics.

Civil, mechanical, and aerospace engineers frequently deal with structures or machines that bend and vibrate. For such applications they need a model of the vibration. The following example illustrates a common problem—the estimation of the deflection characteristics of a cantilever support beam.

EXAMPLE 5.5–1 A Cantilever Beam Deflection Model

The deflection of a cantilever beam is the distance its end moves in response to a force applied at the end (Figure 5.5–3). The following table gives the deflection x that was produced in a particular beam by the given applied force f. Is there a set of axes (rectilinear, semilog, or log-log) with which the data plot is a straight line? If so, use that plot to find a functional relation between f and x.

Force f (lb)	Deflection x (in.)
0	0
100	0.09
200	0.18
300	0.28
400	0.37
500	0.46
600	0.55
700	0.65
800	0.74

■ Solution

The following MATLAB script file generates two plots on rectilinear axes. The data is entered in the arrays `deflection` and `force`.

```
% Enter the data.
deflection = [0,0.09,0.18,0.28,0.37,0.46,0.55,0.65,0.74];
force = [0:100:800];
%
% Plot the data on rectilinear scales.
subplot(2,1,1)
plot(force,deflection,'o'), ...
xlabel('Applied Force (lb)'),ylabel('Deflection (in.)'),...
axis([0 800 0 0.8])
```

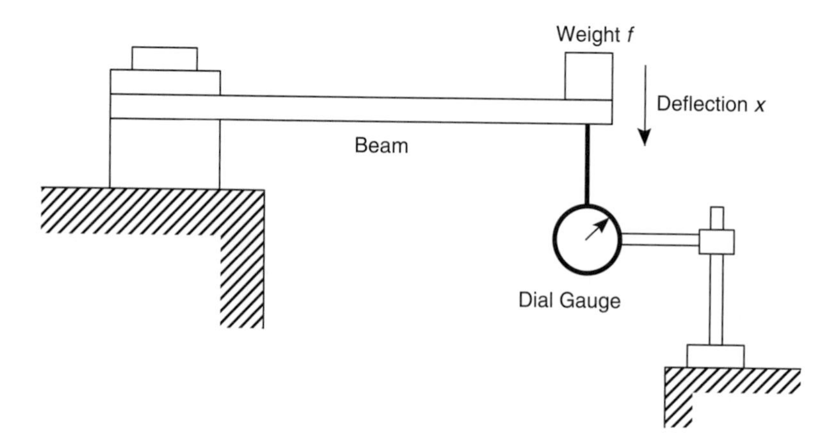

Figure 5.5–3 An experiment to measure force and deflection in a cantilever beam.

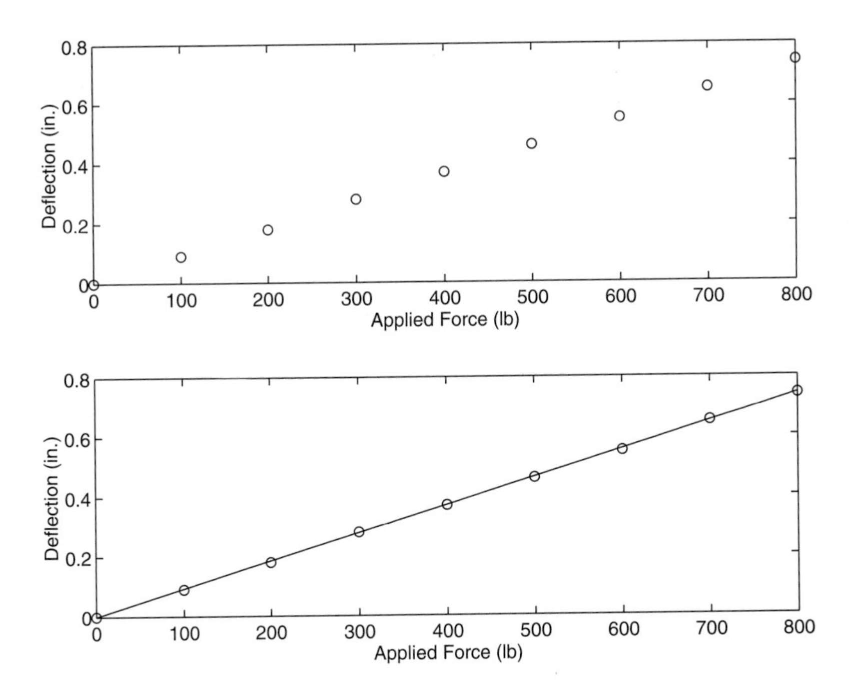

Figure 5.5–4 Plots for the cantilever beam example.

The plot appears in the first subplot in Figure 5.5–4. The data points appear to lie on a straight line that can be described by the relation $f = kx$, where k is called the beam's *spring constant.* We can find the value of k by using the `polyfit` command as shown in the following script file which is a continuation of the preceding script file.

```
% Fit a straight line to the data.
p = polyfit(force,deflection,1);
k = 1/p(1)
% Plot the fitted line and the data.
f = [0:2:800];
x = f/k;
subplot(2,1,2)
plot(f,x,force,deflection,'o'),...
xlabel('Applied Force (lb)'),ylabel('Deflection (in.)'),...
axis([0 800 0 0.8])
```

This file computes the value of the spring constant to be $k = 1079$ lb/in. Thus the force is related to the deflection by $f = 1079x$. The second subplot in Figure 5.5–4 shows the data and the line $x = f/k$, which fits the data well.

Heat Transfer

Civil, mechanical, and chemical engineers are often required to predict the temperatures that will occur in buildings and various industrial processes. Bio-engineers and environmental engineers must develop models of temperature distribution and heat loss in living things and the environment. This area of study is called *heat transfer*. The next example illustrates how we can use function discovery to predict the temperature dynamics of a cooling object.

EXAMPLE 5.5–2 Temperature Dynamics

The temperature of coffee cooling in a porcelain mug at room temperature (68°F) was measured at various times. The data follows.

Time t (sec)	Temperature T (°F)
0	145
620	130
2266	103
3482	90

Develop a model of the coffee's temperature as a function of time and use the model to estimate how long it will take the temperature to reach 120°F.

■ Solution

Because $T(0)$ is finite but nonzero, the power function cannot describe this data, so we do not bother to plot the data on log-log axes.

Common sense tells us that the coffee will cool and its temperature will eventually equal the room temperature. So we subtract the room temperature from the data and plot the relative temperature, $T - 68$, versus time. If the relative temperature is a linear function of time, the model is

$$T - 68 = mt + b$$

If the relative temperature is an exponential function of time, the model is

$$T - 68 = b(10)^{mt}$$

Figure 5.5–5 shows the plots used to solve the problem. The following MATLAB script file generates the top two plots. The time data is entered in the array `time`, and the temperature data is entered in `temp`.

```
% Enter the data.
time = [0,620,2266,3482];
temp = [145,130,103,90];
%Subtract the room temperature.
temp = temp - 68;
% Plot the data on rectilinear scales.
subplot(2,2,1)
plot(time,temp,time,temp,'o'),xlabel('Time (sec)'),...
```

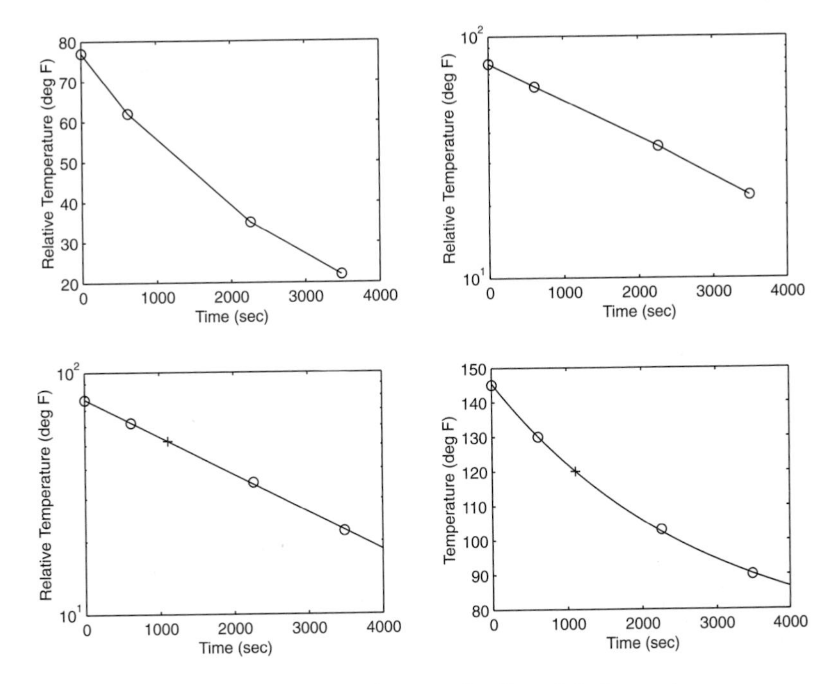

Figure 5.5–5 Temperature of a cooling cup of coffee, plotted on various coordinates.

```
ylabel('Relative Temperature (deg F)')
%
% Plot the data on semilog scales.
subplot(2,2,2)
semilogy(time,temp,time,temp,'o'),xlabel('Time (sec)'),...
ylabel('Relative Temperature (deg F)')
```

The data forms a straight line on the semilog plot only (the top right plot). Thus it can be described with the exponential function $T = 68 + b(10)^{mt}$. Using the `polyfit` command, the following lines can be added to the script file.

```
% Fit a straight line to the transformed data.
p = polyfit(time,log10(temp),1);
m = p(1)
b = 10^p(2)
```

The computed values are $m = -1.5557 \times 10^{-4}$ and $b = 77.4469$. Thus our derived model is $T = 68 + b(10)^{mt}$. To estimate how long it will take for the coffee to cool to 120°F, we must solve the equation $120 = 68 + b(10)^{mt}$ for t. The solution is $t = [(\log_{10}(120 - 68) - \log_{10} b)]/m$. The MATLAB command for this calculation is shown in the following script file, which is a continuation of the previous script and produces the bottom two subplots shown in Figure 5.5–5.

```
% Compute the time to reach 120 degrees.
t_120 = (log10(120-68)-log10(b))/m
% Show the derived curve and estimated point on semilog scales.
t = [0:10:4000];
T = 68+b*10.^(m*t);
subplot(2,2,3)
semilogy(t,T-68,time,temp,'o',t_120,120-68,'+'),
xlabel('Time (sec)'),...
ylabel('Relative Temperature (deg F)')
%
% Show the derived curve and estimated point on rectilinear scales.
subplot(2,2,4)
plot(t,T,time,temp+68,'o',t_120,120,'+'),xlabel('Time (sec)'),...
ylabel('Temperature (deg F)')
```

The computed value of `t_120` is 1112. Thus the time to reach 120° F is 1112 sec. The plot of the model, along with the data and the estimated point (1112, 120) marked with a + sign, is shown in the bottom two subplots in Figure 5.5–5. Because the graph of our model lies near the data points, we can treat its prediction of 1112 sec with some confidence.

Interpolation and Extrapolation

After we discover a functional relation that describes the data, we can use it to predict conditions that lie *within* the range of the original data. This process is called *interpolation*. For example, we can use the coffee cup model to estimate how long it takes for the coffee to cool to 120°F because we have data below and above 120°F. We can be fairly confident of this prediction because our model describes the temperature data very well.

Extrapolation is the process of using the model to predict conditions that lie *outside* the original data range. Extrapolation might be used in the beam example to predict how much force would be required to bend the beam 1.2 in. (The predicted value is found from the model to be $f = 1079(1.2) = 1295$ lb.) We must be careful when using extrapolation because we often have no reason to believe that the mathematical model is valid beyond the range of the original data. For example, if we continue to bend the beam, eventually the force is no longer proportional to the deflection and becomes much greater than that predicted by the linear model $f = kx$.

Extrapolation has a use in making tentative predictions, which must be backed up by testing. The Example 5.5-3 describes an application of extrapolation.

Hydraulic Engineering

Engineers in many fields, including civil, mechanical, bio, nuclear, chemical, and aerospace engineers, often need models to predict the flow rate of fluids

Figure 5.5–6 An experiment to verify Torricelli's principle.

under pressure. *Torricelli's principle of hydraulic resistance* states that the volume flow rate f of a liquid through a restriction—such as an opening or a valve—is proportional to the square root of the pressure drop p across the restriction; that is,

$$f = c\sqrt{p} \tag{5.5–7}$$

where c is a constant. In many applications the weight of liquid in a tank causes the pressure drop (see Figure 5.5–6). In such situations Torricelli's principle states that the flow rate is proportional to the square root of the volume V of liquid in the tank. Thus

$$f = r\sqrt{V}$$

where r is a constant.

Torricelli's principle is widely used to design valves and piping systems for many applications, including water-supply engineering, hydraulically powered machinery, and chemical-processing systems. Here we apply it to a familiar item, a coffee pot.

Hydraulic Resistance — EXAMPLE 5.5–3

A 15-cup coffee pot (see Figure 5.5–6) was placed under a water faucet and filled to the 15-cup line. With the outlet valve open, the faucet's flow rate was adjusted until the water level remained constant at 15 cups, and the time for one cup to flow out of the pot was measured. This experiment was repeated with the pot filled to the various levels shown in the following table:

Liquid volume V (cups)	Time to fill one cup t (sec)
15	6
12	7
9	8
6	9

(a) Use the preceding data to verify Torricelli's principle for the coffee pot and to obtain a relation between the flow rate and the number of cups in the pot. (b) The manufacturer wants to make a 36-cup pot using the same outlet valve but is concerned that a cup will fill too quickly, causing spills. Extrapolate the relation developed in part (a) and predict how long it will take to fill one cup when the pot contains 36 cups.

■ Solution

(a) Torricelli's principle in equation form is $f = rV^{1/2}$, where f is the flow rate through the outlet valve in cups per second, V is the volume of liquid in the pot in cups, and r is a constant whose value is to be found. We see that this relation is a power function where the exponent is 0.5. Thus if we plot $\log_{10}(f)$ versus $\log_{10}(V)$, we should obtain a straight line. The values for f are obtained from the reciprocals of the given data for t. That is, $f = 1/t$ cups per second.

The MATLAB script file follows. The resulting plots appear in Figure 5.5–7. The volume data is entered in the array `cups`, and the time data is entered in `meas_times`.

```
% Data for the problem.
cups = [6,9,12,15];
meas_times = [9,8,7,6];
meas_flow = 1./meas_times;
%
% Fit a straight line to the transformed data.
p = polyfit(log10(cups),log10(meas_flow),1);
coeffs = [p(1),10^p(2)];
m = coeffs(1)
b = coeffs(2)
%
% Plot the data and the fitted line on a loglog plot to see
% how well the line fits the data.
x = [6:0.01:40];
y = b*x.^m;
subplot(2,1,1)
loglog(x,y,cups,meas_flow,'o'),grid,xlabel('Volume (cups)'),...
ylabel('Flow Rate (cups/sec)'),axis([5 15 0.1 0.3])
```

The computed values are $m = 0.433$ and $b = 0.0499$, and our derived relation is $f = 0.0499V^{0.433}$. Because the exponent is 0.433, not 0.5, our model does not agree exactly

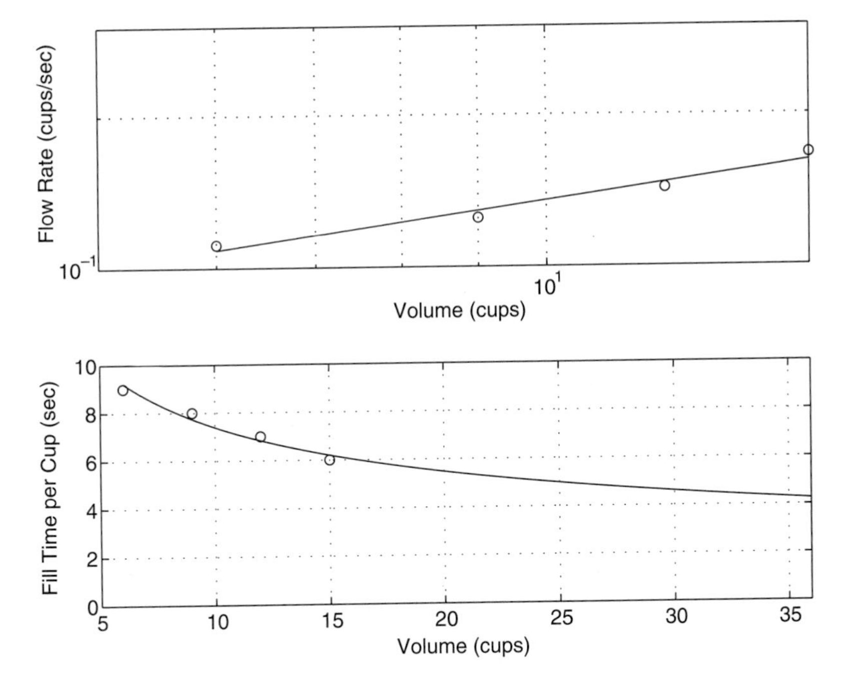

Figure 5.5–7 Flow rate and fill time for a coffee pot.

with Torricelli's principle, but it is close. Note that the first plot in Figure 5.5–7 shows that the data points do not lie exactly on the fitted straight line. In this application it is difficult to measure the time to fill one cup with an accuracy greater than an integer second, so this inaccuracy could have caused our result to disagree with that predicted by Torricelli.

(b) Note that the fill time is $1/f$, the reciprocal of the flow rate. The remainder of the MATLAB script uses the derived flow rate relation $f = 0.0499V^{0.433}$ to plot the extrapolated fill-time curve $1/f$ versus t.

```
% Plot the fill time curve extrapolated to 36 cups.
subplot(2,1,2)
plot(x,1./y,cups,meas_times,'o'),grid,xlabel('Volume(cups)'),...
ylabel('Fill Time per Cup (sec)'),axis([5 36 0 10])
%
% Compute the fill time for V = 36 cups.
V = 36;
f_36 = b*V^m
```

The predicted fill time for one cup is 4.2 sec. The manufacturer must now decide if this time is sufficient for the user to avoid overfilling. (In fact, the manufacturer did construct a 36-cup pot, and the fill time is approximately 4 sec, which agrees with our prediction.)

5.6 Regression

We can distinguish between two types of analysis in experiments involving two variables, say, x and y. In the first type, called *correlation analysis,* both variables are random, and we are interested in finding the relation between them. An example is the analysis of the relation, if any, between the chemistry grades x and physics grades y of a group of students. We will not deal with correlation analysis in this text, because it is an advanced topic in statistics.

With the second type of analysis, called *regression,* one of the variables, say, x, is regarded as an ordinary variable because we can measure it without error or assign it whatever values we want. The variable x is the *independent* or *controlled* variable. The variable y is a random variable. Regression analysis deals with the relationship between y and x. An example is the dependence of the outdoor temperature y on the time of day x.

In the previous section we used the MATLAB function `polyfit` to perform regression analysis with functions that are linear or could be converted to linear form by a logarithmic or other transformation. The `polyfit` function is based on the least squares method, which we now discuss. We also show how to use this function to develop polynomial and other types of functions.

The Least Squares Method

Suppose we have the three data points given in the following table, and we need to determine the coefficients of the straight line $y = mx + b$ that best fit the following data in the least squares sense.

x	y
0	2
5	6
10	11

RESIDUALS

According to the least squares criterion, the line that gives the best fit is the one that minimizes J, the sum of the squares of the vertical differences between the line and the data points (see Figure 5.6–1). These differences are called the *residuals*. Here there are three data points, and J is given by

$$J = \sum_{i=1}^{3} (mx_i + b - y_i)^2$$

Substituting the data values (x_i, y_i) given in the table, we obtain

$$J = (0m + b - 2)^2 + (5m + b - 6)^2 + (10m + b - 11)^2$$

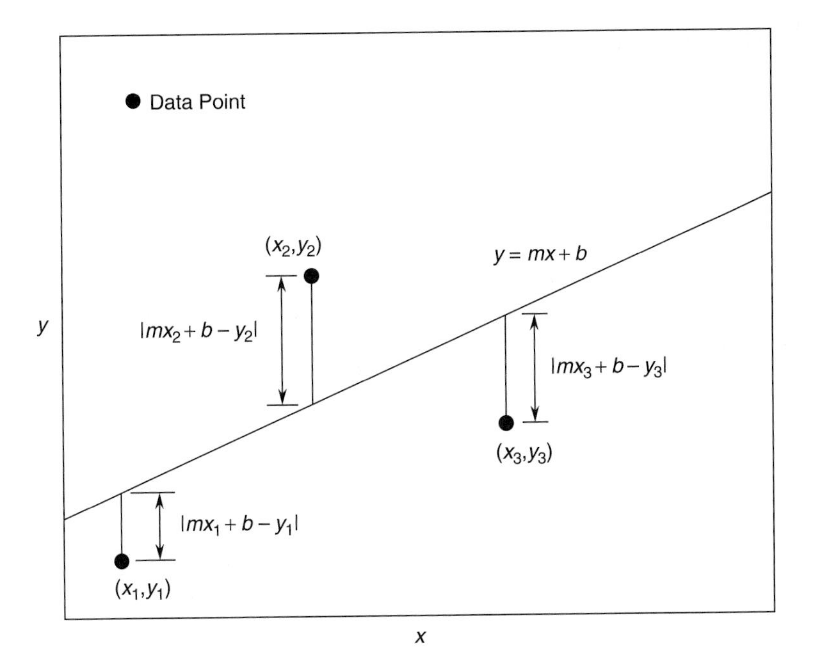

Figure 5.6–1 Illustration of the least squares criterion.

If you are familiar with calculus, you know that the values of m and b that minimize J are found by setting the partial derivatives $\partial J/\partial m$ and $\partial J/\partial b$ equal to zero.

$$\frac{\partial J}{\partial m} = 2(5m + b - 6)(5) + 2(10m + b - 11)(10) = 250m + 30b - 280 = 0$$

$$\frac{\partial J}{\partial b} = 2(b - 2) + 2(5m + b - 6) + 2(10m + b - 11) = 30m + 6b - 38 = 0$$

These conditions give the following equations that must be solved for the two unknowns m and b.

$$250m + 30b = 280$$

$$30m + 6b = 38$$

The solution is $m = 0.9$ and $b = 11/6$. The best straight line in the least squares sense is $y = 0.9x + 11/6$. If we evaluate this equation at the data values $x = 0, 5$, and 10, we obtain the values $y = 1.833, 6.333, 10.8333$. These values are different from the given data values $y = 2$, 6, and 11 because the line is not a perfect fit to the data. The value of J is $J = (1.833 - 2)^2 + (6.333 - 6)^2 + (10.8333 - 11)^2 = 0.16656689$. No other straight line will give a lower value of J for this data. This line is shown in Figure 5.6–2.

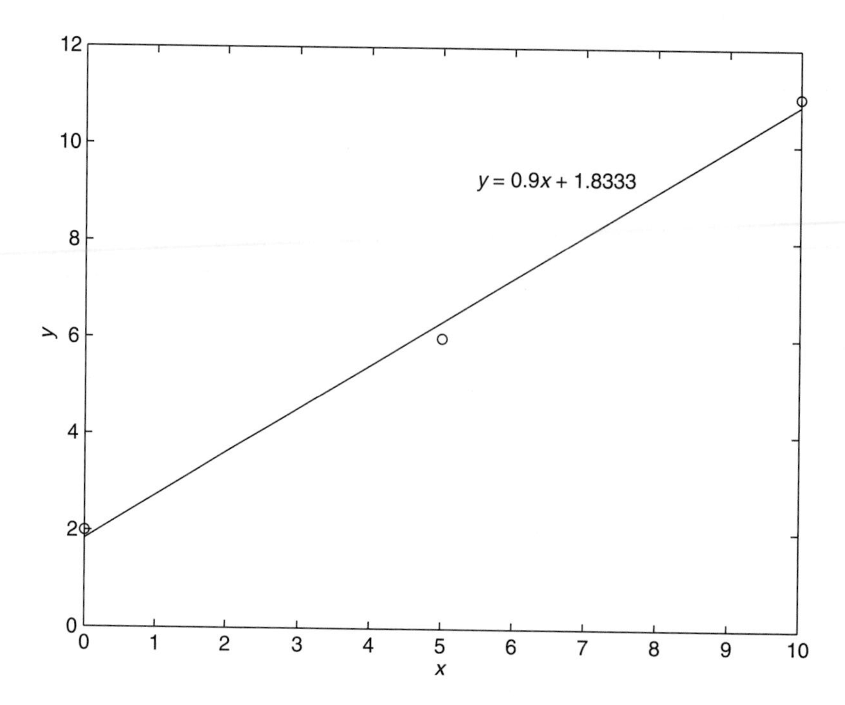

Figure 5.6–2 The least squares fit for the example data.

Now suppose we want to fit a quadratic function $y = a_1x^2 + a_2x + a_3$ to a set of m data points. Then the expression for J is

$$J = \sum_{i=1}^{m} \left(a_1x_i^2 + a_2x_i + a_3 - y_i\right)^2$$

Using calculus (find the derivatives $\partial J/\partial a_1$, $\partial J/\partial a_2$, and $\partial J/\partial a_3$ and set them equal to 0), we can obtain the following equations that must be solved for a_1, a_2, and a_3:

$$a_1 \sum_{i=1}^{m} x_i^4 + a_2 \sum_{i=1}^{m} x_i^3 + a_3 \sum_{i=1}^{m} x_i^2 = \sum_{i=1}^{m} x_i^2 y_i$$

$$a_1 \sum_{i=1}^{m} x_i^3 + a_2 \sum_{i=1}^{m} x_i^2 + a_3 \sum_{i=1}^{m} x_i = \sum_{i=1}^{m} x_i y_i$$

$$a_1 \sum_{i=1}^{m} x_i^2 + a_2 \sum_{i=1}^{m} x_i + ma_3 = \sum_{i=1}^{m} y_i$$

These three linear equations are in terms of the three unknowns a_1, a_2, and a_3, and they can be solved easily in MATLAB.

Table 5.6–1 Functions for polynomial regression

Command	Description
`p = polyfit(x,y,n)`	Fits a polynomial of degree `n` to data described by the vectors `x` and `y`, where `x` is the independent variable. Returns a row vector `p` of length `n+1` that contains the polynomial coefficients in order of descending powers.
`[p,s, mu] = polyfit(x,y,n)`	Fits a polynomial of degree `n` to data described by the vectors `x` and `y`, where `x` is the independent variable. Returns a row vector `p` of length `n+1` that contains the polynomial coefficients in order of descending powers and a structure `s` for use with `polyval` to obtain error estimates for predictions. The optional output variable `mu` is a two-element vector containing the mean and standard deviation of `x`.
`[y,delta] = polyval(p,x,s,mu)`	Uses the optional output structure `s` generated by `[p,s,mu] = polyfit(x,y,n)` to generate error estimates. If the errors in the data used with `polyfit` are independent and normally distributed with constant variance, at least 50 percent of the data will lie within the band `y ± delta`.

In general, for the polynomial $a_1x^n + a_2x^{n-1} + \cdots + a_nx + a_{n+1}$, the sum of the squares of the residuals for m data points is

$$J = \sum_{i=1}^{m} \left(a_1x^n + a_2x^{n-1} + \cdots + a_nx + a_{n+1} - y_i\right)^2$$

The values of the $n+1$ coefficients a_i that minimize J can be found by solving a set of $n+1$ linear equations. The `polyfit` function provides this solution. Its syntax is `p = polyfit(x,y,n)`. The function fits a polynomial of degree n to data described by the vectors `x` and `y`, where x is the independent variable. The result `p` is the row vector of length $n+1$ that contains the polynomial coefficients in order of descending powers. Table 5.6–1 summarizes the `polyfit` and `polyval` functions.

To illustrate the use of the `polyfit` function, consider the data set where $x = 1, 2, 3, \ldots, 9$ and $y = 5, 6, 10, 20, 28, 33, 34, 36, 42$. The following script file finds and plots the first- through fourth-degree polynomials for this data and evaluates J for each polynomial.

```
x = [1:9];
y = [5,6,10,20,28,33,34,36,42];
xp = [1:0.01:9];
for k = 1:4
   coeff = polyfit(x,y,k)
   yp(k,:) = polyval(coeff,xp);
   J(k) = sum((polyval(coeff,x)-y).^2);
end
subplot(2,2,1)
plot(xp,yp(1,:),x,y,'o'),axis([0 10 0 50])
```

```
subplot(2,2,2)
plot(xp,yp(2,:),x,y,'o'),axis([0 10 0 50])
subplot(2,2,3)
plot(xp,yp(3,:),x,y,'o'),axis([0 10 0 50])
subplot(2,2,4)
plot(xp,yp(4,:),x,y,'o'),axis([0 10 0 50])
disp(J)
```

The plots are shown in Figure 5.6–3, and the J values are, to two significant figures, 72, 57, 42, and 4.7. Thus the value of J decreases as the polynomial degree is increased, as we would expect. The figure shows why the fourth-degree polynomial can fit the data better than the lower-degree polynomials. Because it has more coefficients, the fourth-degree polynomial can follow the "bends" in the data more easily than the other polynomials can. The first-degree polynomial (a straight line) cannot bend at all, the quadratic polynomial has one bend, the cubic has two bends, and the quartic has three bends.

The polynomial coefficients in the preceding script file are contained in the vector `polyfit(x,y,k)`. If you need the polynomial coefficients, say, for the cubic polynomial, type `polyfit(x,y,3)` after the program has been run.

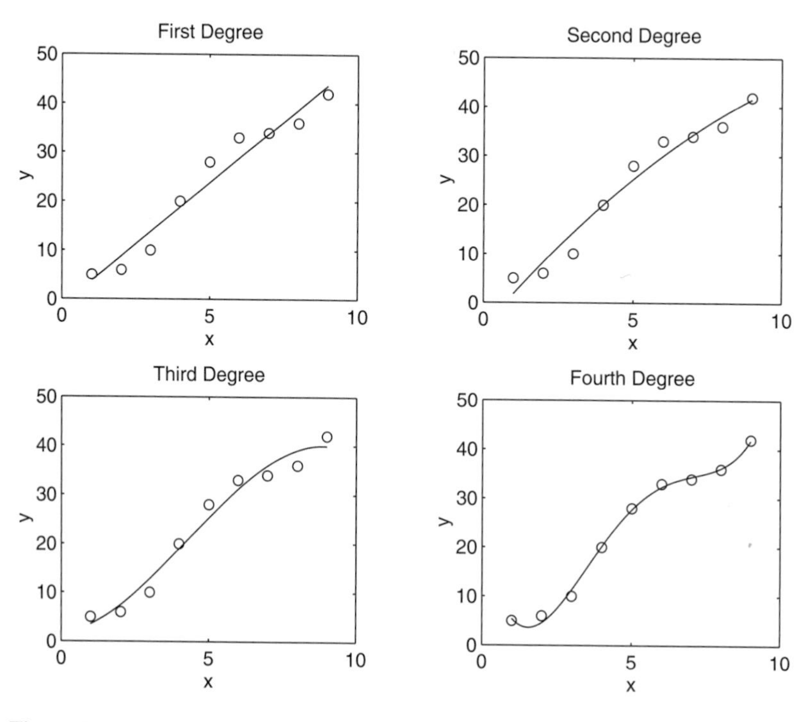

Figure 5.6–3 Regression using polynomials of first through fourth degree.

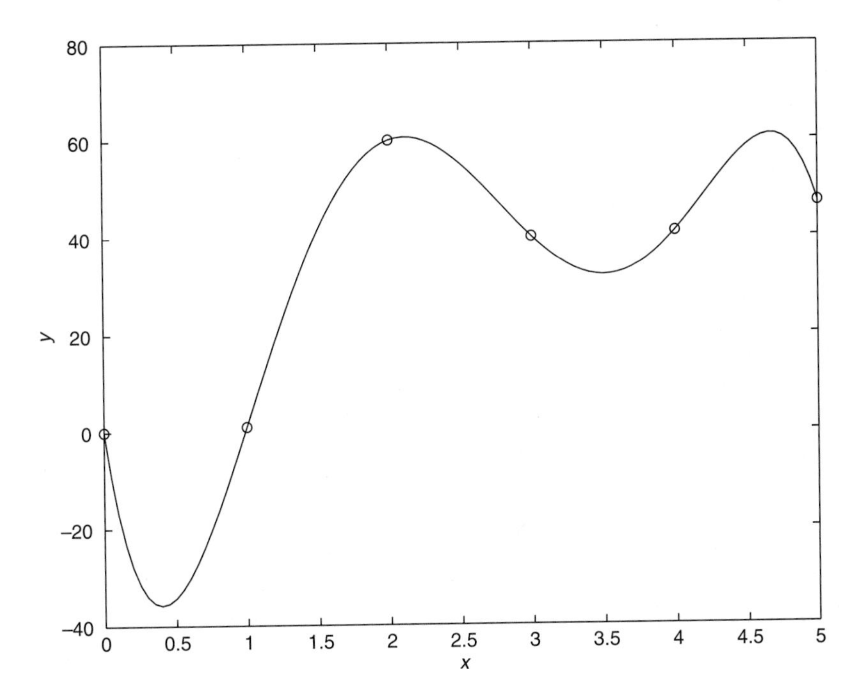

Figure 5.6–4 An example of a fifth-degree polynomial that passes through all six data points but exhibits large excursions between points.

The result is -0.1019, 1.3081, 0.7433, and 1.5556. These values correspond to the polynomial $-0.1019x^3 + 1.3081x^2 + 0.7433x + 1.5556$.

It is tempting to use a high-degree polynomial to obtain the best possible fit. However, there are two dangers in using high-degree polynomials. Figure 5.6–4 illustrates one of these dangers. The data values are $x = 0, 1, \ldots, 5$, and $y = 0$, $1, 60, 40, 41$, and 47. Because there are six data points, a fifth-degree polynomial (which has six coefficients) can be found that passes through all six points and thus its J value will be 0. This polynomial, which can be found using the `polyfit` function, is plotted in Figure 5.6–4 along with the data. Note how the polynomial makes large excursions between the first and the last pairs of data points. Thus we could get large errors if we were to use this polynomial to estimate y values for $0 < x < 1$ and for $4 < x < 5$. High-degree polynomials often exhibit large excursions between the data points and thus should be avoided if possible.

The second danger with using high-degree polynomials is that they can produce large errors if their coefficients are not represented with a large number of significant figures. We examine this issue later in this section.

In some cases it might not be possible to fit the data with a low-degree polynomial. In such cases we might be able to use several cubic polynomials. This method, called cubic splines, is covered in Chapter 7.

Test Your Understanding

T5.6–1 Obtain and plot the first-through fourth-degree polynomials for the following data: $x = 0, 1, \ldots, 5$ and $y = 0, 1, 60, 40, 41$, and 47. Find the coefficients and the J values.
(Answer: The polynomials are $9.5714x + 7.5714$; $-3.6964x^2 + 28.0536x - 4.7500$; $0.3241x^3 - 6.1270x^2 + 32.4934x - 5.7222$; and $2.5208x^4 - 24.8843x^3 + 71.2986x^2 - 39.5304x - 1.4008$. The corresponding J values are 1534, 1024, 1017, and 495, respectively.)

Fitting Other Functions

In the previous section we used the `polyfit` function to fit power and exponential functions. We found the power function $y = bx^m$ that fits the data by typing `p = polyfit(log10(x), log10(y),1)`. The first element p_1 of the vector `p` will be m, and the second element p_2 will be $\log_{10} b$. We can find b from $b = 10^{p_2}$. We can find the exponential function $y = b(10)^{mx}$ that fits the data by typing `p = polyfit(x,log10(y),1)`. The first element p_1 of the vector `p` will be m, and the second element p_2 will be $\log_{10} b$. We can find b from $b = 10^{p_2}$.

Any function is a candidate for fitting data. However, it might be difficult to obtain and solve equations for the function's coefficients. Polynomials are used because their curves can take on many shapes and because they result in a set of linear equations that can be easily solved for the coefficients. As we have just seen, the power and exponential functions can be converted to first-degree polynomials by logarithmic transformations. Other functions can be converted to polynomials by suitable transformations.

Given the data (y, z), the logarithmic function

$$y = m \ln z + b$$

can be converted to a first-degree polynomial by transforming the z values into x values by the transformation $x = \ln z$. The resulting function is $y = mx + b$.

Given the data (y, z), the function

$$y = b(10)^{m/z}$$

can be converted to an exponential function by transforming the z values by the transformation $x = 1/z$. In MATLAB we can type `p = polyfit(1./z, log10(y),1)`. The first element p_1 of the vector `p` will be m, and the second element p_2 will be $\log_{10} b$. We can find b from $b = 10^{p_2}$.

Given the data (v, x), the function

$$v = \frac{1}{mx + b}$$

can be converted to a first-degree polynomial by transforming the v data values with the transformation $y = 1/v$. The resulting function is $y = mx + b$.

The Quality of a Curve Fit

In general, if the arbitrary function $y = f(x)$ is used to represent the data, then the error in the representation is given by $e_i = f(x_i) - y_i$, for $i = 1, 2, 3, \ldots, m$. The error (or residual) e_i is the difference between the data value y_i and the value of y obtained from the function, that is, $f(x_i)$. The least squares criterion used to fit a function $f(x)$ is the sum of the squares of the residuals J. It is defined as

$$J = \sum_{i=1}^{m} [f(x_i) - y_i]^2 \tag{5.6–1}$$

We can use this criterion to compare the quality of the curve fit for two or more functions used to describe the same data. The function that gives the smallest J value gives the best fit.

We denote the sum of the squares of the deviation of the y values from their mean $\overline{y}$ by S, which can be computed from

$$S = \sum_{i=1}^{m} (y_i - \overline{y})^2 \tag{5.6–2}$$

COEFFICIENT OF DETERMINATION

This formula can be used to compute another measure of the quality of the curve fit, the *coefficient of determination,* also known as the *r-squared value.* It is defined as

$$r^2 = 1 - \frac{J}{S} \tag{5.6–3}$$

For a perfect fit, $J = 0$ and thus $r^2 = 1$. Thus the closer r^2 is to 1, the better the fit. The largest r^2 can be is 1. The value of S indicates how much the data is spread around the mean, and the value of J indicates how much of the data spread is unaccounted for by the model. Thus the ratio J/S indicates the fractional variation unaccounted for by the model. It is possible for J to be larger than S, and thus it is possible for r^2 to be negative. Such cases, however, are indicative of a very poor model that should not be used. As a rule of thumb, a good fit accounts for at least 99 percent of the data variation. This value corresponds to $r^2 \geq 0.99$.

For example, the following table gives the values of J, S, and r^2 for the first- through fourth-degree polynomials used to fit the data $x = 1, 2, 3, \ldots, 9$ and $y = 5, 6, 10, 20, 28, 33, 34, 36, 42$.

Degree n	J	S	r^2
1	72	1562	0.9542
2	57	1562	0.9637
3	42	1562	0.9732
4	4.7	1562	0.9970

Because the fourth-degree polynomial has the largest r^2 value, it represents the data better than the representation from first- through third-degree polynomials, according to the r^2 criterion.

To calculate the values of S and r^2, add the following lines to the end of the script file shown on pages 315 to 316.

```
mu = mean(y);
for k=1:4
   S(k) = sum((y-mu).^2);
   r2(k) = 1 - J(k)/S(k);
end
S
r2
```

Regression and Numerical Accuracy

We mentioned that there are two dangers in using high-degree polynomials. The first danger is that high-degree polynomials often exhibit large excursions between the data points, and thus we could get large errors if we were to use a high-degree polynomial to estimate y values between the data points.

The second danger with using high-degree polynomials is that their coefficients often require a large number of significant figures to be represented accurately. Thus if you calculate these coefficients using the `polyfit` function and you intend to include these coefficients in a report, for example, then you should display them with the `format long` or `format long e` commands. As an example, consider the data:

x	y	x	y	x	y
0	0.1	14	2.022	28	0.4308
2	1.884	16	1.65	30	0.203
4	2.732	18	1.5838	32	0.1652
6	3.388	20	1.35	34	−0.073
8	3.346	22	1.0082	36	−0.002
10	3	24	0.718	38	−0.1122
12	2.644	26	0.689	40	0.106

Using the `format long` command, a sixth-degree polynomial fit, `polyfit(x,y,6)`, gives the result

$$
\begin{aligned}
y = & -6.33551 \times 10^{-9}x^6 + 2.09690135 \times 10^{-6}x^5 \\
& - 1.9208956532 \times 10^{-4}x^4 + 7.77770991616 \times 10^{-3}x^3 \\
& - 0.15178006527153x^2 + 1.20369642390774x \\
& + 0.0577394277217
\end{aligned} \tag{5.6–4}
$$

It is plotted along with the data in the top graph in Figure 5.6–5. The J and r^2 values are $J = 0.1859$ and $r^2 = 0.9935$.

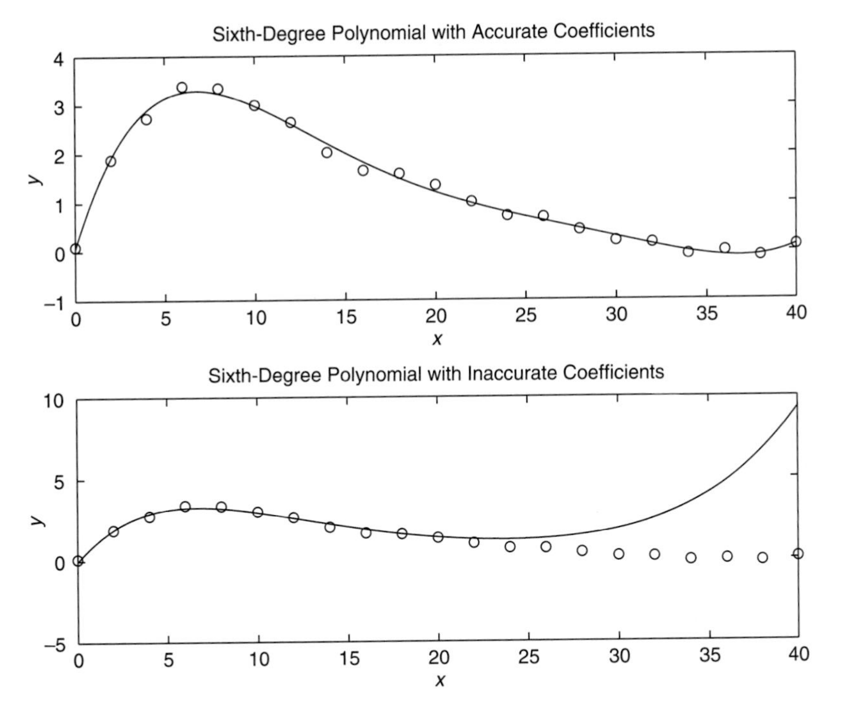

Figure 5.6–5 Effect of coefficient accuracy on a sixth-degree polynomial. The top graph shows the effect of 14 decimal-place accuracy. The bottom graph shows the effect of 8 decimal-place accuracy.

Now suppose we keep only the first eight decimal places of the coefficients. The resulting polynomial is

$$y = 0x^6 + 2.1 \times 10^{-6}x^5 - 1.9209 \times 10^{-4}x^4 + 7.7777 \times 10^{-3}x^3 - 0.15178007x^2 + 1.20369642x + 0.05773943 \tag{5.6–5}$$

This polynomial is plotted in the bottom graph of Figure 5.6–5. Obviously, the polynomial deviates greatly from the data for the larger values of x. The fit is so poor as to be useless for values of x greater than 26 approximately. Polynomials whose coefficients have not been specified accurately enough are prone to large errors at large values of x.

High-degree polynomials have coefficients that not only must be displayed and stored with great accuracy but also are harder to compute accurately. As the polynomial degree increases, the number of linear equations to be solved also increases, and inaccuracies in the numerical solution of these equations become more significant.

An alternative to using a high-degree polynomial is to fit two or more functions to the data. To illustrate this approach, we will fit two cubics because a cubic has greater flexibility than a quadratic but is less susceptible to numerical inaccuracies than higher-degree polynomials. Examining the data plot in

Figure 5.6–5, we see that the data has a bend near $x = 5$ and another bend near $x = 38$. The transition between the convex and concave parts of the data appears to be around $x = 15$. Thus we will fit one cubic over $0 \le x \le 15$ and another over $15 < x \le 40$. The necessary script file follows.

```
x1 = [0:2:14];x2 = [16:2:40];
y1 = [0.1, 1.884, 2.732, 3.388, 3.346, 3, 2.644, 2.022]
y2 = [1.65, 1.5838, 1.35, 1.0082, 0.718, 0.689, 0.4308,
0.203, 0.1652, -0.073, -0.002, -0.1122, 0.106];
x = [x1,x2];
y = [y1,y2];
% Create variables z1 and z2 to generate the curve.
z1 = [0:0.01:15]; z2=[15:0.01:40];
% Fit two cubics.
w1 = polyfit(x1,y1,3);
w2 = polyfit(x2,y2,3);
% Plot the results.
plot(z1,polyval(w1,z1),z2,polyval(w2,z2),x,y,'o'),...
xlabel('x'),ylabel('y')
% Compute the coefficient of determination.
mu1 = mean(y1);
mu2 = mean(y2);
S = sum((y1-mu1).^2) + sum((y2-mu2).^2)
J = sum((polyval(w1,x1)-y1).^2) + sum((polyval(w2,x2)-y2).^2)
r2 = 1 - J/S
```

The values are $S = 12.8618$, $J = 0.0868$, and $r^2 = 0.9932$, which indicates a very good fit. The plot is shown in Figure 5.6–6. The curves are not tangent, but they need not be for this example. However, some applications require the curves to be tangent, and in Section 7.4 we develop a method for fitting cubics that are tangent. Here we estimated the transition point $x = 15$ at which to separate the two cubics. You can experiment with other values to improve the fit, although the fit we achieved is very good.

If we want to estimate a value of y for $0 \le x \le 15$, we use the first cubic, which is found from `w1` and is

$$y = 0.00249494949495x^3 - 0.09924512987013x^2 + 1.03759920634921x + 0.11742424242424$$

To estimate a value of y for $15 < x \le 40$, we use the second cubic, which is found from `w2` and is

$$y = 0.000203642191 14x^3 - 0.01381168831169x^2 + 0.19723598068598x + 1.24452447552445$$

Note that MATLAB reports the coefficients to 14 decimal places.

To demonstrate the robustness of the cubic polynomials, their coefficients were rounded off to eight decimal places. For these rounded coefficients,

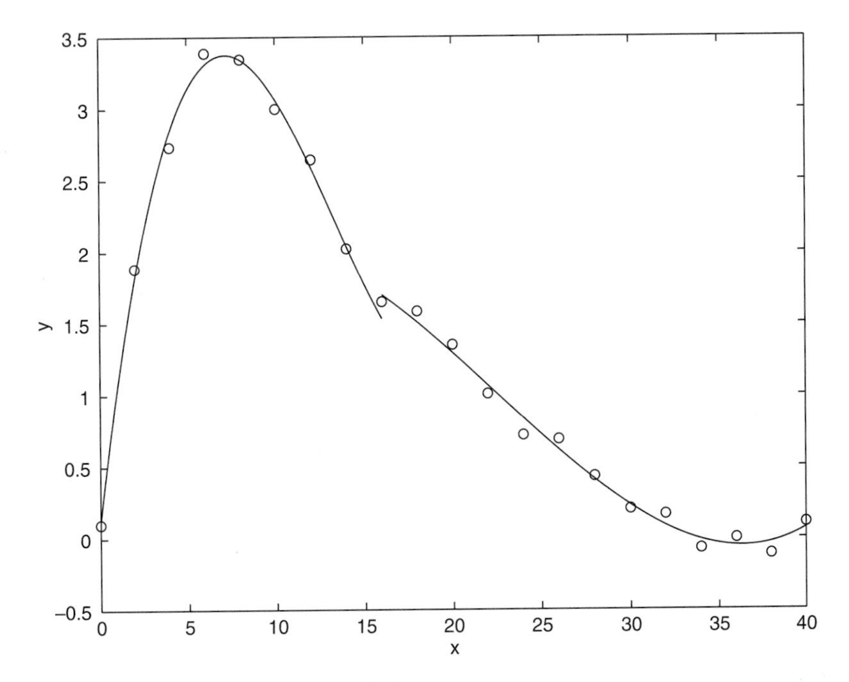

Figure 5.6–6 Use of two cubics to fit data.

$S = 12.8618$, $J = 0.0868$, and $r^2 = 0.9932$; these values are identical to the results obtained with the more accurate coefficients.

Scaling the Data

The effect of computational errors in computing the coefficients can be lessened by properly scaling the x values. When the function `polyfit(x,y,n)` is executed, it will issue a warning message if the polynomial degree `n` is greater than or equal to the number of data points (because there will not be enough equations for MATLAB to solve for the coefficients), or if the vector `x` has repeated, or nearly repeated, points, or if the vector `x` needs centering and/or scaling. The alternate syntax `[p, s, mu] = polyfit (x,y,n)` finds the coefficients `p` of a polynomial of degree `n` in terms of the variable

$$\hat{x} = (x - \mu_x)/\sigma_x$$

The output variable `mu` is a two-element vector, $[\mu_x, \sigma_x]$, where μ_x is the mean of x, and σ_x is the standard deviation of x (the standard deviation is discussed in Chapter 7).

You can scale the data yourself before using `polyfit`. Some common scaling methods are

$$\hat{x} = x - x_{\text{min}} \quad \text{or} \quad \hat{x} = x - \mu_x$$

if the range of x is small, or

$$\hat{x} = \frac{x}{x_{\text{max}}} \quad \text{or} \quad \hat{x} = \frac{x}{x_{\text{mean}}}$$

if the range of x is large.

EXAMPLE 5.6–1 Estimation of Traffic Flow

Civil and transportation engineers must often estimate the future traffic flow on roads and bridges to plan for maintenance or possible future expansion. The following data gives the number of vehicles (in millions) crossing a bridge each year for 10 years. Fit a cubic polynomial to the data and use the fit to estimate the flow in the year 2000.

Year	1990	1991	1992	1993	1994	1995	1996	1997	1998	1999
Vehicle flow (millions)	2.1	3.4	4.5	5.3	6.2	6.6	6.8	7	7.4	7.8

■ Solution

If we attempt to fit a cubic to this data, as in the following session, we get a warning message.

```
>>Year = [1990:1999];
>>Veh_Flow = [2.1,3.4,4.5,5.3,6.2,6.6,6.8,7,7.4,7.8];
>>p = polyfit(Year,Veh_Flow,3)
Warning: Polynomial is badly conditioned.
```

The problem is caused by the large values of the independent variable `Year`. Because their range is small, we can simply subtract 1990 from each value. Continue the session as follows.

```
>>x = Year-1990;
>>p = polyfit(x,Veh_Flow,3)
p =
   0.0087      -0.1851     1.5991    2.0362
>>J = sum((polyval(p3,x)-y).^2);
>>S = sum((y-mean(y)).^2);
>>r2 = 1 - J/S
r2 =
   0.9972
```

Thus the polynomial fit is good because the coefficient of determination is 0.9972. The corresponding polynomial is

$$f = 0.0087(t - 1990)^3 - 0.1851(t - 1990)^2 + 1.5991(t - 1990) + 2.0362$$

where f is the traffic flow in millions of vehicles, and t is the time in years measured from 0. We can use this equation to estimate the flow at the year 2000 by substituting $t = 2000$, or by typing in MATLAB `polyval(p,10)`. Rounded to one decimal place, the answer is 8.2 million vehicles.

Using Residuals

In some previous examples we used the value of the coefficient of determination, r^2, as an indication of the success of the fit. Here we will show how to use the residuals as a guide to choosing an appropriate function to describe the data. In general, if you see a pattern in the plot of the residuals, it indicates that another function can be found to describe the data better.

Modeling Bacteria Growth — EXAMPLE 5.6–2

Bacteria have beneficial uses, such as in the production of foods, beverages, and medicines. On the other hand, some bacteria are important indicators of poor environmental quality. Engineers in the food, environmental, and chemical industries often must understand and be able to model the growth of bacteria. The following table gives data on the growth of a certain bacteria population with time. Fit an equation to this data.

Time (min)	Bacteria (ppm)	Time (min)	Bacteria (ppm)
0	6	10	350
1	13	11	440
2	23	12	557
3	33	13	685
4	54	14	815
5	83	15	990
6	118	16	1170
7	156	17	1350
8	210	18	1575
9	282	19	1830

■ Solution

There is no simple equation that describes bacterial growth under a variety of conditions, so we do not have a predetermined mathematical function to use. The exponential growth law, $y = be^{mt}$ or its equivalent form $y = b(10)^{mt}$, sometimes fits the data, and we will see if it works here. There are other functions that could be tried, but for brevity, here we will try three polynomial fits (linear, quadratic, and cubic), and an exponential fit. We will examine their residuals to determine which best fits the data. The script file is given below. Note that we can write the exponential form as $y = b(10)^{mt} = 10^{mt+a}$, where $b = 10^a$. The coefficients a and m will be obtained with the `polyfit` function.

```
% Time data
x = [0:19];
% Population data
y = [6,13,23,33,54,83,118,156,210,282,...
350,440,557,685,815,990,1170,1350,1575,1830];
% Linear fit
p1 = polyfit(x,y,1);
% Quadratic fit
p2 = polyfit(x,y,2);
% Cubic fit
```

```
p3 = polyfit(x,y,3);
% Exponential fit
p4 = polyfit(x,log10(y),1);
% Residuals
res1 = polyval(p1,x)-y;
res2 = polyval(p2,x)-y;
res3 = polyval(p3,x)-y;
res4 = 10.^polyval(p4,x)-y;
```

You can then plot the residuals as shown in Figure 5.6–7. Note that there is a definite pattern in the residuals of the linear fit. This indicates that the linear function cannot match the curvature of the data. The residuals of the quadratic fit are much smaller, but there is still a pattern, with a random component. This indicates that the quadratic function also cannot match the curvature of the data. The residuals of the cubic fit are even smaller, with no strong pattern and a large random component. This indicates that a polynomial degree higher than three will not be able to match the data curvature any better than the cubic. The residuals for the exponential are the largest of all, and indicate a poor fit. Note also how the residuals systematically increase with t, indicating that the exponential cannot describe the data's behavior after a certain time.

Thus the cubic is the best fit of the four models considered. Its coefficient of determination is $r^2 = 0.9999$. The model is

$$y = 0.1916t^3 + 1.2082t^2 + 3.607t + 7.7307$$

where y is the bacteria population in ppm and t is time in minutes.

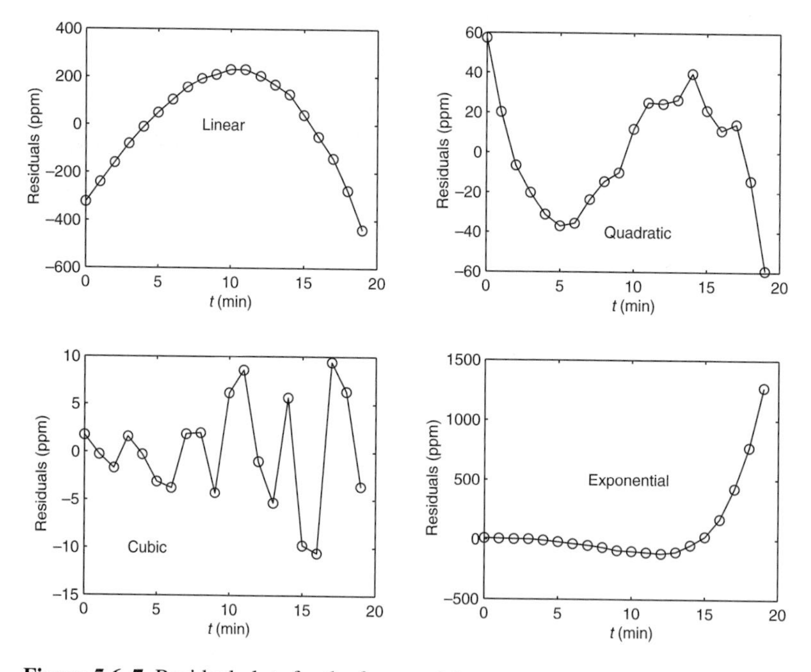

Figure 5.6–7 Residual plots for the four models.

Constraining Models to Pass through a Given Point

Many applications require a model whose form is dictated by physical principles. For example, the force-extension model of a spring must pass through the origin (0, 0) because the spring exerts no force when it is unstretched. Thus a linear spring model $f = a_1 x + a_2$ must have a zero value for a_2. However, in general the `polyfit` function will give a nonzero value for a_2 because of the scatter or measurement error that is usually present in the data.

Linear Model To obtain a zero-intercept model of the form $y = a_1 x$, we must derive the equation for a_1 from basic principles. The sum of the squared residuals in this case is

$$J = \sum_{i=1}^{m} (a_1 x_i - y_i)^2$$

Computing the derivative $\partial J/\partial a_1$ and setting it equal to 0 gives the result

$$a_1 \sum_{i=1}^{m} x_i^2 = \sum_{i=1}^{m} x_i y_i \tag{5.6–6}$$

which can be easily solved for a_1.

Quadratic Model For the quadratic model, $y = a_1 x^2 + a_2 x + a_3$, the coefficient a_3 must be 0 for the curve to pass through the origin. The sum of the squared residuals in this case is

$$J = \sum_{i=1}^{m} \left(a_1 x_i^2 + a_2 x_i - y_i\right)^2$$

Computing the derivatives $\partial J/\partial a_1$ and $\partial J/\partial a_2$ and setting them equal to 0 gives the equations:

$$a_1 \sum_{i=1}^{m} x_i^4 + a_2 \sum_{i=1}^{m} x_i^3 = \sum_{i=1}^{m} y_i x_i^2 \tag{5.6–7}$$

$$a_1 \sum_{i=1}^{m} x_i^3 + a_2 \sum_{i=1}^{m} x_i^2 = \sum_{i=1}^{m} y_i x_i \tag{5.6–8}$$

These can be solved for a_1 and a_2.

If the model is required to pass through a point not at the origin, say the point (x_0, y_0), subtract x_0 from all the x values, subtract y_0 from all the y values, and then use the above equations to find the coefficients. The resulting equations will be of the form

$$y = a_1(x - x_0) + y_0 \tag{5.6–9}$$

and

$$y = a_1(x - x_0)^2 + a_2(x - x_0) + y_0 \tag{5.6–10}$$

You can derive equations for other functions in a similar manner.

Multiple Linear Regression

Suppose that y is a linear function of two or more variables, $x_1, x_2, \ldots$. For example,

$$y = a_0 + a_1 x_1 + a_2 x_2 \tag{5.6–11}$$

To find the coefficient values a_0, a_1, and a_2 to fit a set of data (y, x_1, x_2) in the least squares sense, we can make use of the fact that the left-division method for solving linear equations uses the least squares method when the equation set is overdetermined (see Chapter 6, Section 6.5). To use this method, let n be the number of data points and write the linear equation in matrix form as follows.

$$\mathbf{Xa} = \mathbf{y} \tag{5.6–12}$$

where

$$\mathbf{a} = \begin{bmatrix} a_0 \\ a_1 \\ a_2 \end{bmatrix} \tag{5.6–13}$$

$$\mathbf{X} = \begin{bmatrix} 1 & x_{11} & x_{21} \\ 1 & x_{12} & x_{22} \\ 1 & x_{13} & x_{23} \\ & \cdots & \\ 1 & x_{1n} & x_{2n} \end{bmatrix} \tag{5.6–14}$$

$$\mathbf{y} = \begin{bmatrix} y_1 \\ y_2 \\ y_3 \\ \cdots \\ y_n \end{bmatrix} \tag{5.6–15}$$

where x_{1i}, x_{2i}, and y_i are the data, $i = 1, \ldots, n$. The solution for the coefficients is given by `a = X\y`.

EXAMPLE 5.6–3 Breaking Strength and Alloy Composition

Chemical, civil, mechanical, aerospace, and biomedical engineers need to predict the strength of metal parts as a function of their alloy composition. The tension force y required to break a steel bar is a function of the percentage x_1 and x_2 of each of two alloying elements present in the metal. The following table gives some pertinent data. Obtain a linear model $y = a_0 + a_1 x_1 + a_2 x_2$ to describe the relationship.

Breaking strength (kN) y	% of element 1 x_1	% of element 2 x_2
7.1	0	5
19.2	1	7
31	2	8
45	3	11

■ **Solution**

The script file is as follows:

```
x1 = [0:3]';x2 = [5,7,8,11]';
y = [7.1,19.2,31,45]';
X = [ones(size(x1)), x1, x2];
a = X\y
yp = X*a;
Max_Percent_Error = 100*max(abs((yp-y)./y))
```

The vector `yp` is the vector of breaking-strength values predicted by the model. The scalar `Max_Percent_Error` is the maximum percent error in the four predictions. The results are `a = [0.8000, 10.2429, 1.2143]'` and `Max_Percent_Error = 3.2193`. Thus the model is $y = 0.8 + 10.2429x_1 + 1.2143x_2$. The maximum percent error of the model's predictions, as compared to the given data, is 3.2193 percent.

Linear-in-the-Parameters Regression

Sometimes we want to fit an expression that is neither a polynomial nor a function that can be converted to linear form by a logarithmic or other transformation. In some cases we can still do a least squares fit if the function is a linear expression in terms of its parameters. The following example illustrates the method.

Response of a Biomedical Instrument — EXAMPLE 5.6–4

Biomedical instrumentation is an important engineering field. These devices are used to measure many quantities, such as body temperature, blood oxygen level, heart rate, and so forth. Engineers developing such devices often need to obtain a *response* curve that describes how fast the instrument can make measurements. The theory of instrumentation shows that often the response can be described by one of the following equations, where v is the voltage output, and t is time. In both models, the voltage reaches a steady-state constant value as $t \to \infty$, and T is the time required for the voltage to equal 95 percent of the steady-state value.

$$v(t) = a_1 + a_2 e^{-3t/T} \qquad \text{(First-order model)}$$

$$v(t) = a_1 + a_2 e^{-3t/T} + a_3 t e^{-3t/T} \qquad \text{(Second-order model)}$$

The following data gives the output voltage of a certain device as a function of time. Obtain a function that describes this data.

t (s)	0	0.3	0.8	1.1	1.6	2.3	3
v (V)	0	0.6	1.28	1.5	1.7	1.75	1.8

■ Solution

Plotting the data we estimate that it takes approximately 3 seconds for the voltage to become constant. Thus we estimate that $T = 3$. The first-order model written for each of the n data points results in n equations, which can be expressed as follows:

$$\begin{bmatrix} 1 & e^{-t_1} \\ 1 & e^{-t_2} \\ \cdots & \cdots \\ 1 & e^{-t_n} \end{bmatrix} \begin{bmatrix} a_1 \\ a_2 \end{bmatrix} = \begin{bmatrix} y_1 \\ y_2 \\ \cdots \\ y_n \end{bmatrix}$$

or, in matrix form,

$$\mathbf{Xa} = \mathbf{y}'$$

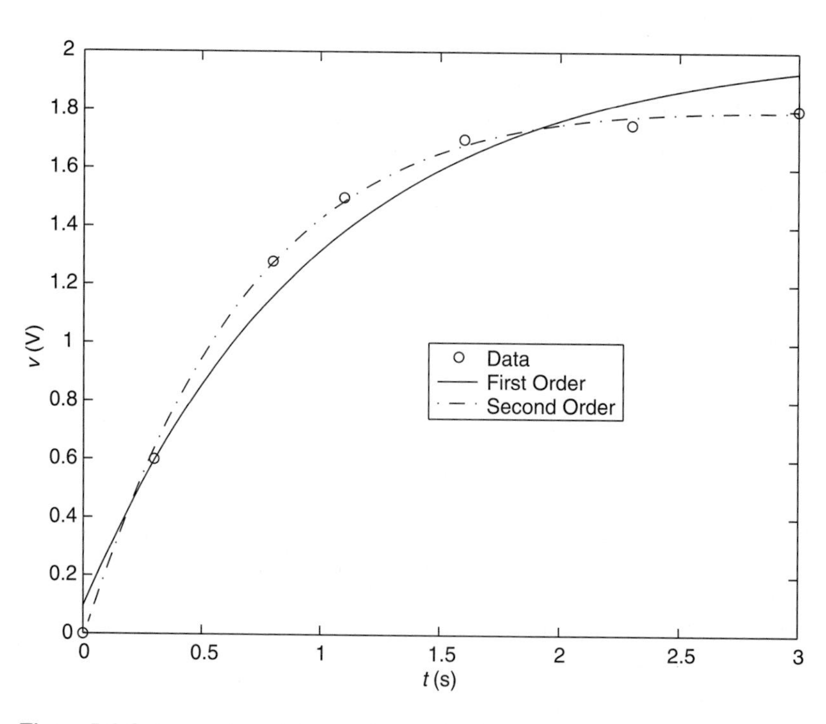

Figure 5.6–8 Comparison of first- and second-order model fits.

which can be solved for the coefficient vector **a** using left division. The following MATLAB script solves the problem.

```
t = [0,0.3,0.8,1.1,1.6,2.3,3];
y = [0,0.6,1.28,1.5,1.7,1.75,1.8];
X = [ones(size(t));exp(-t)]';
a = X\y'
```

The answer is $a_1 = 2.0258$ and $a_2 = -1.9307$.

A similar procedure can be followed for the second-order model.

$$\begin{bmatrix} 1 & e^{-t_1} & t_1e^{-t_1} \\ 1 & e^{-t_2} & t_2e^{-t_2} \\ \cdots & \cdots & \\ 1 & e^{-t_n} & t_ne^{-t_n} \end{bmatrix} \begin{bmatrix} a_1 \\ a_2 \\ a_3 \end{bmatrix} = \begin{bmatrix} y_1 \\ y_2 \\ \cdots \\ y_n \end{bmatrix}$$

Continue the previous script as follows.

```
X = [ones(size(t));exp(-t);t.*exp(-t)]';
a = X\y'
```

The answer is $a_1 = 1.7496$, $a_2 = -1.7682$, and $a_3 = 0.8885$. The two models are plotted with the data in Figure 5.6–8. Clearly the second-order model gives the better fit.

5.7 The Basic Fitting Interface

MATLAB supports curve fitting through the Basic Fitting interface. Using this interface, you can quickly perform basic curve fitting tasks within the same easy-to-use environment. The interface is designed so that you can:

- Fit data using a cubic spline or a polynomial up to degree 10.
- Plot multiple fits simultaneously for a given data set.
- Plot the residuals.
- Examine the numerical results of a fit.
- Interpolate or extrapolate a fit.
- Annotate the plot with the numerical fit results and the norm of residuals.
- Save the fit and evaluated results to the MATLAB workspace.

Depending on your specific curve fitting application, you can use the Basic Fitting interface, the command line functions, or both. Note: you can use the Basic Fitting interface only with two-dimensional data. However, if you plot multiple data sets as a subplot, and at least one data set is two-dimensional, then the interface is enabled.

Two panes of the Basic Fitting interface are shown in Figure 5.7–1. To reproduce this state:

1. Plot some data.
2. Select **Basic Fitting** from the **Tools** menu.
3. Click the right arrow button once.

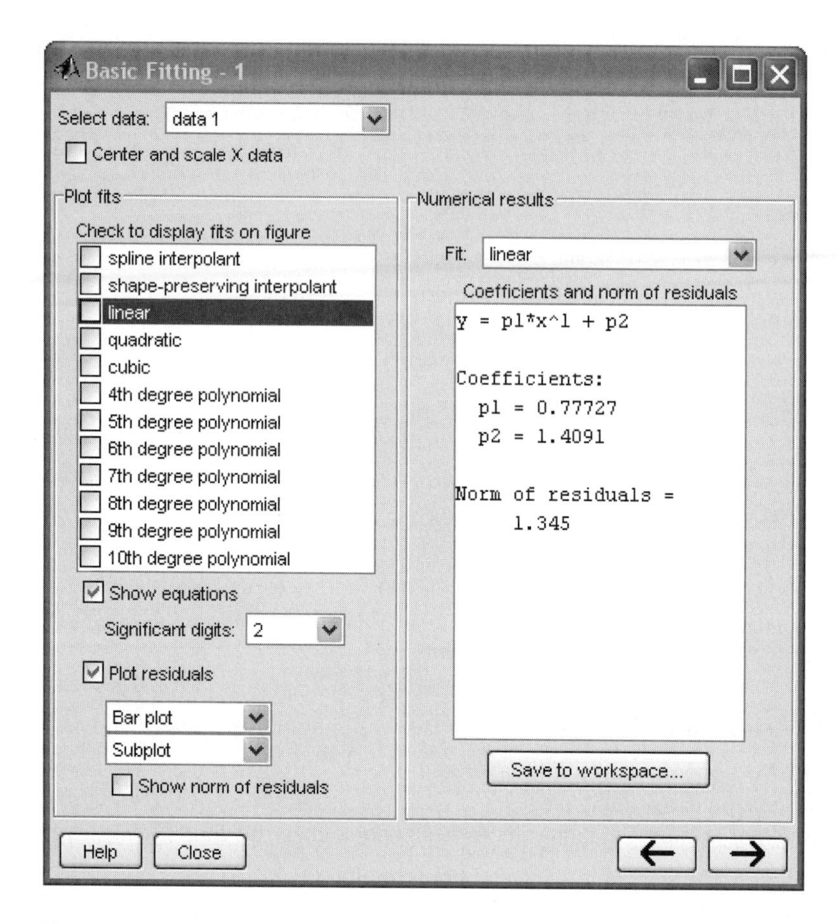

Figure 5.7–1 The Basic Fitting interface.

The third pane is used for interpolating or extrapolating a fit. It appears when you click the right arrow button a second time.

At the top of the first pane is the **Select data** window which contains the names of all the data sets you display in the Figure window associated with the Basic Fitting interface. Use this menu to select the data set to be fit. You can perform multiple fits for the current data set. Use the Plot Editor to change the name of a data set. The remaining items on the first pane are used as follows.

- **Center and scale X data.** If checked, the data is centered at zero mean and scaled to unit standard deviation. You may need to center and scale your data to improve the accuracy of the subsequent numerical computations. As described in the previous section, a warning is returned to the Command window if a fit produces results that may be inaccurate.
- **Plot fits.** This panel allows you to visually explore one or more fits to the current data set.

- **Check to display fits on figure.** Select the fits you want to display for the current data set. You can choose as many fits for a given data set as you want. However, if your data set has n points, then you should use polynomials with, at most, n coefficients. If you fit using polynomials with more than n coefficients, the interface will automatically set a sufficient number of coefficients to 0 during the calculation so that a solution can be obtained.
- **Show equations.** If checked, the fit equation is displayed on the plot.
- **Significant digits.** Select the significant digits associated with the fit coefficient display.
- **Plot residuals.** If checked, the residuals are displayed. You can display the residuals as a bar plot, a scatter plot, a line plot using either the same figure window as the data or using a separate figure window. If you plot multiple data sets as a subplot, then residuals can be plotted only in a separate figure window. See Figure 5.7–2.
- **Show norm of residuals.** If checked, the norm of residuals is displayed. The norm of residuals is a measure of the goodness of fit, where a smaller value indicates a better fit. The norm is the square root of the sum of the squares of the residuals.

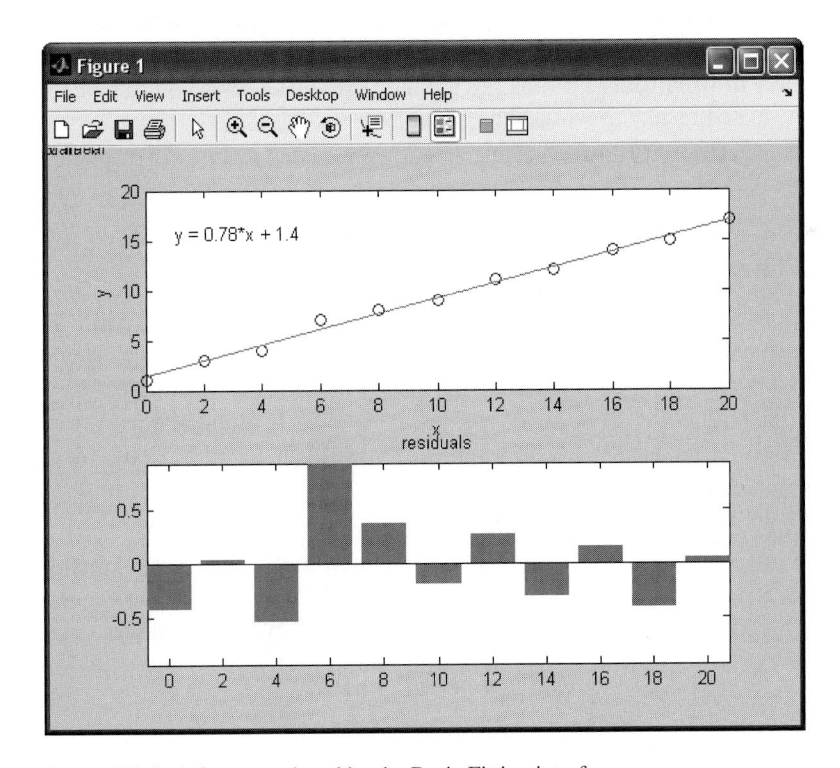

Figure 5.7–2 A figure produced by the Basic Fitting interface.

The second pane of the Basic Fitting Interface is labeled *Numerical Results.* This pane enables you to explore the numerical results of a single fit to the current data set without plotting the fit. It contains three items.

- **Fit.** Use this menu to select an equation to fit to the current data set. The fit results are displayed in the box below the menu. Note that selecting an equation in this menu does not affect the state of the **Plot fits** selection. Therefore, if you want to display the fit in the data plot, you may need to check the relevant check box in **Plot fits**.
- **Coefficients and norm of residuals.** Displays the numerical results for the equation selected in **Fit.** Note that when you first open the **Numerical Results** panel, the results of the last fit you selected in **Plot fits** are displayed.
- **Save to workspace.** Launches a dialog box that allows you to save the fit results to workspace variables.

The third pane of the Basic Fitting interface contains three items.

- **Find $Y = f(X)$.** Use this to interpolate or extrapolate the current fit. Enter a scalar or a vector of values corresponding to the independent variable (`X`). The current fit is evaluated after you click on the **Evaluate** button, and the results are displayed in the associated window. The current fit is displayed in the **Fit** window.
- **Save to workspace.** Launches a dialog box that allows you to save the evaluated results to workspace variables.
- **Plot evaluated results.** If checked, the evaluated results are displayed on the data plot.

5.8 Three-Dimensional Plots

Functions of two variables are sometimes difficult to visualize with a two-dimensional plot. Fortunately, MATLAB provides many functions for creating three-dimensional plots. Here we will summarize the basic functions to create three types of plots: line plots, surface plots, and contour plots. Information about the related functions is available in MATLAB help (category `graph 3d`).

Three-Dimensional Line Plots

Lines in three-dimensional space can be plotted with the `plot3` function. Its syntax is `plot3(x,y,z)`. For example, the following equations generate a three-dimensional curve as the parameter t is varied over some range:

$$x = e^{-0.05t} \sin t$$

$$y = e^{-0.05t} \cos t$$

$$z = t$$

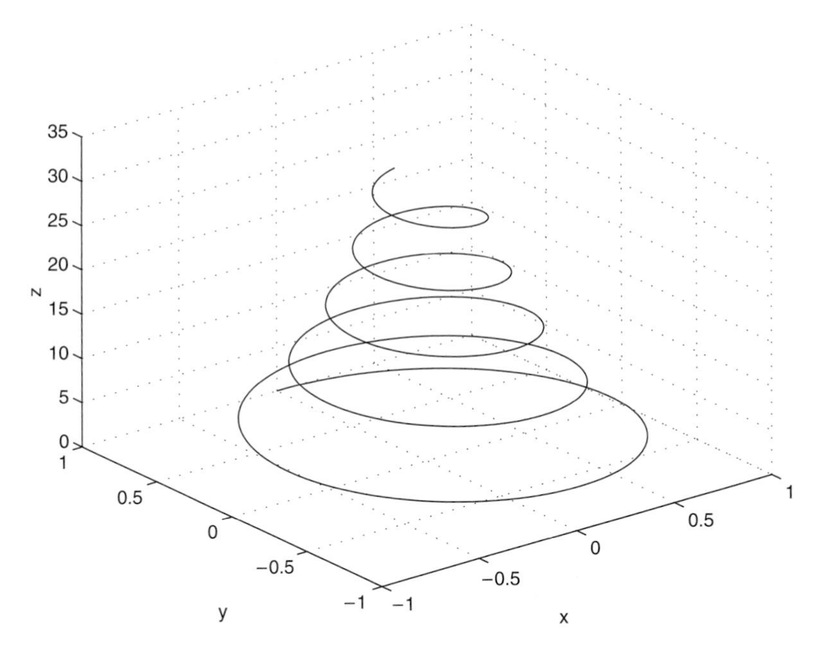

Figure 5.8–1 The curve $x = e^{-0.05t} \sin t$, $y = e^{-0.05t} \cos t$, $z = t$ plotted with the `plot3` function.

If we let t vary from $t = 0$ to $t = 10\pi$, the sin and cos functions will vary through five cycles, while the absolute values of x and y become smaller as t increases. This process results in the spiral curve shown in Figure 5.8–1, which was produced with the following session.

```
>>t = [0:pi/50:10*pi];
>>plot3(exp(-0.05*t).*sin(t),exp(-0.05*t).*cos(t),t),...
xlabel('x'),ylabel('y'),zlabel('z'),grid
```

Note that the `grid` and label functions work with the `plot3` function, and that we can label the z-axis by using the `zlabel` function, which we have seen for the first time. Similarly, we can use the other plot-enhancement functions discussed in Sections 5.1 and 5.2 to add a title and text and to specify line type and color.

Surface Mesh Plots

The function $z = f(x, y)$ represents a surface when plotted on xyz axes, and the `mesh` function provides the means to generate a surface plot. Before you can use this function, you must generate a grid of points in the xy plane, and then evaluate the function $f(x, y)$ at these points. The `meshgrid` function generates the grid. Its syntax is `[X,Y] = meshgrid(x,y)`. If `x = [xmin:xspacing:xmax]` and `y = [ymin:yspacing:ymax]`, then this function will generate the coordinates of a rectangular grid with one corner at $(xmin, ymin)$ and the opposite corner at (*xmax*, *ymax*). Each rectangular panel in the grid will have a width equal to *xspacing* and a depth equal to *yspacing*. The resulting matrices *X* and *Y* contain the coordinate pairs of every point in the grid. These pairs are then used to evaluate the function.

The function `[X,Y] = meshgrid(x)` is equivalent to `[X,Y] = meshgrid(x,x)` and can be used if x and y have the same minimum values, the same maximum values, and the same spacing. Using this form, you can type `[X,Y] = meshgrid (min:spacing:max)`, where `min` and `max` specify the minimum and maximum values of both x and y and `spacing` is the desired spacing of the x and y values.

After the grid is computed, you create the surface plot with the `mesh` function. Its syntax is `mesh(x,y,z)`. The grid, label, and text functions can be used with the `mesh` function. The following session shows how to generate the surface plot of the function $z = xe^{-[(x-y^2)^2+y^2]}$, for $-2 \le x \le 2$ and $-2 \le y \le 2$, with a spacing of 0.1. This plot appears in Figure 5.8–2.

```
>>[X,Y] = meshgrid(-2:0.1:2);
>>Z = X.*exp(-((X-Y.^2).^2+Y.^2));
>>mesh(X,Y,Z),xlabel('x'),ylabel('y'),zlabel('z')
```

Be careful not to select too small a spacing for the x and y values for two reasons: (1) Small spacing creates small grid panels, which make the surface difficult to visualize, and (2) the matrices X and Y can become too large.

The `surf` and `surfc` functions are similar to `mesh` and `meshc` except that the former create a shaded surface plot. You can use the Camera toolbar and some menu items in the Figure window to change the view and lighting of the figure.

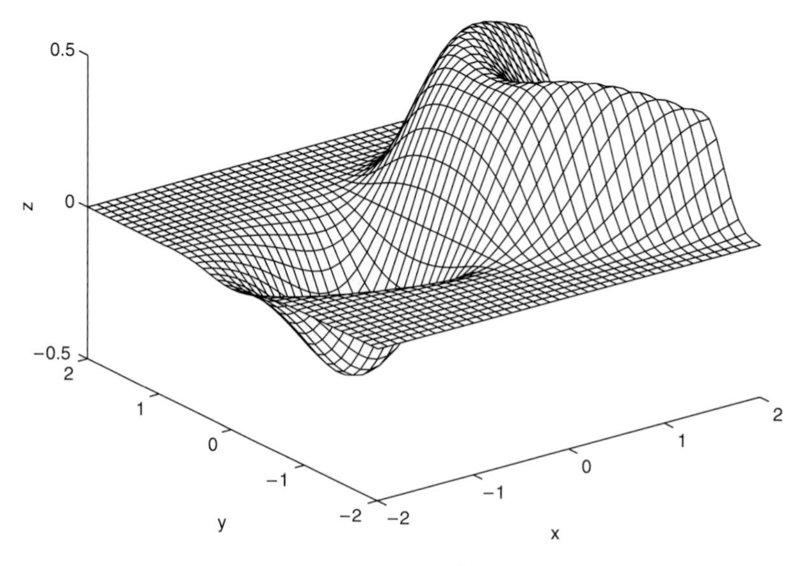

Figure 5.8–2 A plot of the surface $z = xe^{-[(x-y^2)^2+y^2]}$ created with the `mesh` function.

Contour Plots

Topographic plots show the contours of the land by means of constant elevation lines. These lines are also called *contour lines,* and such a plot is called a *contour plot*. If you walk along a contour line, you remain at the same elevation. Contour plots can help you visualize the shape of a function. They can be created with the `contour` function, whose syntax is `contour(X,Y,Z)`. You use this function the same way you use the `mesh` function; that is, first use the `meshgrid` function to generate the grid and then generate the function values. The following session generates the contour plot of the function whose surface plot is shown in Figure 5.8–2; namely, $z = xe^{-[(x-y^2)^2+y^2]}$, for $-2 \le x \le 2$ and $-2 \le y \le 2$, with a spacing of 0.1. This plot appears in Figure 5.8–3.

```
>>[X,Y] = meshgrid(-2:0.1:2);
>>Z = X.*exp(-((X- Y.^2).^2+Y.^2));
>>contour(X,Y,Z),xlabel('x'),ylabel('y')
```

Contour plots and surface plots can be used together to clarify the function. For example, unless the elevations are labeled on contour lines, you cannot tell whether there is a minimum or a maximum point. However, a glance at the surface plot will make this easy to determine. On the other hand, accurate measurements are not possible on a surface plot; these can be done on the contour plot because no

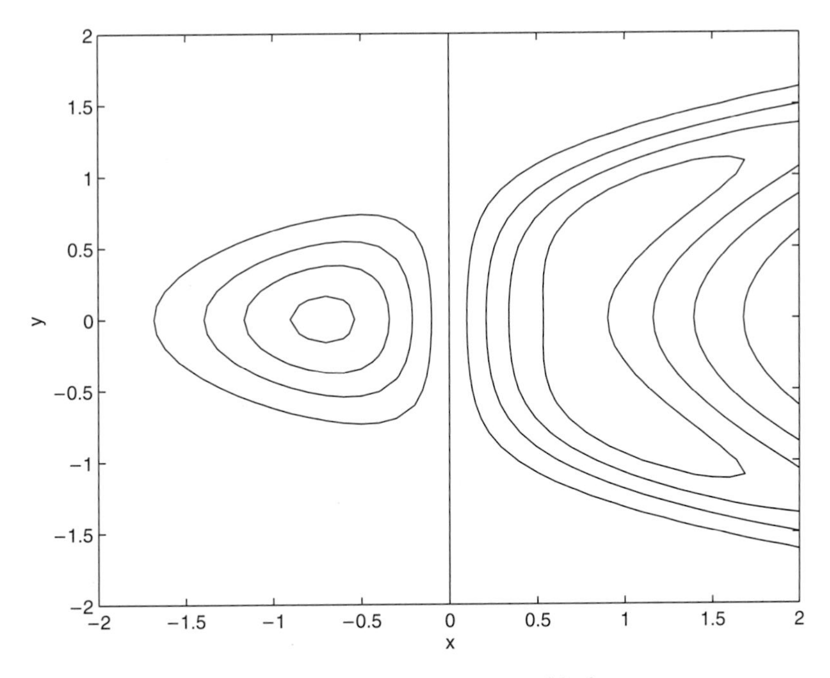

Figure 5.8–3 A contour plot of the surface $z = xe^{-[(x-y^2)^2+y^2]}$ created with the `contour` function.

distortion is involved. Thus a useful function is `meshc`, which shows the contour lines beneath the surface plot. The `meshz` function draws a series of vertical lines under the surface plot, while the `waterfall` function draws mesh lines in one direction only. The results of these functions are shown in Figure 5.8–4 for the function $z = xe^{-(x^2+y^2)}$.

Table 5.8–1 summarizes the functions introduced in this section. For other 3D plot types, type `help specgraph`.

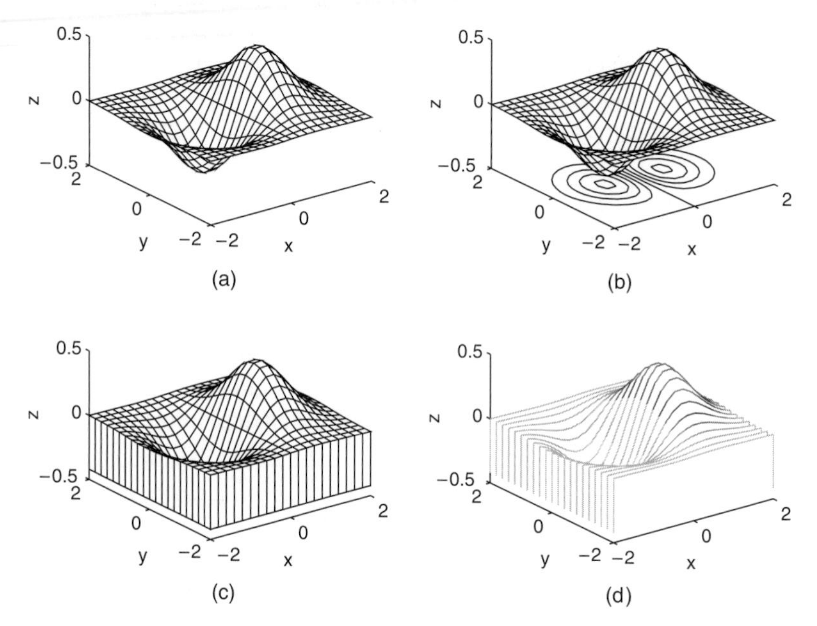

Figure 5.8–4 Plots of the surface $z = xe^{-(x^2+y^2)}$ created with the `mesh` function and its variant forms: `meshc`, `meshz`, and `waterfall`. a) `mesh`, b) `meshc`, c) `meshz`, d) `waterfall`.

Table 5.8–1 Three-dimensional plotting functions

Function	Description
`contour(x,y,z)`	Creates a contour plot.
`mesh(x,y,z)`	Creates a three-dimensional mesh surface plot.
`meshc(x,y,z)`	Same as `mesh` but draws a contour plot under the surface.
`meshz(x,y,z)`	Same as `mesh` but draws a series of vertical reference lines under the surface.
`surf(x,y,z)`	Creates a shaded three-dimensional mesh surface plot.
`surfc(x,y,z)`	Same as `surf` but draws a contour plot under the surface.
`[X,Y] = meshgrid(x,y)`	Creates the matrices `X` and `Y` from the vectors `x` and `y` to define a rectangular grid.
`[X,Y] = meshgrid(x)`	Same as `[X,Y]= meshgrid(x,x)`.
`waterfall(x,y,z)`	Same as `mesh` but draws mesh lines in one direction only.

Test Your Understanding

T5.8–1 Create a surface plot and a contour plot of the function $z = (x - 2)^2 + 2xy + y^2$.

5.9 Summary

This chapter explained how to use the powerful MATLAB commands to create effective and pleasing two-dimensional and three-dimensional plots. You learned an important application of plotting—function discovery—which is the technique for using data plots to obtain a mathematical function that describes the data. Regression can be used to develop a model for cases where there is considerable scatter in the data. These techniques are widely used in engineering applications because engineers frequently need to use mathematical models to predict how their proposed designs will work.

The following guidelines will help you create plots that effectively convey the desired information:

- Label each axis with the name of the quantity being plotted *and its units!*
- Use regularly spaced tick marks at convenient intervals along each axis.
- If you are plotting more than one curve or data set, label each on its plot or use a legend to distinguish them.
- If you are preparing multiple plots of a similar type or if the axes' labels cannot convey enough information, use a title.
- If you are plotting measured data, plot each data point in a given set with the same symbol, such as a circle, square, or cross.
- If you are plotting points generated by evaluating a function (as opposed to measured data), do *not* use a symbol to plot the points. Instead, connect the points with solid lines.

Table 5.9–1 is a guide to the MATLAB commands introduced in this chapter.

Table 5.9–1 Guide to MATLAB commands introduced in Chapter 5

Basic xy plotting commands	Table 5.1–1
Specifiers for data markers, line types, and colors	Table 5.2–1
Plot enhancement commands	Table 5.2–2
Specialized plot commands	Table 5.3–1
The `set` command	Table 5.3–2
The `polyfit` function	Table 5.5–1
Functions for polynomial regression	Table 5.6–1
Three-dimensional plotting functions	Table 5.8–1

Key Terms with Page References

Problems

You can find the answers to problems marked with an asterisk at the end of the text.

Be sure to label and format properly any plots required by the following problems. Label each axis properly. Use a legend, data markers, or different line types as needed. Choose proper axis scaling and tick-mark spacing. Use a title, a grid, or both if they help to interpret the plot.

Section 5.1

1.* *Breakeven analysis* determines the production volume at which the total production cost is equal to the total revenue. At the breakeven point, there is neither profit nor loss. In general, production costs consist of fixed costs and variable costs. Fixed costs include salaries of those not directly involved with production, factory maintenance costs, insurance costs, and so on. Variable costs depend on production volume and include material costs, labor costs, and energy costs. In the following analysis, assume that we produce only what we can sell; thus the production quantity equals the sales. Let the production quantity be Q, in gallons per year.

Consider the following costs for a certain chemical product:
Fixed cost: $3 million per year.
Variable cost: 2.5 cents per gallon of product.
The selling price is 5.5 cents per gallon.

Use this data to plot the total cost and the revenue versus Q, and graphically determine the breakeven point. Fully label the plot and mark the breakeven point. For what range of Q is production profitable? For what value of Q is the profit a maximum?

2. Consider the following costs for a certain chemical product:
Fixed cost: $2.045 million/year.
Variable costs:
Material cost: 62 cents per gallon of product.
Energy cost: 24 cents per gallon of product.
Labor cost: 16 cents per gallon of product.

Assume that we produce only what we sell. Let P be the selling price in dollars per gallon. Suppose that the selling price and the sales quantity Q are interrelated as follows: $Q = 6 \times 10^6 - 1.1 \times 10^6 P$. Accordingly, if we raise the price, the product becomes less competitive and sales drop.

Use this information to plot the fixed and total variable costs versus Q, and graphically determine the breakeven point(s). Fully label the plot and mark the breakeven points. For what range of Q is the production profitable? For what value of Q is the profit a maximum?

3.* Roots of polynomials appear in many engineering applications, such as electrical circuit design and structural vibrations. Find the real roots of the polynomial equation

$$4x^5 + 3x^4 - 95x^3 + 5x^2 - 10x + 80 = 0$$

in the range $-10 \le x \le 10$ by plotting the polynomial.

4. To compute the forces in structures, engineers sometimes must solve equations similar to the following. Use the `fplot` function to find all the positive roots of this equation:

$$x \tan x = 7$$

5.* Cables are used to suspend bridge decks and other structures. If a heavy uniform cable hangs suspended from its two endpoints, it takes the shape of a *catenary* curve whose equation is

$$y = a \cosh\left(\frac{x}{a}\right)$$

where a is the height of the lowest point on the chain above some horizontal reference line, x is the horizontal coordinate measured to the right from the lowest point, and y is the vertical coordinate measured up from the reference line.

Let $a = 10$ m. Plot the catenary curve for $-20 \le x \le 30$ m. How high is each endpoint?

6. Using estimates of rainfall, evaporation, and water consumption, the town engineer developed the following model of the water volume in the reservoir as a function of time.

$$V(t) = 10^9 + 10^8(1 - e^{-t/100}) - 10^7 t$$

where V is the water volume in liters, and t is time in days. Plot $V(t)$ versus t. Use the plot to estimate how many days it will take before the water volume in the reservoir is 50 percent of its initial volume of 10^9 L.

7. It is known that the following Leibniz series converges to the value $\pi/4$ as $n \to \infty$.

$$S(n) = \sum_{k=0}^{n} (-1)^k \frac{1}{2k+1}$$

Plot the difference between $\pi/4$ and the sum $S(n)$ versus n for $0 \le n \le 200$.

8. A certain fishing vessel is initially located in a horizontal plane at $x = 0$ and $y = 10$ mi. It moves on a path for 10 hr such that $x = t$ and $y = 0.5t^2 + 10$, where t is in hours. An international fishing boundary is described by the line $y = 2x + 6$.

a. Plot and label the path of the vessel and the boundary.

b. The perpendicular distance of the point (x_1, y_1) from the line $Ax + By + C = 0$ is given by

$$d = \frac{Ax_1 + By_1 + C}{\pm\sqrt{A^2 + B^2}}$$

where the sign is chosen to make $d \ge 0$. Use this result to plot the distance of the fishing vessel from the fishing boundary as a function of time for $0 \le t \le 10$ hr.

Sections 5.2 and 5.4

9. Plot columns 2 and 3 of the following matrix **A** versus column 1. The data in column 1 is time (seconds). The data in columns 2 and 3 is force (newtons).

$$\mathbf{A} = \begin{bmatrix} 0 & -8 & 6 \\ 5 & -4 & 3 \\ 10 & -1 & 1 \\ 15 & 1 & 0 \\ 20 & 2 & -1 \end{bmatrix}$$

10.* Many engineering applications use the following "small angle" approximation for the sine to obtain a simpler model that is easy to understand and analyze. This approximation states that $\sin x \approx x$, where x must be in radians. Investigate the accuracy of this approximation by creating three plots. For the first, plot $\sin x$ and x versus x for $0 \le x \le 1$. For the second, plot the approximation error $\sin x - x$ versus x for $0 \le x \le 1$. For the third, plot the relative error $[\sin(x) - x]/\sin(x)$ versus x for $0 \le x \le 1$. How small must x be for the approximation to be accurate within 5 percent?

11. You can use trigonometric identities to simplify the equations that appear in many engineering applications. Confirm the identity $\tan(2x) = 2\tan x/(1 - \tan^2 x)$ by plotting both the left and the right sides versus x over the range $0 \le x \le 2\pi$.

12. The complex number identity $e^{ix} = \cos x + i \sin x$ is often used to convert the solutions of engineering design equations into a form that is relatively easy to visualize. Confirm this identity by plotting the imaginary part versus the real part for both the left and right sides over the range $0 \le x \le 2\pi$.

13. Use a plot over the range $0 \leq x \leq 5$ to confirm that $\sin(ix) = i \sinh x$.

14.* The function $y(t) = 1 - e^{-bt}$, where t is time and $b > 0$, describes many engineering processes, such as the height of liquid in a tank as it is being filled and the temperature of an object being heated. Investigate the effect of the parameter b on $y(t)$. To do this, plot y versus t for several values of b on the same plot. How long will it take for $y(t)$ to reach 98 percent of its steady-state value?

15. The following functions describe the oscillations in electrical circuits and the vibrations of machines and structures. Plot these functions on the same plot. Because they are similar, decide how best to plot and label them to avoid confusion.

$$x(t) = 10e^{-0.5t} \sin(3t + 2)$$

$$y(t) = 7e^{-0.4t} \cos(5t - 3)$$

16. The data for a tension test on a steel bar appears in the following table. The *elongation* is the change in the bar's length. The bar was stretched beyond its *elastic limit* so that a permanent elongation remained after the tension force was removed. Plot the tension force versus the elongation. Be sure to label the parts of the curve that correspond to increasing and decreasing tension.

Elongation (in. × 10^{-3})	Increasing tension force (lb)	Decreasing tension force (lb)
0	0	—
1	3500	0
2	6300	3000
3	9200	6000
4	11,500	8800
5	13,000	11,100
6	13,500	12,300
7	13,900	13,500
8	14,100	14,000
9	14,300	14,300
10	14,500	14,500

17. In certain kinds of structural vibrations, a periodic force acting on the structure will cause the vibration amplitude to repeatedly increase and decrease with time. This phenomenon, called *beating,* also occurs in musical sounds. A particular structure's displacement is described by

$$y(t) = \frac{1}{f_1^2 - f_2^2}[\cos(f_2 t) - \cos(f_1 t)]$$

where y is the displacement in inches and t is the time in seconds. Plot y versus t over the range $0 \leq t \leq 20$ for $f_1 = 8$ rad/sec and $f_2 = 1$ rad/sec. Be sure to choose enough points to obtain an accurate plot.

18.* The height $h(t)$ and horizontal distance $x(t)$ traveled by a ball thrown at an angle A with a speed v are given by

$$h(t) = vt \sin A - \frac{1}{2}gt^2$$

$$x(t) = vt \cos A$$

At Earth's surface the acceleration due to gravity is $g = 9.81$ m/s^2.

a. Suppose the ball is thrown with a velocity $v = 10$ m/s at an angle of 35°. Use MATLAB to compute how high the ball will go, how far it will go, and how long it will take to hit the ground.

b. Use the values of v and A given in part a to plot the ball's *trajectory;* that is, plot h versus x for positive values of h.

c. Plot the trajectories for $v = 10$ m/s corresponding to five values of the angle A: 20°, 30°, 45°, 60°, and 70°.

d. Plot the trajectories for $A = 45°$ corresponding to five values of the initial velocity v: 10, 12, 14, 16, and 18 m/s.

19. The perfect gas law relates the pressure p, absolute temperature T, mass m, and volume V of a gas. It states that

$$pV = mRT$$

The constant R is the *gas constant*. The value of R for air is 286.7 N · m/kg · K. Suppose air is contained in a chamber at room temperature (20°C = 293 K). Create a plot having three curves of the gas pressure in N/m^2 versus the container volume V in m^3 for $20 \leq V \leq 100$. The three curves correspond to the following masses of air in the container: $m = 1$ kg; $m = 3$ kg; and $m = 7$ kg.

20. Oscillations in mechanical structures and electric circuits can often be described by the function

$$y(t) = e^{-t/\tau} \sin(\omega t + \phi)$$

where t is time and ω is the oscillation frequency in radians per unit time. The oscillations have a period of $2\pi/\omega$, and their amplitudes decay in time at a rate determined by τ, which is called the *time constant*. The smaller τ is, the faster the oscillations die out.

a. Use these facts to develop a criterion for choosing the spacing of the t values and the upper limit on t to obtain an accurate plot of $y(t)$. (Hint: Consider two cases: $4\tau > 2\pi/\omega$ and $4\tau < 2\pi/\omega$.)

b. Apply your criterion, and plot $y(t)$ for $\tau = 10$, $\omega = \pi$, and $\phi = 2$.

c. Apply your criterion, and plot $y(t)$ for $\tau = 0.1$, $\omega = 8\pi$, and $\phi = 2$.

21. When a constant voltage was applied to a certain motor initially at rest, its rotational speed $s(t)$ versus time was measured. The data appears in the following table:

Time (sec)	1	2	3	4	5	6	7	8	10
Speed (rpm)	1210	1866	2301	2564	2724	2881	2879	2915	3010

Determine whether the following function can describe the data. If so, find the values of the constants b and c.

$$s(t) = b(1 - e^{ct})$$

Section 5.3

22. The following table shows the average temperature for each year in a certain city. Plot the data as a stem plot, a bar plot, and a stairs plot.

Year	1990	1991	1992	1993	1994
Temperature (°C)	18	19	21	17	20

23. \$10,000 invested at 5 percent interest compounded annually will grow according to the formula

$$y(k) = 10^4(1.05)^k$$

where k is the number of years ($k = 0, 1, 2 \ldots$). Plot the amount of money in the account for a 10-year period. Do this problem with four types of plots: the xy plot, the stem plot, the stairs plot, and the bar plot.

24. The volume V and surface area A of a sphere of radius r are given by

$$V = \frac{4}{3}\pi r^3 \qquad A = 4\pi r^2$$

a. Plot V and A versus r in two subplots, for $0.1 \leq r \leq 100$ m. Choose axes that will result in straight-line graphs for both V and A.

b. Plot V and r versus A in two subplots, for $1 \leq A \leq 10^4$ m^2. Choose axes that will result in straight-line graphs for both V and r.

25. The current amount A of a principal P invested in a savings account paying an annual interest rate r is given by

$$A = P\left(1 + \frac{r}{n}\right)^{nt}$$

where n is the number of times per year the interest is compounded. For continuous compounding, $A = Pe^{rt}$. Suppose \$10,000 is initially invested at 3.5 percent ($r = 0.035$).

a. Plot A versus t for $0 \leq t \leq 20$ years for four cases: continuous compounding, annual compounding ($n = 1$), quarterly compounding $n = 4$), and monthly compounding ($n = 12$). Show all four cases on the same subplot and label each curve. On a second subplot, plot the difference between the amount obtained from continuous compounding and the other three cases.

b. Redo part *a* but plot A versus t on log-log and semilog plots. Which plot gives a straight line?

26. The grades of 80 students were distributed as follows.

Letter grade	Number
A	23
B	32
C	19
D	6
Total	80

Use the pie chart function `pie` to plot the grade distribution. Add the title "Grade Distribution" to the chart. Use the `gtext` function or the Plot Editor to add the letter grades to the sections of the pie chart.

27. If we apply a sinusoidal voltage v_i to the circuit shown in Figure P27, then eventually the output voltage v_o will be sinusoidal also, with the same frequency, but with a different amplitude and shifted in time relative to the input voltage. Specifically, if $v_i = A_i \sin \omega t$, then $v_o = A_o \sin(\omega t + \phi)$. The frequency-response plot is a plot of A_o/A_i versus frequency ω. This ratio depends on ω as follows:

$$\frac{A_o}{A_i} = \left| \frac{RCs}{RCs + 1} \right|$$

where $s = \omega i$. For $RC = 0.1$ s, obtain the log-log plot of $|A_o/A_i|$ versus ω over the range of frequencies $1 \le \omega \le 1000$ rad/s. Compare the plot with Figure 5.3–5, which is for a similar circuit.

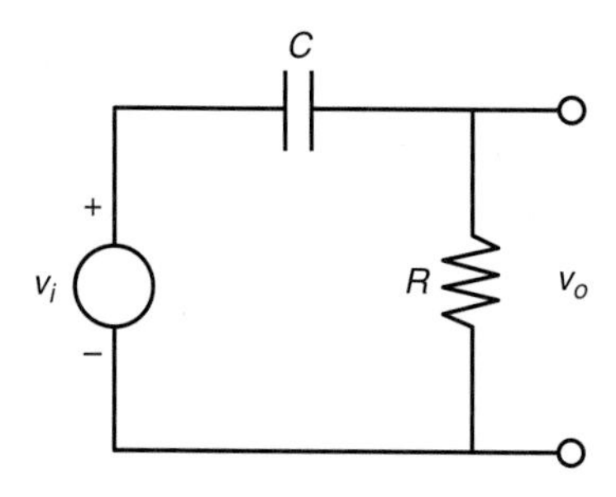

Figure P27

28. If we apply a sinusoidal voltage v_i to the circuit shown in Figure P28, then eventually the output voltage v_o will be sinusoidal also, with the same frequency, but with a different amplitude and shifted in time relative to the input voltage. Specifically, if $v_i = A_i \sin \omega t$, then $v_o = A_o \sin(\omega t + \phi)$. The frequency-response plot is a plot of A_o/A_i versus frequency ω. This ratio depends on ω as follows:

$$\frac{A_o}{A_i} = \left| \frac{1}{LCs^2 + RCs + 1} \right|$$

where $s = \omega i$. For $R = 6\ \Omega$, $L = 3.6 \times 10^{-3}$ H, and $C = 10^{-6}$ F, plot $|A_o/A_i|$ versus ω on rectilinear axes and on log-log axes over the range of frequencies $10^3 \leq \omega \leq 10^6$ rad/s. Is there an advantage to using log-log axes?

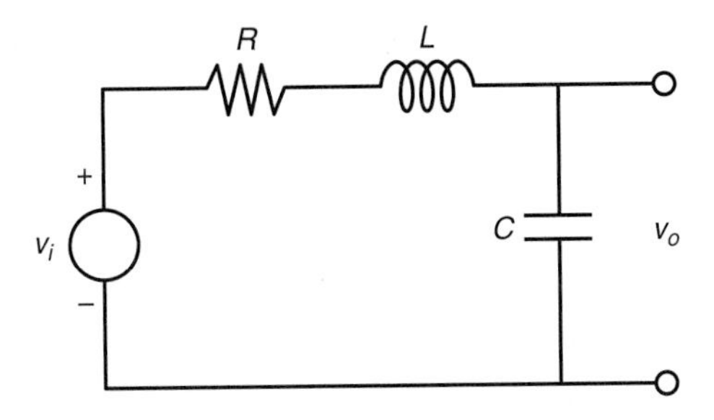

Figure P28

Section 5.5

29. The distance a spring stretches from its "free length" is a function of how much tension force is applied to it. The following table gives the spring length y that the given applied force f produced in a particular spring. The spring's free length is 4.7 in. Find a functional relation between f and x, the extension from the free length ($x = y - 4.7$).

Force f (lb)	Spring length y (in.)
0	4.7
0.47	7.2
1.15	10.6
1.64	12.9

30.* In each of the following problems, determine the best function $y(x)$ (linear, exponential, or power function) to describe the data. Plot the function on the same plot with the data. Label and format the plots appropriately.

a.

x	25	30	35	40	45
y	5	260	480	745	1100

b.

x	2.5	3	3.5	4	4.5	5	5.5	6	7	8	9	10
y	1500	1220	1050	915	810	745	690	620	520	480	410	390

c.

x	550	600	650	700	750
y	41.2	18.62	8.62	3.92	1.86

31. The population data for a certain country is

Year	1990	1991	1992	1993	1994	1995
Population (millions)	10	10.8	11.7	12.7	13.8	14.9

Obtain a function that describes this data. Plot the function and the data on the same plot. Estimate when the population will be double its 1990 size.

32.* The *half-life* of a radioactive substance is the time it takes to decay by half. The half-life of carbon 14, which is used for dating previously living things, is 5500 years. When an organism dies, it stops accumulating carbon 14. The carbon 14 present at the time of death decays with time. Let $C(t)/C(0)$ be the fraction of carbon 14 remaining at time t. In radioactive carbon dating, scientists usually assume that the remaining fraction decays exponentially according to the following formula:

$$\frac{C(t)}{C(0)} = e^{-bt}$$

a. Use the half-life of carbon 14 to find the value of the parameter b, and plot the function.

b. If 90 percent of the original carbon 14 remains, estimate how long ago the organism died.

c. Suppose our estimate of b is off by $\pm$ 1 percent. How does this error affect the age estimate in b?

33. *Quenching* is the process of immersing a hot metal object in a bath for a specified time to obtain certain properties such as hardness. A copper sphere 25 mm in diameter, initially at 300°C, is immersed in a bath at 0°C. The following table gives measurements of the sphere's temperature versus time. Find a functional description of this data. Plot the function and the data on the same plot.

Time (s)	0	1	2	3	4	5	6
Temperature (°C)	300	150	75	35	12	5	2

34. The useful life of a machine bearing depends on its operating temperature, as the following data shows. Obtain a functional description of this data. Plot the function and the data on the same plot. Estimate a bearing's life if it operates at 150°F.

Temperature (°F)	100	120	140	160	180	200	220
Bearing life (hours $\times 10^3$)	28	21	15	11	8	6	4

35. A certain electric circuit has a resistor and a capacitor. The capacitor is initially charged to 100 V. When the power supply is detached, the capacitor voltage decays with time, as the following data table shows. Find a functional description of the capacitor voltage v as a function of time t. Plot the function and the data on the same plot.

Time (s)	0	0.5	1	1.5	2	2.5	3	3.5	4
Voltage (V)	100	62	38	21	13	7	4	2	3

Sections 5.6 and 5.7

36.* The distance a spring stretches from its "free length" is a function of how much tension force is applied to it. The following table gives the spring length y that was produced in a particular spring by the given applied force f. The spring's free length is 4.7 in. Find a functional relation between f and x, the extension from the free length ($x = y - 4.7$). The function must pass through the origin ($x = 0$, $f = 0$).

Force f (lb)	Spring length y (in.)
0	4.7
0.47	7.2
1.15	10.6
1.64	12.9

37. The following data gives the drying time T of a certain paint as a function of the amount of a certain additive A.

a. Find the first-, second-, third-, and fourth-degree polynomials that fit the data and plot each polynomial with the data. Determine the quality of the curve fit for each by computing J, S, and r^2.

b. Use the polynomial giving the best fit to estimate the amount of additive that minimizes the drying time.

A (oz)	0	1	2	3	4	5	6	7	8	9
T (min)	130	115	110	90	89	89	95	100	110	125

38.* The following data gives the stopping distance d as a function of initial speed v, for a certain car model. Find a quadratic polynomial that fits the data. Determine the quality of the curve fit by computing J, S, and r^2.

v (mi/hr)	20	30	40	50	60	70
d (ft)	45	80	130	185	250	330

39. If the acceleration a is constant, Newton's law predicts that the distance travelled versus time is a quadratic function: $d = \frac{a}{2}t^2 + bt$. The following data was taken for a cyclist. Use the data to estimate the cyclist's acceleration a.

t (sec)	0	1	1.5	2	2.5	3	3.5	4
d (ft)	0	4	10	11	26	38	51	65

40.* The number of twists y required to break a certain rod is a function of the percentage x_1 and x_2 of each of two alloying elements present in the rod. The following table gives some pertinent data. Use linear multiple regression to obtain a model $y = a_0 + a_1x_1 + a_2x_2$ of the relationship between the number of twists and the alloy percentages. In addition, find the maximum percent error in the predictions.

Number of twists y	Percentage of element 1 x_1	Percentage of element 2 x_2
40	1	1
51	2	1
65	3	1
72	4	1
38	1	2
46	2	2
53	3	2
67	4	2
31	1	3
39	2	3
48	3	3
56	4	3

41. The following represents pressure samples, in pounds per square inch (psi), taken in a fuel line once every second for 10 sec.

Time (sec)	Pressure (psi)	Time (sec)	Pressure (psi)
1	26.1	6	30.6
2	27.0	7	31.1
3	28.2	8	31.3
4	29.0	9	31.0
5	29.8	10	30.5

a. Fit a first-degree polynomial, a second-degree polynomial, and a third-degree polynomial to this data. Plot the curve fits along with the data points.

b. Use the results from part *a* to predict the pressure at $t = 11$ sec. Explain which curve fit gives the most reliable prediction. Consider the coefficients of determination and the residuals for each fit in making your decision.

42. A liquid boils when its vapor pressure equals the external pressure acting on the surface of the liquid. This is the reason why water boils at a lower temperature at higher altitudes. This information is important for chemical, nuclear, and other engineers who must design processes utilizing boiling liquids. Data on the vapor pressure P of water as a function of temperature T is given in the following table. From theory we know that $\ln P$ is proportional to $1/T$. Obtain a curve fit for $P(T)$ from this data. Use the fit to estimate the vapor pressure at 285 K and at 300 K.

T (degrees K)	P (torr)
273	4.579
278	6.543
283	9.209
288	12.788
293	17.535
298	23.756

43. The salt content of water in the environment affects living organisms and causes corrosion. Environmental and ocean engineers must be aware of these effects. The solubility of salt in water is a function of the water temperature. Let S represent the solubility of NaCl (sodium chloride) as grams of salt in 100 g of water. Let T be temperature in °C. Use the following data to obtain a curve fit for S as a function of T. Use the fit to estimate S when $T = 25$°C.

T (°) C	S (g $NaCl$/100 g H_2O)
10	35
20	35.6
30	36.25
40	36.9
50	37.5
60	38.1
70	38.8
80	39.4
90	40

44. The amount of dissolved oxygen in water affects living organisms and chemical processes. Environmental and chemical engineers must be aware of these effects. The solubility of oxygen in water is a function of the water temperature. Let S represent the solubility of O_2 as millimoles of O_2 per liter of water. Let T be temperature in °C. Use the following data to obtain a curve fit for S as a function of T. Use the fit to estimate S when $T = 8$°C and $T = 50$°C.

T (°) C	S (millimoles O_2/L H_2O)
5	1.95
10	1.7
15	1.55
20	1.40
25	1.30
30	1.15
35	1.05
40	1.00
45	0.95

45. The following function is linear in the parameters a_1 and a_2.

$$y(x) = a_1 + a_2 \ln x$$

Use least squares regression with the following data to estimate the values of a_1 and a_2. Use the curve fit to estimate the values of y at $x = 1.5$ and at $x = 11$.

x	1	2	3	4	5	6	7	8	9	10
y	10	14	16	18	19	20	21	22	23	23

46. Chemical, environmental, and nuclear engineers must be able to predict the changes in chemical concentration in a reaction. A model used for many single reactant processes is:

$$\text{Rate of change of concentration} = -kC^n$$

where C is the chemical concentration and k is the rate constant. The order of the reaction is the value of the exponent n. Solution methods for differential equations (which are discussed in Chapter 8) can show that the solution for a first-order reaction ($n = 1$) is

$$C(t) = C(0)e^{-kt}$$

The following data describes the reaction

$$(CH_3)_3CBr + H_2O \rightarrow (CH_3)_3COH + HBr$$

Use this data to obtain a least squares fit to estimate the value of k.

Time t (h)	C (mol of $(CH_3)_3CBr$/L)
0	0.1039
3.15	0.0896
6.20	0.0776
10.0	0.0639
18.3	0.0353
30.8	0.0207
43.8	0.0101

47. Chemical, environmental, and nuclear engineers must be able to predict the changes in chemical concentration in a reaction. A model used for many single reactant processes is:

$$\text{Rate of change of concentration} = -kC^n$$

where C is the chemical concentration and k is the rate constant. The order of the reaction is the value of the exponent n. Solution methods for differential equations (which are discussed in Chapter 8) can show that the solution for a first-order reaction ($n = 1$) is

$$C(t) = C(0)e^{-kt}$$

and the solution for a second-order reaction ($n = 2$) is

$$\frac{1}{C(t)} = \frac{1}{C(0)} + kt$$

The following data from [Brown, 1994] describes the gas-phase decomposition of nitrogen dioxide at 300°C.

$$2NO_2 \rightarrow 2NO + O_2$$

Time t (s)	C (mol NO_2/L)
0	0.0100
50	0.0079
100	0.0065
200	0.0048
300	0.0038

Determine whether this is a first-order or second-order reaction, and estimate the value of the rate constant k.

48. Chemical, environmental, and nuclear engineers must be able to predict the changes in chemical concentration in a reaction. A model used for many single reactant processes is:

$$\text{Rate of change of concentration} = -kC^n$$

where C is the chemical concentration and k is the rate constant. The order of the reaction is the value of the exponent n. Solution methods for differential equations (which are discussed in Chapter 8) can show that the solution for a first-order reaction ($n = 1$) is

$$C(t) = C(0)e^{-kt}$$

The solution for a second-order reaction ($n = 2$) is

$$\frac{1}{C(t)} = \frac{1}{C(0)} + kt$$

and the solution for a third-order reaction ($n = 3$) is

$$\frac{1}{2C^2(t)} = \frac{1}{2C^2(0)} + kt$$

The following data describes a certain reaction. By examining the residuals, determine whether this is a first-order, second-order, or third-order reaction, and estimate the value of the rate constant k.

Time t (min)	C (mol of reactant/L)
5	0.3575
10	0.3010
15	0.2505
20	0.2095
25	0.1800
30	0.1500
35	0.1245
40	0.1070
45	0.0865

Section 5.8

49. The popular amusement ride known as the corkscrew has a helical shape. The parametric equations for a circular helix are

$$x = a \cos t$$
$$y = a \sin t$$
$$z = bt$$

where a is the radius of the helical path and b is a constant that determines the "tightness" of the path. In addition, if $b > 0$, the helix has the shape of a right-handed screw; if $b < 0$, the helix is left-handed.

Obtain the three-dimensional plot of the helix for the following three cases and compare their appearance with one another. Use $0 \le t \le 10\pi$ and $a = 1$.

a. $b = 0.1$
b. $b = 0.2$
c. $b = -0.1$

50. A robot rotates about its base at two revolutions per minute while lowering its arm and extending its hand. It lowers its arm at the rate of 120° per minute and extends its hand at the rate of 5 m/min. The arm is 0.5 m long. The xyz coordinates of the hand are given by

$$x = (0.5 + 5t)\sin\left(\frac{2\pi}{3}t\right)\cos(4\pi t)$$

$$y = (0.5 + 5t)\sin\left(\frac{2\pi}{3}t\right)\sin(4\pi t)$$

$$z = (0.5 + 5t)\cos\left(\frac{2\pi}{3}t\right)$$

where t is time in minutes.

Obtain the three-dimensional plot of the path of the hand for $0 \le t \le 0.2$ min.

51. Obtain the surface and contour plots for the function $z = x^2 - 2xy + 4y^2$, showing the minimum at $x = y = 0$.

52. Obtain the surface and contour plots for the function $z = -x^2 + 2xy + 3y^2$. This surface has the shape of a saddle. At its saddlepoint at $x = y = 0$, the surface has zero slope, but this point does not correspond to either a minimum or a maximum. What type of contour lines correspond to a saddlepoint?

53. Obtain the surface and contour plots for the function $z = (x - y^2)(x - 3y^2)$. This surface has a singular point at $x = y = 0$, where the surface has zero slope, but this point does not correspond to either a minimum or a maximum. What type of contour lines correspond to a singular point?

54. A square metal plate is heated to 80°C at the corner corresponding to $x = y = 1$. The temperature distribution in the plate is described by

$$T = 80e^{-(x-1)^2}e^{-3(y-1)^2}$$

Obtain the surface and contour plots for the temperature. Label each axis. What is the temperature at the corner corresponding to $x = y = 0$?

55. The following function describes oscillations in some mechanical structures and electric circuits:

$$z(t) = e^{-t/\tau}\sin(\omega t + \phi)$$

In this function t is time, and ω is the oscillation frequency in radians per unit time. The oscillations have a period of $2\pi/\omega$, and their amplitudes decay in time at a rate determined by τ, which is called the *time constant.* The smaller τ is, the faster the oscillations die out.

Suppose that $\phi = 0$, $\omega = 2$, and τ can have values in the range $0.5 \le \tau \le 10$ sec. Then the preceding equation becomes

$$z(t) = e^{-t/\tau}\sin(2t)$$

Obtain a surface plot and a contour plot of this function to help visualize the effect of τ for $0 \le t \le 15$ sec. Let the x variable be time t and the y variable be τ.

56. Many applications require us to know the temperature distribution in an object. For example, this information is important for controlling the material properties such as hardness, when cooling an object formed from molten metal. In a heat transfer course the following description of the temperature distribution in a flat rectangular metal plate is often derived. The temperature on three sides is held constant at T_1, and at T_2 on the fourth side (see Figure P56). The temperature $T(x, y)$ as a function of the xy coordinates shown is given by

$$T(x, y) = (T_2 - T_1)w(x, y) + T_1$$

where

$$w(x, y) = \frac{2}{\pi} \sum_{n \text{ odd}}^{\infty} \frac{2}{n} \sin\left(\frac{n\pi x}{L}\right) \frac{\sinh(n\pi y/L)}{\sinh(n\pi W/L)}$$

The given data for this problem are: $T_1 = 70°\text{F}$, $T_2 = 200°\text{F}$, and $W = L = 2$ ft.

Using a spacing of 0.2 for both x and y, generate a surface mesh plot and a contour plot of the temperature distribution.

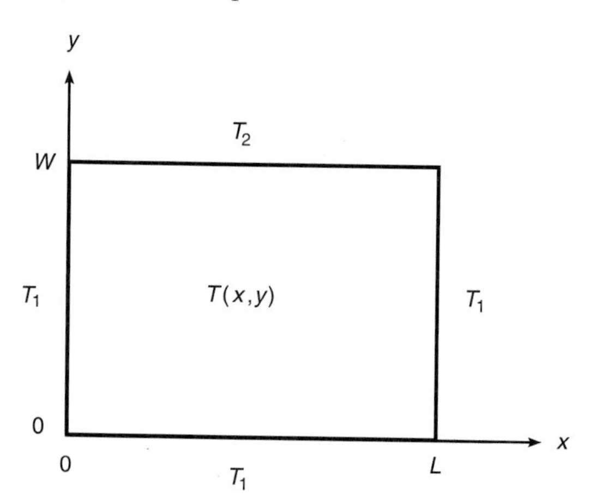

Figure P56

57. The electric potential field V at a point, due to two charged particles, is given by

$$V = \frac{1}{4\pi\epsilon_o}\left(\frac{q_1}{r_1} + \frac{q_2}{r_2}\right)$$

where q_1 and q_2 are the charges of the particles in Coulombs (C), r_1 and r_2 are the distances of the charges from the point (in meters), and ϵ_o is the permittivity of free space, whose value is

$$\epsilon_o = 8.854 \times 10^{-12}\ \text{C}^2/\text{N} \cdot \text{m}^2$$

Suppose the charges are $q_1 = 2 \times 10^{-10}$ C and $q_2 = 4 \times 10^{-10}$ C. Their respective locations in the xy plane are (0.3, 0) and (−0.3, 0) m. Plot the electric potential field on a 3D surface plot with V plotted on the z-axis over the ranges $-0.25 \le x \le 0.25$ and $-0.25 \le y \le 0.25$. Create the plot two ways: a. by using the `surf` function and b. by using the `meshc` function.

58. The grades of 80 students were distributed as follows.

Letter grade	Number
A	23
B	32
C	19
D	6
Total	80

Use the 3D pie chart function `pie3` to plot the grade distribution. Add the title "Grade Distribution" to the chart. Use the Plot Editor to add the letter grades to the sections of the pie chart.

59. Refer to Problem 22 of Chapter 4. Use the function file created for that problem to generate a surface mesh plot and a contour plot of x versus h and W for $0 \le W \le 500$ N and for $0 \le h \le 2$ m. Use the values: $k_1 = 10^4$ N/m; $k_2 = 1.5 \times 10^4$ N/m; and $d = 0.1$ m.

60. Refer to Problem 25 of Chapter 4. To see how sensitive the cost is to location of the distribution center, obtain a surface plot and a contour plot of the total cost as a function of the x and y coordinates of the distribution center location. How much would the cost increase if we located the center 1 mi in any direction from the optimal location?

61. Refer to Example 3.2-2 of Chapter 3. Use a surface plot and a contour plot of the perimeter length L as a function of d and θ over the ranges $1 \le d \le 30$ ft and $0.1 \le \theta \le 1.5$ rad. Are there valleys other than the one corresponding to $d = 7.5984$ and $\theta = 1.0472$? Are there any saddle points?

Engineering in the 21st Century...

Virtual Prototyping

To many people, computer-aided design (CAD) or computer-aided engineering (CAE) means creating engineering drawings. However, it means much more. Engineers can use computers to determine the forces, voltages, currents, and so on that might occur in a proposed design. Then they can use this information to make sure the hardware can withstand the predicted forces or supply the required voltages or currents. Engineers are just beginning to use the full potential of CAE.

The normal stages in the development of a new vehicle, such as an aircraft, formerly consisted of aerodynamic testing a scale model; building a full-size wooden *mock-up* to check for pipe, cable, and structural interferences; and finally building and testing a *prototype,* the first complete vehicle. CAE is changing the traditional development cycle. The new Boeing 777 shown above is the first aircraft to be designed and built using CAE, without the extra time and expense of building a mock-up. The design teams responsible for the various subsystems, such as aerodynamics, structures, hydraulics, and electrical systems, all had access to the same computer database that described the aircraft. Thus when one team made a design change, the database was updated, allowing the other teams to see whether the change affected their subsystem. This process of designing and testing with a computer model has been called *virtual prototyping.* Virtual prototyping reduced engineering changes, errors, and rework by 50 percent, and greatly enhanced the manufacturability of the airplane. When production began, the parts went together easily.

MATLAB is a powerful tool for many CAE applications. It complements geometric modeling packages because it can do advanced calculations that such packages cannot do. ■

CHAPTER 6

Linear Algebraic Equations

OUTLINE

Linear algebraic equations such as

$$5x - 2y = 13$$
$$7x + 3y = 24$$

occur in many engineering applications. For example, electrical engineers use them to predict the power requirements for circuits; civil, mechanical, and aerospace engineers use them to design structures and machines; chemical engineers use them to compute material balances in chemical processes; and industrial engineers apply them to design schedules and operations. The examples and homework problems in this chapter explore some of these applications.

Linear algebraic equations can be solved "by hand" using pencil and paper, by calculator, or with software such as MATLAB. The choice depends on the circumstances. For equations with only two unknown variables, hand solution is easy and adequate. Some calculators can solve equation sets that have many variables. However, the greatest power and flexibility is obtained by using software.

For example, MATLAB can obtain and plot equation solutions as we vary one or more parameters.

Without giving a formal definition of the term *linear algebraic equations,* let us simply say that their unknown variables never appear raised to a power other than unity and never appear as products, ratios, or in transcendental functions such as $\ln(x)$, e^x, and $\cos x$. The simplest linear equation is $ax = b$, which has the solution $x = b/a$ if $a \neq 0$.

In contrast, the following equations are nonlinear:

$$x^2 = 3$$

which has the solutions $x = \pm\sqrt{3}$, and

$$\sin x = 0.5$$

which has the solutions $x = 30^\circ, 150^\circ, 390^\circ, 510^\circ, \ldots$. In contrast to most nonlinear equations, these particular nonlinear equations are easy to solve. For example, we cannot solve the equation $x + 2e^{-x} - 3 = 0$ in "closed form"; that is, we cannot express the solution as a function. We must obtain this solution numerically, as explained in Section 3.2. The equation has two solutions: $x = -0.5831$ and $x = 2.8887$ to four decimal places.

Sets of equations are linear if all the equations are linear. They are nonlinear if at least one of the equations is nonlinear. For example, the set

$$8x - 3y = 1$$
$$6x + 4y = 32$$

is linear because both equations are linear, whereas the set

$$6xy - 2x = 44$$
$$5x - 3y = -2$$

is nonlinear because of the product term xy.

Systematic solution methods have been developed for sets of linear equations. However, no systematic methods are available for nonlinear equations because the nonlinear category covers such a wide range of equations. In this chapter we first review methods for solving linear equations by hand, and we use these methods to develop an understanding of the potential pitfalls that can occur when solving linear equations. Then we introduce some matrix notation that is required for use with MATLAB and that is also useful for expressing solution methods in a compact way. The conditions for the existence and uniqueness of solutions are then introduced. Methods using MATLAB are then treated in four sections: Section 6.2 covers equation sets that have unique solutions; Section 6.3 covers Cramer's method; Sections 6.4 and 6.5 explain how to determine whether a set has a unique solution, multiple solutions, or no solution at all.

6.1 Elementary Solution Methods

You are sure to encounter situations in which MATLAB is not available (such as on a test!), and thus you should become familiar with the hand-solution methods. In addition, understanding these methods will help you understand the MATLAB responses and the pitfalls that can occur when obtaining a computer solution. Finally, hand solutions are sometimes needed when the numerical values of one or more coefficients are unspecified. In this section we cover hand-solution methods; later in the chapter we introduce the MATLAB methods for solving linear equations.

Several methods are available for solving linear algebraic equations by hand. The appropriate choice depends on user preference, on the number of equations, and on the structure of the equations to be solved. We demonstrate two methods: (1) successive elimination of variables and (2) Cramer's method (in Section 6.3). The MATLAB method is based on the successive elimination technique, but Cramer's method gives us some insight into the existence and uniqueness of solutions and into the effects of numerical inaccuracy.

Successive Elimination of Variables

An efficient way to eliminate variables is to multiply one equation by a suitable factor and then add or subtract the resulting equation from another equation in the set. If the factor is chosen properly, the new equation so obtained will contain fewer variables. This process is continued with the remaining equations until only one unknown and one equation remain. A systematic method of doing this is *Gauss elimination.* With this method you multiply the first equation (called the *pivot* equation) by a suitable factor and add the result to one of the other equations in the set to cancel one variable. Repeat the process with the other equations in the set, using the same pivot equation. This step generates a new set of equations, with one less variable. Select the new pivot to be the first equation in this new set and repeat the process until only one variable and one equation remain. This method is suitable for computer implementation, and it forms the basis for many computer methods for solving linear equations. (It is the method used by MATLAB.)

GAUSS ELIMINATION

EXAMPLE 6.1–1 Gauss Elimination

Solve the following set using Gauss elimination:

$$-x + y + 2z = 2 \quad (6.1\text{–}1)$$

$$3x - y + z = 6 \quad (6.1\text{–}2)$$

$$-x + 3y + 4z = 4 \quad (6.1\text{–}3)$$

■ **Solution**

The solution proceeds as follows:

1. Equation (6.1–1) is the pivot equation. Multiply it by -1 and add the result to (6.1–3) to obtain $2y + 2z = 2$, which is equivalent to $y + z = 1$. Next multiply

(6.1–1) by 3 and add the result to (6.1–2) to obtain $2y + 7z = 12$. Thus we have a new set of two equations in two unknowns:

$$y + z = 1 \tag{6.1–4}$$

$$2y + 7z = 12 \tag{6.1–5}$$

2. Equation (6.1–4) is the new pivot equation. Multiply it by -2 and add the result to (6.1–5) to obtain $5z = 10$, or $z = 2$. Substitute this value into (6.1–4) to obtain $y + 2 = 1$, or $y = -1$. Then substitute the values of y and z into (6.1–1) to obtain $-x - 1 + 4 = 2$, or $x = 1$.

Test Your Understanding

T6.1–1 Solve the following equations using Gauss elimination:

$$6x - 3y + 4z = 41$$

$$12x + 5y - 7z = -26$$

$$-5x + 2y + 6z = 14$$

(Answer: $x = 2$, $y = -3$, $z = 5$.)

Singular and Ill-Conditioned Problems

Figure 6.1–1 shows the graphs of the following equations:

$$3x - 4y = 5$$

$$6x - 10y = 2$$

Note that the two lines intersect, and therefore the equations have a solution, which is given by the intersection point: $x = 7$, $y = 4$. A *singular* problem refers to a set of equations having either no unique solution or no solution at all. For example, the set

$$3x - 4y = 5$$

$$6x - 8y = 10$$

is singular and has no unique solution because the second equation is identical to the first equation, multiplied by 2. The graphs of these two equations are identical. All we can say is that the solution must satisfy $y = (3x - 5)/4$, which describes an infinite number of solutions.

On the other hand, the set

$$3x - 4y = 5 \tag{6.1–6}$$

$$6x - 8y = 3 \tag{6.1–7}$$

is singular but has no solution. The graphs of these two equations are distinct but *parallel* (see Figure 6.1–2). Because they do not intersect, no solution exists.

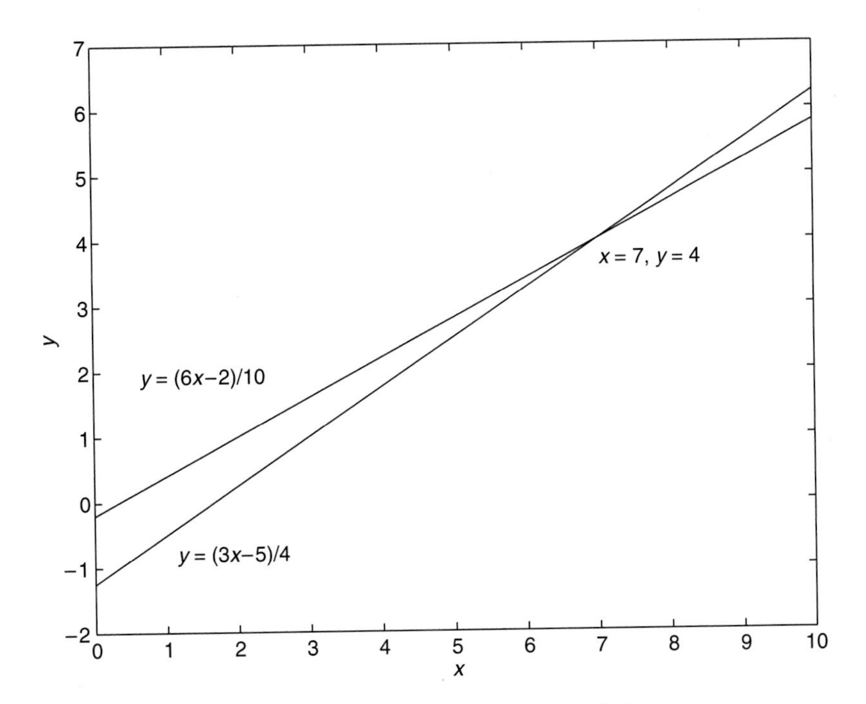

Figure 6.1–1 The graphs of two equations intersect at the solution.

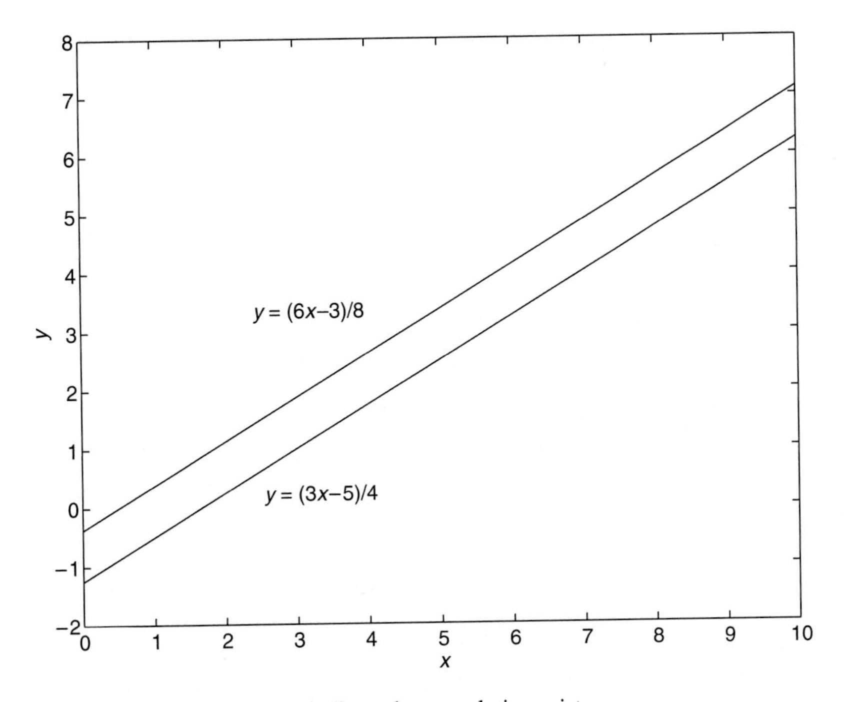

Figure 6.1–2 Parallel graphs indicate that no solution exists.

Homogeneous Equations

As another example, consider the following set of *homogeneous equations* (which means that their right sides are all zero)

$$6x + ay = 0 \tag{6.1–8}$$

$$2x + 4y = 0 \tag{6.1–9}$$

where a is a parameter. Multiply the second equation by 3 and subtract the result from the first equation to obtain

$$(a - 12)y = 0 \tag{6.1–10}$$

The solution is $y = 0$ *only if* $a \neq 12$; substituting $y = 0$ into either (6.1–8) or (6.1–9) shows that $x = 0$. However, if $a = 12$, (6.1–10) implies that $0y = 0$, which is satisfied for any finite value of y; in this case both (6.1–8) and (6.1–9) give $x = -2y$. Thus if $a = 12$, there are an infinite number of solutions for x and y, where $x = -2y$.

Ill-Conditioned Equations

An *ill-conditioned* set of equations is a set that is close to being singular (for example, two equations whose graphs are close to being parallel). The following set would be considered an ill-conditioned set if we carry only two significant figures in our calculations:

$$3x - 4y = 5$$

$$6x - 8.002y = 3$$

To see why, solve the first equation for y to obtain

$$y = \frac{3x - 5}{4} \tag{6.1–11}$$

and solve the second equation to obtain

$$y = \frac{6x - 3}{8.002} = \frac{3x - 1.5}{4.001} \tag{6.1–12}$$

The slope of (6.1–11) is 3/4, whereas the slope of (6.1–12) is 3/4.001. If we had carried only two significant figures, we would have rounded the denominator of the latter expression to 4.0, and thus the two expressions for y would have the same slope and their graphs would be parallel. Thus we see that the ill-conditioned status depends on the accuracy with which the solution calculations are made. Of course, MATLAB uses more than two significant figures in its calculations. However, no computer can represent a number with infinitely many significant figures, and so a given set of equations can appear to be singular if the accuracy required to solve them is greater than the number of significant figures used by the software. If we carry four significant figures in our calculations, we would find that the solution is $x = 4668$ and $y = 3500$.

Test Your Understanding

T6.1–2 Show that the following set has no solution.

$$-4x + 5y = 10$$
$$12x - 15y = 8$$

T6.1–3 For what value of b will the following set have a solution in which both x and y are nonzero? Find the relation between x and y.

$$4x - by = 0$$
$$-3x + 6y = 0$$

(Answer: If $b = 8$, $x = 2y$. If $b \neq 8$, $x = y = 0$.)

6.2 Matrix Methods for Linear Equations

Sets of linear algebraic equations can be expressed in matrix notation, a standard and compact method that is useful for expressing solutions and for developing software applications with an arbitrary number of variables. This section describes the use of matrix notation.

As you saw in Chapter 2, a *matrix* is an ordered array of rows and columns containing numbers, variables, or expressions. A *vector* is a special case of a matrix that has either one row or one column. A *row* vector has one row. A *column* vector has one column. In this chapter a vector is taken to be a column vector unless otherwise specified. Usually, when printed in text, lowercase boldface letters denote a vector, and uppercase boldface letters denote a matrix.

Matrix notation enables us to represent multiple equations as a single matrix equation. For example, consider the following set:

$$2x_1 + 9x_2 = 5 \qquad (6.2\text{–}1)$$

$$3x_1 - 4x_2 = 7 \qquad (6.2\text{–}2)$$

This set can be expressed in vector-matrix form as

$$\begin{bmatrix} 2 & 9 \\ 3 & -4 \end{bmatrix} \begin{bmatrix} x_1 \\ x_2 \end{bmatrix} = \begin{bmatrix} 5 \\ 7 \end{bmatrix}$$

which can be represented in the following compact form

$$\mathbf{Ax} = \mathbf{b} \qquad (6.2\text{–}3)$$

where we have defined the following matrices and vectors:

$$\mathbf{A} = \begin{bmatrix} 2 & 9 \\ 3 & -4 \end{bmatrix} \qquad \mathbf{x} = \begin{bmatrix} x_1 \\ x_2 \end{bmatrix} \qquad \mathbf{b} = \begin{bmatrix} 5 \\ 7 \end{bmatrix}$$

The matrix **A** corresponds in an ordered fashion to the coefficients of x_1 and x_2 in (6.2–1) and (6.2–2). Note that the first row in **A** consists of the coefficients of

x_1 and x_2 on the left side of (6.2–1), and the second row contains the coefficients on the left side of (6.2–2). The vector **x** contains the variables x_1 and x_2, and the vector **b** contains the right sides of (6.2–1) and (6.2–2).

In general, the set of m equations in n unknowns

$$\begin{aligned} a_{11}x_1 + a_{12}x_2 + \cdots + a_{1n}x_n &= b_1 \\ a_{21}x_1 + a_{22}x_2 + \cdots + a_{2n}x_n &= b_2 \\ &\;\cdots\cdots \\ a_{m1}x_1 + a_{m2}x_2 + \cdots + a_{mn}x_n &= b_m \end{aligned} \tag{6.2–4}$$

can be written in the form (6.2–3), where

$$\mathbf{A} = \begin{bmatrix} a_{11} & a_{12} & \cdots & a_{1n} \\ a_{21} & a_{22} & \cdots & a_{2n} \\ \cdot & \cdot & \cdots & \cdot \\ a_{m1} & a_{m2} & \cdots & a_{mn} \end{bmatrix} \tag{6.2–5}$$

$$\mathbf{x} = \begin{bmatrix} x_1 \\ x_2 \\ \vdots \\ x_n \end{bmatrix} \tag{6.2–6}$$

$$\mathbf{b} = \begin{bmatrix} b_1 \\ b_2 \\ \vdots \\ b_m \end{bmatrix} \tag{6.2–7}$$

The matrix **A** has m rows and n columns, so its dimension is expressed as $m \times n$.

Determinants

Determinants are useful for finding out whether a set of equations has a solution. A determinant is a special square array that, unlike a matrix, can be reduced to a *single* number. Vertical bars are used to denote a determinant, whereas square brackets denote a matrix. A determinant having two rows and two columns is a 2×2 determinant. The rule for reducing a 2×2 determinant to a single number is shown below

$$D = \begin{vmatrix} a_{11} & a_{12} \\ a_{21} & a_{22} \end{vmatrix} = a_{11}a_{22} - a_{12}a_{21} \tag{6.2–8}$$

Rules exist to evaluate $n \times n$ determinants by hand, but we can use MATLAB to do this. First enter the determinant as an array. Then use the `det` function to evaluate the determinant. For example, a MATLAB session to compute the determinant

$$A = \begin{vmatrix} 3 & -4 & 1 \\ 6 & 10 & 2 \\ 9 & -7 & 8 \end{vmatrix}$$

would look like:

```
>>A = [3,-4,1;6,10,2;9,-7,8];
>>det(A)
ans =
     8
```

As we have seen, a determinant is not the same as a matrix, but a determinant can be found from a matrix. In the previous MATLAB session, MATLAB can treat the array `A` as matrix. When it executes the function `det(A)`, MATLAB obtains a determinant from the matrix `A`. The determinant obtained from the matrix **A** is expressed as $|\mathbf{A}|$.

Determinants and Singular Problems

We saw in Section 6.1 that a singular problem refers to a set of equations that has either no unique solution or no solution at all. We can use the matrix **A** in the equation set $\mathbf{Ax} = \mathbf{b}$ to determine whether or not the set is singular. For example, in Section 6.1, we saw that the set

$$3x - 4y = 5$$
$$6x - 8y = 10$$

has no unique solution, because the second equation is identical to the first equation, multiplied by 2. The matrix **A** and the vector **b** for this set are

$$\mathbf{A} = \begin{bmatrix} 3 & -4 \\ 6 & -8 \end{bmatrix} \qquad \mathbf{b} = \begin{bmatrix} 5 \\ 10 \end{bmatrix}$$

The determinant of **A** is

$$|\mathbf{A}| = \begin{vmatrix} 3 & -4 \\ 6 & -8 \end{vmatrix} = 3(-8) - (-4)(6) = 0$$

The fact that $|\mathbf{A}| = 0$ indicates that the equation set is singular. We have not proved this statement, but it can be proved.

Consider another example from Section 6.1.

$$3x - 4y = 5$$
$$6x - 8y = 3$$

This set has no solution. The matrix **A** is the same as for the previous set, but the vector **b** is different.

$$\mathbf{b} = \begin{bmatrix} 5 \\ 3 \end{bmatrix}$$

Because $|\mathbf{A}| = 0$, this equation set is also singular. These two examples show that if $|\mathbf{A}| = 0$, the set has either no unique solution or no solution at all.

Now consider another example from Section 6.1, a set of homogeneous equations (the right-hand sides are all zero):

$$6x + ay = 0$$

$$2x + 4y = 0$$

where a is a parameter. As we saw in Section 6.1, this set has the solution $x = y = 0$ unless $a = 12$, in which case there are an infinite number of solutions of the form $x = -2y$. The matrix $\mathbf{A}$ and the vector $\mathbf{b}$ for this set are

$$\mathbf{A} = \begin{bmatrix} 6 & a \\ 2 & 4 \end{bmatrix} \qquad \mathbf{b} = \begin{bmatrix} 0 \\ 0 \end{bmatrix}$$

The determinant of $\mathbf{A}$ is

$$|\mathbf{A}| = \begin{vmatrix} 6 & a \\ 2 & 4 \end{vmatrix} = 6(4) - 2a = 24 - 2a$$

Thus if $a = 12$, $|\mathbf{A}| = 0$ and the equation set is singular.

These examples indicate that for the equation set $\mathbf{Ax} = \mathbf{b}$, if $|\mathbf{A}| = 0$, then there is no unique solution. Depending on the values in the vector $\mathbf{b}$, there may be no solution at all, or an infinite number of solutions.

The Left-Division Method

MATLAB provides the *left-division* method for solving the equation set $\mathbf{Ax} = \mathbf{b}$. The left-division method is based on Gauss elimination. To use the left-division method to solve for $\mathbf{x}$, type `x = A\b`. This method also works in some cases where the number of unknowns does not equal the number of equations. However, this section focuses on problems in which the number of equations equals the number of unknowns. In Sections 6.4 and 6.5, we examine other cases.

If the number of equations equals the number of unknowns and if $|\mathbf{A}| \neq 0$, then the equation set has a solution and it is unique. If $|\mathbf{A}| = 0$ or if the number of equations does not equal the number of unknowns, then you must use the methods presented in Section 6.4.

EXAMPLE 6.2–1 Left-Division Method with Three Unknowns

Use the left-division method to solve the following set:

$$3x + 2y - 9z = -65$$

$$-9x - 5y + 2z = 16$$

$$6x + 7y + 3z = 5$$

■ Solution

The matrix **A** is

$$\mathbf{A} = \begin{bmatrix} 3 & 2 & -9 \\ -9 & -5 & 2 \\ 6 & 7 & 3 \end{bmatrix}$$

We can use MATLAB to check the determinant of **A** to see whether the problem is singular. The session looks like this:

```
>>A = [3,2,-9;-9,-5,2;6,7,3];
>>det(A)
ans =
     288
```

Because $|\mathbf{A}| \neq 0$, a unique solution exists. It is obtained as follows:

```
>>b = [-65;16;5];
>>A\b
ans =
     2.0000
     -4.0000
     7.0000
```

This answer gives the vector **x**, which corresponds to the solution $x = 2$, $y = -4$, $z = 7$. It can be checked by determining whether **Ax** gives the vector **b**, by typing

```
>>A*ans
ans =
     -65.0000
     16.0000
     5.0000
```

which is the vector **b**. Thus the answer is correct.

The backward slash (\) is used for left division. Be careful to distinguish between the *backward* slash (\) and the forward slash (/) which is used for *right* division. Sometimes equation sets are written as $\mathbf{xC} = \mathbf{d}$, where **x** and **d** are *row* vectors. In that case you can use right division to solve the set $\mathbf{xC} = \mathbf{d}$ for **x** by typing `x = d/C`, or you can convert the equations to the form $\mathbf{Ax} = \mathbf{b}$. For example, the matrix equation

$$[x_1 \quad x_2] \begin{bmatrix} 6 & 2 \\ 3 & 5 \end{bmatrix} = [3 \quad -19]$$

corresponds to the equations

$$6x_1 + 3x_2 = 3$$
$$2x_1 + 5x_2 = -19$$

These equations can be written as

$$\begin{bmatrix} 6 & 3 \\ 2 & 5 \end{bmatrix} \begin{bmatrix} x_1 \\ x_2 \end{bmatrix} = \begin{bmatrix} 3 \\ -19 \end{bmatrix}$$

which is in the form $\mathbf{Ax} = \mathbf{b}$.

Linear equations are useful in many engineering fields. Electrical circuits are a common source of linear equation models. The circuit designer must be able to solve them to predict the currents that will exist in the circuit. This information is often needed to determine the power supply requirements, among other things.

EXAMPLE 6.2–2 An Electrical-Resistance Network

The circuit shown in Figure 6.2–1 has five resistances and two applied voltages. Assuming that the positive directions of current flow are in the directions shown in the figure, Kirchhoff's voltage law applied to each loop in the circuit gives

$$-v_1 + R_1 i_1 + R_4 i_4 = 0$$
$$-R_4 i_4 + R_2 i_2 + R_5 i_5 = 0$$
$$-R_5 i_5 + R_3 i_3 + v_2 = 0$$

Conservation of charge applied at each node in the circuit gives

$$i_1 = i_2 + i_4$$
$$i_2 = i_3 + i_5$$

You can use these two equations to eliminate i_4 and i_5 from the first three equations. The result is:

$$(R_1 + R_4)i_1 - R_4 i_2 = v_1$$
$$-R_4 i_1 + (R_2 + R_4 + R_5)i_2 - R_5 i_3 = 0$$
$$R_5 i_2 - (R_3 + R_5)i_3 = v_2$$

Thus we have three equations in three unknowns: i_1, i_2, and i_3.

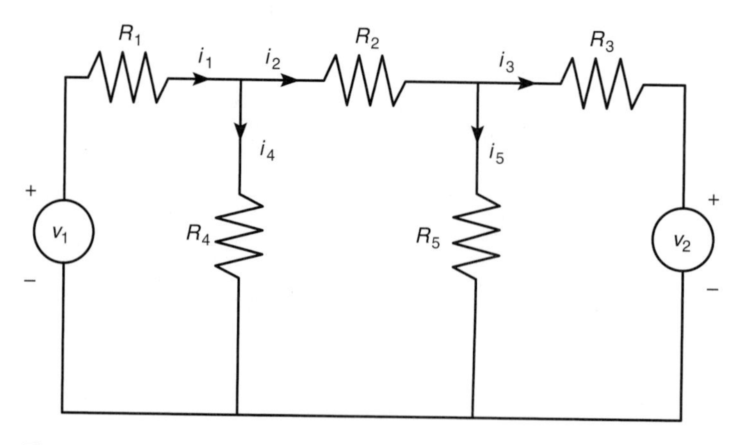

Figure 6.2–1 An electrical-resistance network.

Write a MATLAB script file that uses given values of the applied voltages v_1 and v_2 and given values of the five resistances to solve for the currents i_1, i_2, and i_3. Use the program to find the currents for the case $R_1 = 5$, $R_2 = 100$, $R_3 = 200$, $R_4 = 150$, $R_5 = 250$ kΩ, $v_1 = 100$, and $v_2 = 50$ V. (Note that 1 kΩ = 1000 Ω.)

■ Solution

Because there are as many unknowns as equations, there will be a unique solution if $|\mathbf{A}| \neq 0$; in addition, the left-division method will generate an error message if $|\mathbf{A}| = 0$. The following script file, named `resist.m`, uses the left-division method to solve the three equations for i_1, i_2, and i_3.

```
% File resist.m
% Solves for the currents i_1, i_2, i_3
R = [5,100,200,150,250]*1000;
v1 = 100; v2 = 50;
A1 = [R(1) + R(4), -R(4), 0];
A2 = [-R(4), R(2) + R(4) + R(5), -R(5)];
A3 = [0, R(5), -(R(3) + R(5))];
A = [A1; A2; A3];
b=[v1; 0; v2];
current = A\b;
disp('The currents are:')
disp(current)
```

The row vectors `A1`, `A2`, and `A3` were defined to avoid typing the lengthy expression for `A` in one line. This script is executed from the command prompt as follows:

```
>>resist
The currents are:
   1.0e-003*
   0.9544
   0.3195
   0.0664
```

Because MATLAB did not generate an error message, the solution is unique. The currents are $i_1 = 0.9544$, $i_2 = 0.3195$, and $i_3 = 0.0664$ mA, where 1 mA = 1 milliampere = 0.001 A.

Ethanol Production — EXAMPLE 6.2–3

Engineers in the food and chemical industries use fermentation in many processes. The following equation describes Baker's yeast fermentation.

$$a(C_6H_{12}O_6) + b(O_2) + c(NH_3)$$

$$\rightarrow C_6H_{10}NO_3 + d(H_2O) + e(CO_2) + f(C_2H_6O)$$

The variables $a, b, \ldots, f$ represent the masses of the products involved in the reaction. In this formula $C_6H_{12}O_6$ represents glucose, $C_6H_{10}NO_3$ represents yeast, and C_2H_6O

represents ethanol. This reaction produces ethanol, in addition to water and carbon dioxide. We want to determine the amount of ethanol f produced. The number of C, O, N, and H atoms on the left must balance those on the right side of the equation. This gives four equations:

$$\begin{aligned} 6a &= 6 + e + 2f \\ 6a + 2b &= 3 + d + 2e + f \\ c &= 1 \\ 12a + 3c &= 10 + 2d + 6f \end{aligned}$$

The fermentor is equipped with an oxygen sensor and a carbon dioxide sensor. These enable us to compute the respiratory quotient R:

$$R = \frac{\mathrm{CO_2}}{\mathrm{O_2}} = \frac{e}{b}$$

Thus the fifth equation is $Rb - e = 0$. The yeast yield Y (grams of yeast produced per gram of glucose consumed) is related to a as follows.

$$Y = \frac{144}{180a}$$

where 144 is the molecular weight of yeast and 180 is the molecular weight of glucose. By measuring the yeast yield Y we can compute a as follows: $a = 144/180Y$. This is the sixth equation.

Write a user-defined function that computes f, the amount of ethanol produced, with R and Y as the function's arguments. Test your function for two cases where Y is measured to be 0.5: (a) $R = 1.1$ and (b) $R = 1.05$.

■ Solution

First note that there are only four unknowns because the third equation directly gives $c = 1$, and the sixth equation directly gives $a = 144/180Y$. To write these equations in matrix form, let $x_1 = b, x_2 = d, x_3 = e$, and $x_4 = f$. Then the equations can be written as

$$\begin{aligned} -x_3 - 2x_4 &= 6 - 6(144/180Y) \\ 2x_1 - x_2 - 2x_3 - x_4 &= 3 - 6(144/180Y) \\ -2x_2 - 6x_4 &= 7 - 12(144/180Y) \\ Rx_1 - x_3 &= 0 \end{aligned}$$

In matrix form these become

$$\begin{bmatrix} 0 & 0 & -1 & -2 \\ 2 & -1 & -2 & -1 \\ 0 & -2 & 0 & -6 \\ R & 0 & -1 & 0 \end{bmatrix} \begin{bmatrix} x_1 \\ x_2 \\ x_3 \\ x_4 \end{bmatrix} = \begin{bmatrix} 6 - 6(144/180Y) \\ 3 - 6(144/180Y) \\ 7 - 12(144/180Y) \\ 0 \end{bmatrix}$$

The function file is shown below.

```
function E = ethanol(R,Y)
% Computes ethanol produced from yeast reaction.
A = [0,0,-1,-2;2,-1,-2,-1;...
0,-2,0,-6;R,0,-1,0];
b = [6-6*(144./(180*Y));3-6*(144./(180*Y));...
7-12*(144./(180*Y));0];
x = A\b;
E = x(4);
```

The session is as follows:

```
>>ethanol(1.1,0.5)
ans =
      0.0654
>>ethanol(1.05,0.5)
ans =
      -0.0717
```

The negative value for `E` in the second case indicates that ethanol is being consumed rather than produced.

Matrix Inverse

The solution of the scalar equation $ax = b$ is $x = b/a$ if $a \neq 0$. The division operation of scalar algebra has an analogous operation in matrix algebra. For example, to solve the matrix equation

$$\mathbf{Ax} = \mathbf{b} \tag{6.2–9}$$

for **x**, we must somehow "divide" **b** by **A**. This procedure is developed from the concept of a *matrix inverse*. The inverse of a matrix **A** is defined only if **A** is square and nonsingular. It is denoted by $\mathbf{A}^{-1}$ and has the property that

$$\mathbf{A}^{-1}\mathbf{A} = \mathbf{A}\mathbf{A}^{-1} = \mathbf{I} \tag{6.2–10}$$

where **I** is the identity matrix. Using this property, we multiply both sides of (6.2–9) from the left by $\mathbf{A}^{-1}$ to obtain

$$\mathbf{A}^{-1}\mathbf{Ax} = \mathbf{A}^{-1}\mathbf{b}$$

Because $\mathbf{A}^{-1}\mathbf{Ax} = \mathbf{Ix} = \mathbf{x}$, we obtain

$$\mathbf{x} = \mathbf{A}^{-1}\mathbf{b} \tag{6.2–11}$$

The solution form (6.2–11), $\mathbf{x} = \mathbf{A}^{-1}\mathbf{b}$, is rarely applied in practice to obtain numerical solutions, because calculation of the matrix inverse is subject to numerical inaccuracy, especially for large matrices. However, the equation $\mathbf{x} = \mathbf{A}^{-1}\mathbf{b}$ is a concise representation of the solution and therefore is useful for developing symbolic solutions to problems (for example, such problems are encountered in the solutions of differential equations; see Chapters 8 and 10).

Linearity

The matrix equation $\mathbf{Ax} = \mathbf{b}$ possesses the *linearity* property. The solution $\mathbf{x}$ is $\mathbf{x} = \mathbf{A}^{-1}\mathbf{b}$, and thus $\mathbf{x}$ is proportional to the vector $\mathbf{b}$. We can use this fact to obtain a more generally useful algebraic solution in cases where the right sides are all multiplied by the same scalar. For example, suppose the matrix equation is $\mathbf{Ay} = \mathbf{b}c$, where c is a scalar. The solution is $\mathbf{y} = \mathbf{A}^{-1}\mathbf{b}c = \mathbf{x}c$. Thus if we obtain the solution to $\mathbf{Ax} = \mathbf{b}$, the solution to $\mathbf{Ay} = \mathbf{b}c$ is given by $\mathbf{y} = \mathbf{x}c$. We demonstrate the usefulness of this fact in Example 6.2–4.

When designing structures, engineers must be able to predict how much force will be exerted on each part of the structure so that they can properly select the part's size and material to make it strong enough. The engineers often must solve linear equations to determine these forces. These equations are obtained by applying the principles of statics, which state that the vector sums of forces and moments must be zero if the structure does not move.

EXAMPLE 6.2–4 Calculation of Cable Tension

A mass m is suspended by three cables attached at the three points B, C, and D, as shown in Figure 6.2–2. Let T_1, T_2, and T_3 be the tensions in the three cables AB, AC, and AD, respectively. If the mass m is stationary, the sum of the tension components in the x, in the y, and in the z directions must each be zero. This requirement gives the following three

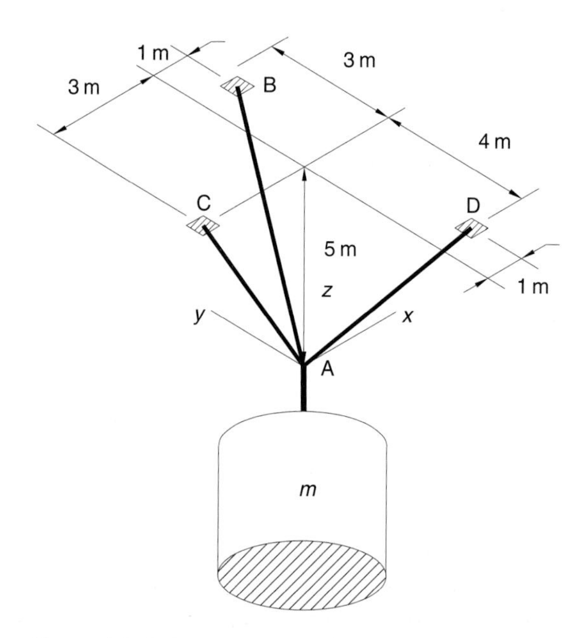

Figure 6.2–2 A mass suspended by three cables.

equations:

$$\frac{T_1}{\sqrt{35}} - \frac{3T_2}{\sqrt{34}} + \frac{T_3}{\sqrt{42}} = 0$$

$$\frac{3T_1}{\sqrt{35}} - \frac{4T_3}{\sqrt{42}} = 0$$

$$\frac{5T_1}{\sqrt{35}} + \frac{5T_2}{\sqrt{34}} + \frac{5T_3}{\sqrt{42}} - mg = 0$$

Use MATLAB to find T_1, T_2, and T_3 in terms of an unspecified value of the weight mg.

■ Solution

If we set $mg = 1$, the equations have the form $\mathbf{Ax} = \mathbf{b}$ where

$$\mathbf{A} = \begin{bmatrix} \frac{1}{\sqrt{35}} & -\frac{3}{\sqrt{34}} & \frac{1}{\sqrt{42}} \\ \frac{3}{\sqrt{35}} & 0 & \frac{-4}{\sqrt{42}} \\ \frac{5}{\sqrt{35}} & \frac{5}{\sqrt{34}} & \frac{5}{\sqrt{42}} \end{bmatrix} \qquad \mathbf{x} = \begin{bmatrix} T_1 \\ T_2 \\ T_3 \end{bmatrix} \qquad \mathbf{b} = \begin{bmatrix} 0 \\ 0 \\ 1 \end{bmatrix}$$

We can use the following MATLAB script file to solve this system for **x** and then multiply the result by mg to obtain the desired result.

```
% File cable.m
% Computes the tensions in three cables.
A1 = [1/sqrt(35), -3/sqrt(34), 1/sqrt(42)];
A2 = [3/sqrt(35), 0, -4/sqrt(42)];
A3 = [5/sqrt(35), 5/sqrt(34), 5/sqrt(42)];
A = [A1; A2; A3];
b = [0; 0; 1];
x = A\b;
disp('The tension T_1 is:')
disp(x(1))
disp('The tension T_2 is:')
disp(x(2))
disp('The tension T_3 is:')
disp(x(3))
```

When this file is executed by typing `cable`, the result is stored in the array `x`, which gives the values $T_1 = 0.5071$, $T_2 = 0.2915$, and $T_3 = 0.4166$. Because MATLAB does not generate an error message when the file is executed, the solution is unique. Using the linearity property, we multiply these results by mg and obtain the following solution to the set $\mathbf{Ay} = \mathbf{b}mg$: $T_1 = 0.5071\, mg$, $T_2 = 0.2915\, mg$, and $T_3 = 0.4166\, mg$.

Calculating a matrix inverse by hand is tedious. The inverse of a 3×3 matrix requires us to evaluate nine 2×2 determinants. We do not give the general procedure here because we will soon explain how to use MATLAB to compute a matrix inverse. The details of computing a matrix inverse can be found in many texts; for example, see [Kreyzig, 1998]. However, the inverse of a 2×2 matrix is easy to find. If $\mathbf{A}$ is given by

$$\mathbf{A} = \begin{bmatrix} a_{11} & a_{12} \\ a_{21} & a_{22} \end{bmatrix}$$

its inverse is given by

$$\mathbf{A}^{-1} = \frac{1}{|\mathbf{A}|} \begin{bmatrix} a_{22} & -a_{12} \\ -a_{21} & a_{11} \end{bmatrix} \tag{6.2–12}$$

Calculation of $\mathbf{A}^{-1}$ can be checked by determining whether $\mathbf{A}^{-1}\mathbf{A} = \mathbf{I}$. Note that the preceding formula shows that $\mathbf{A}^{-1}$ does not exist if $|\mathbf{A}| = 0$ (that is, if $\mathbf{A}$ is singular).

EXAMPLE 6.2–5

The Matrix Inverse Method

Solve the following equations using the matrix inverse:

$$2x + 9y = 5$$
$$3x - 4y = 7$$

■ **Solution**

The matrix $\mathbf{A}$ is

$$\mathbf{A} = \begin{bmatrix} 2 & 9 \\ 3 & -4 \end{bmatrix}$$

Its determinant is $|\mathbf{A}| = 2(-4) - 9(3) = -35$, and its inverse is

$$\mathbf{A}^{-1} = \frac{1}{-35} \begin{bmatrix} -4 & -9 \\ -3 & 2 \end{bmatrix} = \frac{1}{35} \begin{bmatrix} 4 & 9 \\ 3 & -2 \end{bmatrix}$$

The solution is

$$\mathbf{x} = \mathbf{A}^{-1}\mathbf{b} = \frac{1}{35} \begin{bmatrix} 4 & 9 \\ 3 & -2 \end{bmatrix} \begin{bmatrix} 5 \\ 7 \end{bmatrix} = \frac{1}{35} \begin{bmatrix} 83 \\ 1 \end{bmatrix}$$

or $x = 83/35 = 2.3714$ and $y = 1/35 = 0.0286$.

The Matrix Inverse in MATLAB

The MATLAB command `inv(A)` computes the inverse of the matrix $\mathbf{A}$. The following MATLAB session solves the equations given in Example 6.2–5 using MATLAB.

```
>>A = [2,9;3,-4];
>>b = [5;7]
>>x = inv(A)*b
x =
   2.3714
   0.0286
```

If you attempt to solve a singular problem using the `inv` command, MATLAB displays an error message.

Test Your Understanding

T6.2–1 Use the matrix inverse method to solve the following set by hand and by using MATLAB:

$$3x - 4y = 5$$
$$6x - 10y = 2$$

(Answer: $x = 7$, $y = 4$.)

T6.2–2 Use the matrix inverse method to solve the following set by hand and by using MATLAB:

$$3x - 4y = 5$$
$$6x - 8y = 2$$

(Answer: no solution.)

6.3 Cramer's Method

Cramer's method is a systematic method for solving equations, but it is not used as a basis for computer packages because it can be slow and very sensitive to numerical round-off error, especially for a large number of equations. We introduce it here to gain insight into the requirements for a set of equations to have a solution. In addition, Cramer's method provides a systematic method for obtaining solutions of linear equations in symbolic form. This has useful applications in the solution of differential equations (see Chapters 8 and 10). We will use the following set of two equations to illustrate Cramer's method:

$$a_{11}x + a_{12}y = b_1 \tag{6.3–1}$$

$$a_{21}x + a_{22}y = b_2 \tag{6.3–2}$$

To solve these equations, we can multiply the first equation by a_{22} and the second equation by $-a_{12}$ to obtain

$$a_{22}(a_{11}x + a_{12}y) = a_{22}b_1$$
$$-a_{12}(a_{21}x + a_{22}y) = -a_{12}b_2$$

When these two equations are added, the y terms cancel and we obtain the solution for x:

$$x = \frac{b_1a_{22} - b_2a_{12}}{a_{22}a_{11} - a_{12}a_{21}} \tag{6.3–3}$$

We can cancel the x terms in a similar way and obtain the following solution for y:

$$y = \frac{a_{11}b_2 - a_{21}b_1}{a_{22}a_{11} - a_{12}a_{21}} \tag{6.3–4}$$

Note that both solutions have the same denominator, which we denote by $D = a_{22}a_{11} - a_{12}a_{21}$. If this denominator is zero, the above solutions are not valid because we cannot divide by zero. In that case all we can say is that

$$0x = b_1 a_{22} - b_2 a_{12}$$

$$0y = a_{11} b_2 - a_{21} b_1$$

So if $D = 0$, but $b_1 a_{22} - b_2 a_{12} \neq 0$, x is undefined. If $D = 0$ and $b_1 a_{22} - b_2 a_{12} = 0$, there are infinitely many solutions for x (because any finite value of x will satisfy the equation $0x = 0$).

Similarly, if $D = 0$, but $a_{11} b_2 - a_{21} b_1 \neq 0$, y is undefined, and if $a_{11} b_2 - a_{21} b_1 = 0$, there are infinitely many solutions for y.

Cramer's method expresses the above solutions in terms of determinants. The determinant D (called *Cramer's determinant*) formed from the coefficients of equations (6.3–1) and (6.3–2) is as follows:

$$D = \begin{vmatrix} a_{11} & a_{12} \\ a_{21} & a_{22} \end{vmatrix} = a_{11} a_{22} - a_{12} a_{21} \tag{6.3–5}$$

Note that this expression is identical to the denominator of the solutions for x and y given by (6.3–3) and (6.3–4).

If we form a determinant D_1 by replacing the first column of D with the coefficients on the right side of the equation set (6.3–1) and (6.3–2), we obtain

$$D_1 = \begin{vmatrix} b_1 & a_{12} \\ b_2 & a_{22} \end{vmatrix} = b_1 a_{22} - b_2 a_{12}$$

This expression is identical to the numerator of the solution (6.3–3). Thus the solution can be expressed as the ratio of the two determinants $x = D_1/D$.

Next form the determinant D_2 by replacing the second column of D with the coefficients on the right side of the equation set. Thus

$$D_2 = \begin{vmatrix} a_{11} & b_1 \\ a_{21} & b_2 \end{vmatrix} = a_{11} b_2 - a_{21} b_1$$

This expression is identical to the numerator of the solution (6.3–4). Thus $y = D_2/D$.

Cramer's method expresses the solutions as ratios of determinants, and thus it can be extended to equations with more than two variables by using determinants having the appropriate dimension. Before using Cramer's method, be sure the variables are lined up in a consistent order (for example, x, y, z) in each equation and move all constants to the right side. Equations (6.1–1) through (6.1–3) from Example 6.1–1 illustrate this process.

$$-x + y + 2z = 2$$

$$3x - y + z = 6$$

$$-x + 3y + 4z = 4$$

Cramer's determinant for this set is

$$D = \begin{vmatrix} -1 & 1 & 2 \\ 3 & -1 & 1 \\ -1 & 3 & 4 \end{vmatrix} = 10$$

One advantage of Cramer's method is that you can find only one of the unknowns directly if that is all you want. For example, the first unknown is found from $x = D_1/D$, where D_1 is the determinant formed by replacing the first column in the determinant D with the coefficients on the right side of the equation set:

$$D_1 = \begin{vmatrix} 2 & 1 & 2 \\ 6 & -1 & 1 \\ 4 & 3 & 4 \end{vmatrix}$$

This determinant has the value $D_1 = 10$, and thus $x = D_1/D = 10/10 = 1$. Similarly, $y = D_2/D = -10/10 = -1$ and $z = D_3/D = 20/10 = 2$, where

$$D_2 = \begin{vmatrix} -1 & 2 & 2 \\ 3 & 6 & 1 \\ -1 & 4 & 4 \end{vmatrix} \qquad D_3 = \begin{vmatrix} -1 & 1 & 2 \\ 3 & -1 & 6 \\ -1 & 3 & 4 \end{vmatrix}$$

Cramer's Determinant and Singular Problems

When the number of variables equals the number of equations, a singular problem can be identified by computing Cramer's determinant D. If the determinant is zero, the equations are singular because D appears in the denominator of the solutions. For example, for the set

$$3x - 4y = 5$$

$$6x - 8y = 3$$

Cramer's determinant is

$$D = \begin{vmatrix} 3 & -4 \\ 6 & -8 \end{vmatrix} = 3(-8) - 6(-4) = 0$$

Thus the equation set is singular.

Another example is given by the following homogeneous set:

$$6x + ay = 0$$

$$2x + 4y = 0$$

We saw earlier that any finite values of x and y, such that $x = -2y$, are solutions of this set if $a = 12$. If $a \neq 12$, the only solution is $x = y = 0$. Cramer's determinant is

$$D = \begin{vmatrix} 6 & a \\ 2 & 4 \end{vmatrix} = 6(4) - 2a = 24 - 2a$$

and $D = 0$ if $a = 12$. Thus the set is singular if $a = 12$.

In general, for a set of *homogeneous* linear algebraic equations that contains the *same* number of equations as unknowns, a *nonzero* solution exists only if the set is singular; that is, if Cramer's determinant is *zero;* furthermore, the solution is not unique. If Cramer's determinant is not zero, the homogeneous set has a zero solution; that is, all the unknowns are zero.

Cramer's determinant gives some insight into ill-conditioned problems, which are close to being singular. A Cramer's determinant close to zero indicates an ill-conditioned problem.

Test Your Understanding

T6.3–1 Use Cramer's method to solve for x and y in terms of the parameter b. For what value of b is the set singular?

$$4x - by = 5$$
$$-3x + 6y = 3$$

(Answer: $x = (10 + b)/(8 - b)$, $y = 9/(8 - b)$ unless $b = 8$.)

T6.3–2 Use Cramer's method to solve for y. Use MATLAB to evaluate the determinants.

$$2x + y + 2z = 17$$
$$3y + z = 6$$
$$2x - 3y + 4z = 19$$

(Answer: $y = 1$.)

6.4 Underdetermined Systems

You have seen how to use the matrix inverse method $\mathbf{x} = \mathbf{A}^{-1}\mathbf{b}$ to solve the equation set $\mathbf{Ax} = \mathbf{b}$. However, this method works only if the matrix $\mathbf{A}$ is square; that is, if the number of unknowns equals the number of equations. Even if $\mathbf{A}$ is square, the method will not work if $|\mathbf{A}| = 0$ because the matrix inverse $\mathbf{A}^{-1}$ does not exist. The same limitation applies to Cramer's method; it cannot solve equation sets where the number of unknowns does not equal the number of equations.

This section explains how to use MATLAB to solve problems in which the matrix $\mathbf{A}$ is square but $|\mathbf{A}| = 0$, and problems in which $\mathbf{A}$ is not square. The left-division method works for square and nonsquare $\mathbf{A}$ matrices. However, as you will see, if $\mathbf{A}$ is not square, the left-division method can give answers that might be misinterpreted. We explain how to interpret MATLAB results correctly.

An *underdetermined system* does not contain enough information to solve for all of the unknown variables, usually because it has fewer equations than unknowns. Thus an infinite number of solutions can exist, with one or more of the unknowns dependent on the remaining unknowns. For such systems the matrix inverse method and Cramer's method will not work. When there are more equations than unknowns, the left-division method will give a solution with some of the unknowns set equal to zero. A simple example is given by the equation $x + 3y = 6$. All we can do is solve for one of the unknowns in terms of the other; for example, $x = 6 - 3y$. An infinite number of solutions satisfy this equation. The left-division method gives one of these solutions, the one with x set equal to zero: $x = 0$, $y = 2$.

```
>>A = [1,3];
>>b = 6;
>>x = A\b
x =
    0
    2
```

An infinite number of solutions might exist even when the number of equations equals the number of unknowns. This situation can occur when $|\mathbf{A}| = 0$. For such systems the matrix inverse method and Cramer's method will not work, and the left-division method generates an error message warning us that the matrix **A** is singular. In such cases the *pseudoinverse method* `x = pinv(A)*b` gives one solution, the *minimum norm solution.* In cases that have an infinite number of solutions, some of the unknowns can be expressed in terms of the remaining unknowns, whose values are arbitrary. We can use the `rref` command to find these relations. We introduce these commands in this section and give examples showing how to interpret their results.

PSEUDOINVERSE METHOD

MINIMUM NORM SOLUTION

An equation set can be underdetermined even though it has as many equations as unknowns. For example, the set

$$2x - 4y + 5z = -4$$

$$-4x - 2y + 3z = 4$$

$$2x + 6y - 8z = 0$$

has three unknowns and three equations, but it is underdetermined and has infinitely many solutions. This condition occurs because the set has only two independent equations; the third equation can be obtained from the first two. To obtain the third equation, add the first and second equations to obtain $-2x - 6y + 8z = 0$, which is equivalent to the third equation.

Determining whether all the equations are independent might not be easy, especially if the set has many equations. For this reason we now introduce a method that enables us to determine easily whether or not an equation set has a solution and whether or not it is unique. The method requires an understanding of the concept of the *rank* of a matrix.

Matrix Rank

Consider the following 3×3 determinant:

$$\begin{vmatrix} 3 & -4 & 1 \\ 6 & 10 & 2 \\ 9 & -7 & 3 \end{vmatrix}$$

If we eliminate one row and one column in the determinant, we are left with a 2×2 determinant. Depending on which row and column we eliminate, we can obtain any of nine possible 2×2 determinants. These elements are called *subdeterminants*. For example, if we eliminate row 1 and column 1, we obtain

SUBDETERMINANT

$$\begin{vmatrix} 10 & 2 \\ -7 & 3 \end{vmatrix} = 10(3) - 2(-7) = 44$$

If we eliminate row 2 and column 3, we obtain

$$\begin{vmatrix} 3 & -4 \\ 9 & -7 \end{vmatrix} = 3(-7) - 9(-4) = 15$$

Subdeterminants can be used to define the *rank* of a matrix, which provides useful information concerning the existence and nature of solutions. The definition of *matrix rank* is as follows:

Matrix rank. An $m \times n$ matrix $\mathbf{A}$ has a rank $r \geq 1$ if and only if $|\mathbf{A}|$ contains a nonzero $r \times r$ determinant and every square subdeterminant with $r + 1$ or more rows is zero.

For example, the rank of

$$\mathbf{A} = \begin{bmatrix} 3 & -4 & 1 \\ 6 & 10 & 2 \\ 9 & -7 & 3 \end{bmatrix} \tag{6.4–1}$$

is 2 because $|\mathbf{A}| = 0$ whereas $\mathbf{A}$ contains at least one nonzero 2×2 subdeterminant. For example, the subdeterminant obtained by eliminating row 1 and column 1 is nonzero and has the value 44.

MATLAB provides an easy way to determine the rank of a matrix. First define the matrix $\mathbf{A}$ as an array in the usual way. Then type `rank(A)`. For example, the following MATLAB session determines the rank of the matrix given by (6.4–1).

```
>>A = [3,-4,1;6,10,2;9,-7,3];
>>rank(A)
ans =
     2
```

Existence and Uniqueness of Solutions

AUGMENTED MATRIX

The following test determines whether a solution exists and whether or not it is unique. The test requires that we first form the so-called *augmented matrix* [**A b**]. The first n columns of the augmented matrix are the columns of $\mathbf{A}$. The last column of the augmented matrix is the column vector **b**. For example, if

$$\mathbf{A} = \begin{bmatrix} 5 & 3 & -9 \\ -2 & 6 & 8 \end{bmatrix} \qquad \mathbf{b} = \begin{bmatrix} 7 \\ -10 \end{bmatrix}$$

then the augmented matrix is

$$[\mathbf{A} \quad \mathbf{b}] = \begin{bmatrix} 5 & 3 & -9 & 7 \\ -2 & 6 & 8 & -10 \end{bmatrix}$$

The solution test can be stated as follows [Kreyzig, 1998]:

Existence and uniqueness of solutions. The set $\mathbf{Ax} = \mathbf{b}$ with m equations and n unknowns has solutions if and only if rank[**A**] = rank[**A** **b**] (1). Let r = rank[**A**]. If condition (1) is satisfied and if $r = n$, then the solution is unique. If condition (1) is satisfied but $r < n$, an infinite number of solutions exists and r unknown variables can be expressed as linear combinations of the other $n - r$ unknown variables, whose values are arbitrary.

Homogeneous case. The homogeneous set $\mathbf{Ax} = \mathbf{0}$ is a special case in which $\mathbf{b} = \mathbf{0}$. For this case rank[**A**] = rank[**A** **b**] always, and thus the set always has the trivial solution $\mathbf{x} = \mathbf{0}$. A nonzero solution, in which at least one unknown is nonzero, exists if and only if rank[**A**] $< n$. If $m < n$, the homogeneous set always has a nonzero solution.

Recall that if $|\mathbf{A}| = 0$, the equation set is singular. If you try to solve a singular set using MATLAB, it prints a message warning that the matrix is singular and does not try to solve the problem. An ill-conditioned set of equations is a set that is close to being singular. The ill-conditioned status depends on the accuracy with which the solution calculations are made. When the internal numerical accuracy used by MATLAB is insufficient to obtain a solution, MATLAB prints a message to warn you that the matrix is close to singular and that the results might be inaccurate.

A Set Having a Unique Solution — EXAMPLE 6.4-1

Determine whether the following set has a unique solution, and if so, find it:

$$3x - 2y + 8z = 48$$

$$-6x + 5y + z = -12$$

$$9x + 4y + 2z = 24$$

■ Solution

The matrices **A**, **b**, and **x** are

$$\mathbf{A} = \begin{bmatrix} 3 & -2 & 8 \\ -6 & 5 & 1 \\ 9 & 4 & 2 \end{bmatrix} \qquad \mathbf{b} = \begin{bmatrix} 48 \\ -12 \\ 24 \end{bmatrix} \qquad \mathbf{x} = \begin{bmatrix} x \\ y \\ z \end{bmatrix}$$

The following MATLAB session checks the ranks of **A** and [**A** **b**] and finds the solution.

```
>>A = [3,-2,8;-6,5,1;9,4,2];
>>b = [48;-12;24];
>>rank(A)
```

```
ans =
     3
>>rank([A  b])
ans =
     3
>>x = A\b
x =
   2
   -1
   5
```

Because **A** and [**A** **b**] have the same rank, a solution exists. Because this rank equals the number of unknowns (which is three), the solution is unique. The left-division method gives this solution, which is $x = 2$, $y = -1$, $z = 5$.

Test Your Understanding

T6.4–1 Use MATLAB to show that the following set has a unique solution and then find the solution:

$$3x + 12y - 7z = 5$$
$$5x - 6y - 5z = -8$$
$$-2x + 7y + 9z = 5$$

(Answer: The unique solution is $x = -1.0204$, $y = 0.5940$, $z = -0.1332$.)

EXAMPLE 6.4–2

An Underdetermined Set

Show that the following set does not have a unique solution. How many of the unknowns will be undetermined? Interpret the results given by the left-division method.

$$2x - 4y + 5z = -4$$
$$-4x - 2y + 3z = 4$$
$$2x + 6y - 8z = 0$$

■ Solution

A MATLAB session to check the ranks looks like

```
>>A = [2,-4,5;-4,-2,3;2,6,-8];
>>b = [-4;4;0];
>>rank(A)
ans =
     2
>>rank([A  b])
ans =
     2
```

Because the ranks of **A** and [**A** **b**] are equal, a solution exists. However, because the number of unknowns is three, and is one greater than the rank of **A**, one of the unknowns will be undetermined. An infinite number of solutions exists, and we can solve for only two of the unknowns in terms of the third unknown. We will obtain these solutions in Example 6.4.–4.

Note that even though the number of equations equals the number of unknowns here, the matrix **A** is singular. (We know this because its rank is less than three.) Thus we cannot use the matrix inverse method or Cramer's method for this problem.

If we use the left-division method, MATLAB returns a message warning that the problem is singular, rather than producing an answer.

The `pinv` Command and the Euclidean Norm

The `pinv` command can obtain a solution of an underdetermined set. To solve the equation set $\mathbf{Ax} = \mathbf{b}$ using the `pinv` command, type `x = pinv(A)*b`. Underdetermined sets have an infinite number of solutions, and the `pinv` command produces a solution that gives the minimum value of the *Euclidean norm,* which is the magnitude of the solution vector **x**. The magnitude of a vector **v** in three-dimensional space, having components x, y, z, is $\sqrt{x^2 + y^2 + z^2}$. It can be computed using matrix multiplication and the transpose as follows:

$$\sqrt{\mathbf{v}^T\mathbf{v}} = \sqrt{[x \quad y \quad z]^T \begin{bmatrix} x \\ y \\ z \end{bmatrix}} = \sqrt{x^2 + y^2 + z^2}$$

The generalization of this formula to an n-dimensional vector **v** gives the magnitude of the vector and is the Euclidean norm N. Thus

$$N = \sqrt{\mathbf{v}^T\mathbf{v}} \qquad (6.4\text{–}2)$$

Example 6.4–3 shows how to apply the `pinv` command.

A Statically Indeterminate Problem

EXAMPLE 6.4–3

Determine the forces in the three equally spaced supports that hold up a light fixture. The supports are 5 ft apart. The fixture weighs 400 lb, and its mass center is 4 ft from the right end. (a) Solve the problem by hand. (b) Obtain the solution using the MATLAB left-division method and the pseudoinverse method.

■ Solution

(a) Figure 6.4–1 shows the fixture and the free-body diagram, where T_1, T_2, and T_3 are the tension forces in the supports. For the fixture to be in equilibrium, the vertical forces must cancel, and the total moments about an arbitrary fixed point—say, the right endpoint—must be zero. These conditions give the two equations:

$$T_1 + T_2 + T_3 - 400 = 0$$

$$400(4) - 10T_1 - 5T_2 = 0$$

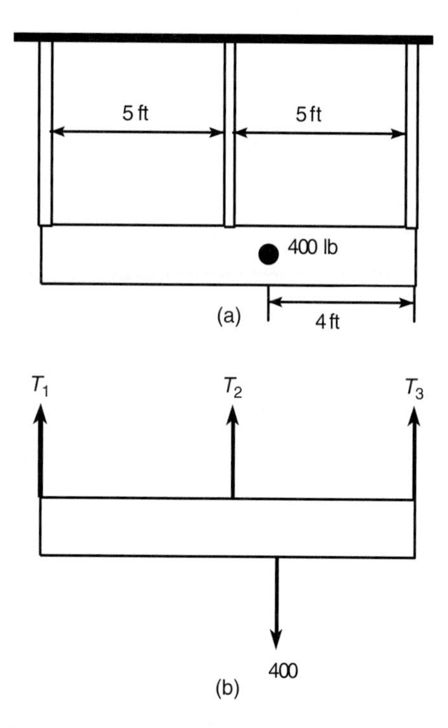

Figure 6.4–1 A light fixture and its free-body diagram.

or

$$T_1 + T_2 + T_3 = 400 \tag{6.4–3}$$

$$10T_1 + 5T_2 + 0T_3 = 1600 \tag{6.4–4}$$

Because there are more unknowns than equations, the set is underdetermined. These equations can be written in the matrix form $\mathbf{Ax} = \mathbf{b}$ as follows:

$$\begin{bmatrix} 1 & 1 & 1 \\ 10 & 5 & 0 \end{bmatrix} \begin{bmatrix} T_1 \\ T_2 \\ T_3 \end{bmatrix} = \begin{bmatrix} 400 \\ 1600 \end{bmatrix}$$

where

$$\mathbf{A} = \begin{bmatrix} 1 & 1 & 1 \\ 10 & 5 & 0 \end{bmatrix} \qquad \mathbf{b} = \begin{bmatrix} 400 \\ 1600 \end{bmatrix} \qquad \mathbf{x} = \begin{bmatrix} T_1 \\ T_2 \\ T_3 \end{bmatrix}$$

$$[\mathbf{A} \quad \mathbf{b}] = \begin{bmatrix} 1 & 1 & 1 & 400 \\ 10 & 5 & 0 & 1600 \end{bmatrix}$$

Because we can find a nonzero 2×2 determinant in both $\mathbf{A}$ and $[\mathbf{A} \quad \mathbf{b}]$, the ranks of $\mathbf{A}$ and $[\mathbf{A} \quad \mathbf{b}]$ are both 2; thus a solution exists. Because the number of unknowns is three, and is one greater than the rank of $\mathbf{A}$, an infinite number of solutions exists and we can solve for only two of the forces in terms of the third force. Equation (6.4–4) gives

$$T_2 = \frac{1600 - 10T_1}{5} = 320 - 2T_1$$

Substitute this expression into (6.4–3) and solve for T_1 to find that $T_1 = T_3 - 80$. Thus the solution in terms of T_3 is

$$T_1 = T_3 - 80$$

$$T_2 = 320 - 2T_1 = 320 - 2(T_3 - 80) = 480 - 2T_3$$

We cannot determine numerical values for any of the forces. Such a problem, in which the equations of statics do not give enough equations to find all of the unknowns, is called *statically indeterminate*.

STATICALLY INDETERMINATE

(b) A MATLAB session to check the ranks and to solve this problem using left division looks like

```
>>A = [1,1,1;10,5,0];
>>b = [400;1600];
>>rank(A)
ans =
     2
>>rank([A  b])
ans =
     2
>>A\b
ans =
      160.0000
      0
      240.0000
```

The answer corresponds to $T_1 = 160$, $T_2 = 0$, and $T_3 = 240$ lb. This example illustrates how the MATLAB left-division operator produces a solution with one or more variables set to zero, for underdetermined sets having more unknowns than equations.

To use the pseudoinverse operator, type the command `pinv(A)*b`. The result is $T_1 = 93.3333$, $T_2 = 133.3333$, and $T_3 = 173.3333$ lb. This answer is the minimum norm solution for real values of the variables. The minimum norm solution consists of the real values of T_1, T_2, and T_3 that minimize

$$\begin{aligned} N &= \sqrt{T_1^2 + T_2^2 + T_3^2} \\ &= \sqrt{(T_3 - 80)^2 + (480 - 2T_3)^2 + T_3^2} \\ &= \sqrt{6T_3^2 - 2080T_3 + 236{,}800} \end{aligned}$$

The smallest value N can have is zero. This result occurs when $T_3 = 173 \pm 97i$, which corresponds to $T_1 = 93 \pm 97i$ and $T_2 = 827 \pm 194i$. This result is a valid solution of the original equations, but is not the minimum norm solution where T_1, T_2, and T_3 are restricted to real values (we know that the forces cannot be complex).

We can find the real value of T_3 that minimizes N by plotting N versus T_3 or by using calculus. The answer is $T_3 = 173.3333$, which gives $T_1 = 93.3333$ and $T_2 = 133.3333$. These values are the minimum norm solution given by the pseudoinverse method.

We must decide whether or not the solutions given by the left-division and the pseudoinverse methods are useful for applications that have an infinite number of solutions, and we must do so in the context of the specific application. For example, in the light-fixture application discussed in Example 6.4–3, only two supports are required, and the left-division solution (the solution with $T_2 = 0$) shows that if the middle support is eliminated, the forces in the end supports will be $T_1 = 160$ and $T_3 = 240$ lb. Suppose we want to use three supports to reduce the load carried by each support. The pseudoinverse solution ($T_1 = 93$, $T_2 = 133$, $T_3 = 173$) is the solution that minimizes the sum of the squares of the support forces.

Many problems are statically indeterminate because the engineer has included more supports than necessary, usually for safety in case one support fails. In practice, when engineers are confronted with a statically indeterminate problem, they supplement the equations of statics with equations that describe the deformations of the supports as functions of the applied forces and moments. These additional equations allow the forces and moments within the structure to be determined unambiguously.

Test Your Understanding

T6.4–2 Use MATLAB to find two solutions to the following set:

$$x + 3y + 2z = 2$$

$$x + y + z = 4$$

(Answer: Minimum norm solution: $x = 4.33$, $y = -1.67$, $z = 1.34$. Left-division solution: $x = 5$, $y = -1$, $z = 0$.)

The Reduced Row Echelon Form

We can express some of the unknowns in an underdetermined set as functions of the remaining unknowns. For example, in the statically indeterminate case of Example 6.4–3, we wrote the solutions for two of the forces in terms of the third:

$$T_1 = T_3 - 80$$

$$T_2 = 480 - 2T_3$$

These two equations are equivalent to

$$T_1 - T_3 = -80$$

$$T_2 + 2T_3 = 480$$

In matrix form these are

$$\begin{bmatrix} 1 & 0 & -1 \\ 0 & 1 & 2 \end{bmatrix} \begin{bmatrix} T_1 \\ T_2 \\ T_3 \end{bmatrix} = \begin{bmatrix} -80 \\ 480 \end{bmatrix}$$

The augmented matrix for the preceding set is

$$\begin{bmatrix} 1 & 0 & -1 & -80 \\ 0 & 1 & 2 & 480 \end{bmatrix}$$

Note that the first two columns form a 2×2 identity matrix. Therefore, the corresponding equations can be solved directly for T_1 and T_2 in terms of T_3.

We can always reduce an underdetermined set to such a form by multiplying the set's equations by suitable factors and adding the resulting equations to eliminate an unknown variable. The MATLAB `rref` command provides a procedure to reduce an equation set to this form, which is called the *reduced row echelon form.* Its syntax is `rref([A b])`. Its output is the augmented matrix [**C d**] that corresponds to the equation set $\mathbf{Cx} = \mathbf{d}$. This set is in reduced row echelon form.

A Singular Set — EXAMPLE 6.4–4

The following underdetermined equation set was analyzed in Example 6.4–2. There it was shown that an infinite number of solutions exists. Use the `pinv` and the `rref` commands to obtain solutions.

$$2x - 4y + 5z = -4$$
$$-4x - 2y + 3z = 4$$
$$2x + 6y - 8z = 0$$

■ Solution

First use the `pinv` command. The MATLAB session follows.

```
>>A = [2,-4,5;-4,-2,3;2,6,-8];
>>b = [-4;4;0];
>>x = pinv(A)*b
x =
   -1.2148
   0.2074
   -0.1481
```

Thus the pseudoinverse method gives the solution: $x = -1.2148$, $y = 0.2074$, $z = -0.1481$. This solution is valid, but it is not the general solution.

To obtain the general solution, we can use the `rref` command. The current MATLAB session continues as follows.

```
>>rref([A b])
ans =
   1     0     -0.1     -1.2000
   0     1     -1.3     0.4000
   0     0        0          0
```

The answer corresponds to the augmented matrix [**C** **d**], where

$$[\mathbf{C} \quad \mathbf{d}] = \begin{bmatrix} 1 & 0 & -0.1 & -1.2 \\ 0 & 1 & -1.3 & 0.4 \\ 0 & 0 & 0 & 0 \end{bmatrix}$$

This matrix corresponds to the matrix equation $\mathbf{Cx} = \mathbf{d}$, or

$$x + 0y - 0.1z = -1.2$$

$$0x + y - 1.3z = 0.4$$

$$0x + 0y + 0z = 0$$

We can easily solve these expressions for x and y in terms of z as follows: $x = 0.1z - 1.2$ and $y = 1.3z + 0.4$. This result is the general solution to the problem, where z is taken to be the arbitrary variable.

Supplementing Underdetermined Systems

Often the linear equations describing the application are underdetermined because not enough information has been specified to determine unique values of the unknowns. In such cases we might be able to include additional information, objectives, or constraints to find a unique solution. We can use the `rref` command to reduce the number of unknown variables in the problem, as illustrated in the next two examples.

EXAMPLE 6.4–5

Production Planning

The following table shows how many hours reactors A and B need to produce 1 ton each of the chemical products 1, 2, and 3. The two reactors are available for 40 hours and 30 hours per week, respectively. Determine how many tons of each product can be produced each week.

Hours	Product 1	Product 2	Product 3
Reactor A	5	3	3
Reactor B	3	3	4

■ Solution

Let x, y, and z be the number of tons each of products 1, 2, and 3 that can be produced in one week. Using the data for reactor A, the equation for its usage in one week is

$$5x + 3y + 3z = 40$$

The data for reactor B gives

$$3x + 3y + 4z = 30$$

This system is underdetermined. The matrices for the equation $\mathbf{Ax} = \mathbf{b}$ are

$$\mathbf{A} = \begin{bmatrix} 5 & 3 & 3 \\ 3 & 3 & 4 \end{bmatrix} \qquad \mathbf{b} = \begin{bmatrix} 40 \\ 30 \end{bmatrix} \qquad \mathbf{x} = \begin{bmatrix} x \\ y \\ z \end{bmatrix}$$

Here the rank($\mathbf{A}$) = rank([$\mathbf{A}$ $\mathbf{b}$]) = 2, which is less than the number of unknowns. Thus an infinite number of solutions exists, and we can determine two of the variables in terms of the third.

Using the `rref` command `rref([A b])`, where `A = [5,3,3;3,3,4]` and `b = [40;30]`, we obtain the following reduced echelon augmented matrix:

$$\begin{bmatrix} 1 & 0 & -0.5 & 5 \\ 0 & 1 & 1.8333 & 5 \end{bmatrix}$$

This matrix gives the reduced system

$$x - 0.5z = 5$$

$$y + 1.8333z = 5$$

which can be easily solved as follows:

$$x = 5 + 0.5z \tag{6.4–5}$$

$$y = 5 - 1.8333z \tag{6.4–6}$$

where z is arbitrary. However, z cannot be completely arbitrary if the solution is to be meaningful. For example, negative values of the variables have no meaning here; thus we require that $x \geq 0$, $y \geq 0$, and $z \geq 0$. Equation (6.4–5) shows that $x \geq 0$ if $z \geq -10$. From (6.4–6), $y \geq 0$ implies that $z \leq 5/1.8333 = 2.727$. Thus valid solutions are those given by (6.4–5) and (6.4–6), where $0 \leq z \leq 2.737$ tons. The choice of z within this range must be made on some other basis, such as profit.

For example, suppose we make a profit of \$400, \$600, and \$100 per ton for products 1, 2, and 3, respectively. Then our total profit P is

$$\begin{aligned} P &= 400x + 600y + 100z \\ &= 400(5 + 0.5z) + 600(5 - 1.8333z) + 100z \\ &= 5000 - 800z \end{aligned}$$

Thus to maximize profit, we should choose z to be the smallest possible value; namely, $z = 0$. This choice gives $x = y = 5$ tons.

However, if the profits for each product were \$3000, \$600, and \$100, the total profit would be $P = 18{,}000 + 500z$. Thus we should choose z to be its maximum; namely, $z = 2.727$ tons. From (6.4–5) and (6.4–6), we obtain $x = 6.36$ and $y = 0$ tons.

EXAMPLE 6.4–6 Traffic Engineering

A traffic engineer wants to know whether measurements of traffic flow entering and leaving a road network are sufficient to predict the traffic flow on each street in the network. For example, consider the network of one-way streets shown in Figure 6.4–2. The numbers in the figure give the measured traffic flows in vehicles per hour. Assume that no vehicles park anywhere within the network. If possible, calculate the traffic flows f_1, f_2, f_3, and f_4. If this is not possible, suggest how to obtain the necessary information.

■ Solution

The flow *into* intersection 1 must equal the flow *out* of the intersection, which gives us

$$100 + 200 = f_1 + f_4$$

Similarly, for the other three intersections, we have

$$f_1 + f_2 = 300 + 200$$

$$600 + 400 = f_2 + f_3$$

$$f_3 + f_4 = 300 + 500$$

Putting these expressions in the matrix form $\mathbf{Ax} = \mathbf{b}$, we obtain

$$\mathbf{A} = \begin{bmatrix} 1 & 0 & 0 & 1 \\ 1 & 1 & 0 & 0 \\ 0 & 1 & 1 & 0 \\ 0 & 0 & 1 & 1 \end{bmatrix} \qquad \mathbf{b} = \begin{bmatrix} 300 \\ 500 \\ 1000 \\ 800 \end{bmatrix} \qquad \mathbf{x} = \begin{bmatrix} f_1 \\ f_2 \\ f_3 \\ f_4 \end{bmatrix}$$

First check the ranks of $\mathbf{A}$ and $[\mathbf{A}\ \ \mathbf{b}]$ using the MATLAB `rank` command. Both have a rank of three, which is less than the number of unknowns, so we can determine three of the unknowns in terms of the fourth. Thus we cannot determine the traffic flows based on the given measurements. This example shows that it is not always possible to find a unique, exact solution even when the number of equations equals the number of unknowns.

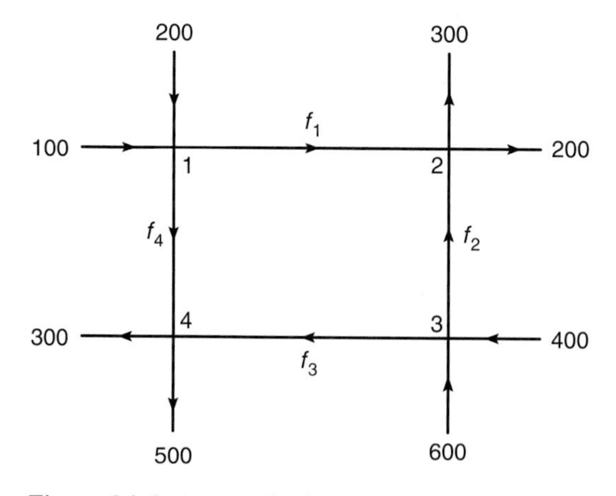

Figure 6.4–2 A network of one-way streets.

Using the `rref([A b])` command produces the reduced augmented matrix

$$\begin{bmatrix} 1 & 0 & 0 & 1 & 300 \\ 0 & 1 & 0 & -1 & 200 \\ 0 & 0 & 1 & 1 & 800 \\ 0 & 0 & 0 & 0 & 0 \end{bmatrix}$$

which corresponds to the following reduced system:

$$f_1 + f_4 = 300$$

$$f_2 - f_4 = 200$$

$$f_3 + f_4 = 800$$

We can easily solve this system as follows: $f_1 = 300 - f_4$, $f_2 = 200 + f_4$, and $f_3 = 800 - f_4$.

If we could measure the flow on one of the internal roads, say, f_4, then we could compute the other flows. So we recommend that the engineer arrange to make this additional measurement.

Test Your Understanding

T6.4–3 Use the `rref` and `pinv` commands and the left-division method to solve the following set:

$$3x + 5y + 6z = 6$$

$$8x - y + 2z = 1$$

$$5x - 6y - 4z = -5$$

(Answer: The set has an infinite number of solutions. The result obtained with the `rref` command is $x = 0.2558 - 0.3721z$, $y = 1.0465 - 0.9767z$, z arbitrary. The `pinv` command gives $x = 0.0571$, $y = 0.5249$, $z = 0.5340$. The left-division method generates an error message.)

T6.4–4 Use the `rref` and `pinv` commands and the left-division method to solve the following set:

$$3x + 5y + 6z = 4$$

$$x - 2y - 3z = 10$$

(Answer: The set has an infinite number of solutions. The result obtained with the `rref` command is $x = 0.2727z + 5.2727$, $y = -1.3636z - 2.3636$, z arbitrary. The solution obtained with left division is $x = 4.8000$, $y = 0$, $z = -1.7333$. The pseudoinverse method gives $x = 4.8394$, $y = -0.1972$, $z = -1.5887$.)

6.5 Overdetermined Systems

An *overdetermined system* is a set of equations that has more independent equations than unknowns. For such a system the matrix inverse method and Cramer's method will not work because the **A** matrix is not square. However, some overdetermined systems have exact solutions, and they can be obtained with the left-division method `x = A\b`. For other overdetermined systems, no exact solution exists. In some of these cases, the left-division method does not yield an answer, while in other cases the left-division method gives an answer that satisfies the equation set only in a "least squares" sense, as explained in Example 6.5–1. When MATLAB gives an answer to an overdetermined set, it does not tell us whether the answer is the exact solution. We must determine this information ourselves, as shown in Example 6.5–2.

EXAMPLE 6.5–1

The Least Squares Method

Suppose we have the following three data points, and we want to find the straight line $y = mx + b$ that best fits the data in some sense.

x	y
0	2
5	6
10	11

(a) Find the coefficients m and b by using the least squares criterion. (b) Find the coefficients by using MATLAB to solve the three equations (one for each data point) for the two unknowns m and b. Compare the answers from (a) and (b).

■ Solution

(a) Because two points define a straight line, unless we are extremely lucky, our data points will not lie on the same straight line. A common criterion for obtaining the straight line that best fits the data is the *least squares* criterion. According to this criterion, the line that minimizes J, the sum of the squares of the vertical differences between the line and the data points, is the "best" fit (see Figure 6.5–1). Here J is

$$J = \sum_{i=1}^{i=3} (mx_i + b - y_i)^2$$

Substituting the data values (x_i, y_i), this expression becomes

$$J = (0m + b - 2)^2 + (5m + b - 6)^2 + (10m + b - 11)^2$$

You can use the `fminsearch` command to find the values of m and b that minimize J. On the other hand, if you are familiar with calculus, you know that the values of m and b that minimize J are found by setting the partial derivatives $\partial J/\partial m$ and $\partial J/\partial b$

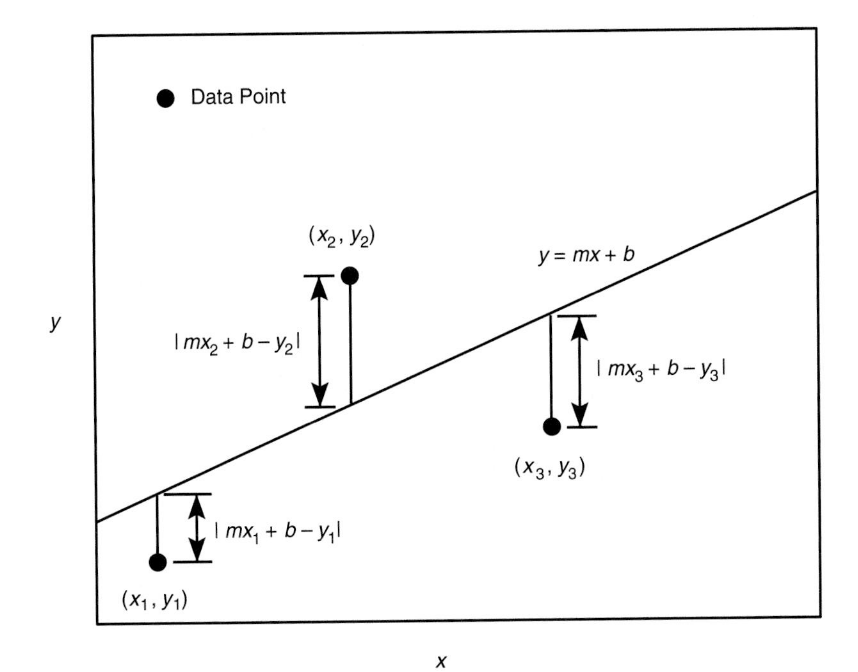

Figure 6.5–1 Illustration of the least squares criterion.

equal to zero:

$$\frac{\partial J}{\partial m} = 2(5m + b - 6)(5) + 2(10m + b - 11)(10)$$

$$= 250m + 30b - 280 = 0$$

$$\frac{\partial J}{\partial b} = 2(b - 2) + 2(5m + b - 6) + 2(10m + b - 11)$$

$$= 30m + 6b - 38 = 0$$

These give the following equations for the two unknowns m and b:

$$250m + 30b = 280$$

$$30m + 6b = 38$$

The solution is $m = 0.9$ and $b = 11/6$. The best straight line in the least squares sense is $y = 0.9x + 11/6 = 0.9x + 1.8333$. It appears in Figure 6.5–2, along with the data points.

(b) Evaluating the equation $y = mx + b$ at each data point gives the following three equations:

$$0m + b = 2 \qquad (6.5\text{–}1)$$

$$5m + b = 6 \qquad (6.5\text{–}2)$$

$$10m + b = 11 \qquad (6.5\text{–}3)$$

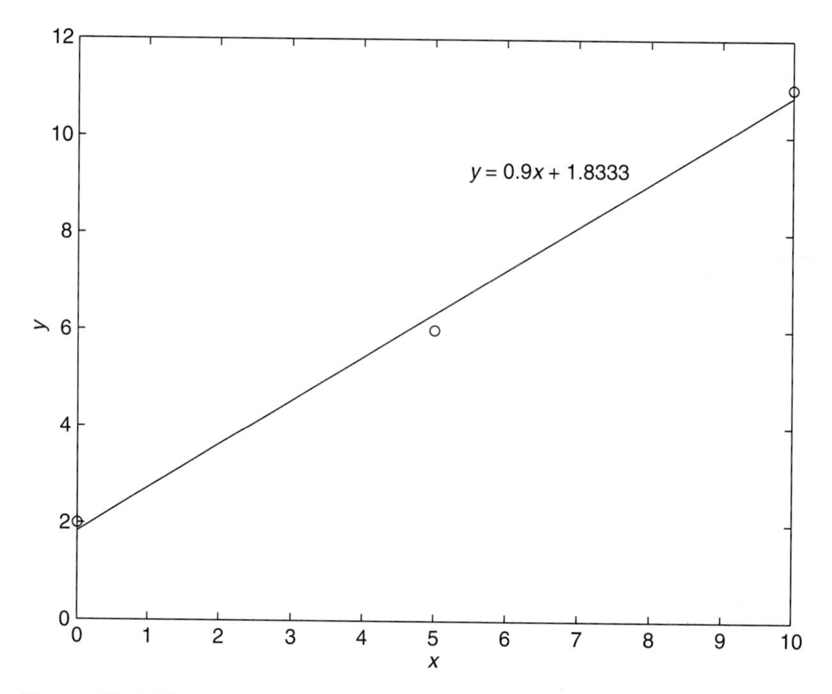

Figure 6.5–2 The least squares fit for the example data.

This set is overdetermined because it has more equations than unknowns. These equations can be written in the matrix form $\mathbf{Ax} = \mathbf{b}$ as follows:

$$\begin{bmatrix} 0 & 1 \\ 5 & 1 \\ 10 & 1 \end{bmatrix} \begin{bmatrix} m \\ b \end{bmatrix} = \begin{bmatrix} 2 \\ 6 \\ 11 \end{bmatrix}$$

where

$$\mathbf{A} = \begin{bmatrix} 0 & 1 \\ 5 & 1 \\ 10 & 1 \end{bmatrix} \qquad \mathbf{x} = \begin{bmatrix} m \\ b \end{bmatrix} \qquad \mathbf{b} = \begin{bmatrix} 2 \\ 6 \\ 11 \end{bmatrix}$$

$$[\mathbf{A} \quad \mathbf{b}] = \begin{bmatrix} 0 & 1 & 2 \\ 5 & 1 & 6 \\ 10 & 1 & 11 \end{bmatrix}$$

Because we can find a nonzero 2×2 determinant in $\mathbf{A}$, its rank is two. However $|\mathbf{A} \quad \mathbf{b}| = -5 \neq 0$, so its rank is three. Thus no exact solution exists for m and b. The following MATLAB session uses left division.

```
>>A = [0,1;5,1;10,1];
>>b = [2;6;11];
>>rank(A)
ans =
     2
```

```
>>rank([A b])
ans =
     3
>>A\b
ans =
      0.9000
      1.8333
```

This result agrees with the least squares solution obtained previously: $m = 0.9$, $b = 11/6 = 1.8333$.

If we now type `A*ans`, MATLAB yields this result:

```
ans =
      1.833
      6.333
      10.8333
```

These values are the y values generated by the line $y = 0.9x + 1.8333$ at the x data values $x = 0, 5, 10$. These values are different from the right sides of the original three equations (6.5–1) through (6.5–3). This result is not unexpected because the least squares solution is not an exact solution of the equations.

Some overdetermined systems have an exact solution. The left-division method sometimes gives an answer for overdetermined systems, but it does not indicate whether the answer is the exact solution. We need to check the ranks of **A** and [**A b**] to know whether the answer is the exact solution. The next example illustrates this situation.

An Overdetermined Set — EXAMPLE 6.5–2

(a) Solve the following equations by hand and (b) solve them using MATLAB. Discuss the solution for two cases: $c = 9$ and $c = 10$.

$$x + y = 1$$

$$x + 2y = 3$$

$$x + 5y = c$$

■ **Solution**

(a) To solve these equations by hand, subtract the first equation from the second to obtain $y = 2$. Substitute this value into the first equation to obtain $x = -1$. Substituting these values into the third equation gives $-1 + 10 = c$, which is satisfied only if $c = 9$. Thus a solution exists if and only if $c = 9$.

(b) The coefficient matrix and the augmented matrix for this problem are

$$\mathbf{A} = \begin{bmatrix} 1 & 1 \\ 1 & 2 \\ 1 & 5 \end{bmatrix} \qquad [\mathbf{A} \quad \mathbf{b}] = \begin{bmatrix} 1 & 1 & 1 \\ 1 & 2 & 3 \\ 1 & 5 & c \end{bmatrix}$$

In MATLAB, enter the array `A = [1,1;1,2;1,5]`. For $c = 9$, type `b = [1;3;9];` the `rank(A)` and `rank([A b])` commands give the result that rank(**A**) = rank([**A** **b**]) = 2. Thus the system has a solution and, because the number of unknowns (two) equals the rank of **A**, the solution is unique. The left-division method `A\b` gives this solution, which is $x = -1$ and $y = 2$.

For $c = 10$, type `b = [1; 3; 10];` the `rank(A)` and `rank ([A\b])` commands give the result that rank(**A**) = 2, but rank([**A** **b**]) = 3. Because rank(**A**) $\neq$ rank([**A** **b**]), no solution exists. However, the left-division method `A\b` gives $x = -1.3846$ and $y = 2.2692$, which is *not* a solution! This conclusion can be verified by substituting these values into the original equation set. This answer is the solution to the equation set in a least squares sense. That is, these values are the values of x and y that minimize J, the sum of the squares of the differences between the equations' left and right sides.

$$J = (x + y - 1)^2 + (x + 2y - 3)^2 + (x + 5y - 10)^2$$

The MATLAB left-division operator sometimes gives the least squares solution when we use the operator to solve problems for which there is no exact solution. A solution exists when $c = 9$, but no solution exists when $c = 10$. The left-division method gives the exact solution when $c = 9$ but gives the least squares solution when $c = 10$.

To interpret MATLAB answers correctly for an overdetermined system, first check the ranks of **A** and [**A** **b**] to see whether an exact solution exists; if one does not exist, then you know that the left-division answer is a least squares solution.

Test Your Understanding

T6.5–1 Use MATLAB to solve the following set:

$$x - 3y = 2$$
$$3x + 5y = 7$$
$$70x - 28y = 153$$

(Answer: The unique solution, $x = 2.2143$, $y = 0.0714$, is given by the left-division method.)

T6.5–2 Use MATLAB to solve the following set:

$$x - 3y = 2$$
$$3x + 5y = 7$$
$$5x - 2y = -4$$

(Answer: No exact solution.)

6.6 Summary

Once you have finished this chapter, you should be able to solve by hand systems of linear algebraic equations that have few variables and use MATLAB to solve systems that have many variables. If the number of equations in the set *equals* the number of unknown variables, the matrix **A** is square and MATLAB provides two ways of solving the equation set **Ax** = **b**:

1. The matrix inverse method; solve for **x** by typing `x = inv(A)*b`.
2. The matrix left-division method; solve for **x** by typing `x = A\b`.

If **A** is square and if MATLAB does not generate an error message when you use one of these methods, then the set has a unique solution, which is given by the left-division method. You can always check the solution for `x` by typing `A*x` to see if the result is the same as `b`. If so, the solution is correct. If you receive an error message, the set is underdetermined, and either it does not have a solution or it has more than one solution. In such a case, if you need more information, you must use the following procedures.

For underdetermined and overdetermined sets, MATLAB provides three ways of dealing with the equation set **Ax** = **b**. (Note that the matrix inverse method will never work with such sets.)

1. The matrix left-division method; solve for **x** by typing `x = A\b`.
2. The pseudoinverse method; solve for **x** by typing `x = pinv(A)*b`.
3. The reduced row echelon form (RREF) method. This method uses the MATLAB command `rref` to obtain a solution.

Table 6.6–1 summarizes the appropriate commands. You should be able to determine whether a unique solution, an infinite number of solutions, or no solution exists. You can get this information by testing for existence and uniqueness of solutions using the following test.

> **Existence and uniqueness of solutions.** The set **Ax** = **b** with m equations and n unknowns has solutions if and only if rank[**A**] = rank[**A b**] (1). Let r = rank[**A**]. If condition (1) is satisfied and if $r = n$, then the solution is unique. If condition (1) is satisfied but $r < n$, an infinite number of solutions exists and r unknown variables can be expressed as linear combinations of the other $n - r$ unknown variables, whose values are arbitrary.
>
> **Homogeneous case.** The homogeneous set **Ax** = **0** is a special case in which **b** = **0**. For this case rank[**A**] = rank[**A b**] always, and thus the set always has the trivial solution **x** = **0**. A nonzero solution, in which at least one unknown is nonzero, exists

Table 6.6–1 Matrix commands for solving linear equations

Command	Description
`det(A)`	Computes the determinant of the array **A**.
`inv(A)`	Computes the inverse of the matrix **A**.
`pinv(A)`	Computes the pseudoinverse of the matrix **A**.
`rank(A)`	Computes the rank of the matrix **A**.
`rref([A b])`	Computes the reduced row echelon form corresponding to the augmented matrix [**A b**].
`x = inv(A)*b`	Solves the matrix equation **Ax** = **b**, using the matrix inverse.
`x = A\b`	Solves the matrix equation **Ax** = **b**, using left division.
`x = d/C`	Solves the matrix equation **xC** = **d**, using right division.

if and only if rank[**A**] $< n$. If $m < n$, the homogeneous set always has a nonzero solution.

Underdetermined Systems

In an *underdetermined* system not enough information is given to determine the values of all the unknown variables.

- An infinite number of solutions might exist in which one or more of the unknowns are dependent on the remaining unknowns.
- For such systems Cramer's method and the matrix inverse method will not work because either **A** is not square or because $|\mathbf{A}| = 0$.
- The left-division method will give a solution with some of the unknowns arbitrarily set equal to zero, but this solution is not the general solution.
- An infinite number of solutions might exist even when the number of equations equals the number of unknowns. The left-division method fails to give a solution in such cases.
- In cases that have an infinite number of solutions, some of the unknowns can be expressed in terms of the remaining unknowns, whose values are arbitrary. The `rref` command can be used to find these relations.

Overdetermined Systems

An *overdetermined* system is a set of equations that has more independent equations than unknowns.

- For such a system Cramer's method and the matrix inverse method will not work because the **A** matrix is not square.
- Some overdetermined systems have exact solutions, which can be obtained with the left-division method `A\b`.
- For overdetermined systems that have no exact solution, the answer given by the left-division method satisfies the equation set only in a least squares sense.
- When we use MATLAB to solve an overdetermined set, the program does not tell us whether the solution is exact. We must determine this information ourselves. The first step is to check the ranks of **A** and [**A b**] to see whether a solution exists; if no solution exists, then we know that the left-division solution is a least squares answer.

Programming Application

In this chapter you saw that the set of linear algebraic equations $\mathbf{Ax} = \mathbf{b}$ with m equations and n unknowns has solutions if and only if (1) rank[**A**] = rank[**A b**]. Let r = rank[**A**]. If condition (1) is satisfied and if $r = n$, then the solution is unique. If condition (1) is satisfied but $r < n$, an infinite number of solutions exists; in addition, r unknown variables can be expressed as linear combinations of the other $n - r$ unknown variables, whose values are arbitrary. In this case we can use the `rref` command to find the relations between the variables. The pseudocode in Table 6.6–2 can be used to outline an equation solver program before writing it.

A condensed flowchart appears in Figure 6.6–1. From this chart or the pseudocode, we can develop the script file shown in Table 6.6–3. The program uses

Table 6.6–2 Pseudocode for the linear equation solver

If the rank of **A** equals the rank of [**A b**], then
  determine whether the rank of **A** equals the number of unknowns. If so, there is a unique solution, which can be computed using left division. Display the results and stop.
  Otherwise, there is an infinite number of solutions, which can be found from the augmented matrix. Display the results and stop.
Otherwise (if the rank of **A** does not equal the rank of [**A b**]), then there are no solutions. Display this message and stop.

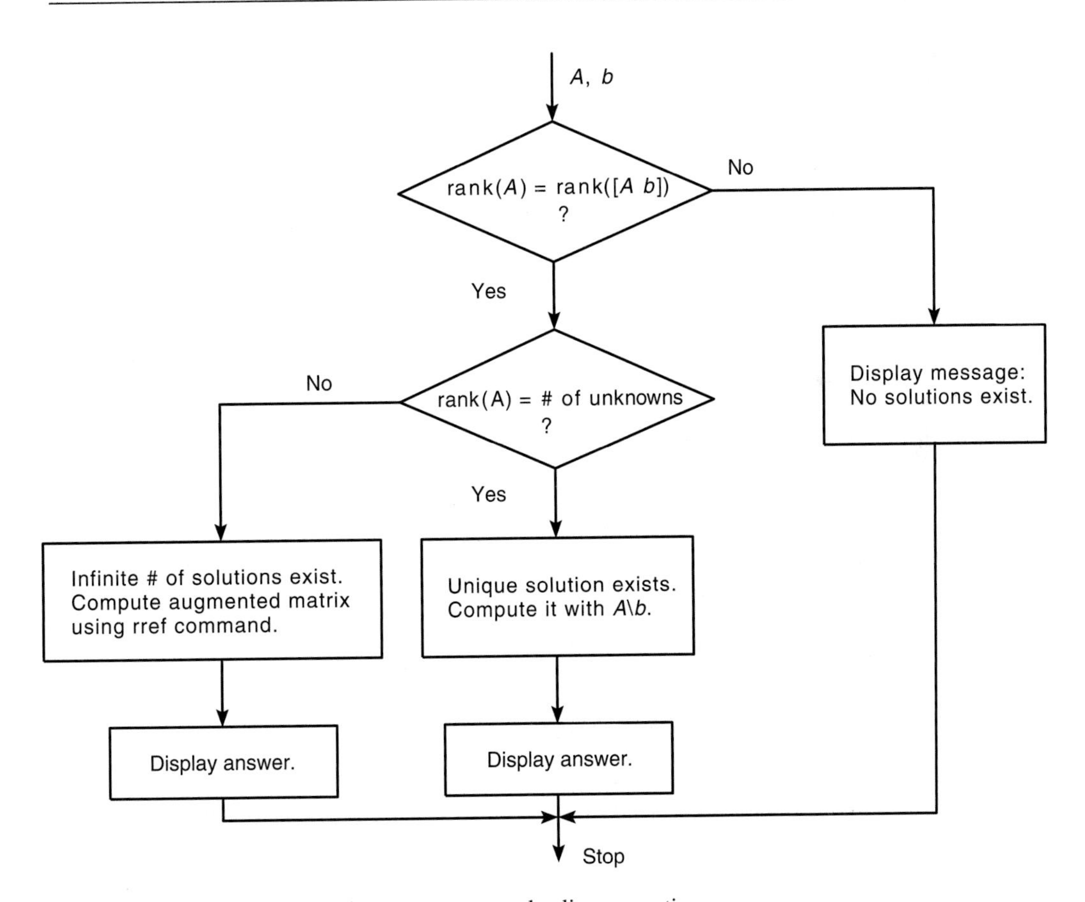

Figure 6.6–1 Flowchart illustrating a program to solve linear equations.

Table 6.6–3 MATLAB program to solve linear equations

```
% Script file lineq.m
% Solves the set Ax = b, given A and b.
% Check the ranks of A and [A  b].
if rank(A) == rank([A b])
   % The ranks are equal.
   size_A = size(A);
   % Does the rank of A equal the number of unknowns?
   if rank(A) == size_A(2)
      % Yes. Rank of A equals the number of unknowns.
      disp('There is a unique solution, which is:')
      x = A\b % Solve using left division.
   else
      % Rank of A does not equal the number of unknowns.
      disp('There is an infinite number of solutions.')
      disp('The augmented matrix of the reduced system is:')
      rref([A b]) % Compute the augmented matrix.
   end
else
   % The ranks of A and [A b] are not equal.
   disp('There are no solutions.')
end
```

the given arrays `A` and `b` to check the rank conditions, the left-division method to obtain the solution, if one exists, and the `rref` method if there is an infinite number of solutions. Note that the number of unknowns equals the number of columns in `A`, which is given by `size_A(2)`, the second element in `size_A`. Note also that the rank of **A** cannot exceed the number of columns in **A**.

Test Your Understanding

T6.6–1 Type in the script file `lineq.m` given in Table 6.6–3 and run it for the following cases. Hand check the answers.

a. `A = [1,-1;1,1],b = [3;5]`
b. `A = [1,-1;2,-2],b = [3;6]`
c. `A = [1,-1;2,-2],b = [3;5]`

Key Terms with Page References

Problems

You can find the answers to problems marked with an asterisk at the end of the text.

Section 6.1

1. Solve the following problems by hand. For each of the following problems, find the unique solution if one exists. If a unique solution does not exist, determine whether no solution exists or a nonunique solution exists.

a. $-5x + y = -6$

$x + y = 6$

b. $-2x + y = -5$

$-2x + y = 3$

c. $-2x + y = 3$

$-8x + 4y = 12$

d. $-2x + y = -5$

$-2x + y = -5.00001$

2. Use elimination of variables to solve the following problem by hand:

$$12x - 5y = 11$$

$$-3x + 4y + 7z = -3$$

$$6x + 2y + 3z = 22$$

3.* Use elimination of variables to solve the following problem by hand:

$$6x - 3y + 4z = 41$$

$$12x + 5y - 7z = -26$$

$$-5x + 2y + 6z = 14$$

4. *a.* Solve the following problem by hand for x, y, and z in terms of the parameter r.

b. For what value of r will a solution *not* exist?

$$3x + 2y - rz = 1$$

$$-x + 3y + 2z = 1$$

$$x - y - z = 1$$

Section 6.2

5. Use the left-division method to solve the following problems. Check your solutions by computing **Ax**.

   a. $$2x + y = 5$$
   $$3x - 9y = 2$$

   b. $$-8x - 5y = 4$$
   $$-2x + 7y = 10$$

6. Use the left-division method to solve the following problems. Check your solutions by computing **Ax**.

   a. $$12x - 5y = 11$$
   $$-3x + 4y + 7z = -3$$
   $$6x + 2y + 3z = 22$$

   b. $$6x - 3y + 4z = 41$$
   $$12x + 5y - 7z = -26$$
   $$-5x + 2y + 6z = 14$$

7. Use MATLAB to solve the following problems:

   a. $$-2x + y = -5$$
   $$-2x + y = 3$$

   b. $$-2x + y = 3$$
   $$-8x + 4y = 12$$

8. Use MATLAB to solve the following problems:

   a. $$-2x + y = -5$$
   $$-2x + y = -5.00001$$

   b. $$x_1 + 5x_2 - x_3 + 6x_4 = 19$$
   $$2x_1 - x_2 + x_3 - 2x_4 = 7$$
   $$-x_1 + 4x_2 - x_3 + 3x_4 = 20$$
   $$3x_1 - 7x_2 - 2x_3 + x_4 = -75$$

9. The circuit shown in Figure P9 has five resistances and one applied voltage. Kirchhoff's voltage law applied to each loop in the circuit shown gives

$$v - R_2 i_2 - R_4 i_4 = 0$$
$$-R_2 i_2 + R_1 i_1 + R_3 i_3 = 0$$
$$-R_4 i_4 - R_3 i_3 + R_5 i_5 = 0$$

Conservation of charge applied at each node in the circuit gives

$$i_6 = i_1 + i_2$$
$$i_2 + i_3 = i_4$$
$$i_1 = i_3 + i_5$$
$$i_4 + i_5 = i_6$$

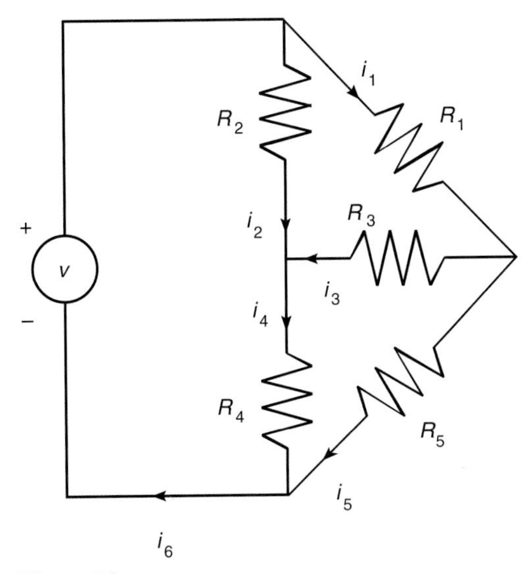

Figure P9

a. Write a MATLAB script file that uses given values of the applied voltage v and the values of the five resistances to solve for the six currents.

b. Use the program developed in part *a* to find the currents for the case: $R_1 = 1$, $R_2 = 5$, $R_3 = 2$, $R_4 = 10$, $R_5 = 5$ kΩ, and $v = 100$ V (1 kΩ = 1000 Ω).

10. Fluid flows in pipe networks can be analyzed in a manner similar to that used for electric-resistance networks. Figure P10 shows a network with three pipes. The volume flow rates in the pipes are q_1, q_2, and q_3. The pressures at the pipe ends are p_a, p_b, and p_c. The pressure at the junction is p_1. Under certain conditions, the pressure-flow rate relation in a pipe has the same form as the voltage-current relation in a resistor. Thus for the three pipes, we have

$$q_1 = \frac{1}{R_1}(p_a - p_1)$$

$$q_2 = \frac{1}{R_2}(p_1 - p_b)$$

$$q_3 = \frac{1}{R_3}(p_1 - p_c)$$

where the R_i are the pipe resistances. From conservation of mass, $q_1 = q_2 + q_3$.

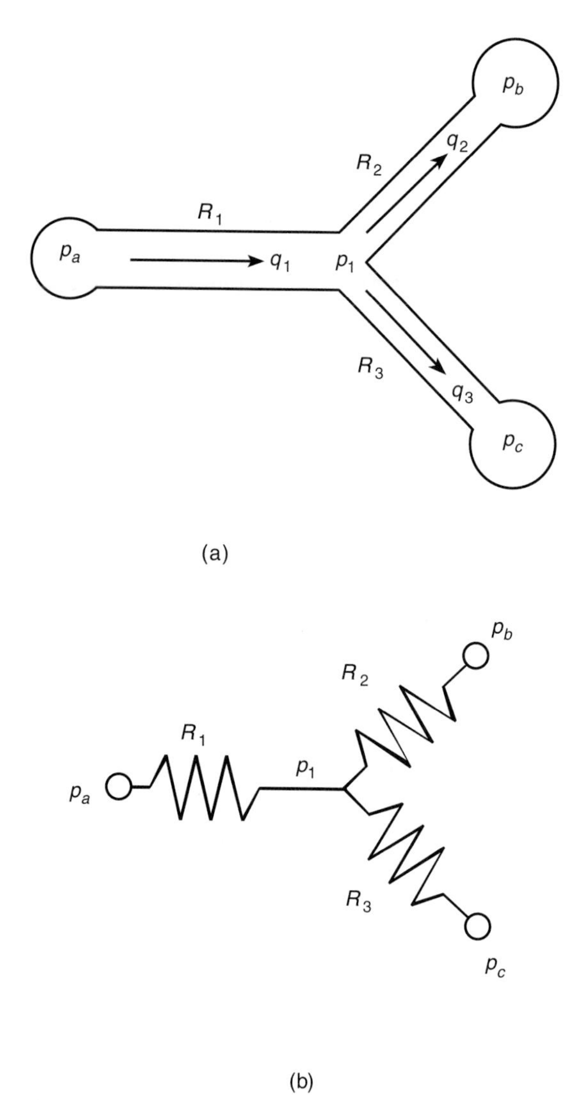

Figure P10

a. Set up these equations in a matrix form $\mathbf{Ax} = \mathbf{b}$ suitable for solving for the three flow rates q_1, q_2, and q_3, and the pressure p_1, given the values of the pressures p_a, p_b, and p_c, and the values of the resistances R_1, R_2, and R_3. Find the expressions for $\mathbf{A}$ and $\mathbf{b}$.

b. Use MATLAB to solve the matrix equations obtained in part *a* for the case: $p_a = 4320$ lb/ft^2, $p_b = 3600$ lb/ft^2, and $p_c = 2880$ lb/ft^2. These correspond to 30, 25, and 20 psi, respectively (1 psi $= 1$ lb/in.^{2}; atmospheric pressure is 14.7 psi). Use the resistance values $R_1 = 10{,}000$; $R_2 = R_3 = 14{,}000$ lb-sec/ft^5. These values correspond to fuel oil flowing through pipes 2 ft long, with 2-in. and 1.4-in. diameters, respectively. The units of the answers will be ft^3/sec for the flow rates and lb/ft^2 for pressure.

11. Figure P11 illustrates a robot arm that has two "links" connected by two "joints": a shoulder, or base, joint and an elbow joint. There is a motor at each joint. The joint angles are θ_1 and θ_2. The (x, y) coordinates of the hand at the end of the arm are given by

$$x = L_1 \cos\theta_1 + L_2 \cos(\theta_1 + \theta_2)$$

$$y = L_1 \sin\theta_1 + L_2 \sin(\theta_1 + \theta_2)$$

where L_1 and L_2 are the lengths of the links.

Polynomials are used for controlling the motion of robots. If we start the arm from rest with zero velocity and acceleration, the following polynomials are used to generate commands to be sent to the joint motor controllers.

$$\theta_1(t) = \theta_1(0) + a_1 t^3 + a_2 t^4 + a_3 t^5$$

$$\theta_2(t) = \theta_2(0) + b_1 t^3 + b_2 t^4 + b_3 t^5$$

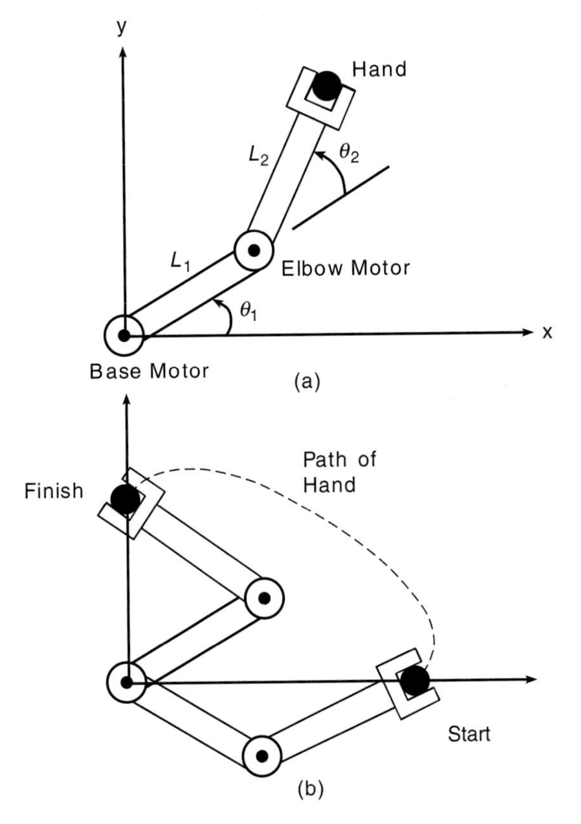

Figure P11

where $\theta_1(0)$ and $\theta_2(0)$ are the starting values at time $t = 0$. The angles $\theta_1(t_f)$ and $\theta_2(t_f)$ are the joint angles corresponding to the desired destination of the arm at time t_f. The values of $\theta_1(0)$, $\theta_2(0)$, $\theta_1(t_f)$, and $\theta_2(t_f)$ can be found using trigonometry if the starting and ending (x, y) coordinates of the hand are specified.

a. Set up a matrix equation to be solved for the coefficients a_1, a_2, and a_3, given values for $\theta_1(0)$, $\theta_1(t_f)$, and t_f. Obtain a similar equation to be solved for the coefficients b_1, b_2, and b_3, given values for $\theta_2(0)$, $\theta_2(t_f)$, and t_f.

b. Use MATLAB to solve for the polynomial coefficients given the values $t_f = 2$ sec, $\theta_1(0) = -19°$, $\theta_2(0) = 44°$, $\theta_1(t_f) = 43°$, and $\theta_2(t_f) = 151°$. (These values correspond to a starting hand location of $x = 6.5$, $y = 0$ ft and a destination location of $x = 0$, $y = 2$ ft for $L_1 = 4$ and $L_2 = 3$ ft.)

c. Use the results of part *b* to plot the path of the hand.

12.* Engineers use the concept of *thermal resistance* R to predict the rate of heat loss through a building wall in order to determine the heating system's requirements. This concept relates the heat flow rate q through a material to the temperature difference ΔT across the material: $q = \Delta T/R$. This relation is like the voltage-current relation for an electrical resistor: $i = v/R$. So the heat flow rate plays the role of electrical current, and the temperature difference plays the role of the voltage difference v. The SI unit for q is W/m^2. A watt (W) is 1 joule/second.

The wall shown in Figure P12 consists of four layers: an inner layer of plaster/lathe 10 mm thick, a layer of fiberglass insulation 125 mm thick, a layer of wood 60 mm thick, and an outer layer of brick 50 mm thick. If we assume that the inner and outer temperatures T_i and T_o have remained constant for some time, then the heat energy stored in the layers is constant; thus the heat flow rate through each layer is the same. Applying conservation of energy gives the following equations:

$$q = \frac{1}{R_1}(T_i - T_1) = \frac{1}{R_2}(T_1 - T_2) = \frac{1}{R_3}(T_2 - T_3) = \frac{1}{R_4}(T_3 - T_o)$$

The thermal resistance of a solid material is given by $R = D/k$ where D is the material's thickness and k is the material's *thermal conductivity.* For the given materials, the resistances for a wall area of 1 m^2 are $R_1 = 0.036$, $R_2 = 4.01$, $R_3 = 0.408$, and $R_4 = 0.038$ °K/W.

Suppose that $T_i = 20$ and $T_o = -10$°C. Find the other three temperatures and the heat loss rate q in W/m^2. Compute the total heat loss rate in watts if the wall's area is 10 m^2.

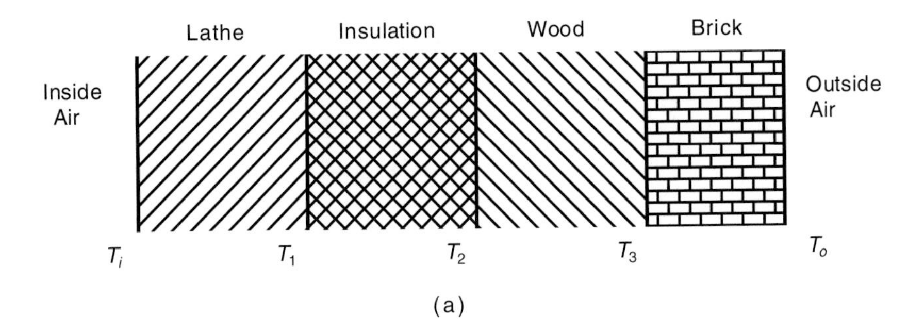

(a)

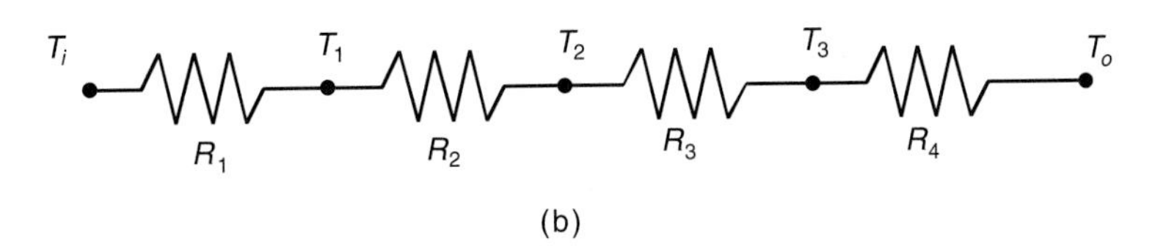

(b)

Figure P12

13. The concept of thermal resistance described in Problem 12 can be used to find the temperature distribution in the flat square plate shown in Figure P13a. The plate's edges are insulated so that no heat can escape, except at two points where the edge temperature is heated to T_a and T_b, respectively. The temperature varies through the plate, so no single point can describe the plate's temperature. One way to estimate the temperature distribution is to imagine that the plate consists of four subsquares and to compute the temperature in each subsquare. Let R be the thermal resistance of the material between the centers of adjacent subsquares. Then we can think of the problem as a network of electrical resistors, as shown in Figure P13b. Let q_{ij} be the heat flow rate between the points whose temperatures are T_i and T_j. If T_a and T_b remain constant for some time, then both the heat energy stored in each subsquare and the heat flow rate between each subsquare are constant. Under these conditions conservation of energy says that the heat flow into a subsquare equals the heat flow out. Applying this principle to each subsquare gives the following equations:

$$q_{a1} = q_{12} + q_{13}$$

$$q_{12} = q_{24}$$

$$q_{13} = q_{34}$$

$$q_{34} + q_{24} = q_{4b}$$

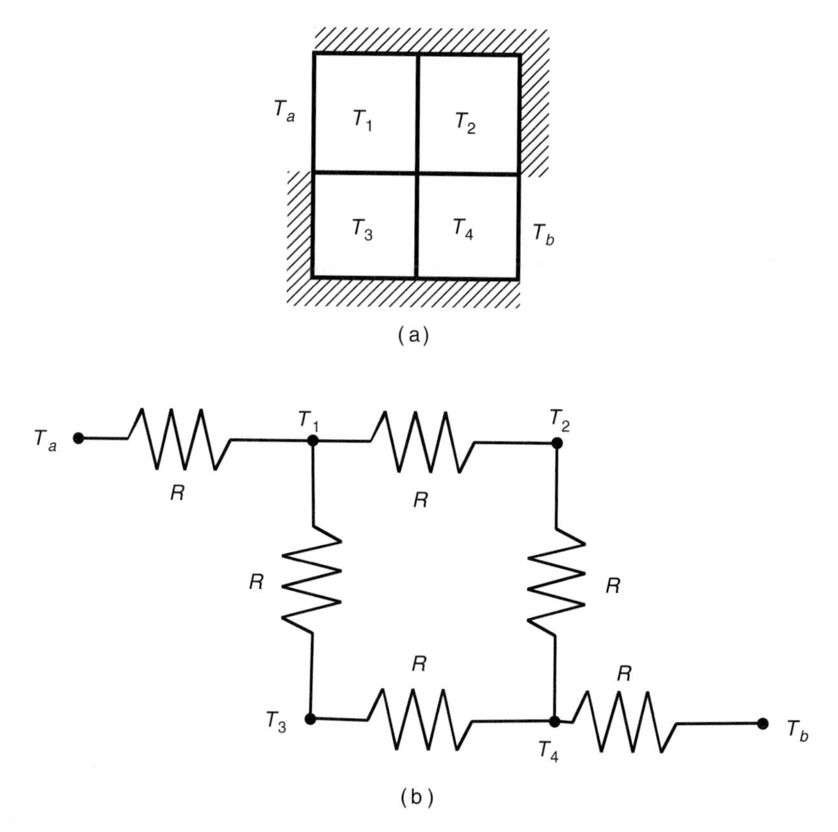

Figure P13

Substituting $q_{ij} = (T_i - T_j)/R$, we find that R can be canceled out of every equation, and the equations can be rearranged as follows:

$$T_1 = \frac{1}{3}(T_a + T_2 + T_3)$$

$$T_2 = \frac{1}{2}(T_1 + T_4)$$

$$T_3 = \frac{1}{2}(T_1 + T_4)$$

$$T_4 = \frac{1}{3}(T_2 + T_3 + T_b)$$

These equations tell us that the temperature of each subsquare is the average of the temperatures in the adjacent subsquares!

Solve these equations for the case where $T_a = 150°\text{C}$ and $T_b = 20°\text{C}$.

14. Use the averaging principle developed in Problem 13 to find the temperature distribution of the plate shown in Figure P14, using the 3×3 grid and the given values $T_a = 150°\text{C}$ and $T_b = 20°\text{C}$.

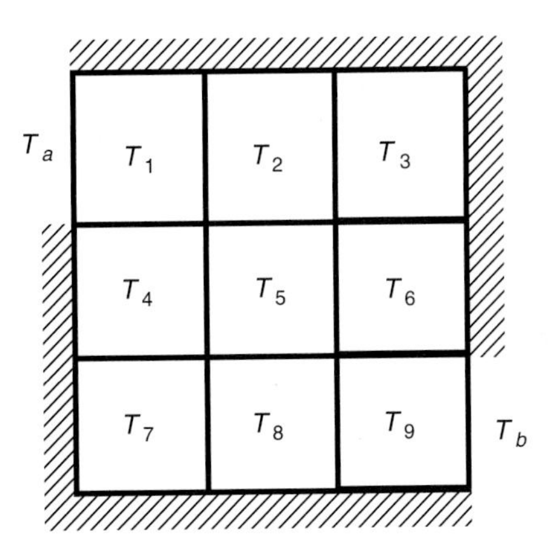

Figure P14

15. A convenient way to draw curves is to select a set of points and slopes on the curve, and find a polynomial whose graph satisfies these conditions. Design the shape of a ski jump with the following specifications. The jump starts at a height of $y = 100$ ft and finishes at a height of 10 ft. From the start at $x = 0$ to the launch point, the jump extends a horizontal distance of 120 ft. A skier using the jump will begin horizontally and will fly off the end at a 30° angle from the horizontal. Develop and solve a set of linear equations for the four coefficients of a cubic polynomial $y = a_3x^3 + a_2x^2 + a_1x + a_0$ for the side view of the jump. Plot the ski jump profile and examine its slope for any undesired behavior. Its slope is given by $s = 3a_3x^2 + 2a_2x + a_1$.

16. A mixture of benzene and toluene with p percent benzene and $(1 - p)$ percent toluene at 10°C is fed continuously to a vessel in which the mixture is heated to 50°C. The liquid product is 40 mole percent benzene and the vapor product is 68.4 mole percent benzene. Using conservation of the overall mass, we obtain the equation

$$V + L = 1$$

where V is the vapor mass and L is the liquid mass. Using conservation of benzene mass, we obtain the equation

$$0.684V + 0.4L = 0.01p$$

Write a user-defined function that uses the left-division method to solve for V and L with p as the function's input argument. Test your function for the case where $p = 50$ percent.

17. Use the matrix inverse method to solve Problems 5*a* and 5*b* by hand.

18. Use the matrix inverse method with MATLAB to solve Problems 6*a* and 6*b*.

19.* *a.* Solve the following matrix equation for the matrix **C**:

$$\mathbf{A}(\mathbf{BC} + \mathbf{A}) = \mathbf{B}$$

b. Evaluate the solution obtained in part *a* for the case:

$$\mathbf{A} = \begin{bmatrix} 3 & 9 \\ -2 & 4 \end{bmatrix} \qquad \mathbf{B} = \begin{bmatrix} 2 & -3 \\ 7 & 6 \end{bmatrix}$$

20.* *a.* Use MATLAB to solve the following equations for x, y, and z as functions of the parameter c:

$$x - 5y - 2z = 11c$$

$$6x + 3y + z = 13c$$

$$7x + 3y - 5z = 10c$$

b. Plot the solutions for x, y, and z versus c on the same plot for $-10 \le c \le 10$.

Section 6.3

21. Use Cramer's method to solve Problems 1*a* and 1*b*.

22. Use Cramer's method to solve Problems 5*a* and 5*b*.

23. For the following equations, what values of the parameter b allow a nonzero solution to exist for x and y? Find that solution.

$$bx - y = 0$$

$$10x + (b + 7)y = 0$$

24. The lightweight rigid rod shown in Figure P24a has a weight W at its end, and it is supported by the two cables and by the pin joint at point O. Denote the tension forces in the cables by T_1 and T_2, and let R_x and R_y be the reaction forces of the pin joint on the beam. The free-body diagram is shown in part *b* of the figure. The values of H, L_1, L_2, and W are given.

Summing moments about point A gives $HR_x - WL_2 = 0$, or $R_x = WL_2/H$. Summing moments about point O gives

$$(L_1 \sin\phi)T_1 + (L_2 \sin\theta)T_2 = WL_2 \tag{1}$$

Summing forces in the x and y directions gives

$$(\cos\phi)T_1 + (\cos\theta)T_2 = \frac{WL_2}{H} \tag{2}$$

$$(\sin\phi)T_1 + (\sin\theta)T_2 + R_y = W \tag{3}$$

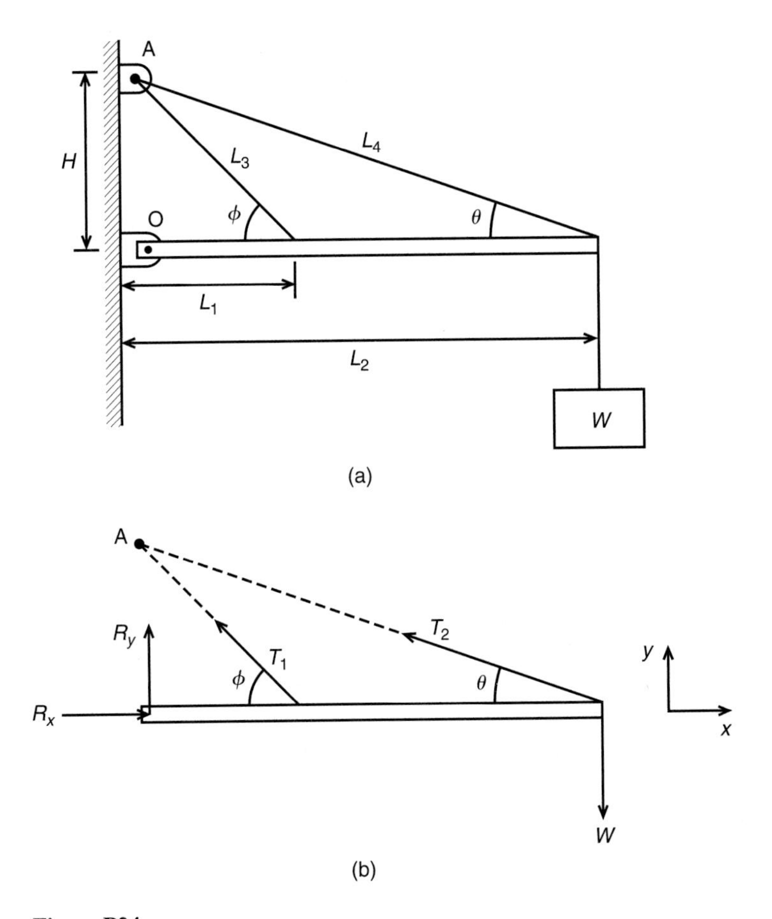

Figure P24

a. Use Cramer's method to solve equations (1), (2), and (3) for T_1, T_2, and R_y in terms of Cramer's determinant D.

b. Use the fact that $\sin\phi = H/L_3$, $\sin\theta = H/L_4$, $\cos\phi = L_1/L_3$, and $\cos\theta = L_2/L_3$ to evaluate D. What does this result say about the solutions for T_1, T_2, and R_y?

Section 6.4

25.* Use MATLAB to solve the following problem:

$$7x + 9y - 9z = 22$$
$$3x + 2y - 4z = 12$$
$$x + 5y - z = -2$$

26. The following table shows how many hours reactors A and B need to produce 1 ton each of the chemical products 1, 2, and 3. The two reactors are available for 35 hours and 40 hours per week, respectively.

Hours	Product 1	Product 2	Product 3
Reactor A	6	2	10
Reactor B	3	5	2

Let x, y, and z be the number of tons each of products 1, 2, and 3 that can be produced in one week.

a. Use the data in the table to write two equations in terms of x, y, and z. Determine whether or not a unique solution exists. If not, use MATLAB to find the relations between x, y, and z.

b. Note that negative values of x, y, and z have no meaning here. Find the allowable ranges for x, y, and z.

c. Suppose the profits for each product are \$200, \$300, and \$100 for products 1, 2, and 3, respectively. Find the values of x, y, and z to maximize the profit.

d. Suppose the profits for each product are \$200, \$500, and \$100 for products 1, 2, and 3, respectively. Find the values of x, y, and z to maximize the profit.

27. See Figure P27. Assume that no vehicles stop within the network. A traffic engineer wants to know whether the traffic flows $f_1, f_2, \ldots, f_7$ (in vehicles per hour) can be computed given the measured flows shown in the figure. If not, then determine how many more traffic sensors need

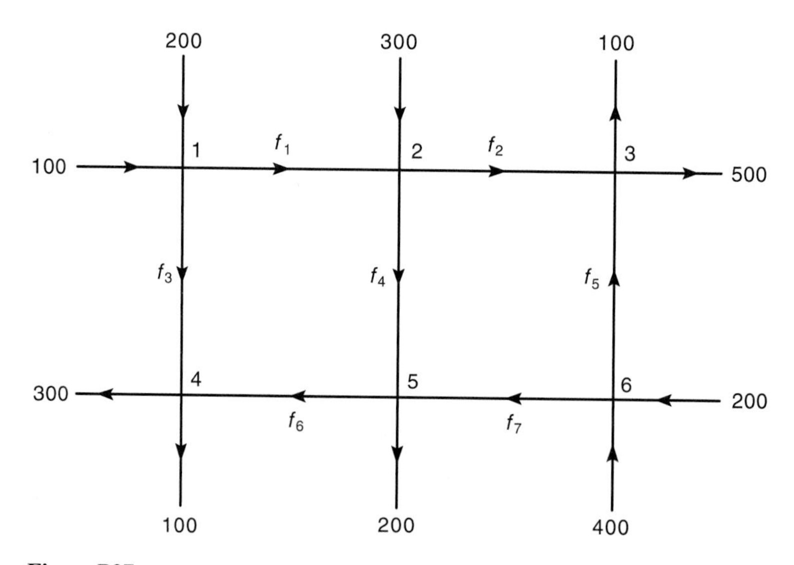

Figure P27

to be installed and obtain the expressions for the other traffic flows in terms of the measured quantities.

Section 6.5

28.* Use MATLAB to solve the following problem:

$$x - 3y = 2$$
$$x + 5y = 18$$
$$4x - 6y = 20$$

29.* Use MATLAB to solve the following problem:

$$x - 3y = 2$$
$$x + 5y = 18$$
$$4x - 6y = 10$$

30. Use MATLAB to find the coefficients of the quadratic polynomial $y = ax^2 + bx + c$ that passes through these three points: $(x, y) = (1, 4)$, $(4, 73)$, $(5, 120)$.

31. Use MATLAB to find the coefficients of the cubic polynomial $y = ax^3 + bx^2 + cx + d$ that passes through the three points given in Problem 30.

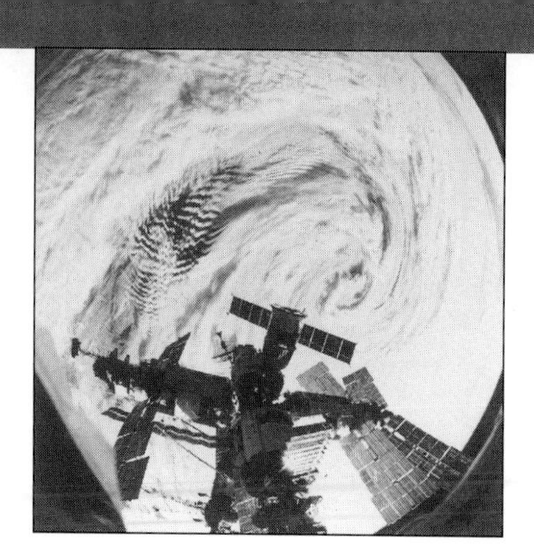

Courtesy of NASA.

Engineering in the 21st Century...

International Engineering

The end of the Cold War has resulted in increased international cooperation. A vivid example of this is the cooperation between the Russian and American space agencies. The photo shows the American space shuttle Atlantis docked with the Russian space station Mir while orbiting over a typhoon. Now Russia, the United States, and several other countries are collaborating on a more ambitious space project: the International Space Station (ISS). The launch of the first element of the station, a Russian-built cargo block, occurred in 1998. The first permanent crew arrived in 2000. The station will have a crew of seven, will weigh over 1,000,000 pounds, and will be 290 feet long and 356 feet wide. To assemble it will require 28 U.S. shuttle flights and 41 Russian flights over a period of 5 years.

The new era of international cooperation has resulted in increased integration of national economies into one global economy. For example, many components for cars assembled in the United States are supplied by companies in other countries. In the 21st century improved communications and computer networks will enable engineering teams in different countries to work simultaneously on different aspects of the product's design, employing a common design database. To prepare for such a future in international engineering, students need to become familiar with other cultures and languages, as well as proficient in computer-aided design.

MATLAB is available in international editions and has been widely used in Europe for many years. Thus engineers familiar with MATLAB will already have a head start in preparing for a career in international engineering. ■

CHAPTER 7

Probability, Statistics, and Interpolation

OUTLINE

This chapter is unlike the first six chapters in that it covers relatively few MATLAB functions. These functions are easy to use and have widespread and important uses in statistics and data analysis; however, their proper application requires informed judgment and experience. Here we present enough of the mathematical foundations of these methods to allow you to apply them properly.

We begin with an introduction to basic statistics and probability in Section 7.1. You will see how to obtain and interpret *histograms,* which are specialized plots for displaying statistical results. The *normal distribution,* commonly called the *bell-shaped curve,* forms the basis of many statistical methods and is covered in Section 7.2. As you saw in Chapter 4, MATLAB is useful for programming simulations. In Section 7.3 you will see how to include random processes in your simulation programs. In Section 7.4 you will see how to use interpolation with data tables to estimate values that are not in the table. An interesting application of this method is in machine control, and we give an example of controlling a robot arm.

When you have finished this chapter, you should be able to use MATLAB to do the following:

- Solve basic problems in statistics and probability.
- Create simulations incorporating random processes.
- Apply interpolation techniques.

7.1 Statistics, Histograms, and Probability

MEAN

MEDIAN

In all likelihood you have computed an *average,* for example, the average of all your test scores in a course. To find your average, you add your scores and divide by the number of tests. The mathematical term for this average is the *mean.* On the other hand, the *median* is the value in the middle of the data if the number of data points is odd. For example, if the test scores on a particular test in a class of 27 students have a median of 74, then 13 students scored below 74, 13 scored above 74, and one student obtained a grade of 74. If the number of data points is even, the median is the mean of the two values closest to the middle. The mean need not be the same as the median. For example, for the data 60, 65, 68, 74, 88, 95, the mean is 75, whereas the median is the mean of 68 and 74, or 71.

MATLAB provides the `mean(x)` and `median(x)` functions to perform these computations. If `x` is a vector, the mean (or median) value of the vector's values is returned. However, if `x` is a matrix, a row vector is returned containing the mean (or median) value of each column of `x`. These functions do not require the elements in `x` to be sorted in ascending or descending order.

In many applications, the mean and the median do not adequately describe a data set. Two data sets can have the same mean (or the same median) yet be very different. For example, the test scores 60, 65, 68, 74, 88, 95 have the same mean as the scores 71, 72, 73, 77, 78, 79, but the two sets describe very different test outcomes. The first set of scores vary over a large range, whereas in the second set the scores are tightly grouped about the mean.

The way the data are spread around the mean can be described by a *histogram* plot. A *histogram* is a plot of the frequency of occurrence of data values versus the values themselves. For example, suppose that in a class of 20 students the 20 scores on the first test were

```
61 61 65 67 69 72 74 74 76 77
83 83 85 88 89 92 93 93 95 98
```

On this test there are five scores in the 60–69 range, five in the 70–79 range, five in the 80–89 range, and five in the 90–100 range. The histogram for these scores is shown in the top graph in Figure 7.1–1. It is a bar plot of the number of scores that occur within each range, with the bar centered in the middle of the range (for example, the bar for the range 60–69 is centered at 64.5, and the asterisk on the plot's abscissa shows the bar's center).

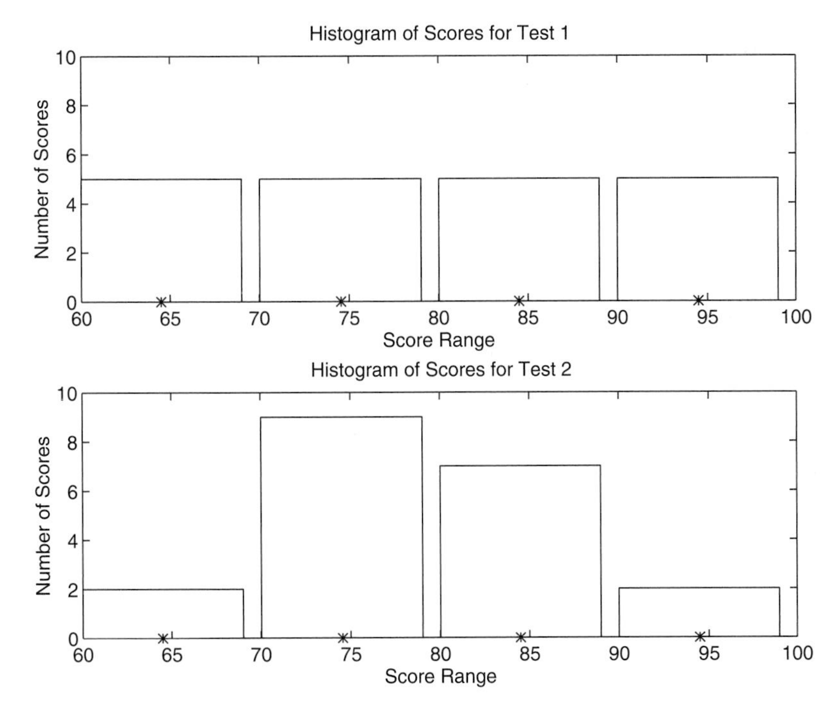

Figure 7.1–1 Histograms of test scores for 20 students.

Suppose that on the second test the following 20 scores were achieved:

66	69	72	74	75	76	77	78	78	79
79	80	81	83	84	85	87	88	90	94

On this test there are two scores in the 60–69 range, nine in the 70–79 range, seven in the 80–89 range, and two in the 90–100 range. The histogram for these scores is shown in the bottom graph in Figure 7.1–1. The mean on both tests is identical and is 79.75. However, the distribution of the scores is very different. On the first test we say that the scores are evenly, or "uniformly," distributed between 60 and 100, whereas on the second test the scores are more clustered around the mean.

BINS

To plot a histogram, you must group the data into subranges, called *bins.* In this example the four bins are the ranges 60–69, 70–79, 80–89, and 90–100. The choice of the bin width and bin center can drastically change the shape of the histogram. If the number of data values is relatively small, the bin width cannot be small because some of the bins will contain no data and the resulting histogram might not usefully illustrate the distribution of the data.

To obtain a histogram, first sort the data if it has not yet been sorted (you can use the `sort` function here). Then choose the bin ranges and bin centers and count the number of values in each bin. Use the `bar` function to plot the number of values in each bin versus the bin centers as a bar chart. The function `bar(x,y)` creates a bar chart of `y` versus `x`. The MATLAB script file that generates Figure 7.1–1 follows. We have selected the bin centers to be in the middle of the ranges 60–69, 70–79, 80–89, 90–99.

```
% Collect the scores in each bin.
test1 = [5,5,5,5];
test2 = [2,9,7,2];
% Specify the bin centers.
x = [64.5,74.5,84.5,94.5];
subplot(2,1,1)
bar(x,test1),axis([60 100 0 10]),...
title('Histogram of Scores for Test 1'),...
xlabel('Score Range'),ylabel('Number of Scores')
subplot(2,1,2)
bar(x,test2),axis([60 100 0 10]),...
title('Histogram of Scores for Test 2'),...
xlabel('Score Range'),ylabel('Number of Scores')
```

MATLAB provides the `hist` command to generate a histogram. This command has several forms. Its basic form is `hist(y)`, where `y` is a vector containing the data. This form aggregates the data into 10 bins evenly spaced between the minimum and maximum values in `y`. The second form is `hist(y,n)`, where `n` is a user-specified scalar indicating the number of bins. The third form is `hist(y,x)`, where `x` is a user-specified vector that determines the location of the bin centers; the bin widths are the distances between the centers.

EXAMPLE 7.1–1

Breaking Strength of Thread

To ensure proper quality control, a thread manufacturer selects samples and tests them for breaking strength. Suppose that 20 thread samples are pulled until they break, and the breaking force is measured in newtons rounded off to integer values. The breaking force values recorded were 92, 94, 93, 96, 93, 94, 95, 96, 91, 93, 95, 95, 95, 92, 93, 94, 91, 94, 92, and 93. Plot the histogram of the data.

■ Solution

Store the data in the vector `y`, which is shown in the following script file. Because there are six outcomes (91, 92, 93, 94, 95, 96 N), we choose six bins. However, if you use `hist(y,6)`, the bins will not be centered at 91, 92, 93, 94, 95, and 96. So use the form `hist(y,x)`, where `x = [91:96]`. The following script file generates the histogram shown in Figure 7.1–2.

```
% Thread breaking strength data for 20 tests.
y = [92,94,93,96,93,94,95,96,91,93,...
95,95,95,92,93,94,91,94,92,93];
% The six possible outcomes are 91,92,93,94,95,96.
x = [91:96];
hist(y,x),axis([90 97 0 6]),ylabel('Absolute Frequency'),...
xlabel('Thread Strength (N)'),...
title('Absolute Frequency Histogram for 20 Tests')
```

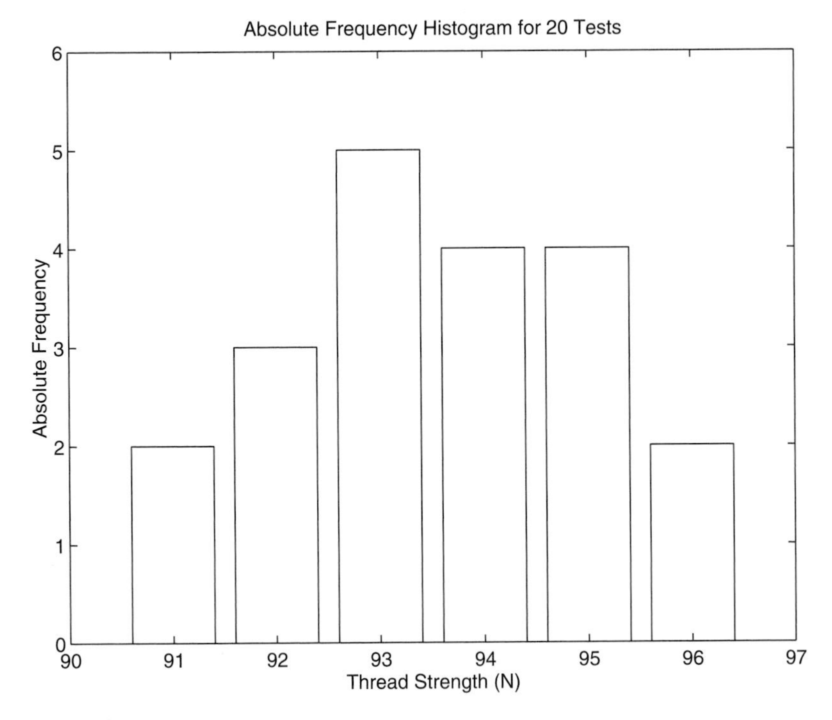

Figure 7.1–2 Histograms for 20 tests of thread strength.

The *absolute frequency* is the number of times a particular outcome occurs. For example, in 20 tests this data shows that a 95 occurred four times. The absolute frequency is 4, and its *relative frequency* is 4/20, or 20 percent of the time.

ABSOLUTE FREQUENCY

RELATIVE FREQUENCY

When there is a large amount of data, you can avoid typing in every data value by first aggregating the data. The following example shows how this is done using the `ones` function. The following data was generated by testing 100 thread samples. The number of times 91, 92, 93, 94, 95, or 96 N was measured is 13, 15, 22, 19, 17, and 14, respectively.

```
% Thread strength data for 20 tests.
y = [91*ones(1,13),92*ones(1,15),93*ones(1,22),...
94*ones(1,19),95*ones(1,17),96*ones(1,14)];
x = [91:96];
hist(y,x),ylabel('Absolute Frequency'),...
xlabel('Thread Strength (N)'),...
title('Absolute Frequency Histogram for 100 Tests')
```

The result appears in Figure 7.1–3.

The `hist` function is somewhat limited in its ability to produce useful histograms. Unless all the outcome values are the same as the bin centers (as is the case with the thread examples), the graph produced by the `hist` function

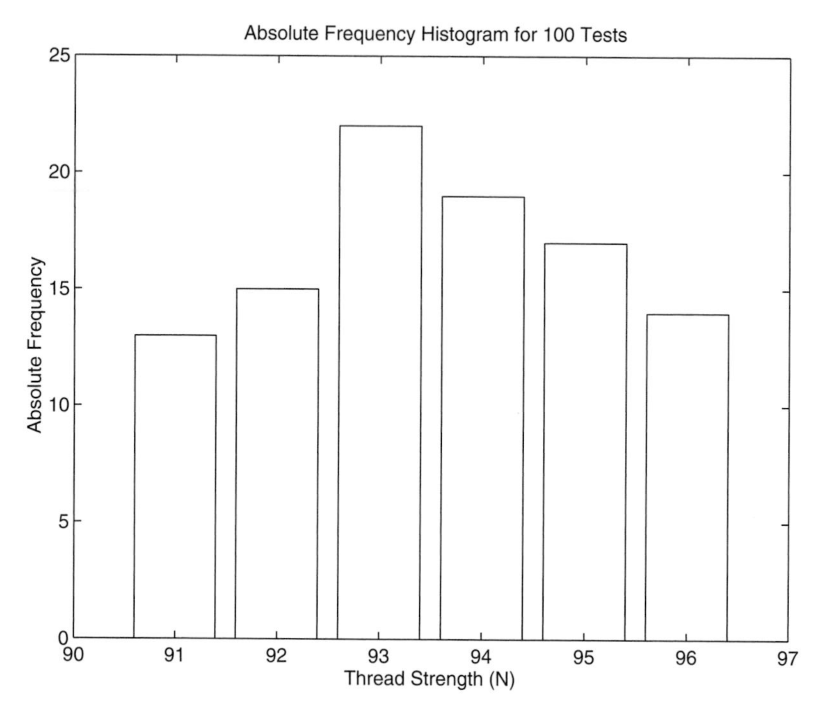

Figure 7.1–3 Absolute frequency histogram for 100 thread tests.

will not be satisfactory. This case occurs when you want to obtain a *relative* frequency histogram. In such cases you can use the bar function to generate the histogram. The following script file generates the relative frequency histogram for the 100 thread tests. Note that if you use the bar function, you must aggregate the data first.

```
% Relative frequency histogram using the bar function.
tests = 100;
y = [13,15,22,19,17,14]/tests;
x = [91:96];
bar(x,y),ylabel('Relative Frequency'),...
xlabel('Thread Strength (N)'),...
title('Relative Frequency Histogram for 100 Tests')
```

The result appears in Figure 7.1–4.

The fourth, fifth, and sixth forms of the hist function do not generate a plot, but are used to compute the frequency counts and bin locations. The bar function can then be used to plot the histogram. The syntax of the fourth form is [z,x] = hist(y), where z is the returned vector containing the frequency count and x is the returned vector containing the bin locations. The fifth and sixth forms are [z,x] = hist(y,n) and [z,x] = hist(y,x). In the latter

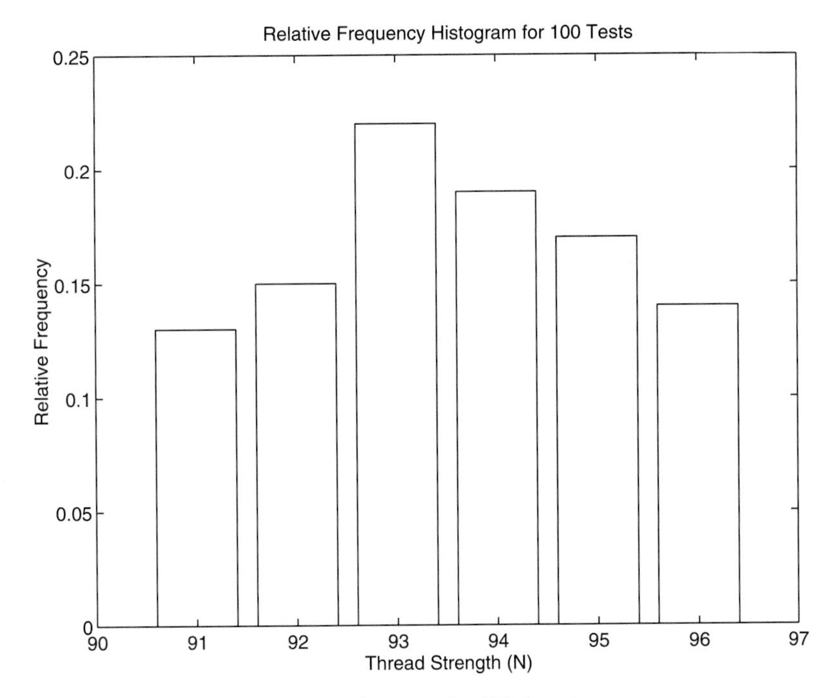

Figure 7.1–4 Relative frequency histogram for 100 thread tests.

case the returned vector x is the same as the user-supplied vector. The following script file shows how the sixth form can be used to generate a relative frequency histogram for the thread example with 100 tests.

```
tests = 100;
y = [91*ones(1,13),92*ones(1,15),93*ones(1,22),...
94*ones(1,19),95*ones(1,17),96*ones(1,14)];
x = [91:96];
[z,x] = hist(y,x);bar(x,z/tests),...
ylabel('Relative Frequency'),xlabel('Thread Strength (N)'),...
title('Relative Frequency Histogram for 100 Tests')
```

The plot generated by this M-file will be identical to that shown in Figure 7.1–4. These commands are summarized in Table 7.1–1.

Test Your Understanding

T7.1–1 In 50 tests of thread, the number of times 91, 92, 93, 94, 95, or 96 N was measured was 7, 8, 10, 6, 12, and 7, respectively. Obtain the absolute and relative frequency histograms.

Table 7.1–1 Histogram functions

Command	Description
`bar(x,y)`	Creates a bar chart of `y` versus `x`.
`hist(y)`	Aggregates the data in the vector `y` into 10 bins evenly spaced between the minimum and maximum values in `y`.
`hist(y,n)`	Aggregates the data in the vector `y` into `n` bins evenly spaced between the minimum and maximum values in `y`.
`hist(y,x)`	Aggregates the data in the vector `y` into bins whose center locations are specified by the vector `x`. The bin widths are the distances between the centers.
`[z,x] = hist(y)`	Same as `hist(y)` but returns two vectors `z` and `x` that contain the frequency count and the bin locations.
`[z,x] = hist(y,n)`	Same as `hist(y,n)` but returns two vectors `z` and `x` that contain the frequency count and the bin locations.
`[z,x] = hist(y,x)`	Same as `hist(y,x)` but returns two vectors `z` and `x` that contain the frequency count and the bin locations. The returned vector `x` is the same as the user-supplied vector `x`.

The Data Statistics Tool

With the Data Statistics tool you can calculate statistics for data and add plots of the statistics to a graph of the data. The tool is accessed from the Figure window after you plot the data. Click on the **Tools** menu, then select **Data Statistics.** The menu appears as shown in Figure 7.1–5. To plot the mean of the dependent variable (y), click the box in the row labeled `mean` under the column labeled `Y`, as shown in the figure. You can plot other statistics as well; these are shown in the figure. You can save the statistics to the workspace as a structure by clicking on the **Save to Workspace** button. This opens a dialog box that prompts you for a name for the structure containing the x data, and a name for the y data structure.

Probability

Probability is expressed as a number between 0 and 1 or as a percentage between 0 percent and 100 percent. For example, because there are six possible outcomes from rolling a single die, the probability of obtaining a specific number on one roll is $1/6$, or 16.67 percent. Thus if you roll the die a large number of times, you expect to obtain a 2 one-sixth of the time. Figure 7.1–6 shows the theoretical uniform probabilities for rolling a single die, and the relative frequency histogram for the data from 100 die rolls. The number of times a 1, 2, 3, 4, 5, or 6 occurred was 21, 14, 18, 16, 19, and 12 respectively. The plots of the theory and the data are very similar, but not identical. In general, if you had rolled the die 1000 times instead of 100 times, the histogram would look even more like the theoretical probability plot.

If you roll two balanced dice, each roll has 36 possible outcomes because each die can produce six numbers. There is only one way to obtain a sum of 2, but there are two ways to obtain a sum of 3, and so on. Thus the probability of rolling a sum of 2 is $1/36$, and the probability of rolling a sum of 3 is $1/36 + 1/36 = 2/36$.

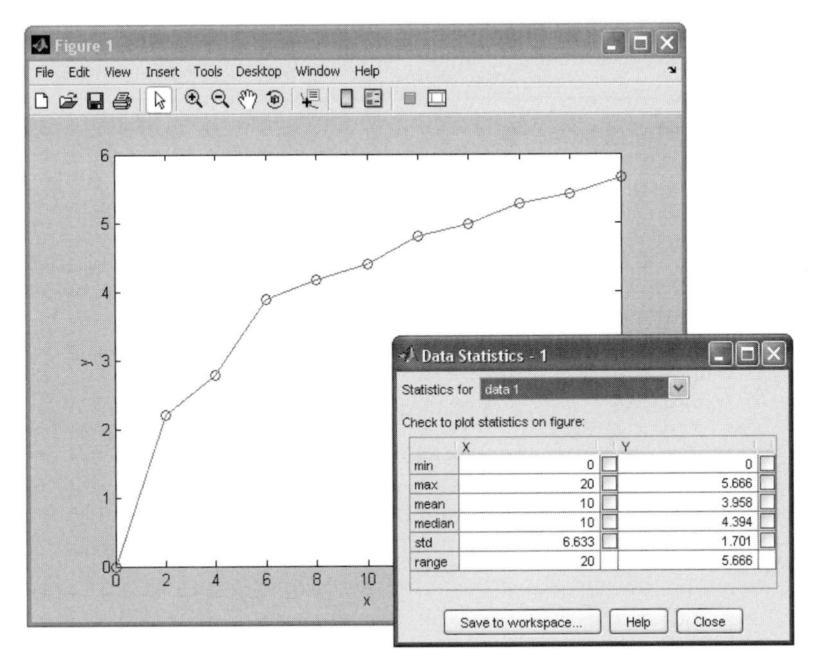

Figure 7.1–5 The Data Statistics tool.

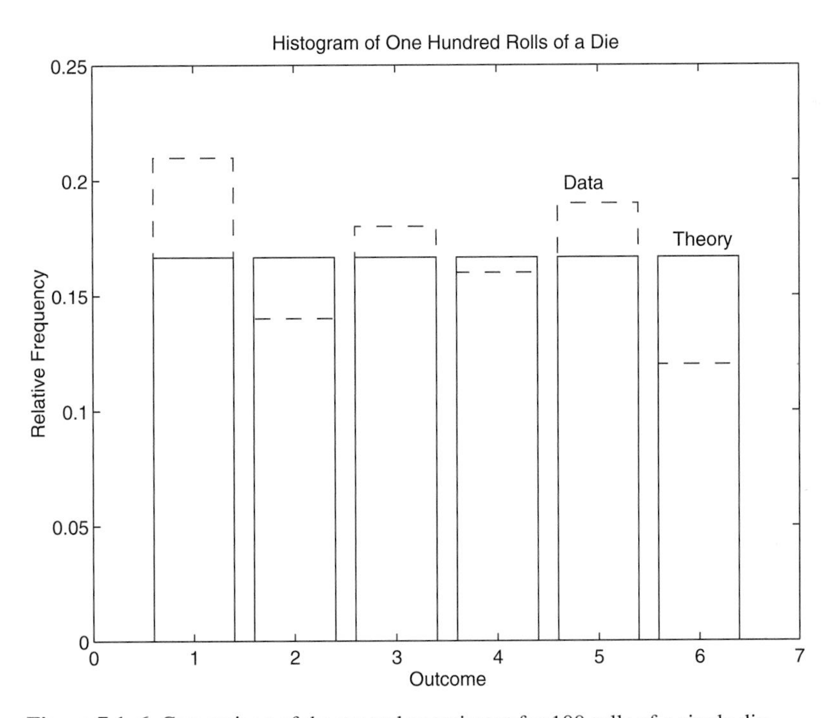

Figure 7.1–6 Comparison of theory and experiment for 100 rolls of a single die.

Continuing this line of reasoning, you can obtain the theoretical probabilities for the sum of two dice, as shown in the following table.

Probabilities for the sum of two dice

Sum	2	3	4	5	6	7	8	9	10	11	12
Probability ($\times$ 36)	1	2	3	4	5	6	5	4	3	2	1

An experiment was performed by rolling two dice 100 times and recording the sums. The data follows.

Data for two dice

Sum	2	3	4	5	6	7	8	9	10	11	12
Frequency	5	5	8	11	20	10	8	12	7	10	4

Figure 7.1–7 shows the relative frequency histogram and the theoretical probabilities on the same plot. If you had collected more data, the histogram would have been closer to the theoretical probabilities.

The theoretical probabilities can be used to predict the outcome of an experiment. Note that the sum of the theoretical probabilities for two dice equals 1, because it is 100 percent certain to obtain a sum between 2 and 12. The sum of the probabilities corresponding to the outcomes 3, 4, and 5 is $2/36 + 3/36 + 4/36 = 1/4$. This result corresponds to a probability of 25 percent. Thus if you roll two dice many times, 25 percent of the time you would expect to obtain a sum of either 3, 4, or 5.

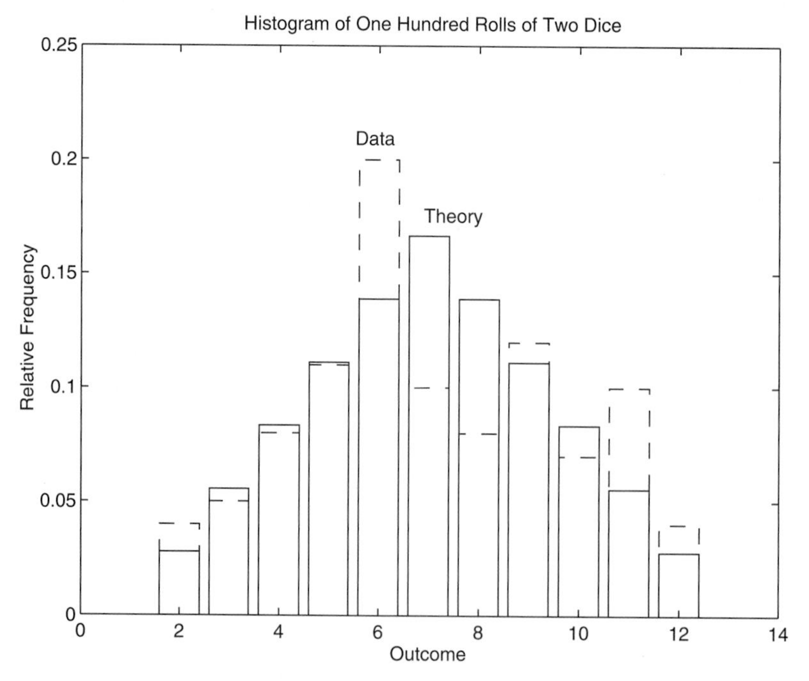

Figure 7.1–7 Comparison of theory and experiment for 100 rolls of two dice.

In many applications the theoretical probabilities are not available because the underlying causes of the process are not understood well enough. In such applications you can use the histogram to make predictions. For example, if you did not have the theoretical probabilities for the sum of two dice, you could use the data to estimate the probability. Using the previously given data from 100 rolls, you can estimate the probability of obtaining a sum of either 3, 4, or 5 by summing the relative frequencies of these three outcomes. This sum is $(5+8+11)/100 = 0.24$, or 24 percent. Thus on the basis of the data from 100 rolls, 24 percent of the time you can estimate that you would obtain a sum of either 3, 4, or 5. The accuracy of the estimates so obtained is highly dependent on the number of trials used to collect the data; the more trials, the better. Many sophisticated statistical methods are available to assess the accuracy of such predictions; these methods are covered in advanced courses.

Test Your Understanding

T7.1–2 If you roll a pair of balanced dice 200 times, how many times would you expect to obtain a sum of 7? How many times would you expect to obtain a sum of either 9, 10, or 11? How many times would you expect to obtain a sum less than 7?
(Answer: 33 times, 50 times, and 83 times.)

7.2 The Normal Distribution

Rolling a die is an example of a process whose possible outcomes are a limited set of numbers; namely, the integers from 1 to 6. For such processes the probability is a function of a discrete-valued variable, that is, a variable having a limited number of values. For example, the following table gives the measured heights of 100 men 20 years of age. The heights were recorded to the nearest 1/2 in., so the height variable is discrete valued.

Table 7.2–1 Height data for men 20 years of age

Height (in.)	Frequency	Height (in.)	Frequency
64	1	70	9
64.5	0	70.5	8
65	0	71	7
65.5	0	71.5	5
66	2	72	4
66.5	4	72.5	4
67	5	73	3
67.5	4	73.5	1
68	8	74	1
68.5	11	74.5	0
69	12	75	1
69.5	10		

Scaled Frequency Histogram

You can plot the data as a histogram using either the absolute or relative frequencies. However, another useful histogram uses data scaled so that the total area under the histogram's rectangles is 1. This *scaled frequency histogram* is the absolute frequency histogram divided by the total area of that histogram. The area of each rectangle on the absolute frequency histogram equals the bin width times the absolute frequency for that bin. Because all the rectangles have the same width, the total area is the bin width times the sum of the absolute frequencies. The following M-file produces the scaled histogram shown in Figure 7.2–1.

```
% Absolute frequency data.
y_abs=[1,0,0,0,2,4,5,4,8,11,12,10,9,8,7,5,4,4,3,1,1,0,1];
binwidth = 0.5;
% Compute scaled frequency data.
area = binwidth*sum(y_abs);
y_scaled = y_abs/area;
% Define the bins.
bins = [64:binwidth:75];
% Plot the scaled histogram.
bar(bins,y_scaled),...
ylabel('Scaled Frequency'),xlabel('Height (in.)')
```

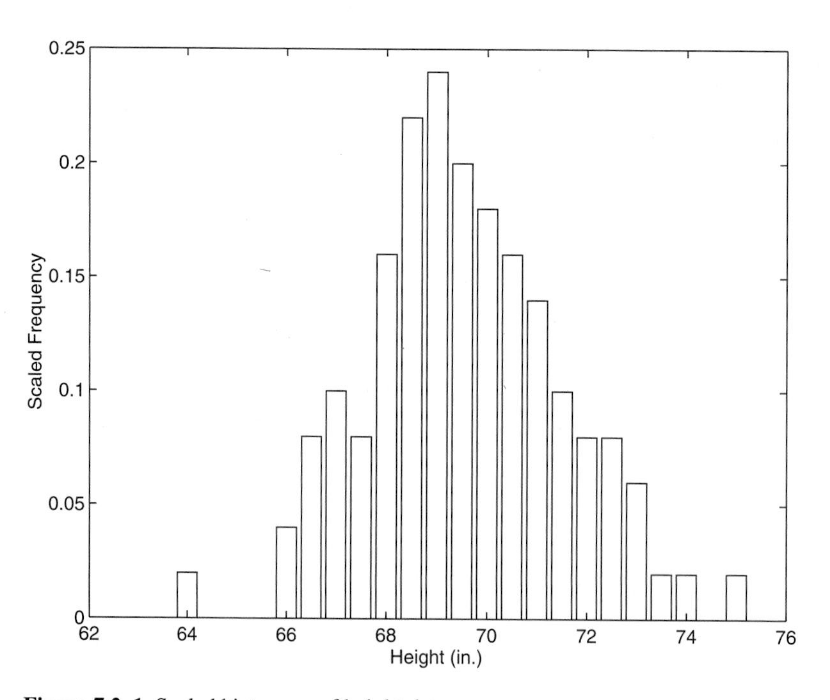

Figure 7.2–1 Scaled histogram of height data.

Because the total area under the scaled histogram is 1, the fractional area corresponding to a range of heights gives the probability that a randomly selected 20-year-old man will have a height in that range. For example, the heights of the scaled histogram rectangles corresponding to heights of 67 through 69 in. are 0.1, 0.08, 0.16, 0.22, and 0.24. Because the bin width is 0.5, the total area corresponding to these rectangles is $(0.1+0.08+0.16+0.22+0.24)(0.5) = 0.4$. Thus 40 percent of the heights lie between 67 and 69 in.

You can use the `cumsum` function to calculate areas under the scaled frequency histogram, and therefore calculate probabilities. If `x` is a vector, `cumsum(x)` returns a vector the same length as `x`, whose elements are the sum of the previous elements. For example, if `x = [2, 5, 3, 8]`, `cumsum(x) = [2, 7, 10, 18]`. If `A` is a matrix, `cumsum(A)` computes the cumulative sum of each row. The result is a matrix the same size as `A`.

After running the previous script, the last element of `cumsum(y_scaled)*binwidth` is 1, which is the area under the scaled frequency histogram. To compute the probability of a height lying between 67 and 69 in. (that is, above the 6th value up to the 11th value, type)

```
>>prob = cumsum(y_scaled)*binwidth;
>>prob67_69 = prob(11)-prob(6)
```

The result is `prob67_69 = 0.4000`, which agrees with our previous calculation of 40 percent.

Continuous Approximation to the Scaled Histogram

In the height data given previously, there was a limited number of possible outcomes because the heights were measured to within 1/2 in. That is, if a particular man's height is between 66 and 66.5, we would measure and record his height as either 66 or 66.5. The number of possible outcomes is doubled if we were to measure the heights to within 1/4 in. Other processes can have an infinite set of possible outcomes. For example, you could obtain an infinite number of possible height measurements in the human population if you could measure a person's height to enough decimal places. For example, an infinite number of values exist between 66 inches and 66.5 in.

Figure 7.2–2 shows the scaled histogram for very many height measurements taken to within 1/4 in. For many processes, as we decrease the bin width and increase the number of measurements, the tops of the rectangles in the scaled histogram often form a smooth bell-shaped curve such as the one shown in Figure 7.2–2.

For processes having an infinite number of possible outcomes, the probability is a function of a *continuous* variable and is plotted as a curve rather than as rectangles. It is based on the same concept as the scaled histogram; that is, the total area under the curve is 1, and the fractional area gives the probability of occurrence of a specific range of outcomes. A probability function that describes many processes is the *normal* or *Gaussian* function, which is shown in Figure 7.2–3.

NORMAL OR GAUSSIAN FUNCTION

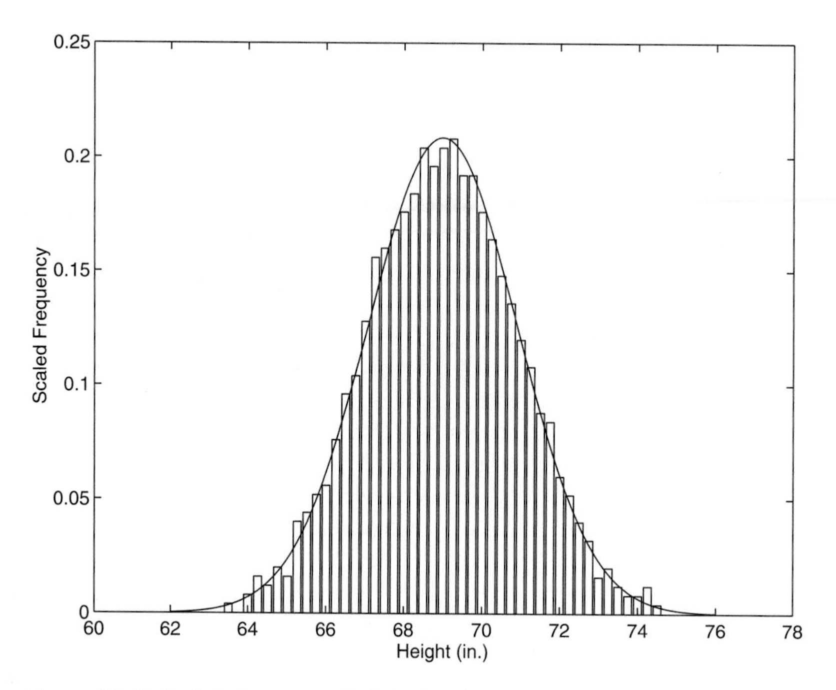

Figure 7.2–2 Scaled histogram of height data for very many measurements.

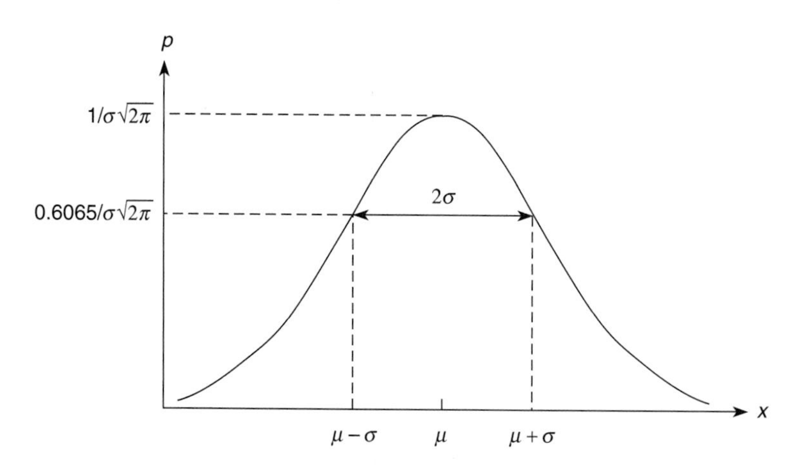

Figure 7.2–3 The basic shape of the normal distribution curve.

NORMALLY DISTRIBUTED

STANDARD DEVIATION

This function is also known as the "bell-shaped curve." Outcomes that can be described by this function are said to be "normally distributed." The normal probability function is a two-parameter function; one parameter, μ, is the mean of the outcomes, and the other parameter, σ, is the *standard deviation.* The mean μ locates the peak of the curve and is the most likely value to occur. The width, or spread, of the curve is described by the parameter σ. Sometimes the term

VARIANCE

variance is used to describe the spread of the curve. The variance is the square of the standard deviation σ.

The normal probability function is described by the following equation:

$$p(x) = \frac{1}{\sigma\sqrt{2\pi}} e^{-(x-\mu)^2/2\sigma^2} \tag{7.2–1}$$

Figure 7.2–4 is a plot of this function for three cases having the same mean, $\mu = 10$, but different standard deviations: $\sigma = 1$, 2, and 3. Note how the peak height decreases as σ is increased. The reason is that the area under the curve must equal 1 (because the value of the random variable x must certainly lie between $-\infty$ and $+\infty$).

Recall that the fractional area under a scaled histogram gives the probability that a range of outcomes will occur. The fractional area under a probability function curve also gives this probability. It can be shown that 68.3 percent, or approximately 68 percent, of the area lies between the limits of $\mu - \sigma \leq x \leq \mu + \sigma$. Consequently, if a variable is normally distributed, there is a 68 percent chance that a randomly selected sample will lie within one standard deviation of the mean. In addition, 95.5 percent, or approximately 96 percent, of the area lies between the limits of $\mu - 2\sigma \leq x \leq \mu + 2\sigma$, and 99.7 percent, or practically 100 percent, of the area lies between the limits of $\mu - 3\sigma \leq x \leq \mu + 3\sigma$. So there is a 96 percent chance that a randomly selected sample will lie within two

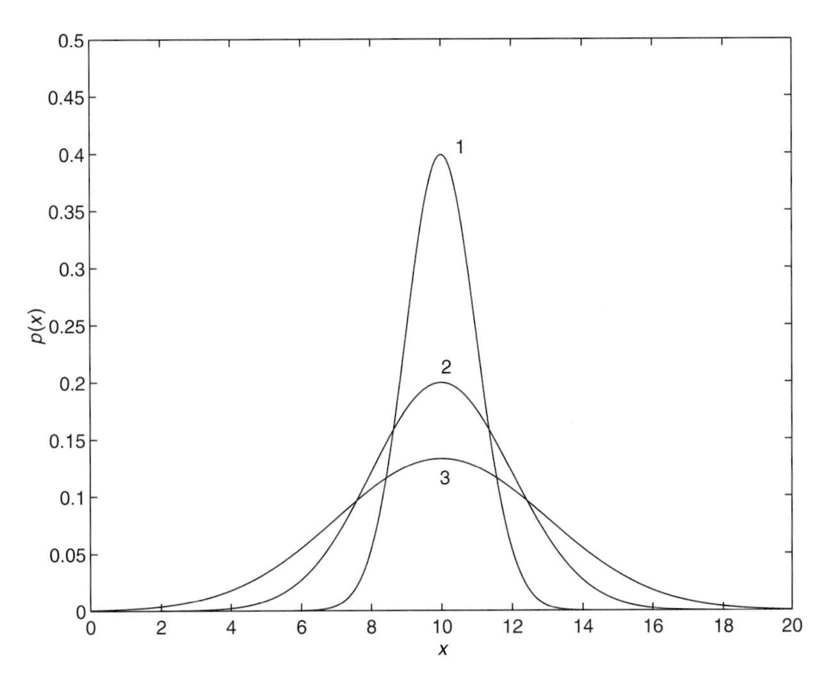

Figure 7.2–4 The effect on the normal distribution curve of increasing σ. For this case $\mu = 10$, and the three curves correspond to $\sigma = 1$, $\sigma = 2$, and $\sigma = 3$.

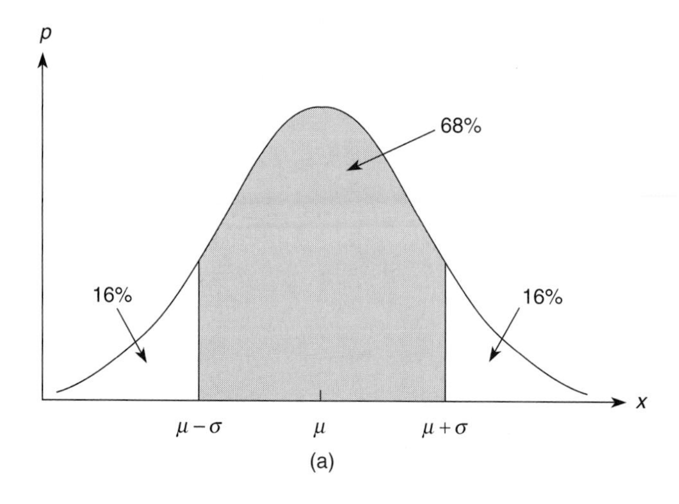

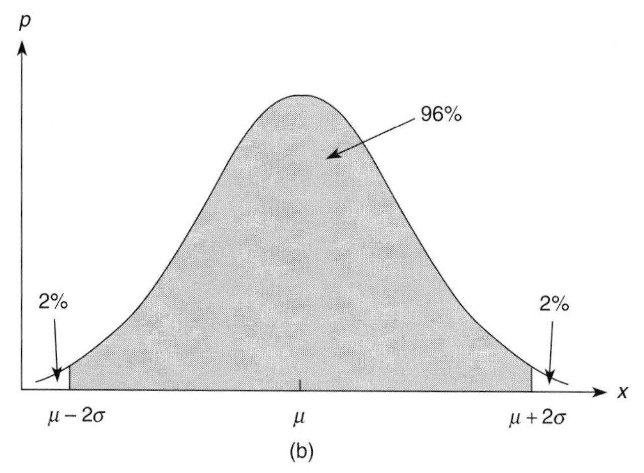

Figure 7.2–5 Probability interpretation of (a) the $\mu \pm \sigma$ and (b) the $\mu \pm 2\sigma$ limits.

standard deviations of the mean, and a 99.7 percent chance that a randomly selected sample will lie within three standard deviations of the mean. Figure 7.2–5 illustrates the areas associated with the $\mu \pm \sigma$ and $\mu \pm 2\sigma$ limits. For example, if the variable is normally distributed with a mean equal to 20 and a standard deviation equal to 2, there is a 68 percent chance that a randomly selected sample will lie between 18 and 22, a 96 percent chance that it will lie between 16 and 24, and a 99.7 percent chance that it will lie between 14 and 26.

Estimating the Mean and Standard Deviation

In most applications you do not know the mean or variance of the distribution of possible outcomes, but must estimate them from experimental data. An estimate

Table 7.2–1 Statistical functions

Command	Description
`cumsum(x)`	Creates a vector the same size as x, containing the cumulative sum of the elements of x.
`mean(x)`	Calculates the mean of the data stored in the vector x.
`median(x)`	Calculates the median of the data stored in the vector x.
`std(x)`	Uses equation (7.2–3) to calculate the standard deviation of the data stored in the vector x.

of the mean μ is denoted by $\bar{x}$ and is found in the same way you compute an average, namely,

$$\bar{x} = \frac{1}{n}(x_1 + x_2 + x_3 + \cdots + x_n) = \frac{1}{n}\sum_{i=1}^{n} x_i \tag{7.2–2}$$

where the n data values are $x_1, x_2, \ldots, x_n$. The variance of a set of data values is the average of their squared deviations from their mean $\bar{x}$. Thus the standard deviation σ is computed from a set of n data values as follows:

$$\sigma = \sqrt{\frac{\sum_{i=1}^{n}(x_i - \bar{x})^2}{n-1}} \tag{7.2–3}$$

You would expect that the divisor should be n rather than $n - 1$. However, using $n - 1$ gives a better estimate of the standard deviation when the number of data points n is small. The MATLAB function `mean(x)` uses (7.2–2) to calculate the mean of the data stored in the vector `x`. The function `std(x)` uses (7.2–3) to calculate the standard deviation. Table 7.2–1 summarizes these functions.

Mean and Standard Deviation of Heights

EXAMPLE 7.2–1

Statistical analysis of data on human proportions is required in many engineering applications. For example, designers of submarine crew quarters need to know how small they can make bunk lengths without eliminating a large percentage of prospective crew members. Use MATLAB to estimate the mean and standard deviation for the height data given in Table 7.2–1.

■ Solution

The script file follows. The data given in Table 7.2–1 is the absolute frequency data and is stored in the vector `y_abs`. A bin width of 1/2 in. is used because the heights were measured to the nearest 1/2 in. The vector `bins` contains the heights in 1/2 in. increments.

To compute the mean and standard deviation, reconstruct the original (raw) height data from the absolute frequency data. Note that this data has some zero entries. For example, none of the 100 men had a height of 65 in. Thus to reconstruct the raw data, start with a null vector `y_raw` and fill it with the height data obtained from the absolute frequencies. The `for` loop checks to see whether the absolute frequency for a particular bin is nonzero. If it is nonzero, append the appropriate number of data values to the vector y_raw. If the particular bin frequency is 0, y_raw is left unchanged.

```
% Absolute frequency data.
y_abs = [1,0,0,0,2,4,5,4,8,11,12,10,9,8,7,5,4,4,3,1,1,0,1];
binwidth = 0.5;
% Define the bins.
bins = [64:binwidth:75];
% Fill the vector y_raw with the raw data.
% Start with a null vector.
y_raw = [];
for i = 1:length(y_abs)
   if y_abs(i)>0
      new = bins(i)*ones(1,y_abs(i));
   else
      new = [];
   end
y_raw = [y_raw,new];
end
% Compute the mean and standard deviation.
mu = mean(y_raw),sigma = std(y_raw)
```

When you run this program, you will find that the mean is $\mu = 69.6$ in. and the standard deviation is $\sigma = 1.96$ in.

As discussed earlier, you can use the 1σ, 2σ, and 3σ points to estimate the 68.3 percent, 95.5 percent, and 99.7 percent probabilities, respectively. Thus for the preceding height data, 68.3 percent of 20-year-old men will be between $\mu - \sigma = 67.3$ and $\mu + \sigma = 71.3$ in. tall.

If you need to compute the probability at other points, you can use the erf function. Typing erf(x) returns the area to the left of the value $t = x$ under the curve of the function $2e^{-t^2}/\sqrt{\pi}$. This area, which is a function of x, is known as the *error function,* and is written as $\text{erf}(x)$. The probability that the random variable x is less than or equal to b is written as $P(x \leq b)$ if the outcomes are normally distributed. This probability can be computed from the error function as follows [Kreyzig, 1998]:

ERROR FUNCTION

$$P(x \leq b) = \frac{1}{2}\left[1 + \text{erf}\left(\frac{b-\mu}{\sigma\sqrt{2}}\right)\right] \tag{7.2–4}$$

The probability that the random variable x is no less than a and no greater than b is written as $P(a \leq x \leq b)$. It can be computed as follows:

$$P(a \leq x \leq b) = \frac{1}{2}\left[\text{erf}\left(\frac{b-\mu}{\sigma\sqrt{2}}\right) - \text{erf}\left(\frac{a-\mu}{\sigma\sqrt{2}}\right)\right] \tag{7.2–5}$$

These equations are useful for computing probabilities of outcomes for which the data is scarce or missing altogether.

Estimation of Height Distribution

EXAMPLE 7.2–2

Use the results of Example 7.2–1 to estimate how many 20-year-old men are no taller than 68 in. How many are within 3 in. of the mean?

■ Solution

In Example 7.2–1 the mean and standard deviation were found to be $\mu = 69.3$ in. and $\sigma = 1.96$ in. In Table 7.2–1, note that few data points are available for heights less than 68 in. However, if you assume that the heights are normally distributed, you can use equation (7.2–4) to estimate how many men are shorter than 68 in. Use (7.2–4) with $b = 68$; that is,

$$P(x \le 68) = \frac{1}{2}\left[1 + \text{erf}\left(\frac{68 - 69.3}{1.96\sqrt{2}}\right)\right]$$

To determine how many men are within 3 in. of the mean, use (7.2–5) with $a = \mu - 3 = 66.3$ and $b = \mu + 3 = 72.3$; that is,

$$P(66.3 \le x \le 72.3) = \frac{1}{2}\left[\text{erf}\left(\frac{3}{1.96\sqrt{2}}\right) - \text{erf}\left(\frac{-3}{1.96\sqrt{2}}\right)\right]$$

In MATLAB these expressions are computed in a script file as follows:

```
mu = 69.3;
sigma = 1.96;
% How many are no taller than 68 inches?
b1 = 68;
P1 = (1+erf((b1-mu)/(sigma*sqrt(2))))/2
% How many are within 3 inches of the mean?
a2 = 66.3;
b2 = 72.3;
P2 = (erf((b2-mu)/(sigma*sqrt(2)))-erf((a2-mu)/(sigma*sqrt(2))))/2
```

When you run this program, you obtain the results `P1 = 0.2536` and `P2 = 0.8741`. Thus 25 percent of 20-year-old men are estimated to be 68 inches or less in height, and 87 percent are estimated to be between 66.3 and 72.3 inches tall.

Test Your Understanding

T7.2–1 Suppose that 10 more height measurements are obtained so that the following numbers must be *added* to Table 7.2–1.

Height (in.)	Additional data
64.5	1
65	2
66	1
67.5	2
70	2
73	1
74	1

(a) Plot the scaled frequency histogram. (b) Find the mean and standard deviation. (c) Use the mean and standard deviation to estimate how many 20-year-old men are no taller than 69 in. (d) Estimate how many are between 68 and 72 in. tall.
(Answers: (b) mean = 69.4 in., standard deviation = 2.14 in.; (c) 43 percent; (d) 63 percent.)

Sums and Differences of Random Variables

It can be proved that the mean of the sum (or difference) of two independent normally distributed random variables equals the sum (or difference) of their means, but the variance is always the sum of the two variances. That is, if x and y are normally distributed with means μ_x and μ_y, and variances σ_x^2 and σ_y^2, and if $u = x + y$ and $v = x - y$, then

$$\mu_u = \mu_x + \mu_y \tag{7.2–6}$$

$$\mu_v = \mu_x - \mu_y \tag{7.2–7}$$

$$\sigma_u^2 = \sigma_v^2 = \sigma_x^2 + \sigma_y^2 \tag{7.2–8}$$

These properties are applied in some of the homework problems.

7.3 Random Number Generation

We often do not have a simple probability distribution to describe the distribution of outcomes in many engineering applications. For example, the probability that a circuit consisting of many components will fail is a function of the number and the age of the components, but we often cannot obtain a function to describe the failure probability. In such cases engineers often resort to simulation to make predictions. The simulation program is executed many times, using a random set of numbers to represent the failure of one or more components, and the results are used to estimate the desired probability.

Uniformly Distributed Numbers

In a sequence of *uniformly distributed* random numbers, all values within a given interval are equally likely to occur. The MATLAB function `rand` generates random numbers uniformly distributed over the interval [0,1]. Type `rand` to obtain a single random number in the interval [0, 1]. Typing `rand` again generates a different number because the MATLAB algorithm used for the `rand` function requires a "state" to start. MATLAB obtains this state from the computer's CPU clock. Thus every time the `rand` function is used, a different result will be obtained. For example,

```
rand
ans =
     0.6161
rand
ans =
     0.5184
```

Type `rand(n)` to obtain an $n \times n$ matrix of uniformly distributed random numbers in the interval [0, 1]. Type `rand(m,n)` to obtain an $m \times n$ matrix of random numbers. For example, to create a 1×100 vector `y` having 100 random values in the interval [0, 1], type `y = rand(1,100)`. Using the `rand` function this way is equivalent to typing `rand` 100 times. Even though there is a single call to the `rand` function, the `rand` function's calculation has the effect of using a different state to obtain each of the 100 numbers so that they will be random.

Use `Y = rand(m,n,p,...)` to generate a multidimensional array `Y` having random elements. Typing `rand(size(A))` produces an array of random entries that is the same size as `A`.

For example, the following script makes a random choice between two equally probable alternatives.

```
if rand < 0.5
   disp('heads')
else
  disp('tails')
end
```

In order to compare the results of two or more simulations, you sometimes will need to generate the same sequence of random numbers each time the simulation runs. To generate the same sequence, you must use the same state each time. The current state `s` of the uniform number generator can be obtained by typing `s = rand('state')`. This returns a 35-element vector containing the current state of the uniform generator. To set the state of the generator to `s`, type `rand('state',s)`. Typing `rand('state',0)` resets the generator to its initial state. Typing `rand('state',j)`, for integer `j`, resets the generator to state `j`. Typing `rand('state',sum(100*clock))` resets the generator to a different state each time. Table 7.3–1 summarizes these functions.

The following session shows how to obtain the same sequence every time `rand` is called.

```
>>rand('state',0)
>>rand
ans =
     0.9501
```

Table 7.3–1 Random number functions

Command	Description
`rand`	Generates a single uniformly distributed random number between 0 and 1.
`rand(n)`	Generates an $n \times n$ matrix containing uniformly distributed random numbers between 0 and 1.
`rand(m,n)`	Generates an $m \times n$ matrix containing uniformly distributed random numbers between 0 and 1.
`s = rand('state')`	Returns a 35-element vector `s` containing the current state of the uniformly distributed generator.
`rand('state',s)`	Sets the state of the uniformly distributed generator to `s`.
`rand('state',0)`	Resets the uniformly distributed generator to its initial state.
`rand('state',j)`	Resets the uniformly distributed generator to state `j`, for integer `j`.
`rand('state',sum(100*clock))`	Resets the uniformly distributed generator to a different state each time it is executed.
`randn`	Generates a single normally distributed random number having a mean of 0 and a standard deviation of 1.
`randn(n)`	Generates an $n \times n$ matrix containing normally distributed random numbers having a mean of 0 and a standard deviation of 1.
`randn(m,n)`	Generates an $m \times n$ matrix containing normally distributed random numbers having a mean of 0 and a standard deviation of 1.
`s = randn('state')`	Like `rand('state')` but for the normally distributed generator.
`randn('state',s)`	Like `rand('state',s)` but for the normally distributed generator.
`randn('state',0)`	Like `rand('state',0)` but for the normally distributed generator.
`randn('state',j)`	Like `rand('state',j)` but for the normally distributed generator.
`randn('state',sum(100*clock))`	Like `rand('state',sum(100*clock))` but for the normally distributed generator.
`randperm(n)`	Generates a random permutation of the integers from 1 to `n`.

```
>>rand
ans =
     0.2311
>>rand('state',0)
>>rand
ans =
     0.9501
>>rand
ans =
     0.2311
```

You need not start with the initial state in order to generate the same sequence. To show this, continue the above session as follows.

```
>>s = rand('state');
>>rand('state',s)
>>rand
ans =
     0.6068
>>rand('state',s)
>>rand
ans =
     0.6068
```

You can use the `rand` function to generate random numbers in an interval other than [0, 1]. For example, to generate values in the interval [2, 10], first generate a random number between 0 and 1, multiply it by 8 (the difference between the upper and lower bounds), and then add the lower bound (2). The result is a value that is uniformly distributed in the interval [2, 10]. The general formula for generating a uniformly distributed random number y in the interval $[a, b]$ is

$$y = (b - a)x + a \qquad (7.3\text{–}1)$$

where x is a random number uniformly distributed in the interval [0, 1]. For example, to generate a vector `y` containing 1000 uniformly distributed random numbers in the interval [2, 10], you type `y = 8*rand(1,1000) + 2`. You can check the results with the `mean`, `min`, and `max` functions. You should obtain values close to 6, 2, and 10, respectively.

You can use `rand` to generate random results for games involving dice, for example, but you must use it to create integers. An easier way is to use the `randperm(n)` function, which generates a random permutation of the integers from 1 to `n`. For example, `randperm(6)` might generate the vector `[3 2 6 4 1 5]`, or some other permutation of the numbers from 1 to 6. Note that `randperm` calls `rand` and therefore changes the state of the generator.

Test Your Understanding

T7.3–1 Use MATLAB to generate a vector `y` containing 1500 uniformly distributed random numbers in the interval [−5, 15]. Check your results with the `mean`, `min`, and `max` functions.

Simulations Using Random Numbers

Simulations using random number generators can be used to analyze and find solutions to problems that are difficult or impossible to solve using standard mathematics. The following example illustrates this method.

Optimal Production Quantity — EXAMPLE 7.3–1

You have recently taken a position as the engineer in charge of your company's seasonal product, which is manufactured during the off-season. You want to determine the optimum production level. If you produce more units than you can sell, your profit will not be as large as it could be; if you produce too many units, the unsold units at the end of the season will hurt profits. A review of the company's records shows that the fixed cost of production is $30,000 per season, no matter how many units are made, and that it costs $2000 above the fixed cost to make one unit. In addition, past sales have fluctuated between 25 and 50 units per season, with no increasing or decreasing trend in the sales data. Your sales force estimates that you cannot raise the price above $4000 because of the competition.

Determine the optimal number of units to produce for each season. Assume a selling price of $4000 during the season. Assume also that at the end of the season all unsold units can be sold at $1000 each.

■ Solution

Given that the previous sales were always between 25 and 50 units per season, with no trend in the data, the most reasonable model for the sales is a uniform distribution over the interval [25, 50]. In the following script file, the vector `cost` is the cost as a function of production level (the number of units produced). The loop over `k` computes the profit for each of the 26 production levels starting with 25 and ending with 50. The loop over `m` does a random number simulation to compute the profit for a particular production level. The variable `demand` represents the demand for the product; it is a uniformly distributed number in the interval [25, 50]. If the demand is greater than the number of units produced, then all the units will be sold and the income is $4000 times the number produced. If the demand is less than the number of units produced, then the income is $4000 times the number sold plus $1000 times the number unsold, which is the number produced minus the demand. For each random demand, the profit is computed as the difference between the income and the cost. This amount is the scalar `profit`. The cumulative profit `cum_profit` is the sum of profits for all the random simulations for a particular production level. The expected profit `expected_profit` for a particular production level is the profit for that level averaged over all the random simulations.

```
n = 5000; % number of random simulations
level = [25:50]; % initialize the production level vector
cost = 30000 + 2000*level;
for k = 1:26
   cum_profit = 0;
   for m = 1:n
      demand = floor(rand*(50-25)+26);
      if demand >= level(k)
         income = 4000*level(k);
      else
         income = 4000*demand+1000*(level(k)-demand);
      end
      profit = income-cost(k);
      cum_profit = cum_profit+profit;
   end
   expected_profit = cum_profit/n;
   p(k,1) = level(k);
   p(k,2) = expected_profit;
end
plot(p(:,1),p(:,2),'o',p(:,1),p(:,2),'-'),...
xlabel('No. of Units'),ylabel('Profit ($)')
```

The choice of 5000 simulations was a somewhat arbitrary compromise between accuracy and the amount of time required to do the calculations. The more simulations, the more accurate the results, but the program will take longer to finish. The results are shown in Figure 7.3–1. The maximum profit is attained for a production level in the range of 39 to

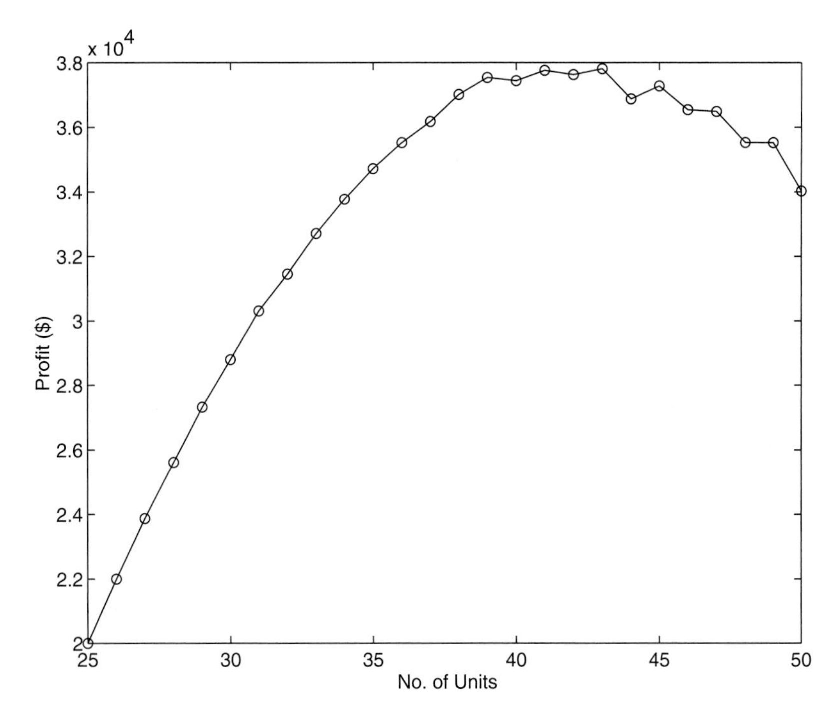

Figure 7.3–1 Profit versus quantity plot for 5000 simulations.

43 units per season. The profit is nearly $38,000 per season, which is much more than the profit obtained by a conservative production level of 25 units per season.

You should try different values of `n` and compare the results.

Test Your Understanding

T7.3–2 In Example 7.3–1, what is the optimum production level and resulting profit if we sell unsold units at the end of the season at cost ($2000), instead of below cost?
(Answer: Optimum production level = 50 units per season; profit = $46,000.)

Normally Distributed Random Numbers

In a sequence of normally distributed random numbers, the values near the mean are more likely to occur. We have noted that the outcomes of many processes can be described by the normal distribution. Although a uniformly distributed random variable has definite upper and lower bounds, a normally distributed random variable does not.

The MATLAB function `randn` will generate a single number that is normally distributed with a mean equal to 0 and a standard deviation equal to 1. Type `randn(n)` to obtain an $n \times n$ matrix of such numbers. Type `randn(m,n)` to obtain an $m \times n$ matrix of random numbers.

The functions for retrieving and specifying the state of the normally distributed random number generator are identical to those for the uniformly distributed generator, except that `randn(...)` replaces `rand(...)` in the syntax. These functions are summarized in Table 7.3–1.

You can generate a sequence of normally distributed numbers having a mean μ and standard deviation σ from a normally distributed sequence having a mean of 0 and a standard deviation of 1. You do this by multiplying the values by σ and adding μ to each result. Thus if x is a random number with a mean of 0 and a standard deviation of 1, use the following equation to generate a new random number y having a standard deviation of σ and a mean of μ.

$$y = \sigma x + \mu \tag{7.3–2}$$

For example, to generate a vector `y` containing 2000 random numbers normally distributed with a mean of 5 and a standard deviation of 3, you type `y = 3*randn(1,2000) + 5`. You can check the results with the `mean` and `std` functions. You should obtain values close to 5 and 3, respectively.

Test Your Understanding

T7.3–3 Use MATLAB to generate a vector `y` containing 1800 random numbers normally distributed with a mean of 7 and a standard deviation of 10. Check your results with the `mean` and `std` functions. Why can't you use the `min` and `max` functions to check your results?

Functions of Random Variables If y and x are linearly related, as

$$y = bx + c \tag{7.3–3}$$

and if x is normally distributed with a mean μ_x and standard deviation σ_x, it can be shown that the mean and standard deviation of y are given by

$$\mu_y = b\mu_x + c \tag{7.3–4}$$

$$\sigma_y = |b|\sigma_x \tag{7.3–5}$$

However, it is easy to see that the means and standard deviations do not combine in a straightforward fashion when the variables are related by a nonlinear function. For example, if x is normally distributed with a mean of 0, and if $y = x^2$, it is easy to see that the mean of y is not 0, but is positive. In addition, y is not normally distributed.

Some advanced methods are available for deriving a formula for the mean and variance of $y = f(x)$, but for our purposes, the simplest way is to use random number simulation.

It was noted in the previous section that the mean of the sum (or difference) of two independent normally distributed random variables equals the sum (or difference) of their means, but the variance is always the sum of the two variances. However, if z is a nonlinear function of x and y, then the mean and variance of z cannot be found with a simple formula. In fact, the distribution of z will not even be normal. This outcome is illustrated by the following example.

Statistical Analysis and Manufacturing Tolerances

EXAMPLE 7.3–2

Suppose you must cut a triangular piece off the corner of a square plate by measuring the distances x and y from the corner (see Figure 7.3–2). The desired value of x is 10 in., and the desired value of θ is 20°. This requires that $y = 3.64$ in. We are told that measurements of x and y are normally distributed with means of 10 and 3.64, respectively, with a standard deviation equal to 0.05 in. Determine the standard deviation of θ and plot the relative frequency histogram for θ.

■ Solution

From Figure 7.3–2, we see that the angle θ is determined by $\theta = \tan^{-1}(y/x)$. We can find the statistical distribution of θ by creating random variables x and y that have means of 10 and 3.64, respectively, with a standard deviation of 0.05. The random variable θ is then found by calculating $\theta = \tan^{-1}(y/x)$ for each random pair (x, y). The following script file shows this procedure.

```
s = 0.05; % standard deviation of x and y
n = 8000; % number of random simulations
x = 10 + s*randn(1,n);
y = 3.64 + s*randn(1,n);
theta = (180/pi)*atan(y./x);
mean_theta = mean(theta)
sigma_theta = std(theta)
xp = [19:0.1:21];
z = hist(theta,xp);
yp = z/n;
bar(xp,yp),xlabel('Theta (degrees)'),ylabel('Relative Frequency')
```

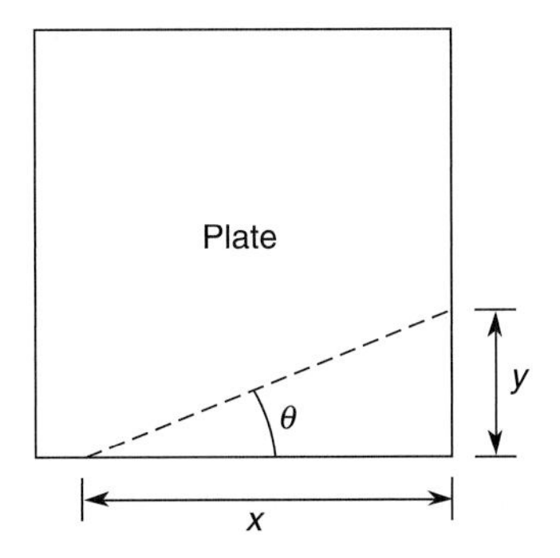

Figure 7.3–2 Dimensions of a triangular cut.

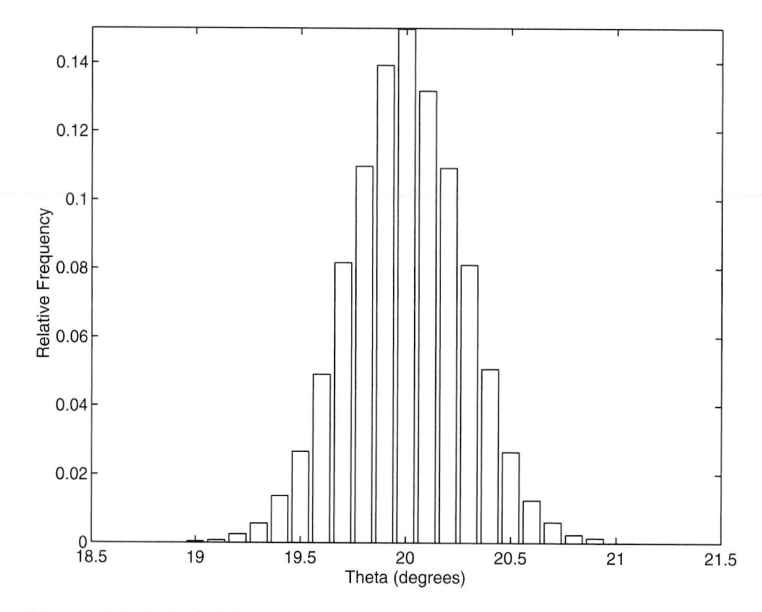

Figure 7.3–3 Scaled histogram of the angle θ.

The choice of 8000 simulations was a compromise between accuracy and the amount of time required to do the calculations. You should try different values of `n` and compare the results. The results gave a mean of 19.9993° for θ with a standard deviation of 0.2730°. The histogram is shown in Figure 7.3–3. Although the plot resembles the normal distribution, the values of θ are not distributed normally. From the histogram we can calculate that approximately 65 percent of the values of θ lie between 19.8 and 20.2. This range corresponds to a standard deviation of 0.2°, not 0.273° as calculated from the simulation data. Thus the curve is not a normal distribution.

This example shows that the interaction of two of more normally distributed variables does not produce a result that is normally distributed. In general, the result is normally distributed if and only if the result is a linear combination of the variables.

7.4 Interpolation

Engineering problems often require the analysis of data pairs. For example, the paired data might represent a *cause and effect,* or *input-output relationship,* such as the current produced in a resistor as a result of an applied voltage, or a *time history,* such as the temperature of an object as a function of time. Another type of paired data represents a *profile,* such as a road profile (which shows the height of the road along its length). In some applications we want to estimate a variable's value between the data points. This process is called *interpolation*. In other cases we might need to estimate the variable's value outside of the given data range. This process is *extrapolation*. Interpolation and extrapolation are greatly aided by plotting the data. Such plots, some perhaps using logarithmic axes, often help to discover a functional description of the data.

Suppose that x represents the independent variable in the data (such as the applied voltage in the preceding example), and y represents the dependent variable (such as the resistor current). In some applications the data set will contain only one value of y for each value of x. In other cases there will be several measured values of y for a particular value of x. This condition is called *data scatter*. For example, suppose we apply 10 V to a resistor, and measure 3.1 mA of current. Then, repeating the experiment, suppose we measure 3.3 mA the second time. If we average the two results, the resulting data point will be $x = 10$ V, $y = 3.2$ mA, which is an example of *aggregating* the data. In this section we assume that the data have been aggregated if necessary, so only one value of y corresponds to a specific value of x. You can use the methods of Sections 7.1 and 7.2 to aggregate the data by computing its mean. The data's standard deviation indicates how much the data is spread around the aggregated point.

Suppose we have the following temperature measurements, taken once an hour starting at 7:00 A.M. The measurements at 8 A.M. and 10 A.M. are missing for some reason, perhaps because of equipment malfunction.

Time	7 A.M.	9 A.M.	11 A.M.	12 noon
Temperature (°F)	49	57	71	75

A plot of this data is shown in Figure 7.4–1 with the data points connected by dashed lines. If we need to estimate the temperature at 10 A.M., we can read the value from the dashed line that connects the data points at 9 A.M. and 11 A.M. From the plot we thus estimate the temperature at 8 A.M. to be 53°F and at 10 A.M. to be 64°F. We have just performed *linear interpolation* on the data to obtain an *estimate* of the missing data. Linear interpolation is so named because it is equivalent to connecting the data points with a linear function (a straight line).

Of course we have no reason to believe that the temperature follows the straight lines shown in the plot, and our estimate of 64°F will most likely be incorrect, but it might be close enough to be useful. When using interpolation, we must always keep in mind that our results will be approximate and should be used with caution. In general, the more closely spaced the data, the more accurate the interpolation. Plotting the data sometimes helps to judge the accuracy of the interpolation.

Using straight lines to connect the data points is the simplest form of interpolation. Another function could be used if we have a good reason to do so. Later in this section we use polynomial functions to do the interpolation.

Linear interpolation in MATLAB is obtained with the `interp1` and `interp2` functions. Suppose that `x` is a vector containing the independent variable data and that `y` is a vector containing the dependent variable data. If `x_int` is a vector containing the value or values of the independent variable at which we wish to estimate the dependent variable, then typing `interp1(x,y,x_int)`

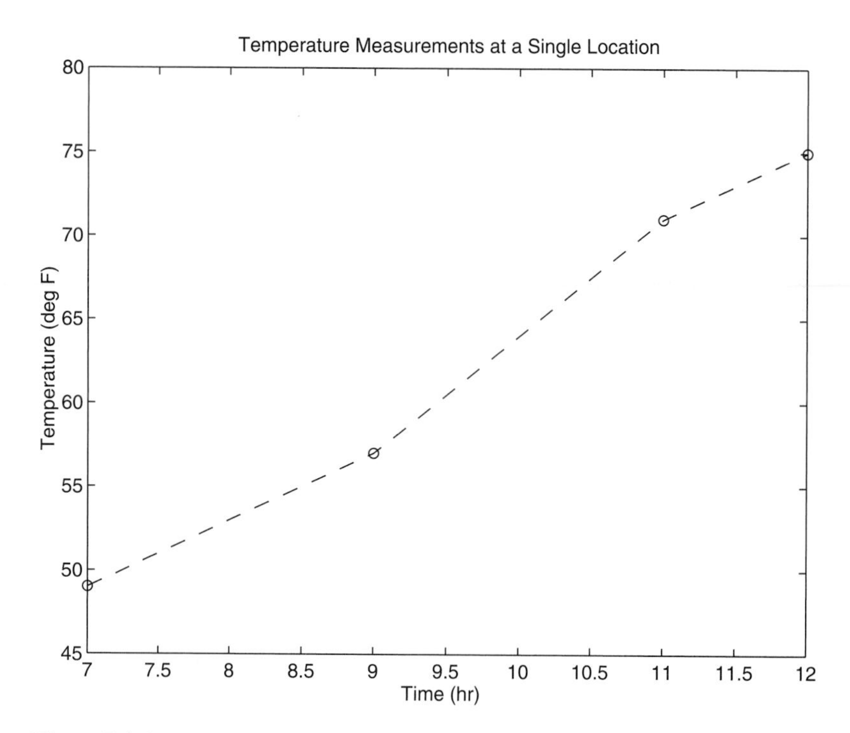

Figure 7.4–1 A plot of temperature data versus time.

produces a vector the same size as `x_int` containing the interpolated values of y that correspond to `x_int`. For example, the following session produces an estimate of the temperatures at 8 A.M. and 10 A.M. from the preceding data. The vectors `x` and `y` contain the times and temperatures, respectively.

```
>>x = [7, 9, 11, 12];
>>y = [49, 57, 71, 75];
>>x_int = [8, 10];
>>interp1(x,y,x_int)
ans =
     53
     64
```

You must keep in mind two restrictions when using the `interp1` function. The values of the independent variable in the vector `x` must be in ascending order, and the values in the interpolation vector `x_int` must lie within the range of the values in `x`. Thus we cannot use the `interp1` function to estimate the temperature at 6 A.M., for example.

The `interp1` function can be used to interpolate in a table of values by defining `y` to be a matrix instead of a vector. For example, suppose we now have temperature measurements at three locations and that the measurements at 8 A.M. and 10 A.M. are missing for all three locations. The data is

	Temperatures (°F)		
Time	Location 1	Location 2	Location 3
7 A.M.	49	52	54
9 A.M.	57	60	61
11 A.M.	71	73	75
12 noon	75	79	81

We define x as before, but now we define y to be a matrix whose three columns contain the second, third, and fourth columns of the preceding table. The following session produces an estimate of the temperatures at 8 A.M. and 10 A.M. at each location.

```
>>x = [7, 9, 11, 12]';
>>y(:,1) = [49, 57, 71, 75]';
>>y(:,2) = [52, 60, 73, 79]';
>>y(:,3) = [54, 61, 75, 81]';
>>x_int = [8, 10]';
>>interp1(x,y,x_int)
ans =
     53.0000      56.0000      57.5000
     64.0000      65.5000      68.0000
```

Thus the estimated temperatures at 8 A.M. at each location are 53, 56, and 57.5°F, respectively. At 10 A.M. the estimated temperatures are 64, 65.5, and 68. From this example we see that if the first argument x in the interp1(x,y,x_int) function is a *vector* and the second argument y is a *matrix,* the function interpolates between the rows of y and computes a matrix having the same number of columns as y and the same number of rows as the number of values in x_int.

Note that we need not define two separate vectors x and y. Rather, we can define a single matrix that contains the entire table. For example, by defining the matrix temp to be the preceding table, the session would look like this:

```
>>temp(:,1) = [7, 9, 11, 12]';
>>temp(:,2) = [49, 57, 71, 75]';
>>temp(:,3) = [52, 60, 73, 79]';
>>temp(:,4) = [54, 61, 75, 81]';
>>x_int = [8, 10]';
>>interp1(temp(:,1),temp(:,2:4),x_int)
ans =
     53.0000      56.0000      57.5000
     64.0000      65.5000      68.0000
```

Two-Dimensional Interpolation

Now suppose that we have temperature measurements at four locations at 7 A.M. These locations are at the corners of a rectangle 1 mi wide and 2 mi long. Assigning a coordinate system origin (0, 0) to the first location, the coordinates of the other locations are (1, 0), (1, 2), and (0, 2); see Figure 7.4–2. The temperature measurements are shown in the figure. The temperature is a function of two variables, the coordinates x and y. MATLAB provides the `interp2` function to interpolate functions of two variables. If the function is written as $z = f(x, y)$ and we wish to estimate the value of z for $x = x_i$ and $y = y_i$, the syntax is `interp2(x,y,z,x_i,y_i)`.

Suppose we want to estimate the temperature at the point whose coordinates are (0.6, 1.5). Put the x coordinates in the vector `x` and the y coordinates in the vector `y`. Then put the temperature measurements in a matrix `z` such that going across a row represents an increase in x and going down a column represents an increase in y. The session to do this is as follows:

```
>>x = [0,1];
>>y = [0,2];
>>z = [49,54;53,57]
z =
   49    54
   53    57
>>interp2(x,y,z,0.6,1.5)
ans =
      54.5500
```

Thus the estimated temperature is 54.55°.

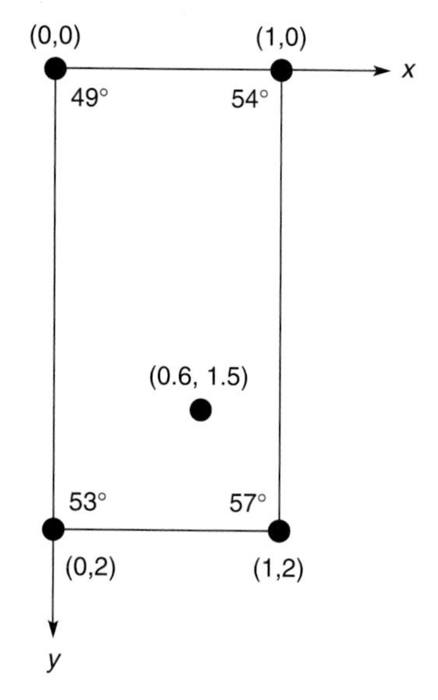

Figure 7.4–2 Temperature measurements at four locations.

Table 7.4–1 Linear interpolation functions

Command	Description
`y_int=interp1(x,y,x_int)`	Used to linearly interpolate a function of one variable: $y = f(x)$. Returns a linearly interpolated vector `y_int` at the specified value `x_int`, using data stored in `x` and `y`.
`z_int=interp2(x,y,z,x_,y_int)`	Used to linearly interpolate a function of two variables: $y = f(x, y)$. Returns a linearly interpolated vector `z_int` at the specified values `x_int` and `y_int`, using data stored in `x`, `y`, and `z`.

The syntax of the `interp1` and `interp2` functions is summarized in Table 7.4–1. MATLAB also provides the `interpn` function for interpolating multidimensional arrays.

Cubic-Spline Interpolation

We have seen that the use of high-order polynomials can exhibit undesired behavior between the data points, and this behavior can make high-order polynomials unsuitable for interpolation. An alternative procedure that is widely used is to fit the data points using a lower-order polynomial between *each pair* of adjacent data points. This method is called *spline* interpolation and is so named for the splines used by illustrators to draw a smooth curve through a set of points. Such a device can be constructed from a length of lead coated with rubber or flexible plastic. It can be bent to provide a drawing guide passing through the data points.

CUBIC SPLINES

Spline interpolation obtains an exact fit that is also smooth. The most common procedure uses cubic polynomials, called *cubic splines,* and thus is called *cubic-spline interpolation.* If the data is given as n pairs of (x, y) values, then $n - 1$ cubic polynomials are used. Each has the form

$$y_i(x) = a_i(x - x_i)^3 + b_i(x - x_i)^2 + c_i(x - x_i) + d_i$$

for $x_i \leq x \leq x_{i+1}$ and $i = 1, 2, \ldots, n - 1$. The coefficients a_i, b_i, c_i, and d_i for each polynomial are determined so that the following three conditions are satisfied for each polynomial:

1. The polynomial must pass through the data points at its endpoints at x_i and x_{i+1}.
2. The slopes of adjacent polynomials must be equal at their common data point.
3. The curvatures of adjacent polynomials must be equal at their common data point.

For example, a set of cubic splines for the temperature data given earlier follows (y represents the temperature values, and x represents the hourly values). The data is repeated here.

x	7	9	11	12
y	49	57	71	75

We will shortly see how to use MATLAB to obtain these polynomials. For $7 \leq x \leq 9$,

$$y_1(x) = -0.35(x-7)^3 + 2.85(x-7)^2 - 0.3(x-7) + 49$$

For $9 \leq x \leq 11$,

$$y_2(x) = -0.35(x-9)^3 + 0.75(x-9)^2 + 6.9(x-9) + 57$$

For $11 \leq x \leq 12$,

$$y_3(x) = -0.35(x-11)^3 - 1.35(x-11)^2 + 5.7(x-11) + 71$$

MATLAB provides the `spline` command to obtain a cubic-spline interpolation. Its syntax is `y_int = spline(x,y,x_int)`, where `x` and `y` are vectors containing the data and `x_int` is a vector containing the values of the independent variable x at which we wish to estimate the dependent variable y. The result `y_int` is a vector the same size as `x_int` containing the interpolated values of y that correspond to `x_int`. The spline fit can be plotted by plotting the vectors `x_int` and `y_int`. For example, the following session produces and plots a cubic-spline fit to the preceding data, using an increment of 0.01 in the x values.

```
>>x = [7,9,11,12];
>>y = [49,57,71,75];
>>x_int = [7:0.01:12];
>>y_int = spline(x,y,x_int);
>>plot(x,y,'o',x,y,'-- ',x_int,y_int),...
 xlabel('Time (hr)'),ylabel('Temperature (deg F)',...
 title('Temperature Measurements at a Single Location'),...
 axis([7 12 45 80])
```

The plot is shown in Figure 7.4–3. The dashed lines represent linear interpolation, and the solid curve is the cubic spline. Note that if we use the spline plot to estimate the temperature at 8 A.M., we obtain approximately 51°F. If we evaluate the spline polynomial at $x = 8$, we obtain $y(8) = 51.2$°F. In either case the estimate is different from the 53°F estimate obtained from linear interpolation. It is impossible to say which estimate is more accurate without having more understanding of the temperature dynamics.

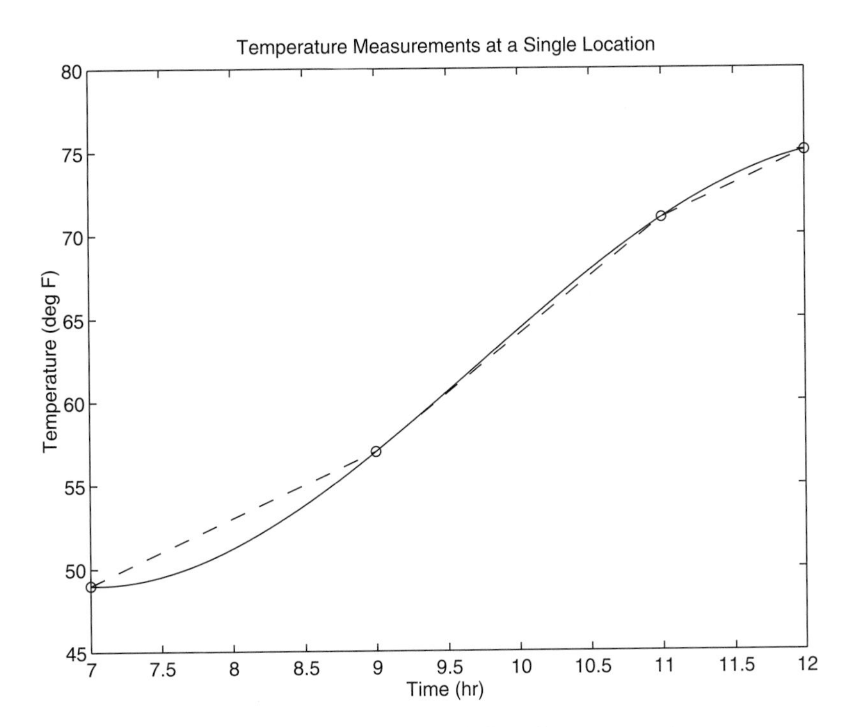

Figure 7.4–3 Linear and cubic-spline interpolation of temperature data.

Instead of plotting the cubic spline and estimating a point on the plot, we can obtain a more accurate calculation, more quickly, by using the following variation of the `interp1` function.

```
y_est = interp1(x,y,x_est,'spline')
```

In this form the function returns a column vector `y_est` that contains the estimated values of y that correspond to the x values specified in the vector `x_est`, using cubic-spline interpolation. For example, to estimate the value of y at $x = 8$ for the preceding data, we would type `interp1(x,y,8,'spline')`. MATLAB returns the answer 51.2. To estimate the value of y at two different points, say, $x = 8$ and $x = 10$, we would type `interp1(x,y,[8, 10],'spline')`. MATLAB returns the answers 51.2 and 64.3.

In some applications it is helpful to know the polynomial coefficients, but we cannot obtain the spline coefficients from the `interp1` function. However, we can use the form

```
[breaks, coeffs, m, n] = unmkpp(spline(x,y))
```

to obtain the coefficients of the cubic polynomials. The vector `breaks` contains the x values of the data, and the matrix `coeffs` is an $m \times n$ matrix containing the coefficients of the polynomials. The scalars `m` and `n` give the dimensions of the matrix `coeffs`; `m` is the number of polynomials, and `n` is the number of coefficients for each polynomial (MATLAB will fit a lower-order polynomial if possible, so there can be fewer than four coefficients). For example, using the same data, the following session produces the coefficients of the polynomials given earlier:

```
>>x = [7,9,11,12];
>>y = [49,57,71,75];
>>[breaks, coeffs, m, n] = unmkpp(spline(x,y))
breaks =
   7    9   11   12
coeffs =
   -0.3500  2.8500 -0.3000  49.0000
   -0.3500  0.7500   6.900  57.0000
   -0.3500 -1.3500  5.7000  71.0000
m =
   3
n =
   4
```

The first row of the matrix `coeffs` contains the coefficients of the first polynomial, and so on. These functions are summarized in Table 7.4–2. The Basic Fitting interface, which is available on the Tools menu of the Figure window, can be used for cubic-spline interpolation. See Chapter 5, Section 5.7 for instructions for using the interface.

Table 7.4–2 Polynomial interpolation functions

Command	Description
`y_est = interp1(x,y,x_est,'spline')`	Returns a column vector `y_est` that contains the estimated values of *y* that correspond to the *x* values specified in the vector `x_est`, using cubic-spline interpolation.
`y_int = spline(x,y,x_int)`	Computes a cubic-spline interpolation where `x` and `y` are vectors containing the data and `x_int` is a vector containing the values of the independent variable *x* at which we wish to estimate the dependent variable *y*. The result `y_int` is a vector the same size as `x_int` containing the interpolated values of *y* that correspond to `x_int`.
`[breaks, coeffs, m, n] = unmkpp(spline(x,y))`	Computes the coefficients of the cubic-spline polynomials for the data in `x` and `y`. The vector `breaks` contains the *x* values, and the matrix `coeffs` is an $m \times n$ matrix containing the polynomial coefficients. The scalars `m` and `n` give the dimensions of the matrix `coeffs`; `m` is the number of polynomials, and `n` is the number of coefficients for each polynomial.

Application to Robot Control

Cubic splines are used to control robotic devices such as the arm shown in Figure 7.4–4. This particular arm is a simple one that moves in a two-dimensional plane. It has a motor at its "shoulder" and a motor at its "elbow." When these two motors rotate through the proper angles, the arm can place its hand at a desired point in the plane. Robots have a supervisory computer that accepts user input, such as the desired location of the hand. The computer then calculates how much rotation each motor must produce to move the hand from its starting point to its desired final position. Then, at regular time intervals, the supervisory computer feeds the angular motion commands to the two smaller computers that control each motor. In many devices the supervisory computer uses cubic-spline interpolation to generate the angular motion commands. The advantage of a cubic spline is that it results in smooth robot motion. We now show how the splines are computed.

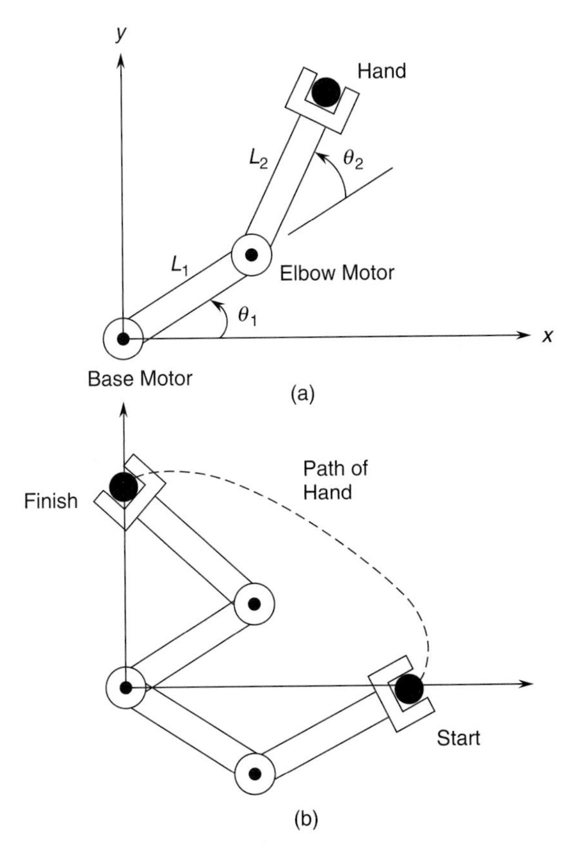

Figure 7.4–4 A robot arm having two joints. (a) Dimensions and arm angles. (b) The hand's path is not a straight line, but a complicated curve.

We can use trigonometry to obtain the following expressions for the coordinates of the elbow and the hand. For the elbow, $x_{\text{elbow}} = L_1 \cos\theta_1$ and $y_{\text{elbow}} = L_1 \sin\theta_1$. The hand's coordinates (x, y) are given by

$$x = x_{\text{elbow}} + L_2\cos(\theta_1 + \theta_2) = L_1 \cos\,\theta_1 + L_2\cos(\theta_1 + \theta_2) \quad (7.4\text{–}1)$$

$$y = y_{\text{elbow}} + L_2\sin(\theta_1 + \theta_2) = L_1 \sin\,\theta_1 + L_2\sin(\theta_1 + \theta_2) \quad (7.4\text{–}2)$$

These two equations can be solved for θ_1 and θ_2 in terms of x and y so that we can determine the elbow and arm angles necessary to place the hand at a specified position given by x and y. We omit the details of the solution and just state the results.

$$R^2 = x^2 + y^2 \quad (7.4\text{–}3)$$

$$\cos\theta_2 = \frac{R^2 - L_1^2 - L_2^2}{2L_1L_2} \quad (7.4\text{–}4)$$

$$\cos\beta = \frac{R^2 + L_1^2 - L_2^2}{2L_1R} \quad (7.4\text{–}5)$$

$$\alpha = \arctan\frac{y}{x} \quad (7.4\text{–}6)$$

$$\theta_1 = \begin{cases} \alpha + \beta & \text{if } \theta_2 < 0 \\ \alpha - \beta & \text{if } \theta_2 \geq 0 \end{cases} \quad (7.4\text{–}7)$$

If we use a cubic polynomial to express the joint angles as functions of time for $0 \leq t \leq T$, then we can specify four conditions for each angle. Two of these conditions should require the polynomial to pass through the starting value $\theta(0)$ and the ending value $\theta(T)$. The other two conditions require that the slope of the polynomial should be 0 at the start and finish because we want the robot to start from rest and to finish at rest. The cubic polynomial that satisfies these conditions is

$$\theta(t) = at^3 + bt^2 + \theta(0) \quad (7.4\text{–}8)$$

where $a = 2[\theta(0) - \theta(T)]/T^3$ and $b = -3[\theta(0) - \theta(T)]/T^2$.

For example, suppose the robot arm shown in Figure 7.4–4a has the lengths $L_1 = 4$ and $L_2 = 3$ ft. Suppose also that in 2 sec we want the hand to move from $x(0) = 6.5$, $y(0) = 0$ to $x(2) = 0$, $y(2) = 6.5$ ft. We obtain the following angle solutions from equations (7.4–3) through (7.4–7): $\theta_1(0) = -18.717°$, $\theta_1(2) = 71.283°$, and $\theta_2(0) = \theta_2(2) = 44.049°$. The polynomials are $\theta_1(t) = -22.5t^3 + 67.5t^2 - 18.717$ and $\theta_2(t) = 44.049$. If the robot controller uses these polynomials to generate commands to the motors, then the path of the hand looks like that shown in Figure 7.4–4b. Note that the hand's path is not a straight line between the starting and stopping points, and this motion might result in a collision with nearby objects.

In many applications we want to control the hand's path more precisely. For example, the robot might be used for welding along a specific path. Suppose we want the path to be a straight line. Then we must specify a series of intermediate locations for the hand to pass through. These locations must lie on the straight line connecting the starting and ending hand positions. These points are called *knot points.* To obtain smooth motion, we want the hand to pass through these points without stopping or changing speed, so we use cubic splines to interpolate between these points. If the hand is initially at the point (x_0, y_0) and we want it to move to (x_f, y_f) in a straight line, the line's equation is

KNOT POINT

$$y = \frac{y_f - y_0}{x_f - x_0}(x - x_0) + y_0 \tag{7.4–9}$$

Robot Path Control Using Three Knot Points

EXAMPLE 7.4–1

Write a MATLAB program to compute the arm angle solution for three knot points. Write another program to compute the splines required to generate three knot points and to plot the path of the robot's hand. Do this problem for the case where $L_1 = 4$, $L_2 = 3$, $(x_0, y_0) = (6, 0)$, $(x_f, y_f) = (0, 4)$

■ Solution

The following script file performs these calculations. It solves the specific case where $L_1 = 4$, $L_2 = 3$, $(x_0, y_0) = (6, 0)$, $(x_f, y_f) = (0, 4)$, using three equally spaced knot points along the straight-line path.

```
% Solution for joint angles using n knot points.
% Enter the arm lengths below.
L1 = 4;L2 = 3;
% Enter the specific values of the starting point (x0,y0),
% the stopping point (xf,yf), and the number of knot points n.
x0 = 6;y0 = 0;xf = 0;yf = 4;n = 3;
% Obtain the points on a straight line from (x0,y0) to (xf,yf).
x = linspace(x0,xf,n+1);y = ((yf-y0)/(xf-x0))*(x-x0)+y0;
% Begin the angle solution. The answers will be in degrees.
R = sqrt(x.^2+y.^2);
theta2 = acos((R.^2-L1^2-L2^2)/(2*L1*L2))*(180/pi);
beta = acos((R.^2+L1^2-L2^2)/(2*L1*R))*(180/pi);
alpha = atan2(y,x)*(180/pi);
if theta2 < 0
     theta1 = alpha+beta;
else
     theta1 = alpha-beta;
end
```

The results are `theta1 = [-18.7170, -17.2129, 4.8980, 35.9172, 71.2830]` and `theta2 = [44.0486, 86.6409, 99.2916, 86.6409, 44.0486]`.

The second program uses these angles to compute the splines and plot the hand's path. The plot is shown in Figure 7.4–5. The hand's path is very close to a straight line, as desired.

```
% Example of robot path using splines with three knot points.
% Define the joint angles and times along the straight line path.
theta1 = [-18.717, -17.2129, 4.8980, 35.9172, 71.2830];
```

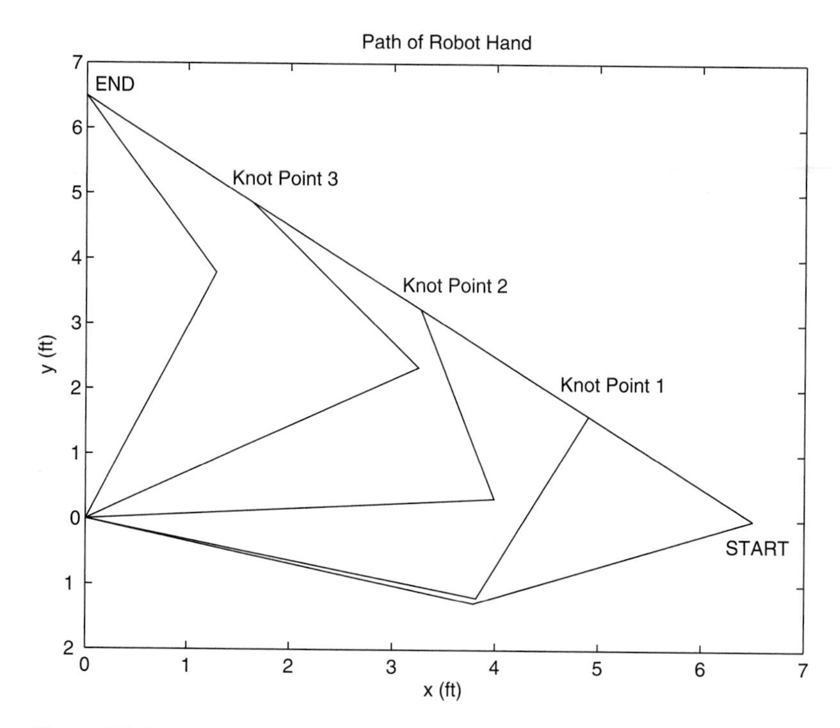

Figure 7.4–5 Path of a robot hand using three knot points.

```
theta2 = [44.0486, 86.6409, 99.2916, 86.6409, 44.0486];
t = [0, 0.5, 1, 1.5, 2];
% Use 200 points for interpolation.
t_i = linspace(0,2,200);
theta1_i = spline(t,theta1,t_i);
theta2_i = spline(t,theta2,t_i);
% Compute the (x,y) coordinates of the hand.
x = 4*cos(theta1_i*(pi/180))+3*cos((theta1_i+theta2_i)*(pi/180));
y = 4*sin(theta1_i*(pi/180))+3*sin((theta1_i+theta2_i)*(pi/180));
% Compute the (x,y) coordinates of the elbow.
x_elbow = [4*cos(theta1_i*(pi/180))];
y_elbow = [4*sin(theta1_i*(pi/180))];
% Compute the coordinates to display the arm's position at 5 points.
x1 = [0,x_elbow(1),x(1)];y1 = [0,y_elbow(1),y(1)];
x2 = [0,x_elbow(50),x(50)];y2 = [0, y_elbow(50),y(50)];
x3 = [0,x_elbow(100),x(100)];y3 = [0,y_elbow(100),y(100)];
x4 = [0,x_elbow(150),x(150)];y4 = [0,y_elbow(150),y(150)];
x5 = [0,x_elbow(200),x(200)];y5 = [0,y_elbow(200),y(200)];
% Plot the results.
plot(x,y,x1,y1,x2,y2,x3,y3,x4,y4,x5,y5),...
axis([0 7 -2 7]),title('Path of Robot Hand'),...
xlabel('x (ft)'),ylabel('y (ft)')
```

Table 7.5–1 Guide to MATLAB functions introduced in Chapter 7

Histogram functions	See Table 7.1–1
Statistical functions	See Table 7.2–1
Random number functions	See Table 7.3–1
Linear interpolation functions	See Table 7.4–1
Polynomial interpolation functions	See Table 7.4–2

Miscellaneous functions	
Command	**Description**
`erf(x)`	Computes the error function erf(x).

7.5 Summary

This chapter introduces MATLAB functions that have widespread and important uses in statistics and data analysis; however, their proper application requires informed judgment and experience. Section 7.1 gives an introduction to basic statistics and probability, including histograms, which are specialized plots for displaying statistical results. The normal distribution that forms the basis of many statistical methods is covered in Section 7.2. Section 7.3 covers random number generators and their use in simulation programs. Section 7.4 covers interpolation methods, including linear and spline interpolation. Table 7.5–1 is a guide to the functions introduced in this chapter.

Now that you have finished this chapter, you should be able to use MATLAB to

- Solve basic problems in statistics and probability.
- Create simulations incorporating random processes.
- Apply interpolation to data.

Key Terms with Page References

Absolute frequency, 421
Bins, 419
Cubic splines, 449
Error function, 434
Gaussian function, 429
Histogram, 418
Interpolation, 444
Knot point, 455
Mean, 418
Median, 418
Normally distributed, 430
Normal function, 429
Relative frequency, 421
Scaled frequency histogram, 428
Standard deviation, 430
Uniformly distributed, 436
Variance, 431

Problems

You can find the answers to problems marked with an asterisk at the end of the text.

Section 7.1

1. The following list gives the measured gas mileage in miles per gallon for 22 cars of the same model. Plot the absolute frequency histogram and the relative frequency histogram.

23	25	26	25	27	25	24	22	23	25	26
26	24	24	22	25	26	24	24	24	27	23

2. Thirty pieces of structural timber of the same dimensions were subjected to an increasing lateral force until they broke. The measured force in pounds required to break them is given in the following list. Plot the absolute frequency histogram. Try bin widths of 50, 100, and 200 lb. Which gives the most meaningful histogram? Try to find a better value for the bin width.

243	236	389	628	143	417	205
404	464	605	137	123	372	439
497	500	535	577	441	231	675
132	196	217	660	569	865	725
457	347					

3. The following list gives the measured breaking force in newtons for a sample of 60 pieces of certain type of cord. Plot the absolute frequency histogram. Try bin widths of 10, 30, and 50 N. Which gives the most meaningful histogram? Try to find a better value for the bin width.

311	138	340	199	270	255	332	279	231	296	198	269
257	236	313	281	288	225	216	250	259	323	280	205
279	159	276	354	278	221	192	281	204	361	321	282
254	273	334	172	240	327	261	282	208	213	299	318
356	269	355	232	275	234	267	240	331	222	370	226

4.* If you roll a balanced pair of dice 300 times, how many times would you expect to obtain a sum of 8? A sum of either 3, 4, or 5? A sum of less than 9?

Section 7.2

5. For the data given in Problem 1,
   a. Plot the scaled frequency histogram.
   b. Compute the mean and standard deviation and use them to estimate the lower and upper limits of gas mileage corresponding to 68 percent of cars of this model. Compare these limits with those of the data.

6. For the data given in Problem 2,

a. Plot the scaled frequency histogram.
b. Compute the mean and standard deviation and use them to estimate the lower and upper limits of strength corresponding to 68 percent and to 96 percent of such timber pieces. Compare these limits with those of the data.

7. For the data given in Problem 3,

a. Plot the scaled frequency histogram.
b. Compute the mean and standard deviation, and use them to estimate the lower and upper limits of breaking force corresponding to 68 percent and 96 percent of cord pieces of this type. Compare these limits with those of the data.

8.* Data analysis of the breaking strength of a certain fabric shows that it is normally distributed with a mean of 200 lb and a variance of 9.

a. Estimate the percentage of fabric samples that will have a breaking strength no less than 194 lb.
b. Estimate the percentage of fabric samples that will have a breaking strength no less than 197 lb and no greater than 203 lb.

9. Data from service records shows that the time to repair a certain machine is normally distributed with a mean of 50 min and a standard deviation of 5 min. Estimate how often it will take more than 60 min to repair a machine.

10. Measurements of a number of fittings show that the pitch diameter of the thread is normally distributed with a mean of 5.007 mm and a standard deviation of 0.005 mm. The design specifications require that the pitch diameter must be 5 ± 0.01 mm. Estimate the percentage of fittings that will be within tolerance.

11. A certain product requires that a shaft be inserted into a bearing. Measurements show that the diameter d_1 of the cylindrical hole in the bearing is normally distributed with a mean of 3 cm with a variance of 0.0064. The diameter d_2 of the shaft is normally distributed with a mean of 2.96 cm and a variance of 0.0036.

a. Compute the mean and the variance of the clearance $c = d_1 - d_2$.
b. Find the probability that a given shaft will not fit into the bearing. (Hint: Find the probability that the clearance is negative.)

12.* A shipping pallet holds 10 boxes. Each box holds 300 parts of different types. The part weight is normally distributed with a mean of 1 lb and a standard deviation of 0.2 lb.

a. Compute the mean and standard deviation of the pallet weight.
b. Compute the probability that the pallet weight will exceed 3015 lb.

13. A certain product is assembled by placing three components end to end. The components' lengths are L_1, L_2, and L_3. Each component is manufactured on a different machine, so the random variation in their lengths is independent of each other. The lengths are normally distributed with means of 1, 2, and 1.5 ft and variances of 0.00014, 0.0002, and 0.0003, respectively.

a. Compute the mean and variance of the length of the assembled product.

b. Estimate what percentage of assembled products will be no less than 4.48 and no more than 4.52 ft in length.

Section 7.3

14. Use a random number generator to produce 1000 uniformly distributed numbers with a mean of 10, a minimum of 2, and a maximum of 18. Obtain the mean and the histogram of these numbers and discuss whether or not they appear uniformly distributed with the desired mean.

15. Use a random number generator to produce 1000 normally distributed numbers with a mean of 20 and a variance of 4. Obtain the mean, variance, and the histogram of these numbers and discuss whether or not they appear normally distributed with the desired mean and variance.

16. The mean of the sum (or difference) of two independent random variables equals the sum (or difference) of their means, but the variance is always the sum of the two variances. Use random number generation to verify this statement for the case where $z = x + y$, where x and y are independent and normally distributed random variables. The mean and variance of x are $\mu_x = 10$ and $\sigma_x^2 = 2$. The mean and variance of y are $\mu_y = 15$ and $\sigma_y^2 = 3$. Find the mean and variance of z by simulation and compare the results with the theoretical prediction. Do this for 100, 1000, and 5000 trials.

17. Suppose that $z = xy$, where x and y are independent and normally distributed random variables. The mean and variance of x are $\mu_x = 10$ and $\sigma_x^2 = 2$. The mean and variance of y are $\mu_y = 15$ and $\sigma_y^2 = 3$. Find the mean and variance of z by simulation. Does $\mu_z = \mu_x \mu_y$? Does $\sigma_z^2 = \sigma_x^2 \sigma_y^2$? Do this for 100, 1000, and 5000 trials.

18. Suppose that $y = x^2$, where x is a normally distributed random variable with a mean and variance of $\mu_x = 0$ and $\sigma_x^2 = 4$. Find the mean and variance of y by simulation. Does $\mu_y = \mu_x^2$? Does $\sigma_y = \sigma_x^2$? Do this for 100, 1000, and 5000 trials.

19.* Suppose you have analyzed the price behavior of a certain stock by plotting the scaled frequency histogram of the price over a number of months. Suppose that the histogram indicates that the price is normally distributed with a mean $100 and a standard deviation of $5. Write a

MATLAB program to simulate the effects of buying 50 shares of this stock whenever the price is below the $100 mean, and selling all your shares whenever the price is above $105. Analyze the outcome of this strategy over 250 days (the approximate number of business days in a year). Define the profit as the yearly income from selling stock plus the value of the stocks you own at year's end, minus the yearly cost of buying stock. Compute the mean yearly profit you would expect to make, the minimum expected yearly profit, the maximum expected yearly profit, and the standard deviation of the yearly profit. The broker charges 6 cents per share bought or sold with a minimum fee of $40 per transaction. Assume you make only one transaction per day.

20. Suppose that data shows that a certain stock price is normally distributed with a mean of $150 and a variance of 100. Create a simulation to compare the results of the following two strategies over 250 days. You start the year with 1000 shares. With the first strategy, every day the price is below $140 you buy 100 shares, and every day the price is above $160 you sell all the shares you own. With the second strategy, every day the price is below $150 you buy 100 shares, and every day the price is above $160 you sell all the shares you own. The broker charges 5 cents per share traded with a minimum of $35 per transaction.

21. Write a script file to simulate 100 plays of a game in which you flip two coins. You win the game if you get two heads, lose if you get two tails, and you flip again if you get one head and one tail. Create three user-defined functions to use in the script. Function `flip` simulates the flip of one coin, with the state `s` of the random number generator as the input argument, and the new state `s` and the result of the flip (0 for a tail and 1 for a head) as the outputs. Function `flips` simulates the flipping of two coins, and calls `flip`. The input of `flips` is the state `s`, and the outputs are the new state `s` and the result (0 for two tails, 1 for a head and a tail, and 2 for two heads). Function `match` simulates a turn at the game. Its input is the state `s`, and its outputs are the result (1 for win, 0 for lose) and the new state `s`. The script should first reset the random number generator to its initial state, compute the state `s`, and then pass this state to the user-defined functions.

22. Write a script file to play a simple number guessing game as follows. The script should generate a random integer in the range 1, 2, 3, . . . , 14, 15. It should provide for the player to make repeated guesses of the number, and should indicate if the player has won or give the player a hint after each wrong guess. The responses and hints are:

- "You won," and then stop the game.
- "Very close," if the guess is within 1 of the correct number.
- "Getting close," if the guess is within 2 or 3 of the correct number.
- "Not close," if the guess is not within 3 of the correct number.

Section 7.4

23.* Interpolation is useful when one or more data points are missing. This situation often occurs with environmental measurements, such as temperature, because of the difficulty of making measurements around the clock. The following table of temperature versus time data is missing readings at 5 and 9 hours. Use linear interpolation with MATLAB to estimate the temperature at those times.

Time (hours, P.M.)	1	2	3	4	5	6	7	8	9	10	11	12
Temperature (°C)	10	12	18	24	?	21	20	18	?	15	13	8

24. The following table gives temperature data in °C as a function of time of day and day of the week at a specific location. Data is missing for the entries marked with a question mark (?). Use linear interpolation with MATLAB to estimate the temperature at the missing points.

	Day				
Hour	**Mon**	**Tues**	**Wed**	**Thurs**	**Fri**
1	17	15	12	16	16
2	13	?	8	11	12
3	14	14	9	?	15
4	17	15	14	15	19
5	23	18	17	20	24

25. Consider the robot arm shown in Figure 7.4–4 on page 453. Suppose the arm lengths are $L_1 = 4$ and $L_2 = 3$ ft. Suppose we want the arm to start from rest at the location $x(0) = 6$, $y(0) = 0$ and stop three seconds later at the location $x(3) = 0$, $y(3) = 4$. Write a MATLAB program to compute the arm angle solution using two knot points. Write another program to compute the splines required to generate the two knot points and to plot the path of the robot's hand. Discuss whether the hand's path deviates much from a straight line.

26. Computer-controlled machines are used to cut and to form metal and other materials when manufacturing products. These machines often use cubic splines to specify the path to be cut or the contour of the part to be shaped. The following coodinates specify the shape of a certain car's front fender. Fit a series of cubic splines to the coordinates and plot the splines along with the coordinate points.

x (ft)	0	0.25	0.75	1.25	1.5	1.75	1.875	2	2.125	2.25
y (ft)	1.2	1.18	1.1	1	0.92	0.8	0.7	0.55	0.35	0

27. The following data is the measured temperature T of water flowing from a hot water faucet after it is turned on at time $t = 0$.

t (sec)	T (°F)	t (sec)	T (°F)
0	72.5	6	109.3
1	78.1	7	110.2
2	86.4	8	110.5
3	92.3	9	109.9
4	110.6	10	110.2
5	111.5		

a. Plot the data first connecting them with straight lines, and then with a cubic spline.

b. Estimate the temperature values at the following times using linear interpolation and then cubic-spline interpolation: $t = 0.6, 2.5, 4.7, 8.9$.

c. Use both the linear and cubic-spline interpolations to estimate the time it will take for the temperature to equal the following values: $T = 75, 85, 90, 105$.

Courtesy Ford Motor Company.

Engineering in the 21st Century...

Energy-Efficient Transportation

Western societies have become very dependent on transportation powered by gasoline and diesel fuel. There is some disagreement about how long it will take to exhaust these fuel resources, but it will certainly happen. Novel engineering developments in both personal and mass transportation will be needed to reduce our dependence on such fuels. These developments will be required in a number of areas such as engine design, electric motor and battery technology, lightweight materials, and aerodynamics.

A number of such initiatives are underway. The photo shows the Synergy 2010 concept car developed by Ford to explore new technologies for a circa 2010 family car. Designed for six passengers, the car is one-third lighter and 40 percent more aerodynamic than today's sleekest cars. It is an example of a hybrid-electric vehicle that has two power sources. Three devices—an internal combustion engine, a gas turbine, and a fuel cell—are candidates for the primary power source. The secondary power source might be a flywheel or a battery. Ford is presently investigating a small, 1.0-L, direct-injection, compression-ignited engine to power a generator to produce electricity for motors located at each wheel. A flywheel would collect excess engine and braking energy, to be released to supplement the engine for quick acceleration or to climb hills.

The weight reduction is to be achieved with all-aluminum unibody construction and by improved design of the engine, flywheel, radiator, and brakes to make use of advanced materials such as composites and magnesium. Other manufacturers are investigating plastic bodies made from recycled materials. The fin-shaped vertical front fenders create an air extractor for the cooling system and control the airflow along the sides to reduce drag.

Further research on the design and technology concepts and the required manufacturing processes is required to make the car affordable. MATLAB can assist engineers performing such research. ■

CHAPTER 8

Numerical Calculus and Differential Equations

OUTLINE

This chapter covers numerical methods for computing integrals and derivatives and for solving ordinary differential equations. Some integrals cannot be evaluated analytically, and we need to compute them numerically with an approximate method. In addition, it is often necessary to use data to estimate rates of change, and this process requires a numerical estimate of the derivative. Finally, many differential equations cannot be solved analytically, and so we must solve them using appropriate numerical techniques.

When you have finished this chapter, you should be able to

- Use MATLAB to numerically evaluate integrals.
- Use numerical methods with MATLAB to estimate derivatives.
- Use the analytical expressions for simple integrals and derivatives to check the accuracy of numerical methods.
- Use MATLAB's numerical differential equation solvers to obtain solutions.
- Use the analytical solutions of simple differential equations to check the accuracy of numerical methods.

8.1 Review of Integration and Differentiation

The *integral* of a function $f(x)$ for $a \leq x \leq b$ can be interpreted as the area between the $f(x)$ curve and the x-axis, bounded by the limits $x = a$ and $x = b$. Figure 8.1–1 illustrates this area. If we denote this area by A, then we can write A as

$$A = \int_a^b f(x)\,dx \tag{8.1–1}$$

INTEGRAND

The *integrand* is $f(x)$. The *lower* and *upper limits of integration* here are a and b. The symbol x here is the *variable of integration.*

Integrals are often encountered in engineering applications. Here are some examples:

Acceleration and velocity: An object's velocity $v(b)$ at time $t = b$, starting with velocity $v(0)$ at $t = 0$, is the integral of its acceleration $a(t)$:

$$v(b) = \int_0^b a(t)\,dt + v(0) \tag{8.1–2}$$

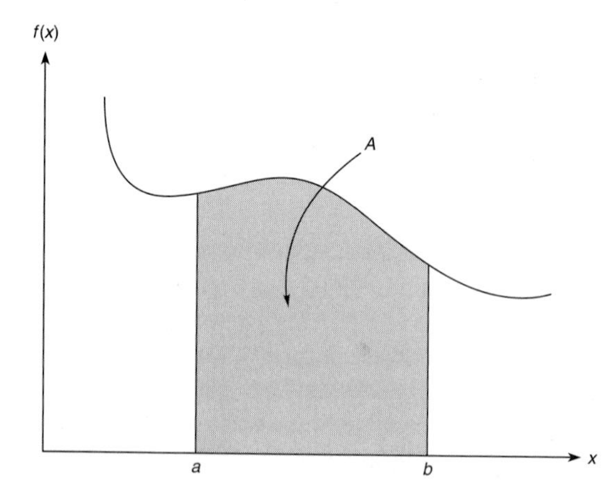

Figure 8.1–1 The area under the curve of $f(x)$ from $x = a$ to $x = b$.

Velocity and distance: The position of an object moving at velocity $v(t)$ from the time $t = a$ to the time $t = b$ and starting at position $x(a)$ at $t = a$ is

$$x(b) = \int_a^b v(t)\,dt + x(a) \tag{8.1–3}$$

Capacitor voltage and current: The charge Q across a capacitor is the integral of the current i applied to the capacitor. If a capacitor initially holds a charge $Q(a)$, then if the current $i(t)$ is applied from time $t = a$ to time $t = b$, the capacitor voltage v^- is

$$v(b) = \frac{1}{C}\left[\int_a^b i(t)\,dt + Q(a)\right] \tag{8.1–4}$$

where C is the capacitance of the capacitor.

Work expended: The mechanical work done in pushing an object a distance d is the integral

$$W = \int_0^d f(x)\,dx \tag{8.1–5}$$

where $f(x)$ is the force as a function of position x. For a linear spring $f(x) = kx$, where k is the *spring constant,* and the work done against the spring is

$$W = \int_0^d kx\,dx \tag{8.1–6}$$

In many applications one or both integral limits are variables. For example, an object's velocity at time t is the integral of its acceleration. If the object is moving at time $t = 0$ with a velocity $v(0)$, its velocity is given by

$$v(t) = \int_0^t a(t)\,dt + v(0) \tag{8.1–7}$$

Integrals have the *linearity* property. If c and d are not functions of x, then

$$\int_a^b [cf(x) + dg(x)]\,dx = c\int_a^b f(x)\,dx + d\int_a^b g(x)\,dx \tag{8.1–8}$$

Another useful property of integrals is the following:

$$\int_a^b f(x)\,dx = \int_a^c f(x)\,dx + \int_c^b f(x)\,dx \tag{8.1–9}$$

The integrals of many functions can be evaluated analytically. We can obtain a formula for the integral, and we are said to have obtained the answer in "closed form." Here are some common examples that you have probably seen before:

$$\int_a^b x^n\,dx = \left.\frac{x^{n+1}}{n+1}\right|_{x=a}^{x=b} = \frac{b^{n+1}}{n+1} - \frac{a^{n+1}}{n+1} \quad n \neq -1 \qquad (8.1\text{–}10)$$

$$\int_a^b \frac{1}{x}\,dx = \ln x\Big|_{x=a}^{x=b} = \ln b - \ln a \qquad (8.1\text{–}11)$$

$$\int_\pi^{2\pi} \sin x\,dx = -\cos x\Big|_{x=\pi}^{x=2\pi} = -[\cos 2\pi - \cos \pi] = -2 \qquad (8.1\text{–}12)$$

DEFINITE INTEGRAL

INDEFINITE INTEGRAL

The integrals we have seen thus far are called *definite* integrals because they have specified limits of integration. *Indefinite* integrals do not have the limits specified. For example, the following integral is indefinite:

$$\int \cos x\,dx = \sin x \qquad (8.1\text{–}13)$$

Not every function can be integrated analytically. For example, the following is *Fresnel's cosine integral,* which has not been evaluated analytically:

$$\int \cos x^2\,dx \qquad (8.1\text{–}14)$$

In such cases we must compute the definite integral using numerical methods. These are treated in Section 8.2.

Improper Integrals and Singularities

Some integrals have infinite values, depending on their integration limits. These are called *improper integrals*. For example, the following integral can be found in most integral tables:

$$\int \frac{1}{x-1}\,dx = \ln|x-1|$$

However, it is an improper integral if the integration limits include the point $x = 1$. To see why, consider Figure 8.1–2, which is a plot of the function $y = 1/(x-1)$. The function becomes undefined as x approaches 1 from the left or the right. To show that the integral is improper, take the limit of the integral as the upper integration limit h approaches 1 from the left.

$$\begin{aligned}\lim_{h\to 1-} \int_0^h \frac{1}{x-1}\,dx &= \lim_{h\to 1-} \ln|x-1|\big|_0^h \\ &= \lim_{h\to 1-} (\ln|h-1| - \ln|-1|) \\ &= \lim_{h\to 1-} \ln(1-h) = -\infty\end{aligned}$$

because $\ln 0 = -\infty$. So even though an integral can be found in an integral table, you should examine the integrand to check for points at which the integrand is undefined. These points are called *singularities.* If singularities lie on or within the integration limits, you need to check the limit of the integral to see whether it

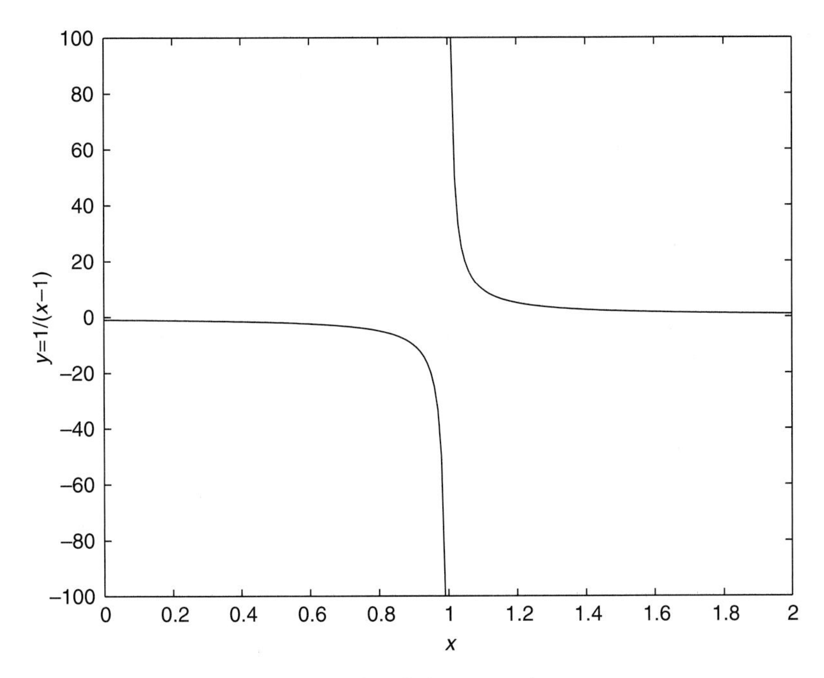

Figure 8.1–2 A function having a singularity at $x = 1$.

is improper. The same warning applies when using numerical methods to evaluate integrals. The points at which the integrand is undefined can cause problems for the numerical method.

Test Your Understanding

T8.1–1 Compute the area under the $\sin(x)$ curve from $x = 0$ to $x = \pi$. (Answer: 2.)

T8.1–2 If a rocket is launched from rest with an acceleration of 5g, *a*. how fast will it be going, and *b*. how high will it be 10 s after launch? (1g = 9.81 m/s^2) (Answer: (a) 490.5 m/s; (b) 2452.5 m.)

Derivatives

Differentiation and integration are complementary operations. For example, if

$$g(x) = \int f(x)\,dx$$

then $f(x)$ is the derivative of $g(x)$ with respect to x. This relation is written as

$$f(x) = \frac{dg}{dx} \tag{8.1–15}$$

The derivative of $f(x)$ can be interpreted geometrically as the slope of $f(x)$. Thus the derivative is sometimes called the "slope function."

A few basic rules and a short table of derivatives enable us to compute derivatives of complicated functions. The *product* rule says that if $h(x) = f(x)g(x)$, then

$$\frac{dh}{dx} = f(x)\frac{dg}{dx} + g(x)\frac{df}{dx} \tag{8.1–16}$$

The *quotient* rule states that if $h(x) = f(x)/g(x)$, then

$$\frac{dh}{dx} = \frac{g(x)\frac{df}{dx} - f(x)\frac{dg}{dx}}{g^2} \tag{8.1–17}$$

The following *chain rule* enables us to obtain derivatives by decomposing functions into basic functions whose derivatives are given in a table. If $z = f(y)$ and if $y = g(x)$, then z is indirectly a function of x, and the chain rule says that

$$\frac{dz}{dx} = \frac{dz}{dy}\frac{dy}{dx} \tag{8.1–18}$$

Here are a few examples of derivatives:

$$\frac{dx^n}{dx} = nx^{n-1} \tag{8.1–19}$$

$$\frac{d\ln x}{dx} = \frac{1}{x} \tag{8.1–20}$$

$$\frac{d\sin x}{dx} = \cos x \tag{8.1–21}$$

$$\frac{d\cos x}{dx} = -\sin x \tag{8.1–22}$$

Using this short table of derivatives and the product rule, we can show that

$$\frac{d(x^2\sin x)}{dx} = x^2\cos x + 2x\sin x$$

Use the chain rule with $z = y^2$ and $y = \sin x$ to obtain the following derivative:

$$\frac{d(\sin^2 x)}{dx} = 2y\frac{dy}{dx} = 2\sin x\cos x$$

Test Your Understanding

T8.1–3 Derive the expressions for the derivatives of the following functions with respect to x:
a. $\sin(3x)$; *b*. $\cos^2 x$; *c*. $x^3\ln x$; *d*. $(\sin x)/x^2$.
(Answers: *a*. $3\cos(3x)$; *b*. $-2\cos x\sin x$; *c*. $x^2(3\ln x + 1)$; *d*. $(x\cos x - 2\sin x)/x^3$.)

8.2 Numerical Integration

This section shows how to use MATLAB to calculate values of definite integrals using approximate methods. In Chapter 10 we show how to use MATLAB to obtain the closed-form solution of some integrals.

Trapezoidal Integration

The simplest way to find the area under a curve is to split the area into rectangles (Figure 8.2–1a). If the widths of the rectangles are small enough, the sum of their areas gives the approximate value of the integral. A more sophisticated method is to use trapezoidal elements (Figure 8.2–1b). Each trapezoid is called a *panel*. It is not necessary to use panels of the same width; to increase the method's accuracy you can use narrow panels where the function is changing rapidly. When the widths are adjusted according to the function's behavior, the method is said to be *adaptive*. MATLAB implements *trapezoidal integration* with the `trapz` function. Its syntax is `trapz(x,y)`, where the array `y` contains the function values at the points contained in the array `x`. If you want the integral of a single function, then `y` is a vector. To integrate more than one function, place their values in a matrix `y`; typing `trapz(x,y)` will compute the integral of each column of `y`.

PANEL

You cannot directly specify a function to integrate with the `trapz` function; you must first compute and store the function's values ahead of time in an array. Later we discuss two other integration functions, the `quad` and `quadl` functions, that can accept functions directly. However, they cannot handle arrays of values. So the functions complement one another. These functions are summarized in Table 8.2–1.

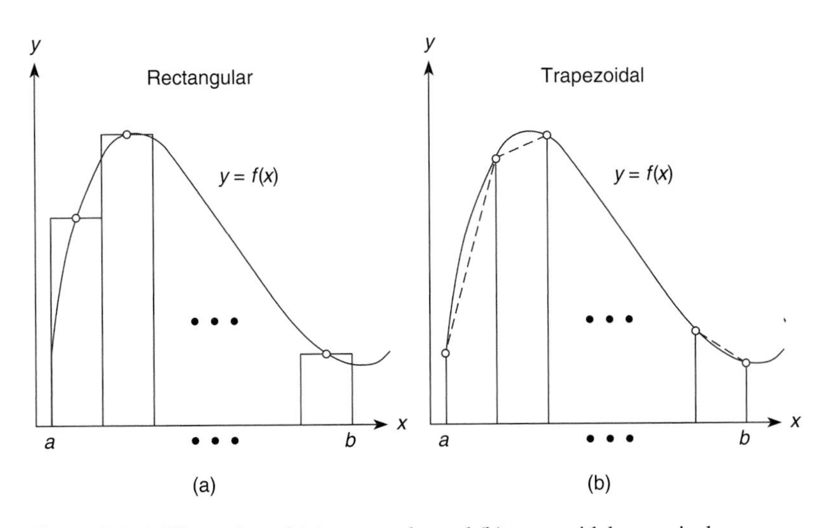

Figure 8.2–1 Illustration of (a) rectangular and (b) trapezoidal numerical integration.

Table 8.2–1 Numerical integration functions

Command	Description
`quad('function',a,b,tol)`	Uses an adaptive Simpson's rule to compute the integral of the function `'function'` with `a` as the lower integration limit and `b` as the upper limit. The parameter `tol` is optional. `tol` indicates the specified error tolerance.
`quadl('function',a,b,tol)`	Uses Lobatto quadrature to compute the integral of the function `'function'`. The rest of the syntax is identical to `quad`.
`trapz(x,y)`	Uses trapezoidal integration to compute the integral of `y` with respect to `x`, where the array `y` contains the function values at the points contained in the array `x`.

As a simple example of the use of the `trapz` function, let us compute the integral

$$\int_0^{\pi} \sin x \, dx$$

The exact answer is

$$\int_0^{\pi} \sin x \, dx = -\cos x \Big|_0^{\pi} = \cos 0 - \cos \pi = 2$$

To investigate the effect of panel width, let us first use 10 panels with equal widths of $\pi/10$. The script file is

```
x = linspace(0,pi,10);
y = sin(x);
trapz(x,y)
```

The answer is 1.9797, which gives a relative error of $100(2 - 1.9797)/2) = 1\%$. Now try 100 panels of equal width; replace the array `x` with `x = linspace (0,pi,100)`. The answer is 1.9998 for a relative error of $100(2 - 1.9998)/2 = 0.01\%$. If we examine the plot of the integrand $\sin x$, we would see that the function is changing faster near $x = 0$ and $x = \pi$ than near $x = \pi/2$. Thus we could achieve the same accuracy using fewer panels if narrower panels are used near $x = 0$ and $x = \pi$.

When numerical integration was done by hand (before World War II), it was important to use as few panels as necessary to achieve the desired accuracy. However, with the speed of modern computers, it is not difficult to find a reasonable number of panels of uniform width to achieve the required accuracy for many problems. The following method usually works: Compute the integral with a reasonable number of panels (say, 100). Then double the number of panels and compare the answers. If they are close, you have the solution. If not, continue increasing the number of panels until the answers converge to a common value. This method does not always give accurate results, but it can be tried before using variable panel widths or methods more sophisticated than trapezoidal integration.

We can use numerical integration to find the velocity when either the acceleration function $a(t)$ cannot be integrated or the acceleration is given as a table of values. The following example illustrates the latter case.

Velocity from an Accelerometer

EXAMPLE 8.2–1

An *accelerometer* measures acceleration and is used in aircraft, rockets, and other vehicles to estimate the vehicle's velocity and displacement. The accelerometer integrates the acceleration signal to produce an estimate of the velocity, and it integrates the velocity estimate to produce an estimate of displacement. Suppose the vehicle starts from rest at time $t = 0$ and that its measured acceleration is given in the following table.

(a) Estimate the velocity after 10 s.

(b) Estimate the velocity at the times $t = 1, 2, \ldots, 10$ s.

(c) Check your program by using a case that can be solved analytically.

Time (s)	0	1	2	3	4	5	6	7	8	9	10
Acceleration (m/s^2)	0	2	4	7	11	17	24	32	41	48	51

■ Solution

(a) We must use the `trapz` function here, because the acceleration is given as a table of values. We cannot use the `quad` or `quadl` functions. The relation between velocity and acceleration is

$$v(10) = \int_0^{10} a(t)\,dt + v(0) = \int_0^{10} a(t)\,dt$$

The script file follows.

```
t = [0:10];
a = [0,2,4,7,11,17,24,32,41,48,51];
vf = trapz(t,a)
```

The answer for the final velocity is given by `vf` and is 211.5 m/s.

(b) To find the velocity at the times $t = 1, 2, \ldots, 10$, we can use the fact that $v(t_1) = 0$ and write (8.1–2) as

$$v(t_k) = \int_{t_1}^{t_k} a(t)\,dt + v(t_1) = \int_{t_1}^{t_k} a(t)\,dt \tag{8.2–1}$$

for $k = 2, 3, \ldots, 11$. (Note that there are 11 values in the sequence $t = 0, 1, 2, \ldots, 10$. Thus t_{11} corresponds to $t = 10$.)

```
t = [0:10];
a = [0,2,4,7,11,17,24,32,41,48,51];
v(1) = 0;
for k = [2:11]
   v(k) = trapz(t(1:k),a(1:k));
end
disp([t',v'])
```

The preceding method uses more calculations than necessary because it does not take advantage of the velocity value calculated in the previous pass through the `for` loop. The following script file is thus more efficient. It is based on (8.1–2), which can be written as follows by substituting t_k for 0 and t_{k+1} for b.

$$v(t_{k+1}) = \int_{t_k}^{t_{k+1}} a(t)\,dt + v(t_k)$$

```
t = [0:10];
a = [0,2,4,7,11,17,24,32,41,48,51];
v(1) = 0;
for k = [1:10]
   v(k+1) = trapz(t(k:k+1),a(k:k+1))+v(k);
end
disp([t',v'])
```

For either method the answers are those given in the following table:

Time (s)	0	1	2	3	4	5	6	7	8	9	10
Velocity (m/s)	0	1	4	9.5	18.5	32.5	53	81	117	162	211.5

(c) Because the trapezoidal rule uses straight-line segments to connect the data points, it will give the exact solution for any panel width when the integrand is a linear function. So we can use the linear acceleration function $a(t) = t$ to test the program. The velocity is given by

$$v(t) = \int_0^t t\,dt = \left.\frac{t^2}{2}\right|_0^t = \frac{t^2}{2}$$

The velocity at $t = 10$ is 50. The following script file can be used to check the method used in part (b):

```
t = [0:10];
a = t;
v(1) = 0;
for k = [1:10]
   v(k+1) = trapz(t(k:k+1),a(k:k+1))+v(k);
end
disp(v(11))
```

When this file is run it produces the answer `v(11) = 50`, which is correct.

Test Your Understanding

T8.2–1 Modify the above script file to estimate the displacement at the times $t = 1, 2, \ldots, 10$ s.
(Partial answer: The displacement after 10 s is 584.25 m.)

Quadrature Functions

As we have just seen, when the integrand is a linear function (one having a plot that is a straight line), trapezoidal integration gives the exact answer. However, if the integrand is not a linear function, then the trapezoidal representation will be inexact. We can represent the function's curve by quadratic functions to obtain more accuracy. This approach is taken with *Simpson's rule,* which divides the integration range $b - a$ into an even number of sections and uses a different quadratic for each pair of adjacent panels. A quadratic function has three parameters, and Simpson's rule computes these parameters by requiring the quadratic to pass through the function's three points corresponding to the two adjacent panels. To obtain more accuracy, we can use polynomials of a degree higher than two.

MATLAB function `quad` implements an adaptive version of Simpson's rule, while the `quadl` function is based on an adaptive Lobatto integration algorithm. The term *quad* is an abbreviation of *quadrature,* which is an old term for the process of measuring areas. The syntax of both functions is identical and is summarized in Table 8.2–1. In its basic form, the syntax is `quad('function',a,b)`, where `'function'` is the name of the function representing the integrand, `a` is the lower integration limit, and `b` is the upper limit. To illustrate, let us compute an integral we already know.

$$\int_0^{\pi} \sin x \, dx = 2$$

The session consists of one command:

```
>>A = quad('sin',0,pi)
```

The answer given by MATLAB is 2, which is correct. We use `quadl` the same way; namely, `A = quadl('sin',0,pi)`.

Although the `quad` and `quadl` functions are more accurate than `trapz`, they are restricted to computing the integrals of functions and cannot be used when the integrand is specified by a set of points. Thus we could not have used `quad` and `quadl` in Example 8.2–1.

Slope Function Singularities

In addition to singularities of the integrand, another condition can cause problems for numerical integration methods. This occurs when the slope of the integrand becomes infinite either on or within the integration limits; that is, the *slope function has a singularity.* A simple example of this is the square root function $y = \sqrt{x}$, shown in the top graph of Figure 8.2–2. The slope of this function is its derivative, which is $dy/dx = 0.5/\sqrt{x}$. This slope is plotted in the bottom graph of the figure. Note that it becomes infinite at $x = 0$.

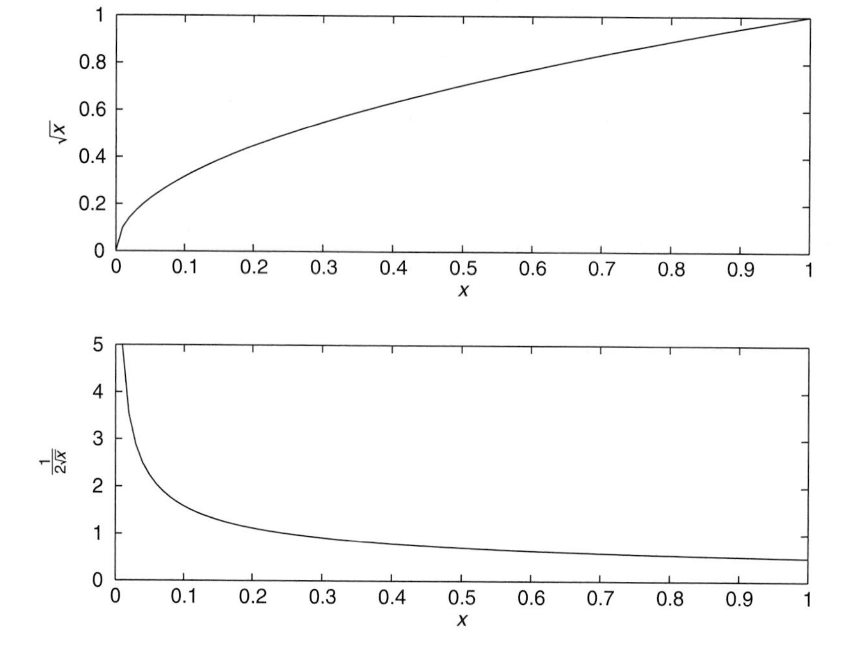

Figure 8.2–2 A function having a singularity in its slope function. The top graph shows the function $y = \sqrt{x}$. The bottom graph shows the derivative of $y = \sqrt{x}$. The slope has a singularity at $x = 0$.

Let us use all three integration functions and compare their performance in computing the following integral:

$$\int_0^1 \sqrt{x}\,dx$$

To use the trapz function, we must evaluate the integrand at a chosen number of points. The number of points determines the panel width. Here we will use a spacing of 0.01, which requires 100 panels. The script file follows.

```
A1 = quad('sqrt',0,1)
A2 = quadl('sqrt',0,1)
x = [0:0.01:1];
y = sqrt(x);
A3 = trapz(x,y)
```

The answers are $A_1 = A_2 = 0.6667$ and $A_3 = 0.6665$. The correct answer, which can be obtained analytically, is $2/3 = 0.6667$ to four decimal places. The quad and quadl functions can produce a message warning that a singularity might be present; however, they do not produce such a message here. This indicates that these functions are capable of dealing with this singularity. The trapz is not affected by the slope singularity here, and it gives a reasonably accurate answer.

You can use quad and quadl to integrate user-defined functions, as shown in the following example. Because the quad and quadl functions call the integrand function using vector arguments, you must always use array operations when defining the function. The following example shows how this is done.

Evaluation of Fresnel's Cosine Integral

EXAMPLE 8.2–2

Some simple looking integrals cannot be evaluated in closed form. An example is Fresnel's cosine integral:

$$\int_0^b \cos x^2 \, dx \tag{8.2–2}$$

Compute the integral when the upper limit is $b = \sqrt{2\pi}$.

■ Solution

Plot $\cos x^2$ and its slope versus x for $0 \le x \le \sqrt{2\pi}$ and use the plots to check for any singularities that might cause problems for the integration function. Use a small enough step size to identify such points. Using 300 plotting points, the plot shows no problems.

Use the quad function to evaluate the integral. Define the integrand with a user-defined function as shown by the following function file:

```
function c2 = cossq(x)
% cosine squared function.
c2 = cos(x.^2);
```

Note that we must use array exponentiation. The quad function is called as follows: `quad('cossq',0,sqrt(2*pi))`. The result is 0.6119.

The quad functions have some optional arguments. For example, one syntax of the quad function is

```
quad('function',a,b,tol)
```

where tol indicates the specified error tolerance. The function iterates until the relative error is less than tol. The default value of tol is 15*eps.

Test Your Understanding

T8.2–2 Use both the quad and quadl functions to compute the integral

$$\int_2^5 \frac{1}{x} \, dx$$

and compare the answers with that obtained from the closed-form solution.

T8.2–3 Use a tolerance of 0.001 to integrate the square root function from 0 to 1 by typing `quad('sqrt',0,1,0.001)`. Do the same using the quadl function, and compare with the results obtained with the default tolerance.

8.3 Numerical Differentiation

As we have seen, the derivative of a function can be interpreted graphically as the slope of the function. This interpretation leads to methods for computing the derivative numerically. Numerical differentiation must be performed when we do not have the function represented as a formula that can be differentiated using the rules presented in Section 8.1. Two major types of applications require numerical differentiation. In the first type, data has been collected and must be analyzed afterward using *postprocessing* to find rates of change. In the second type, the rates must be estimated in *real time* as the measurements are made. This application occurs in control systems. For example, an aircraft autopilot needs to estimate the rate of change of pitch angle to control the aircraft properly. Numerical differentiation with postprocessing can use all of the data and need not be particularly fast. However, real-time numerical differentiation requires a fast algorithm that can use only the data measured up to the current time. These two requirements place a heavier demand on real-time algorithms as compared to postprocessing algorithms.

Here we will introduce some simple algorithms for computing the derivative numerically. Consider Figure 8.3–1, which shows three data points that represent a function $y(x)$. Recall that the definition of the derivative is

$$\frac{dy}{dx} = \lim_{\Delta x \to 0} \frac{\Delta y}{\Delta x} \tag{8.3–1}$$

The success of numerical differentiation depends heavily on two factors: the spacing of the data points and the scatter present in the data due to measurement error. The greater the spacing, the more difficult it is to estimate the derivative. We assume here that the spacing between the measurements is regular; that is, $x_3 - x_2 = x_2 - x_1 = \Delta x$. Suppose we want to estimate the derivative dy/dx at

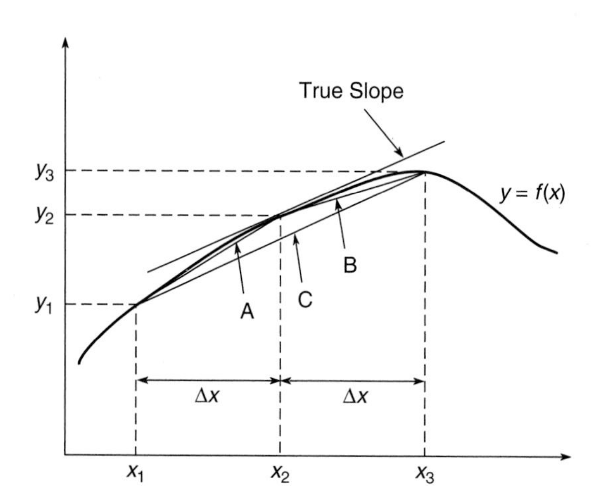

Figure 8.3–1 Illustration of methods for estimating the derivative dy/dx.

the point x_2. The correct answer is the slope of the straight line passing through the point (x_2, y_2), but we do not have a second point on that line, so we cannot find its slope. Therefore, we must estimate the slope by using nearby data points. One estimate can be obtained from the straight line labeled A in the figure. Its slope is

$$m_A = \frac{y_2 - y_1}{x_2 - x_1} = \frac{y_2 - y_1}{\Delta x} \tag{8.3–2}$$

BACKWARD DIFFERENCE

This estimate of the derivative is called the *backward difference* estimate and is actually a better estimate of the derivative at $x = x_1 + (\Delta x)/2$, rather than at $x = x_2$. Another estimate can be obtained from the straight line labeled B. Its slope is

$$m_B = \frac{y_3 - y_2}{x_3 - x_2} = \frac{y_3 - y_2}{\Delta x} \tag{8.3–3}$$

FORWARD DIFFERENCE

This estimate is called the *forward difference* estimate and is a better estimate of the derivative at $x = x_2 + (\Delta x)/2$, rather than at $x = x_2$. Examining the plot, you might think that the average of these two slopes would provide a better estimate of the derivative at $x = x_2$ because the average tends to cancel out the effects of measurement error. The average of m_A and m_B is

$$m_C = \frac{m_A + m_B}{2} = \frac{1}{2}\left(\frac{y_2 - y_1}{\Delta x} + \frac{y_3 - y_2}{\Delta x}\right) = \frac{y_3 - y_1}{2\Delta x} \tag{8.3–4}$$

CENTRAL DIFFERENCE

This is the slope of the line labeled C, which connects the first and third data points. This estimate of the derivative is called the *central difference* estimate.

The diff Function

MATLAB provides the `diff` function to use for computing derivative estimates. Its syntax is `d = diff(x)`, where `x` is a vector of values, and the result is a vector `d` containing the differences between adjacent elements in `x`. That is, if `x` has n elements, `d` will have $n - 1$ elements, where $d = [x(2) - x(1), x(3) - x(2), \ldots, x(n) - x(n-1)]$. For example, if `x = [5, 7, 12, -20]`, then `diff(x)` returns the vector `[2, 5, -32]`.

Let us compare the backward difference and central difference methods by considering a sinusoidal signal that is measured 51 times during one half-period. The measurements are in error by a uniformly distributed error between −0.025 and 0.025. Figure 8.3–2 shows the data and the underlying sine curve. The following script file implements the two methods. The results are shown in Figure 8.3–3. Clearly the central difference method does better in this example.

```
% Comparison of numerical derivative algorithms.
x = [0:pi/50:pi];
n = length(x);
% true derivative
td = cos(x);
% generating function with +/-0.025 random error.
y = sin(x) + 0.05*(rand(1,51)-0.5);
```

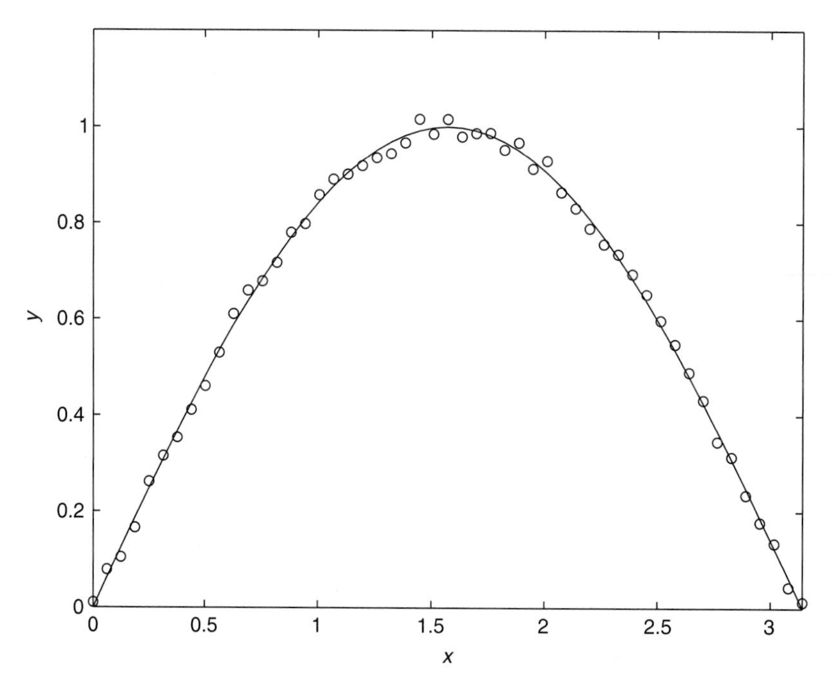

Figure 8.3–2 Measurements of a sine function containing uniformly distributed random errors between −0.025 and 0.025.

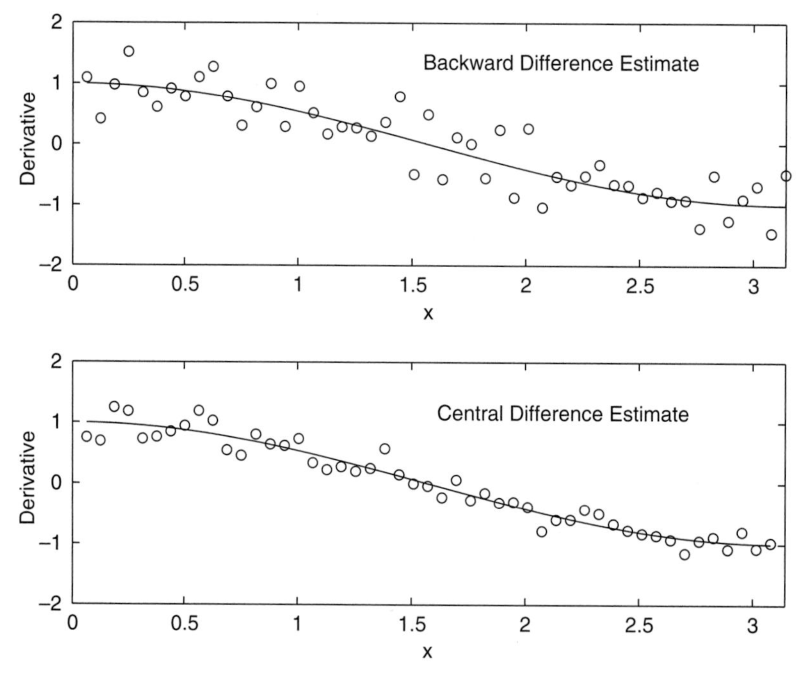

Figure 8.3–3 Comparison of backward difference and central difference methods for the data shown in Figure 8.3–2.

```
% backward difference
d1 = diff(y)./diff(x);
subplot(2,1,1)
plot(x(2:n),td(2:n),x(2:n),d1,'o'),xlabel('x'),...
ylabel('Derivative'),axis([0 pi -2 2]),...
gtext('Backward Difference Estimate')
% central difference
d2 = (y(3:n)-y(1:n-2))./(x(3:n)-x(1:n-2));
subplot(2,1,2)
plot(x(2:n-1),td(2:n-1),x(2:n-1),d2,'o'),xlabel('x'),...
ylabel('Derivative'),axis([0 pi -2 2]),...
gtext('Central Difference Estimate')
```

Many more-advanced numerical differentiation procedures have been developed; for example, a central difference method using four points instead of two is commonly used. Some algorithms are suitable only for postprocessing, whereas others have been developed specifically for real-time applications. Postprocessing algorithms are covered in advanced texts dealing with data analysis. Real-time algorithms are covered in signal-processing texts.

Test Your Understanding

T8.3–1 Modify the previous program to use the forward difference method to estimate the derivative. Plot the results and compare with the results from the backward and central difference methods.

Polynomial Derivatives

MATLAB provides the `polyder` function to compute the derivative of a polynomial. The derivative of

$$f(x) = a_1x^n + a_2x^{n-1} + a_3x^{n-2} + \cdots + a_{n-1}x + a_n$$

is

$$\frac{df}{dx} = na_1x^{n-1} + (n-1)a_2x^{n-2} + \cdots + a_{n-1}$$

$$= b_1x^{n-1} + b_2x^{n-2} + \cdots + b_{n-1}$$

Because polynomial derivatives can be obtained from the preceding formula, the `polyder` function is technically not a numerical differentiation operation. Its syntax has several forms. The basic form is

```
b = polyder(p)
```

where `p` is a vector whose elements are the coefficients of the polynomial arranged in descending powers; that is, $p = [a_1, a_2, \ldots, a_n]$. The output of `polyder` is the vector `b` containing the coefficients of the derivative; that is,

$b = [b_1, b_2, \ldots, b_{n-1}]$. The second syntax form is

```
b = polyder(p1,p2)
```

This form computes the derivative of the *product* of the two polynomials `p1` and `p2`. The third form is

```
[num, den] = polyder(p2,p1)
```

This form computes the derivative of the *quotient* p_2/p_1. The vector of coefficients of the numerator of the derivative is given by `num`. The denominator is given by `den`.

Here are some examples of the use of `polyder`. Let $p_1 = 5x + 2$ and $p_2 = 10x^2 + 4x - 3$. Then

$$\frac{dp_2}{dx} = 20x + 4$$

$$p_1 p_2 = 50x^3 + 40x^2 - 7x - 6$$

$$\frac{d(p_1 p_2)}{dx} = 150x^2 + 80 - 7$$

$$\frac{d(p_2/p_1)}{dx} = \frac{50x^2 + 40x + 18}{25x^2 + 20x + 4}$$

These results can be obtained as follows:

```
p1 = [5, 2];p2 = [10, 4, -3];
der2 = polyder(p2)
prod = polyder(p1,p2)
[num, den] = polyder(p2,p1)
```

The results are `der2 = [20, 4]`,`prod = [150, 80, -7]`,`num = [50, 40, 23]`, and `den = [25, 20, 4]`. The numerical differentiation functions are summarized in Table 8.3–1.

Table 8.3–1 Numerical differentiation functions

Command	Description
`d = diff(x)`	Returns a vector `d` containing the differences between adjacent elements in the vector `x`.
`b = polyder(p)`	Returns a vector `b` containing the coefficients of the derivative of the polynomial represented by the vector `p`.
`b = polyder(p1,p2)`	Returns a vector `b` containing the coefficients of the polynomial that is the derivative of the product of the polynomials represented by `p1` and `p2`.
`[num, den] = polyder(p2,p1)`	Returns the vectors `num` and `den` containing the coefficients of the numerator and denominator polynomials of the derivative of the quotient p_2/p_1, where `p1` and `p2` are polynomials.

8.4 Analytical Solutions to Differential Equations

In this section we introduce some important concepts and terminology associated with differential equations, and we develop analytical solutions to some differential equations commonly found in engineering applications. These solutions will give us insight into the proper use of numerical methods for solving differential equations. They also give us some test cases to use to check our programs.

Solution by Direct Integration

An *ordinary differential equation (ODE)* is an equation containing ordinary derivatives of the dependent variable. An equation containing partial derivatives with respect to two or more independent variables is a partial differential equation (PDE). Solution methods for PDEs are an advanced topic, and we will not treat them in this text.

A simple example of an ODE is the equation

$$\frac{dy}{dt} = t^2 \tag{8.4–1}$$

Here the dependent variable is y, and t is the independent variable. We can solve for y by integrating both sides of the equation with respect to the independent variable t.

$$\int_0^t \frac{dy}{dt}\,dt = \int_0^t t^2\,dt = \left.\frac{t^3}{3}\right|_0^t = \frac{t^3}{3}$$

The solution is

$$y(t) = y(0) + \frac{t^3}{3} \tag{8.4–2}$$

You can always check your answer by substituting it into the differential equation and evaluating the solution at $t = 0$. Try this method for the preceding solution.

It will be convenient to use the following abbreviated "dot" notation for derivatives.

$$\dot{y}(t) = \frac{dy}{dt} \tag{8.4–3}$$

$$\ddot{y}(t) = \frac{d^2y}{dt^2} \tag{8.4–4}$$

Oscillatory Forcing Function

Now consider the following equation:

$$\frac{dy}{dt} = f(t) \tag{8.4–5}$$

The function $f(t)$ is sometimes called the *forcing function* because it "forces" the solution to behave with a certain pattern. Let us see what this pattern is if the forcing function is sinusoidal: $f(t) = \sin \omega t$. We can solve this case with the technique used earlier:

$$\int_0^t \frac{dy}{dt}\, dt = \int_0^t \sin \omega t \, dt = -\left.\frac{\cos \omega t}{\omega}\right|_0^t = \frac{1 - \cos \omega t}{\omega}$$

or

$$y(t)\Big|_0^t = y(t) - y(0) = \frac{1 - \cos \omega t}{\omega}$$

The solution is

$$y(t) = y(0) + \frac{1 - \cos \omega t}{\omega} \qquad (8.4\text{–}6)$$

A Second-Order Equation

The *order* of a differential equation is the order of its highest derivative. Thus (8.4–1) is a first-order equation. The following is a second-order equation:

$$\frac{d^2y}{dt^2} = t^3 \qquad (8.4\text{–}7)$$

To solve it we must integrate twice. Integrating once gives

$$\int_0^t \frac{d^2y}{dt^2}\, dt = \frac{dy}{dt} - \dot{y}(0) = \int_0^t t^3\, dt = \frac{t^4}{4}$$

or

$$\frac{dy}{dt} = \frac{t^4}{4} + \dot{y}(0)$$

Integrating once more gives

$$\int_0^t \frac{dy}{dt}\, dt = y(t) - y(0) = \int_0^t \left[\frac{t^4}{4} + \dot{y}(0)\right] dt = \frac{t^5}{20} + t\dot{y}(0)$$

The solution is $y(t) = t^5/20 + t\dot{y}(0) + y(0)$. Note that because the ODE is second order, we need to specify two initial condition values to complete the solution; one of these is the value of the derivative at $t = 0$.

Substitution Method for First-Order Equations

Consider the differential equation

$$\tau \frac{dy}{dt} + y = f(t) \qquad (8.4\text{–}8)$$

where τ is a constant and $f(t)$ is a given function. Linear equations can often be solved with the trial solution form $y(t) = Ae^{st}$. Note that $dy/dt = sAe^{st}$. Substitute this form into the differential equation with $f(t) = 0$ to obtain

$$\tau \frac{dy}{dt} + y = \tau s A e^{st} + A e^{st} = 0$$

For the solution to be general, Ae^{st} cannot be 0 and thus we can cancel it out of the equation to obtain $\tau s + 1 = 0$. This equation is called the *characteristic equation,* and its root $s = -1/\tau$ is the *characteristic root*. To find A, we evaluate the solution form at $t = 0$. This evaluation gives $y(0) = Ae^0 = A$. Thus the solution is

CHARACTERISTIC ROOT

$$y(t) = y(0)e^{-t/\tau} \tag{8.4–9}$$

This solution is called the *free response* because it describes the behavior or response of the process when the *forcing function* $f(t)$ is 0; that is, when the process is "free" of the influence of $f(t)$. The solution $y(t)$ decays with time if $\tau > 0$. It starts at $y(0)$ when $t = 0$, equals $0.02y(0)$ at $t = 4\tau$, and equals $0.01y(0)$ at $t = 5\tau$. Thus τ gives an indication of how fast $y(t)$ decays, and τ is called the *time constant.*

FREE RESPONSE

TIME CONSTANT

Now suppose that $f(t) = 0$ for $t < 0$ and suddenly increases to the constant value M at $t = 0$. Such a function is called a *step* function because its plot looks like a single stair step. The height of the step is M. The solution form for this case is $y(t) = Ae^{st} + B$. The initial condition gives $B = y(0) - A$, and thus $y(t) = Ae^{st} + y(0) - A$. Substituting this into the differential equation, we find that $\tau s + 1 = 0$ and $A = y(0) - M$. The solution for $y(t)$ is

$$y(t) = y(0)e^{-t/\tau} + M\left(1 - e^{-t/\tau}\right) \tag{8.4–10}$$

The *forced response* is the term $M\left(1 - e^{-t/\tau}\right)$, which is due to the forcing function. Thus we see that the *total response* for this equation is the sum of the free and the forced responses. Figure 8.4–1 shows the free and total response for the case where $\tau = 0.1$, $y(0) = 2$, and $M = 10$. Note that the free response is essentially 0 (1 percent of its initial value) for $t > 5\tau = 0.5$. Note also that the total response is essentially constant for $t > 5\tau = 0.5$. Thus the time constant tells us how long it takes for the free response to disappear and how long it takes for the total response to reach steady state.

FORCED RESPONSE

TOTAL RESPONSE

Nonlinear Equations

Nonlinear ODEs can be recognized by the fact that the dependent variable or its derivatives appear raised to a power or in a transcendental function. For example, the following equations are nonlinear:

$$y\ddot{y} + 5\dot{y} + y = 0$$

$$\dot{y} + \sin y = 0$$

$$\dot{y} + \sqrt{y} = 0$$

Because of the great variety of possible nonlinear equation forms, no general solution method exists for them. Each class must be treated separately.

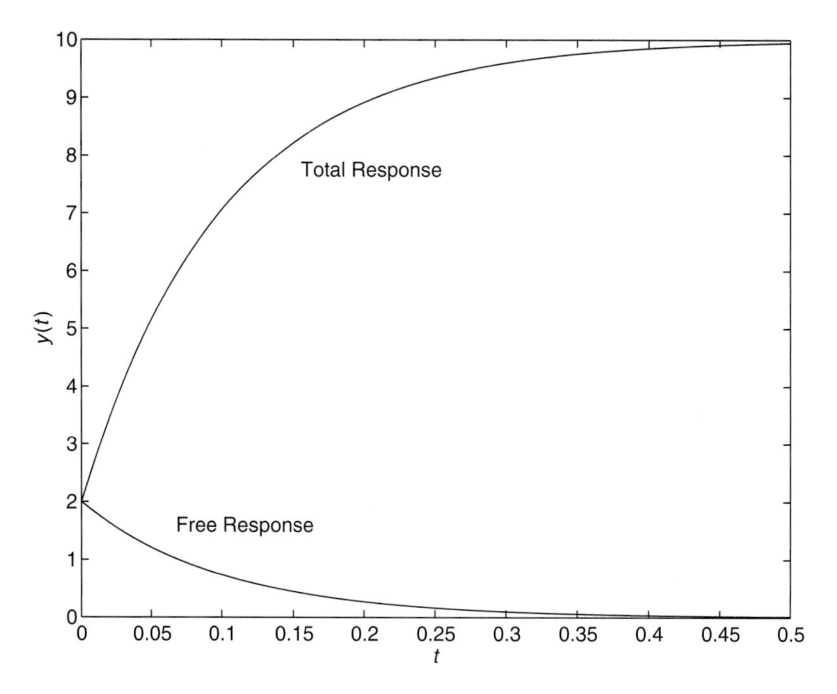

Figure 8.4–1 Free and total step response of the equation $0.1\dot{y} + y = 10$, $y(0) = 2$.

Substitution Method for Second-Order Equations

Consider the second-order equation

$$\ddot{y} - c^2 y = 0 \tag{8.4–11}$$

If we substitute $y(t) = Ae^{st}$ into this equation, we obtain

$$\left(s^2 - c^2\right) Ae^{st} = 0$$

which is satisfied for all values of t only if $s^2 - c^2 = 0$. This gives two values for the unknown constant s; namely, $s = \pm c$. The general solution is

$$y(t) = A_1 e^{ct} + A_2 e^{-ct} \tag{8.4–12}$$

UNSTABLE EQUATION

STABLE EQUATION

Note that the solution becomes infinite as $t \to \infty$ regardless of whether c is positive or negative. If the free response becomes infinite, the equation is said to be *unstable*. If the free response "dies out" (becomes 0), the equation is said to be *stable*. We need two initial conditions to determine the coefficients A_1 and A_2. For example, suppose that $c = 2$, and we are told that $y(0) = 6$ and $\dot{y}(0) = 4$. Then from (8.4–12), $y(0) = 6 = A_1 + A_2$, and $\dot{y}(0) = 2A_1 - 2A_2 = 4$. These two equations have the solutions $A_1 = 4$, $A_2 = 2$.

The following second-order equation is similar to (8.4–11) except that the coefficient of y is positive.

$$\ddot{y} + \omega^2 y = 0 \tag{8.4–13}$$

Substituting $y(t) = Ae^{st}$ into this equation, we find that the general solution is

$$y(t) = A_1 e^{i\omega t} + A_2 e^{-i\omega t} \tag{8.4–14}$$

This solution is difficult to interpret until we use *Euler's identities*

$$e^{\pm i\omega t} = \cos \omega t \pm i \sin \omega t \tag{8.4–15}$$

If we substitute these two identities into (8.4–14) and collect terms, we would find that the solution has the form

$$y(t) = B_1 \sin \omega t + B_2 \cos \omega t$$

where B_1 and B_2 are constants that depend on the initial conditions and are $B_1 = \dot{y}(0)/\omega$ and $B_2 = y(0)$. The solution is

$$y(t) = \frac{\dot{y}(0)}{\omega} \sin \omega t + y(0) \cos \omega t \tag{8.4–16}$$

The solution oscillates with constant amplitude and a frequency of ω radians per unit time. The *period P* of the oscillation is the time between adjacent peaks and is related to the frequency as follows:

$$P = \frac{2\pi}{\omega} \tag{8.4–17}$$

The frequency f in cycles per unit time is given by $f = 1/P$.

The following equation is often used as a model of structural vibrations and some types of electric circuits.

$$m\ddot{y} + c\dot{y} + ky = f(t) \tag{8.4–18}$$

Suppose for now that $f(t) = 0$. Substituting $y(t) = Ae^{st}$, we obtain

$$\left(ms^2 + cs + k\right) Ae^{st} = 0$$

which is satisfied for all values of t only if $s^2 + cs + k = 0$. The characteristic roots here can fall into one of the following three categories:

1. Real and distinct: s_1 and s_2.
2. Real and equal: s_1.
3. Complex conjugates: $s = a \pm i\omega$.

In the first case, the solution form is

$$y(t) = A_1 e^{s_1 t} + A_2 e^{s_2 t} \tag{8.4–19}$$

For the second case,

$$y(t) = (A_1 + A_2 t)\, e^{s_1 t} \tag{8.4–20}$$

For the third case,

$$y(t) = B_1 e^{at} \sin \omega t + B_2 e^{at} \cos \omega t \tag{8.4–21}$$

These solutions can be obtained with the same methods used to solve the earlier equations. The values of the constants A_i and B_i depend on the initial conditions. Let us look at four specific cases.

1. *Real, distinct roots:* Suppose that $m = 1$, $c = 8$, and $k = 15$. The characteristic roots are $s = -3, -5$. The form of the free response is

$$y(t) = A_1 e^{-3t} + A_2 e^{-5t}$$

   The equation has two time constants, which are the negative reciprocals of the roots. They are $\tau_1 = 1/3$ and $\tau_2 = 1/5$. Note that the term e^{-5t} disappears first and that it corresponds to the smallest time constant. The solution is essentially 0 after the term e^{-3t} disappears. This term corresponds to the largest time constant. The time constant of this term is $\tau = 1/3$, so for most practical purposes the solution will be 0 after $5\tau = 5/3$.

2. *Complex roots:* Suppose that $m = 1$, $c = 10$, and $k = 601$. The characteristic roots are $s = -5 \pm 24i$. The form of the free response is

$$y(t) = B_1 e^{-5t} \sin 24t + B_2 e^{-5t} \cos 24t \tag{8.4–22}$$

   This solution will oscillate at a frequency of 24 radians per unit time, which corresponds to a period $P = 2\pi/24 = \pi/12$. The oscillations will disappear when the term e^{-5t} disappears. The time constant of this term is $\tau = 1/5$, so for most practical purposes the oscillations will disappear after $5\tau = 5/5 = 1$. Thus we should see approximately $(5/5)/(\pi/12) \approx 4$ cycles of the oscillations before they die out.

3. *Unstable case, complex roots:* Suppose that $m = 1$, $c = -4$, and $k = 20$. The characteristic roots are $s = 2 \pm 4i$. The form of the free response is

$$y(t) = B_1 e^{2t} \sin 4t + B_2 e^{2t} \cos 4t \tag{8.4–23}$$

   This solution will oscillate at a frequency of 4 radians per unit time, which corresponds to a period $P = 2\pi/4 = \pi/2$. Because the term e^{2t} increases with time, the oscillation amplitude also increases. Because the free response continues to increase, this case is said to be "unstable." You can recognize an unstable linear equation by the fact that at least one of its characteristic roots will have a positive real part.

4. *Unstable case, real roots:* Suppose that $m = 1$, $c = 3$, and $k = -10$. The characteristic roots are $s = 2$ and $s = -5$. The form of the free response is

$$y(t) = A_1 e^{2t} + A_2 e^{-5t}$$

   The solution will become infinite as $t \to \infty$ because of the e^{2t} term. Thus the equation is unstable.

Figure 8.4–2 shows the response for each case the initial conditions $y(0) = 1$, $\dot{y}(0) = 0$. In the next section we will show how to obtain these plots.

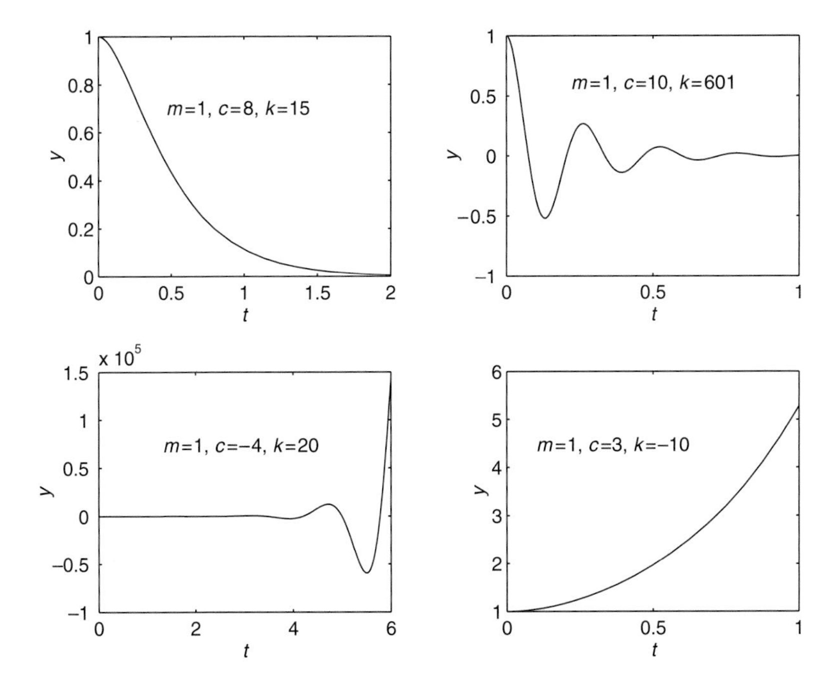

Figure 8.4–2 The free responses for the four cases discussed in the text.

Test Your Understanding

T8.4–1 Find the form of the free response of the following equations:

a. $\ddot{y} + 11\dot{y} + 28y = 0$
b. $\ddot{y} + 6\dot{y} + 34y = 0$
c. $\ddot{y} - 2\dot{y} - 15y = 0$
d. $\ddot{y} + 6\dot{y} - 40y = 0$

Summary

The solutions obtained in this section can be used to check the results of a numerical solution technique. In addition, these solutions have also pointed out the following facts that will be helpful for properly using the numerical techniques presented in the next section.

1. For certain types of differential equations, called linear equations, the characteristic polynomial can be found by making the substitution $y(t) = Ae^{st}$.
2. If any of the characteristic roots has a positive real part, the equation is unstable. If all the roots have negative real parts, the equation is stable.
3. If the equation is stable, the time constants can be found from the negative reciprocal of the real parts of the characteristic roots.

4. The equation's largest time constant indicates how long the solution takes to reach steady state.
5. The equation's smallest time constant indicates how fast the solution changes with t.
6. The frequency of oscillation of the free response can be found from the imaginary parts of the characteristic roots.
7. The rate of change of the forcing function affects the rate of change of the solution. In particular, if the forcing function oscillates, the solution of a linear equation will also oscillate and at the same frequency.
8. The number of initial conditions needed to obtain the solution equals the order of the equation.

8.5 Numerical Methods for Differential Equations

It is not always possible to obtain the closed-form solution of a differential equation. In this section we introduce numerical methods for solving differential equations. First we treat first-order equations, and in the next section we show how to extend the techniques to higher-order equations. The essence of a numerical method is to convert the differential equation into a difference equation that can be programmed on a calculator or digital computer. Numerical algorithms differ partly as a result of the specific procedure used to obtain the difference equations. In general, as the accuracy of the approximation is increased, so is the complexity of the programming involved. Understanding the concept of *step size* and its effects on solution accuracy is important. To provide a simple introduction to these issues, we begin with the simplest numerical method, the *Euler method.*

The Euler Method

The *Euler method* is the simplest algorithm for numerical solution of a differential equation. It usually gives the least accurate results but provides a basis for understanding more sophisticated methods. Consider the equation

$$\frac{dy}{dt} = r(t)y \tag{8.5–1}$$

where $r(t)$ is a known function. From the definition of the derivative,

$$\frac{dy}{dt} = \lim_{\Delta t \to 0} \frac{y(t + \Delta t) - y(t)}{\Delta t}$$

If the time increment Δt is small enough, the derivative can be replaced by the approximate expression

$$\frac{dy}{dt} \approx \frac{y(t + \Delta t) - y(t)}{\Delta t} \tag{8.5–2}$$

Use (8.5–2) to replace (8.5–1) by the following approximation:

$$\frac{y(t + \Delta t) - y(t)}{\Delta t} = r(t)y(t)$$

or

$$y(t + \Delta t) = y(t) + r(t)y(t)\Delta t \tag{8.5–3}$$

Assume that the right side of (8.5–1) remains constant over the time interval $(t, t + \Delta t)$. Then equation (8.5–3) can be written in more convenient form as follows:

$$y(t_{k+1}) = y(t_k) + r(t_k)y(t_k)\Delta t \tag{8.5–4}$$

where $t_{k+1} = t_k + \Delta t$. The smaller Δt is, the more accurate are our two assumptions leading to (8.5–4). This technique for replacing a differential equation with a difference equation is the Euler method. The increment Δt is called the *step size*. The Euler method for the general first-order equation $\dot{y} = f(t, y)$ is

STEP SIZE

$$y(t_{k+1}) = y(t_k) + \Delta t f[t_k, y(t_k)] \tag{8.5–5}$$

This equation can be applied successively at the times t_k by putting it in a `for` loop. For example, the following script file solves the differential equation $\dot{y} = ry$ and plots the solution over the range $0 \le t \le 0.5$ for the case where $r = -10$ and the initial condition is $y(0) = 2$. The time constant is $\tau = -1/r = 0.1$, and the true solution is $y(t) = 2e^{-10t}$. To illustrate the effect of the step size on the solution's accuracy, we use a step size $\Delta t = 0.02$, which is 20 percent of the time constant.

```
r = -10; delta = 0.02; y(1) = 2;
k = 0;
for time = [delta:delta:0.5]
   k = k + 1;
   y(k+1) = y(k) + r*y(k)*delta;
end
t = [0:delta:0.5];
y_true = 2*exp(-10*t);
plot(t,y,'o',t,y_true),xlabel('t'),ylabel('y')
```

Figure 8.5–1 shows the results. The numerical solution is shown by the small circles. The true solution is shown by the solid line. There is some noticeable error. If we had used a step size equal to 5 percent of the time constant, the error would not be noticeable on the plot.

Numerical methods have their greatest errors when trying to obtain solutions that are rapidly changing. Rapid changes can be due to small time constants or to oscillations. To illustrate the difficulties caused by an oscillating solution, consider the following equation

$$\dot{y} = \sin t \tag{8.5–6}$$

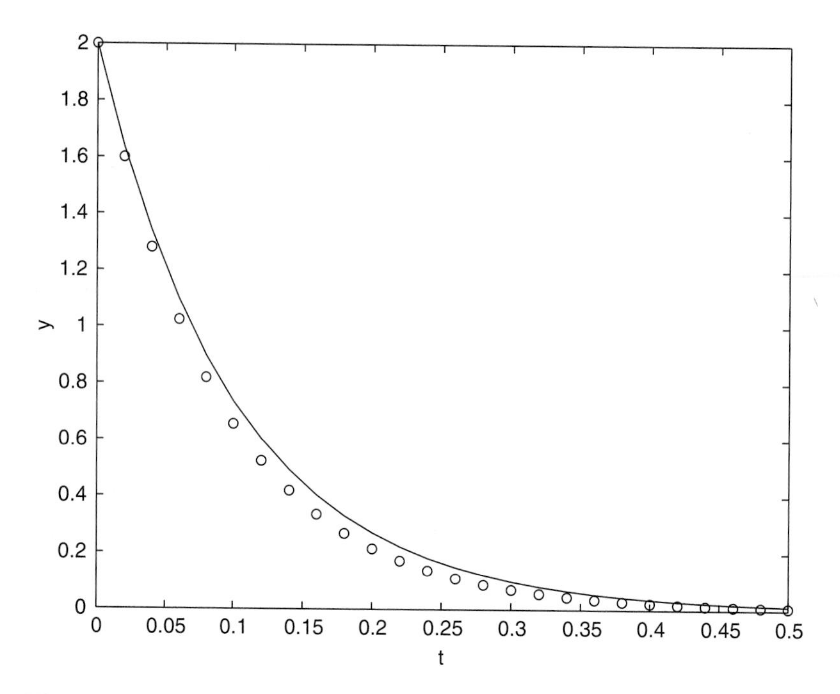

Figure 8.5–1 Euler method solution for the free response of $\dot{y} = -10y$, $y(0) = 2$.

for $y(0) = 0$ and $0 \le t \le 4\pi$. The true solution is $y(t) = 1 - \cos t$, and its period is 2π. To compare the results with those obtained from the `ode23` function later on, we will use a step size equal to $1/13$ of the period, or $\Delta t = 2\pi/13$. The script file is

```
delta = 2*pi/13; y(1) = 0;
k = 0;
for time = [delta:delta:4*pi]
   k = k + 1;
   y(k+1) = y(k) + sin(time-delta)*delta;
end
t_true = [0:delta/10:4*pi];
y_true = 1 - cos(t_true);
t = [0:delta:4*pi];
plot(t,y,'o',t_true,y_true),xlabel('t'),ylabel('y')
```

The results are shown in Figure 8.5–2, where the numerical solution is shown by the small circles. There is noticeable error, especially near the peaks and valleys, where the solution is rapidly changing.

The accuracy of the Euler method can be improved by using a smaller step size. However, very small step sizes require longer runtimes and can result in a large accumulated error because of round-off effects. So we seek better algorithms to use for more challenging applications.

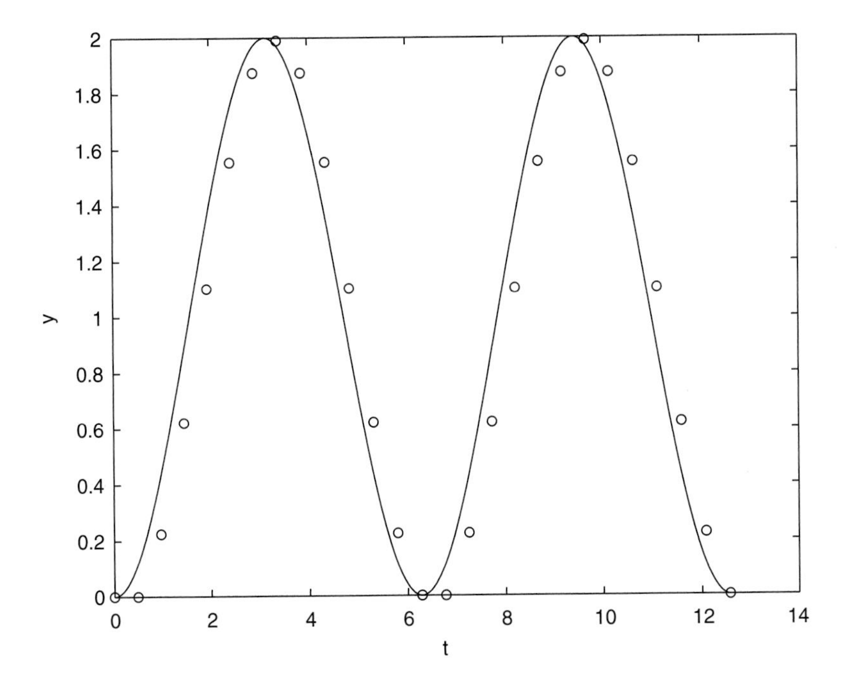

Figure 8.5–2 Euler method solution of $\dot{y} = \sin t$, $y(0) = 0$.

The Predictor-Corrector Method

We now consider two methods that are more powerful than the Euler method. The first is a *predictor-corrector* method. The second method is the *Runge-Kutta* family of algorithms. In this section we apply these techniques to first-order equations. We then extend them to higher-order equations in the next section.

The Euler method can have a serious deficiency in problems where the variables are rapidly changing because the method assumes the variables are constant over the time interval Δt. One way of improving the method is to use a better approximation to the right side of the equation

$$\frac{dy}{dt} = f(t, y) \tag{8.5–7}$$

The Euler approximation is

$$y(t_{k+1}) = y(t_k) + \Delta t f[t_k, y(t_k)] \tag{8.5–8}$$

Suppose instead we use the average of the right side of (8.5–7) on the interval (t_k, t_{k+1}). This gives

$$y(t_{k+1}) = y(t_k) + \frac{\Delta t}{2}(f_k + f_{k+1}) \tag{8.5–9}$$

where

$$f_k = f[t_k, y(t_k)] \tag{8.5–10}$$

with a similar definition for f_{k+1}. Equation (8.5–9) is equivalent to integrating (8.5–7) with the trapezoidal rule.

The difficulty with (8.5–9) is that f_{k+1} cannot be evaluated until $y(t_{k+1})$ is known, but this is precisely the quantity being sought. A way out of this difficulty is by using the Euler formula (8.5–8) to obtain a preliminary estimate of $y(t_{k+1})$. This estimate is then used to compute f_{k+1} for use in (8.5–9) to obtain the required value of $y(t_{k+1})$.

The notation can be changed to clarify the method. Let $h = \Delta t$ and $y_k = y(t_k)$, and let x_{k+1} be the estimate of $y(t_{k+1})$ obtained from the Euler formula (8.5–8). Then, by omitting the t_k notation from the other equations, we obtain the following description of the predictor-corrector process:

$$\text{Euler predictor: } x_{k+1} = y_k + hf(t_k, y_k) \tag{8.5–11}$$

$$\text{Trapezoidal corrector: } y_{k+1} = y_k + \frac{h}{2}[f(t_k, y_k) + f(t_{k+1}, x_{k+1})] \tag{8.5–12}$$

MODIFIED EULER METHOD

This algorithm is sometimes called the *modified Euler method*. However, note that any algorithm can be tried as a predictor or a corrector. Thus many other methods can be classified as predictor-corrector. For purposes of comparison with the Runge-Kutta methods, we can express the modified Euler method as

$$g_1 = hf(t_k, y_k) \tag{8.5–13}$$

$$g_2 = hf(t_k + h, y_k + g_1) \tag{8.5–14}$$

$$y_{k+1} = y_k + \tfrac{1}{2}(g_1 + g_2) \tag{8.5–15}$$

For example, the following script file solves the differential equation $\dot{y} = ry$ and plots the solution over the range $0 \le t \le 0.5$ for the case where $r = -10$ and the initial condition is $y(0) = 2$. The time constant is $\tau = -1/r = 0.1$, and the true solution is $y(t) = 2e^{-10t}$. To illustrate the effect of the step size on the solution's accuracy, we use a step size $\Delta t = 0.02$, which is 20 percent of the time constant.

```
r = -10; delta = 0.02; y(1) = 2;
k = 0;
for time = [delta:delta:0.5]
   k = k + 1;
   x(k+1) = y(k) + delta*r*y(k);
   y(k+1) = y(k) + (delta/2)*(r*y(k) + r*x(k+1));
end
t = [0:delta:0.5];
y_true = 2*exp(-10*t);
plot(t,y,'o',t,y_true),xlabel('t'),ylabel('y')
```

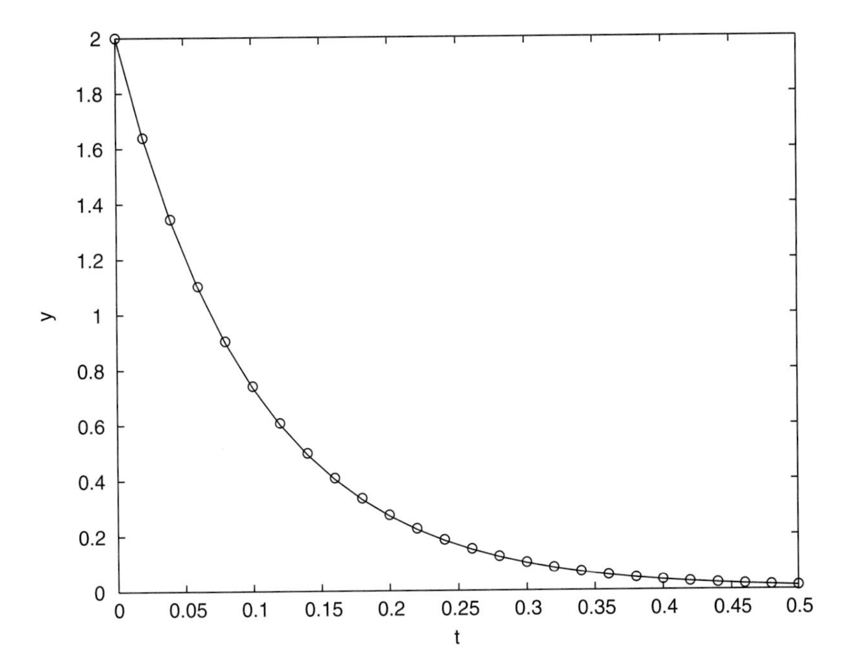

Figure 8.5–3 Modified Euler solution of $\dot{y} = -10y$, $y(0) = 2$.

Figure 8.5–3 shows the results, with the numerical solution shown by the small circles and the true solution by the solid line. There is less error than with the Euler method using the same step size.

To illustrate how the modified Euler method works with an oscillating solution, consider the equation $\dot{y} = \sin t$ for $y(0) = 0$ and $0 \le t \le 4\pi$. To compare the results with those obtained with the Euler method, we will use a step size $\Delta t = 2\pi/13$. Note that because the right side of the ODE is not a function of y, we do not need the Euler predictor (8.5–11). The script file is

```
delta = 2*pi/13; y(1)=0;
k = 0;
for time = [delta:delta:4*pi]
   k = k + 1;
   y(k+1) = y(k) + (delta/2)*(sin(time-delta) + sin(time));
end
t_true = [0:delta/10:4*pi];
y_true = 1 - cos(t_true);
t = [0:delta:4*pi];
plot(t,y,'o',t_true,y_true),xlabel('t'),ylabel('y')
```

The results are shown in Figure 8.5–4. The error is less than with the Euler method, but there is still some noticeable error near the peaks, where the solution is rapidly changing. This error can be decreased by decreasing the step size.

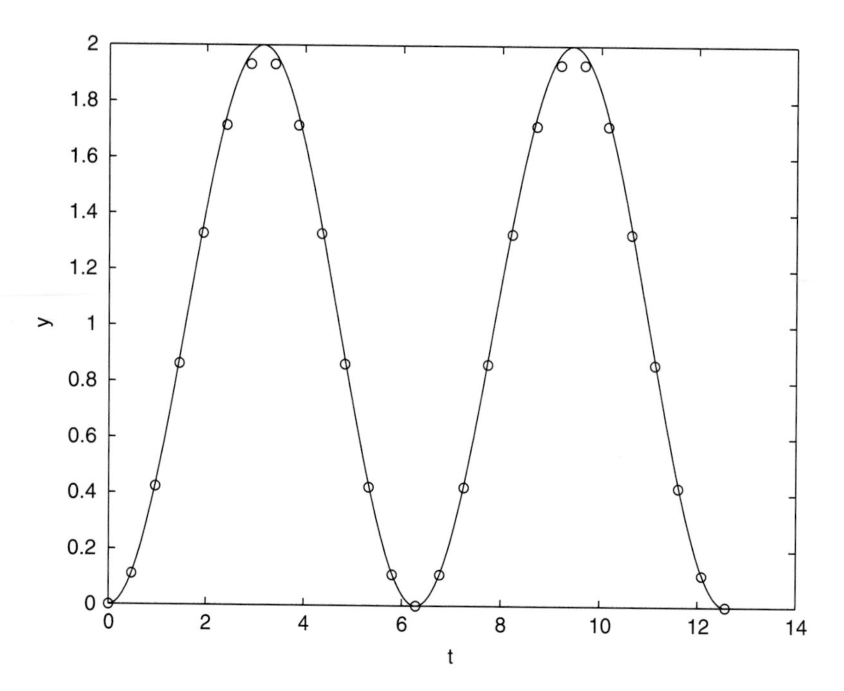

Figure 8.5–4 Modified Euler solution of $\dot{y} = \sin t$, $y(0) = 0$.

Test Your Understanding

T8.5–1 Use MATLAB to compare the solutions for $0 \leq t \leq 2$ obtained with the Euler and modified Euler methods for the following equation:

$$10\frac{dy}{dt} + y = te^{-2t} \quad y(0) = 2$$

Compare the results with the analytical solution $y(t) = (732e^{-0.1t} - 19te^{-2t} - 10e^{-2t})/361$.

Runge-Kutta Methods

The Taylor series representation forms the basis of several methods for solving differential equations, including the *Runge-Kutta methods.* The Taylor series may be used to represent the solution $y(t+h)$ in terms of $y(t)$ and its derivatives as follows.

$$y(t+h) = y(t) + h\dot{y}(t) + \tfrac{1}{2}h^2\ddot{y}(t) + \cdots \qquad (8.5\text{–}16)$$

The number of terms kept in the series determines its accuracy. The required derivatives are calculated from the differential equation. If these derivatives can be found, (8.5–16) can be used to march forward in time. In practice, the high-order derivatives can be difficult to calculate, and the series (8.5–16) is truncated at some term. The Runge-Kutta methods were developed because of the difficulty in computing the derivatives. These methods use several evaluations of the function $f(t, y)$ in a way that approximates the Taylor series. The number of terms in the series that is duplicated determines the order of the Runge-Kutta method. Thus a fourth-order Runge-Kutta algorithm duplicates the Taylor series through the term involving h^4.

The second-order Runge-Kutta methods express y_{k+1} as

$$y_{k+1} = y_k + w_1 g_1 + w_2 g_2 \tag{8.5–17}$$

where w_1 and w_2 are constant weighting factors and

$$g_1 = hf(t_k, y_k) \tag{8.5–18}$$

$$g_2 = hf(t_k + \alpha h, y_k + \beta h f_k) \tag{8.5–19}$$

The family of second-order Runge-Kutta algorithms is categorized by the parameters α, β, w_1, and w_2. To duplicate the Taylor series through the h^2 term, these coefficients must satisfy the following:

$$w_1 + w_2 = 1 \tag{8.5–20}$$

$$w_1 \alpha = \frac{1}{2} \tag{8.5–21}$$

$$w_2 \beta = \frac{1}{2} \tag{8.5–22}$$

Thus one of the parameters can be chosen independently.

The family of fourth-order Runge-Kutta algorithms expresses y_{k+1} as

$$y_{k+1} = y_k + w_1 g_1 + w_2 g_2 + w_3 g_3 + w_4 g_4 \tag{8.5–23}$$

$$\begin{aligned} g_1 &= hf(t_k, y_k) \\ g_2 &= hf(t_k + \alpha_1 h, y_k + \alpha_1 g_1) \\ g_3 &= hf[t_k + \alpha_2 h, y_k + \beta_2 g_2 + (\alpha_2 - \beta_2) g_1] \\ g_4 &= hf[t_k + \alpha_3 h, y_k + \beta_3 g_2 + \gamma_3 g_3 + (\alpha_3 - \beta_3 - \gamma_3) g_1] \end{aligned} \tag{8.5–24}$$

Comparison with the Taylor series yields eight equations for the 10 parameters. Thus two parameters can be chosen in light of other considerations. A number of different choices have been used. For example, the classical method, which reduces to Simpson's rule for integration if $f(t, y)$ is a function of only t, uses the following set of parameters:

$$\begin{aligned} w_1 &= w_4 = 1/6 \\ w_2 &= w_3 = 1/3 \\ \alpha_1 &= \alpha_2 = 1/2 \\ \beta_2 &= 1/2 \\ \gamma_3 &= \alpha_3 = 1 \\ \beta_3 &= 0 \end{aligned} \tag{8.5–25}$$

MATLAB ODE Solvers `ode23` and `ode45`

In addition to the many variations of the predictor-corrector and Runge-Kutta algorithms that have been developed, some more-advanced algorithms use a variable step size. These algorithms use larger step sizes when the solution is changing more slowly. MATLAB provides functions, called *solvers,* that implement Runge-Kutta methods with variable step size. These are the `ode23`, `ode45`, and `ode113` functions. The `ode23` function uses a combination of second- and third-order Runge-Kutta methods, whereas `ode45` uses a combination of fourth- and fifth-order methods. In general, `ode45` is faster and more accurate, but it uses larger step sizes that can produce a solution plot that is not as smooth as the plot produced with `ode23`. These solvers are classified as low order and medium order, respectively. The solver `ode113` is based on a variable-order algorithm.

MATLAB contains four additional solvers; these are `ode23t`, `ode15s`, `ode23s`, and `ode23tb`. These and the other solvers are categorized in Table 8.5–1. Some of these solvers are classified as "stiff." A stiff solver is one that is well-suited for solving *stiff equations,* which are described in Section 8.8.

Solver Syntax

In this section we limit our coverage to first-order equations. Solution of higher-order equations, where **y** is a vector, is covered in Section 8.6. When used to solve the equation $\dot{y} = f(t, y)$, the basic syntax is (using `ode23` as the example):

```
[t,y] = ode23('ydot', tspan, y0)
```

where `ydot` is the name of the function file whose inputs must be t and y and whose output must be a column vector representing dy/dt; that is, $f(t, y)$. The number of rows in this column vector must equal the order of the equation. The syntax for `ode23`, `ode45`, and `ode113` is identical. The vector `tspan` contains the starting and ending values of the independent variable t, and optionally, any intermediate values of t where the solution is desired. For example, if no intermediate values are specified, `tspan` is `[t0, tf]`, where `t0` and `tf` are the desired starting and ending values of the independent parameter t. As another example, using `tspan = [0, 5, 10]` tells MATLAB to find the solution at $t = 5$ and at $t = 10$. You can solve equations backward in time by specifying `t0` to be greater than `tf`. The parameter `y0` is the value $y(t_0)$. The function file

Table 8.5–1 ODE solvers

Solver name	Description
`ode23`	Nonstiff, low-order solver.
`ode45`	Nonstiff, medium-order solver.
`ode113`	Nonstiff, variable-order solver.
`ode23s`	Stiff, low-order solver.
`ode23t`	Moderately stiff, trapezoidal-rule solver.
`ode23tb`	Stiff, low-order solver.
`ode15s`	Stiff, variable-order solver.

Table 8.5–2 Basic syntax of ODE solvers

Command	Description
`[t, y] = ode23('ydot', tspan, y0)`	Solves the vector differential equation $\dot{\mathbf{y}} = \mathbf{f}(t, \mathbf{y})$ specified in the function file `ydot`, whose inputs must be t and y and whose output must be a *column* vector representing $d\mathbf{y}/dt$; that is, $\mathbf{f}(t, \mathbf{y})$. The number of rows in this column vector must equal the order of the equation. The vector `tspan` contains the starting and ending values of the independent variable t, and optionally, any intermediate values of t where the solution is desired. The vector `y0` contains $\mathbf{y}(t_0)$. The function file must have two input arguments `t` and `y` even for equations where $\mathbf{f}(t, \mathbf{y})$ is not a function of t. The syntax is identical for the other solvers.

must have two input arguments `t` and `y` even for equations where $f(t, y)$ is not a function of t. You need not use array operations in the function file because the ODE solvers call the file with scalar values for the arguments. The basic syntax of ODE solvers is summarized in Table 8.5–2.

As a first example of using a solver, let us solve an equation whose solution is known in closed form so that we can make sure we are using the method correctly.

Response of an RC Circuit — EXAMPLE 8.5–1

The model of the RC circuit shown in Figure 8.5–5 can be found from Kirchhoff's voltage law and conservation of charge.

$$RC\frac{dy}{dt} + y = v(t)$$

Suppose the value of RC is 0.1 s. Use a numerical method to find the free response for the case where the applied voltage v is 0 and the initial capacitor voltage is $y(0) = 2$ V. Compare the results with the analytical solution.

■ Solution

The equation for the circuit becomes

$$0.1\dot{y} + y = 0$$

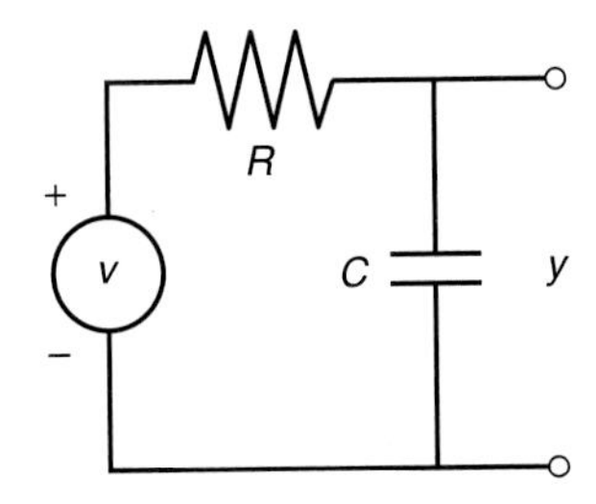

Figure 8.5–5 An RC circuit.

First solve this for $\dot{y}$:

$$\dot{y} = -10y$$

Next define the following function file. Note that the order of the input arguments must be t and y.

```
function ydot = rccirc(t,y)
% An RC circuit model with no applied voltage.
ydot = -10*y;
```

The initial time is $t = 0$, so set `t0` to be 0. The time constant is 0.1, so the response will be 2 percent of its initial value at $t = 4(0.1) = 0.4$ s and 1 percent at $t = 5(0.1) = 0.5$ s. So we can choose `tf` to be between 0.4 and 0.5 s, depending on how much of the response we wish to see. The analytical solution is $y(t) = 2e^{-10t}$. The function is called as follows, and the solution plotted along with the analytical solution `y_true`.

```
[t, y] = ode45('rccirc', [0, 0.4], 2);
y_true = 2*exp(-10*t);
plot(t,y,'o',t,y_true), xlabel('Time(s)'),...
ylabel('Capacitor Voltage')
```

Note that we need not generate the array `t` to evaluate `y_true`, because `t` is generated by the `ode45` function. The plot is shown in Figure 8.5–6. The numerical solution is marked by the circles, and the analytical solution is indicated by the solid line. The numerical solution clearly gives an accurate answer. Note that the step size automatically selected by the `ode45` function is 0.02, which is the same step size we used with the Euler

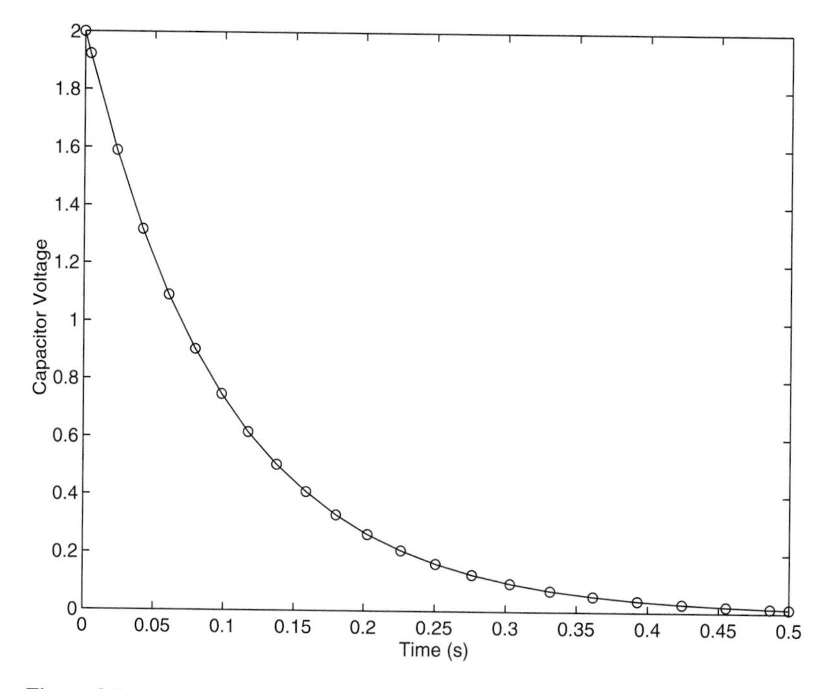

Figure 8.5–6 Free response of an RC circuit.

method to obtain Figure 8.5–1. Comparing the two plots demonstrates the accuracy of the Runge-Kutta algorithm relative to the Euler method.

Effect of Step Size

The spacing used by ode23 is smaller than that used by ode45 because ode45 has less truncation error than ode23 and thus can use a larger step size. When the solution changes rapidly, such as with an oscillatory solution, the circles representing the numerical solution do not always yield a smooth curve. Thus ode23 is sometimes more useful for plotting the solution because it often gives a smoother curve. For example, consider the problem

$$\frac{dy}{dt} = \sin t, \quad y(0) = 0$$

whose analytical solution is $y(t) = 1 - \cos t$. To solve the differential equation numerically, define the following function file:

```
function ydot = sinefn(t,y)
ydot = sin(t);
```

Use the following script file to compute the solution for $y(0) = 0$:

```
t_true = [0:0.01:4*pi];
subplot(2,1,1)
[t, y] = ode45('sinefn', [0, 4*pi], 0);
y1_true = 1 - cos(t_true);
plot(t,y,'o',t_true,y1_true), xlabel('t'), ylabel('y'),...
axis([0 4*pi -0.5 2.5]),gtext('ode45')
subplot(2,1,2)
[t, y] = ode23('sinefn', [0, 4*pi], 0);
y2_true = 1 - cos(t_true);
plot(t,y,'o',t_true,y2_true), xlabel('t'),ylabel ('y'),...
axis([0 4*pi -0.5 2.5]),gtext('ode23')
```

Figure 8.5–7 shows the solution generated by ode45 (the top graph) and ode23 (the bottom graph). Note the difference in step sizes.

Numerical Methods and Linear Equations

Even though there are general methods available for finding the analytical solutions of linear differential equations, it is nevertheless sometimes more convenient to use a numerical method to find the solution. Examples of such situations are when the forcing function is a complicated function or when the order of the differential equation is higher than two. In such cases the algebra involved in obtaining the analytical solution might not be worth the effort, especially if the main objective is to obtain a plot of the solution. In such cases we can still use the characteristic roots to check the validity of the numerical results. Example 8.5–2 demonstrates this method for an equation with a complicated forcing function.

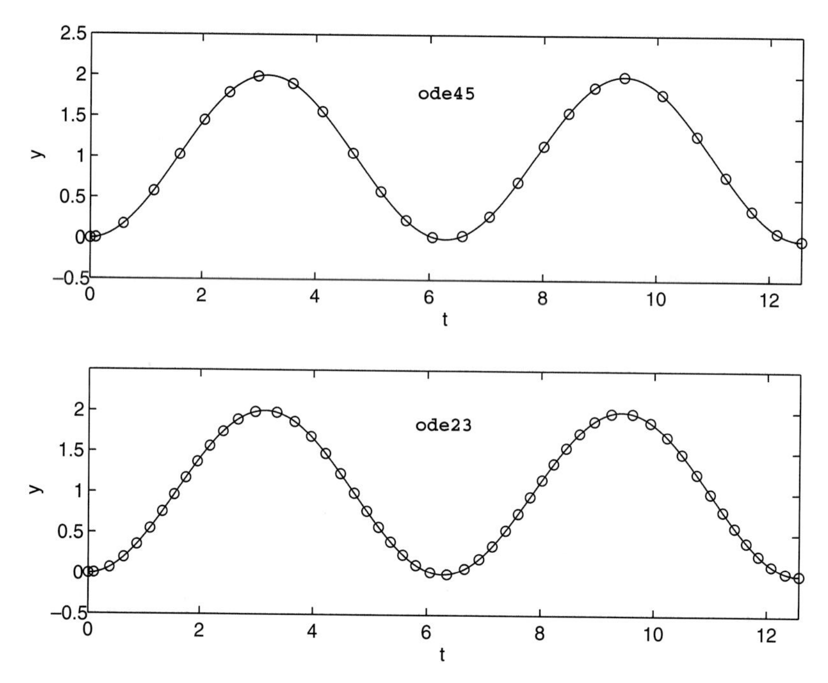

Figure 8.5–7 Numerical solutions of the equation $\dot{y} = \sin t$, $y(0) = 0$.

Use of Global Parameters

The function file to be used with the `ode` functions must have only the two arguments, t and $\mathbf{y}$. Therefore, any additional parameters needed to describe the differential equation cannot be passed through the function call. The `global x y z` command allows all functions and files using that command to share the values of the variables `x`, `y`, and `z`. Use of the `global` command avoids the necessity of changing the values of certain parameters in every file. Example 8.5–2 demonstrates the use of this command.

EXAMPLE 8.5–2

Decaying Sine Voltage Applied to an RC Circuit

Consider the RC circuit shown in Figure 8.5–5 where $RC = 0.1$ s. Now suppose that the applied voltage oscillates sinusoidally with a period P and a decaying amplitude. The voltage is given by

$$v(t) = 10e^{-t/\tau_1}\sin(2\pi t/P) \tag{8.5–26}$$

Thus the voltage amplitude will be less than 2 percent of its initial value, or essentially 0 for $t \geq 4\tau_1$. Find and plot the output voltage $y(t)$ for three cases:

1. $\tau_1 = 0.3$ s, $P = 2$ s.
2. $\tau_1 = 0.05$ s, $P = 0.03$ s.
3. $\tau_1 = 0.3$ s, $P = 0.03$ s.

Interpret the results in light of the circuit's characteristic root. In all three cases, take the initial voltage $y(0)$ to be 0.

■ Solution

The circuit's equation was given earlier. It is

$$RC\frac{dy}{dt} + y = v(t) \tag{8.5–27}$$

Solving for the derivative and using $RC = 0.1$, we obtain

$$\frac{dy}{dt} = 10[v(t) - y] \tag{8.5–28}$$

In this problem we will be examining the effects of changing the values of the parameters τ_1, P, and the final time t_f. Therefore, we will declare these parameters as global. The function file for the derivative is

```
function ydot = circuit(t,y)
% RC circuit model with a decaying sine input.
global tau_1 P
v = 10*exp(-t/tau_1)*sin((2*pi/P)*t);
ydot = 10*(v-y);
```

The `ode23` solver is called with the script file `circplot.m`:

```
% file circplot
% Solves the circuit equation and plots the response.
global  tau_1 P
tau_1 = 0.3;
P = 2;
tf = 2;
[t, y] = ode23('circuit', [0, tf], 0);
subplot(2,1,1)
tp = [0:tf/300:tf];
v = 10*exp(-tp/tau_1).*sin((2*pi/P)*tp);
plot(tp,v),xlabel('Time (s)'),...
ylabel('Applied Voltage (V)')
subplot(2,1,2)
plot(t,y),xlabel('Time (s)'),...
ylabel('Capacitor Voltage (V)')
```

The simulation is performed by entering the desired values of τ_1, P, and t_f in the file `circplot`, saving the file, and typing `circplot` at the prompt. Note that you need not change the file `circuit`. Note also that the variable `tf` need not be declared global in the file `circuit` because that file does not use `tf`.

The circuit's time constant is $\tau = RC = 0.1$ s. Therefore, if the applied voltage v is constant, the output voltage y will reach a constant value after approximately $4\tau = 0.4$ s. So the "speed of response" of the circuit is about 0.4 s. The applied voltage will be essentially 0 after $t = 4\tau_1$. We must estimate the final time `tf` to use. This time is determined by the larger of the times $4\tau = 0.4$ and $4\tau_1$. Now let us look at each case.

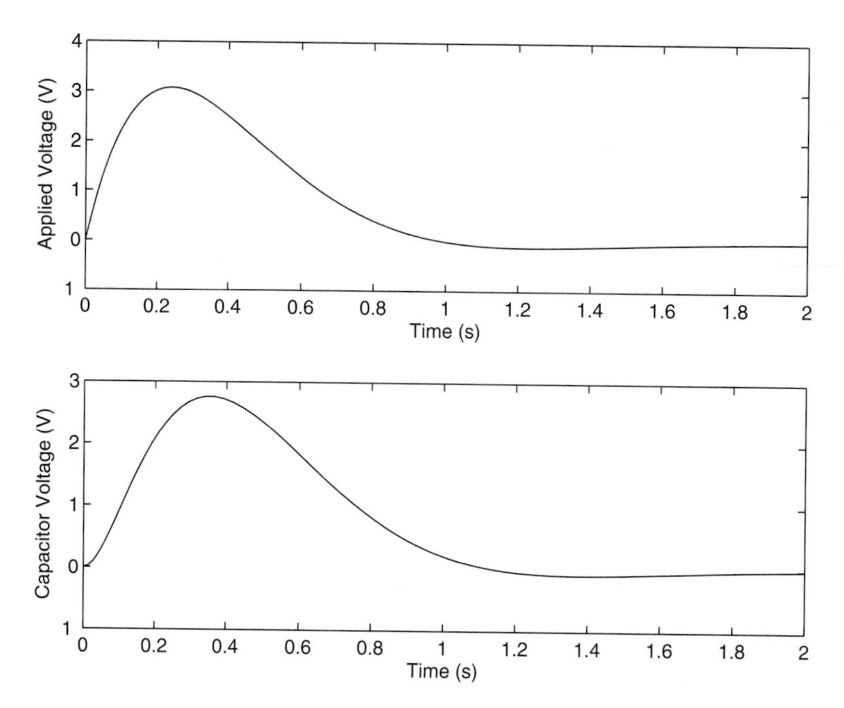

Figure 8.5–8 Plots of the applied voltage and the capacitor voltage when the applied voltage is $v(t) = 10e^{-t/0.3}\sin(\pi t)$.

Case 1: $\tau_1 = 0.3$, $P = 2$. In this case the applied voltage will be essentially 0 after $t = 4\tau_1 = 1.2$ s, which is longer than the response time of the circuit. So we choose a final time t_f no less than 1.2. However, because the period of the applied voltage is 2 s, we select $t_f = 2$ s to see the effects of one full period. The resulting plot is shown in Figure 8.5–8 along with a plot of the applied voltage for reference. Because the period is greater than 4τ, we do not see oscillations in the capacitor voltage; that is, the circuit's response has died out before the applied voltage can complete one cycle. The plot shows no discrepancies from the behavior predicted on the basis the circuit's time constant and the time constant and period of the applied voltage.

Case 2: $\tau_1 = 0.05$, $P = 0.03$. In this case the applied voltage will be essentially 0 after $t = 4\tau = 0.2$ s, and its period is 0.03, both of which are shorter than the circuit's response time (0.4 s). So the proper final time to use is determined by the response time of the circuit and thus we choose a final time of $t_f = 0.4$. The results are shown in Figure 8.5–9. Because the period is shorter than both 4τ and $4\tau_1$, we should see oscillations in the capacitor voltage, and we do. The capacitor voltage is essentially 0 after 0.4 s, as predicted.

Case 3: $\tau_1 = 0.3$, $P = 0.03$. In this case the applied voltage will be essentially 0 after $t = 4\tau_1 = 1.2$ s, which is longer than the period and the response time of the circuit. So we expect to see oscillations in the solution; the proper final time is determined by the decay time of the applied voltage and should be at least 1.2. The results are shown in Figure 8.5–10. Because the period is shorter than both 4τ and $4\tau_1$, we should see oscillations in the capacitor voltage, and we do.

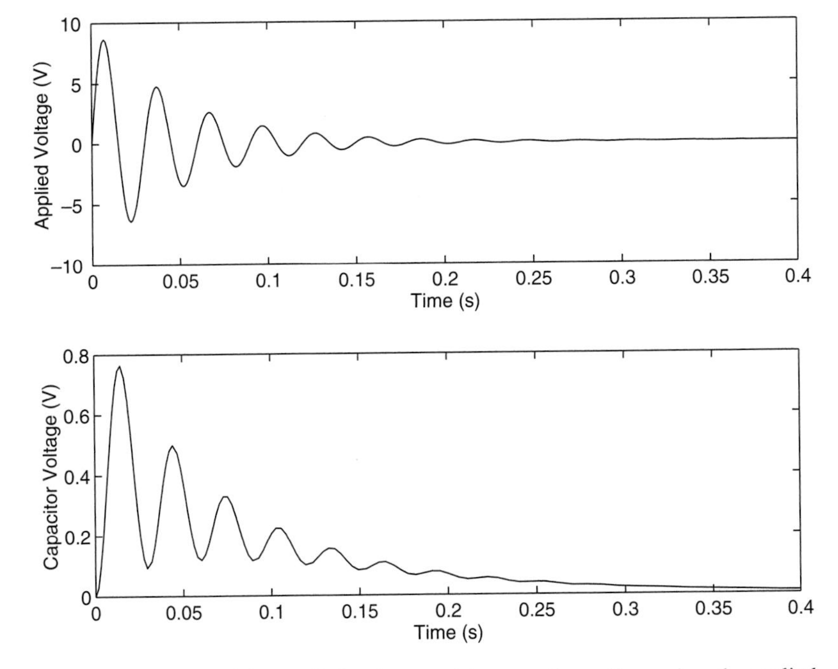

Figure 8.5–9 Plots of the applied voltage and the capacitor voltage when the applied voltage is $v(t) = 10e^{-t/0.05} \sin(2\pi t/0.03)$.

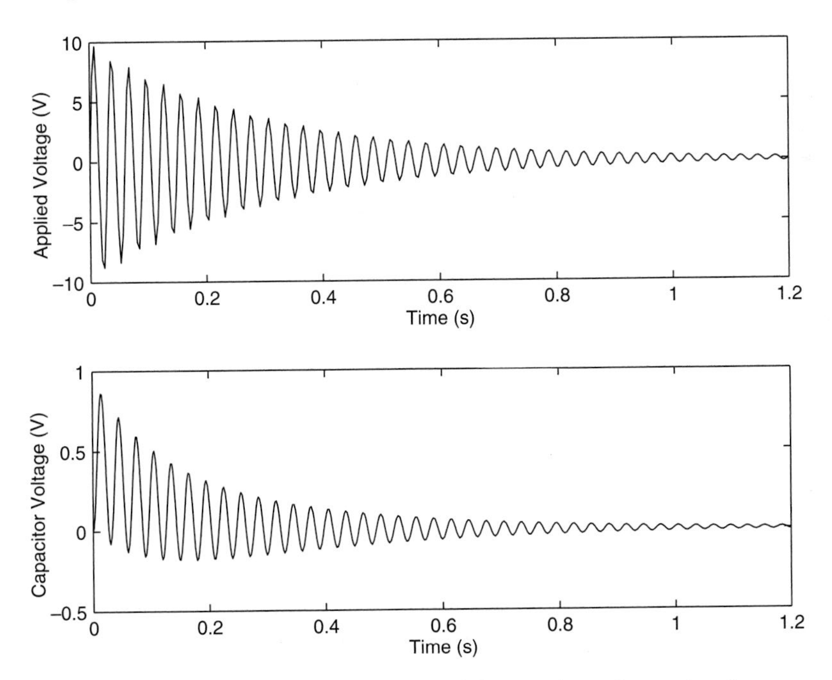

Figure 8.5–10 Plots of the applied voltage and the capacitor voltage when the applied voltage is $v(t) = 10e^{-t/0.3} \sin(2\pi t/0.03)$.

Test Your Understanding

T8.5–2 Suppose you want to run several simulations of the equation

$$\tau \frac{dy}{dt} + y = a + b \sin ct, \quad y(0) = y_0$$

for several values of τ, a, b, c, and y_0. Write a program to do this using the `global` command.

Solution of Nonlinear Equations

When the differential equation is nonlinear, we often have no analytical solution to use for checking our numerical results. In such cases we can use our physical insight to guard against grossly incorrect results. We can also check the equation for singularities that might affect the numerical procedure. Finally, we can sometimes use an approximation to replace the nonlinear equation with a linear one that can be solved analytically. Although the linear approximation does not give the exact answer, its solution can be used to see whether our numerical answer is "in the ballpark." Example 8.5–3 illustrates this approach.

EXAMPLE 8.5–3 Liquid Height in a Spherical Tank

Figure 8.5–11 shows a spherical tank for storing water. The tank is filled through a hole in the top and drained through a hole in the bottom. If the tank's radius is r, you can use integration to show that the volume of water in the tank as a function of its height h is given by

$$V(h) = \pi r h^2 - \pi \frac{h^3}{3} \tag{8.5–29}$$

Torricelli's principle (see Chapter 5) states that the liquid flow rate through the hole is proportional to the square root of the height h. Further studies in fluid mechanics have identified the relation more precisely, and the result is that the volume flow rate through

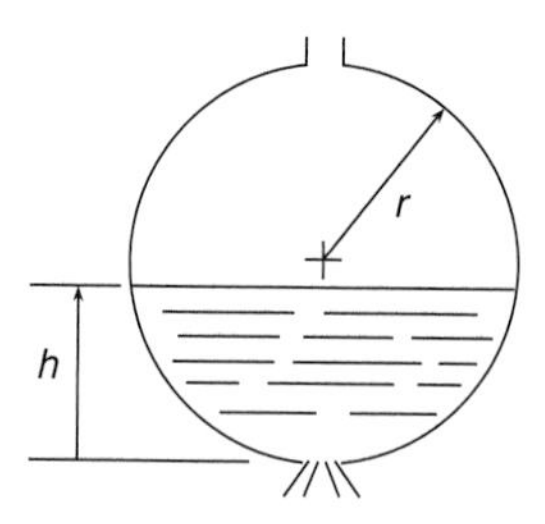

Figure 8.5–11 Draining of a spherical tank.

the hole is given by

$$q = C_d A \sqrt{2gh} \tag{8.5–30}$$

where A is the area of the hole, g is the acceleration due to gravity, and C_d is an experimentally determined value that depends partly on the type of liquid. For water, $C_d = 0.6$ is a common value. We can use the principle of conservation of mass to obtain a differential equation for the height h. Applied to this tank, the principle says that the rate of change of liquid volume in the tank must equal the volume flow rate out of the tank; that is

$$\frac{dV}{dt} = -q \tag{8.5–31}$$

From (8.5–29),

$$\frac{dV}{dt} = 2\pi r h \frac{dh}{dt} - \pi h^2 \frac{dh}{dt} = \pi h (2r - \pi h) \frac{dh}{dt}$$

Substituting this and (8.5–30) into (8.5–31) gives the required equation for h.

$$\pi (2rh - h^2) \frac{dh}{dt} = -C_d A \sqrt{2gh} \tag{8.5–32}$$

Use MATLAB to solve this equation to determine how long it will take for the tank to empty if the initial height is 9 ft. The tank has a radius of $r = 5$ ft and has a 1-in. diameter hole in the bottom. Use $g = 32.2$ ft/sec^2. Discuss how to check the solution.

■ Solution

With $C_d = 0.6$, $r = 5$, $g = 32.2$, and $A = \pi(1/24)^2$, (8.5–32) becomes

$$\frac{dh}{dt} = -\frac{0.0334\sqrt{h}}{10h - h^2} \tag{8.5–33}$$

Because this is a nonlinear equation, we have no analytical solution to use for checking our numerical results. We can use our physical insight to guard against grossly incorrect results. We can also check the above expression for dh/dt for singularities. The denominator does not become 0 unless $h = 0$ or $h = 10$, which correspond to a completely empty and a completely full tank. So we will avoid singularities if $0 < h(0) < 10$. Finally, we can use the following approximation to estimate the time to empty. Replace h on the right side of (8.5–33) with its average value, namely, $(9-0)/2 = 4.5$ ft. This gives $dh/dt = -0.00286$, whose solution is $h(t) = h(0) - 0.00286t = 9 - 0.00286t$. According to this equation, if $h(0) = 9$, the tank will be empty at $t = 9/0.00286 = 3147$ sec, or 52 min. We will use this value as a reality check on our answer.

The function file based on equation (8.5–33) is

```
function hdot = height(t,h)
hdot = -(0.0334*sqrt(h))/(10*h-h^2);
```

The file is called as follows, using the ode45 solver.

```
[t, h] = ode45('height', [0, 2475], 9);
plot(t,h),xlabel('Time (sec)'),ylabel('Height (ft)')
```

The resulting plot is shown in Figure 8.5–12. Note how the height changes more rapidly when the tank is nearly full or nearly empty. This condition is to be expected because

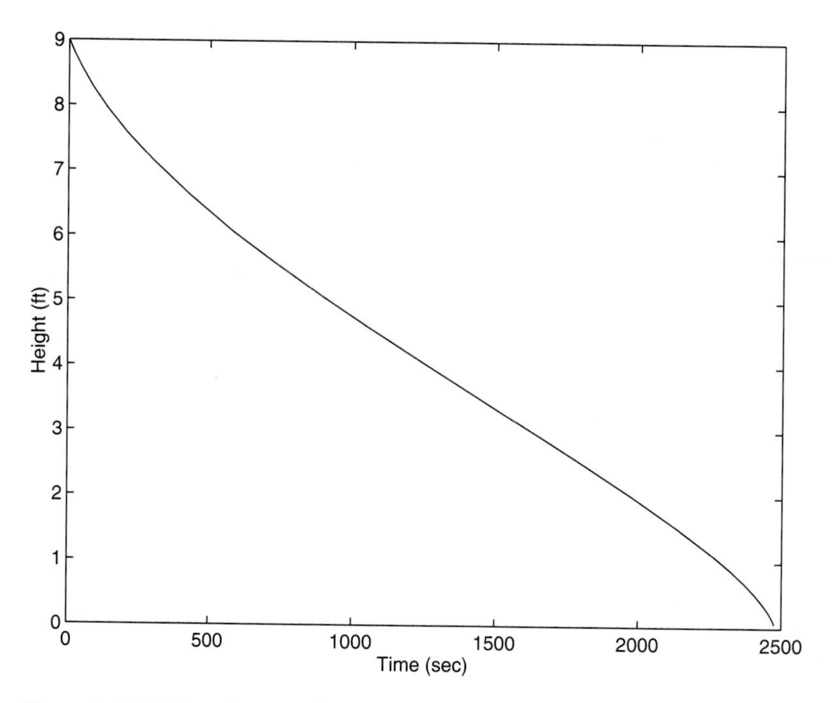

Figure 8.5–12 Plot of water height in a spherical tank.

of the effects of the tank's curvature. The tank empties in 2475 sec, or 41 min. This value is not grossly different from our rough estimate of 52 min, so we should feel comfortable accepting the numerical results.

The value of the final time of 2475 sec was found by increasing the final time until the plot showed that the height became 0. You could use a `while` loop to do this, by increasing the final time in the loop while calling `ode45` repeatedly. However, the advanced syntax for ODE solvers can be used for detecting when the height becomes 0. We will examine these capabilities in Section 8.7.

8.6 Extension to Higher-Order Equations

To use the ODE solvers to solve an equation higher than order 2, you must first write the equation as a set of first-order equations. This is easily done. Consider the second-order equation

$$5\ddot{y} + 7\dot{y} + 4y = f(t) \tag{8.6–1}$$

Solve it for the highest derivative:

$$\ddot{y} = \frac{1}{5}f(t) - \frac{4}{5}y - \frac{7}{5}\dot{y} \tag{8.6–2}$$

Define two new variables x_1 and x_2 to be y and its derivative $\dot{y}$. That is, define

$$x_1 = y$$
$$x_2 = \dot{y}$$

These definitions imply that

$$\dot{x}_1 = x_2$$
$$\dot{x}_2 = \frac{1}{5}f(t) - \frac{4}{5}x_1 - \frac{7}{5}x_2$$

This form is sometimes called the *Cauchy form* or the *state-variable form.*

CAUCHY FORM

STATE-VARIABLE FORM

Now write a function file that computes the values of $\dot{x}_1$ and $\dot{x}_2$ and stores them in a *column* vector. To do so, we must first have a function specified for $f(t)$. Suppose that $f(t) = \sin t$. Then the required file is

```
function xdot = example1(t,x)
% Computes derivatives of two equations
xdot(1) = x(2);
xdot(2) = (1/5)*(sin(t)-4*x(1)-7*x(2));
xdot = [xdot(1); xdot(2)];
```

Note that `xdot(1)` represents $\dot{x}_1$, `xdot(2)` represents $\dot{x}_2$, `x(1)` represents x_1, and `x(2)` represents x_2. Once you become familiar with the notation for the state-variable form, you will see that the preceding code could be replaced with the following shorter form:

```
function xdot = example1(t,x)
% Computes derivatives of two equations
xdot = [x(2); (1/5)*(sin(t)-4*x(1)-7*x(2))];
```

Suppose we want to solve (8.6–1) for $0 \le t \le 6$ with the initial conditions $y(0) = 3$, $\dot{y}(0) = 9$. Then the initial condition for the *vector* **x** is `[3, 9]`. To use `ode45`, you type

```
[t, x] = ode45('example1', [0, 6], [3, 9]);
```

Each row in the matrix `x` corresponds to a time returned in the column vector `t`. If you type `plot(t,x)`, you will obtain a plot of both x_1 and x_2 versus t. Note that `x` is a matrix with two columns; the first column contains the values of x_1 at the various times generated by the solver. The second column contains the values of x_2. Thus to plot only x_1, type `plot(t,x(:,1))`.

Solution of Nonlinear Equations

We mentioned earlier that when solving nonlinear equations, sometimes it is possible to check the numerical results by using an approximation that reduces the equation to a linear one. The following example illustrates such an approach with a second-order equation.

EXAMPLE 8.6–1 A Nonlinear Pendulum Model

By studying the dynamics of a pendulum like that shown in Figure 8.6–1, we can better understand the dynamics of machines such as a robot arm. The pendulum shown consists of a concentrated mass m attached to a rod whose mass is small compared to m. The rod's length is L. The equation of motion for this pendulum is

$$\ddot{\theta} + \frac{g}{L}\sin\theta = 0 \tag{8.6–3}$$

Suppose that $L = 1$ m and $g = 9.81$ m/s^2. Use MATLAB to solve this equation for $\theta(t)$ for two cases: $\theta(0) = 0.5$ rad and $\theta(0) = 0.8\pi$ rad. In both cases $\dot{\theta}(0) = 0$. Discuss how to check the accuracy of the results.

■ Solution

If we use the small angle approximation $\sin\theta \approx \theta$, the equation becomes

$$\ddot{\theta} + \frac{g}{L}\theta = 0 \tag{8.6–4}$$

which is linear and has a solution given by (8.4–16):

$$\theta(t) = \theta(0)\cos\sqrt{\frac{g}{L}}t \tag{8.6–5}$$

Thus the amplitude of oscillation is $\theta(0)$ and the period is $P = 2\pi/\sqrt{g/L} = 2$ s. We can use this information to select a final time and to check our numerical results.

First rewrite the pendulum equation (8.6–3) as two first-order equations. To do this, let

$$x_1 = \theta$$
$$x_2 = \dot{\theta}$$

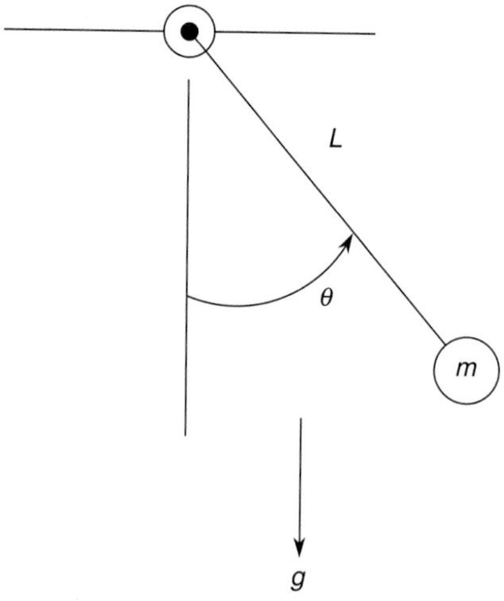

Figure 8.6–1 A pendulum.

Thus

$$\dot{x}_1 = \dot{\theta} = x_2$$

$$\dot{x}_2 = \ddot{\theta} = -\frac{g}{L} \sin x_1$$

The following function file is based on the last two equations. Remember that the output xdot must be a *column* vector.

```
function xdot = pendul(t,x)
global g L
xdot = [x(2); -(g/L)*sin(x(1))];
```

It is called as follows. The vectors ta and xa contain the results for the case where $\theta(0) = 0.5$. The vectors tb and xb contain the results for $\theta(0) = 0.8\pi$.

```
global g L
g = 9.81;L = 1;
[ta, xa] = ode45('pendul', [0, 5], [0.5, 0]);
[tb, xb] = ode45('pendul', [0, 5], [0.8*pi, 0]);
plot(ta,xa(:,1),tb,xb(:,1)),xlabel('Time (s)'),...
ylabel('Angle (rad)'),gtext('Case 1'),gtext('Case 2')
```

The results are shown in Figure 8.6–2. The amplitude remains constant, as predicted by the small angle analysis, and the period for the case where $\theta(0) = 0.5$ is a little longer

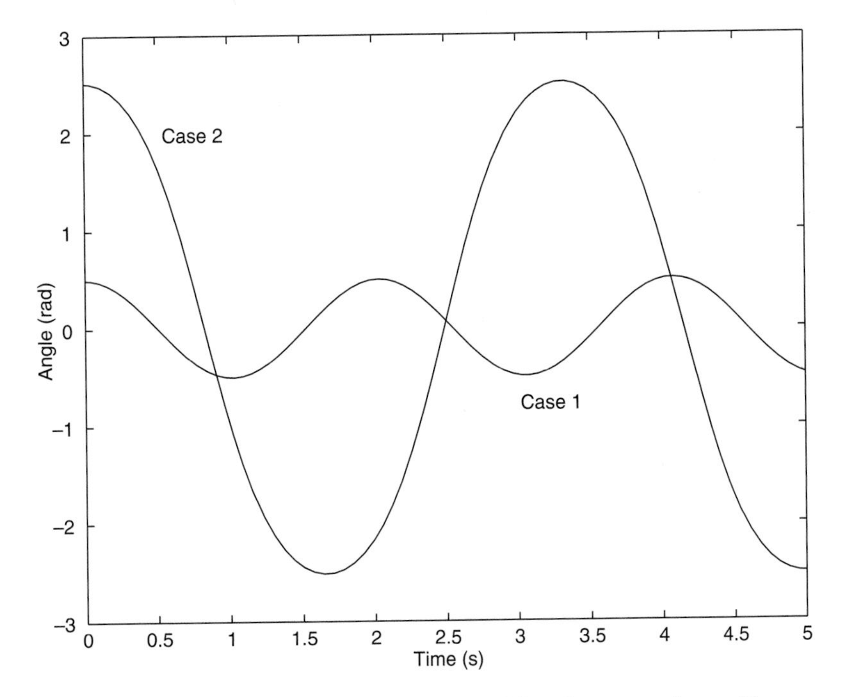

Figure 8.6–2 The pendulum angle as a function of time for two starting positions.

than 2 s, the value predicted by the small angle analysis. Therefore, we can place some confidence in the numerical procedure. For the case where $\theta(0) = 0.8\pi$, the period of the numerical solution is about 3.3 s. This longer period illustrates an important property of nonlinear differential equations. The free response of a linear equation has the same period for any initial conditions; however, the form of the free response of a nonlinear equation oftens depends on the particular values of the initial conditions.

Matrix Methods

We can use matrix operations to reduce the number of lines to be typed in the derivative function file. For example, the following equation describes the motion of a mass connected to a spring with viscous friction acting between the mass and the surface. Another force $f(t)$ also acts on the mass (see Figure 8.6–3).

$$m\ddot{y} + c\dot{y} + ky = f(t) \tag{8.6–6}$$

This equation can be put into Cauchy form by letting $x_1 = y$ and $x_2 = \dot{y}$, which gives

$$\dot{x}_1 = x_2$$

$$\dot{x}_2 = \frac{1}{m}f(t) - \frac{k}{m}x_1 - \frac{c}{m}x_2$$

These two equations can be written as one matrix equation as follows:

$$\begin{bmatrix} \dot{x}_1 \\ \dot{x}_2 \end{bmatrix} = \begin{bmatrix} 0 & 1 \\ -\dfrac{k}{m} & -\dfrac{c}{m} \end{bmatrix} \begin{bmatrix} x_1 \\ x_2 \end{bmatrix} + \begin{bmatrix} 0 \\ \dfrac{1}{m} \end{bmatrix} f(t)$$

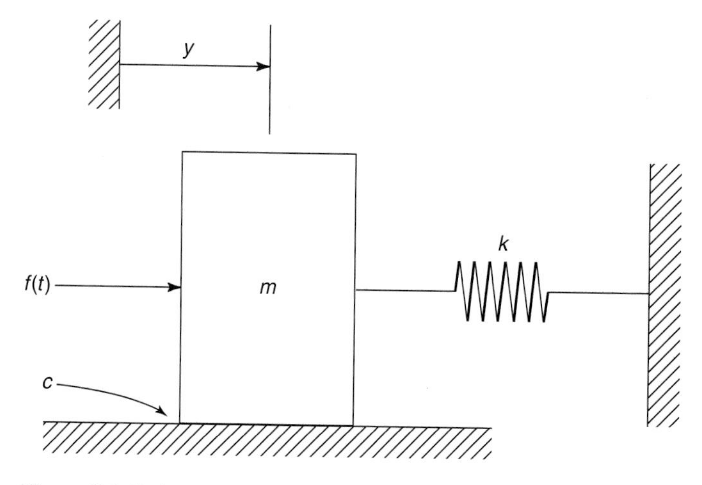

Figure 8.6–3 A mass and spring with viscous surface friction.

In compact form this is

$$\dot{\mathbf{x}} = \mathbf{A}\mathbf{x} + \mathbf{B}f(t) \tag{8.6–7}$$

where

$$\mathbf{A} = \begin{bmatrix} 0 & 1 \\ -\dfrac{k}{m} & -\dfrac{c}{m} \end{bmatrix}$$

$$\mathbf{B} = \begin{bmatrix} 0 \\ \dfrac{1}{m} \end{bmatrix}$$

and

$$\mathbf{x} = \begin{bmatrix} x_1 \\ x_2 \end{bmatrix}$$

The following function file shows how to use matrix operations. In this example $m = 1$, $c = 2$, $k = 5$, and the applied force f is a constant equal to 10.

```
function xdot = msd(t,x)
% function file for mass with spring and damping.
% position is first variable, velocity is second variable.
global c f k m
A = [0, 1;-k/m, -c/m];
B = [0; 1/m];
xdot = A*x + B*f;
```

Note that the output `xdot` will be a column vector because of the definition of matrix-vector multiplication. The characteristic roots are the roots of $ms^2 + cs + k = s^2 + 2s + 5 = 0$ and are $s = -1 \pm 2i$. The time constant is 1, and the steady-state response will thus be reached after $t = 4$. The period of oscillation will be π. Thus if we choose a final time of 5, we will see the entire response. Using the initial conditions $x_1(0) = 0$, $x_2(0) = 0$, the solver is called as follows:

```
global c f k m
m = 1;c = 2;k = 5;
f = 10;
[t, x] = ode23('msd', [0, 5], [0, 0]);
plot(t,x),xlabel('Time (s)'),...
ylabel('Displacement (m) and Velocity (m/s)'),...
gtext('Displacement'), gtext('Velocity')
```

The result is shown in Figure 8.6–4.

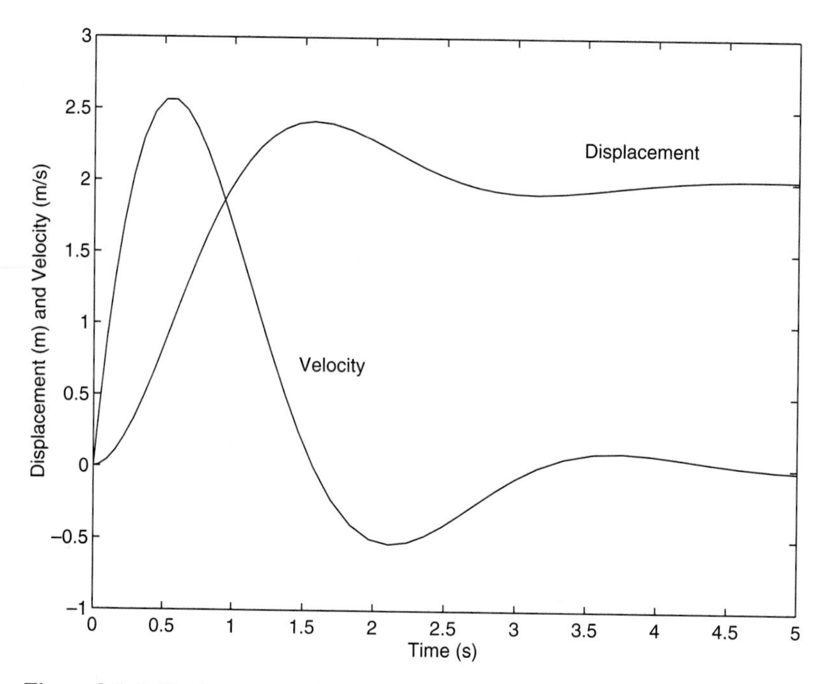

Figure 8.6–4 Displacement and velocity of the mass as a function of time.

Test Your Understanding

T8.6–1 Plot the position and velocity of a mass with a spring and damping, having the parameter values $m = 2$, $c = 3$, and $k = 7$. The applied force is $f = 35$, the initial position is $y(0) = 2$, and the initial velocity is $\dot{y}(0) = -3$.

Characteristic Roots from the `eig` Function

We saw in Section 8.4 that you can obtain the characteristic polynomial and roots for a linear ODE by substituting $y(t) = Ae^{st}$ for the dependent variable $y(t)$. However, when the ODE is in state-variable form, there is more than one dependent variable. In such cases you must substitute $x_1(t) = A_1e^{st}, x_2(t) = A_2e^{st}, \ldots$ for the dependent variables $x_1(t), x_2(t), \ldots$. For example, making these substitutions into the following

$$\dot{x}_1 = -3x_1 + x_2 \tag{8.6–8}$$

$$\dot{x}_2 = -x_1 - 7x_2 \tag{8.6–9}$$

gives

$$sA_1e^{st} = -3A_1e^{st} + A_2e^{st}$$

$$sA_2e^{st} = -A_1e^{st} - 7A_2e^{st}$$

Collect like terms and cancel the e^{st} terms to obtain

$$(s+3)A_1 - A_2 = 0$$
$$A_1 + (s+7)A_2 = 0$$

As we saw in Chapter 6, a nonzero solution will exist for A_1 and A_2 if and only if the determinant is zero. This requirement leads to

$$\begin{vmatrix} s+3 & -1 \\ 1 & s+7 \end{vmatrix} = s^2 + 10s + 22 = 0$$

This is the characteristic equation, and its roots are $s = -6.7321$ and $s = -3.2679$.

MATLAB provides the `eig` function to compute the characteristic roots when the model is given in the state-variable form (8.6–7). Its syntax is `eig(A)`, where `A` is the matrix that appears in (8.6–7). (The function's name is an abbreviation of *eigenvalue,* which is another name for characteristic root.) For example, the matrix **A** for the equations (8.6–8) and (8.6–9) is

EIGENVALUE

$$\begin{bmatrix} -3 & 1 \\ -1 & -7 \end{bmatrix}$$

To find the characteristic roots, type

```
A = [-3, 1;-1, -7];
r = eig(A)
```

The answer so obtained is `r = [-6.7321, -3.2679]`, which agrees with the answer we found by hand. To find the time constants, which are the negative reciprocals of the real parts of the roots, you type `tau = -1./real(r)`. The time constants are 0.1485 and 0.3060.

Programming Detailed Forcing Functions

As a final example of higher-order equations, we now show how to program detailed forcing functions within the derivative function file. We now use a dc motor as the application. The equations for an armature-controlled dc motor (such as a permanent magnet motor) shown in Figure 8.6–5 are the following. They result from Kirchhoff's voltage law and Newton's law applied to a rotating

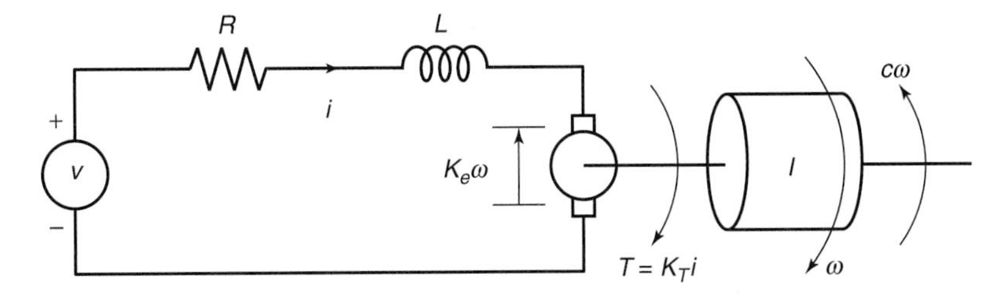

Figure 8.6–5 An armature-controlled dc motor.

inertia. The motor's current is i, and its rotational velocity is ω.

$$L\frac{di}{dt} = -Ri - K_e\omega + v(t) \tag{8.6–10}$$

$$I\frac{d\omega}{dt} = K_T i - c\omega \tag{8.6–11}$$

L, R, and I are the motor's inductance, resistance, and inertia; K_T and K_e are the torque constant and back emf constant; c is a viscous damping constant; and $v(t)$ is the applied voltage. These equations can be put into matrix form as follows, where $x_1 = i$ and $x_2 = \omega$.

$$\begin{bmatrix} \dot{x}_1 \\ \dot{x}_2 \end{bmatrix} = \begin{bmatrix} -\dfrac{R}{L} & -\dfrac{K_e}{L} \\ \dfrac{K_T}{I} & -\dfrac{c}{I} \end{bmatrix} \begin{bmatrix} x_1 \\ x_2 \end{bmatrix} + \begin{bmatrix} \dfrac{1}{L} \\ 0 \end{bmatrix} v(t)$$

EXAMPLE 8.6–2 Trapezoidal Profile for a dc Motor

In many applications we want to accelerate the motor to a desired speed and allow it to run at that speed for some time before decelerating to a stop. Investigate whether an applied voltage having a trapezoidal profile will accomplish this. Use the values $R = 0.6\ \Omega$, $L = 0.002$ H, $K_T = 0.04$ N · m/A, $K_e = 0.04$ V · s/rad, $c = 0$, and $I = 6 \times 10^{-5}$ kg · m^2. The applied voltage in volts is given by

$$v(t) = \begin{cases} 100t & 0 \le t < 0.1 \\ 10 & 0.1 \le t \le 0.4 \\ -100(t - 0.4) + 10 & 0.4 < t \le 0.5 \\ 0 & t > 0.5 \end{cases}$$

This function is shown in the top graph in Figure 8.6–6.

■ Solution

First find the time constants using the `eig` function. Use the following script file:

```
R = 0.6;L = 0.002;c = 0;
K_T = 0.04;K_e = 0.04;I = 6e-5;
A = [-R/L, -K_e/L; K_T/I, -c/I];
% compute the characteristic roots and time constants.
disp('The characteristic roots are:')
eig(A)
disp('The time constants are:')
-1./real(eig(A))
```

The roots are $s = -245.7427$ and $s = -54.2573$. The time constants are $\tau_1 = 0.0041$ and $\tau_2 = 0.0184$ s. The largest time constant indicates that the motor's response time is approximately $4(0.0184) = 0.0736$ s. Because this time is less than the time needed for the applied voltage to reach 10 V, the motor should be able to follow the desired trapezoidal profile reasonably well. To know for certain, we must solve the motor's differential

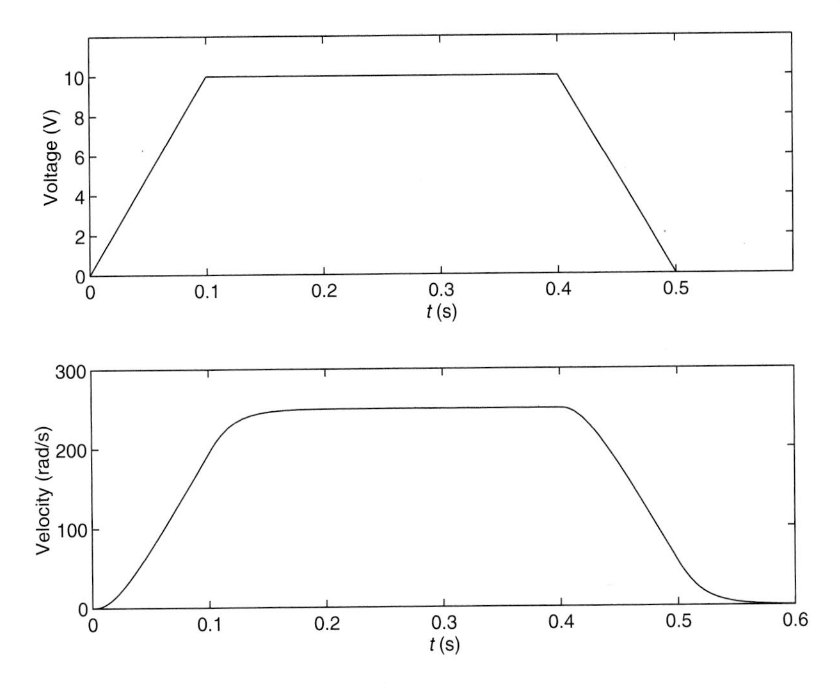

Figure 8.6–6 Voltage input and resulting velocity response of a dc motor.

equations. Use the following derivative function file:

```
function xdot = dcmotor(t,x)
% dc motor model with trapezoidal voltage profile.
% First variable is current; second is velocity.
global c I K_T K_e L R
A = [-R/L, -K_e/L; K_T/I, -c/I];
B = [1/L; 0];
if t < 0.1
   v = 100*t;
elseif t <= 0.3
   v = 10;
elseif t <= 0.4
   v = -100*(t - 0.4) + 10;
else
   v = 0;
end
xdot = A*x + B*v;
```

Using the initial conditions $x_1(0) = 0$, $x_2(0) = 0$, the solver is called as follows:

```
global c I K_T K_e L R
R = 0.6;L = 0.002;c = 0;
K_T = 0.04;K_e = 0.04;I = 6e-5;
[t, x] = ode23('dcmotor', [0, 0.5], [0, 0]);
```

The results are plotted in Figure 8.6–6. The motor's velocity follows a trapezoidal profile as expected, although there is some slight deviation because of its electrical resistance and inductance and its mechanical inertia.

8.7 ODE Solvers in the Control System Toolbox

Many of the functions of the Control System toolbox can be used to solve linear, time-invariant (constant-coefficient) differential equations. They are sometimes more convenient to use and more powerful than the ODE solvers discussed thus far, because general solutions can be found for linear, time-invariant equations. Here we discuss several of these functions. These are summarized in Table 8.7–1. The other features of the Control System toolbox require advanced methods and will not be covered here.

Model Forms

LTI OBJECT

The ODE solvers in the Control System toolbox can accept various descriptions of the equations to be solved. Version 4 of the Control System toolbox introduced the *LTI object,* which describes a linear, time-invariant equation, or sets of equations, here referred to as the *system.* An LTI object can be created from different descriptions of the system; it can be analyzed with several functions; and it can be accessed to provide alternative descriptions of the system. For example, the equation

$$2\ddot{x} + 3\dot{x} + 5x = f(t) \tag{8.7–1}$$

REDUCED FORM

is one description of a particular system. This description is called the *reduced form.* The standard arrangement of the reduced form has the forcing function on the right-hand side of the equals sign, and all the functions of the dependent variable on the left-hand side. This distinction is important when using the `tf` function.

The following is a state-model description of the same system:

$$\dot{\mathbf{x}} = \mathbf{A}\mathbf{x} + \mathbf{B}u \tag{8.7–2}$$

Table 8.7–1 LTI object functions

Command	Description
`sys = ss(A, B, C, D)`	Creates an LTI object in state-space form, where the matrices `A`, `B`, `C`, and `D` correspond to those in the model $\dot{\mathbf{x}} = \mathbf{Ax} + \mathbf{Bu}$, $\mathbf{y} = \mathbf{Cx} + \mathbf{Du}$.
`[A, B, C, D] = ssdata(sys)`	Extracts the matrices `A`, `B`, `C`, and `D` corresponding to those in the model $\dot{\mathbf{x}} = \mathbf{Ax} + \mathbf{Bu}$, $\mathbf{y} = \mathbf{Cx} + \mathbf{Du}$.
`sys = tf(right,left)`	Creates an LTI object in transfer-function form, where the vector `right` is the vector of coefficients of the right-hand side of the equation, arranged in descending derivative order, and `left` is the vector of coefficients of the left-hand side of the equation, also arranged in descending derivative order.
`[right, left] = tfdata(sys)`	Extracts the coefficients on the right- and left-hand sides of the reduced-form model.

where $x_1 = x$, $x_2 = \dot{x}$, $u = f(t)$,

$$\mathbf{A} = \begin{bmatrix} 0 & 1 \\ -\frac{5}{2} & -\frac{3}{2} \end{bmatrix} \tag{8.7–3}$$

$$\mathbf{B} = \begin{bmatrix} 0 \\ \frac{1}{2} \end{bmatrix} \tag{8.7–4}$$

and

$$\mathbf{x} = \begin{bmatrix} x_1 \\ x_2 \end{bmatrix} \tag{8.7–5}$$

Both model forms contain the same information. However, each form has its own advantages, depending on the purpose of the analysis.

Because there are two or more state variables in a state model, we need to be able to specify which state variable, or combination of variables, constitutes the output of the simulation. For example, model (8.7–2) through (8.7–5) can represent the motion of a mass, with x_1 the position, and x_2 the velocity of the mass. We need to be able to specify whether we want to see a plot of the position, or the velocity, or both. This specification of the output, denoted by the vector **y**, is done in general with the matrices **C** and **D**, which are defined as follows:

$$\mathbf{y} = \mathbf{Cx} + \mathbf{Du} \tag{8.7–6}$$

Continuing the previous example, if we want the output to be the position $x = x_1$, then $y = x_1$; and we would select

$$\mathbf{C} = [1 \quad 0]$$
$$\mathbf{D} = 0$$

Thus, in this case, (8.7–6) reduces to $y = x_1$.

To create an LTI object from the reduced form (8.7–1), you use the `tf(right,left)` function and type

```
>>sys1 = tf(1, [2, 3, 5]);
```

where the vector `right` is the vector of coefficients of the right side of the equation, arranged in descending derivative order, and `left` is the vector of coefficients of the left side of the equation, also arranged in descending derivative order. The result `sys1` is the LTI object that describes the system in the reduced form, also called the *transfer-function form*. (The name of the function `tf` stands for *transfer function*, which is an equivalent way of describing the coefficients on the left and right side of the equation.)

TRANSFER-FUNCTION FORM

Here is another example. The LTI object `sys2` in transfer-function form for the equation

$$6\frac{d^3x}{dt^3} - 4\frac{d^2x}{dt^2} + 7\frac{dx}{dt} + 5x = 3\frac{d^2f}{dt^2} + 9\frac{df}{dt} + 2f \tag{8.7–7}$$

is created by typing

```
>>sys2 = tf([3, 9, 2],[6, -4, 7, 5]);
```

To create an LTI object from a state model, you use the `ss(A,B,C,D)` function, where `ss` stands for *state space.* The matrix arguments of the function are the matrices in the following standard form of a state model:

$$\dot{\mathbf{x}} = \mathbf{Ax} + \mathbf{Bu} \tag{8.7–8}$$

$$\mathbf{y} = \mathbf{Cx} + \mathbf{Du} \tag{8.7–9}$$

where $\mathbf{x}$ is the vector of state variables, $\mathbf{u}$ is the vector of input functions, and $\mathbf{y}$ is the vector of output variables. For example, to create an LTI object in state-model form for the system described by (8.7–2) through (8.7–5), you type

```
>>A = [0, 1; -5/2, -3/2];
>>B = [0;1/2];
>>C = [1, 0];
>>D = 0;
>>sys3 = ss(A,B,C,D);
```

An LTI object defined using the `tf` function can obtain an equivalent state-model description of the system. To create a state model for the system described by the LTI object `sys1` created previously in transfer-function form you type `ss(sys1)`. You will then see the resulting **A**, **B**, **C**, and **D** matrices on the screen. To extract and save the matrices, use the `ssdata` function as follows:

```
>>[A1, B1, C1, D1] = ssdata(sys1);
```

The results are

$$\mathbf{A1} = \begin{bmatrix} -1.5 & -1.25 \\ 2 & 0 \end{bmatrix}$$

$$\mathbf{B1} = \begin{bmatrix} 0.5 \\ 0 \end{bmatrix}$$

$$\mathbf{C1} = [0 \quad 0.5]$$

$$\mathbf{D1} = [0]$$

When using `ssdata` to convert a transfer-function form into a state model, note that the output y will be a scalar that is identical to the solution variable of the reduced form; in this case the solution variable of (8.7–1) is the variable x. To interpret the state model, we need to relate its state variables x_1 and x_2 to x. The values of the matrices **C1** and **D1** tell us that the output variable is $y = 0.5x_2$. Because the output y is the same as x, we then see that $x_2 = 2x$. The other state variable x_1 is related to x_2 by $\dot{x}_2 = 2x_1$. Thus $x_1 = \dot{x}$.

To create a transfer-function description of the system `sys3`, previously created from the state model, you type `tfsys3 = tf(sys3);`. To extract and save the coefficients of the reduced form, use the `tfdata` function as follows:

```
>>[right, left] = tfdata(sys3, 'v');
```

For this example, the vectors returned are `right = 1` and `left = [1, 1.5, 2.5]`. The optional parameter `'v'` tells MATLAB to return the coefficients as vectors; otherwise, they are returned as cell arrays. These functions are summarized in Table 8.7–1.

Test Your Understanding

T8.7–1 Obtain the state model for the reduced-form model

$$5\ddot{x} + 7\dot{x} + 4x = f(t)$$

Then convert the state model back to reduced form and see whether you get the original reduced-form model.

ODE Solvers

The Control System toolbox provides several solvers for linear models. These solvers are categorized by the type of input function they can accept: zero input, impulse input, step input, and a general input function. These are summarized in Table 8.7–2.

The `initial` Function: The `initial` function computes and plots the free response of a state model. This response is sometimes called the *initial condition response* or the *undriven response* in the MATLAB documentation. The basic syntax is `initial(sys,x0)`, where `sys` is the LTI object in state-model form and `x0` is the initial condition vector. The time span and number of solution points are chosen automatically. For example, to find the free response of the state model (8.7–2) through (8.7–5), for $x_1(0) = 5$ and $x_2(0) = -2$, first

Table 8.7–2 Basic syntax of the LTI ODE solvers

Command	Description
`impulse(sys)`	Computes and plots the unit-impulse response of the LTI object `sys`.
`initial(sys,x0)`	Computes and plots the free response of the LTI object `sys` given in state-model form, for the initial conditions specified in the vector `x0`.
`lsim(sys,u,t)`	Computes and plots the response of the LTI object `sys` to the input specified by the vector `u`, at the times specified by the vector `t`.
`step(sys)`	Computes and plots the unit-step response of the LTI object `sys`.

(See the text for description of extended syntax.)

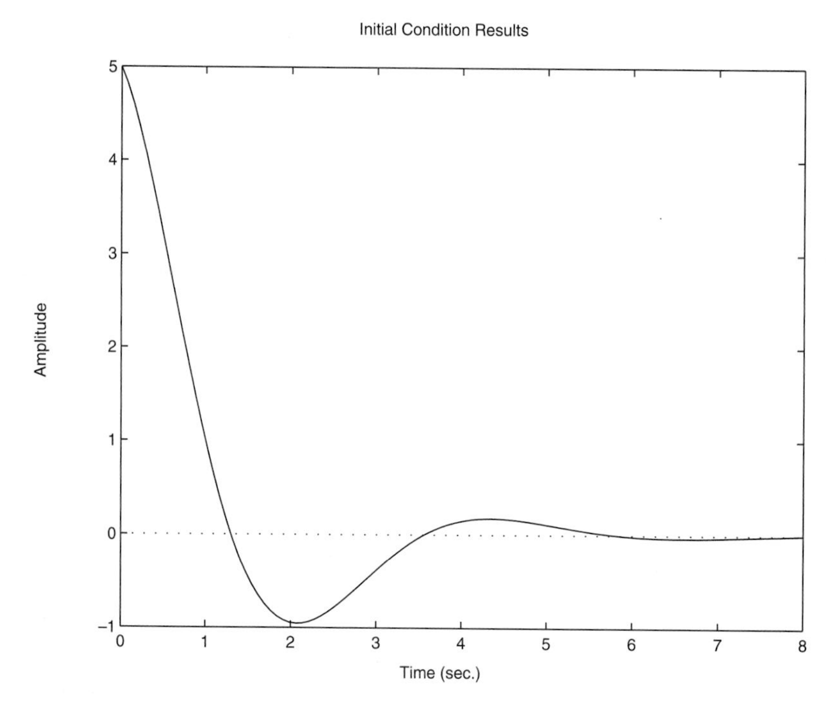

Figure 8.7–1 Free response of the model given by (8.7–2) through (8.7–5) for $x_1(0) = 5$ and $x_2(0) = -2$.

define it in state-model form. This was done previously to obtain the system `sys3`. Then use the `initial` function as follows:

```
>>initial(sys3, [5, -2])
```

The plot shown in Figure 8.7–1 will be displayed on the screen.

To specify the final time `tf`, use the syntax `initial(sys,x0, tf)`. To specify a vector of times of the form `t = [0:dt:tf]`, at which to obtain the solution, use the syntax `initial(sys,x0,t)`. When called with left-hand arguments, as `[y, t, x] = initial(sys,x0, ...)`, the function returns the output response `y`, the time vector `t` used for the simulation, and the state vector `x` evaluated at those times. The columns of the matrices `y` and `x` are the outputs and the states, respectively. The number of rows in `y` and `x` equals `length(t)`. No plot is drawn. The syntax `initial(sys1,sys2, ...,x0,t)` plots the free response of multiple LTI systems on a single plot. The time vector `t` is optional. You can specify line color, line style, and marker for each system; for example, `initial (sys1,'r',sys2,'y--',sys3, 'gx',x0)`.

The `impulse` **Function:** The `impulse` function plots the unit-impulse response for each input-output pair of the system, assuming that the initial conditions are zero. (If you are unfamiliar with the impulse input, see Chapter 10,

Section 10.5.) The basic syntax is `impulse(sys)`, where `sys` is the LTI object. Unlike the `initial` function, the `impulse` function can be used with either a state model or a transfer-function model. The time span and number of solution points are chosen automatically. For example, the impulse response of (8.7–1) is found as follows:

```
>>sys1 = tf(1,[2, 3, 5]);
>>impulse(sys1)
```

To specify the final time `tf`, use the syntax `impulse(sys,tf)`. To specify a vector of times of the form `t = [0:dt:tf]`, at which to obtain the solution, use the syntax `impulse(sys,t)`. When called with left-hand arguments, as `[y, t] = impulse(sys, ...)`, the function returns the output response `y` and the time vector `t` used for the simulation. No plot is drawn. The array `y` is $p \times q \times m$, where p is `length(t)`, q is the number of outputs, and m is the number of inputs. To obtain the state vector solution, use the syntax `[y, t, x] = impulse(sys, ...)`.

The syntax `impulse(sys1,sys2, ...,t)` plots the impulse response of multiple LTI systems on a single plot. The time vector `t` is optional. You can specify line color, line style, and marker for each system; for example, `impulse(sys1,'r',sys2,'y--',sys3,'gx')`.

The `step` Function: The `step` function plots the unit-step response for each input-output pair of the system, assuming that the initial conditions are zero. (If you are unfamiliar with the step function, see Chapter 10, Section 10.5.) The basic syntax is `step(sys)` where `sys` is the LTI object. The `step` function can be used with either a state model or a transfer-function model. The time span and number of solution points are chosen automatically. To specify the final time `tf`, use the syntax `step(sys,tf)`. To specify a vector of times of the form `t = [0:dt:tf]`, at which to obtain the solution, use the syntax `step(sys,t)`. When called with left-hand arguments, as `[y, t] = step(sys, ...)`, the function returns the output response `y` and the time vector `t` used for the simulation. No plot is drawn. The array `y` is $p \times q \times m$, where p is `length(t)`, q is the number of outputs, and m is the number of inputs. To obtain the state vector solution for state-space models, use the syntax `[y, t, x] = step(sys, ...)`.

The syntax `step(sys1,sys2, ...,t)` plots the step response of multiple LTI systems on a single plot. The time vector `t` is optional. You can specify line color, line style, and marker for each system; for example, `step(sys1,'r', sys2,'y--',sys3,'gx')`. To find the step response, for zero initial conditions, of the state model (8.7–2) through (8.7–5), and the reduced-form model

$$5\ddot{x} + 7\dot{x} + 5x = 5\dot{f} + f \qquad (8.7\text{–}10)$$

the session is (assuming `sys3` is still available):

```
>>sys4 = tf([5, 1], [5, 7, 5]);
>>step(sys3,'b',sys4,'--')
```

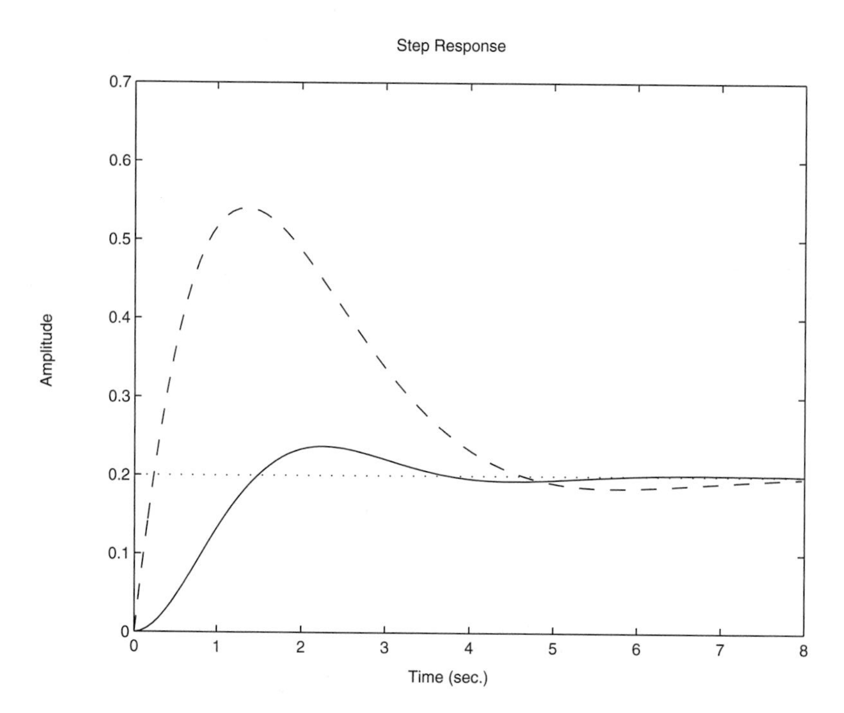

Figure 8.7–2 Step response of the model given by (8.7–2) through (8.7–5) and the model (8.7–10), for zero initial conditions.

The result is shown in Figure 8.7–2. The steady-state response is indicated by the horizontal dotted line. Note how the steady-state response and the time to reach that state are automatically determined.

The `lsim` Function: The `lsim` function plots the response of the system to an arbitrary input. The basic syntax for zero initial conditions is `lsim(sys, u,t)`, where `sys` is the LTI object; `t` is a time vector having regular spacing, as `t = [0:dt:tf]`; and `u` is a matrix with as many columns as inputs and whose *i*th row specifies the value of the input at time `t(i)`. To specify nonzero initial conditions for a state-space model, use the syntax `lsim(sys,u,t,x0)`.

When called with left-hand arguments, as `[y, t] = lsim(sys, u,...)`, the function returns the output response `y` and the time vector `t` used for the simulation. The columns of the matrix `y` are the outputs, and the number of its rows equals `length(t)`. No plot is drawn. To obtain the state vector solution for state-space models, use the syntax `[y, t, x] = lsim(sys,u,...)`. The syntax `lsim(sys1,sys2,...,u,t,x0)` plots the free response of multiple LTI systems on a single plot. The initial condition vector `x0` is needed only if the initial conditions are nonzero. You can specify line color, line style, and marker for each system; for example, `lsim(sys1,'r',sys2,'y-- ', sys3,'gx',u,t)`.

We will see an example of the `lsim` function shortly.

LTI Viewer: The Control System toolbox has the LTI Viewer to assist in the analysis of LTI systems. It provides an interactive user interface that allows you to switch between different types of response plots and between the analysis of different systems. The viewer is invoked by typing `ltiview`. See MATLAB help for more information.

Predefined Input Functions

You can always create any complicated input function to use with the ODE solvers `odexxx` or `lsim` by defining a vector containing the input function's values at specified times. However, MATLAB provides several predefined functions that you might find easier to use. These are described in the following paragraphs, and are summarized in Table 8.7–3.

The function `stepfun(t,t0)` returns a vector the same length as `t` with zeros where `t` < `t0` and ones where `t` ≥ `t0`. The vector `t` must have monotonically increasing elements.

The function `square(t)` returns a vector the same length as `t` corresponding to a square wave of period 2π. The function `square(t)` is similar to `sin(t)`, with extreme points at ± 1, except that the wave is square. The syntax `square(t,duty)` generates a square wave with a duty cycle specified by `duty`, which is the percentage of the period over which the function is positive.

The function `sawtooth(t)` returns a vector the same length as `t` corresponding to a sawtooth wave of period 2π. The function `sawtooth(t)` is similar to `sin(t)`, with extreme points at ± 1, except that the wave shape is a

Table 8.7–3 Predefined input functions

Command	Description
`[u, t] = gensig(type,period,tf,dt)`	Generates a periodic input of a specified type `type`, having a period `period`. The following types are available: sine wave (`type` = `sin`), square wave (`type` = `square`), and narrow-width periodic pulse (`type` = `pulse`). The vector `t` contains the times, and the vector `u` contains the input values at those times. All generated inputs have unit amplitudes. The optional parameters `tf` and `dt` specify the time duration `tf` of the input and the spacing `dt` between the time instants.
`sawtooth(t,width)`	Returns a vector the same length as `t` corresponding to a sawtooth wave of period 2π. The optional parameter `width` generates a modified sawtooth wave where `width` determines the fraction between 0 and 2π at which the maximum occurs (0 ≤ `width` ≤ 1).
`square(t,duty)`	Returns a vector the same length as `t` corresponding to a square wave of period 2π. The optional parameter `duty` generates a square wave with a duty cycle specified by `duty`, which is the percentage of the period over which the function is positive.
`stepfun(t,t0)`	Returns a vector the same length as `t` with zeros where `t` < `t0` and ones where `t` ≥ `t0`.

triangle. The syntax `sawtooth(t, width)` generates a modified sawtooth wave where `width` determines the fraction between 0 and 2π at which the maximum occurs ($0 \le$ `width` ≤ 1).

The function `[u, t] = gensig(type,period)` generates a periodic input of a specified type `type`, having a period `period`. The following types are available: sine wave (`type = 'sin'`), square wave (`type = 'square'`), and narrow-width periodic pulse (`type = 'pulse'`). The vector `t` contains the times, and the vector `u` contains the input values at those times. All generated inputs have unit amplitudes. The syntax `[u, t] = gensig(type,period, tf,dt)` specifies the time duration `tf` of the input and the spacing `dt` between the time instants.

For example, suppose a square wave with period 5 is applied to the following reduced-form model.

$$\ddot{x} + 2\dot{x} + 4x = 4f \tag{8.7–11}$$

To find the response for zero initial conditions, over the interval $0 \le t \le 10$ and using a step size of 0.01, the session is

```
>>sys4 = tf(4,[1,2,4]);
>>[u, t] = gensig('square',5,10,0.01);
>>[y, t] = lsim(sys4,u,t);plot(t,y,t,u),axis([0 10 -0.5 1.5]),...
 xlabel('Time'),ylabel('Response')
```

The result is shown in Figure 8.7–3.

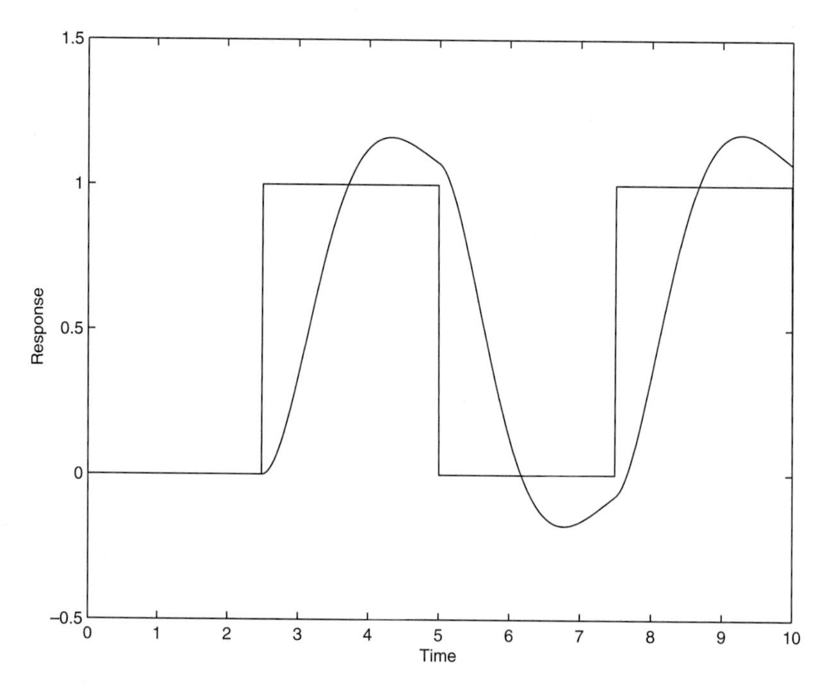

Figure 8.7–3 Square-wave response of the model $\ddot{x} + 2\dot{x} + 4x = 4f$.

8.8 Advanced Solver Syntax

The complete syntax of the ODE solver is as follows and is summarized in Table 8.8–1.

```
[t, y] = ode23('ydot', tspan, y0, options, p1, p2, ...)
```

where the `options` argument is created with the new `odeset` function, and `p1, p2, ...` are optional parameters that can be passed to the function file `ydot` every time it is called. If these optional parameters are used, but no `options` are set, use `options = [ ]` as a placeholder.

The function file `ydot` can now take additional input arguments. It has the form `ydot(t, y, flag, p1, p2, ...)`, where `flag` is an argument that notifies the function `ydot` that the solver is expecting a specific kind of information. We will see an example of the `flag` argument shortly.

The `odeset` Function

The `odeset` function creates an options structure to be supplied to the solver. Its syntax is

```
options = odeset('name1', 'value1' 'name2', 'value2',...)
```

Table 8.8–1 Complete syntax of ODE solvers

Command	Description
`[t, y] = ode23('ydot', tspan, y0, options, p1, p2, ...)`	Solves the differential equation $\dot{\mathbf{y}} = \mathbf{f}(t, \mathbf{y})$ specified in the function file `ydot`, whose inputs must be t and $\mathbf{y}$ and whose output must be a *column* vector representing $d\mathbf{y}/dt$; that is, $\mathbf{f}(t, \mathbf{y})$. The number of rows in this column vector must equal the order of the equation. The vector `tspan` contains the starting and ending values of the independent variable t, and optionally, any intermediate values of t where the solution is desired. The vector `y0` contains $\mathbf{y}(t_0)$. The function file must have two input arguments `t` and `y` even for equations where $\mathbf{f}(t, \mathbf{y})$ is not a function of t. The `options` argument is created with the `odeset` function, and `p1, p2, ...` are optional parameters that can be passed to the function file `ydot` every time it is called. If these optional parameters are used, but no `options` are set, use `options = [ ]` as a placeholder. The function file `ydot` can take additional input arguments. It has the form `ydot(t, y, flag, p1, p2, ...)`, where `flag` is an argument that notifies the function `ydot` that the solver is expecting a specific kind of information. The syntax for all the solvers is identical to that of `ode23`.
`options = odeset('name1', 'value1', 'name2', 'value2',...)`	Creates an integrator options structure `options` to be used with the ODE solver, in which the named properties have the specified values, where `name` is the name of a *property* and `value` is the value to be assigned to the property. Any unspecified properties have default values. `odeset` with no input arguments displays all property names and their possible values.

where `name` is the name of a *property* and `value` is the value to be assigned to the property.

A simple example will clarify things. The `Refine` property is used to increase the number of output points from the solver by an integer factor n. The default value of n is 1 for all the solvers except `ode45`, whose default value is 4 because of the solver's large step sizes. Suppose we want to solve the equation

$$\frac{dy}{dt} = \sin t$$

for $0 \leq t \leq 4\pi$ with $y(0) = 0$. Define the following function file:

```
function ydot = sinefn(t,y)
ydot = sin(t);
```

Then use the `odeset` function to set the value of `Refine` to $n = 8$, and call the `ode45` solver, as shown here:

```
options = odeset('Refine',8);
[t, y] = ode45('sinefn', [0, 4*pi], 0, options);
```

Many properties can be set with the `odeset` function. To see a list of these, type `odeset`.

Another property is the `Events` property, which has two possible values: `on` and `off`. It can be used to locate transitions to, from, or through zeros of a user-defined function. This property can be used to detect when the ODE solution reaches a certain value. To use this property, you must write the ODE file `ydot` to have three outputs, as follows:

```
[value, isterminal, direction] = ydot(t,y,flag)
```

The vector `value` should be programmed to contain the vector describing the event when the flag equals `'events'`. The vector `value` should be programmed to contain the derivatives when the flag is not set to `'events'` or is empty. `value` is evaluated at the beginning and end of each step, and if any of its elements make transitions to, from, or through zero, the solver determines the time when the transition occurred. The direction of the transition is specified in the vector `direction`. A `1` indicates positive direction, a `-1` indicates negative direction, and a `0` indicates "don't care."

The `isterminal` vector is a logical vector of `1`s and `0`s that specifies whether a zero crossing of the corresponding element in `value` is a terminal event. A `1` corresponds to a terminal event and halts the solver; a `0` corresponds to a nonterminal event.

For example, suppose we want to simulate a dropped ball bouncing up from the floor. The equation of motion of the ball in free flight is

$$m\ddot{y} = -mg$$

or $\ddot{y} = -g$, where y is the ball's height above the floor. Put this into state-variable form by defining $y_1 = y$ and $y_2 = \dot{y}$. The equations are

$$\dot{y}_1 = y_2$$

$$\dot{y}_2 = -g$$

Define the following equation file to use with the solver. We use $g = 9.81$ m/s^2 and an initial height of 10 m. The `value` vector here consists of the state vector. The "event" here is a bounce, which occurs when the height is 0.

```
function [value, isterminal, direction] = ballode(t, y, flag)
if (nargin<3)|isempty(flag)
   value = [y(2); -9.81];
elseif flag == 'events'
   % Returns height and velocity
   value = y;
   %Instructs to stop when first
   %variable (height) is zero
   isterminal = [1; 0];
   % Instructs to detect an event only when the first
   % variable (height) is decreasing,
   % regardless of velocity
   direction = [-1; 0];
else
   error(['unknown flag''''flag'''.']);
end
```

We assume that the ball loses 10 percent of its speed each time it bounces. The script file to find the ball's motion up to the second bounce follows. It calls the solver before and after the bounce.

```
options = odeset('Events','on')
[t1, y1] = ode45('ballode', [0, 10], [10, 0], options);
[t2, y2] = ode45('ballode',[t1(length(t1)), 10],...
[0, -0.9*y1(length(t1),2)],options);
t = [t1; t2];y = [y1; y2];
plot(t,y(:,1)),xlabel('Time (s)'),...
ylabel('Ball Height (m)'),axis([0 5 0 10])
```

The resulting plot of height versus time is shown in Figure 8.8–1.

Stiff Differential Equations

A *stiff* differential equation is one whose response changes rapidly over a time scale that is short compared to the time scale over which we are interested in the solution. Stiff equations present a challenge to solve numerically. A small step size is needed to solve for the rapid changes, but many steps are needed to obtain the solution over the longer time interval, and thus a large error might

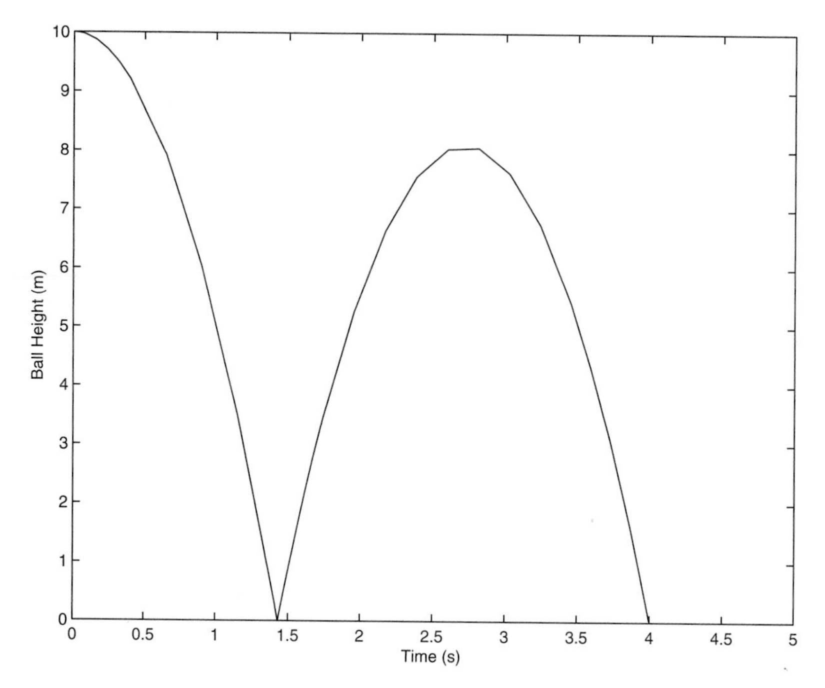

Figure 8.8–1 Height of a bouncing ball as a function of time.

accumulate. For example, the following equation might be considered "stiff":

$$\ddot{y} + 1001\dot{y} + 1000y = 0 \tag{8.8–1}$$

The characteristic roots are $s = -1$ and $s = -1000$. If the initial conditions are $y(0) = 1$ and $\dot{y}(0) = 0$, the closed-form solution is

$$y(t) = \frac{1}{999}\left(1000e^{-t} - e^{-1000t}\right) \tag{8.8–2}$$

That part of the response due to the term e^{-1000t} disappears after approximately $t = 4/1000 = 0.004$, but the term due to e^{-t} does not disappear until after $t = 4$. Thus it would be difficult for a plot to show the solution accurately, and a numerical solver would need a small step size to compute the rapid changes due to the e^{-1000t} term, much smaller than that required to compute the longer-term response due to the e^{-t} term. The result can be a large accumulated error because of the small step size combined with the large number of steps required to obtain the full solution.

All of the solvers in MATLAB do rather well with moderately stiff equations, but if you have trouble solving an equation with one of these solvers, try one of the four solvers specifically designed to handle stiff equations. These are `ode15s`, a variable-order method; `ode23s`, a low-order method; `ode23tb`, another low-order method; and ode23t, a trapezoidal method. Their syntax is identical to the ode23 and ode45 solvers.

We have not covered all of the new solver capabilities provided in MATLAB. For more information and additional examples, consult online help.

8.9 Summary

This chapter covered numerical methods for computing integrals and derivatives and for solving ordinary differential equations. Some integrals and many differential equations cannot be evaluated analytically, and we need to compute them numerically with an approximate method. In addition, it is often necessary to use data to estimate rates of change, and this process requires a numerical estimate of the derivative.

Now that you have finished this chapter, you should be able to the following:

- Use the trapz, quad, and quadl functions to numerically evaluate integrals. Note that trapz must be supplied with numerical values for the integrand, whereas quad and quadl can and must use a function.
- Use numerical methods with the diff function to estimate derivatives.
- Use the analytical expressions for simple integrals and derivatives to check the accuracy of numerical methods.
- Use MATLAB ODE solvers to solve differential equations.
- Use the analytical solutions of simple differential equations to check the accuracy of numerical methods.
- Apply approximations to check the validity of numerical solutions of nonlinear differential equations.

Table 8.9–1 summarizes the MATLAB commands introduced in this chapter.

Table 8.9–1 Guide to MATLAB functions introduced in Chapter 8

Numerical integration functions	Table 8.2–1
Numerical differentiation functions	Table 8.3–1
ODE solvers	Table 8.5–1
Basic syntax of ODE solvers	Table 8.5–2
LTI object functions	Table 8.7–1
Basic syntax of the LTI ODE solvers	Table 8.7–2
Predefined input functions	Table 8.7–3
Complete syntax of ODE solvers	Table 8.8–1

Miscellaneous command	
Command	**Description**
eig(A)	Computes the eigenvalues of the matrix A, which are the characteristic roots of the vector-matrix differential equation $\dot{\mathbf{x}} = \mathbf{A}\mathbf{x}$.

Key Terms with Page References

Problems

You can find the answers to problems marked with an asterisk at the end of the text.

Note: Unless otherwise directed by your instructor, for each of the following problems select the easiest method (analytical or numerical) to solve the problem. Discuss the reasons for your choice. Your instructor might require you to use a numerical method and might require you to check the numerical solution with the analytical solution if possible.

Sections 8.1, 8.2, and 8.3

1.* An object moves at a velocity $v(t) = 5 + 7t^2$ starting from the position $x(2) = 5$ at $t = 2$. Determine its position at $t = 10$.

2. The total distance D traveled by an object moving at velocity $v(t)$ from the time $t = a$ to the time $t = b$ is

$$D = \int_a^b |v(t)|\, dt$$

The absolute value $|v(t)|$ is used to account for the possibility that $v(t)$ might be negative. Suppose an object moves with a velocity of $v(t) = \cos(\pi t)$ for $0 \le t \le 1$. Find the total distance traveled and the object's location $x(1)$ at $t = 1$ if $x(0) = 2$.

3. An object starts with an initial velocity of 3 m/s at $t = 0$ and accelerates with an acceleration of $a(t) = 5t$ m/s^2. Plot its velocity as a function of time for $0 \le t \le 5$ and find the total distance the object travels in 5 s.

4. The equation for the voltage $v(t)$ across a capacitor as a function of time is

$$v(t) = \frac{1}{C}\left(\int_0^t i(t)\,dt + Q_0\right)$$

where $i(t)$ is the applied current and Q_0 is the initial charge. A certain capacitor initially holds no charge. Its capacitance is $C = 10^{-5}$ F. If a current $i(t) = 2[1 + \sin(5t)]10^{-4}$ A is applied to the capacitor, plot the voltage $v(t)$ as a function of time for $0 \le t \le 1.2$ s.

5.* Two electrons a distance x apart repel each other with a force k/x^2, where k is a constant. Let one electron be fixed at $x = 0$. Determine the work done by the force of repulsion in moving the second charge from $x = 1$ to $x = 5$.

6. A certain object's position as a function of time is given by $x(t) = 6t \sin 5t$. Plot its velocity and acceleration as functions of time for $0 \le t \le 5$.

7.* A ball was thrown vertically with a velocity $v(0)$ m/s. Its measured height as a function of time was determined to be $h(t) = 6t - 4.9t^2$ m. Determine its initial velocity.

8. The volume of liquid in a spherical tank of radius r as a function of the liquid's height h above the tank bottom is given by

$$V(h) = \pi r h^2 - \pi \frac{h^3}{3}$$

a. Determine the volume rate of change dV/dh with respect to height.
b. Determine the volume rate of change dV/dt with respect to time.

9. A certain object moves with the velocity $v(t)$ given in the following table. Determine the object's position $x(t)$ at $t = 10$ s if $x(0) = 3$.

Time (s)	0	1	2	3	4	5	6	7	8	9	10
Velocity (m/s)	0	2	5	7	9	12	15	18	22	20	17

10.* A tank having vertical sides and a bottom area of 100 ft^2 stores water. To fill the tank, water is pumped into the top at the rate given in the following table. Determine the water height $h(t)$ at $t = 10$ min.

Time (min)	0	1	2	3	4	5	6	7	8	9	10
Flow rate (ft^3/min)	0	80	130	150	150	160	165	170	160	140	120

11. A cone-shaped paper drinking cup (the kind used at water fountains) has a radius R and a height H. If the water height in the cup is h, the water volume is given by

$$V = \frac{1}{3}\pi\left(\frac{R}{H}\right)^2 h^3$$

Suppose that the cup's dimensions are $R = 1.5$ in. and $H = 4$ in.

a. If the flow rate from the fountain into the cup is 2 in.^{3}/sec, how long will it take to fill the cup to the brim?

b. If the flow rate from the fountain into the cup is given by $2(1 - e^{-2t})$ in.^{3}/sec, how long will it take to fill the cup to the brim?

12. A certain object has a mass of 100 kg and is acted on by a force $f(t) = 500[2 - e^{-t}\sin(5\pi t)]$ N. The mass is at rest at $t = 0$. Determine the object's velocity at $t = 5$ s.

13.* A rocket's mass decreases as it burns fuel. The equation of motion for a rocket in vertical flight can be obtained from Newton's law and is

$$m(t)\frac{dv}{dt} = T - m(t)g$$

where T is the rocket's thrust and its mass as a function of time is given by $m(t) = m_0(1 - rt/b)$. The rocket's initial mass is m_0, the burn time is b, and r is the fraction of the total mass accounted for by the fuel.

Use the values $T = 48{,}000$ N, $m_0 = 2200$ kg, $r = 0.8$, $g = 9.81$ m/s^2, and $b = 40$ s. Determine the rocket's velocity at burnout.

14. The equation for the voltage $v(t)$ across a capacitor as a function of time is

$$v(t) = \frac{1}{C}\left(\int_0^t i(t)\,dt + Q_0\right)$$

where $i(t)$ is the applied current and Q_0 is the initial charge. Suppose that $C = 10^{-6}$ F and that $Q_0 = 0$. Suppose the applied current is $i(t) = [0.01 + 0.3e^{-5t}\sin(25\pi t)]10^{-3}$ A. Plot the voltage $v(t)$ for $0 \le t \le 0.3$ s.

15. Plot the estimate of the derivative dy/dx from the following data. Do this using forward, backward, and central differences. Compare the results.

x	0	1	2	3	4	5	6	7	8	9	10
y	0	2	5	7	9	12	15	18	22	20	17

16. At a relative maximum of a curve $y(x)$, the slope dy/dx is 0. Use the following data to estimate the values of x and y that correspond to a maximum point.

x	0	1	2	3	4	5	6	7	8	9	10
y	0	2	5	7	9	10	8	7	6	8	10

17. Compare the performance of the forward, backward, and central difference methods for estimating the derivative of the following function: $y(x) = e^{-x}\sin(3x)$. Use 101 points from $x = 0$ to $x = 4$. Use a random additive error of ± 0.01.

Sections 8.4 and 8.5

18. Plot the free and total response of the equation

$$5\dot{y} + y = f(t)$$

if $f(t) = 0$ for $t < 0$ and $f(t) = 10$ for $t \geq 0$. The initial condition is $y(0) = 5$.

19. The equation for the voltage y across the capacitor of an RC circuit is

$$RC\frac{dy}{dt} + y = v(t)$$

where $v(t)$ is the applied voltage. Suppose that $RC = 0.2$ s and that the capacitor voltage is initially 2 V. Suppose also that the applied voltage goes from 0 to 10 V at $t = 0$. Plot the voltage $y(t)$ for $0 \leq t \leq 1$ s.

20. The following equation describes the temperature $T(t)$ of a certain object immersed in a liquid bath of constant temperature T_b.

$$10\frac{dT}{dt} + T = T_b$$

Suppose the object's temperature is initially $T(0) = 70°$F and the bath temperature is $T_b = 170°$C.

a. How long will it take for the object's temperature T to reach the bath temperature?
b. How long will it take for the object's temperature T to reach 168°F?
c. Plot the object's temperature $T(t)$ as a function of time.

21.* The equation of motion of a rocket-propelled sled is, from Newton's law,

$$m\dot{v} = f - cv$$

where m is the sled mass, f is the rocket thrust, and c is a air resistance coefficient. Suppose that $m = 1000$ kg and $c = 500$ N · s/m. Suppose $v(0) = 0$ and f is constant for $t \geq 0$.

a. What is the form of the step response $v(t)$?
b. Determine the final speed the sled will reach as a function of f. How long will it take to reach that speed?

22.* The following equation describes the motion of a mass connected to a spring with viscous friction on the surface.

$$m\ddot{y} + c\dot{y} + ky = f(t)$$

where $f(t)$ is an applied force.

a. What is the form of the free response if $m = 3$, $c = 18$, and $k = 102$?
b. What is the form of the free response if $m = 3$, $c = 39$, and $k = 120$?

23. The equation for the voltage y across the capacitor of an RC circuit is

$$RC\frac{dy}{dt} + y = v(t)$$

where $v(t)$ is the applied voltage. Suppose that $RC = 0.2$ s and that the capacitor voltage is initially 2 V. Suppose also that the applied voltage is $v(t) = 10[2 - e^{-t}\sin(5\pi t)]$. Plot the voltage $y(t)$ for $0 \leq t \leq 5$ s. Interpret the results in terms of the circuit's time constant and the behavior of the applied voltage.

24. The equation describing the water height h in a spherical tank with a circular drain of area A at the bottom is

$$\pi\left(2rh - h^2\right)\frac{dh}{dt} = -C_d A\sqrt{2gh}$$

Suppose the tank's radius is $r = 3$ m and that the circular drain hole has a radius of 2 cm. Assume that $C_d = 0.5$ and that the initial water height is $h(0) = 5$ m. Use $g = 9.81$ m/s^2.

a. Use an approximation to estimate how long it takes for the tank to empty.

b. Plot the water height as a function of time until $h(t) = 0$.

25. The following equation describes a certain dilution process, where $y(t)$ is the concentration of salt in a tank of fresh water to which salt brine is being added.

$$\frac{dy}{dt} + \frac{2}{10 + 2t}y = 4$$

Suppose that $y(0) = 0$.

a. Plot $y(t)$ for $0 \leq t \leq 10$.

b. Check your results by using an approximation that converts the differential equation into one having constant coefficients.

Sections 8.6, 8.7, and 8.8

26. The following equation describes the motion of a certain mass connected to a spring with viscous friction on the surface

$$3\ddot{y} + 18\dot{y} + 102y = f(t)$$

where $f(t)$ is an applied force. Suppose that $f(t) = 0$ for $t < 0$ and $f(t) = 10$ for $t \geq 0$.

a. Plot $y(t)$ for $y(0) = \dot{y}(0) = 0$.

b. Plot $y(t)$ for $y(0) = 0$ and $\dot{y}(0) = 10$. Discuss the effect of the nonzero initial velocity.

27. The following equation describes the motion of a certain mass connected to a spring with viscous friction on the surface

$$3\ddot{y} + 39\dot{y} + 120y = f(t)$$

where $f(t)$ is an applied force. Suppose that $f(t) = 0$ for $t < 0$ and $f(t) = 10$ for $t \geq 0$.

a. Plot $y(t)$ for $y(0) = \dot{y}(0) = 0$.
b. Plot $y(t)$ for $y(0) = 0$ and $\dot{y}(0) = 10$. Discuss the effect of the nonzero initial velocity.

28. The following equation describes the motion of a certain mass connected to a spring with no friction

$$3\ddot{y} + 75y = f(t)$$

where $f(t)$ is an applied force. The equation's characteristic roots are $s = \pm 5i$, so the system's natural frequency of oscillation is 5 rad/s. Suppose the applied force is sinusoidal with a frequency of ω rad/s and an amplitude of 10 N: $f(t) = 10 \sin(\omega t)$.

Suppose that the initial conditions are $y(0) = \dot{y}(0) = 0$. Plot $y(t)$ for $0 \leq t \leq 20$ s. Do this for the following three cases. Compare the results of each case.

a. $\omega = 1$ rad/s.
b. $\omega = 5.1$ rad/s.
c. $\omega = 10$ rad/s.

29. Van der Pol's equation has been used to describe many oscillatory processes. It is

$$\ddot{y} - \mu(1 - y^2)\dot{y} + y = 0$$

Plot $y(t)$ for $\mu = 1$ and $0 \leq t \leq 20$, using the initial conditions $y(0) = 2$, $\dot{y}(0) = 0$.

30. The equation of motion for a pendulum whose base is accelerating horizontally with an acceleration $a(t)$ is

$$L\ddot{\theta} + g \sin\theta = a(t)\cos\theta$$

Suppose that $g = 9.81$ m/s^2, $L = 1$ m, and $\dot{\theta}(0) = 0$. Plot $\theta(t)$ for $0 \leq t \leq 10$ s for the following three cases:

a. The acceleration is constant: $a = 5$ m/s^2 and $\theta(0) = 0.5$ rad.
b. The acceleration is constant: $a = 5$ m/s^2 and $\theta(0) = 3$ rad.
c. The acceleration is linear with time: $a = 0.5t$ m/s^2 and $\theta(0) = 3$ rad.

31. The equations for an armature-controlled dc motor are the following. The motor's current is i, and its rotational velocity is ω.

$$L\frac{di}{dt} = -Ri - K_e\omega + v(t) \qquad (8.6\text{–}10)$$

$$I\frac{d\omega}{dt} = K_T i - c\omega \qquad (8.6\text{–}11)$$

where L, R, and I are the motor's inductance, resistance, and inertia; K_T and K_e are the torque constant and back emf constant; c is a viscous damping constant; and $v(t)$ is the applied voltage.

Use the values $R = 0.8\ \Omega$, $L = 0.003$ H, $K_T = 0.05$ N $\cdot$ m/A, $K_e = 0.05$ V $\cdot$ s/rad, $c = 0$, and $I = 8 \times 10^{-5}$ kg $\cdot$ m^2.

a. Find the motor's characteristic roots and time constants. If the applied voltage is constant, approximately how long will it take for the motor to reach a constant speed?

b. Suppose the applied voltage is 20 V. Plot the motor's speed and current versus time. Choose a final time large enough to show the motor's speed becoming constant.

c. Suppose the applied voltage is trapezoidal as given here:

$$v(t) = \begin{cases} 400t & 0 \le t < 0.05 \\ 20 & 0.05 \le t \le 0.2 \\ -400(t - 0.2) + 20 & 0.2 < t \le 0.25 \\ 0 & t > 0.25 \end{cases}$$

Plot the motor's speed versus time for $0 \le t \le 0.3$ s. Also plot the applied voltage versus time. Does the motor speed follow a trapezoidal profile?

32.* (Control System toolbox) Find the state-space form of the following model:

$$10\ddot{y} + 3\dot{y} + 7y = f(t)$$

33. (Control System toolbox) Find the state-space form of the following model:

$$10\ddot{y} + 6\dot{y} + 2y = f + 3\dot{f}$$

34.* (Control System toolbox) Find the reduced form of the following state model in terms of x_1.

$$\begin{bmatrix} \dot{x}_1 \\ \dot{x}_2 \end{bmatrix} = \begin{bmatrix} -4 & -1 \\ 2 & -3 \end{bmatrix} \begin{bmatrix} x_1 \\ x_2 \end{bmatrix} + \begin{bmatrix} 2 \\ 5 \end{bmatrix} u(t)$$

35. (Control System toolbox) The following state model describes the motion of a certain mass connected to a spring with viscous friction on the surface, where $m = 1$, $c = 2$, and $k = 5$.

$$\begin{bmatrix} \dot{x}_1 \\ \dot{x}_2 \end{bmatrix} = \begin{bmatrix} 0 & 1 \\ -5 & -2 \end{bmatrix} \begin{bmatrix} x_1 \\ x_2 \end{bmatrix} + \begin{bmatrix} 0 \\ 1 \end{bmatrix} f(t)$$

a. Use the `initial` function to plot the position x_1 of the mass if the initial position is 5 and the initial velocity is 3.

b. Use the `step` function to plot the step response of the position and velocity for zero initial conditions, where the magnitude of the step input is 10. Compare your plot with that shown in Figure 8.6–4.

36. Consider the following equation:

$$5\ddot{y} + 2\dot{y} + 10y = f(t)$$

a. Plot the free response for the initial conditions $y(0) = 10$, $\dot{y}(0) = -5$.
b. Plot the unit-step response (for zero initial conditions).
c. The *total response* to a step input is the sum of the free response and the step response. Demonstrate this fact for this equation by plotting the sum of the solutions found in parts *a* and *b* and comparing the plot with that generated by solving for the total response with $y(0) = 10$, $\dot{y}(0) = -5$.

37. (Control System toolbox) Use the `lsim` function to solve the dc motor problem given in Example 8.6–2. Compare your results with those shown in Figure 8.6–6.

38. (Control System toolbox) The model for the RC circuit shown in Figure P38 is

$$RC\frac{dv_o}{dt} + v_o = v_i$$

For $RC = 0.1$ s, plot the voltage response $v_o(t)$ for the case where the applied voltage is a single square pulse of height 10 V and duration 0.2 s, starting at $t = 0$. The initial capacitor voltage is 0.

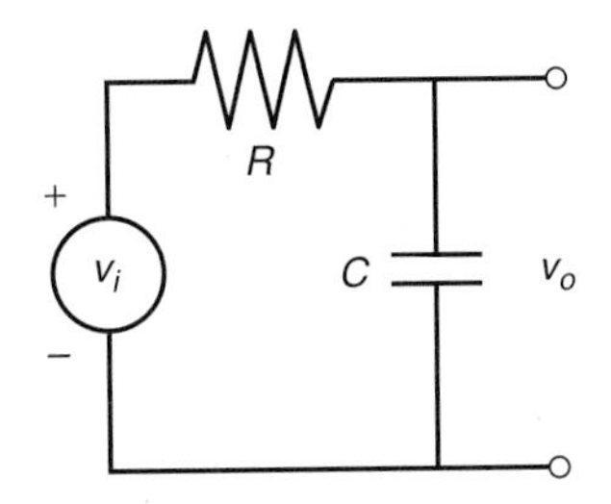

Figure P38

39. Van der Pol's equation is

$$\ddot{y} - \mu(1 - y^2)\dot{y} + y = 0$$

This equation is stiff for large values of the parameter μ. Compare the performance of `ode45` and `ode23s` for this equation. Use $\mu = 1000$ and $0 \le t \le 3000$ with the initial conditions $y(0) = 2$, $\dot{y}(0) = 0$. Plot $y(t)$ versus t.

40. Use MATLAB to plot the trajectory of a ball thrown at an angle of 30° to the horizontal with a speed of 30 m/s. The ball bounces off the horizontal surface and loses 20 percent of its vertical speed with each bounce. Plot the trajectory showing three bounces.

Engineering in the 21st Century...

Embedded Control Systems

An embedded control system is a microprocessor and sensor suite designed to be an integral part of a product. The aerospace and automotive industries have used embedded controllers for some time, but the decreasing cost of components now make embedded controllers feasible for more consumer and biomedical applications.

For example, embedded controllers can greatly increase the performance of orthopedic devices. One model of an artificial leg now uses sensors to measure in real time the walking speed, the knee joint angle, and the loading due to the foot and ankle. These measurements are used by the controller to adjust the hydraulic resistance of a piston to produce a more stable, natural, and efficient gait. The controller algorithms are adaptive in that they can be tuned to an individual's characteristics and their settings changed to accommodate different physical activities.

Engines incorporate embedded controllers to improve efficiency. Embedded controllers in new active suspensions use actuators to improve on the performance of traditional passive systems consisting only of springs and dampers. One design phase of such systems is *hardware-in-the-loop testing,* in which the controlled object (the engine or vehicle suspension) is replaced with a real-time simulation of its behavior. This enables the embedded system hardware and software to be tested faster and less expensively than with the physical prototype, and perhaps even before the prototype is available.

Simulink is often used to create the simulation model for hardware-in-the-loop testing. The Control Systems and the Signal Processing toolboxes, and the DSP and Fixed Point blocksets, are also useful for such applications. ■

CHAPTER 9

Simulink

OUTLINE

Simulink is built on top of MATLAB, so you must have MATLAB to use Simulink. It is included in the Student Edition of MATLAB, and is also available separately from The MathWorks, Inc. The popularity of Simulink has greatly increased in the last few years, as evidenced by the increasing number of short courses offered at meetings sponsored by professional organizations such as the American Society of Engineering Education.

Simulink provides a graphical user interface that uses various types of elements called *blocks* to create a simulation of a dynamic system—that is, a system that can be modeled with differential or difference equations whose independent variable is time. For example, one block type is a multiplier, another performs a sum, and another is an integrator. The Simulink graphical interface enables you to position the blocks, resize them, label them, specify block parameters, and interconnect the blocks to describe complicated systems for simulation.

This chapter starts with simulations of simple systems that require few blocks. Gradually, through a series of examples, more block types are introduced. The chosen applications require only a basic knowledge of physics and thus can be appreciated by readers from any engineering discipline. By the time you have finished this chapter you will have seen the block types needed to simulate a large variety of common engineering applications.

9.1 Simulation Diagrams

BLOCK DIAGRAM

You develop Simulink models by constructing a diagram that shows the elements of the problem to be solved. Such diagrams are called *simulation diagrams* or *block diagrams*. Consider the equation $\dot{y} = 10f(t)$. Its solution can be represented symbolically as

$$y(t) = \int 10f(t)\,dt$$

which can be thought of as two steps, using an intermediate variable x:

$$x(t) = 10f(t) \qquad \text{and} \qquad y(t) = \int x(t)\,dt$$

GAIN BLOCK

INTEGRATOR BLOCK

This solution can be represented graphically by the simulation diagram shown in Figure 9.1–1a. The arrows represent the variables y, x, and f. The blocks represent cause-and-effect processes. Thus, the block containing the number 10 represents the process $x(t) = 10f(t)$, where $f(t)$ is the cause (the *input*) and $x(t)$ represents the effect (the *output*). This type of block is called a *multiplier* or *gain* block.

The block containing the integral sign $\int$ represents the integration process $y(t) = \int x(t)\,dt$, where $x(t)$ is the cause (the *input*) and $y(t)$ represents the effect (the *output*). This type of block is called an *integrator* block.

There is some variation in the notation and symbols used in simulation diagrams. Figure 9.1–1b shows one variation. Instead of being represented by a box, the multiplication process is now represented by a triangle like that used to represent an electrical amplifier, hence the name *gain* block.

In addition, the integration symbol in the integrator block has been replaced by the operator symbol $1/s$, which derives from the notation used for the Laplace transform (see Section 10.5 for a discussion of this transform). Thus the equation $\dot{y} = 10f(t)$ is represented by $sy = 10f$, and the solution is represented as

$$y = \frac{10f}{s}$$

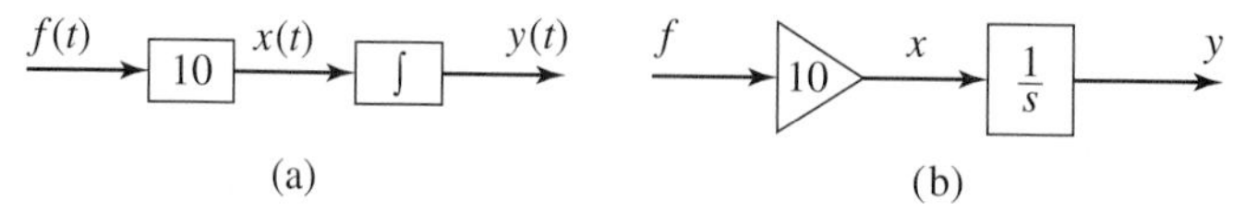

Figure 9.1–1 Simulation diagrams for $\dot{y} = 10f(t)$.

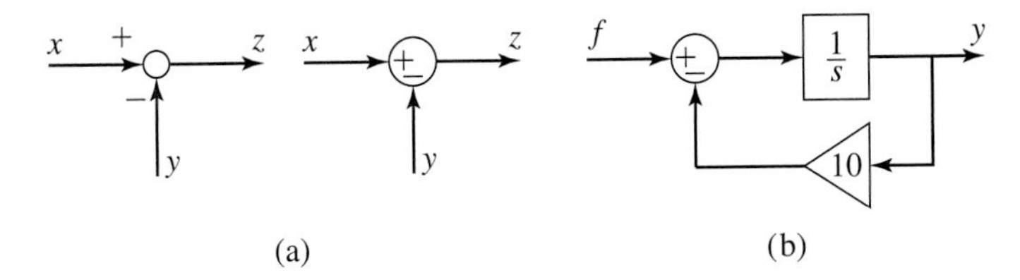

Figure 9.1–2 (a) The summer element. (b) Simulation diagram for $\dot{y} = f(t) - 10y$.

or as the two equations:

$$x = 10f \qquad \text{and} \qquad y = \frac{1}{s}x$$

SUMMER

Another element used in simulation diagrams is the *summer* that, despite its name, is used to subtract as well as to sum variables. Two versions of its symbol are shown in Figure 9.1–2a. In each case the symbol represents the equation $z = x - y$. Note that a plus or minus sign is required for each input arrow.

The summer symbol can be used to represent the equation $\dot{y} = f(t) - 10y$, which can be expressed as

$$y(t) = \int [f(t) - 10y]\,dt$$

or as

$$y = \frac{1}{s}(f - 10y)$$

You should study the simulation diagram shown in Figure 9.1–2b to confirm that it represents this equation. This figure forms the basis for developing a Simulink model to solve the equation.

9.2 Introduction to Simulink

LIBRARY BROWSER

Type `simulink` in the MATLAB Command window to start Simulink. The Simulink Library Browser window opens. See Figure 9.2–1. The Simulink blocks are located in "libraries." These libraries are displayed under the Simulink heading in Figure 9.2–1. Depending on what other Mathworks products are installed, you might see additional items in this window, such as the Control System Toolbox and Stateflow. These provide additional Simulink blocks, which can be displayed by clicking on the plus sign to the left of the item. As Simulink evolves through new versions, some libraries are renamed and some blocks are moved to different libraries, so the library we specify here might change in later releases. The best way to locate a block, given its name, is to type its name in the Find pane at the top of the Simulink Library Browser. When you press Enter, Simulink will take you to the block location and will display a brief description of the block in the pane below the Find pane.

To create a new model, click on the icon that resembles a clean sheet of paper, or select **New** from the **File** menu in the Browser. A new **Untitled** window opens for you to create the model. To select a block from the Library Browser, double-click on the appropriate library and a list of blocks within that library then

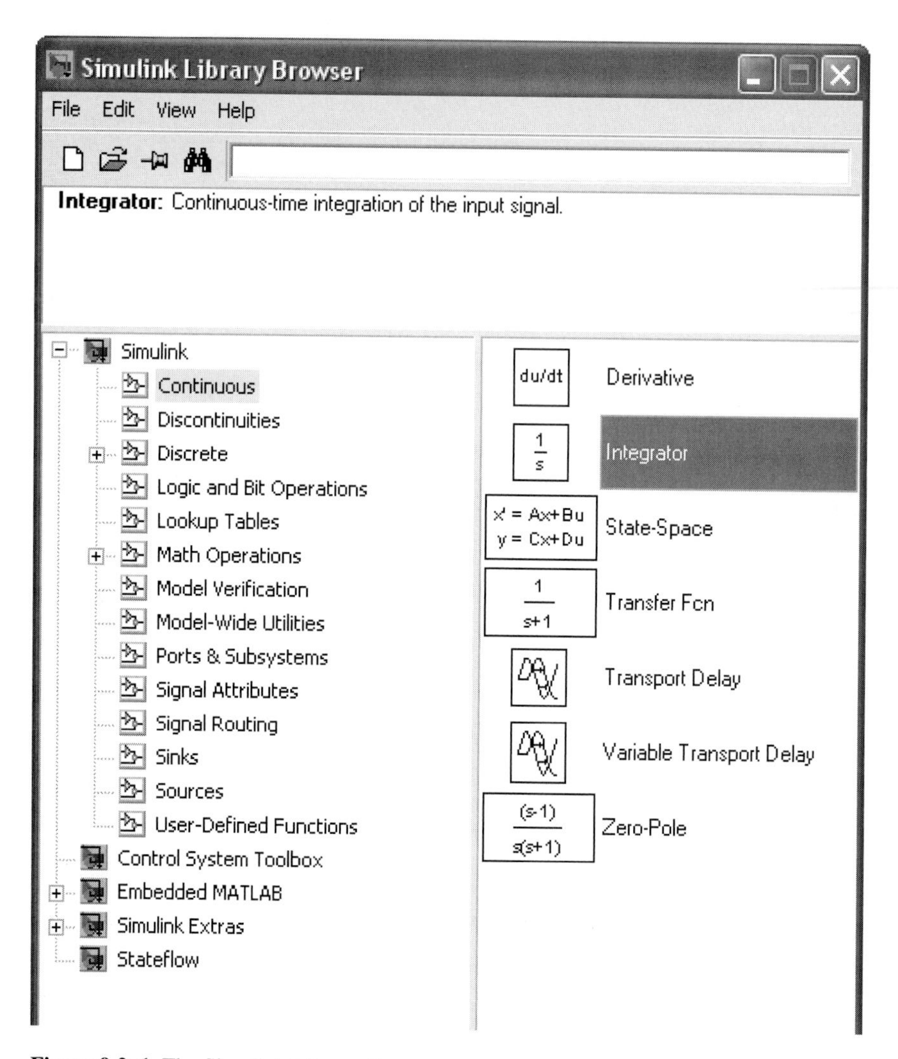

Figure 9.2–1 The Simulink Library Browser.

appears as shown in Figure 9.2–1. This figure shows the result of double-clicking on the Continuous library, then clicking on the Integrator block.

Click on the block name or icon, hold the mouse button down, drag the block to the new model window, and release the button. Note that when you click on the block name in the Library Browser, a brief description of the block's function appears at the top of the Browser. You can access help for that block by right-clicking on its name or icon, and selecting **Help** from the drop-down menu.

Simulink model files have the extension `.mdl`. Use the **File** menu in the model window to Open, Close, and Save model files. To print the block diagram of the model, select **Print** on the **File** menu. Use the **Edit** menu to copy, cut and paste blocks. You can also use the mouse for these operations. For example, to delete a block, click on it and press the **Delete** key.

Getting started with Simulink is best done through examples, which we now present.

Simulink Solution of $\dot{y} = 10 \sin t$ EXAMPLE 9.2–1

Use Simulink to solve the following problem for $0 \le t \le 13$.

$$\frac{dy}{dt} = 10 \sin t \qquad y(0) = 0$$

The exact solution is $y(t) = 10(1 - \cos t)$.

■ Solution

To construct the simulation, do the following steps. Refer to Figure 9.2–2.

1. Start Simulink and open a new model window as described previously.
2. Select and place in the new window the Sine Wave block from the Sources library. Double-click on it to open the Block Parameters window, and make sure the Amplitude is set to 1, the Frequency to 1, the Phase to 0, and the Sample time to 0. Then click **OK.**
3. Select and place the Gain block from the Math Operations library, double-click on it, and set the Gain value to 10 in the Block Parameters window. Then click **OK.** Note that the value 10 then appears in the triangle. To make the number more visible, click on the block, and drag one of the corners to expand the block so that all the text is visible.
4. Select and place the Integrator block from the Continuous library, double-click on it to obtain the Block Parameters window, and set the Initial condition to 0 (this is because $y(0) = 0$). Then click **OK.**
5. Select and place the Scope block from the Sinks library.
6. Once the blocks have been placed as shown in Figure 9.2–2, connect the input port on each block to the outport port on the preceding block. To do this, move the cursor to an input port or an output port; the cursor will change to a cross. Hold the mouse button down, and drag the cursor to a port on another block. When you release the mouse button, Simulink will connect them with an arrow pointing at the input port. Your model should now look like that shown in Figure 9.2–2.
7. Click on the **Simulation** menu, and click the **Configuration Parameters** item. Click on the Solver tab, and enter 13 for the Stop time. Make sure the Start time is 0. Then click **OK.**
8. Run the simulation by clicking on the **Simulation** menu, and then clicking the **Start** item. You can also start the simulation by clicking on the **Start** icon on the toolbar (this is the black triangle).

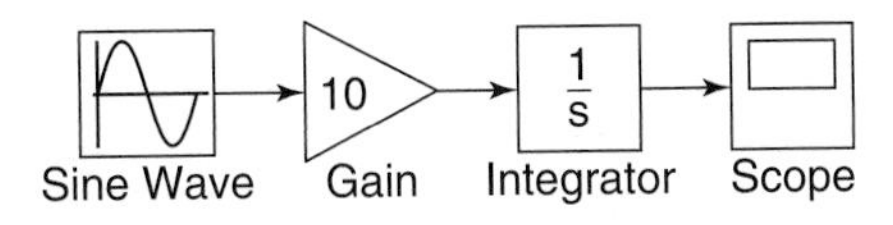

Figure 9.2–2 Simulink model for $\dot{y} = 10 \sin t$.

9. You will hear a bell sound when the simulation is finished. Then double-click on the Scope block and then click on the binoculars icon in the Scope display to enable autoscaling. You should see an oscillating curve with an amplitude of 10 and a period of 2π. The independent variable in the Scope block is time t; the input to the block is the dependent variable y. This completes the simulation.

In the **Configuration Parameters** submenu under the **Simulation** menu, you can select the ODE solver to use by clicking on the Solver tab. The default is `ode45`.

To have Simulink automatically connect two blocks, select the source block, hold down the **Ctrl** key, and left-click on the destination block. Simulink also provides easy ways to connect multiple blocks and lines; see the help for information.

Note that blocks have a Block Parameters window that opens when you double-click on the block. This window contains several items, the number and nature of which depend on the specific type of block. In general, you can use the default values of these parameters, except where we have explicitly indicated that they should be changed. You can always click on **Help** within the Block Parameters window to obtain more information.

When you click on **Apply,** any changes immediately take effect and the window remains open. If you click on **OK,** the changes take effect but the window closes.

Note that most blocks have default labels. You can edit text associated with a block by clicking on the text and making the changes. You can save the Simulink model as an `.mdl` file by selecting **Save** from the **File** menu in Simulink. The model file can then be reloaded at a later time. You can also print the diagram by selecting **Print** on the **File** menu.

The Scope block is useful for examining the solution, but if you want to obtain a labeled and printed plot you can use the To Workspace block, which is described in the next example.

EXAMPLE 9.2–2 Exporting to the MATLAB Workspace

We now demonstrate how to export the results of the simulation to the MATLAB workspace, where they can be plotted or analyzed with any of the MATLAB functions.

■ Solution

Modify the Simulink model constructed in Example 9.2–1 as follows. Refer to Figure 9.2–3.

1. Delete the arrow connecting the Scope block by clicking on it and pressing the Delete key. Delete the Scope block in the same way.
2. Select and place the To Workspace block from the Sinks library and the Clock block from the Sources library.
3. Select and place the Mux block from the Signal Routing library, double-click on it, and set the Number of inputs to 2. Click **OK.** (The name Mux is an

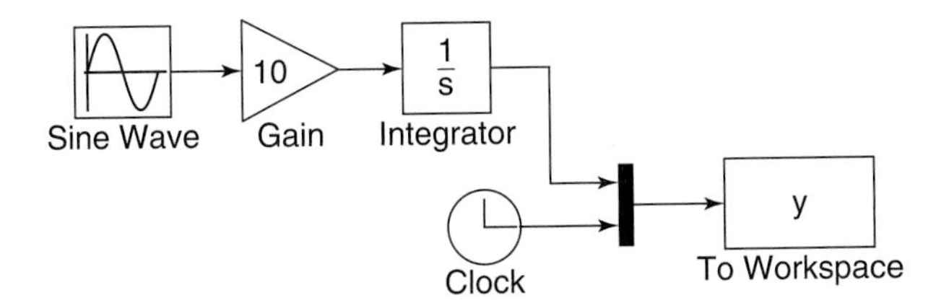

Figure 9.2–3 Simulink model using the Clock and To Workspace blocks.

abbreviation for multiplexer, which is an electrical device for transmitting several signals.)

4. Connect the top input port of the Mux block to the output port of the Integrator block. Then use the same technique to connect the bottom input port of the Mux block to the outport port of the Clock block. Your model should now look like that shown in Figure 9.2–3.
5. Double-click on the To Workspace block. You can specify any variable name you want as the output; the default is `simout`. Change its name to `y`. The output variable `y` will have as many rows as there are simulation time steps, and as many columns as there are inputs to the block. The second column in our simulation will be time, because of the way we have connected the Clock to the second input port of the Mux. Specify the Save format as Array. Use the default values for the other parameters (these should be `inf`, `1`, and `-1` for Maximum number of rows, Decimation, and Sample time, respectively). Click on **OK.**
6. After running the simulation, you can use the MATLAB plotting commands from the Command window to plot the columns of `y` (or `simout` in general). To plot $y(t)$, type in the MATLAB Command window:

```
>>plot(y(:,2),y(:,1)),xlabel('t'),ylabel('y')
```

Simulink can be configured to put the time variable `tout` into the MATLAB workspace automatically when you are using the To Workspace block. This is done with the Data I/O tab under **Configuration Parameters** on the Simulation menu. The alternative is to use the Clock block to put `tout` into the workspace. The Clock block has one parameter, Decimation. If this parameter is set to 1, the Clock block will output the time every time step; if set to 10 for example, the block will output every 10 time steps, and so on.

Simulink Model for $\dot{y} = -10y + f(t)$ — EXAMPLE 9.2–3

Construct a Simulink model to solve

$$\dot{y} = -10y + f(t) \qquad y(0) = 1$$

where $f(t) = 2\sin 4t$, for $0 \leq t \leq 3$.

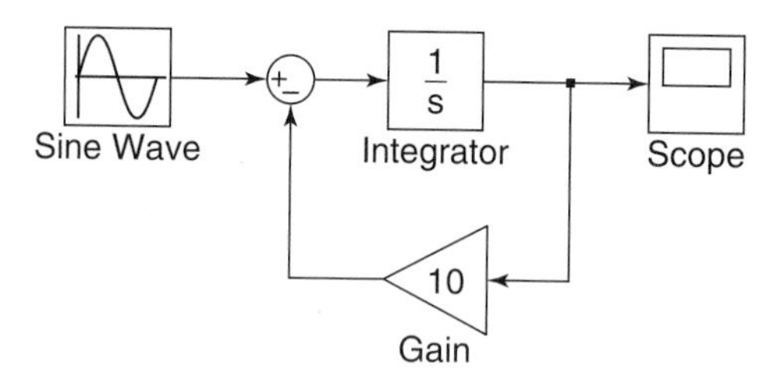

Figure 9.2–4 Simulink model for $\dot{y} = -10y + f(t)$.

■ Solution

To construct the simulation, do the following steps.

1. You can use the model shown in Figure 9.2–2 by rearranging the blocks as shown in Figure 9.2–4. You will need to add a Sum block.
2. Select the Sum block from the Math Operations library and place it as shown in the simulation diagram. Its default setting adds two input signals. To change this, double-click on the block, and in the List of Signs window, type |+-. The signs are ordered counterclockwise from the top. The symbol | is a spacer indicating here that the top port is to be empty.
3. To reverse the direction of the Gain block, right-click on the block, select **Format** from the pop-up menu, and select **Flip Block.**
4. When you connect the negative input port of the Sum block to the output port of the Gain block, Simulink will attempt to draw the shortest line. To obtain the more standard appearance shown in Figure 9.2–4, first extend the line vertically down from the Sum input port. Release the mouse button and then click on the end of the line and attach it to the Gain block. The result will be a line with a right angle. Do the same to connect the input of the Gain to the arrow connecting the Integrator and the Scope. A small dot appears to indicate that the lines have been successfully connected. This point is called a *takeoff point* because it takes the value of the variable represented by the arrow (here, the variable y) and makes that value available to another block.
5. Select **Configuration Parameters** from the **Simulation** menu, and set the Stop time to 3. Then click **OK.**
6. Run the simulation as before and observe the results in the Scope.

9.3 Linear State-Variable Models

State-variable models, unlike transfer-function models, can have more than one input and more than one output. Simulink has the State-Space block that represents the linear state-variable model $\dot{\mathbf{x}} = \mathbf{Ax} + \mathbf{Bu}$, $\mathbf{y} = \mathbf{Cx} + \mathbf{Du}$. (See Section 8.7 for discussion of this model form.) The vector **u** represents the inputs, and the vector **y** represents the outputs. Thus, when you are connecting inputs to the State-Space block, care must be taken to connect them in the proper order. Similar care must be taken when connecting the block's outputs to another block. The following example illustrates how this is done.

Simulink Model of a Two-Mass System

EXAMPLE 9.3–1

Consider the two-mass system shown in Figure 9.3–1. Suppose the parameter values are $m_1 = 5$, $m_2 = 3$, $c_1 = 4$, $c_2 = 8$, $k_1 = 1$, and $k_2 = 4$. The equations of motion are

$$5\ddot{x}_1 + 12\dot{x}_1 + 5x_1 - 8\dot{x}_2 - 4x_2 = 0$$

$$3\ddot{x}_2 + 8\dot{x}_2 + 4x_2 - 8\dot{x}_1 - 4x_1 = f(t)$$

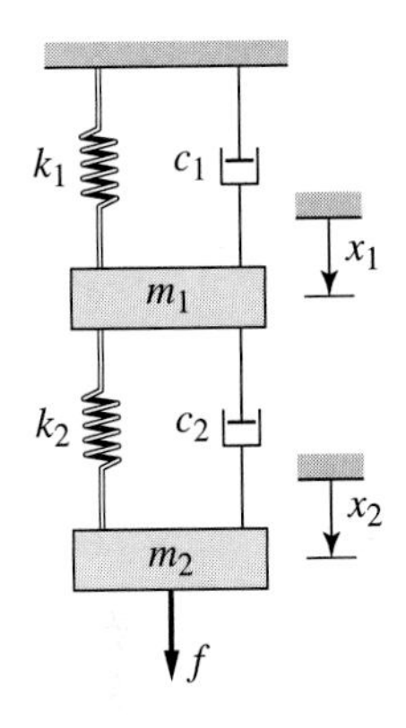

Figure 9.3–1 A vibrating system having two masses.

These equations can be expressed in state-variable form as

$$\dot{z}_1 = z_2 \qquad \dot{z}_2 = \frac{1}{5}(-5z_1 - 12z_2 + 4z_3 + 8z_4)$$

$$\dot{z}_3 = z_4 \qquad \dot{z}_4 = \frac{1}{3}[4z_1 + 8z_2 - 4z_3 - 8z_4 + f(t)]$$

In vector-matrix form these equations are

$$\dot{\mathbf{z}} = \mathbf{Az} + \mathbf{B}f(t)$$

where

$$\mathbf{A} = \begin{bmatrix} 0 & 1 & 0 & 0 \\ -1 & -\frac{12}{5} & \frac{4}{5} & \frac{8}{5} \\ 0 & 0 & 0 & 1 \\ \frac{4}{3} & \frac{8}{3} & -\frac{4}{3} & -\frac{8}{3} \end{bmatrix} \qquad \mathbf{B} = \begin{bmatrix} 0 \\ 0 \\ 0 \\ \frac{1}{3} \end{bmatrix}$$

and

$$\mathbf{z} = \begin{bmatrix} z_1 \\ z_2 \\ z_3 \\ z_4 \end{bmatrix} = \begin{bmatrix} x_1 \\ \dot{x}_1 \\ x_2 \\ \dot{x}_2 \end{bmatrix}$$

Develop a Simulink model to plot the unit-step response of the variables x_1 and x_2 with the initial conditions $x_1(0) = 0.2$, $\dot{x}_1(0) = 0$, $x_2(0) = 0.5$, $\dot{x}_2(0) = 0$.

■ Solution

First select appropriate values for the matrices in the output equation $\mathbf{y} = \mathbf{Cz} + \mathbf{B}f(t)$. Since we want to plot x_1 and x_2, which are z_1 and z_2, we choose **C** and **D** as follows.

$$\mathbf{C} = \begin{bmatrix} 1 & 0 & 0 & 0 \\ 0 & 0 & 1 & 0 \end{bmatrix} \qquad \mathbf{D} = \begin{bmatrix} 0 \\ 0 \end{bmatrix}$$

To create this simulation, obtain a new model window. Then do the following to create the model shown in Figure 9.3–2.

1. Select and place in the new window the Step block. Double-click on it to obtain the Block Parameters window, and set the Step time to 0, the Initial and Final values to 0 and 1, and the Sample time to 0. Click **OK.** The Step time is the time at which the step input begins.

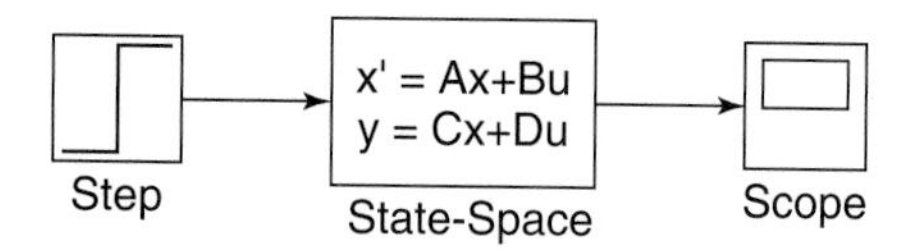

Figure 9.3–2 Simulink model containing the State-Space block and the Step block.

2. Select and place the State-Space block. Double-click on it, and enter `[0, 1, 0, 0; -1, -12/5, 4/5, 8/5; 0, 0, 0, 1; 4/3, 8/3, -4/3, -8/3]` for **A**, `[0; 0; 0; 1/3]` for **B**, `[1, 0, 0, 0; 0, 0, 1, 0]` for **C**, and `[0; 0]` for **D**. Then enter `[0.2; 0; 0.5; 0]` for the initial conditions. Click **OK.** Note that the dimension of the matrix **B** tells Simulink that there is one input. The dimensions of the matrices **C** and **D** tell Simulink that there are two outputs.
3. Select and place the Scope block.
4. Once the blocks have been placed, connect the input port on each block to the outport port on the preceding block as shown in the figure.
5. Experiment with different values of the Stop time until the Scope shows that the steady-state response has been reached. For this application, a Stop time of 25 is satisfactory. The plots of both x_1 and x_2 will appear in the Scope.

9.4 Piecewise-Linear Models

Unlike linear models, closed-form solutions are not available for most nonlinear differential equations, and we must therefore solve such equations numerically. A nonlinear ordinary differential equation can be recognized by the fact that the dependent variable or its derivatives appears raised to a power or in a transcendental function. For example, the following equations are nonlinear.

$$y\ddot{y} + 5\dot{y} + y = 0 \qquad \dot{y} + \sin y = 0 \qquad \dot{y} + \sqrt{y} = 0$$

Piecewise-linear models are actually nonlinear, although they may appear to be linear. They are composed of linear models that take effect when certain conditions are satisfied. The effect of switching back and forth between these linear models makes the overall model nonlinear. An example of such a model is a mass attached to a spring and sliding on a horizontal surface with Coulomb friction. The model is

$$m\ddot{x} + kx = f(t) - \mu mg \quad \text{if } \dot{x} \geq 0$$

$$m\ddot{x} + kx = f(t) + \mu mg \quad \text{if } \dot{x} < 0$$

These two linear equations can be expressed as the single, nonlinear equation

$$m\ddot{x} + kx = f(t) - \mu mg\,\text{sign}(\dot{x}) \quad \text{where} \quad \text{sign}(\dot{x}) = \begin{cases} +1 & \text{if } \dot{x} \geq 0 \\ -1 & \text{if } \dot{x} < 0 \end{cases}$$

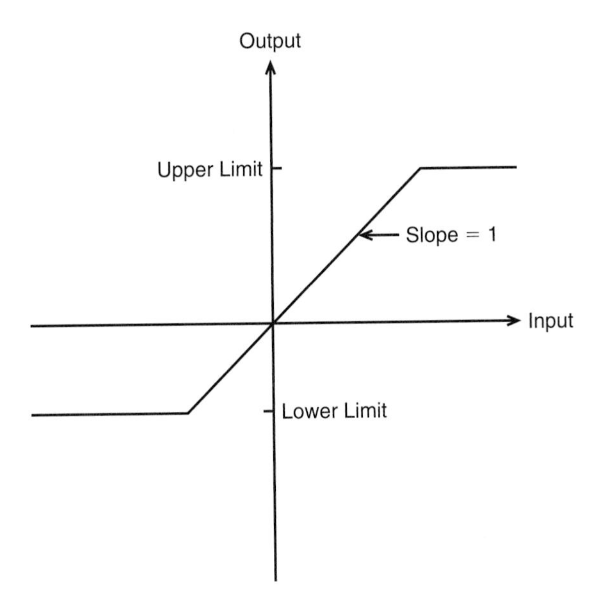

Figure 9.4–1 The saturation nonlinearity.

Solutions of models that contain piecewise-linear functions are very tedious to program. However, Simulink has built-in blocks that represent many of the commonly-found functions such as Coulomb friction. Therefore Simulink is especially useful for such applications. One such block is the Saturation block in the Discontinuities library. The block implements the saturation function shown in Figure 9.4–1.

Simulink Model of a Rocket-Propelled Sled

EXAMPLE 9.4–1

A rocket-propelled sled on a track is represented in Figure 9.4–2 as a mass m with an applied force f that represents the rocket thrust. The rocket thrust initially is horizontal, but the engine accidentally pivots during firing and rotates with an angular acceleration of $\ddot{\theta} = \pi/50$ rad/s. Compute the sled's velocity v for $0 \leq t \leq 6$ if $v(0) = 0$. The rocket thrust is 4000 N and the sled mass is 450 kg.

The sled's equation of motion is

$$450\dot{v} = 4000 \cos \theta(t)$$

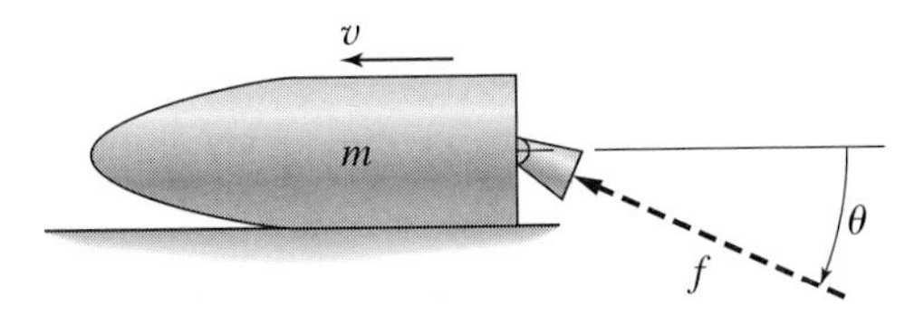

Figure 9.4–2 A rocket-propelled sled.

To obtain $\theta(t)$, note that

$$\dot{\theta} = \int_0^t \ddot{\theta}\, dt = \frac{\pi}{50}t$$

and

$$\theta = \int_0^t \dot{\theta}\, dt = \int_0^t \frac{\pi}{50} t\, dt = \frac{\pi}{100}t^2$$

Thus the equation of motion becomes

$$450\dot{v} = 4000 \cos\left(\frac{\pi}{100}t^2\right)$$

or

$$\dot{v} = \frac{80}{9} \cos\left(\frac{\pi}{100}t^2\right)$$

The solution is formally given by

$$v(t) = \frac{80}{9} \int_0^t \cos\left(\frac{\pi}{100}t^2\right) dt$$

Unfortunately, no closed-form solution is available for the integral, which is called *Fresnel's cosine integral*. The value of the integral has been tabulated numerically, but we will use Simulink to obtain the solution.

(a) Create a Simulink model to solve this problem for $0 \le t \le 10$ s.

(b) Now suppose that the engine angle is limited by a mechanical stop to 60°, which is $60\pi/180$ rad. Create a Simulink model to solve the problem.

■ Solution

(a) There are several ways to create the input function $\theta = (\pi/100)t^2$. Here we note that $\ddot{\theta} = \pi/50$ rad/s and that

$$\dot{\theta} = \int_0^t \ddot{\theta}\, dt$$

and

$$\theta = \int_0^t \dot{\theta}\, dt = \frac{\pi}{100}t^2$$

Thus we can create $\theta(t)$ by integrating the constant $\ddot{\theta} = \pi/50$ twice. The simulation diagram is shown in Figure 9.4–3. This diagram is used to create the corresponding Simulink model shown in Figure 9.4–4.

There are two new blocks in this model. The Constant block is in the Sources library. After placing it, double-click on it and type `pi/50` in its Constant Value window.

The Trigonometric block is in the Math Operations library. After placing it, double-click on it and select `cos` in its Function window.

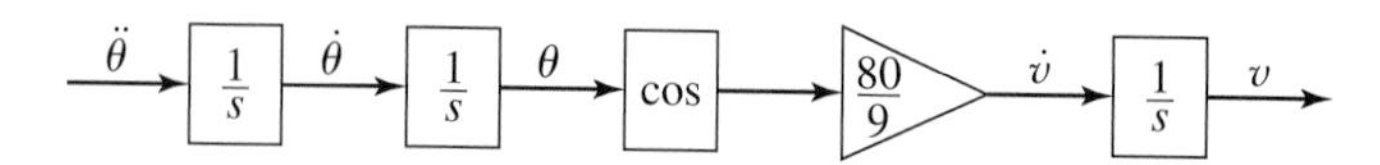

Figure 9.4–3 Simulation diagram for $v = (80/9)\cos(\pi t^2/100)$.

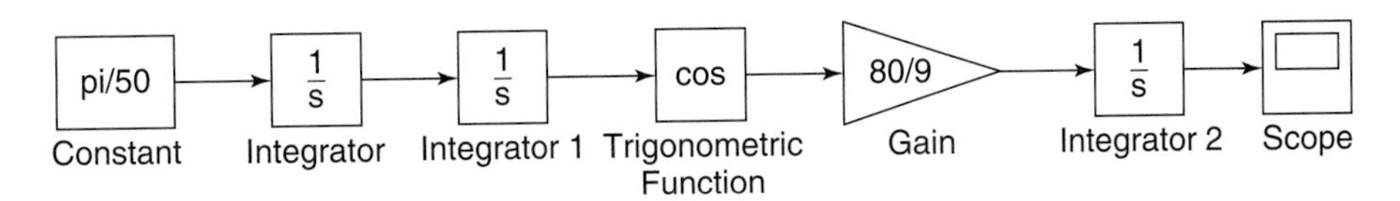

Figure 9.4–4 Simulink model for $v = (80/9)\cos(\pi t^2/100)$.

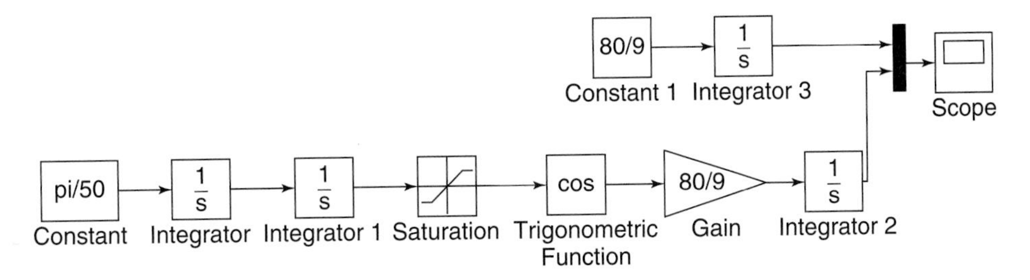

Figure 9.4–5 Simulink model for $v = (80/9)\cos(\pi t^2/100)$ with a Saturation block.

Set the Stop time to 10, run the simulation, and examine the results in the Scope.

(b) Modify the model in Figure 9.4–4 as follows to obtain the model shown in Figure 9.4–5. We use the Saturation block in the Discontinuities library to limit the range of θ to $60\pi/180$ rad. After placing the block as shown in Figure 9.4–5, double-click on it and type `60*pi/180` in its Upper Limit window. Then type 0 in its Lower Limit window.

Enter and connect the remaining elements as shown, and run the simulation. The upper Constant block and Integrator block are used to generate the solution when the engine angle is $\theta = 0$, as a check on our results. (The equation of motion for $\theta = 0$ is $\dot{v} = 80/9$, which gives $v(t) = 80t/9$.)

If you prefer, you can substitute a To Workspace block for the Scope, add a Clock block, and change the number of inputs to the Mux block to three (do this by double-clicking on it). Then you can plot the results in MATLAB. The resulting plot is shown in Figure 9.4–6.

The Relay Block

The Simulink Relay block is an example of something that is tedious to program in MATLAB but is easy to implement in Simulink. Figure 9.4–7a is a graph of the logic of a relay. The relay switches the output between two specified values, named *On* and *Off* in the figure. Simulink calls these values "Output when on" and "Output when off." When the relay output is *On,* it remains *On* until the input drops below the value of the Switch-off point parameter, named *SwOff* in the figure. When the relay output is *Off*, it remains *Off* until the input exceeds the value of the Switch-on point parameter, named *SwOn* in the figure.

The Switch-on point parameter value must be greater than or equal to the Switch-off point value. Note that the value of *Off* need not be zero. Note also

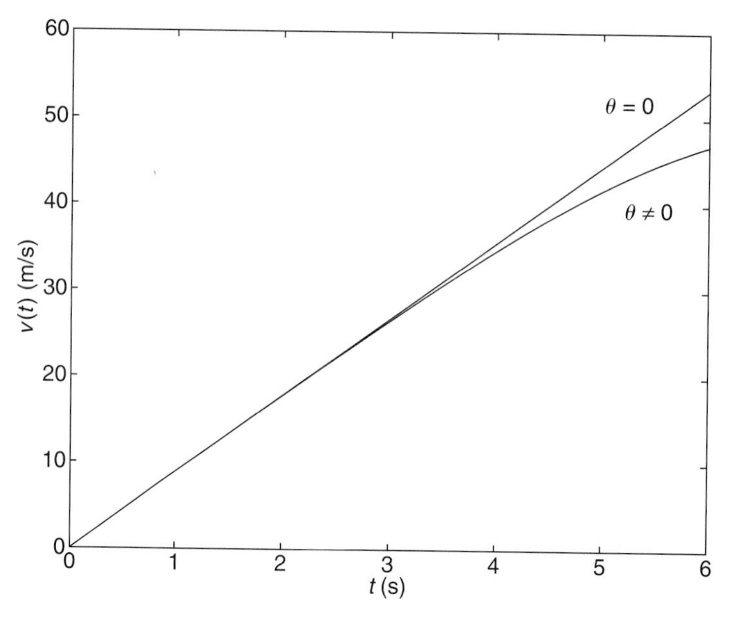

Figure 9.4–6 Speed response of the sled for $\theta = 0$ and $\theta \neq 0$.

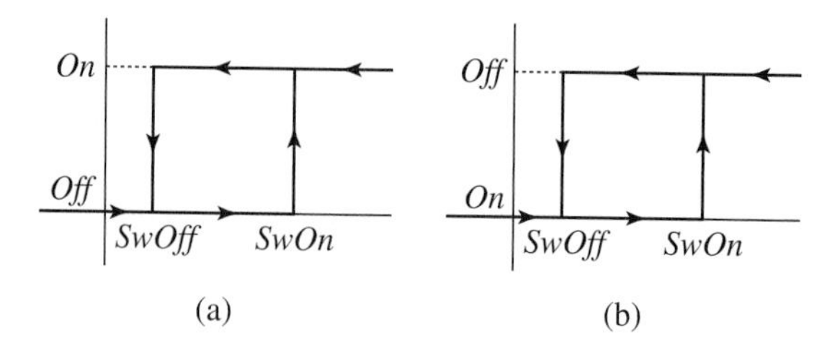

Figure 9.4–7 The relay function. (a) The case where $On > Off$. (b) The case where $On < Off$.

that the value of *Off* need not be less than the value of *On*. The case where $Off > On$ is shown in Figure 9.4–7b. As we will see in the following example, it is sometimes necessary to use this case.

EXAMPLE 9.4–2 Model of a Relay-Controlled Motor

The model of an armature-controlled dc motor was discussed in Section 8.6. See Figure 9.4–8. The model is

$$L\frac{di}{dt} = -Ri - K_e\omega + v(t)$$

$$I\frac{d\omega}{dt} = K_T i - c\omega - T_d(t)$$

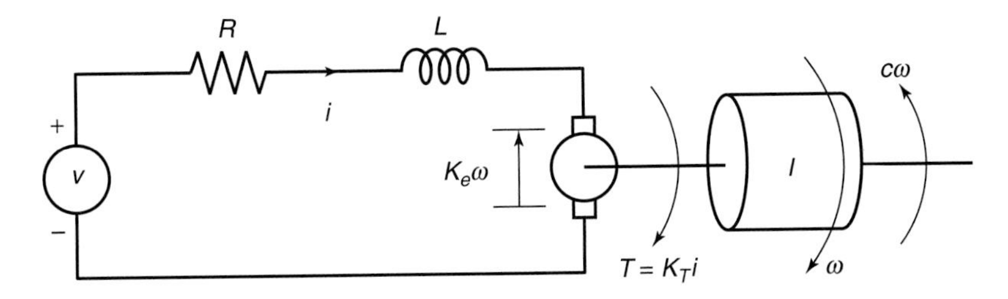

Figure 9.4–8 An armature-controlled dc motor.

where the model now includes a torque $T_d(t)$ acting on the motor shaft, due for example, to some unwanted source such as Coulomb friction or wind gusts. Control system engineers call this a "disturbance." These equations can be put into matrix form as follows, where $x_1 = i$ and $x_2 = \omega$.

$$\begin{bmatrix} \dot{x}_1 \\ \dot{x}_2 \end{bmatrix} = \begin{bmatrix} -\dfrac{R}{L} & -\dfrac{K_e}{L} \\ \dfrac{K_T}{I} & -\dfrac{c}{I} \end{bmatrix} \begin{bmatrix} x_1 \\ x_2 \end{bmatrix} + \begin{bmatrix} \dfrac{1}{L} & 0 \\ 0 & -\dfrac{1}{I} \end{bmatrix} \begin{bmatrix} v(t) \\ T_d(t) \end{bmatrix}$$

Use the values $R = 0.6\ \Omega$, $L = 0.002$ H, $K_T = 0.04$ N · m/A, $K_e = 0.04$ V · s/rad, $c = 0.01$ N · m · s/rad, and $I = 6 \times 10^{-5}$ kg · m^2.

Suppose we have a sensor that measures the motor speed, and we use the sensor's signal to activate a relay to switch the applied voltage $v(t)$ between 0 and 100 V to keep the speed between 250 and 350 rad/s. This corresponds to the relay logic shown in Figure 9.4–7b, with *SwOff* = 250, *SwOn* = 350, *Off* = 100, and *On* = 0. Investigate how well this scheme will work if the disturbance torque is a step function that increases from 0 to 3 N · m, starting at $t = 0.05$ s. Assume that the system starts from rest with $\omega(0) = 0$ and $i(0) = 0$.

■ Solution

For the given parameter values,

$$\mathbf{A} = \begin{bmatrix} -300 & -20 \\ 666.7 & -166.7 \end{bmatrix} \qquad \begin{bmatrix} 500 & 0 \\ 0 & -16667 \end{bmatrix}$$

To examine the speed ω as output, we choose $\mathbf{C} = [0, 1]$ and $\mathbf{D} = [0, 0]$. To create this simulation, first obtain a new model window. Then do the following.

1. Select and place in the new window the Step block from the Sources library. Label it Disturbance Step as shown in Figure 9.4–9. Double-click on it to obtain the Block Parameters window, and set the Step time to 0.05, the Initial and Final values to 0 and 3, and the Sample time to 0. Click **OK.**
2. Select and place the Relay block from the Discontinuities library. Double-click on it, and set the Switch-on and Switch-off points to 350 and 250, and set the Output when on and Output when off to 0 and 100. Click **OK.**
3. Select and place the Mux block from the Signal Routing library. The Mux block combines two or more signals into a vector signal. Double-click on it, and set the

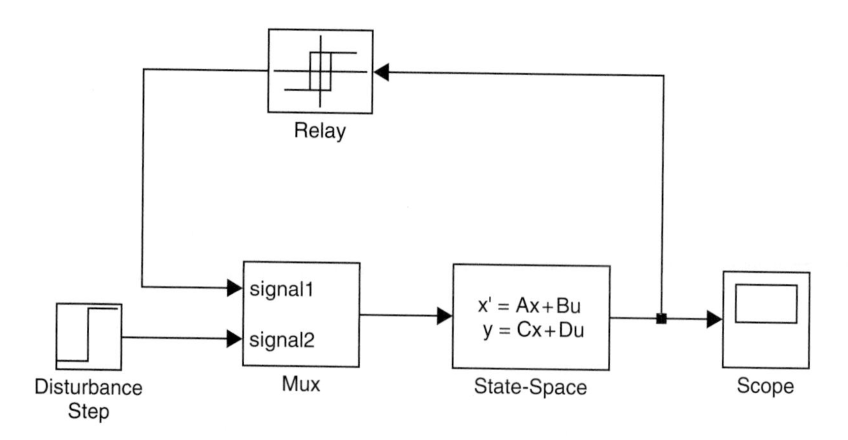

Figure 9.4–9 Simulink model of a relay-controlled motor.

Number of inputs to 2, and the Display option to signals. Click **OK.** Then click on the Mux icon in the model window, and drag one of the corners to expand the box so that all the text is visible.

4. Select and place the State-Space block from the Continuous library. Double-click on it, and enter `[-300, -20; 666.7, -166.7]` for **A**, `[500, 0; 0, -16667]` for **B**, `[0, 1]` for **C**, and `[0, 0]` for **D**. Then enter `[0; 0]` for the initial conditions. Click **OK.** Note that the dimension of the matrix **B** tells Simulink that there are two inputs. The dimensions of the matrices **C** and **D** tell Simulink that there is one output.
5. Select and place the Scope block from the Sinks library.
6. Once the blocks have been placed, connect the input port on each block to the outport port on the preceding block as shown in the figure. It is important to connect the top port of the Mux block (which corresponds to the first input, $v(t)$) to the output of the Relay block, and to connect the bottom port of the Mux block (which corresponds to the second input, $T_d(t)$) to the output of the Disturbance Step block.
7. Set the Stop time to 0.1 (which is simply an estimate of how long is needed to see the complete response), run the simulation, and examine the plot of $\omega(t)$ in the Scope. You should see something like Figure 9.4–10. If you want to examine the current $i(t)$, change the matrix **C** to `[1, 0]`, and run the simulation again.

The results show that the relay logic control scheme keeps the speed within the desired limits of 250 and 350 before the disturbance torque starts to act. The speed oscillates because when the applied voltage is zero, the speed decreases as a result of the back emf and the viscous damping. The speed drops below 250 when the disturbance torque starts to act, because the applied voltage is 0 at that time. As soon as the speed drops below 250, the relay controller switches the voltage to 100, but it now takes longer for the speed to increase because the motor torque must now work against the disturbance.

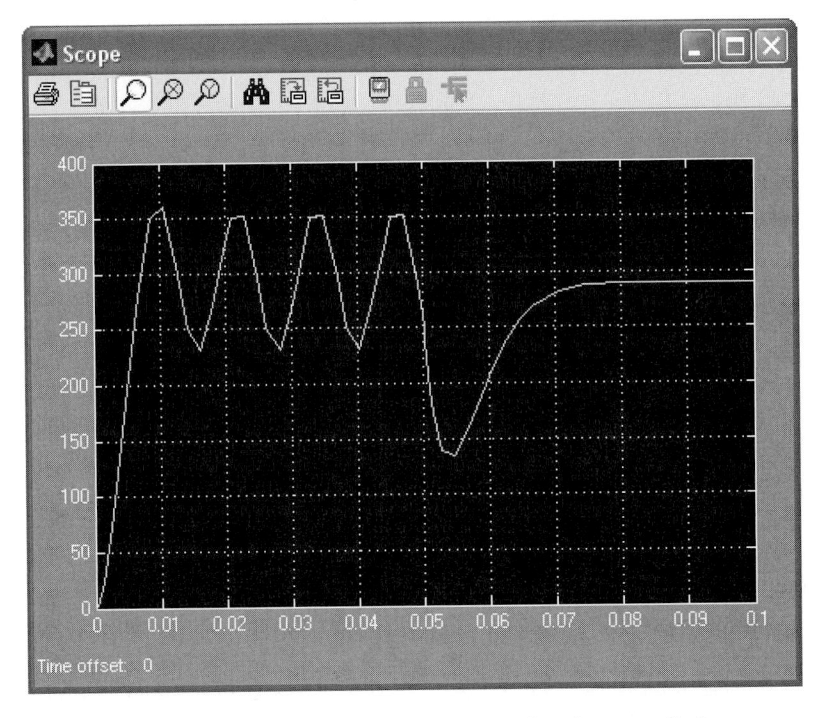

Figure 9.4–10 Scope display of the speed response of a relay-controlled motor.

Note that the speed becomes constant, instead of oscillating. This is because with $v = 100$, the system achieves a steady-state condition in which the motor torque equals the sum of the disturbance torque and the viscous damping torque. Thus the acceleration is zero.

One practical use of this simulation is to determine how long the speed is below the limit of 250. The simulation shows that this time is approximately 0.013 s. Other uses of the simulation include finding the period of the speed's oscillation (about 0.013 s) and the maximum value of the disturbance torque that can be tolerated by the relay controller (it is about 3.7 N · m).

9.5 Transfer-Function Models

The equation of motion of a mass-spring-damper system is

$$m\ddot{y} + c\dot{y} + ky = f(t) \tag{9.5–1}$$

As with the Control System toolbox, Simulink can accepts a system description in transfer-function form and in state-variable form. (See Section 8.7 for a discussion of these forms.) If the mass-spring system is subjected to a sinusoidal forcing function $f(t)$, it is easy to use the MATLAB commands presented thus far to solve and plot the response $y(t)$. However, suppose that the force $f(t)$ is created

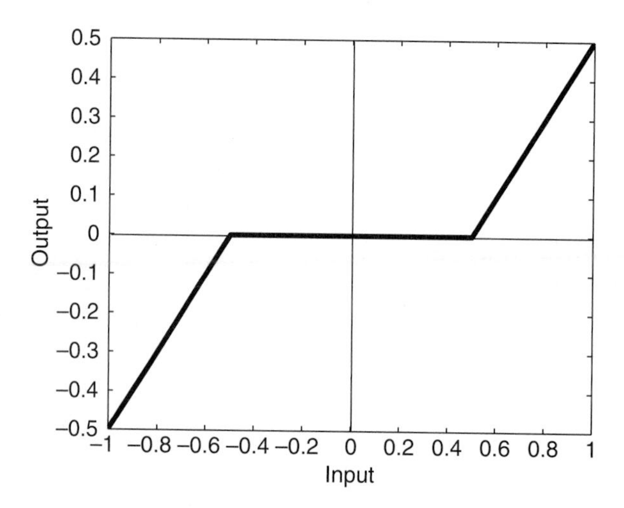

Figure 9.5–1 A dead-zone nonlinearity.

DEAD ZONE

by applying a sinusoidal input voltage to a hydraulic piston that has a *dead-zone* nonlinearity. This means that the piston does not generate a force until the input voltage exceeds a certain magnitude, and thus the system model is piecewise linear.

A graph of a particular dead-zone nonlinearity is shown in Figure 9.5–1. When the input (the independent variable on the graph) is between −0.5 and 0.5, the output is zero. When the input is greater than or equal to the upper limit of 0.5, the output is the input minus the upper limit. When the input is less than or equal to the lower limit of −0.5, the output is the input minus the lower limit. In this example, the dead zone is symmetric about 0, but it need not be in general.

Simulations with dead-zone nonlinearities are somewhat tedious to program in MATLAB, but are easily done in Simulink. The following example illustrates how it is done.

EXAMPLE 9.5–1 Response with a Dead Zone

Create and run a Simulink simulation of a mass-spring-damper model (9.5–1) using the parameter values $m = 1$, $c = 2$, and $k = 4$. The forcing function is the function $f(t) = \sin 1.4t$. The system has the dead-zone nonlinearity shown in Figure 9.5–1.

■ Solution

To construct the simulation, do the following steps.

1. Start Simulink and open a new model window as described previously.
2. Select and place in the new window the Sine Wave block from the Sources library. Double-click on it, and set the Amplitude to 1, the Frequency to 1.4, the Phase to 0, and the Sample time to 0. Click **OK.**

3. Select and place the Dead Zone block from the Discontinuities library, double-click on it, and set the Start of dead zone to -0.5 and the End of dead zone to 0.5. Click **OK.**
4. Select and place the Transfer Fcn block from the Continuous library, double-click on it, and set the Numerator to `[1]` and the Denominator to `[1, 2, 4]`. Click **OK.**
5. Select and place the Scope block from the Sinks library.
6. Once the blocks have been placed, connect the input port on each block to the outport port on the preceding block. Your model should now look like Figure 9.5–2.
7. Click on the **Simulation** menu, then click the **Configuration Parameters** item. Click on the Solver tab, and enter 10 for the Stop time. Make sure the Start time is 0. Then click **OK.**
8. Run the simulation by clicking on the **Simulation** menu, and then clicking the **Start** item. You should see an oscillating curve in the Scope display.

It is informative to plot both the input and the output of the Transfer Fcn block versus time on the same graph. To do this,

1. Delete the arrow connecting the Scope block to the Transfer Fcn block. Do this by clicking on the arrow line and then pressing the **Delete** key.
2. Select and place the Mux block from the Signal Routing library, double-click on it, and set the Number of inputs to 2. Click **OK.**
3. Connect the top input port of the Mux block to the output port of the Transfer Fcn block. Then use the same technique to connect the bottom input port of the Mux block to the arrow from the outport port of the Dead Zone block. Just remember to start with the input port. Simulink will sense the arrow automatically and make the connection. Your model should now look like Figure 9.5–3.
4. Set the Stop time to 10, run the simulation as before, and bring up the Scope display. You should see what is shown in Figure 9.5–4. This plot shows the effect of the dead zone on the sine wave.

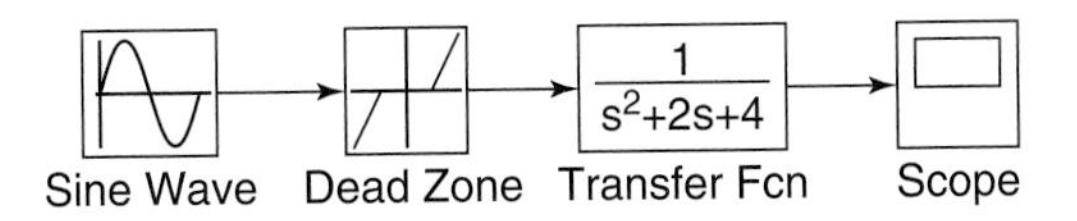

Figure 9.5–2 The Simulink model of dead-zone response.

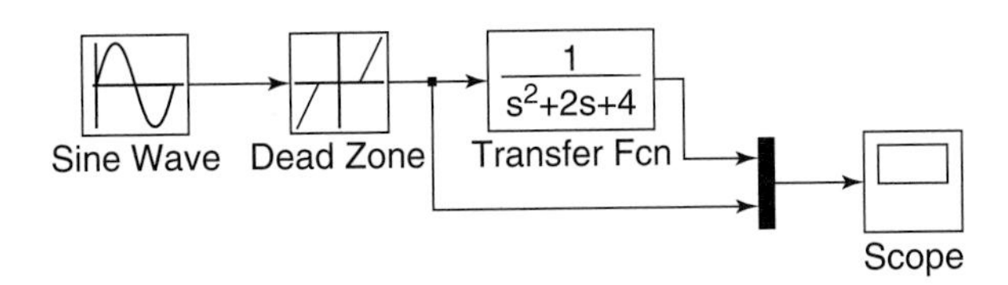

Figure 9.5–3 Modification of the dead-zone model to include a Mux block.

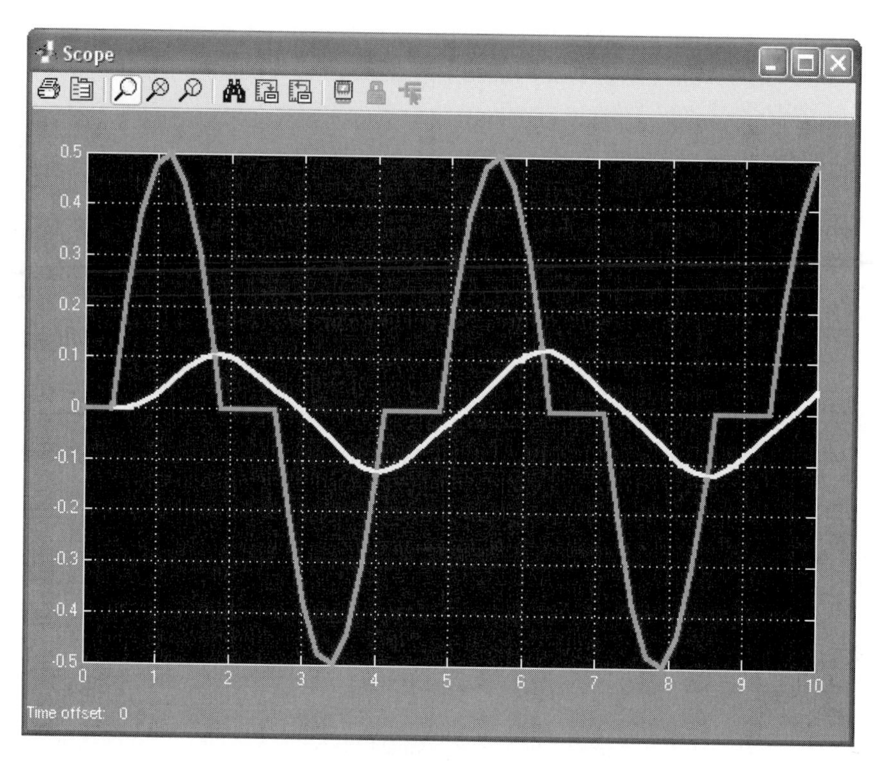

Figure 9.5–4 The response of the dead-zone model.

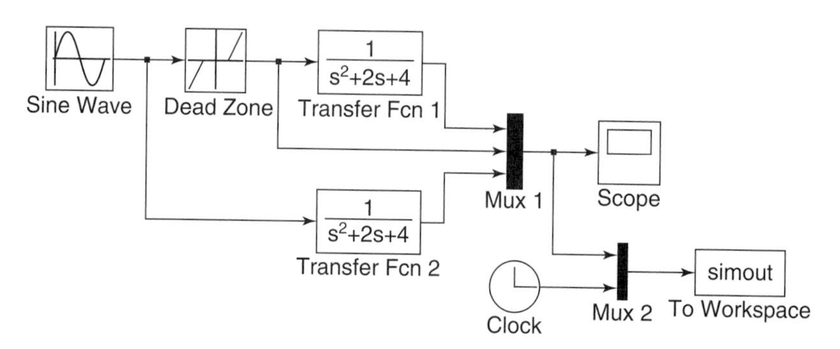

Figure 9.5–5 Modification of the dead-zone model to export variables to the MATLAB workspace.

You can bring the simulation results into the MATLAB workspace by using the To Workspace block. For example, suppose we want to examine the effects of the dead zone by comparing the response of the system with and without a dead zone. We can do this with the model shown in Figure 9.5–5. To create this model,

1. Copy the Transfer Fcn block by right-clicking on it, holding down the mouse button, and dragging the block copy to a new location. Then release the button. Copy the Mux block in the same way.

2. Double-click on the first Mux block and change the number of its inputs to 3.
3. In the usual way, select and place the To Workspace block from the Sinks library and the Clock box from the Sources library. Double-click on the To Workspace block. You can specify any variable name you want as the output; the default is `simout`. Change its name to `y`. The output variable `y` will have as many rows as there are simulation time steps, and as many columns as there are inputs to the block. The fourth column in our simulation will be time, because of the way we have connected the Clock to the second Mux. Specify the Save format as Matrix. Use the default values for the other parameters (these should be `inf`, `1`, and `-1` for Maximum number of rows, Decimation, and Sample time, respectively). Click on **OK.**
4. Connect the blocks as shown, and run the simulation.
5. You can use the MATLAB plotting commands from the Command window to plot the columns of `y`; for example, to plot the response of the two systems and the output of the Dead Zone block versus time, type

```
>>plot(y(:,4),y(:,1),y(:,4),y(:,2),y(:,4),y(:,3))
```

9.6 Nonlinear State-Variable Models

Nonlinear models cannot be put into transfer-function form or the state-variable form $\dot{\mathbf{x}} = \mathbf{Ax} + \mathbf{Bu}$. However, they can be simulated in Simulink. The following example shows how this can be done.

Model of a Nonlinear Pendulum — EXAMPLE 9.6–1

The pendulum shown in Figure 9.6–1 has the following nonlinear equation of motion, if there is viscous friction in the pivot and if there is an applied moment $M(t)$ about the pivot.

$$I\ddot{\theta} + c\dot{\theta} + mgL\sin\theta = M(t)$$

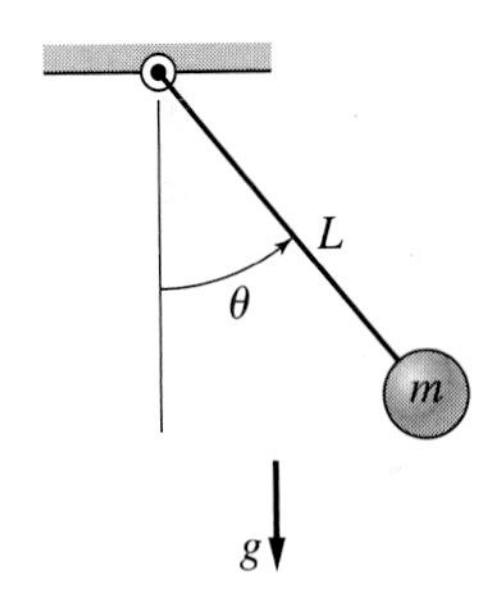

Figure 9.6–1 A pendulum.

where I is the mass moment of inertia about the pivot. Create a Simulink model for this system for the case where $I = 4$, $mgL = 10$, $c = 0.8$, and $M(t)$ is a square wave with an amplitude of 3 and a frequency of 0.5 Hz. Assume that the initial conditions are $\theta(0) = \pi/4$ rad and $\dot{\theta}(0) = 0$.

■ Solution

To simulate this model in Simulink, define a set of variables that lets you rewrite the equation as two first-order equations. Thus let $\omega = \dot{\theta}$. Then the model can be written as

$$\dot{\theta} = \omega$$

$$\dot{\omega} = \frac{1}{I}[-c\omega - mgL\sin\theta + M(t)] = 0.25[-0.8\omega - 10\sin\theta + M(t)]$$

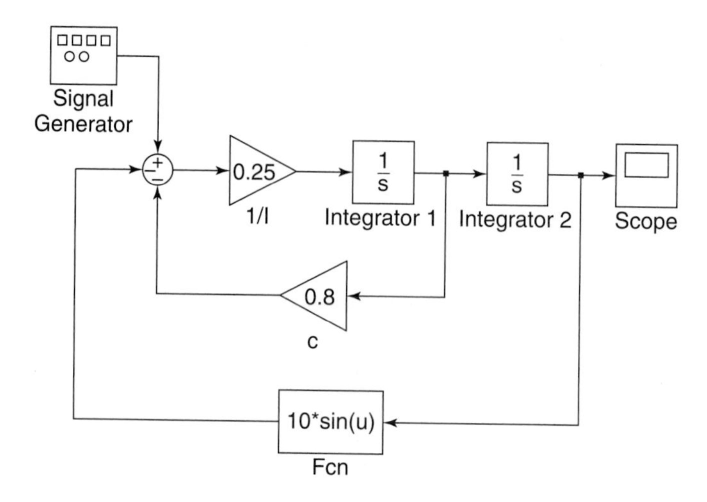

Figure 9.6–2 Simulink model of nonlinear pendulum dynamics.

Integrate both sides of each equation over time to obtain

$$\theta = \int \omega \, dt$$

$$\omega = 0.25 \int [-0.8\omega - 10 \sin\theta + M(t)] \, dt$$

We will introduce four new blocks to create this simulation. Obtain a new model window and do the following.

1. Select and place in the new window the Integrator block from the Continuous library, and change its label to Integrator 1 as shown in Figure 9.6–2. You can edit text associated with a block by clicking on the text and making the changes. Double-click on the block to obtain the Block Parameters window, and set the Initial condition to 0 (this is the initial condition $\dot{\theta}(0) = 0$). Click **OK.**
2. Copy the Integrator block to the location shown and change its label to Integrator 2. Set its initial condition to $\pi/4$ by typing `pi/4` in the Block Parameters window. This is the initial condition $\theta(0) = \pi/4$.
3. Select and place a Gain block from the Math Operations library, double-click on it, and set the Gain value to 0.25. Click **OK.** Change its label to `1/I`. Then click on the block, and drag one of the corners to expand the box so that all the text is visible.
4. Copy the Gain box, change its label to `c`, and place it as shown in Figure 9.6–2. Double-click on it, and set the Gain value to 0.8. Click **OK.** To flip the box left to right, right-click on it, select **Format,** and select **Flip.**
5. Select and place the Scope block from the Sinks library.
6. For the term $10 \sin\theta$, we cannot use the Trig function block in the Math library because we need to multiply the $\sin\theta$ by 10. So we use the Fcn block under the User-Defined Functions library (Fcn stands for function). Select and place this

block as shown. Double-click on it, and type `10*sin(u)` in the expression window. This block uses the variable `u` to represent the input to the block. Click **OK.** Then flip the block.

7. Select and place the Sum block from the Math Operations library. Double-click on it, and select round for the Icon shape. In the List of Signs window, type `+--`. Click **OK.**
8. Select and place the Signal Generator block from the Sources library. Double-click on it, select square wave for the Wave form, 3 for the Amplitude, and 0.5 for the Frequency, and Hertz for the Units. Click **OK.**
9. Once the blocks have been placed, connect arrows as shown in the figure.
10. Set the Stop time to 10, run the simulation, and examine the plot of $\theta(t)$ in the Scope. This completes the simulation.

9.7 Subsystems

One potential disadvantage of a graphical interface such as Simulink is that, to simulate a complex system, the diagram can become rather large and therefore somewhat cumbersome. Simulink, however, provides for the creation of *subsystem blocks,* which play a role analogous to that of subprograms in a programming language. A subsystem block is actually a Simulink program represented by a single block. A subsystem block, once created, can be used in other Simulink programs. We also introduce some other blocks in this section.

To illustrate subsystem blocks we will use a simple hydraulic system whose model is based on the conservation of mass principle familiar to engineers. Because the governing equations are similar to other engineering applications, such as electric circuits and devices, the lessons learned from this example will enable you to use Simulink for other applications.

A Hydraulic System

The working fluid in a *hydraulic* system is an incompressible fluid such as water or a silicon-based oil. (*Pneumatic* systems operate with compressible fluids, such as air.) Consider a hydraulic system composed of a tank of liquid of mass density ρ (Figure 9.7–1). The tank shown in cross section in the figure is cylindrical with a bottom area A. A flow source dumps liquid into the tank at the mass flow rate

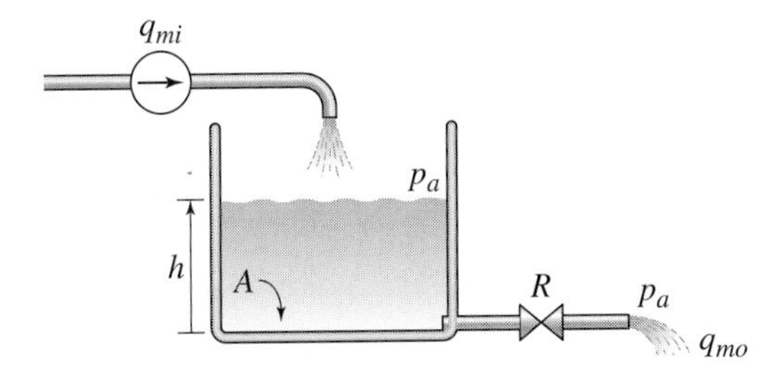

Figure 9.7–1 A hydraulic system with a flow source.

$q_{mi}(t)$. The total mass in the tank is $m = \rho A h$, and from conservation of mass we have

$$\frac{dm}{dt} = \rho A \frac{dh}{dt} = q_{mi} - q_{mo} \tag{9.7–1}$$

since ρ and A are constants.

If the outlet is a pipe that discharges to atmospheric pressure p_a and provides a resistance to flow that is proportional to the pressure difference across its ends, then the outlet flow rate is

$$q_{mo} = \frac{1}{R}[(\rho g h + p_a) - p_a] = \frac{\rho g h}{R}$$

where R is called the *fluid resistance*. Substituting this expression into equation (9.7–1) gives the model:

$$\rho A \frac{dh}{dt} = q_{mi}(t) - \frac{\rho g}{R} h \tag{9.7–2}$$

The transfer function is

$$\frac{H(s)}{Q_{mi}(s)} = \frac{1}{\rho A s + \rho g/R}$$

On the other hand, the outlet may be a valve or other restriction that provides nonlinear resistance to the flow. In such cases, a common model is the signed-square-root relation

$$q_{mo} = \frac{1}{R}\mathrm{SSR}(\Delta p)$$

where q_{mo} is the outlet mass flow rate, R is the resistance, Δp is the pressure difference across the resistance, and

$$\mathrm{SSR}(\Delta p) = \begin{cases} \sqrt{\Delta p} & \text{if } \Delta p \ge 0 \\ -\sqrt{|\Delta p|} & \text{if } \Delta p < 0 \end{cases}$$

Note that we may express the SSR(u) function in MATLAB as follows: `sgn(u)*sqrt(abs(u))`.

Consider the slightly different system shown in Figure 9.7–2, which has a flow source q and two pumps that supply liquid at the pressures p_l and p_r. Suppose

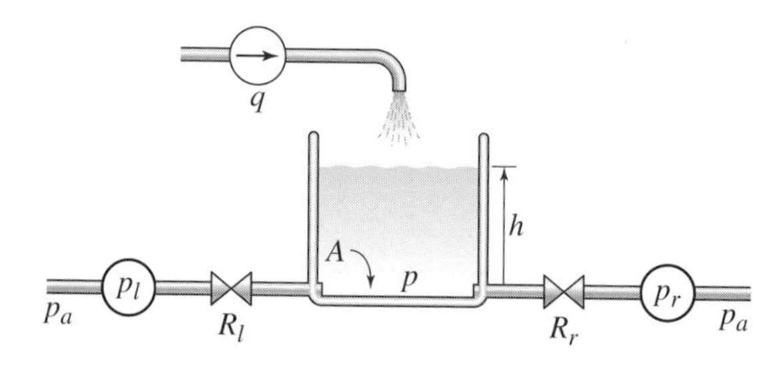

Figure 9.7–2 A hydraulic system with a flow source and pump.

the resistances are nonlinear and obey the signed-square-root relation. Then the model of the system is the following:

$$\rho A\frac{dh}{dt} = q + \frac{1}{R_l}\text{SSR}(p_l - p) - \frac{1}{R_r}\text{SSR}(p - p_r)$$

where A is the bottom area and $p = \rho g h$. The pressures p_l and p_r are the *gage* pressures at the left- and right-hand sides. Gage pressure is the difference between the absolute pressure and atmospheric pressure. Note that the atmospheric pressure p_a cancels out of the model because of the use of gage pressure.

We will use this application to introduce the following Simulink elements:

- Subsystem blocks, and
- Input and Output Ports.

You can create a subsystem block in one of two ways: by dragging the Subsystem block from the library to the model window or by first creating a Simulink model and then "encapsulating" it within a bounding box. We will illustrate the latter method.

We will create a subsystem block for the liquid-level system shown in Figure 9.7–2. First construct the Simulink model shown in Figure 9.7–3. The oval blocks are Input and Output Ports (In 1 and Out 1), which are available in the Ports and Subsystems library. Note that you can use MATLAB variables and expressions when entering the gains in each of the four Gain blocks.

Before running the program we will assign values to these variables in the MATLAB Command window. Enter the gains for the four Gain blocks using the expressions shown in the block. You may also use a variable as the Initial condition of the Integrator block. Name this variable `h0`.

The SSR blocks are examples of the Fcn block, which is in the User-Defined Functions library. Double-click on the block and enter the MATLAB expression

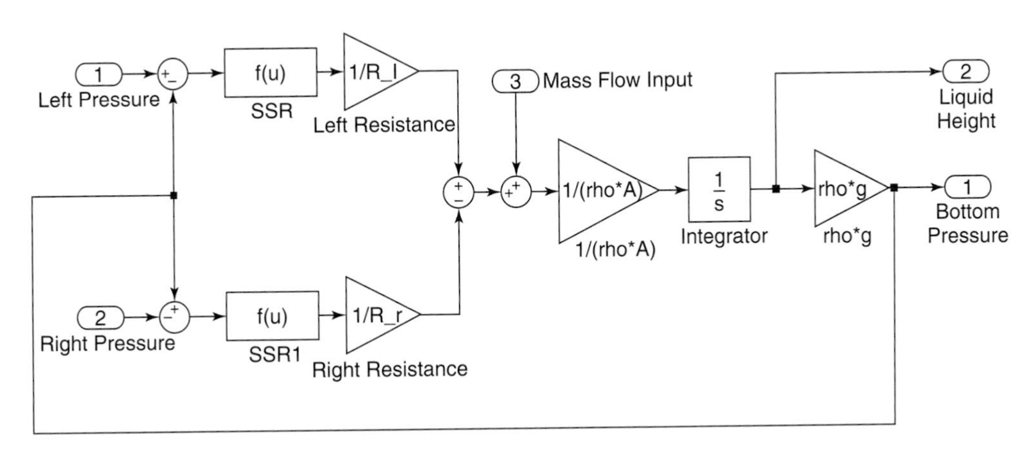

Figure 9.7–3 Simulink model of the system shown in Figure 9.7–2.

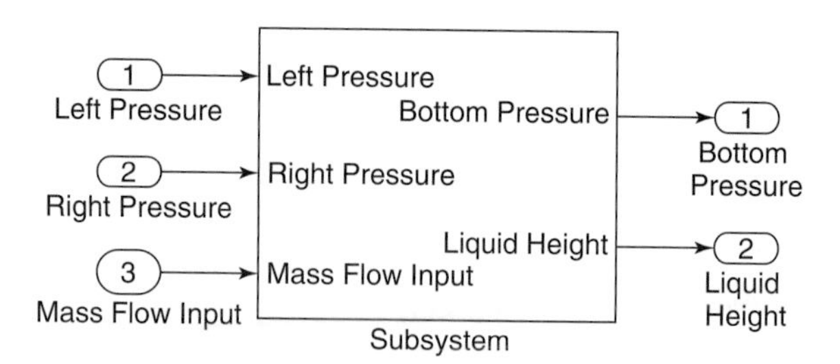

Figure 9.7–4 The Subsystem block.

`sgn(u)*sqrt(abs(u))`. Note that the Fcn block requires you to use the variable `u`. The output of the Fcn block must be a scalar, as is the case here, and you cannot perform matrix operations in the Fcn block, but these are not needed here. (An alternative to the Fcn block is the MATLAB Fcn block to be discussed in Section 9.9.) Save the model and give it a name, such as Tank.

Now create a "bounding box" surrounding the diagram. Do this by placing the mouse cursor in the upper left, holding the mouse button down, and dragging the expanding box to the lower right to enclose the entire diagram. Then choose **Create Subsystem** from the **Edit** menu. Simulink will then replace the diagram with a single block having as many input and output ports as required and will assign default names. You can resize the block to make the labels readable. You may view or edit the subsystem by double-clicking on it. The result is shown in Figure 9.7–4.

Connecting Subsystem Blocks

We now create a simulation of the system shown in Figure 9.7–5, where the mass inflow rate q is a step function. To do this, create the Simulink model shown in Figure 9.7–6. The square blocks are Constant blocks from the Sources library. These give constant inputs (which are not the same as step function inputs).

The larger rectangular blocks are two subsystem blocks of the type just created. To insert them into the model, first open the Tank subsystem model, select **Copy** from the **Edit** menu, then paste it twice into the new model window. Connect

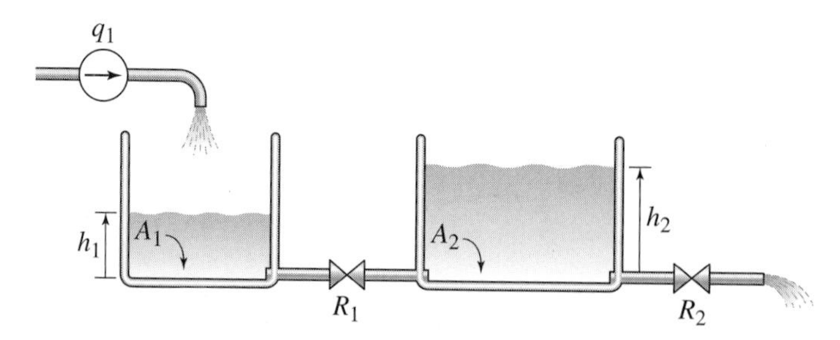

Figure 9.7–5 A hydraulic system with two tanks.

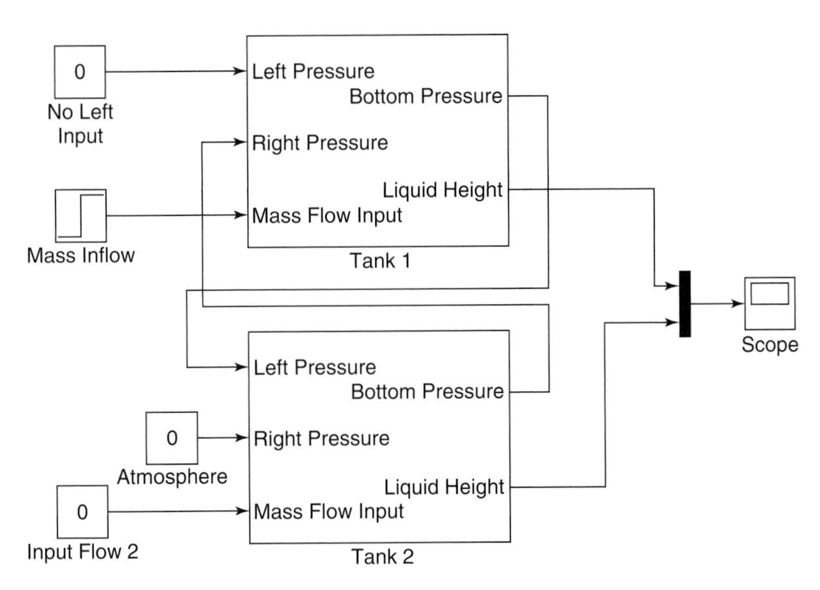

Figure 9.7–6 Simulink model of the system shown in Figure 9.7–5.

the input and output ports and edit the labels as shown. Then double-click on the Tank 1 subsystem block, set the left-side gain `1/R_l` equal to 0, the right-side gain `1/R_r` equal to `1/R_1`, and the gain `1/rho*A` equal to `1/rho*A_1`. Set the Initial condition of the integrator to `h10`. Note that setting the gain `1/R_l` equal to 0 is equivalent to $R\_l = \infty$, which indicates no inlet on the left-hand side.

Then double-click on the Tank 2 subsystem block, set the left-side gain `1/R_l` equal to `1/R_1`, the right-side gain `1/R_r` equal to `1/R_2`, and the gain `1/rho*A` equal to `1/rho*A_2`. Set the Initial condition of the integrator to `h20`. For the Step block, set the Step time to 0, the Initial value to 0, the Final value to the variable `q_1`, and the Sample time to 0. Save the model using a name other than Tank.

Before running the model, in the Command window assign numerical values to the variables. As an example, you may type the following values for water, in U. S. Customary units, in the Command window.

```
>>A_1 = 2;A_2 = 5;rho = 1.94;g = 32.2;
>>R_1 = 20;R_2 = 50;q_1 = 0.3;h10 = 1;h20 = 10;
```

After selecting a simulation Stop time, you may run the simulation. The Scope will display the plots of the heights h_1 and h_2 versus time.

Figures 9.7–7, 9.7–8, and 9.7–9 illustrate some electrical and mechanical systems that are likely candidates for application of subsystem blocks. In Figure 9.7–7, the basic element for the subsystem block is an RC circuit. In Figure 9.7–8, the basic element for the subsystem block is a mass connected to two elastic elements.

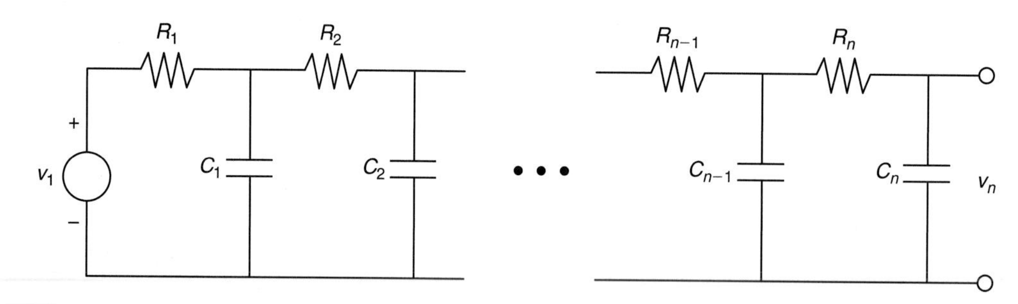

Figure 9.7–7 A network of RC loops.

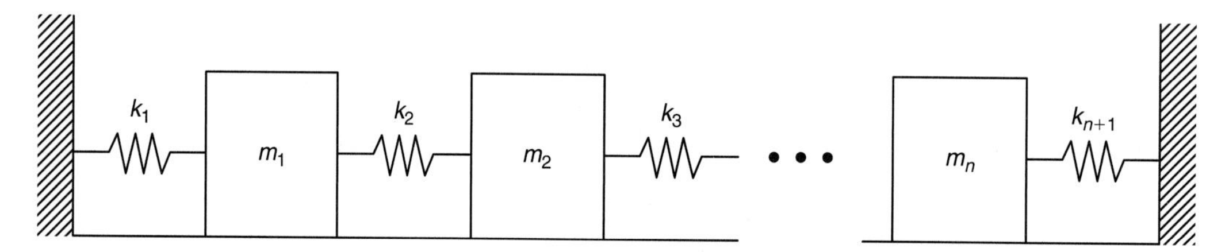

Figure 9.7–8 A vibrating system.

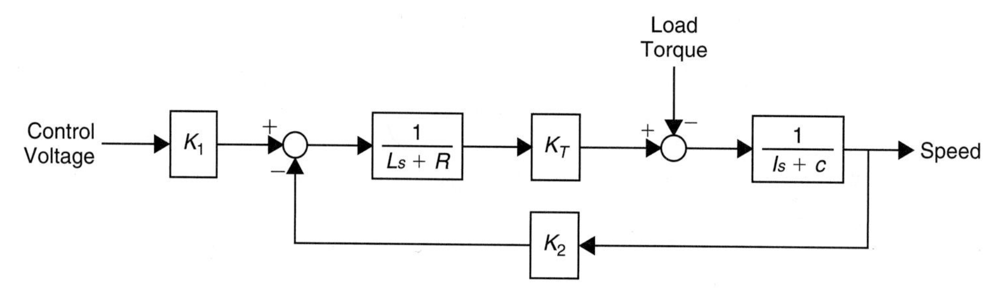

Figure 9.7–9 An armature-controlled dc motor.

Figure 9.7–9 is the block diagram of an armature-controlled dc motor, which may be converted into a subsystem block. The inputs for the block would be the voltage from a controller and a load torque, and the output would be the motor speed. Such a block would be useful in simulating systems containing several motors, such as a robot arm.

9.8 Dead Time in Models

TRANSPORT DELAY

Dead time, also called *transport delay,* is a time delay between an action and its effect. It occurs, for example, when a fluid flows through a conduit. If the fluid velocity v is constant and the conduit length is L, it takes a time $T = L/v$ for the fluid to move from one end to the other. The time T is the dead time.

Let $\Theta_1(t)$ denote the incoming fluid temperature and $\Theta_2(t)$ the temperature of the fluid leaving the conduit. If no heat energy is lost, then $\Theta_2(t) = \Theta_1(t - T)$.

From the shifting property of the Laplace transform,

$$\Theta_2(s) = e^{-Ts}\Theta_1(s)$$

So the transfer function for a dead-time process is e^{-Ts}.

Dead time may be described as a "pure" time delay, in which no response at all occurs for a time T, as opposed to the time lag associated with the time constant of a response, for which $\Theta_2(t) = (1 - e^{-t/\tau})\Theta_1(t)$.

Some systems have an unavoidable time delay in the interaction between components. The delay often results from the physical separation of the components and typically occurs as a delay between a change in the actuator signal and its effect on the system being controlled, or as a delay in the measurement of the output.

Another, perhaps unexpected, source of dead time is the computation time required for digital control computer to calculate the control algorithm. This can be a significant dead time in systems using inexpensive and slower microprocessors.

The presence of dead time means the system does not have a characteristic equation of finite order. In fact, there are an infinite number of characteristic roots for a system with dead time. This can be seen by noting that the term e^{-Ts} can be expanded in an infinite series as

$$e^{-Ts} = \frac{1}{e^{Ts}} = \frac{1}{1 + Ts + T^2s^2/2 + \cdots}$$

The fact that there are an infinite number of characteristic roots means that the analysis of dead-time processes is difficult, and often simulation is the only practical way to study such processes.

Systems having dead-time elements are easily simulated in Simulink. The block implementing the dead-time transfer function e^{-Ts} is called the "Transport Delay" block.

Consider the model of the height h of liquid in a tank, such as that shown in Figure 9.7–1, whose input is a mass flow rate q_i. Suppose that it takes a time T for the change in input flow to reach the tank following a change in the valve opening. Thus, T is a dead time. For specific parameter values, the transfer function has the form

$$\frac{H(s)}{Q_i(s)} = e^{-Ts}\frac{2}{5s + 1}$$

Figure 9.8–1 shows a Simulink model for this system. After placing the Transport Delay block, set the delay to 1.25. Set the Step time to 0 in the Step Function block. We will now discuss the other blocks in the model.

Specifying Initial Conditions with Transfer Functions

The "Transfer Fcn (with initial outputs)" block, so-called to distinguish it from the Transfer Fcn block, enables us to set the initial value of the block output. In our model, this corresponds to the initial liquid height in the tank. This feature thus provides a useful improvement over traditional transfer-function analysis, in which initial conditions are assumed to be zero.

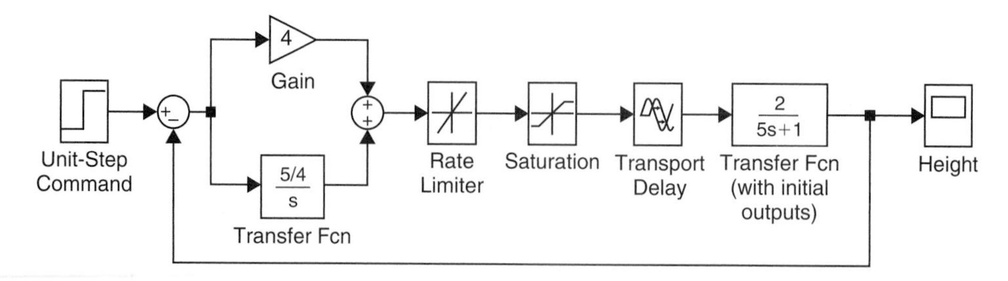

Figure 9.8–1 Simulink model of a hydraulic system with dead time.

The "Transfer Fcn (with initial outputs)" block is equivalent to adding the free response to the block output, with all the block's state variables set to zero except for the output variable. The block also lets you assign an initial value to the block input, but we will not use this feature and so will leave the Initial input set to 0 in the Block Parameters window. Set the Initial output to 0.2 to simulate an initial liquid height of 0.2.

The Saturation and Rate Limiter Blocks

Suppose that the minimum and maximum flow rates available from the input flow valve are 0 and 2. These limits can be simulated with the Saturation block, which was discussed in Section 9.4. After placing the block as shown in Figure 9.8–1, double-click on it and type 2 in its Upper limit window and 0 in the Lower limit window.

In addition to being limited by saturation, some actuators have limits on how fast they can react. This limitation might be due to deliberate restrictions placed on the unit by its manufacturer to avoid damage to the unit. An example is a flow control valve whose rate of opening and closing is controlled by a "rate limiter." Simulink has such a block, and it can be used in series with the Saturation block to model the valve behavior. Place the Rate Limiter block as shown in Figure 9.8–1. Set the Rising slew rate to 1 and the Falling slew rate to -1.

A Control System

PI CONTROLLER

The Simulink model shown in Figure 9.8–1 is for a specific type of control system called a *PI controller*, whose response $f(t)$ to the error signal $e(t)$ is the sum of a term proportional to the error signal and a term proportional to the integral of the error signal. That is,

$$f(t) = K_P e(t) + K_I \int_0^t e(t)\,dt$$

where K_P and K_I are called the proportional and integral gains. Here the error signal $e(t)$ is the difference between the unit-step command representing the desired height and the actual height. In transform notation this expression becomes

$$F(s) = K_P E(s) + \frac{K_I}{s} E(s) = \left(K_P + \frac{K_I}{s} \right) E(s)$$

In Figure 9.8–1, we used the values $K_P = 4$ and $K_I = 5/4$. These values are computed using the methods of control theory (For a discussion of control systems, see, for example, [Palm, 2005]). The simulation is now ready to be run. Set the Stop time to 30 and observe the behavior of the liquid height $h(t)$ in the Scope. Does it reach the desired height of 1?

9.9 Simulation of a Vehicle Suspension

Linear or linearized models are useful for predicting the behavior of dynamic systems because powerful analytical techniques are available for such models, especially when the inputs are relatively simple functions such as the impulse, step, ramp, and sine. Often in the design of an engineering system, however, we must eventually deal with nonlinearities in the system and with more complicated inputs such as trapezoidal functions, and this must often be done with simulation.

In this section we introduce four additional Simulink elements that enable us to model a wide range of nonlinearities and input functions, namely,

- the Derivative block,
- the Signal Builder block,
- the Look-Up Table block, and
- the MATLAB Fcn block.

As our example, we will use the single-mass suspension model shown in Figure 9.9–1, where the spring and damper forces f_s and f_d have the nonlinear models shown in Figures 9.9–2 and 9.9–3. The damper model is unsymmetric and represents a damper whose force during rebound is higher than during jounce (in order to minimize the force transmitted to the passenger compartment when the vehicle strikes a bump). The bump is represented by the trapezoidal function $y(t)$ shown in Figure 9.9–4. This function corresponds approximately to a vehicle traveling at 30 mi/hr over a road surface elevation 0.2 m high and 48 m long.

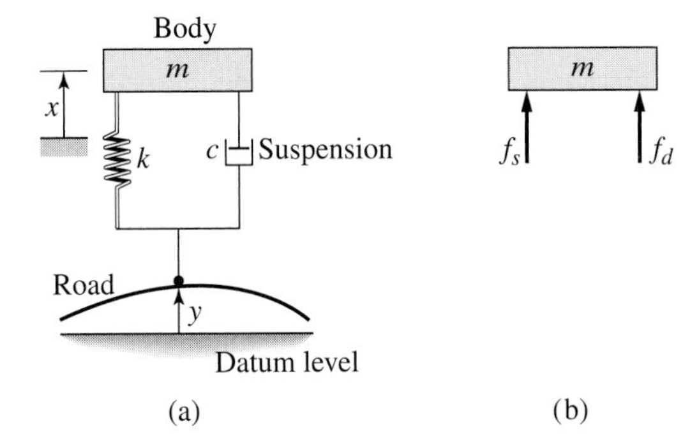

Figure 9.9–1 Single-mass model of a vehicle suspension.

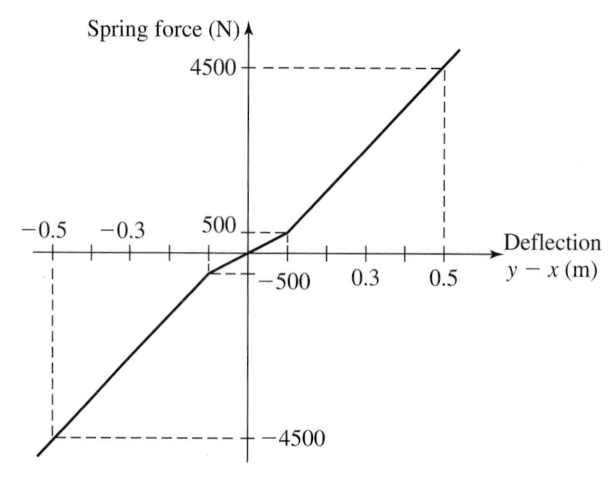

Figure 9.9–2 Nonlinear spring function.

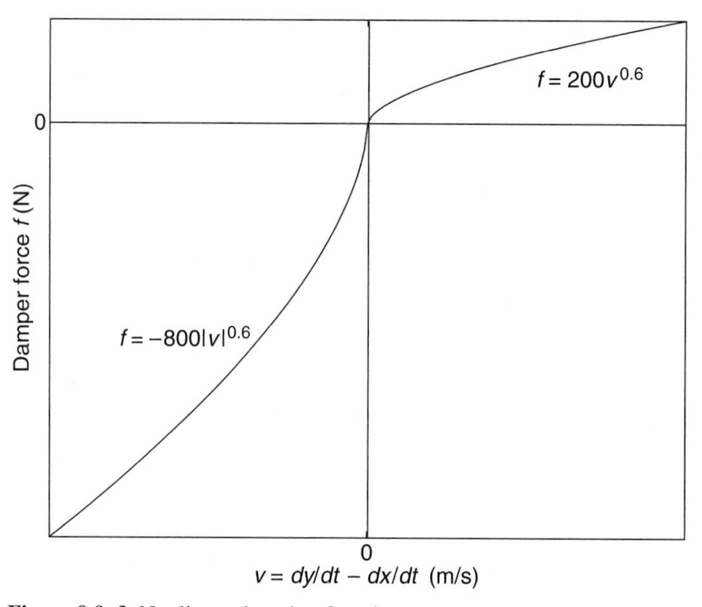

Figure 9.9–3 Nonlinear damping function.

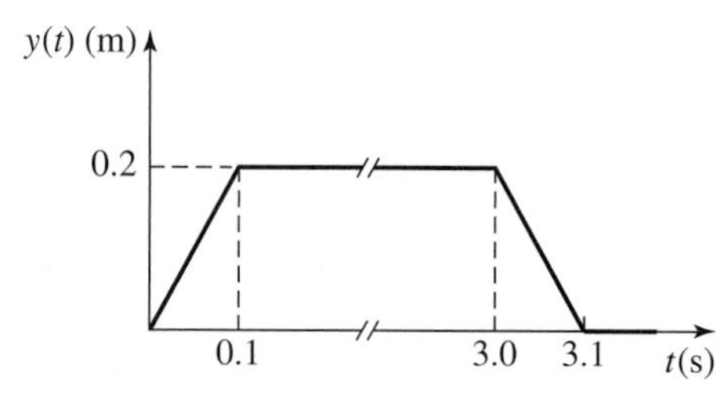

Figure 9.9–4 Road surface profile.

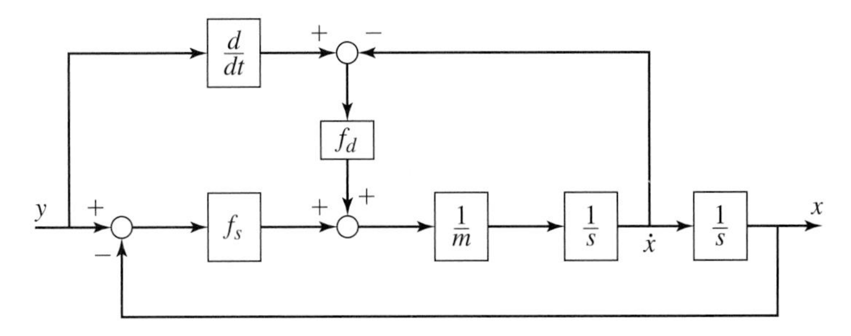

Figure 9.9–5 Simulation diagram of a vehicle suspension model.

The system model from Newton's law is

$$m\ddot{x} = f_s(y - x) + f_d(\dot{y} - \dot{x})$$

where $m = 400$ kg, $f_s(y - x)$ is the nonlinear spring function shown in Figure 9.9–2, and $f_d(\dot{y} - \dot{x})$ is the nonlinear damper function shown in Figure 9.9–3. The corresponding simulation diagram is shown in Figure 9.9–5.

The Derivative and Signal Builder Blocks

The simulation diagram shows that we need to compute $\dot{y}$. Because Simulink uses numerical and not analytical methods, it computes derivatives only approximately, using the Derivative block. We must keep this in mind when using rapidly changing or discontinuous inputs. The Derivative block has no settings, so merely place it in the Simulink diagram as shown in Figure 9.9–6.

Next, place the Signal-Builder block, then double-click on it. A plot window appears that enables you to place points to define the input function. Follow the directions in the window to create the function shown in Figure 9.9–4.

The Look-Up Table Block

The spring function f_s is created with the Look-Up Table block. After placing it as shown, double-click on it and enter `[-0.5, -0.1, 0, 0.1, 0.5]` for the Vector of input values and `[-4500, -500, 0, 500, 4500]` for the Vector of output values. Use the default settings for the remaining parameters.

Place the two integrators as shown, and make sure the initial values are set to 0. Then place the Gain block and set its gain to $1/400$. The To Workspace block and the Clock will enable us to plot $x(t)$ and $y(t) - x(t)$ versus t in the MATLAB Command window.

The MATLAB Fcn Block

In Section 9.7 we used the Fcn block to implement the signed-square-root function. We cannot use this block for the damper function shown in Figure 9.9–3 because

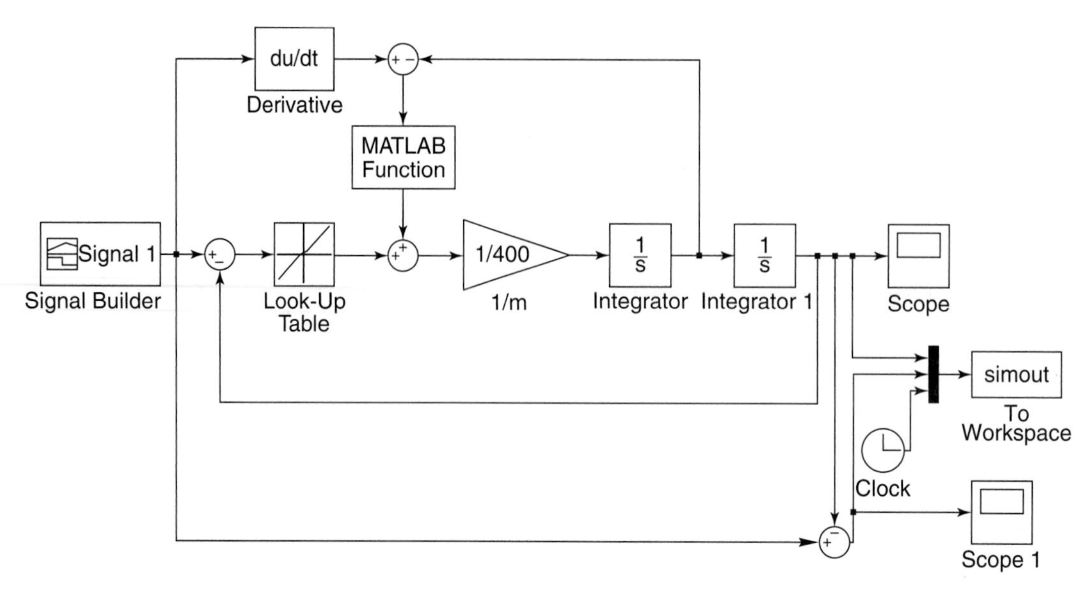

Figure 9.9–6 Simulink model of a vehicle suspension system.

we must write a user-defined function to describe it. This function is as follows.

```
function f = damper(v)
if v <= 0
  f = -800*(abs(v)).^(0.6);
else
  f = 200*v.^(0.6);
end
```

Create and save this function file. After placing the MATLAB Fcn block, double-click on it and enter its name `damper`. Make sure Output dimensions is set to -1 and the Output signal type is set to auto.

The Fcn, MATLAB Fcn, Math Function, and S-Function blocks can be used to implement functions, but each has its advantages and limitations. The Fcn block can contain an expression, but its output must be a scalar, and it cannot call a function file. The MATLAB Fcn block is slower than the Fcn block, but its output can be an array, and it can call a function file.

The Math Function block can produce an array output, but it is limited to a single MATLAB function and cannot use an expression or call a file. The S-Function block provides more advanced features, such as the ability to use C language code.

The Simulink model when completed should look like Figure 9.9–6. You can plot the response $x(t)$ in the Command window as follows:

```
>>x = simout(:,1);
>>t = simout(:,3);
>>plot(t,x),grid,xlabel('t (s)'),ylabel('x (m)')
```

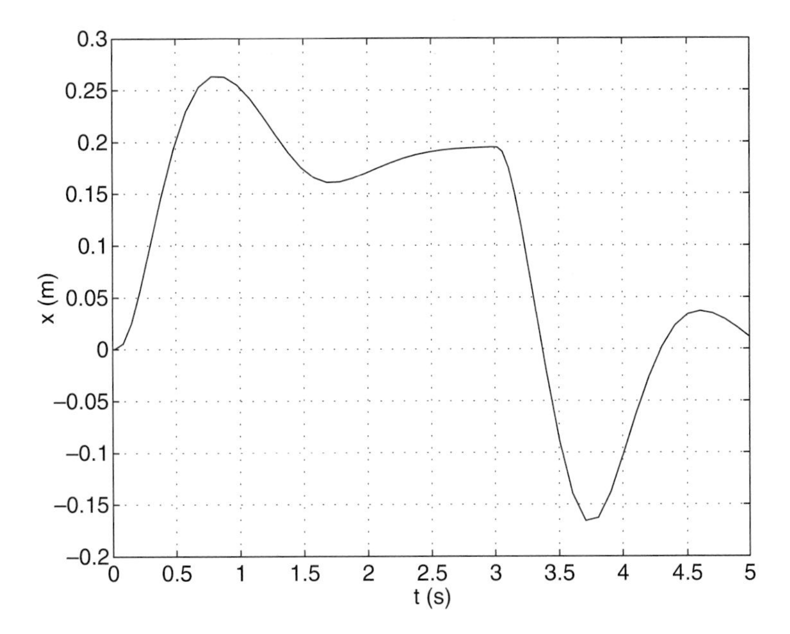

Figure 9.9–7 Output of the Simulink model shown in Figure 9.9–6.

The result is shown in Figure 9.9–7. The maximum overshoot is seen to be $(0.26 - 0.2) = 0.06$ m, but the maximum *under*shoot is seen to be much greater, -0.168 m.

9.10 Summary

The Simulink model window contains menu items we have not discussed. However, the ones we have discussed are the most important ones for getting started. We have introduced just a few of the blocks available within Simulink. Some of the blocks not discussed deal with discrete-time systems (ones modeled with difference, rather than differential, equations), digital logic systems, and other types of mathematical operations. In addition, some blocks have additional properties that we have not mentioned. However, the examples given here will help you get started in exploring the other features of Simulink. Consult the online help for information about these items.

Key Terms with Page References

Problems

Section 9.1

1. Draw a simulation diagram for the following equation.

$$\dot{y} = 5f(t) - 7y$$

2. Draw a simulation diagram for the following equation.

$$5\ddot{y} + 3\dot{y} + 7y = f(t)$$

3. Draw a simulation diagram for the following equation.

$$3\ddot{y} + 5 \sin y = f(t)$$

Section 9.2

4. Create a Simulink model to plot the solution of the following equation for $0 \le t \le 6$.

$$10\ddot{y} = 7 \sin 4t + 5 \cos 3t \qquad y(0) = 4 \qquad \dot{y}(0) = 1$$

5. A projectile is launched with a velocity of 100 m/s at an angle of 30° above the horizontal. Create a Simulink model to solve the projectile's equations of motion where x and y are the horizontal and vertical displacements of the projectile.

$$\ddot{x} = 0 \qquad x(0) = 0 \qquad \dot{x}(0) = 100 \cos 30^\circ$$

$$\ddot{y} = -g \qquad y(0) = 0 \qquad \dot{y}(0) = 100 \sin 30^\circ$$

Use the model to plot the projectile's trajectory y versus x for $0 \le t \le 10$ s.

6. The following equation has no analytical solution even though it is linear.

$$\dot{x} + x = \tan t \qquad x(0) = 0$$

The approximate solution, which is less accurate for large values of t, is

$$x(t) = \frac{1}{3}t^3 - t^2 + 3t - 3 + 3e^{-t}$$

Create a Simulink model to solve this problem and compare its solution with the approximate solution over the range $0 \le t \le 1$.

7. Construct a Simulink model to plot the solution of the following equation for $0 \le t \le 10$.

$$15\dot{x} + 5x = 4u_s(t) - 4u_s(t-2) \qquad x(0) = 2$$

where $u_s(t)$ is a unit-step function (in the Block Parameters window of the Step block, set the Step time to 0, the Initial value to 0, and the Final value to 1).

8. A tank having vertical sides and a bottom area of 100 ft^2 is used to store water. To fill the tank, water is pumped into the top at the rate given in the following table. Use Simulink to solve for and plot the water height $h(t)$ for $0 \le t \le 10$ min.

Time (min)	0	1	2	3	4	5	6	7	8	9	10
Flow Rate (ft^3/min)	0	80	130	150	150	160	165	170	160	140	120

Section 9.3

9. Construct a Simulink model to plot the solution of the following equations for $0 \le t \le 2$.

$$\dot{x}_1 = -6x_1 + 4x_2$$

$$\dot{x}_2 = 5x_1 - 7x_2 + f(t)$$

where $f(t) = 2t$. Use the Ramp block in the Sources library.

10. Construct a Simulink model to plot the solution of the following equations for $0 \le t \le 3$.

$$\dot{x}_1 = -6x_1 + 4x_2 + f_1(t)$$

$$\dot{x}_2 = 5x_1 - 7x_2 + f_2(t)$$

where $f_1(t)$ is a step function of height 3 starting at $t = 0$, and $f_2(t)$ is a step function of height -3 starting at $t = 1$.

Section 9.4

11. Use the Saturation block to create a Simulink model to plot the solution of the following equation for $0 \le t \le 6$.

$$3\dot{y} + y = f(t) \qquad y(0) = 2$$

where

$$f(t) = \begin{cases} 8 & \text{if } 10 \sin 3t > 8 \\ -8 & \text{if } 10 \sin 3t < -8 \\ 10 \sin 3t & \text{otherwise} \end{cases}$$

12. Construct a Simulink model of the following problem.

$$5\dot{x} + \sin x = f(t) \qquad x(0) = 0$$

The forcing function is

$$f(t) = \begin{cases} -5 & \text{if } g(t) \le -5 \\ g(t) & \text{if } -5 \le g(t) \le 5 \\ 5 & \text{if } g(t) \ge 5 \end{cases}$$

where $g(t) = 10 \sin 4t$.

13. If a mass-spring system has Coulomb friction on the surface rather than viscous friction, its equation of motion is

$$m\ddot{y} = -ky + f(t) - \mu mg \qquad \text{if } \dot{y} \ge 0$$

$$m\ddot{y} = -ky + f(t) + \mu mg \qquad \text{if } \dot{y} < 0$$

where μ is the coefficient of friction. Develop a Simulink model for the case where $m = 1$ kg, $k = 5$ N/m, $\mu = 0.4$, and $g = 9.8$ m/s^2. Run the simulation for two cases: (a) the applied force $f(t)$ is a step function with a magnitude of 10 N and (b) the applied force is sinusoidal: $f(t) = 10 \sin 2.5t$. Either the Sign block in the Math Operations library or the Coulomb and Viscous Friction block in the Discontinuities library can be used, but since there is no viscous friction in this problem, the Sign block is easier to use.

14. A certain mass, $m = 2$ kg, moves on a surface inclined at an angle $\phi = 30^\circ$ above the horizontal. Its initial velocity is $v(0) = 3$ m/s up the incline. An external force of $f_1 = 5$ N acts on it parallel to and up the incline. The coefficient of Coulomb friction is $\mu = 0.5$. Use the Sign block and create a Simulink model to solve for the velocity of the mass until the mass comes to rest. Use the model to determine the time at which the mass comes to rest.

15. *a.* Develop a Simulink model of a thermostatic control system in which the temperature model is

$$RC\frac{dT}{dt} + T = Rq + T_a(t)$$

where T is the room air temperature in °F, T_a is the ambient (outside) air temperature in °F, time t is measured in hours, q is the input from the heating system in lb-ft/hr, R is the thermal resistance, and C is the thermal capacitance. The thermostat switches q on at the value q_{max} whenever the temperature drops below 69° and switches q to $q = 0$ whenever the temperature is above 71°. The value of q_{max} indicates the heat output of the heating system.

Run the simulation for the case where $T(0) = 70^\circ$ and $T_a(t) = 50 + 10 \sin(\pi t/12)$. Use the values $R = 5 \times 10^{-5}$ °F-hr/lb-ft and

$C = 4 \times 10^4$ lb-ft/°F. Plot the temperatures T and T_a versus t on the same graph, for $0 \le t \le 24$ hr. Do this for two cases: $q_{max} = 4 \times 10^5$ and $q_{max} = 8 \times 10^5$ lb-ft/hr. Investigate the effectiveness of each case.

b. The integral of q over time is the energy used. Plot $\int q\, dt$ versus t and determine how much energy is used in 24 hr for the case where $q_{max} = 8 \times 10^5$.

16. Refer to Problem 15. Use the simulation with $q = 8 \times 10^5$ to compare the energy consumption and the thermostat cycling frequency for the two temperature bands (69°, 71°) and (68°, 72°).

17. Consider the liquid-level system shown in Figure 9.7–1. The governing equation based on conservation of mass is (9.7–2). Suppose that the height h is controlled by using a relay to switch the input flow rate between the values 0 and 50 kg/s. The flow rate is switched on when the height is less than 4.5 m and is switched off when the height reaches 5.5 m. Create a Simulink model for this application using the values $A = 2$ m^2, $R =$ 400 N · s/(kg · m^2), $\rho = 1000$ kg/m^3, and $h(0) = 1$ m. Obtain a plot of $h(t)$.

Section 9.5

18. Use the Transfer Function block to construct a Simulink model to plot the solution of the following equation for $0 \le t \le 4$.

$$2\ddot{x} + 12\dot{x} + 10x = 5u_s(t) - 5u_s(t-2) \qquad x(0) = \dot{x}(0) = 0$$

19. Use Transfer Function blocks to construct a Simulink model to plot the solution of the following equations for $0 \le t \le 2$.

$$3\ddot{x} + 15\dot{x} + 18x = f(t) \qquad x(0) = \dot{x}(0) = 0$$

$$2\ddot{y} + 16\dot{y} + 50y = x(t) \qquad y(0) = \dot{y}(0) = 0$$

where $f(t) = 50u_s(t)$.

20. Use Transfer Function blocks to construct a Simulink model to plot the solution of the following equations for $0 \le t \le 2$.

$$3\ddot{x} + 15\dot{x} + 18x = f(t) \qquad x(0) = \dot{x}(0) = 0$$

$$2\ddot{y} + 16\dot{y} + 50y = x(t) \qquad y(0) = \dot{y}(0) = 0$$

where $f(t) = 50u_s(t)$. At the output of the first block there is a dead zone for $-1 \le x \le 1$. This limits the input to the second block.

21. Use Transfer Function blocks to construct a Simulink model to plot the solution of the following equations for $0 \le t \le 2$.

$$3\ddot{x} + 15\dot{x} + 18x = f(t) \qquad x(0) = \dot{x}(0) = 0$$

$$2\ddot{y} + 16\dot{y} + 50y = x(t) \qquad y(0) = \dot{y}(0) = 0$$

where $f(t) = 50u_s(t)$. At the output of the first block there is a saturation that limits x be $|x| \le 1$. This limits the input to the second block.

Section 9.6

22. Construct a Simulink model to plot the solution of the following equation for $0 \le t \le 4$.

$$2\ddot{x} + 12\dot{x} + 10x^2 = 5\sin 0.8t \qquad x(0) = \dot{x}(0) = 0$$

23. Create a Simulink model to plot the solution of the following equation for $0 \le t \le 3$.

$$\dot{x} + 10x^2 = 2\sin 4t \qquad x(0) = 1$$

24. Construct a Simulink model of the following problem.

$$10\dot{x} + \sin x = f(t) \qquad x(0) = 0$$

The forcing function is $f(t) = \sin 2t$. The system has the dead-zone nonlinearity shown in Figure 9.5–1.

25. The following model describes a mass supported by a nonlinear, hardening spring. The units are SI. Use $g = 9.81$ m/s^2.

$$5\ddot{y} = 5g - (900y + 1700y^3) \qquad y(0) = 0.5 \qquad \dot{y}(0) = 0$$

Create a Simulink model to plot the solution for $0 \le t \le 2$.

26. Consider the system for lifting a mast shown in Figure P26. The 70-ft long mast weighs 500 lb. The winch applies a force $f = 380$ lb to the cable. The mast is supported initially at an angle of 30°, and the cable at A is initially horizontal. The equation of motion of the mast is

$$25{,}400\,\ddot{\theta} = -17{,}500\cos\theta + \frac{626{,}000}{Q}\sin(1.33 + \theta)$$

where

$$Q = \sqrt{2020 + 1650\cos(1.33 + \theta)}$$

Create and run a Simulink model to solve for and plot $\theta(t)$ for $\theta(t) \le \pi/2$ rad.

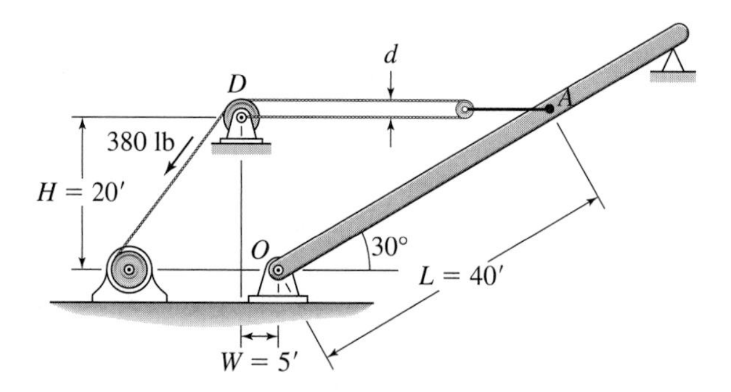

Figure P26

27. The equation describing the water height h in a spherical tank with a drain at the bottom is

$$\pi(2rh - h^2)\frac{dh}{dt} = -C_d A\sqrt{2gh}$$

Suppose the tank's radius is $r = 3$ m and that the circular drain hole of area A has a radius of 2 cm. Assume that $C_d = 0.5$, and that the initial water height is $h(0) = 5$ m. Use $g = 9.81$ m/s^2. Use Simulink to solve the nonlinear equation and plot the water height as a function of time until $h(t) = 0$.

28. A cone-shaped paper drinking cup (like the kind used at water fountains) has a radius R and a height H. If the water height in the cup is h, the water volume is given by

$$V = \frac{1}{3}\pi\left(\frac{R}{H}\right)^2 h^3$$

Suppose that the cup's dimensions are $R = 1.5$ in. and $H = 4$ in.

a. If the flow rate from the fountain into the cup is 2 in.^{3}/sec, use Simulink to determine how long will it take to fill the cup to the brim.

b. If the flow rate from the fountain into the cup is given by $2(1 - e^{-2t})$ in.^{3}/sec, use Simulink to determine how long will it take to fill the cup to the brim.

Section 9.7

29. Refer to Figure 9.7–2. Assume that the resistances obey the linear relation, so that the mass flow q_l through the left-hand resistance is $q_l = (p_l - p)/R_l$, with a similar linear relation for the right-hand resistance.

a. Create a Simulink subsystem block for this element.

b. Use the subsystem block to create a Simulink model of the system shown in Figure 9.7–5. Assume that the mass inflow rate is a step function.

c. Use the Simulink model to obtain plots of $h_1(t)$ and $h_2(t)$ for the following parameter values: $A_1 = 2$ m^2, $A_2 = 5$ m^2, $R_1 = 400$ N$\cdot$ s/(kg$\cdot$m^2), $R_2 = 600$ N$\cdot$s/(kg$\cdot$m^2), $\rho = 1000$ kg/m^3, $q_{mi} = 50$ kg/s, $h_1(0) = 1.5$ m, and $h_2(0) = 0.5$ m.

30. *a.* Use the subsystem block developed in Section 9.7 to construct a Simulink model of the system shown in Figure P30. The mass inflow rate is a step function.

b. Use the Simulink model to obtain plots of $h_1(t)$ and $h_2(t)$ for the following parameter values: $A_1 = 3$ ft^2, $A_2 = 5$ ft^2, $R_1 = 30$ lb-sec/(slug-ft^2), $R_2 = 40$ lb-sec/(slug-ft^2), $\rho = 1.94$ slug/ft^3, $q_{mi} = 0.5$ slug/sec, $h_1(0) = 2$ ft, and $h_2(0) = 5$ ft.

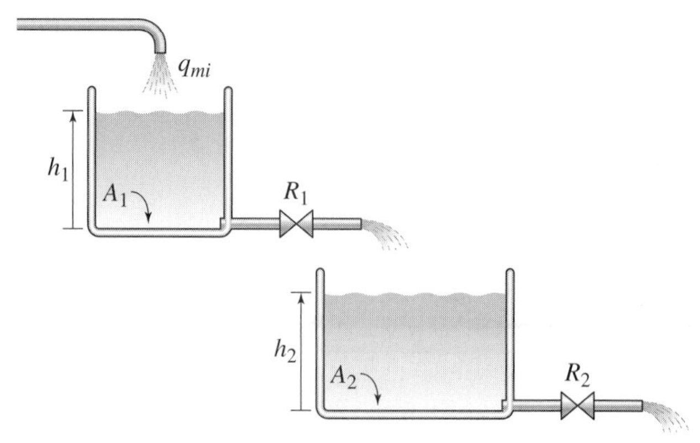

Figure P30

31. Consider Figure 9.7–7 for the case where there are three RC loops with the values $R_1 = R_3 = 10^4\ \Omega$, $R_2 = 5 \times 10^4\ \Omega$, $C_1 = C_3 = 10^{-6}$ F, and $C_2 = 4 \times 10^{-6}$ F.

a. Develop a subsystem block for one RC loop.
b. Use the subsystem block to construct a Simulink model of the entire system of three loops. Plot $v_3(t)$ over $0 \le t \le 3$ for $v_1(t) = 12 \sin 10t$ V.

32. Consider Figure 9.7–8 for the case where there are three masses. Use the values $m_1 = m_3 = 10$ kg, $m_2 = 30$ kg, $k_1 = k_4 = 10^4$ N/m, and $k_2 = k_3 = 2 \times 10^4$ N/m.

a. Develop a subsystem block for one mass.
b. Use the subsystem block to construct a Simulink model of the entire system of three masses. Plot the displacements of the masses over $0 \le t \le 2$ s for if the initial displacement of m_1 is 0.1 m.

Section 9.8

33. Refer to Figure P30. Suppose there is a dead time of 10 sec between the outflow of the top tank and the lower tank. Use the subsystem block developed in Section 9.7 to create a Simulink model of this system. Using the parameters given in problem 30, plot the height h_1 and h_2 versus time.

Section 9.9

34. Redo the Simulink suspension model developed in Section 9.9, using the spring relation and input function shown in Figure P34, and the following damper relation.

$$f_d(v) = \begin{cases} -500|v|^{1.2} & v \le 0 \\ 50v^{1.2} & v > 0 \end{cases}$$

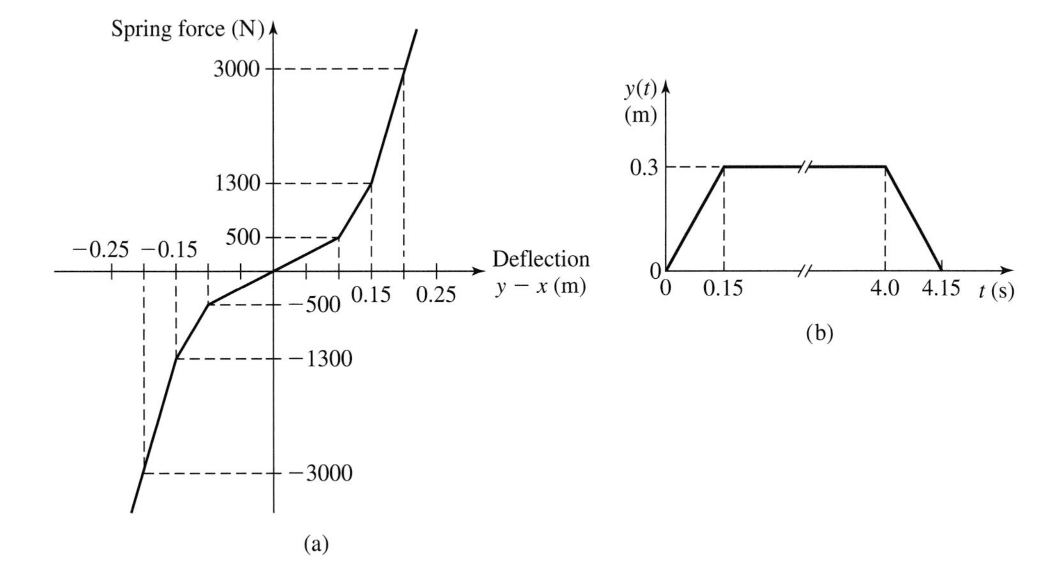

Figure P34

Use the simulation to plot the response. Evaluate the overshoot and undershoot.

35. Consider the system shown in Figure P35. The equations of motion are

$$m_1\ddot{x}_1 + (c_1 + c_2)\dot{x}_1 + (k_1 + k_2)x_1 - c_2\dot{x}_2 - k_2x_2 = 0$$
$$m_2\ddot{x}_2 + c_2\dot{x}_2 + k_2x_2 - c_2\dot{x}_1 - k_2x_1 = f(t)$$

Suppose that $m_1 = m_2 = 1$, $c_1 = 3$, $c_2 = 1$, $k_1 = 1$, and $k_2 = 4$.

a. Develop a Simulink model of this system. In doing this, consider whether to use a state-variable representation or a transfer-function representation of the model.

b. Use the Simulink model to plot the response $x_1(t)$ for the following input. The initial conditions are zero.

$$f(t) = \begin{cases} t & 0 \le t \le 1 \\ 2 - t & 1 < t < 2 \\ 0 & t \ge 2 \end{cases}$$

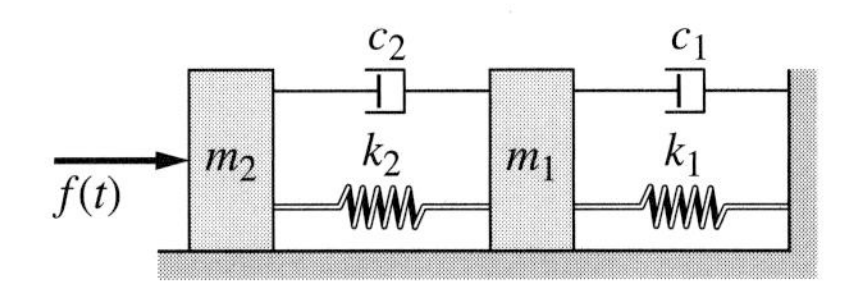

Figure P35

Engineering in the 21st Century...

Rebuilding the Infrastructure

During the Great Depression, many public works projects that improved the nation's infrastructure were undertaken to stimulate the economy and provide employment. These projects included highways, bridges, water supply systems, sewer systems, and electrical power distribution networks. Following World War II another burst of such activity culminated in the construction of the interstate highway system. As we enter the 21st century, much of the infrastructure is 30 to 70 years old and is literally crumbling or not up-to-date. One survey showed that more than 25 percent of the nation's bridges are substandard. These need to be repaired or replaced with new bridges, such as the one under construction in Savannah, Georgia, shown above.

Rebuilding the infrastructure requires engineering methods different from those in the past because labor and material costs are now higher and environmental and social issues have greater importance than before. Infrastructure engineers must take advantage of new materials, inspection technology, construction techniques, and labor-saving machines.

Also, some infrastructure components, such as communications networks, need to be replaced because they are outdated and do not have sufficient capacity or ability to take advantage of new technology. An example is the "information infrastructure," which includes physical facilities to transmit, store, process, and display voice, data, and images. Better communications and computer networking technology will be needed for such improvements. Many of the MATLAB toolboxes provide advanced support for such work, including the Financial, Communications, Image Processing, Signal Processing, PDE, and Wavelet toolboxes. ■

CHAPTER 10

Symbolic Processing with MATLAB

OUTLINE

Up to now we have used MATLAB to perform numerical operations only; that is, our answers have been numbers, not expressions. In this chapter we use MATLAB to perform *symbolic processing* to obtain answers in the form of expressions. Symbolic processing is the term used to describe how a computer performs operations on mathematical expressions in the way, for example, that humans do algebra with pencil and paper. Whenever possible, we wish to obtain solutions in closed form because they give us greater insight into the problem. For example, we often can see how to improve an engineering design by modeling it with mathematical expressions that do not have specific parameter values. Then we can analyze the expressions and decide which parameter values will optimize the design.

SYMBOLIC EXPRESSION

This chapter explains how to define a *symbolic expression* such as $y = \sin x/\cos x$ in MATLAB and how to use MATLAB to simplify expressions wherever possible. For example, the previous function simplifies to $y = \sin x/\cos x = \tan x$. MATLAB can perform operations such as addition and multiplication on mathematical expressions, and we can use MATLAB to obtain symbolic solutions to algebraic equations such as $x^2 + 2x + a = 0$ (the solution for x is $x = -1 \pm \sqrt{1 - a}$). MATLAB can also perform symbolic differentiation and integration and can solve ordinary differential equations in closed form.

To use the methods of this chapter, you must have either the Symbolic Math toolbox or the Student Edition of MATLAB, which contains all the functions of the Symbolic Math toolbox but has limited access to the Maple kernel.

The programs in this chapter are compatible with versions 2 through 3.1 of the toolbox, *although different versions might give slightly different error messages and slightly different displays of expressions.*

The symbolic processing capabilities in MATLAB are based on the Maple V software package, which was developed by Waterloo Maple Software, Inc. The MathWorks has licensed the Maple "engine," that is, the core of Maple. If you have used Maple before, however, or plan to use it in the future, you should be aware that the syntax used by MATLAB differs from that used by the commercially available Maple package.

We cover in this chapter a subset of the capabilities of the Symbolic Math toolbox. Specifically we treat

- Symbolic algebra.
- Symbolic methods for solving algebraic and transcendental equations.
- Symbolic methods for solving ordinary differential equations.
- Symbolic calculus, including integration, differentiation, limits, and series.
- Laplace transforms.
- Selected topics in linear algebra, including symbolic methods for obtaining determinants, matrix inverses, and eigenvalues.

The topic of Laplace transforms is included because they provide one way of solving differential equations and are often covered along with differential equations.

We do not discuss the following features of the Symbolic Math toolbox: canonical forms of symbolic matrices; variable precision arithmetic that allows you to evaluate expressions to a specified numerical accuracy; and special mathematical functions such as Fourier transforms. Details on these capabilities can be found in the online help.

When you have finished this chapter, you should be able to use MATLAB to

- Create symbolic expressions and manipulate them algebraically.
- Obtain symbolic solutions to algebraic and transcendental equations.
- Perform symbolic differentiation and integration.
- Evaluate limits and series symbolically.
- Obtain symbolic solutions to ordinary differential equations.
- Obtain Laplace transforms.
- Perform symbolic linear algebra operations, including obtaining expressions for determinants, matrix inverses, and eigenvalues.

10.1 Symbolic Expressions and Algebra

The `sym` function can be used to create "symbolic objects" in MATLAB. If the input argument to `sym` is a string, the result is a symbolic number or variable. If the input argument is a numeric scalar or matrix, the result is a symbolic representation of the given numeric values. For example, typing `x = sym('x')` creates the symbolic variable with name `x`, and typing `y = sym('y')` creates a symbolic variable named `y`. Typing `x = sym('x','real')` tells MATLAB to assume that `x` is real. Typing `x = sym('x','unreal')` tells MATLAB to assume that `x` is not real.

The `syms` function enables you to combine more than one such statement into a single statement. For example, typing `syms x` is equivalent to typing `x = sym('x')`, and typing `syms x y u v` creates the four symbolic variables `x`, `y`, `u`, and `v`. When used without arguments, `syms` lists the symbolic objects in the workspace. The `syms` command, however, cannot be used to create symbolic constants; you must use `sym` for this purpose.

The `syms` command enables you to specify that certain variables are real. For example,

```
>>syms x y real
```

SYMBOLIC CONSTANT

You can use the `sym` function to create *symbolic constants* by using a numerical value for the argument. For example, typing `pi = sym('pi')`, `fraction = sym('1/3')`, and `sqroot2 = sym('sqrt(2)')` create symbolic constants that avoid the floating-point approximations inherent in the values of π, 1/3, and $\sqrt{2}$. If you create the symbolic constant π this way, it temporarily replaces the built-in numeric constant, and you no longer obtain a numerical value when you type its name. For example,

```
>>pi = sym('pi')
pi =
   pi
>>sqroot2 = sym('sqrt(2)')
sqroot2 =
   sqrt(2)
>>a = 3*sqrt(2)     % This gives a numeric result.
a =
   4.2426
>>b = 3*sqroot2     % This gives a symbolic result.
b =
   3*2^(1/2)
```

The advantage of using symbolic constants is that they need not be evaluated (with the accompanying round-off error) until a numeric answer is required.

Symbolic constants can look like numbers but are actually symbolic expressions. Symbolic expressions can look like character strings but are a different sort of quantity. You can use the `class` function to determine whether or not a quantity is symbolic, numeric, or a character string. We will give examples of the `class` function later.

The effect of round-off error needs to be considered when converting MATLAB floating-point values to symbolic constants in this way. You can use an optional second argument with the `sym` function to specify the technique for converting floating-point numbers. Refer to the online help for more information.

Symbolic Expressions

You can use symbolic variables in expressions and as arguments of functions. You use the operators `+ - * / ^` and the built-in functions just as you use them with numerical calculations. For example, typing

```
>>syms x y
>>s = x + y;
>>r = sqrt(x^2 + y^2);
```

creates the symbolic variables `s` and `r`. The terms `s = x + y` and `r = sqrt(x^2 + y^2)` are examples of symbolic *expressions*. The variables `s` and `r` created this way are *not* the same as user-defined function files. That is, if you later assign `x` and `y` numeric values, typing `r` will not cause MATLAB to evaluate the equation $r = \sqrt{x^2 + y^2}$. We will see later how to evaluate symbolic expressions numerically.

The `syms` command enables you to specify that expressions have certain characteristics. For example, in the following session MATLAB will treat the expression `w` as a nonnegative number:

```
>>syms x y real
>>w = x^2 + y^2;
```

To clear `x` of its real property, type `syms x unreal`. Note that typing `clear x` eliminates `x` from the workspace and does not make `x` a nonreal variable.

The vector and matrix notation used in MATLAB also applies to symbolic variables. For example, you can create a symbolic matrix `A` as follows:

```
>>n = 3;
>>syms x;
>>A = x.^((0:n)'*(0:n))
A =
    [ 1, 1, 1, 1]
    [ 1, x, x^2, x^3]
    [ 1, x^2, x^4, x^6]
    [ 1, x^3, x^6, x^9]
```

Note that it was not necessary to use `sym` or `syms` to declare `A` to be a symbolic variable beforehand. It is recognized as a symbolic variable because it is created with a symbolic expression.

DEFAULT VARIABLE

In MATLAB the variable `x` is the *default* independent variable, but other variables can be specified to be the independent variable. It is important to know which variable is the independent variable in an expression. The function `findsym(E)` can be used to determine the symbolic variable used by MATLAB in a particular expression `E`.

The function `findsym(E)` finds the symbolic variables in a symbolic expression or matrix, where `E` is a scalar or matrix symbolic expression, and returns a string containing all of the symbolic variables appearing in `E`. The variables are returned in alphabetical order and are separated by commas. If no symbolic variables are found, `findsym` returns the empty string.

By contrast, the function `findsym(E,n)` returns the `n` symbolic variables in `E` closest to x, with the tie breaker going to the variable closer to z. The following session shows some examples of its use:

```
>>syms b x1 y
>>findsym(6*b+y)
ans =
     b,y
>>findsym(6*b+y+x) %Note: x has not been declared symbolic.
??? Undefined function or variable 'x'.
>>findsym(6*b+y,1) %Find the one variable closest to x
ans =
     y
>>findsym(6*b+y+x1,1) %Find the one variable closest to x
ans =
     x1
>>findsym(6*b+y*i) %i is not symbolic
ans =
     b, y
```

Manipulating Expressions

The following functions can be used to manipulate expressions by collecting coefficients of like powers, expanding powers, and factoring expressions, for example.

The function `collect(E)` collects coefficients of like powers in the expression `E`. If there is more than one variable, you can use the optional form `collect(E,v)`, which collects all the coefficients with the same power of `v`.

```
>>syms x y
>>E = (x-5)^2+(y-3)^2;
>>collect(E)
ans =
     x^2-10*x+25+(y-3)^2
>>collect(E,y)
ans =
     y^2-6*y+(x-5)^2+9
```

The function expand(E) expands the expression E by carrying out powers. For example,

```
>>syms x y
>>expand((x+y)^2) % applies algebra rules
ans =
     x^2+2*x*y+y^2
>>expand(sin(x+y)) % applies trig identities
ans =
     sin(x)*cos(y)+cos(x)*sin(y)
>>simplify(6*((sin(x))^2+(cos(x))^2)) % applies another trig identity
ans =
     6
```

The function factor(E) factors the expression E. For example,

```
>>syms x y
>>factor(x^2-1)
ans =
     (x-1)*(x+1)
```

The function simplify(E) simplifies the expression E, using Maple's simplification rules. For example,

```
>>syms x y
>>simplify(x*sqrt(x^8*y^2))
ans =
     x*(x^8*y^2)^(1/2)
```

The function simple(E) searches for the shortest form of the expression E in terms of number of characters. When called, the function displays the results of each step of its search. When called without the argument, simple acts on the previous expression. The form [r, how] = simple(E) does not display intermediate steps, but saves those steps in the string how. The shortest form found is stored in r.

You can use the operators + - * / and ^ with symbolic expressions to obtain new expressions. The following session illustrates how this is done.

```
>>syms x y
>>E1 = x^2+5;              % define two expressions
>>E2 = y^3-2;
>>S1 = E1 + E2              % add the expressions
S1 =
     x^2+3+y^3
>>S2 = E1*E2               % multiply the expressions

S2 =
     (x^2+5)*(y^3-2)
```

```
>>expand(S2)            % expand the product
ans =
     x^2*y^3-2*x^2+5*y^3-10
>>E3 = x^3+2*x^2+5*x+10;     % define a third expression
>>S3 = E3/E1            % divide two expressions
      S3 = (x^3+2*x^2+5*x+10)/(x^2+5)
>>simplify(S3)            % see if some terms cancel
ans =
     x+2
```

The function `[num den] = numden(E)` returns two symbolic expressions that represent the numerator `num` and denominator `den` for the rational representation of the expression `E`.

```
>>syms x
>>E1 = x^2+5;
>>E4 = 1/(x+6);
>>[num, den] = numden(E1+E4)
num =
     x^3+6*x^2+5*x+31
den =
     x+6
```

The function `double(E)` converts the expression `E` to numeric form. The expression `E` must not contain any symbolic variables. The term *double* stands for floating-point, *double* precision. For example,

```
>>sqroot2 = sym('sqrt(2)');
>>y = 6*sqroot2
y =
   6*2^(1/2)
z = double(y)
z =
   8.4853
```

The function `poly2sym(p)` converts a coefficient vector `p` to a symbolic polynomial. The form `poly2sym(p, 'v')` generates the polynomial in terms of the variable `v`. For example,

```
>>poly2sym([2,6,4])
ans =
     2*x^2+6*x+4
>>poly2sym([5,-3,7],'y')
ans =
     5*y^2-3*y+7
```

The function `sym2poly(E)` converts the expression `E` to a polynomial coefficient vector.

```
>>syms x
>>sym2poly(9*x^2+4*x+6)
ans =
     9 4 6
```

The function `pretty(E)` displays the expression `E` on the screen in a form that resembles typeset mathematics.

The function `subs(E,old,new)` substitutes `new` for `old` in the expression `E`, where `old` can be a symbolic variable or expression and `new` can be a symbolic variable, expression, or matrix, or a numeric value or matrix. For example,

```
>>syms x y
>>E = x^2+6*x+7;
>>F = subs(E,x,y)
F =
   y^2+6*y+7
```

If `old` and `new` are cell arrays of the same size, each element of `old` is replaced by the corresponding element of `new`. If `E` and `old` are scalars and `new` is an array or cell array, the scalars are expanded to produce an array result.

If you want to tell MATLAB that f is a function of the variable t, type `f = sym('f(t)')`. Thereafter, `f` behaves like a function of `t`, and you can manipulate it with the toolbox commands. For example, to create a new function $g(t) = f(t+2) - f(t)$, the session is

```
>>syms t
>>f = sym('f(t)');
>>g = subs(f,t,t+2)-f
g =
   f(t+2)-f(t)
```

Once a specific function is defined for $f(t)$, the function $g(t)$ will be available. We will use this technique with the Laplace transform in Section 10.5.

MATLAB does not have a symbolic factorial function, but you can use the `sym` and `subs` functions to access the Maple factorial function to compute $(n-1)!$ as follows:

```
>>kfac = sym('k!');
>>syms k n
>>E = subs(kfac,k,n-1)
E =
   (n-1)!
>>expand (E)
ans =
     n!/n
```

To compute a numeric factorial, say, 5!, type `factorial(5)`, or use the `prod` function and type `prod(1:5)`.

To perform multiple substitutions, enclose the new and old elements in braces. For example, to substitute $a = x$ and $b = 2$ into the expression $E = a \sin b$, the session is

```
>>syms a b x
>>E = a*sin(b);
>>F = subs(E,{a, b}, {x, 2})
F =
   x*sin(2)
```

Evaluating Expressions

In most applications we eventually want to obtain numerical values or a plot from the symbolic expression. Use the `subs` and `double` functions to evaluate an expression numerically. Use `subs(E,old,new)` to replace `old` with a numeric value `new` in the expression `E`. The result is of class double. For example,

```
>>syms x
>>E = x^2+6*x+7;
>>G = subs(E,x,2)
G =
   23
>>class(G)
ans =
      double
```

Sometimes MATLAB will display all zeros as the result of evaluating an expression, whereas in fact the value can be nonzero but so small that you need to evaluate the expression with more accuracy to see that it is nonzero. You can use the `digits` and the `vpa` functions to change the number of digits MATLAB uses for calculating and evaluating expressions. The accuracy of individual arithmetic operations in MATLAB is 16 digits, whereas symbolic operations can be carried out to an arbitrary number of digits. The default is 32 digits. Type `digits(d)` to change the number of digits used to `d`. Be aware that larger values of `d` will require more time and computer memory to perform operations. Type `vpa(E)` to compute the expression `E` to the number of digits specified by the default value of 32 or the current setting of `digits`. Type `vpa(E,d)` to compute the expression `E` using `d` digits. (The abbreviation *vpa* stands for "variable precision arithmetic.")

Plotting Expressions

The MATLAB function `ezplot(E)` generates a plot of a symbolic expression `E`, which is a function of one variable. The default range of the independent variable is the interval $[-2\pi, 2\pi]$ unless this interval contains a singularity. The optional form `ezplot(E,[xmin xmax])` generates a plot over the range from `xmin`

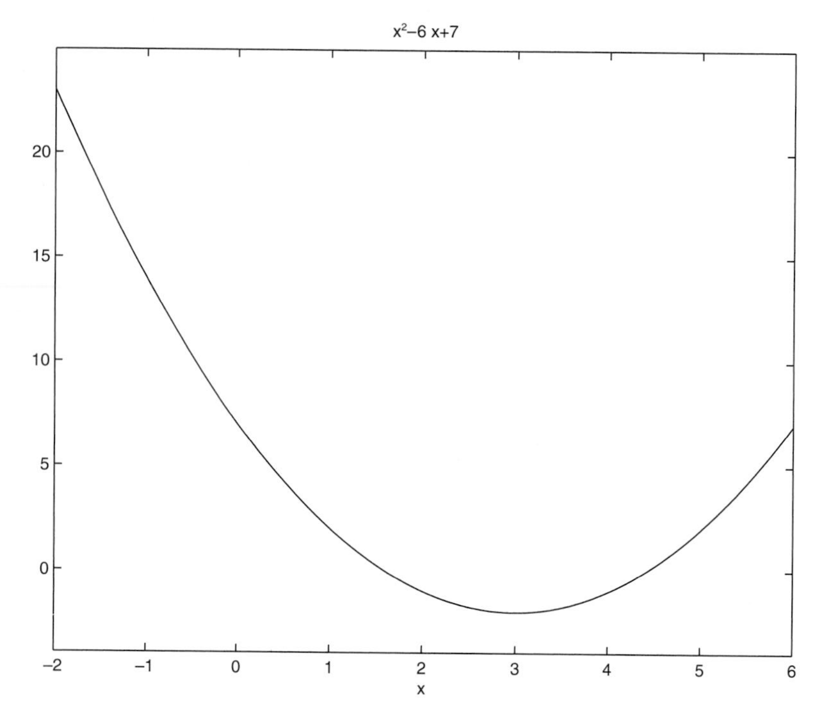

Figure 10.1–1 Plot of the function $E = x^2 - 6x + 7$ generated by the `ezplot` function.

to `xmax`. Of course, you can enhance the plot generated by `ezplot` by using the plot format commands discussed in Chapter 5; for example, the `axis`, `xlabel`, and `ylabel` commands.

For example,

```
>>syms x
>>E = x^2-6*x+7;
>>ezplot(E,[-2 6])
```

The plot is shown in Figure 10.1–1. Note that the expression is automatically placed at the top of the plot and that the axis label for the independent variable is automatically placed. You can use other plot-enhancement functions with `ezplot` to modify its appearance. For example, sometimes the automatic selection of the ordinate scale is not satisfactory. To obtain an ordinate scale from −5 to 25 and to place a label on the ordinate, you would type

```
>>ezplot(E),axis([-2 6 -5 25]),ylabel('E')
```

Order of Precedence

MATLAB does not always arrange expressions in a form that we normally would use. For example, MATLAB might provide an answer in the form `-c+b`, whereas we would normally write `b-c`. The order of precedence used by MATLAB must be constantly kept in mind to avoid misinterpreting the MATLAB output (see pages 9 and 10 for the order of precedence). MATLAB frequently expresses results in the form `1/a*b`, whereas we would normally write `b/a`. MATLAB sometimes writes `x^(1/2)*y^(1/2)` instead of grouping the terms as `(x*y)^(1/2)`, and often fails to cancel negative signs where possible, as in `-a/(-b*c-d)`, instead of `a/(b*c+d)`. For example,

```
>>syms x
>>E = x^2-6*x+7
>>F = -E/3
F =
   -1/3*x^2+2*x-7/3
```

The answer is $F = -(1/3)x^2 + 2x - 7/3$, which we would normally express as $F = -(x^2 - 6x + 7)/3$.

Tables 10.1–1 and 10.1–2 summarize the functions for creating, evaluating, and manipulating symbolic expressions.

Table 10.1–1 Functions for creating and evaluating symbolic expressions

Command	Description
`class(E)`	Returns the class of the expression `E`.
`digits(d)`	Sets the number of decimal digits used to do variable precision arithmetic. The default is 32 digits.
`double(E)`	Converts the expression `E` to numeric form.
`ezplot(E)`	Generates a plot of a symbolic expression `E`, which is a function of one variable. The default range of the independent variable is the interval $[-2\pi, 2\pi]$ unless this interval contains a singularity. The optional form `ezplot(E,[xmin xmax])` generates a plot over the range from `xmin` to `xmax`.
`findsym(E)`	Finds the symbolic variables in a symbolic expression or matrix, where `E` is a scalar or matrix symbolic expression, and returns a string containing all the symbolic variables appearing in `E`. The variables are returned in alphabetical order and are separated by commas. If no symbolic variables are found, `findsym` returns the empty string.
`findsym(E,n)`	Returns the `n` symbolic variables in `E` closest to x, with the tie breaker going to the variable closer to z.
`[num den] = numden(E)`	Returns two symbolic expressions that represent the numerator expression `num` and denominator expression `den` for the rational representation of the expression `E`.
`x = sym('x')`	Creates the symbolic variable with name `x`. Typing `x = sym('x','real')` tells MATLAB to assume that `x` is real. Typing `x = sym('x','unreal')` tells MATLAB to assume that `x` is not real.
`syms x y u v`	Creates the symbolic variables `x`, `y`, `u`, and `v`. When used without arguments, `syms` lists the symbolic objects in the workspace.
`vpa(E,d)`	Sets the number of digits used to evaluate the expression `E` to `d`. Typing `vpa(E)` causes `E` to be evaluated to the number of digits specified by the default value of 32 or by the current setting of digits.

Table 10.1–2 Functions for manipulating symbolic expressions

Command	Description
`collect(E)`	Collects coefficients of like powers in the expression `E`.
`expand(E)`	Expands the expression `E` by carrying out powers.
`factor(E)`	Factors the expression `E`.
`poly2sym(p)`	Converts a polynomial coefficient vector `p` to a symbolic polynomial. The form `poly2sym(p,'v')` generates the polynomial in terms of the variable `v`.
`pretty(E)`	Displays the expression `E` on the screen in a form that resembles typeset mathematics.
`simple(E)`	Searches for the shortest form of the expression `E` in terms of number of characters. When called, the function displays the results of each step of its search. When called without the argument, `simple` acts on the previous expression. The form `[r, how] = simple(E)` does not display intermediate steps, but saves those steps in the string `how`. The shortest form found is stored in `r`.
`simplify(E)`	Simplifies the expression `E` using Maple's simplification rules.
`subs(E,old,new)`	Substitutes `new` for `old` in the expression `E`, where `old` can be a symbolic variable or expression, `new` can be a symbolic variable, expression, or matrix, or a numeric value or matrix.
`sym2poly(E)`	Converts the expression `E` to a polynomial coefficient vector.

Test Your Understanding

T10.1–1 Given the expressions: $E_1 = x^3 - 15x^2 + 75x - 125$ and $E_2 = (x+5)^2 - 20x$, use MATLAB to

a. Find the product $E_1 E_2$ and express it in its simplest form.

b. Find the quotient E_1/E_2 and express it in its simplest form.

c. Evaluate the sum $E_1 + E_2$ at $x = 7.1$ in symbolic form and in numeric form.

(Answers: *a.* $(x-5)^5$; *b.* $x - 5$; *c.* 13,671/1000 in symbolic form, 13.6710 in numeric form.)

10.2 Algebraic and Transcendental Equations

The Symbolic Math toolbox can solve algebraic and transcendental equations, as well as systems of such equations. A *transcendental* equation is one that contains one or more transcendental functions, such as $\sin x$, e^x, or $\log x$. The appropriate function to solve such equations is the `solve` function.

The function `solve(E)` solves a symbolic expression or equation represented by the expression `E`. If `E` represents an *equation,* the equation's expression must be enclosed in single quotes. If `E` represents an expression, then the solution obtained will be the roots of the expression `E`; that is, the solution of the equation $E = 0$. Multiple expressions or equations can be solved by separating them with a comma, as `solve(E1,E2, ..., En)`. Note that you need not declare the symbolic variable with the `sym` or `syms` function before using `solve`.

There are three ways to use the `solve` function. For example, to solve the equation $x + 5 = 0$, one way is

```
>>eq1 = 'x+5=0';
>>solve(eq1)
```

```
ans =
      -5
```

The second way is

```
>>solve('x+5=0')
ans =
      -5
```

The third way is

```
>>syms x
>>solve(x+5)
ans =
      -5
```

You can store the result in a named variable as follows:

```
>>syms x
>>x = solve(x+5)
x =
    -5
```

To solve the equation $e^{2x} + 3e^x = 54$, the session is

```
>>solve('exp(2*x)+3*exp(x)=54')
ans =
    [ log(-9)]
    [ log(6)]
```

Note that `log(6)` is $\ln(6) = 1.7918$, whereas `log(-9)`, which is $\ln(-9)$, is a complex number. To see this, in MATLAB type `log(-9)` to obtain $2.1972 + 3.1416i$. So we obtained two solutions, instead of one, and now we must decide whether both are meaningful. The answer depends on the application that produced the original equation. If the application requires a real number for a solution, then we should choose `log(6)` as the answer.

The following sessions provide some more examples of the use of these functions.

```
>>eq2 = 'y^2+3*y+2=0';
>>solve(eq2)
ans =
      [-2]
      [-1]
>>eq3 = 'x^2+9*y^4=0';
>>solve(eq3) %Note that x is presumed to be the unknown variable
ans =
      [ 3*i*y^2]
      [-3*i*y^2]
```

When more than one variable occurs in the expression, MATLAB assumes that the variable closest to x in the alphabet is the variable to be found. You can specify the solution variable using the syntax `solve(E,'v')`, where `v` is the solution variable. For example,

```
>>solve('b^2+8*c+2*b=0') %solves for c because it is closer to x
ans =
     -1/8*b^2-1/4*b
>>solve('b^2+8*c+2*b=0','b') % solves for b
ans =
    [ -1+(1-8*c)^(1/2)]
    [ -1-(1-8*c)^(1/2)]
```

Thus the solution of $b^2 + 8c + 2b = 0$ for c is $c = -(b^2 + 2b)/8$. The solution for b is $b = -1 \pm \sqrt{1 - 8c}$.

You can save the solutions as vectors by using the form `[x, y] = solve(eq1, eq2)`. Note the difference in the output formats in the following example:

```
>>eq4 = '6*x+2*y=14';
>>eq5 = '3*x+7*y=31';
>>solve(eq4,eq5)
ans =
     x: [1x1 sym]
     y: [1x1 sym]
>>x = ans.x
x =
   1
>>y = ans.y =
   4
>>[x, y] = solve(eq4,eq5)
x =
   1
y =
   4
```

SOLUTION STRUCTURE

You can save the *solution in a structure* with named fields (see Chapter 2, Section 2.7 for a discussion of structures and fields). The individual solutions are saved in the fields. For example, continue the preceding session as follows:

```
>>S = solve(eq4,eq5)
S =
   x: [1x1 sym]
   y: [1x1 sym]
>>S.x
ans =
     1
>>S.y
ans =
     4
```

Test Your Understanding

T10.2–1 Use MATLAB to solve the equation $\sqrt{1-x^2} = x$.
(Answer: $x = \sqrt{2}/2$.)

T10.2–2 Use MATLAB to solve the equation set $x + 6y = a$, $2x - 3y = 9$ for x and y in terms of the parameter a.
(Answer: $x = (a + 18)/5$, $y = (2a - 9)/15$.)

Intersection of Two Circles

EXAMPLE 10.2–1

We want to find the intersection points of two circles. The first circle has a radius of 2 and is centered at $x = 3$, $y = 5$. The second circle has a radius b and is centered at $x = 5$, $y = 3$. See Figure 10.2–1.

(a) Find the (x, y) coordinates of the intersection points in terms of the parameter b.

(b) Evaluate the solution for the case where $b = \sqrt{3}$.

■ Solution

(a) The intersection points are found from the solutions of the two equations for the circles. These equations are

$$(x - 3)^2 + (y - 5)^2 = 4$$

for the first circle, and

$$(x - 5)^2 + (y - 3)^2 = b^2$$

The session to solve these equations follows. Note that the result `x: [2x1 sym]` indicates that there are two solutions for x. Similarly, there are two solutions for y.

```
>>syms x y b
>>S = solve((x-3)^2+(y-5)^2-4,(x-5)^2+(y-3)^2-b^2)
```

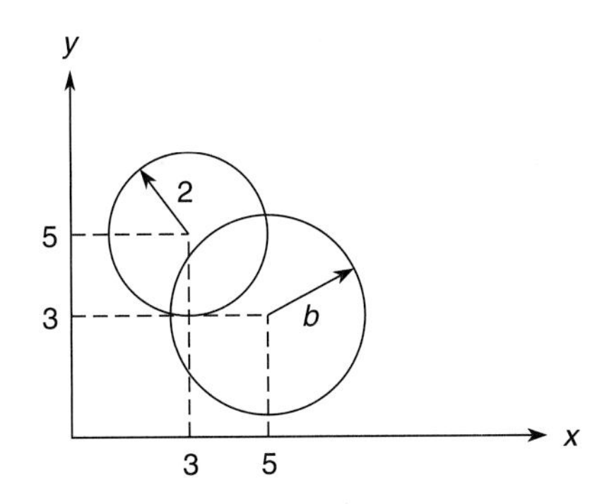

Figure 10.2–1 Intersection points of two circles.

```
S =
   x:   [2x1 sym]
   y:   [2x1 sym]
>>S.x
ans =
      [ 9/2-1/8*b^2+1/8*(-16+24*b^2-b^4)^(1/2)]
      [ 9/2-1/8*b^2-1/8*(-16+24*b^2-b^4)^(1/2)]
```

The solution for the x coordinates of the intersection points is

$$x = \frac{9}{2} - \frac{1}{8}b^2 \pm \frac{1}{8}\sqrt{-16 + 24b^2 - b^4}$$

The solution for the y coordinates can be found in a similar way by typing `S.y`.
(b) Continue the session by substituting $b = \sqrt{3}$ into the expression for x.

```
>>subs(S.x,b,sqrt(3))
ans =
     4.9820
     3.2680
```

Thus the x coordinates of the two intersection points are $x = 4.982$ and $x = 3.268$. The y coordinates can be found in a similar way.

Test Your Understanding

T10.2–3 Find the y coordinates of the intersection points in Example 10.2–1. Use $b = \sqrt{3}$.
(Answer: $y = 4.7320, 3.0180$.)

Equations containing periodic functions can have an infinite number of solutions. In such cases the `solve` function restricts the solution search to solutions near 0. For example, to solve the equation $\sin(2x) - \cos x = 0$, the session is

```
>>solve('sin(2*x)-cos(x)=0')
ans =
      [ 1/2*pi]
      [ -1/2*pi]
      [ 1/6*pi]
      [ 5/6*pi]
```

EXAMPLE 10.2–2 Positioning a Robot Arm

Figure 10.2–2 shows a robot arm having two joints and two links. The angles of rotation of the motors at the joints are θ_1 and θ_2. From trigonometry we can derive the following

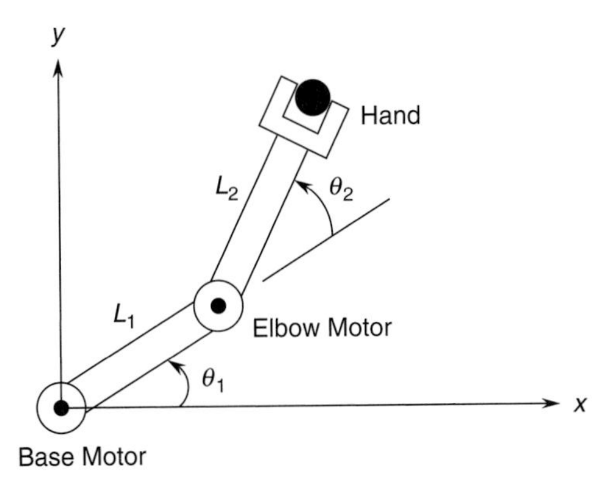

Figure 10.2–2 A robot arm having two joints and two links.

expressions for the (x, y) coordinates of the hand:

$$x = L_1 \cos\theta_1 + L_2 \cos(\theta_1 + \theta_2)$$

$$y = L_1 \sin\theta_1 + L_2 \sin(\theta_1 + \theta_2)$$

Suppose that the link lengths are $L_1 = 4$ ft and $L_2 = 3$ ft.

(a) Compute the motor angles required to position the hand at $x = 6$ ft, $y = 2$ ft.

(b) It is desired to move the hand along the straight line where x is constant at 6 ft and y varies from $y = 0.1$ to $y = 3.6$ ft. Obtain a plot of the required motor angles as a function of y.

■ Solution

(a) Substituting the given values of L_1, L_2, x, and y into the above equations gives

$$6 = 4\cos\theta_1 + 3\cos(\theta_1 + \theta_2)$$

$$2 = 4\sin\theta_1 + 3\sin(\theta_1 + \theta_2)$$

The following session solves these equations. The variables `th1` and `th2` represent θ_1 and θ_2.

```
>>S = solve('4*cos(th1)+3*cos(th1+th2)=6',...
'4*sin(th1)+3*sin(th1+th2)=2')
S =
   th1:[2x1 sym]
   th2:[2x1 sym]
>>double(S.th1)*(180/pi) % convert from radians to degrees
ans =
     -3.2981
     40.1680
>>double(S.th2)*(180/pi) % convert from radians to degrees
ans =
     51.3178
     -51.3178
```

Thus there are two solutions. The first is $\theta_1 = -3.2981°$, $\theta_2 = 51.3178°$. This is called the "elbow-down" solution. The second solution is $\theta_1 = 40.168°, \theta_2 = -51.3178°$.

This is called the "elbow-up" solution. When a problem can be solved numerically, as in this case, the `solve` function will not perform a symbolic solution. In part (b), however, the symbolic solution capabilities of the `solve` function are put to use.

(b) First we find the solutions for the motor angles in terms of the variable y. Then we evaluate the solution for numerical values of y and plot the results. The script file is shown below. Note that because the problem has three symbolic variables, we must tell the `solve` function that we want to solve for θ_1 and θ_2.

```
S = solve('4*cos(th1)+3*cos(th1+th2)=6',...
'4*sin(th1)+3*sin(th1+th2)=y', 'th1','th2');
yr = [1:0.1:3.6];
th1r = [subs(S.th1(1),'y',yr);subs(s.th1(2),'y',yr)];
th2r = [subs(S.th2(1),'y',yr);subs(s.th2(2),'y',yr)];
th1d = (180/pi)*th1r;
th2d = (180/pi)*th2r;
subplot(2,1,1)
plot(yr,th1d,2,-3.2981,'x',2,40.168,'o'),xlabel('y (feet)'),...
ylabel('Theta1 (degrees)')
subplot(2,1,2)
plot(yr,th2d,2,-51.3178,'o',2,51.3178,'x'),xlabel('y (feet)'),...
ylabel('Theta2 (degrees)')
```

The results are shown in Figure 10.2–3, where we have marked the solutions from part (a) to check the validity of the symbolic solutions. The elbow-up solutions are marked

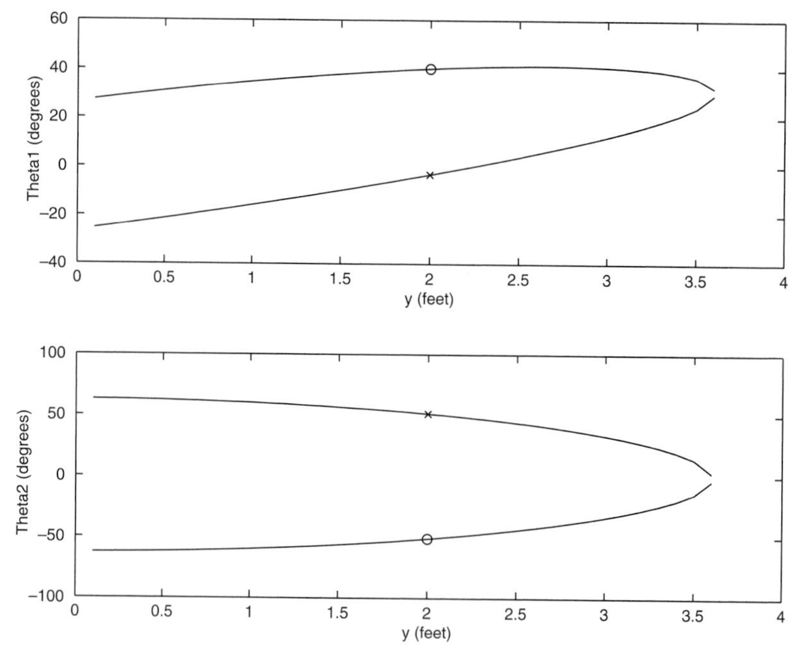

Figure 10.2–3 Plot of the motor angles for the robot hand moving along a vertical line.

with an o, and the elbow-down solutions are marked with an x. We could have printed the expression for the solutions for θ_1 and θ_2 as functions of y, but the expressions are cumbersome and unnecessary if all we want is the plot.

MATLAB is powerful enough to solve the robot arm equations for arbitrary values of the hand coordinates (x, y). However, the resulting expressions for θ_1 and θ_2 are complicated.

Table 10.2–1 summarizes the `solve` function.

Table 10.2–1 Functions for solving algebraic and transcendental equations

Command	Description
`solve(E)`	Solves a symbolic expression or equation represented by the expression `E`. If `E` represents an *equation,* the equation's expression must be enclosed in single quotes. If `E` represents an expression, then the solution obtained will be the roots of the expression `E`; that is, the solution of the equation $E = 0$. You need not declare the symbolic variable with the `sym` or `syms` function before using `solve`.
`solve(E1,...,En)`	Solves multiple expressions or equations.
`S = solve(E)`	Saves the solution in the structure `S`.

10.3 Calculus

In Chapter 8 we discussed techniques for performing numerical differentiation and numerical integration; this section covers differentiation and integration of symbolic expressions to obtain closed form results for the derivatives and integrals.

Differentiation

The `diff` function is used to obtain the symbolic derivative. Although this function has the same name as the function used to compute numerical differences (see Chapter 8), MATLAB detects whether or not a symbolic expression is used in the argument and directs the calculation accordingly. The basic syntax is `diff(E)`, which returns the derivative of the expression `E` with respect to the default independent variable.

For example, the derivatives

$$\frac{dx^n}{dx} = nx^{n-1}$$

$$\frac{d \ln x}{dx} = \frac{1}{x}$$

$$\frac{d \sin^2 x}{dx} = 2 \sin x \cos x$$

$$\frac{d \sin y}{dy} = \cos y$$

are obtained with the following session:

```
>>syms n x y
>>diff(x^n)
ans =
     x^n*n/x
>>simplify(ans)
ans =
     x^(n-1)*n
>>diff(log(x))
ans =
     1/x
>>diff((sin(x))^2)
ans =
     2*sin(x)*cos(x)
>>diff(sin(y))
ans =
     cos(y)
```

If the expression contains more than one variable, the `diff` function operates on the variable x, or the variable closest to x, unless told to do otherwise. When there is more than one variable, the `diff` function computes the *partial* derivative. For example, if

$$f(x, y) = \sin(xy)$$

then

$$\frac{\partial f}{\partial x} = y\cos(xy)$$

The corresponding session is

```
>>diff(sin(x*y))
ans =
     cos(x*y)*y
```

There are three ways to use the `diff` function, but the third way is preferred. Using the function x^2 as an example, the first way is

```
>>E = 'x^2';
>>diff(E)
ans =
     2*x
```

The second way is

```
>>diff('x^2')
ans =
     2*x
```

The third way is

```
>>syms x
>>diff(x^2)
ans =
      2*x
```

Note that the expression to be differentiated need not be placed in quotes if the variable has been declared to be symbolic. This method is preferred because the use of quoted strings bypasses the use of symbolic expressions, which is the whole point of the Symbolic Math toolbox.

There are three other forms of the `diff` function. The function `diff(E,v)` returns the derivative of the expression `E` with respect to the variable `v`. For example,

$$\frac{\partial[x \sin(xy)]}{\partial y} = x^2 \cos(xy)$$

is given by

```
>>syms x y
>>diff(x*sin(x*y),y)
ans =
      x^2*cos(x*y)
```

The function `diff(E,n)` returns the nth derivative of the expression `E` with respect to the default independent variable. For example,

$$\frac{d^2(x^3)}{dx^2} = 6x$$

is given by

```
>>syms x
>>diff(x^3,2)
ans =
      6*x
```

The function `diff(E,v,n)` returns the nth derivative of the expression `E` with respect to the variable `v`. For example,

$$\frac{\partial^2[x \sin(xy)]}{\partial y^2} = -x^3 \sin(xy)$$

is given by

```
>>syms x y
>>diff(x*sin(x*y),y,2)
ans =
      -x^3*sin(x*y)
```

Table 10.3–1 summarizes the differentiation functions.

Table 10.3–1 Symbolic calculus functions

Command	Description
`diff(E)`	Returns the derivative of the expression `E` with respect to the default independent variable.
`diff(E,v)`	Returns the derivative of the expression `E` with respect to the variable `v`.
`diff(E,n)`	Returns the nth derivative of the expression `E` with respect to the default independent variable.
`diff(E,v,n)`	Returns the nth derivative of the expression `E` with respect to the variable `v`.
`int(E)`	Returns the integral of the expression `E` with respect to the default independent variable.
`int(E,v)`	Returns the integral of the expression `E` with respect to the variable `v`.
`int(E,a,b)`	Returns the integral of the expression `E` with respect to the default independent variable over the interval $[a, b]$, where a and b are numeric quantities.
`int(E,v,a,b)`	Returns the integral of the expression `E` with respect to the variable `v` over the interval $[a, b]$, where a and b are numeric quantities.
`int(E,m,n)`	Returns the integral of the expression `E` with respect to the default independent variable over the interval $[m, n]$, where m and n are symbolic expressions.
`limit(E)`	Returns the limit of the expression `E` as the default independent variable goes to 0.
`limit(E,a)`	Returns the limit of the expression `E` as the default independent variable goes to `a`.
`limit(E,v,a)`	Returns the limit of the expression `E` as the variable `v` goes to `a`.
`limit(E,v,a,'d')`	Returns the limit of the expression `E` as the variable `v` goes to `a` from the direction specified by `d`, which may be `right` or `left`.
`symsum(E)`	Returns the symbolic summation of the expression `E`.
`taylor(f,n,a)`	Gives the first `n-1` terms in the Taylor series for the function defined in the expression `f`, evaluated at the point $x = a$. If the parameter `a` is omitted, the function returns the series evaluated at $x = 0$.

Max-Min Problems

The derivative can be used to find the maximum or minimum of a continuous function, say, $f(x)$, over an interval $a \leq x \leq b$. A *local* maximum or local minimum (one that does not occur at one of the boundaries $x = a$ or $x = b$) can occur only at a *critical point,* which is a point where either $df/dx = 0$ or df/dx does not exist. If $d^2f/dx^2 > 0$, the point is a relative minimum; if $d^2f/dx^2 < 0$, the point is a relative maximum. If $d^2f/dx^2 = 0$, the point is neither a minimum nor a maximum, but is an *inflection point.* If multiple candidates exist, you must evaluate the function at each point to determine the *global* maximum and global minimum.

EXAMPLE 10.3–1 Topping the Green Monster

The Green Monster is a wall 37 ft high in left field at Fenway Park in Boston. The wall is 310 ft from home plate down the left-field line. Assuming that the batter hits the ball 4 ft above the ground, and neglecting air resistance, determine the *minimum* speed the batter must give to the ball to hit it over the Green Monster. In addition, find the angle at which the ball must be hit (see Figure 10.3–1).

■ Solution

The equations of motion for a projectile launched with a speed v_0 at an angle θ relative to the horizontal are

$$x(t) = (v_0 \cos\theta)t \qquad y(t) = -\frac{gt^2}{2} + (v_0 \sin\theta)t$$

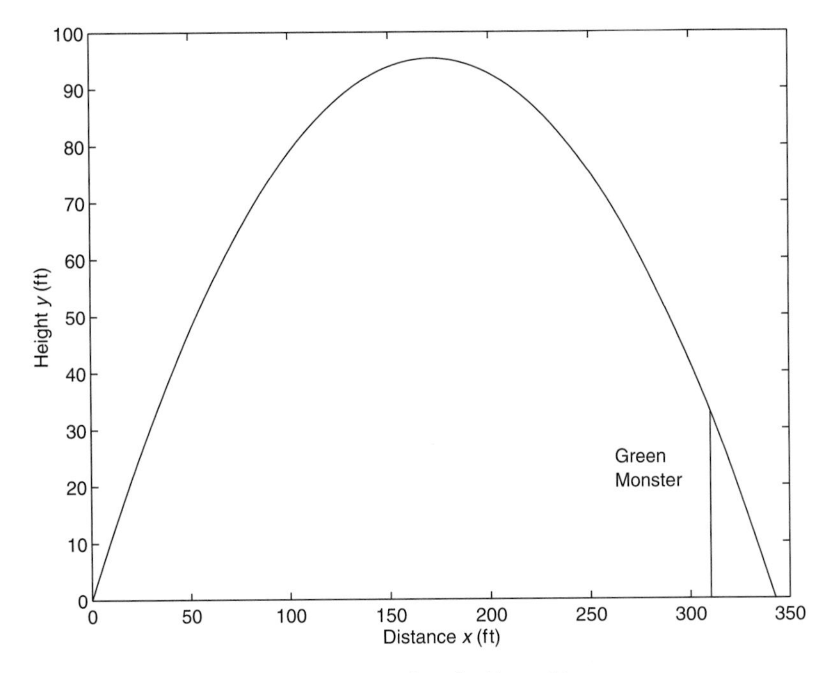

Figure 10.3–1 A baseball trajectory to clear the Green Monster.

where $x = 0$, $y = 0$ is the location of the ball when it is hit. Because we are not concerned with the time of flight in this problem, we can eliminate t and obtain an equation for y in terms of x. To do so, solve the x equation for t and substitute this into the y equation to obtain

$$y(t) = -\frac{g}{2}\frac{x^2(t)}{v_0^2 \cos^2\theta} + x(t)\tan\theta$$

(You could use MATLAB to do this algebra if you wish. We will use MATLAB to do the more difficult task to follow.)

Because the ball is hit 4 ft above the ground, the ball must rise $37-4 = 33$ ft to clear the wall. Let h represent the relative height of the wall (33 ft). Let d represent the distance to the wall (310 ft). Use $g = 32.2$ ft/sec^2. When $x = d$, $y = h$. Thus the previous equation gives

$$h = -\frac{g}{2}\frac{d^2}{v_0^2 \cos^2\theta} + d\tan\theta$$

which can easily be solved for v_0^2 as follows:

$$v_0^2 = \frac{g}{2}\frac{d^2}{\cos^2\theta(d\tan\theta - h)}$$

Because $v_0 > 0$, minimizing v_0^2 is equivalent to minimizing v_0. Note also that $gd^2/2$ is a multiplicative factor in the expression for v_0^2. Thus the minimizing value of θ is independent of g and can be found by minimizing the function

$$f = \frac{1}{\cos^2\theta(d\tan\theta - h)}$$

The session to do this is as follows. The variable `th` represents the angle θ of the ball's velocity vector relative to the horizontal. The first step is to calculate the derivative $df/d\theta$ and solve the equation $df/d\theta = 0$ for θ.

```
>>syms d g h th
>>f = 1/(((cos(th))^2)*(d*tan(th)-h));
>>dfdth = diff(f,th);
>>thmin = solve(dfdth,th);
>>thmin = subs(thmin,{d,h},{310,33})
thmin =
    0.8384
   -0.7324
```

Obviously, the negative angle is not a proper solution, so the only solution candidate is $\theta = 0.8384$ rad, or about 48°. To verify that this angle is a minimum solution, and not a maximum or an inflection point, we can check the second derivative $d^2f/d\theta^2$. If this derivative is positive, the solution represents a minimum. To check this solution and to find the speed required, continue the session as follows:

```
>>second = diff(f,2,th);    % This is the second derivative.
>>second = subs(second,{th,d,h},{thmin(1),310,33})
second =
   0.0321
>>v2 = (g*d^2/2)*f;
>>v2min = subs(v2,{d,h,g},{310,33,32.2});
>>vmin = sqrt(v2min);
>>vmin = double(subs(vmin(1),{th,d,h,g},{thmin(1),310,33,32.2}))
vmin =
   105.3613
```

Because the second derivative is positive, the solution is a minimum. Thus the minimum speed required is 105.3613 ft/sec, or about 72 mi/hr. A ball hit with this speed will clear the wall only if it is hit at an angle of approximately 48°.

Test Your Understanding

T10.3–1 Given that $y = \sinh(3x)\cosh(5x)$, use MATLAB to find dy/dx at $x = 0.2$.
(Answer: 9.2288.)

T10.3–2 Given that $z = 5\cos(2x)\ln(4y)$, use MATLAB to find $\partial z/\partial y$.
(Answer: $5\cos(2x)/y$.)

Integration

The `int(E)` function is used to integrate a symbolic expression `E`. It attempts to find the symbolic expression `I` such that `diff(I)=E`. If the integral does not exist in closed form or MATLAB cannot find the integral even if it exists, the function will return the expression unevaluated.

As with the `diff` function, there are three ways to use the `int` function; two of the ways use quoted strings, but the following way is preferred. Using the function $2x$ as an example, the method is

```
>>syms x
>>int(2*x)
ans =
     x^2
```

The function `int(E)` returns the integral of the expression `E` with respect to the default independent variable. For example, you can obtain the following integrals with the next session:

$$\int x^n \, dx = \frac{x^{n+1}}{n+1}$$

$$\int \frac{1}{x} \, dx = \ln x$$

$$\int \cos x \, dx = \sin x$$

$$\int \sin y \, dy = -\cos y$$

```
>>syms n x y
>>int(x^n)
ans =
     x^(n+1)/(n+1)
>>int(1/x)
ans =
     log(x)
>>int(cos(x))
ans =
     sin(x)
>>int(sin(y))
ans =
     -cos(y)
```

Here are the other forms of the `int` function. The form `int(E,v)` returns the integral of the expression `E` with respect to the variable `v`. For example, the result

$$\int x^n \, dn = \frac{x^n}{\ln x}$$

can be obtained with the session:

```
>>syms n x
>>int(x^n,n)
ans =
     1/log(x)*x^n
```

The form `int(E,a,b)` returns the integral of the expression `E` with respect to the default independent variable evaluated over the interval $[a, b]$, where `a` and `b` are numeric expressions. For example, the result

$$\int_2^5 x^2\,dx = \left.\frac{x^3}{3}\right|_2^5 = 39$$

is obtained as follows:

```
>>syms x
>>int(x^2,2,5)
ans =
     39
```

The form `int(E,v,a,b)` returns the integral of the expression `E` with respect to the variable `v` evaluated over the interval $[a, b]$, where `a` and `b` are numeric quantities. For example, the result

$$\int_0^5 xy^2\,dy = x\left.\frac{y^3}{3}\right|_0^5 = \frac{125}{3}x$$

is obtained from

```
>>syms x y
>>int(xy^2,y,0,5)
ans =
     125/3*x
```

The result

$$\int_a^b x^2\,dx = \frac{b^3}{3} - \frac{a^3}{3}$$

is obtained from

```
>>syms a b x
>>int(x^2,a,b)
ans =
     1/3*b^3-1/3*a^3
```

The form `int(E,m,n)` returns the integral of the expression `E` with respect to the default independent variable evaluated over the interval $[m, n]$, where `m` and `n` are symbolic expressions. For example,

$$\int_1^t x\,dx = \left.\frac{x^2}{2}\right|_1^t = \frac{1}{2}t^2 - \frac{1}{2}$$

$$\int_t^{e^t} \sin x\,dx = -\cos x|_t^{e^t} = -\cos(e^t) + \cos t$$

are given by this session:

```
>>syms t x
>>int(x,1,t)
ans =
     1/2*t^2-1/2
int(sin(x),t,exp(t))
ans =
     -cos(exp(t)) + cos(t)
```

The following session gives an example for which no integral can be found. The indefinite integral exists, but the definite integral does not exist if the limits of integration include the singularity at $x = 1$. The integral is

$$\int \frac{1}{x-1}\,dx = \ln|x-1|$$

The session is

```
>>syms x
>>int(1/(x-1))
ans =
     log(x-1)
>>int(1/(x-1),0,2)
ans =
     NaN
```

Table 10.3–1 summarizes the integration functions.

Test Your Understanding

T10.3–3 Given that $y = x\sin(3x)$, use MATLAB to find $\int y\,dx$. (Answer: $[\sin(3x) - 3x\cos(3x)]/9$.)

T10.3–4 Given that $z = 6y^2\tan(8x)$, use MATLAB to find $\int z\,dy$. (Answer: $2y^3\tan(8x)$.)

T10.3–5 Use MATLAB to evaluate

$$\int_{-2}^{5} x\sin(3x)\,dx$$

(Answer: 0.6672.)

Taylor Series

Taylor's theorem states that a function $f(x)$ can be represented in the vicinity of $x = a$ by the expansion

$$f(x) = f(a) + \left(\frac{df}{dx}\right)\bigg|_{x=a}(x-a) + \frac{1}{2}\left(\frac{d^2 f}{dx^2}\right)\bigg|_{x=a}(x-a)^2 + \cdots + \frac{1}{k!}\left(\frac{d^k f}{dx^k}\right)\bigg|_{x=a}(x-a)^k + \cdots + R_n \tag{10.3–1}$$

The term R_n is the remainder and is given by

$$R_n = \frac{1}{n!}\left(\frac{d^n f}{dx^n}\right)\bigg|_{x=b}(x-a)^n \tag{10.3–2}$$

where b lies between a and x.

These results hold if $f(x)$ has continuous derivatives through order n. If R_n approaches 0 for large n, the expansion is called the *Taylor* series for $f(x)$ about $x = a$. If $a = 0$, the series is sometimes called the *Maclaurin* series.

Some common examples of the Taylor series are

$$\sin x = x - \frac{x^3}{3!} + \frac{x^5}{5!} - \frac{x^7}{7!} + \cdots, \quad -\infty < x < \infty$$

$$\cos x = 1 - \frac{x^2}{2!} + \frac{x^4}{4!} - \frac{x^6}{6!} + \cdots, \quad -\infty < x < \infty$$

$$e^x = 1 + x + \frac{x^2}{2!} + \frac{x^3}{3!} + \frac{x^4}{4!} + \cdots, \quad -\infty < x < \infty$$

where $a = 0$ in all three examples.

The `taylor(f,n,a)` function gives the first `n-1` terms in the Taylor series for the function defined in the expression `f`, evaluated at the point $x = a$. If the parameter `a` is omitted the function returns the series evaluated at $x = 0$. Here are some examples:

```
>>syms x
>>f = exp(x);
>>taylor(f,4)
ans =
     1+x+1/2*x^2+1/6*x^3
>>taylor(f,3,2)
ans =
     exp(2)+exp(2)*(x-2)+1/2*exp(2)*(x-2)^2
```

The latter expression corresponds to

$$e^2\left[1 + (x-2) + \tfrac{1}{2}(x-2)^2\right]$$

Sums

The `symsum(E)` function returns the symbolic summation of the expression `E`; that is

$$\sum_{x=0}^{x-1} E(x) = E(0) + E(1) + E(2) + \cdots + E(x-1)$$

The `symsum(E,a,b)` function returns the sum of the expression `E` as the default symbolic variable varies from `a` to `b`. That is, if the symbolic variable is `x`, then `S = symsum(E,a,b)` returns

$$\sum_{x=a}^{b} E(x) = E(a) + E(a+1) + E(a+2) + \cdots + E(b)$$

Here are some examples. The summations

$$\sum_{k=0}^{10} k = 0 + 1 + 2 + 3 + \cdots + 9 + 10 = 55$$

$$\sum_{k=0}^{n-1} k = 0 + 1 + 2 + 3 + \cdots + n - 1 = \frac{1}{2}n^2 - \frac{1}{2}n$$

$$\sum_{k=1}^{4} k^2 = 1 + 4 + 9 + 16 = 30$$

are given by

```
>>syms k n
>>symsum(k,0,10)
ans =
      55
>>symsum(k^2, 1, 4)
ans =
      30
>>symsum(k,0,n-1)
ans =
      1/2*n^2-1/2*n
>>factor(ans)
ans =
      1/2*n*(n-1)
```

The latter expression is the standard form of the result.

Limits

The function `limit(E,a)` returns the limit

$$\lim_{x \to a} E(x)$$

if x is the symbolic variable. There are several variations of this syntax. The basic form `limit(E)` finds the limit as $x \to 0$. For example

$$\lim_{x\to 0} \frac{\sin(ax)}{x} = a$$

is given by

```
>>syms a x
>>limit(sin(a*x)/x)
ans =
     a
```

The form `limit(E,v,a)` finds the limit as $v \to a$. For example,

$$\lim_{x\to 3} \frac{x-3}{x^2-9} = \frac{1}{6}$$

$$\lim_{x\to 0} \frac{\sin(x+h)-\sin(x)}{h}$$

are given by

```
>>syms h x
>>limit((x-3)/(x^2-9),3)
ans =
     1/6
>>limit((sin(x+h)-sin(x))/h,h,0)
ans =
     cos(x)
```

The forms `limit(E,v,a,'right')` and `limit(E,v,a,'left')` specify the direction of the limit. For example,

$$\lim_{x\to 0-} \frac{1}{x} = -\infty$$

$$\lim_{x\to 0+} \frac{1}{x} = \infty$$

are given by

```
>>syms x
>>limit(1/x,x,0,'left')
ans =
     -inf
>>limit(1/x,x,0,'right')
ans =
     inf
```

Table 10.3–1 summarizes the series and limit functions.

Test Your Understanding

T10.3–6 Use MATLAB to find the first three nonzero terms in the Taylor series for $\cos x$.
(Answer: $1 - x^2/2 + x^4/24$.)

T10.3–7 Use MATLAB to find a formula for the sum

$$\sum_{m=0}^{m-1} m^3$$

(Answer: $m^4/4 - m^3/2 + m^2/4$.)

T10.3–8 Use MATLAB to evaluate

$$\sum_{n=0}^{7} \cos(\pi n)$$

(Answer: 0.)

T10.3–9 Use MATLAB to evaluate

$$\lim_{x \to 5} \frac{2x - 10}{x^3 - 125}$$

(Answer: 2/75.)

10.4 Differential Equations

A first-order ordinary differential equation (ODE) can be written in the form

$$\frac{dy}{dt} = f(t, y)$$

where t is the independent variable and y is a function of t. A solution to such an equation is a function $y = g(t)$ such that $dg/dt = f(t, g)$, and the solution will contain one arbitrary constant. This constant becomes determined when we apply an additional condition of the solution by requiring that the solution have a specified value $y(t_1)$ when $t = t_1$. The chosen value t_1 is often the smallest, or starting value, of t, and if so, the condition is called the *initial condition* (quite often $t_1 = 0$). The general term for such a requirement is a *boundary condition*, and MATLAB lets us specify conditions other than initial conditions. For example, we can specify the value of the dependent variable at $t = t_2$, where $t_2 > t_1$.

INITIAL CONDITION

BOUNDARY CONDITION

Methods for obtaining a numerical solution to differential equations were covered in Chapter 8. However, we prefer to obtain an analytical solution whenever possible because it is more general and thus more useful for designing engineering devices or processes.

A second-order ODE has the following form:

$$\frac{d^2y}{dt^2} = f\left(t, y, \frac{dy}{dt}\right)$$

Its solution will have two arbitrary constants that can be determined once two additional conditions are specified. These conditions are often the specified values of y and dy/dt at $t = 0$. The generalization to third-order and higher equations is straightforward.

We will occasionally use the following abbreviations for the first- and second-order derivatives:

$$\dot{y} = \frac{dy}{dt} \qquad \ddot{y} = \frac{d^2y}{dt^2}$$

MATLAB provides the `dsolve` function for solving ordinary differential equations. Its various forms differ according to whether they are used to solve single equations or sets of equations, whether or not boundary conditions are specified, and whether or not the default independent variable `t` is acceptable. Note that `t` is the default independent variable and not `x` as with the other symbolic functions. The reason is that many ODE models of engineering applications have time t as the independent variable.

Solving a Single Differential Equation

The `dsolve` function's syntax for solving a single equation is `dsolve('eqn')`. The function returns a symbolic solution of the ODE specified by the symbolic expression `eqn`. Use the uppercase letter `D` to represent the first derivative, use `D2` to represent the second derivative, and so on. Any character immediately following the differentiation operator is taken to be the dependent variable. Thus `Dw` represents dw/dt. Because of this syntax, you cannot use uppercase D as symbolic variable when using the `dsolve` function.

The arbitrary constants in the solution are denoted by `C1`, `C2`, and so on. The number of such constants is the same as the order of the ODE. For example, the equation

$$\frac{dy}{dt} + 2y = 12$$

has the solution

$$y(t) = 6 + C_1e^{-2t}$$

The solution can be found with the following session. Note that you need not declare `y` to be symbolic prior to using `dsolve`.

```
>>dsolve('Dy+2*y=12')
ans =
     6+C1*exp(-2*t)
```

There can be symbolic constants in the equation. For example,

$$\frac{dy}{dt} = \sin(at)$$

has the solution

$$y(t) = -\frac{\cos(at)}{a} + C_1$$

It can be found as follows:

```
>>dsolve('Dy=sin(a*t)')
ans =
      (-cos(a*t)+C1*a)/a
```

Here is a second-order example:

$$\frac{d^2y}{dt^2} = c^2y$$

The solution $y(t) = C_1e^{ct} + C_2e^{-ct}$ can be found with the following session:

```
dsolve('D2y=c^2*y')
ans =
      C1*exp(-c*t) + C2*exp(c*t)
```

Solving Sets of Equations

Sets of equations can be solved with `dsolve`. The appropriate syntax is `dsolve('eqn1','eqn2',...)`. The function returns a symbolic solution of the set of equations specified by the symbolic expressions `eqn1` and `eqn2`.

For example, the set

$$\frac{dx}{dt} = 3x + 4y$$

$$\frac{dy}{dt} = -4x + 3y$$

has the solution $x(t) = C_1e^{3t}\cos 4t + C_2e^{3t}\sin 4t$, $y(t) = -C_1e^{3t}\sin 4t + C_2e^{3t}\cos 4t$. The session is

```
>>[x, y] = dsolve('Dx=3*x+4*y','Dy=-4*x+3*y')
   x = C1*exp(3*t)*cos(4*t)+C2*exp(3*t)*sin(4*t)
   y = -C1*exp(3*t)*sin(4*t)+C2*exp(3*t)*cos(4*t)
```

Specifying Initial and Boundary Conditions

Conditions on the solutions at specified values of the independent variable can be handled as follows. The form `dsolve('eqn', 'cond1', 'cond2',...)` returns a symbolic solution of the ODE specified by the symbolic expression `eqn`, subject to the conditions specified in the expressions `cond1`, `cond2`, and so on. If `y` is the dependent variable, these conditions are specified as follows: `y(a) = b`, `Dy(a) = c`, `D2y(a) = d`, and so on. These correspond to $y(a)$, $\dot{y}(a)$, $\ddot{y}(a)$, and so on. If the number of conditions is less than the order of the equation, the returned solution will contain arbitrary constants `C1`, `C2`, and so on.

For example, the problem

$$\frac{dy}{dt} = \sin(bt), \quad y(0) = 0$$

has the solution $y(t) = [1 - \cos(bt)]/b$. It can be found as follows:

```
>>dsolve('Dy=sin(b*t)','y(0)=0')
ans =
      -cos(b*t)/b+1/b
```

The problem

$$\frac{d^2y}{dt^2} = c^2y, \quad y(0) = 1, \quad \dot{y}(0) = 0$$

has the solution $y(t) = (e^{ct} + e^{-ct})/2$. The session is

```
>>dsolve('D2y=c^2*y','y(0)=1','Dy(0)=0')
ans =
      1/2*exp(c*t)+1/2*exp(-c*t)
```

Arbitrary boundary conditions, such as $y(0) = c$, can be used. For example, the solution of the problem

$$\frac{dy}{dt} + ay = b, \quad y(0) = c$$

is

$$y(t) = \frac{b}{a} + \left(c - \frac{b}{a}\right) e^{-at}$$

The session is

```
>>dsolve('Dy+a*y=b','y(0)=c')
ans =
      1/a*b+exp(-a*t)*(-1/a*b+c)
```

Plotting the Solution

The `ezplot` function can be used to plot the solution, just as with any other symbolic expression, provided no undetermined constants such as `C1` are present. For example, the problem

$$\frac{dy}{dt} + 10y = 10 + 4\sin(4t), \qquad y(0) = 0$$

has the solution

$$y(t) = 1 - \tfrac{4}{29}\cos(4t) + \tfrac{10}{29}\sin(4t) - \tfrac{25}{29}e^{-10t}$$

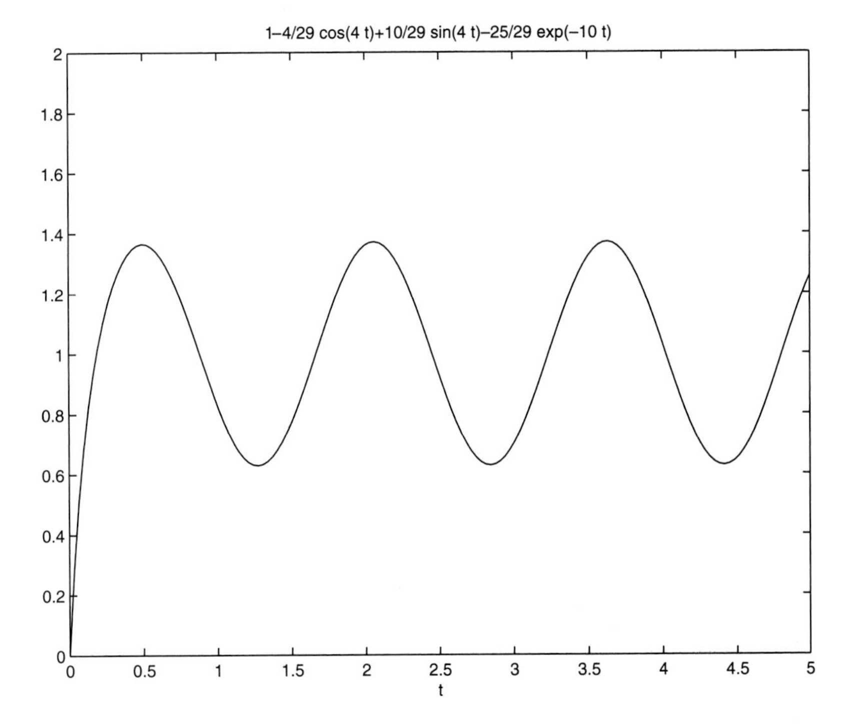

Figure 10.4–1 Plot of the solution of $\dot{y} + 10y = 10 + 4\sin(4t)$, $y(0) = 0$.

The session is

```
>>y = dsolve('Dy+10*y=10+4*sin(4*t)','y(0)=0')
y =
    1-4/29*cos(4*t)+ 10/29*sin(4*t)-25/29*exp(-10*t)
>>ezplot(y),axis([0 5 0 2])
```

The plot is shown in Figure 10.4–1.

Sometimes the `ezplot` function uses too few values of the independent variable and thus does not produce a smooth plot. To override the spacing chosen by the `ezplot` function, you can use the `subs` function to substitute an array of values for the independent variable. Note that you must define `t` to be a symbolic variable. For example, you can continue the previous session as follows:

```
>>syms t
>>x = [0:0.05:5];
>>P = subs(y,t,x);
>>plot(x,P),axis([0 5 0 2]), xlabel('t')
```

Equation Sets with Boundary Conditions

Sets of equations with specified boundary conditions can be solved as follows. The function `dsolve('eqn1','eqn2',...,'cond1','cond2',...)`

returns a symbolic solution of a set of equations specified by the symbolic expressions `eqn1`, `eqn2`, and so on, subject to the initial conditions specified in the expressions `cond1`, `cond2`, and so on.

For example, the problem

$$\frac{dx}{dt} = 3x + 4y, \qquad x(0) = 0$$

$$\frac{dy}{dt} = -4x + 3y, \quad y(0) = 1$$

has the solution

$$x(t) = e^{3t}\sin(4t), \quad y(t) = e^{3t}\cos(4t)$$

The session is

```
>>dsolve('Dx=3*x+4*y','Dy=-4*x+3*y','x(0)=0','y(0)=1')
[x,y] =
     x = exp(3*t)*sin(4*t), y = exp(3*t)*cos(4*t)
```

It is not necessary to specify only initial conditions. The conditions can be specified at different values of t. For example, to solve the problem

$$\frac{d^2y}{dt^2} + 9y = 0, \quad y(0) = 1, \quad \dot{y}(\pi) = 2$$

the session is

```
>>dsolve('D2y+9*y=0','y(0)=1','Dy(pi)=2')
ans =
     -2/3*sin(3*t)+cos(3*t)
```

Using Other Independent Variables

Although the default independent variable is `t`, you can use the following syntax to specify a different independent variable. The function `dsolve('eqn1', 'eqn2', ..., 'cond1','cond2', ..., 'x')` returns a symbolic solution of a set of equations where the independent variable is `x`.

For example, the solution of the equation

$$\frac{dv}{dx} + 2v = 12$$

is $v(x) = 6 + C_1e^{-2x}$. The session is

```
>>dsolve('Dv+2*v=12','x')
ans =
     6+exp(-2*x)*C1
```

Test Your Understanding

T10.4–1 Use MATLAB to solve the equation

$$\frac{d^2y}{dt^2} + b^2y = 0$$

Check the answer by hand or with MATLAB.
(Answer: $y(t) = C_1 \cos(bt) + C_2 \sin(bt)$.)

T10.4–2 Use MATLAB to solve the problem

$$\frac{d^2y}{dt^2} + b^2y = 0, \qquad y(0) = 1, \qquad \dot{y}(0) = 0$$

Check the answer by hand or with MATLAB.
(Answer: $y(t) = \cos(bt)$.)

Solving Nonlinear Equations

MATLAB can solve many nonlinear first-order differential equations. For example, the problem

$$\frac{dy}{dt} = 4 - y^2, \qquad y(0) = 1 \tag{10.4–1}$$

can be solved with the following session

```
>>dsolve('Dy=4-y^2', 'y(0)=1')
ans =
     2*(exp(4*t-log(-1/3))+1)/(-1+exp(4*t-log(-1/3)))
>>simple(ans)
ans =
     2*(3*exp(4*t)-1)/(1+3*exp(4*t))
```

which is equivalent to

$$y(t) = 2\frac{3e^{4t} - 1}{1 + 3e^{4t}}$$

Not all nonlinear equations can be solved in closed form. For example, the following equation is the equation of motion of a specific pendulum.

$$\frac{d^2y}{dt^2} + 9\sin(y) = 0, \qquad y(0) = 1, \qquad \dot{y}(0) = 0$$

The following session generates a message that a solution could not be found.

```
>>dsolve('D2y+9*sin(y)=0','y(0)=1','Dy(0)=0')
```

Table 10.4–1 The `dsolve` function

Command	Description
`dsolve('eqn')`	Returns a symbolic solution of the ODE specified by the symbolic expression `eqn`. Use the uppercase letter `D` to represent the first derivative; use `D2` to represent the second derivative, and so on. Any character immediately following the differentiation operator is taken to be the dependent variable.
`dsolve('eqn1','eqn2', ...)`	Returns a symbolic solution of the set of equations specified by the symbolic expressions `eqn1`, `eqn2`, and so on.
`dsolve('eqn','cond1','cond2', ...)`	Returns a symbolic solution of the ODE specified by the symbolic expression `eqn`, subject to the conditions specified in the expressions `cond1`, `cond2`, and so on. If `y` is the dependent variable, these conditions are specified as follows: `y(a) = b`, `Dy(a) = c`, `D2(a) = d`, and so on.
`dsolve('eqn1','eqn2', ..., 'cond1', 'cond2', ...)`	Returns a symbolic solution of a set of equations specified by the symbolic expressions `eqn1`, `eqn2`, and so on, subject to the initial conditions specified in the expressions `cond1`, `cond2`, and so on.

Thus MATLAB was unable to find a closed-form solution. It is possible, however, that later versions of MATLAB will be able to solve this equation. Try it and see!

Table 10.4–1 summarizes the functions for solving differential equations.

10.5 Laplace Transforms

This section explains how to use the *Laplace transform* with MATLAB to solve some types of differential equations that cannot be solved with `dsolve`. Application of the Laplace transform converts a linear differential equation problem into an algebraic problem. With proper algebraic manipulation of the resulting quantities, the solution of the differential equation can be recovered in an orderly fashion by inverting the transformation process to obtain a function of time. We assume that you are familiar with the fundamentals of differential equations outlined in Chapter 8, Section 8.4.

The Laplace transform $\mathcal{L}[y(t)]$ of a function $y(t)$ is defined to be

$$\mathcal{L}[y(t)] = \int_0^\infty y(t)e^{-st}\,dt \qquad (10.5\text{–}1)$$

The integration removes t as a variable, and the transform is thus a function of only the Laplace variable s, which may be a complex number. The integral exists

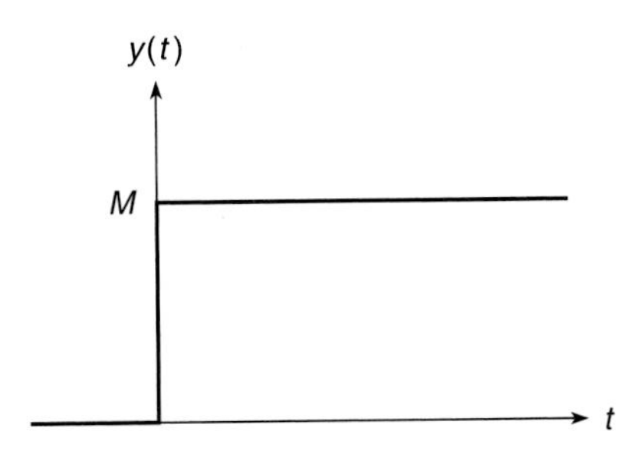

Figure 10.5–1 A step function of magnitude M.

for most of the commonly encountered functions if suitable restrictions are placed on s. An alternative notation is the use of the uppercase symbol to represent the transform of the corresponding lowercase symbol; that is,

$$Y(s) = \mathcal{L}[y(t)]$$

STEP FUNCTION

We will use the *one-sided* transform, which assumes that the variable $y(t)$ is 0 for $t < 0$. For example, the *step function* is such a function. Its name comes from the fact that its graph looks like a stair step (see Figure 10.5–1).

The height of the step is M and is called the *magnitude*. The *unit-step function*, denoted $u_s(t)$, has a height of $M = 1$ and is defined as follows:

$$u_s(t) = \begin{cases} 0 & t < 0 \\ 1 & t > 0 \\ \text{indeterminate} & t = 0 \end{cases}$$

The engineering literature generally uses the term *step* function, whereas the mathematical literature uses the name *Heaviside* function. The Symbolic Math toolbox includes the `Heaviside(t)` function, which produces a unit-step function.

A step function of height M can be written as $y(t) = Mu_s(t)$. Its transform is

$$\mathcal{L}[y(t)] = \int_0^\infty Mu_s(t)e^{-st}dt = M\int_o^\infty e^{-st}dt = -M\left.\frac{e^{-st}}{s}\right|_0^\infty = \frac{M}{s}$$

where we have assumed that the real part of s is greater than 0, so that the limit of e^{-st} exists as $t \to \infty$. Similar considerations of the region of convergence of the integral apply for other functions of time. However, we need not concern ourselves with this topic here, because the transforms of all the common functions have been calculated and tabulated. They can be obtained in MATLAB with the Symbolic Math toolbox by typing `laplace(function)`, where `function` is a symbolic expression representing the function $y(t)$ in (10.5–1). The default independent variable is `t`, and the default result is a function of `s`. The optional form is `laplace (function,x,y)`, where `function` is a function of `x` and `y` is the Laplace variable.

Here is a session with some examples. The functions are t^3, e^{-at}, and $\sin bt$.

```
>>syms b t
>>laplace(t^3)
ans =
     6/s^4
>>laplace(exp(-b*t))
ans =
     1/(s+b)
>>laplace(sin(b*t))
ans =
     b/(s^2+b^2)
```

Because the transform is an integral, it has the properties of integrals. In particular, it has the *linearity property,* which states that if a and b are not functions of t, then

$$\mathcal{L}[af_1(t) + bf_2(t)] = a\mathcal{L}[f_1(t)] + b\mathcal{L}[f_2(t)] \tag{10.5–2}$$

The *inverse Laplace transform* $\mathcal{L}^{-1}[Y(s)]$ is that time function $y(t)$ whose transform is $Y(s)$; that is, $y(t) = \mathcal{L}^{-1}[Y(s)]$. The inverse operation is also linear. For example, the inverse transform of $10/s + 4/(s+3)$ is $10 + 4e^{-3t}$. Inverse transforms can be found using the `ilaplace` function. For example,

```
>>syms b s
>>ilaplace(1/s^4)
ans =
     1/6*t^3
>>ilaplace(1/(s+b))
ans =
     exp(-b*t)
>>ilaplace(b/(s^2+b^2)
ans =
     sin(b*t)
```

The transforms of derivatives are useful for solving differential equations. Applying integration by parts to the definition of the transform, we obtain

$$\mathcal{L}\left(\frac{dy}{dt}\right) = \int_0^\infty \frac{dy}{dt} e^{-st}\,dt = y(t)\,e^{-st}\Big|_0^\infty + s\int_0^\infty y(t)e^{-st}\,dt$$

$$= s\mathcal{L}[y(t)] - y(0) = sY(s) - y(0) \tag{10.5–3}$$

This procedure can be extended to higher derivatives. For example, the result for the second derivative is

$$\mathcal{L}\left(\frac{d^2y}{dt^2}\right) = s^2Y(s) - sy(0) - \dot{y}(0) \tag{10.5–4}$$

The general result for any order derivative is

$$\mathcal{L}\left(\frac{d^n y}{dt^n}\right) = s^n Y(s) - \sum_{k=1}^{n} s^{n-k} g_{k-1} \tag{10.5–5}$$

where

$$g_{k-1} = \left.\frac{d^{k-1} y}{dt^{k-1}}\right|_{t=0} \tag{10.5–6}$$

Application to Differential Equations

The derivative and linearity properties can be used to solve the differential equation

$$a\dot{y} + y = bv(t) \tag{10.5–7}$$

If we multiply both sides of the equation by e^{-st} and then integrate over time from $t = 0$ to $t = \infty$, we obtain

$$\int_0^\infty (a\dot{y} + y)\, e^{-st}\, dt = \int_0^\infty bv(t)\, e^{-st}\, dt$$

or

$$\mathcal{L}(a\dot{y} + y) = \mathcal{L}[bv(t)]$$

or, using the linearity property,

$$a\mathcal{L}(\dot{y}) + \mathcal{L}(y) = b\mathcal{L}[v(t)]$$

Using the derivative property and the alternative transform notation, the preceding equation can be written as

$$a[sY(s) - y(0)] + Y(s) = bV(s)$$

where $V(s)$ is the transform of v. This equation is an algebraic equation for $Y(s)$ in terms of $V(s)$ and $y(0)$. Its solution is

$$Y(s) = \frac{ay(0)}{as + 1} + \frac{b}{as + 1} V(s) \tag{10.5–8}$$

Applying the inverse transform to (10.5–8) gives

$$y(t) = \mathcal{L}^{-1}\left[\frac{ay(0)}{as + 1}\right] + \mathcal{L}^{-1}\left[\frac{b}{as + 1} V(s)\right] \tag{10.5–9}$$

From the transform given earlier, it can be seen that

$$\mathcal{L}^{-1}\left[\frac{ay(0)}{as + 1}\right] = \mathcal{L}^{-1}\left[\frac{y(0)}{s + 1/a}\right] = y(0)e^{-t/a}$$

which is the free response. The forced response is given by

$$\mathcal{L}^{-1}\left[\frac{b}{as + 1} V(s)\right] \tag{10.5–10}$$

This inverse transform cannot be evaluated until $V(s)$ is specified. Suppose $v(t)$ is a unit-step function. Then $V(s) = 1/s$, and (10.5–10) becomes

$$\mathcal{L}^{-1}\left[\frac{b}{s(as+1)}\right]$$

To find the inverse transform, enter

```
>>syms a b s
>>ilaplace(b/(s*(a*s+1)))
ans =
      2*b*exp(-1/2*t/a)*sinh(1/2*t/a)
```

Thus the forced response of (10.5–7) to a unit-step input is $2be^{-t/2a}\sinh(t/2a)$, which is equivalent to $b(1 - e^{-t/a})$.

You can use the `Heaviside` function with the `dsolve` function to find the step response, but the resulting expressions are more complicated than those obtained with the Laplace transform method.

Consider the second-order model

$$\ddot{x} + 1.4\dot{x} + x = f(t) \tag{10.5–11}$$

Transforming this equation gives

$$[s^2X(s) - sx(0) - \dot{x}(0)] + 1.4[sX(s) - x(0)] + X(s) = F(s)$$

Solve for $X(s)$.

$$X(s) = \frac{x(0)s + \dot{x}(0) + 1.4x(0)}{s^2 + 1.4s + 1} + \frac{F(s)}{s^2 + 1.4s + 1}$$

The free response is obtained from

$$x(t) = \mathcal{L}^{-1}\left[\frac{x(0)s + \dot{x}(0) + 1.4x(0)}{s^2 + 1.4s + 1}\right]$$

Suppose the initial conditions are $x(0) = 2$ and $\dot{x}(0) = -3$. Then the free response is obtained from

$$x(t) = \mathcal{L}^{-1}\left[\frac{2s - 0.2}{s^2 + 1.4s + 1}\right] \tag{10.5–12}$$

It can be found by typing

```
>>ilaplace((2*s-0.2)/(s^2+1.4*s+1))
```

The free response thus found is

$$x(t) = e^{-0.7t}\left[2\cos\left(\frac{\sqrt{51}}{10}t\right) - \frac{16\sqrt{51}}{51}\sin\left(\frac{\sqrt{51}}{10}t\right)\right]$$

The forced response is obtained from

$$x(t) = \mathcal{L}^{-1}\left[\frac{F(s)}{s^2 + 1.4s + 1}\right]$$

If $f(t)$ is a unit-step function, $F(s) = 1/s$ and the forced response is

$$x(t) = \mathcal{L}^{-1}\left[\frac{1}{s(s^2 + 1.4s + 1)}\right]$$

To find the forced response, enter

```
>>ilaplace(1/(s*(s^2+1.4*s+1)))
```

The answer obtained is

$$x(t) = 1 - e^{-0.7t}\left[\cos\left(\frac{\sqrt{51}}{10}t\right) + \frac{7\sqrt{51}}{51}\sin\left(\frac{\sqrt{51}}{10}t\right)\right] \quad (10.5\text{–}13)$$

Input Derivatives

Two similar mechanical systems are shown in Figure 10.5–2. In both cases the input is a displacement $y(t)$. Their equations of motion are

$$m\ddot{x} + c\dot{x} + kx = ky + c\dot{y} \quad (10.5\text{–}14)$$

$$m\ddot{x} + c\dot{x} + kx = ky \quad (10.5\text{–}15)$$

The only difference between these systems is that the system in Figure 10.5–2a has an equation of motion containing the derivative of the input function $y(t)$. Both systems are examples of the more general differential equation

$$m\ddot{x} + c\dot{x} + kx = dy + g\dot{y} \quad (10.5\text{–}16)$$

As noted earlier, you can use the `Heaviside` function with the `dsolve` function to find the step response of equations containing derivatives of the input, but the resulting expressions are more complicated than those obtained with the Laplace transform method.

We now demonstrate how to use the Laplace transform to find the step response of equations containing derivatives of the input. Suppose the initial conditions are zero. Then transforming (10.5–16) gives

$$X(s) = \frac{d + gs}{ms^2 + cs + k}Y(s) \quad (10.5\text{–}17)$$

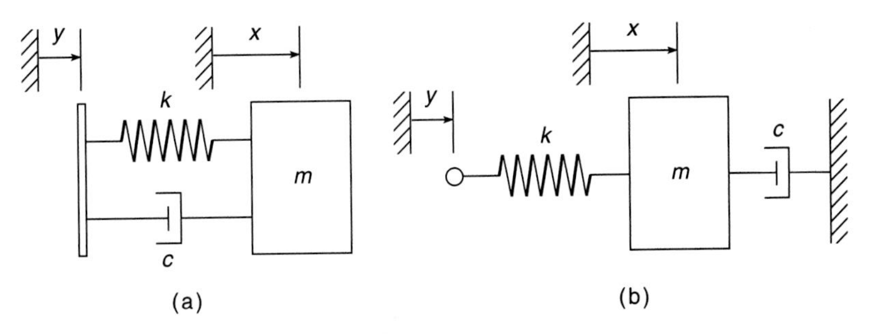

Figure 10.5–2 Two mechanical systems. The model for (a) contains the derivative of the input $y(t)$; the model for (b) does not.

Let us compare the unit-step response of (10.5–16) for two cases using the values $m = 1$, $c = 1.4$, and $k = 1$, with zero initial conditions. The two cases are $g = 0$ and $g = 5$.

With $Y(s) = 1/s$, (10.5–17) gives

$$X(s) = \frac{1 + gs}{s(s^2 + 1.4s + 1)} \qquad (10.5\text{–}18)$$

The response for the case $g = 0$ was found earlier. It is given by (10.5–13). The response for $g = 5$ is found by typing

```
>>syms s
>>ilaplace((1+5*s)/(s*(s^2+1.4*s+1)))
```

The response obtained is

$$x(t) = 1 - e^{0.7t}\left[\cos\left(\frac{\sqrt{51}}{10}t\right) + \frac{43\sqrt{51}}{51}\sin\left(\frac{\sqrt{51}}{10}t\right)\right] \qquad (10.5\text{–}19)$$

Figure 10.5–3 shows the responses given by (10.5–13) and (10.5–19). The effect of differentiating the input is an increase in the response's peak value.

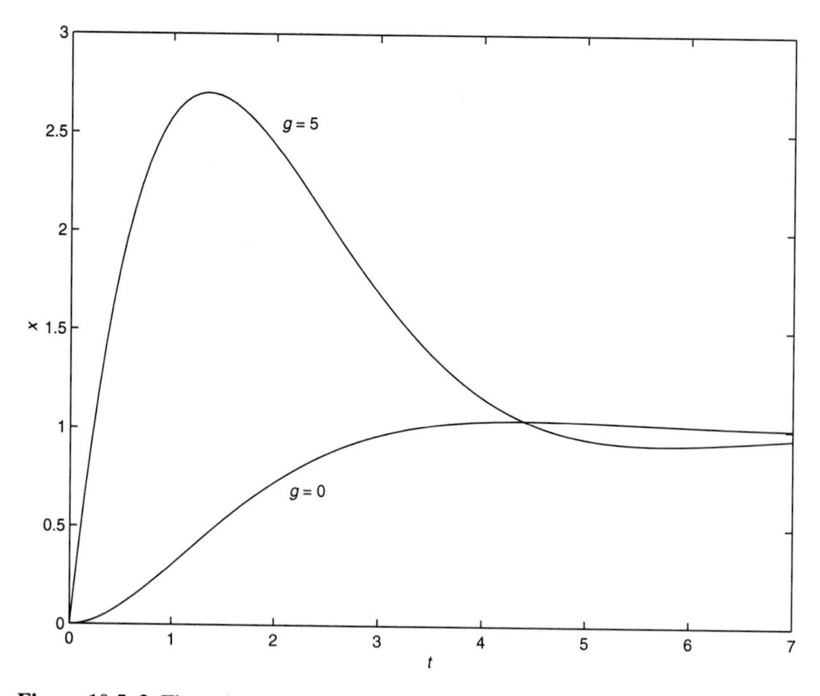

Figure 10.5–3 The unit-step response of the model $\ddot{x} + 1.4\dot{x} + x = u + g\dot{u}$ for $g = 0$ and $g = 5$.

Impulse Response

The area A under the curve of the *pulse function* shown in Figure 10.5–4a is called the *strength* of the pulse. If we let the pulse duration T approach 0 while keeping the area A constant, we obtain the *impulse* function of strength A, represented by Figure 10.5–4b. If the strength is 1, we have a *unit* impulse. The impulse can be thought of as the derivative of the step function and is a mathematical abstraction for convenience in analyzing the response of systems subjected to an input that is applied and removed suddenly, such as the force from a hammer blow.

The engineering literature generally uses the term *impulse* function, whereas the mathematical literature uses the name *Dirac delta* function. The Symbolic Math toolbox includes the `Dirac(t)` function, which returns a unit impulse. You can use the `Dirac` function with the `dsolve` function when the input function is an impulse, but the resulting expressions are more complicated than those obtained with the Laplace transform.

It can be shown that the transform of an impulse of strength A is simply A. So, for example, to find the impulse response of $\ddot{x} + 1.4\dot{x} + x = f(t)$, where $f(t)$ is an impulse of strength A, for zero inital conditions, first obtain the transform.

$$X(s) = \frac{1}{s^2 + 1.4s + 1} F(s) = \frac{A}{s^2 + 1.4s + 1}$$

Then you type

```
>>syms A s
>>ilaplace(A/(s^2+1.4*s+1))
```

The response obtained is

$$x(t) = \frac{10A\sqrt{51}}{51} e^{-0.7t} \sin\left(\frac{\sqrt{51}}{10} t\right)$$

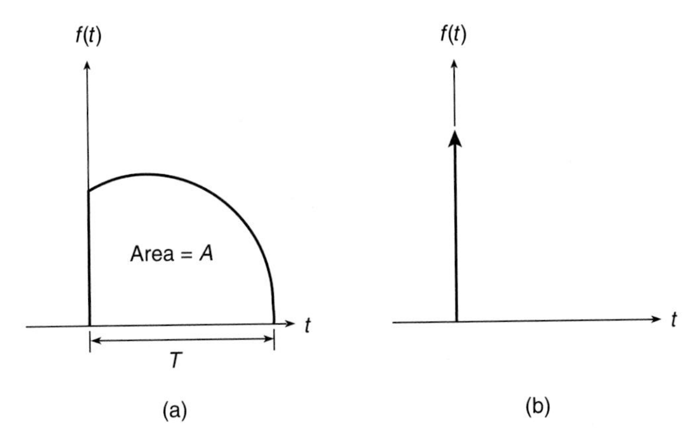

Figure 10.5–4 Pulse and impulse functions.

Direct Method

Instead of performing by hand the algebra required to find the response transform, we could use MATLAB to do the algebra for us. We now demonstrate the most direct way of using MATLAB to solve an equation with the Laplace transform. One advantage of this method is that we are not required to use the transform identities (10.5–3) through (10.5–6) for the derivatives. Let us solve the equation

$$a\frac{dy}{dt} + y = f(t) \tag{10.5–20}$$

with $f(t) = \sin t$, in terms of an unspecified value for $y(0)$. Here is the session:

```
>>syms a L s t
>>y = sym('y(t)');
>>dydt = sym('diff(y(t),t)');
>>f = sin(t);
>>eq = a*dydt+y-f;
>>E = laplace(eq,t,s)
E =
   a*(s*laplace(y(t),t,s)-y(0)) + laplace(y(t),t,s)- 1/(s^2+1)
>>E = subs(E,'laplace(y(t),t,s)',L)
E =
   a*(s*L-y(0))+L-1/(s^2+1)
>>L = solve(E,L)
L =
   (a*y(0)*s^2+a*y(0)+1)/(a*s^3+a*s+s^2+1)
>>I = simplify(ilaplace(L))
I =
   (-a*cos(t)+sin(t)+exp(-t/a)*y(0)+exp(-t/a)*a^2*y(0)+
   exp(-t/a)*a)/(1+a^2)
>>I = collect(I,exp(-t/a))
I =
   1/(1+a^2)*(-a*cos(t)+sin(t))+(a+y(0)+y(0)*a^2)/(1+a^2)*exp(-t/a)
```

The answer is

$$y(t) = \frac{1}{1+a^2}\{\sin t - a\cos t + e^{-t/a}[y(0) + a^2 y(0) + a]\}$$

Note that this session consists of the following steps:

1. Define the symbolic variables, including the derivatives that appear in the equation. Note that $y(t)$ is explicitly expressed as a function of t in these definitions.
2. Move all terms to the left side of the equation and define the left side as a symbolic expression.
3. Apply the Laplace transformation to the differential equation to obtain an algebraic equation.

Table 10.5–1 Laplace transform functions

Command	Description
`ilaplace(function)`	Returns the inverse Laplace transform of `function`.
`laplace(function)`	Returns the Laplace transform of `function`.
`laplace(function,x,y)`	Returns the Laplace transform of `function`, which is a function of `x`, in terms of the Laplace variable `y`.

4. Substitute a symbolic variable, here `L`, for the expression `laplace(y(t),t,s)` in the algebraic equation. Then solve the equation for the variable `L`, which is the transform of the solution.
5. Invert `L` to find the solution as a function of t.

   Note that this procedure can also be used to solve sets of equations.

Test Your Understanding

T10.5–1 Find the Laplace transform of the following functions: $1 - e^{-at}$ and $\cos bt$. Use the `ilaplace` function to check your answers.

T10.5–2 Use the Laplace transform to solve the problem $5\ddot{y} + 20\dot{y} + 15y = 30u - 4\dot{u}$, where $u(t)$ is a unit-step function and $y(0) = 5$, $\dot{y}(0) = 1$. (Answer: $y(t) = -1.6e^{-3t} + 4.6e^{-t} + 2$.)

Table 10.5–1 summarizes the Laplace transform functions.

10.6 Symbolic Linear Algebra

You can perform operations with symbolic matrices in much the same way as with numeric matrices. Here we give examples of finding matrix products, the matrix inverse, eigenvalues, and the characteristic polynomial of a matrix.

Remember that using symbolic matrices avoids numerical imprecision in subsequent operations. You can create a symbolic matrix from a numeric matrix in several ways, as shown in the following session:

```
>>A = sym([3, 5; 2, 7]);    % the most direct method
>>B = [3, 5; 2, 7];
>>C = sym(B);%B is preserved as a numeric matrix
>>D = subs(A,[3, 5; 2, 7]);
```

The first method is the most direct. Use the second method when you want to keep a numeric version of the matrix. The matrices `A` and `C` are symbolic and identical. The matrices `B` and `D` look like `A` and `C` but are numeric of class double.

You can create a symbolic matrix consisting of functions. For example, the relationship between the coordinates (x_2, y_2) of a coordinate system rotated counterclockwise through an angle a relative to the (x_1, y_1) coodinate system is

$$x_2 = x_1 \cos a + y_1 \sin a$$

$$y_2 = y_1 \cos a - x_1 \sin a$$

These equations can be expressed in matrix form as

$$\begin{bmatrix} x_2 \\ y_2 \end{bmatrix} = \begin{bmatrix} \cos a & \sin a \\ -\sin a & \cos a \end{bmatrix} \begin{bmatrix} x_1 \\ y_1 \end{bmatrix} = \mathbf{R} \begin{bmatrix} x_1 \\ y_1 \end{bmatrix}$$

where the rotation matrix $\mathbf{R}(a)$ is defined as

$$\mathbf{R}(a) = \begin{bmatrix} \cos a & \sin a \\ -\sin a & \cos a \end{bmatrix} \tag{10.6–1}$$

The symbolic matrix **R** can be defined in MATLAB as follows:

```
>>syms a
>>R = [cos(a), sin(a); -sin(a), cos(a)]
R =
   [ cos(a),    sin(a) ]
   [ -sin(a),    cos(a) ]
```

If we rotate the coordinate system twice by the same angle to produce a third coordinate system (x_3, y_3), the result is the same as a single rotation with twice the angle. Let us see if MATLAB gives that result. The vector-matrix equation is

$$\begin{bmatrix} x_3 \\ y_3 \end{bmatrix} = \mathbf{R} \begin{bmatrix} x_2 \\ y_2 \end{bmatrix} = \mathbf{RR} \begin{bmatrix} x_1 \\ y_1 \end{bmatrix}$$

Thus $\mathbf{R}(a)\mathbf{R}(a)$ should be the same as $\mathbf{R}(2a)$. Continue the previous session as follows:

```
>>Q = R*R
Q =

   [ cos(a)^2-sin(a)^2,    2*cos(a)*sin(a) ]
   [ -2*cos(a)*sin(a), cos(a)^2-sin(a)^2 ]
>>Q = simple(Q)
Q =
   [ cos(2*a),   sin(2*a) ]
   [ -sin(2*a), cos(2*a) ]
```

The matrix **Q** is the same as $\mathbf{R}(2a)$, as we suspected.

To evaluate a matrix numerically, use the `subs` and `double` functions. For example, for a rotation of $a = \pi/4$ rad (45°),

```
>>R = subs(R,a,pi/4)
```

Characteristic Polynomial and Roots

Sets of first-order differential equations can be expressed in vector-matrix notation as

$$\dot{\mathbf{x}} = \mathbf{A}\mathbf{x} + \mathbf{B}\mathbf{f}(t)$$

where $\mathbf{x}$ is the vector of dependent variables and $\mathbf{f}(t)$ is a vector containing the forcing functions. For example, the equation set

$$\dot{x}_1 = x_2$$
$$\dot{x}_2 = -kx_1 - 2x_2 + f(t)$$

comes from the equation of motion of a mass connected to a spring and sliding on a surface having viscous friction. The term $f(t)$ is the applied force acting on the mass. For this set the vector $\mathbf{x}$ and the matrices $\mathbf{A}$ and $\mathbf{B}$ are

$$\mathbf{x} = \begin{bmatrix} x_1 \\ x_2 \end{bmatrix}$$

$$\mathbf{A} = \begin{bmatrix} 0 & 1 \\ -k & -2 \end{bmatrix} \qquad \mathbf{B} = \begin{bmatrix} 0 \\ 1 \end{bmatrix}$$

The equation $|s\mathbf{I} - \mathbf{A}| = 0$ is the characteristic equation of the model, where s represents the characteristic roots of the model. Use the `poly(A)` function to find this polynomial, and note that MATLAB uses the default symbolic variable `x` to represent the roots. For example, to find the characteristic equation and solve for the roots in terms of the spring constant k, use the following session:

```
>>syms k
>>A = [0 ,1;-k, -2];
>>poly(A)
ans =
      x^2+2*x+k
>>solve(ans)
ans =
      [ -1+(1-k)^(1/2) ]
      [ -1-(1-k)^(1/2) ]
```

Thus the roots are $s = -1 \pm \sqrt{1-k}$.

Use the `eig(A)` function to find the roots directly without finding the characteristic equation (*eig* stands for "eigenvalue," which is another term for "characteristic root"). For example,

```
>>syms k
>>A = [0 ,1;-k, -2];
>>eig(A)
ans =
      [ -1+(1-k)^(1/2) ]
      [ -1-(1-k)^(1/2) ]
```

You can use the inv(A) and det(A) functions to invert and find the determinant of a matrix symbolically. For example, using the same matrix A from the previous session,

```
>>inv(A)
ans =
      [ -2/k,   -1/k ]
      [ 1, 0 ]
>>A*ans     % verify that the inverse is correct
ans =
      [ 1, 0 ]
      [ 0, 1 ]
>>det(A)
ans =
      k
```

Solving Linear Algebraic Equations

You can use matrix methods in MATLAB to solve linear algebraic equations symbolically. You can use the matrix inverse method, if the inverse exists, or the left-division method (see Chapter 6 for a discussion of these methods). For example, to solve the set

$$2x - 3y = 3$$

$$5x + cy = 19$$

using both methods, the session is

```
>>syms c
>>A = sym([2, -3; 5, c]);
>>b = sym([3; 19]);
>>x = inv(A)*b   % the matrix inverse method
x =
    [3*c/(2*c+15)+57/(2*c+15)]
    [23/(2*c+15)]
>>x = A\b   % the left-division method
x =
    [3*(19+c)/(2*c+15)]
    [23/(2*c+15)]
```

Although the results appear to be different, they both reduce to the same solution: $x = 3(19 + c)/(2c + 15)$, $y = 23/(2c + 15)$.

Table 10.6–1 summarizes the functions used in this section. Note that their syntax is identical to the numeric versions used in earlier chapters.

Table 10.6–1 Linear algebra functions

Command	Description
`det(A)`	Returns the determinant of the matrix `A` in symbolic form.
`eig(A)`	Returns the eigenvalues (characteristic roots) of the matrix `A` in symbolic form.
`inv(A)`	Returns the inverse of the matrix `A` in symbolic form.
`poly(A)`	Returns the characteristic polynomial of the matrix `A` in symbolic form.

Test Your Understanding

T10.6–1 Consider three successive coordinate rotations using the same angle a. Show that the product **RRR** of the rotation matrix $\mathbf{R}(a)$ given by (10.6–1) equals $\mathbf{R}(3a)$.

T10.6–2 Find the characteristic polynomial and roots of the following matrix.

$$\mathbf{A} = \begin{bmatrix} -2 & 1 \\ -3k & -5 \end{bmatrix}$$

(Answers: $s^2 + 7s + 10 + 3k$ and $s = (-7 \pm \sqrt{9 - 12k})/2$.)

T10.6–3 Use the matrix inverse and the left-division method to solve the following set.

$$-4x + 6y = -2c$$

$$7x - 4y = 23$$

(Answer: $x = (69 - 4c)/13$, $y = (46 - 7c)/13$.)

10.7 Summary

This chapter covers a subset of the capabilities of the Symbolic Math toolbox, specifically

- Symbolic algebra.
- Symbolic methods for solving algebraic and transcendental equations.
- Symbolic methods for solving ordinary differential equations.
- Symbolic calculus, including integration, differentiation, limits, and series.
- Laplace transforms.
- Selected topics in linear algebra, including symbolic methods for obtaining determinants, matrix inverses, and eigenvalues.

Now that you have finished this chapter, you should be able to use MATLAB to

- Create symbolic expressions and manipulate them algebraically.
- Obtain symbolic solutions to algebraic and transcendental equations.

Table 10.7–1 Guide to MATLAB commands introduced in Chapter 10

Creating and evaluating expressions	Table 10.1–1
Manipulating symbolic expressions	Table 10.1–2
Solving algebraic and transcendental equations	Table 10.2–1
Symbolic calculus functions	Table 10.3–1
The `dsolve` function	Table 10.4–1
Laplace transform functions	Table 10.5–1
Linear algebra functions	Table 10.6–1

Miscellaneous functions	
Command	**Description**
`Dirac(t)`	Dirac delta function (unit-impulse function at $t = 0$).
`Heaviside(t)`	Heaviside function (unit-step function making a transition from 0 to 1 at $t = 0$).

- Perform symbolic differentiation and integration.
- Evaluate limits and series symbolically.
- Obtain symbolic solutions to ordinary differential equations.
- Obtain Laplace transforms.
- Perform symbolic linear algebra operations, including obtaining expressions for determinants, matrix inverses, and eigenvalues.

Table 10.7–1 is a guide by category to the functions introduced in this chapter.

Key Terms with Page References

Boundary condition, 615
Default variable, 589
Impulse function, 629
Initial condition, 615
Laplace transform, 622
Solution structure, 598
Step function, 623
Symbolic constant, 587
Symbolic expression, 585

Problems

You can find the answers to problems marked with an asterisk at the end of the text.

Section 10.1

1. Use MATLAB to prove the following identities:

a. $\sin^2 x + \cos^2 x = 1$
b. $\sin(x + y) = \sin x \cos y + \cos x \sin y$
c. $\sin 2x = 2 \sin x \cos x$
d. $\cosh^2 x - \sinh^2 x = 1$

2. Use MATLAB to express $\cos 5\theta$ as a polynomial in x, where $x = \cos\theta$.

3.* Two polynomials in the variable x are represented by the coefficient vectors `p1 = [6, 2, 7, -3]` and `p2 = [10, -5, 8]`.

a. Use MATLAB to find the product of these two polynomials; express the product in its simplest form.

b. Use MATLAB to find the numeric value of the product if $x = 2$.

4.* The equation of a circle of radius r centered at $x = 0$, $y = 0$ is

$$x^2 + y^2 = r^2$$

Use the `subs` and other MATLAB functions to find the equation of a circle of radius r centered at the point $x = a$, $y = b$. Rearrange the equation into the form $Ax^2 + Bx + Cxy + Dy + Ey^2 = F$ and find the expressions for the coefficients in terms of a, b, and r.

5. The equation for a curve called the "lemniscate" in polar coordinates (r, θ) is

$$r^2 = a^2 \cos(2\theta)$$

Use MATLAB to find the equation for the curve in terms of Cartesian coordinates (x, y), where $x = r\cos\theta$ and $y = r\sin\theta$.

Section 10.2

6.* The law of cosines for a triangle states that $a^2 = b^2 + c^2 - 2bc\cos A$, where a is the length of the side opposite the angle A, and b and c are the lengths of the other sides.

a. Use MATLAB to solve for b.

b. Suppose that $A = 60°$, $a = 5$ m, and $c = 2$ m. Determine b.

7. Use MATLAB to solve the polynomial equation $x^3 + 8x^2 + ax + 10 = 0$ for x in terms of the parameter a, and evaluate your solution for the case $a = 17$. Use MATLAB to check the answer.

8.* The equation for an ellipse centered at the origin of the Cartesian coordinates (x,y) is

$$\frac{x^2}{a^2} + \frac{y^2}{b^2} = 1$$

where a and b are constants that determine the shape of the ellipse.

a. In terms of the parameter b, use MATLAB to find the points of intersection of the two ellipses described by

$$x^2 + \frac{y^2}{b^2} = 1$$

and

$$\frac{x^2}{100} + 4y^2 = 1$$

b. Evaluate the solution obtained in part a for the case $b = 2$.

9. The equation

$$r = \frac{p}{1 - \epsilon \cos \theta}$$

describes the polar coordinates of an orbit with the coordinate origin at the sun. If $\epsilon = 0$, the orbit is circular; if $0 < \epsilon < 1$, the orbit is elliptical. The planets have orbits that are nearly circular; comets have orbits that are highly elongated with ϵ nearer to 1. It is of obvious interest to determine whether or not a comet's or an asteroid's orbit will intersect that of a planet. For each of the following two cases, use MATLAB to determine whether or not orbits A and B intersect. If they do, determine the polar coordinates of the intersection point. The units of distance are AU, where 1 AU is the mean distance of the Earth from the sun.

a. Orbit A: $p = 1, \epsilon = 0.01$. Orbit B: $p = 0.1, \epsilon = 0.9$.
b. Orbit A: $p = 1, \epsilon = 0.01$. Orbit B: $p = 1.1, \epsilon = 0.5$.

10. Figure 10.2–2 on page 601 shows a robot arm having two joints and two links. The angles of rotation of the motors at the joints are θ_1 and θ_2. From trigonometry we can derive the following expressions for the (x, y) coordinates of the hand.

$$x = L_1 \cos \theta_1 + L_2 \cos(\theta_1 + \theta_2)$$

$$y = L_1 \sin \theta_1 + L_2 \sin(\theta_1 + \theta_2)$$

Suppose that the link lengths are $L_1 = 3$ ft and $L_2 = 2$ ft.

a. Compute the motor angles required to position the hand at $x = 3$ ft, $y = 1$ ft. Identify the elbow-up and elbow-down solutions.
b. Suppose you want to move the hand along a straight, horizontal line at $y = 1$ for $2 \le x \le 4$. Plot the required motor angles versus x. Label the elbow-up and elbow-down solutions.

Section 10.3

11. Use MATLAB to find all the values of x where the graph of $y = 3^x - 2x$ has a horizontal tangent line.

12.* Use MATLAB to determine all the local minima and local maxima and all the inflection points where $dy/dx = 0$ of the following function:

$$y = x^4 - \tfrac{16}{3}x^3 + 8x^2 - 4$$

13. The surface area of a sphere of radius r is $S = 4\pi r^2$. Its volume is $V = 4\pi r^3/3$.

a. Use MATLAB to find the expression for dS/dV.
b. A spherical balloon expands as air is pumped into it. What is the rate of increase in the balloon's surface area with volume when its volume is 30 in.^{3}?

14. Use MATLAB to find the point on the line $y = 2 - x/3$ that is closest to the point $x = -3$, $y = 1$.

15. A particular circle is centered at the origin and has a radius of 5. Use MATLAB to find the equation of the line that is tangent to the circle at the point $x = 3$, $y = 4$.

16. Ship A is traveling north at 6 mi/hr, and ship B is traveling west at 12 mi/hr. When ship A was dead ahead of ship B, it was 6 mi away. Use MATLAB to determine how close the ships come to each other.

17. Suppose you have a wire of length L. You cut a length x to make a square, and use the remaining length $L - x$ to make a circle. Use MATLAB to find the length x that maximizes the sum of the areas enclosed by the square and the circle.

18.* A certain spherical street lamp emits light in all directions. It is mounted on a pole of height h (see Figure P18). The brightness B at point P on the sidewalk is directly proportional to $\sin\theta$ and inversely proportional to the square of the distance d from the light to the point. Thus

$$B = \frac{c}{d^2}\sin\theta$$

where c is a constant. Use MATLAB to determine how high h should be to maximize the brightness at point P, which is 30 ft from the base of the pole.

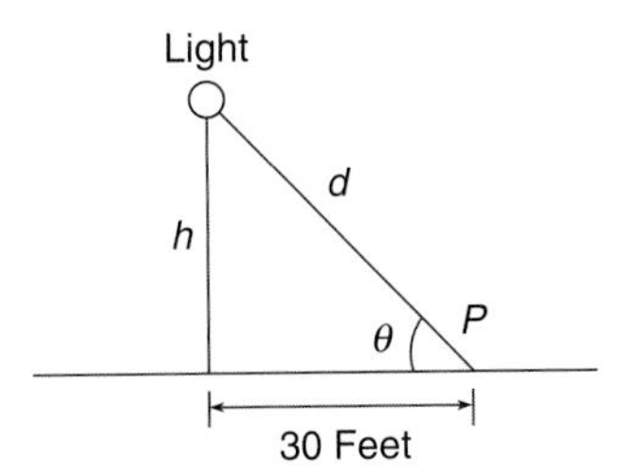

Figure P18

19.* A certain object has a mass $m = 100$ kg and is acted on by a force $f(t) = 500[2 - e^{-t}\sin(5\pi t)]$ N. The mass is at rest at $t = 0$. Use MATLAB to compute the object's velocity v at $t = 5$ s. The equation of motion is $m\dot{v} = f(t)$.

20. A rocket's mass decreases as it burns fuel. The equation of motion for a rocket in vertical flight can be obtained from Newton's law and is

$$m(t)\frac{dv}{dt} = T - m(t)g$$

where T is the rocket's thrust and its mass as a function of time is given by $m(t) = m_0(1 - rt/b)$. The rocket's initial mass is m_0, the burn time is b, and r is the fraction of the total mass accounted for by the fuel. Use the values $T = 48{,}000$ N; $m_0 = 2200$ kg; $r = 0.8$; $g = 9.81$ m/s^2; and $b = 40$ s.

a. Use MATLAB to compute the rocket's velocity as a function of time for $t \le b$.

b. Use MATLAB to compute the rocket's velocity at burnout.

21. The equation for the voltage $v(t)$ across a capacitor as a function of time is

$$v(t) = \frac{1}{C}\left(\int_0^t i(t)\,dt + Q_0\right)$$

where $i(t)$ is the applied current and Q_0 is the initial charge. Suppose that $C = 10^{-6}$ F and that $Q_0 = 0$. If the applied current is $i(t) = [0.01 + 0.3e^{-5t}\sin(25\pi t)]10^{-3}$ A, use MATLAB to compute and plot the voltage $v(t)$ for $0 \le t \le 0.3$ s.

22. The power P dissipated as heat in a resistor R as a function of the current $i(t)$ passing through it is $P = i^2 R$. The energy $E(t)$ lost as a function of time is the time integral of the power. Thus

$$E(t) = \int_0^t P(t)\,dt = R\int_0^t i^2(t)\,dt$$

If the current is measured in amperes, the power is in watts and the energy is in joules (1 W = 1 J/s). Suppose that a current $i(t) = 0.2[1 + \sin(0.2t)]$ A is applied to the resistor.

a. Determine the energy $E(t)$ dissipated as a function of time.

b. Determine the energy dissipated in 1 min if $R = 1000\ \Omega$.

23. The RLC circuit shown in Figure P23 can be used as a *narrow-band filter*. If the input voltage $v_i(t)$ consists of a sum of sinusoidally varying voltages with different frequencies, the narrow-band filter will allow to pass only those voltages whose frequencies lie within a narrow range. These filters are used in tuning circuits, such as those used in AM radios, to allow reception only of the carrier signal of the desired radio station. The magnification ratio M of a circuit is the ratio of the amplitude of the output voltage $v_o(t)$ to the amplitude of the input voltage $v_i(t)$. It is a

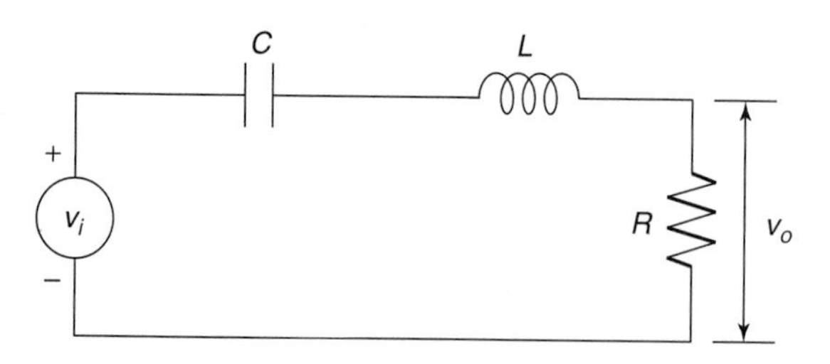

Figure P23

function of the radian frequency ω of the input voltage. Formulas for M are derived in elementary electrical circuits courses. For this particular circuit, M is given by

$$M = \frac{RC\omega}{\sqrt{(1 - LC\omega^2)^2 + (RC\omega)^2}}$$

The frequency at which M is a maximum is the frequency of the desired carrier signal.

a. Determine this frequency as a function of R, C, and L.

b. Plot M versus ω for two cases where $C = 10^{-5}$ F and $L = 5 \times 10^{-3}$ H. For the first case, $R = 1000\ \Omega$. For the second case, $R = 10\ \Omega$. Comment on the filtering capability of each case.

24. The shape of a cable hanging with no load other than its own weight is a *catenary* curve. A particular bridge cable is described by the catenary $y(x) = 10\cosh((x - 20)/10)$ for $0 \le x \le 50$, where x and y are the horizontal and vertical coordinates measured in feet. (See Figure P24.) It is desired to hang plastic sheeting from the cable to protect passersby while the bridge is being repainted. Use MATLAB to determine how many square feet of sheeting are required. Assume that the bottom edge of the sheeting is located along the x-axis at $y = 0$.

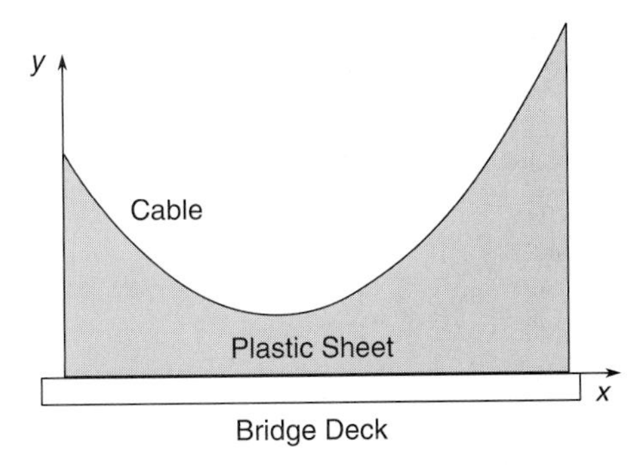

Figure P24

25. The shape of a cable hanging with no load other than its own weight is a *catenary* curve. A particular bridge cable is described by the catenary $y(x) = 10\cosh((x - 20)/10)$ for $0 \le x \le 50$, where x and y are the horizontal and vertical coordinates measured in feet.

The length L of a curve described by $y(x)$ for $a \le x \le b$ can be found from the following integral:

$$L = \int_a^b \sqrt{1 + \left(\frac{dy}{dx}\right)^2}\, dx$$

Determine the length of the cable.

26. Use the first five nonzero terms in the Taylor series for e^{ix}, $\sin x$, and $\cos x$ about $x = 0$ to demonstrate the validity of Euler's formula $e^{ix} = \cos x + i \sin x$.

27. Find the Taylor series for $e^x \sin x$ about $x = 0$ in two ways: *a*. by multiplying the Taylor series for e^x and that for $\sin x$, and *b*. by using the `taylor` function directly on $e^x \sin x$.

28. Integrals that cannot be evaluated in closed form sometimes can be evaluated approximately by using a series representation for the integrand. For example, the following integral is used for some probability calculations (see Chapter 7, Section 7.2):

$$I = \int_0^1 e^{-x^2} dx$$

a. Obtain the Taylor series for e^{-x^2} about $x = 0$ and integrate the first six nonzero terms in the series to find I. Use the seventh term to estimate the error.

b. Compare your answer with that obtained with the MATLAB `erf(t)` function, defined as

$$\text{erf}(t) = \frac{2}{\sqrt{\pi}} \int_0^t e^{-t^2} dt$$

29.* Use MATLAB to compute the following limits:

a. $\displaystyle \lim_{x \to 1} \frac{x^2 - 1}{x^2 - x}$

b. $\displaystyle \lim_{x \to -2} \frac{x^2 - 4}{x^2 + 4}$

c. $\displaystyle \lim_{x \to 0} \frac{x^4 + 2x^2}{x^3 + x}$

30. Use MATLAB to compute the following limits:

a. $\displaystyle \lim_{x \to 0+} x^x$

b. $\displaystyle \lim_{x \to 0+} (\cos x)^{1/\tan x}$

c. $\displaystyle \lim_{x \to 0+} \left(\frac{1}{1 - x}\right)^{-1/x^2}$

d. $\displaystyle \lim_{x \to 0-} \frac{\sin x^2}{x^3}$

e. $\lim_{x\to 5-} \frac{x^2 - 25}{x^2 - 10x + 25}$

f. $\lim_{x\to 1+} \frac{x^2 - 1}{\sin(x - 1)^2}$

31. Use MATLAB to compute the following limits:

a. $\lim_{x\to \infty} \frac{x + 1}{x}$

b. $\lim_{x\to -\infty} \frac{3x^3 - 2x}{2x^3 + 3}$

32. Find the expression for the sum of the geometric series

$$\sum_{k=0}^{n-1} r^k$$

for $r \neq 1$.

33. A particular rubber ball rebounds to one-half its original height when dropped on a floor.

a. If the ball is initially dropped from a height h and is allowed to continue to bounce, find the expression for the total distance traveled by the ball after the ball hits the floor for the nth time.

b. If it is initially dropped from a height of 10 ft, how far will the ball have traveled after it hits the floor for the eighth time?

Section 10.4

34. The equation for the voltage y across the capacitor of an RC circuit is

$$RC\frac{dy}{dt} + y = v(t)$$

where $v(t)$ is the applied voltage. Suppose that $RC = 0.2$ s and that the capacitor voltage is initially 2 V. If the applied voltage goes from 0 to 10 V at $t = 0$, use MATLAB to determine and plot the voltage $y(t)$ for $0 \leq t \leq 1$ s.

35. The following equation describes the temperature $T(t)$ of a certain object immersed in a liquid bath of temperature $T_b(t)$:

$$10\frac{dT}{dt} + T = T_b$$

Suppose the object's temperature is initially $T(0) = 70°$F and the bath temperature is 170°F. Use MATLAB to answer the following questions:

a. Determine $T(t)$.

b. How long will it take for the object's temperature T to reach 168°F?

c. Plot the object's temperature $T(t)$ as a function of time.

36.* This equation describes the motion of a mass connected to a spring with viscous friction on the surface

$$m\ddot{y} + c\dot{y} + ky = f(t)$$

where $f(t)$ is an applied force. The position and velocity of the mass at $t = 0$ are denoted by x_0 and v_0. Use MATLAB to answer the following questions:

a. What is the free response in terms of x_0 and v_0 if $m = 3$, $c = 18$, and $k = 102$?

b. What is the free response in terms of x_0 and v_0 if $m = 3$, $c = 39$, and $k = 120$?

37. The equation for the voltage y across the capacitor of an RC circuit is

$$RC\frac{dy}{dt} + y = v(t)$$

where $v(t)$ is the applied voltage. Suppose that $RC = 0.2$ s and that the capacitor voltage is initially 2 V. If the applied voltage is $v(t) = 10[2 - e^{-t}\sin(5\pi t)]$, use MATLAB to compute and plot the voltage $y(t)$ for $0 \le t \le 5$ s.

38. The following equation describes a certain dilution process, where $y(t)$ is the concentration of salt in a tank of fresh water to which salt brine is being added:

$$\frac{dy}{dt} + \frac{2}{10 + 2t}y = 4$$

Suppose that $y(0) = 0$. Use MATLAB to compute and plot $y(t)$ for $0 \le t \le 10$.

39. This equation describes the motion of a certain mass connected to a spring with viscous friction on the surface

$$3\ddot{y} + 18\dot{y} + 102y = f(t)$$

where $f(t)$ is an applied force. Suppose that $f(t) = 0$ for $t < 0$ and $f(t) = 10$ for $t \ge 0$.

a. Use MATLAB to compute and plot $y(t)$ when $y(0) = \dot{y}(0) = 0$.

b. Use MATLAB to compute and plot $y(t)$ when $y(0) = 0$ and $\dot{y}(0) = 10$.

40. This equation describes the motion of a certain mass connected to a spring with viscous friction on the surface

$$3\ddot{y} + 39\dot{y} + 120y = f(t)$$

where $f(t)$ is an applied force. Suppose that $f(t) = 0$ for $t < 0$ and $f(t) = 10$ for $t \ge 0$.

a. Use MATLAB to compute and plot $y(t)$ when $y(0) = \dot{y}(0) = 0$.
b. Use MATLAB to compute and plot $y(t)$ when $y(0) = 0$ and $\dot{y}(0) = 10$.

41. The equations for an armature-controlled dc motor follow. The motor's current is i and its rotational velocity is ω.

$$L\frac{di}{dt} = -Ri - K_e\omega + v(t)$$

$$I\frac{d\omega}{dt} = K_T i - c\omega$$

L, R, and I are the motor's inductance, resistance, and inertia; K_T and K_e are the torque constant and back emf constant; c is a viscous damping constant; and $v(t)$ is the applied voltage.

Use the values $R = 0.8\ \Omega$, $L = 0.003$ H, $K_T = 0.05$ N · m/A, $K_e = 0.05$ V · s/rad, $c = 0$, and $I = 8 \times 10^{-5}$ N · m · s^2.

Suppose the applied voltage is 20 V. Use MATLAB to compute and plot the motor's speed and current versus time for zero initial conditions. Choose a final time large enough to show the motor's speed becoming constant.

Section 10.5

42. The RLC circuit described in Problem 23 and shown in Figure P23 on page 640 has the following differential equation model:

$$LC\ddot{v}_o + RC\dot{v}_o + v_o = RC\dot{v}_i(t)$$

Use the Laplace transform method to solve for the unit-step response of $v_o(t)$ for zero initial conditions, where $C = 10^{-5}$ F and $L = 5 \times 10^{-3}$ H. For the first case (a broadband filter), $R = 1000\ \Omega$. For the second case (a narrow-band filter), $R = 10\ \Omega$. Compare the step responses of the two cases.

43. The differential equation model for a certain speed control system for a vehicle is

$$\ddot{v} + (1 + K_p)\dot{v} + K_I v = K_p\dot{v}_d + K_I v_d$$

where the actual speed is v, the desired speed is $v_d(t)$, and K_p and K_I are constants called the "control gains." Use the Laplace transform method to find the unit-step response (that is, $v_d(t)$ is a unit-step function). Use zero initial conditions. Compare the response for three cases:

a. $K_p = 9$, $K_I = 50$
b. $K_p = 9$, $K_I = 25$
c. $K_p = 54$, $K_I = 250$

44. The differential equation model for a certain position control system for a metal cutting tool is

$$\frac{d^3x}{dt^3} + (6 + K_D)\frac{d^2x}{dt^2} + (11 + K_p)\frac{dx}{dt} + (6 + K_I)x$$

$$= K_D\frac{d^2x_d}{dt^2} + K_p\frac{dx_d}{dt} + K_I x_d$$

where the actual tool position is x; the desired position is $x_d(t)$; and K_p, K_I, and K_D are constants called the control gains. Use the Laplace transform method to find the unit-step response (that is, $x_d(t)$ is a unit-step function). Use zero initial conditions. Compare the response for three cases:

a. $K_p = 30$, $K_I = K_D = 0$
b. $K_p = 27$, $K_I = 17.18$, $K_D = 0$
c. $K_p = 36$, $K_I = 38.1$, $K_D = 8.52$

45.* The differential equation model for the motor torque $m(t)$ required for a certain speed control system is

$$4\ddot{m} + 4K\dot{m} + K^2m = K^2\dot{v}_d$$

where the desired speed is $v_d(t)$, and K is a constant called the control gain.

a. Use the Laplace transform method to find the unit-step response (that is, $v_d(t)$ is a unit-step function). Use zero initial conditions.
b. Use symbolic manipulation in MATLAB to find the value of the peak torque in terms of the gain K.

Section 10.6

46. Show that $\mathbf{R}^{-1}(a)\mathbf{R}(a) = \mathbf{I}$, where $\mathbf{I}$ is the identity matrix and $\mathbf{R}(a)$ is the rotation matrix given by (10.6–1). This equation shows that the inverse coordinate transformation returns you to the original coordinate system.

47. Show that $\mathbf{R}^{-1}(a) = \mathbf{R}(-a)$. This equation shows that a rotation through a negative angle is equivalent to an inverse transformation.

48.* Find the characteristic polynomial and roots of the following matrix:

$$\mathbf{A} = \begin{bmatrix} -6 & 2 \\ 3k & -7 \end{bmatrix}$$

49.* Use the matrix inverse and the left-division method to solve the following set for x and y in terms of c:

$$4cx + 5y = 43$$

$$3x - 4y = -22$$

50. The currents i_1, i_2, and i_3 in the circuit shown in Figure P50 are described by the following equation set if all the resistances are equal to R.

$$\begin{bmatrix} 2R & -R & 0 \\ -R & 3R & -R \\ 0 & R & -2R \end{bmatrix} \begin{bmatrix} i_1 \\ i_2 \\ i_3 \end{bmatrix} = \begin{bmatrix} v_1 \\ 0 \\ v_2 \end{bmatrix}$$

v_1 and v_2 are applied voltages; the other two currents can be found from $i_4 = i_1 - i_2$ and $i_5 = i_2 - i_3$.

a. Use both the matrix inverse method and the left-division method to solve for the currents in terms of the resistance R and the voltages v_1 and v_2.

b. Find the numerical values for the currents if $R = 1000\ \Omega$, $v_1 = 100$ V, and $v_2 = 25$ V.

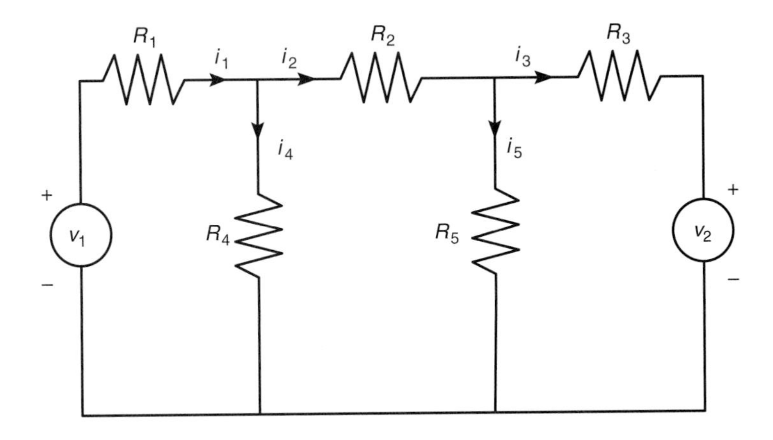

Figure P50

51. The equations for the armature-controlled dc motor shown in Figure P51 follow. The motor's current is i, and its rotational velocity is ω.

$$L\frac{di}{dt} = -Ri - K_e\omega + v(t)$$

$$I\frac{d\omega}{dt} = K_T i - c\omega$$

L, R, and I are the motor's inductance, resistance, and inertia; K_T and K_e are the torque constant and back emf constant; c is a viscous damping constant; and $v(t)$ is the applied voltage.

a. Find the characteristic polynomial and the characteristic roots.

b. Use the values $R = 0.8\ \Omega$, $L = 0.003$ H, $K_T = 0.05$ N · m/A, $K_e = 0.05$ V · s/rad, and $I = 8 \times 10^{-5}$ kg · m^2. The damping constant

c is often difficult to determine with accuracy. For these values find the expressions for the two characteristic roots in terms of c.

c. Using the parameter values in part *b*, determine the roots for the following values of c (in newton meter second): $c = 0$, $c = 0.01$, $c = 0.1$, and $c = 0.2$. For each case, use the roots to estimate how long the motor's speed will take to become constant; also discuss whether or not the speed will oscillate before it becomes constant.

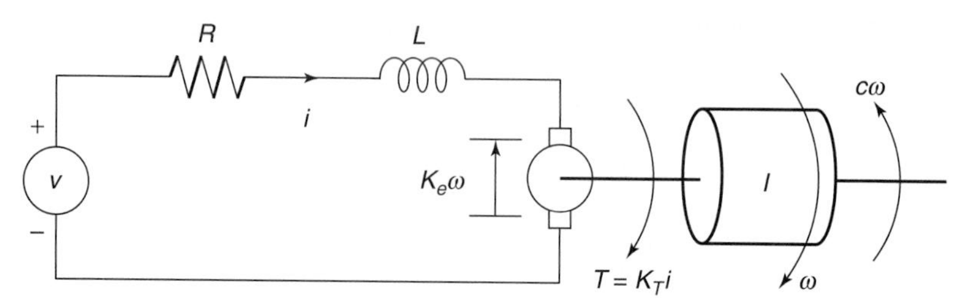

Figure P51

APPENDIX A

Guide to Commands and Functions in This Text

Operators and special characters

Item	Description	Pages
+	Plus; addition operator.	9
-	Minus; subtraction operator.	9
*	Scalar and matrix multiplication operator.	9
.*	Array multiplication operator.	87
^	Scalar and matrix exponentiation operator.	9
.^	Array exponentiation operator.	92
\	Left-division operator.	9
/	Right-division operator.	9
.\	Array left-division operator.	91
./	Array right-division operator.	91
:	Colon; generates regularly spaced elements and represents an entire row or column.	20, 72, 75, 124
( )	Parentheses; encloses function arguments and array indices; overrides precedence.	9, 45
[ ]	Brackets; encloses array elements.	19, 72
{}	Braces; encloses cell elements	113
.	Decimal point.	13
...	Ellipsis; line-continuation operator.	13
,	Comma; separates statements, and elements in a row of an array.	12, 13
;	Semicolon; separates columns in an array, and suppresses display.	12, 74, 124
%	Percent sign; designates a comment, and specifies formatting.	30, 673
'	Quote sign and transpose operator.	25, 72, 124
.'	Nonconjugated transpose operator.	72, 124
=	Assignment (replacement) operator.	11
@	Creates a function handle.	163

Logical and relational operators

Item	Description	Pages
==	Relational operator: equal to.	44, 191
~=	Relational operator: not equal to.	44, 191
<	Relational operator: less than.	44, 191
<=	Relational operator: less than or equal to.	44, 191
>	Relational operator: greater than.	44, 191
>=	Relational operator: greater than or equal to.	44, 191
&	Logical operator: AND.	194
&&	Short-circuit AND.	197
|	Logical operator: OR.	194
||	Short-circuit OR.	197
~	Logical operator: NOT.	194

Special variables and constants

Item	Description	Pages
ans	Most recent answer.	8
eps	Accuracy of floating-point precision.	15
i,j	The imaginary unit $\sqrt{-1}$.	15
Inf	Infinity.	15
NaN	Undefined numerical result (not a number).	15
pi	The number π.	15

Commands for managing a session

Item	Description	Pages
clc	Clears Command window.	13
clear	Removes variables from memory.	13
doc	Displays documentation.	43
exist	Checks for existence of file or variable.	13, 32
global	Declares variables to be global.	153
help	Displays help text in the Command window.	43
helpwin	Displays help text in the Help Browser.	43
lookfor	Searches help entries for a keyword.	41, 43
quit	Stops MATLAB.	13
who	Lists current variables.	13
whos	Lists current variables (long display).	13

System and file commands

Item	Description	Pages
cd	Changes current directory.	24
date	Displays current date.	167
dir	Lists all files in current directory.	24
load	Loads workspace variables from a file.	22, 173
path	Displays search path.	24
pwd	Displays current directory.	24
save	Saves workspace variables in a file.	22
type	Displays contents of a file.	43
what	Lists all MATLAB files.	24
wk1read	Reads .wk1 spreadsheet file.	173
xlsread	Reads .xls spreadsheet file.	173

Input/output commands

Item	Description	Pages
disp	Displays contents of an array or string.	36
format	Controls screen display format.	16, 17
fprintf	Performs formatted writes to screen or file.	36, 672, 673
input	Displays prompts and waits for input.	36, 210
menu	Displays a menu of choices.	36, 37
;	Suppresses screen printing.	12

Numeric display formats

Item	Description	Pages
format short	Four decimal digits (default).	16, 17
format long	16 decimal digits.	16, 17
format short e	Five digits plus exponent.	16, 17
format long e	16 digits plus exponent.	16, 17
format bank	Two decimal digits.	16, 17
format +	Positive, negative, or zero.	16, 17
format rat	Rational approximation.	16, 17
format compact	Suppresses some line feeds.	16, 17
format loose	Resets to less compact display mode.	16, 17

Array functions

Item	Description	Pages
`cat`	Concatenates arrays.	77, 82
`find`	Finds indices of nonzero elements.	45, 77, 78, 198
`length`	Computes number of elements.	77, 79
`linspace`	Creates regularly spaced vector.	73, 77
`logspace`	Creates logarithmically spaced vector.	73, 77
`max`	Returns largest element.	77, 78
`min`	Returns smallest element.	77, 78
`size`	Computes array size.	77, 78, 209
`sort`	Sorts each column.	77, 78
`sum`	Sums each column.	77, 78

Special matrices

Item	Description	Pages
`eye`	Creates an identity matrix.	105
`ones`	Creates an array of ones.	105
`zeros`	Creates an array of zeros.	105

Matrix functions for solving linear equations

Item	Description	Pages
`det`	Computes determinant of an array.	379, 635
`inv`	Computes inverse of a matrix.	376, 635
`pinv`	Computes pseudoinverse of a matrix.	385, 399
`rank`	Computes rank of a matrix.	382, 399
`rref`	Computes reduced row echelon form.	389, 399

Exponential and logarithmic functions

Item	Description	Pages
`exp(x)`	Exponential; e^x.	142
`log(x)`	Natural logarithm; $\ln x$.	142
`log10(x)`	Common (base 10) logarithm; $\log x = \log_{10} x$.	142
`sqrt(x)`	Square root; $\sqrt{x}$.	142

Complex functions

Item	Description	Pages
`abs(x)`	Absolute value; $\|x\|$.	142
`angle(x)`	Angle of a complex number x.	142
`conj(x)`	Complex conjugate of x.	142
`imag(x)`	Imaginary part of a complex number x.	142
`real(x)`	Real part of a complex number x.	142

Numeric functions

Item	Description	Pages
`ceil`	Rounds to the nearest integer toward ∞.	112, 116
`fix`	Rounds to the nearest integer toward zero.	142
`floor`	Rounds to the nearest integer toward $-\infty$.	142
`round`	Rounds toward the nearest integer.	142
`sign`	Signum function.	148

Trigonometric functions

Item	Description	Pages
`acos(x)`	Inverse cosine; $\arccos x = \cos^{-1} x$.	146
`acot(x)`	Inverse cotangent; $\text{arccot}\, x = \cot^{-1} x$.	146
`acsc(x)`	Inverse cosecant; $\text{arccsc}\, x = \csc^{-1} x$.	146
`asec(x)`	Inverse secant; $\text{arcsec}\, x = \sec^{-1} x$.	146
`asin(x)`	Inverse sine; $\arcsin x = \sin^{-1} x$.	146
`atan(x)`	Inverse tangent; $\arctan x = \tan^{-1} x$.	146
`atan2(y,x)`	Four-quadrant inverse tangent.	146
`cos(x)`	Cosine; $\cos x$.	146
`cot(x)`	Cotangent; $\cot x$.	146
`csc(x)`	Cosecant; $\csc x$.	146
`sec(x)`	Secant; $\sec x$.	146
`sin(x)`	Sine; $\sin x$.	146
`tan(x)`	Tangent; $\tan x$.	146

Hyperbolic functions

Item	Description	Pages
`acosh(x)`	Inverse hyperbolic cosine; $\cosh^{-1} x$.	148
`acoth(x)`	Inverse hyperbolic cotangent; $\coth^{-1} x$.	148
`acsch(x)`	Inverse hyperbolic cosecant; $\text{csch}^{-1} x$.	148
`asech(x)`	Inverse hyperbolic secant; $\text{sech}^{-1} x$.	148
`asinh(x)`	Inverse hyperbolic sine; $\sinh^{-1} x$.	148
`atanh(x)`	Inverse hyperbolic tangent; $\tanh^{-1} x$.	148
`cosh(x)`	Hyperbolic cosine; $\cosh x$.	148
`coth(x)`	Hyperbolic cotangent; $\cosh x / \sinh x$.	148
`csch(x)`	Hyperbolic cosecant; $1/\sinh x$.	148
`sech(x)`	Hyperbolic secant; $1/\cosh x$.	148
`sinh(x)`	Hyperbolic sine; $\sinh x$.	148
`tanh(x)`	Hyperbolic tangent; $\sinh x / \cosh x$.	148

Polynomial functions

Item	Description	Pages
`conv`	Computes product of two polynomials.	108, 109
`deconv`	Computes ratio of polynomials.	108, 109
`eig`	Computes the eigenvalues of a matrix.	515, 635
`poly`	Computes polynomial from roots.	107, 108
`polyfit`	Fits a polynomial to data.	302, 303, 315
`polyval`	Evaluates polynomial.	108, 109, 315
`roots`	Computes polynomial roots.	20, 107, 108

String functions

Item	Description	Pages
findstr	Finds occurrences of a string.	210
lower	Converts string to all lowercase.	210
strcmp	Compares strings.	210, 240
upper	Converts string to all uppercase.	210

Logical functions

Item	Description	Pages
any	True if any elements are nonzero.	198
all	True if all elements are nonzero.	198
find	Finds indices of nonzero elements.	45, 77, 78, 198
finite	True if elements are finite.	198
ischar	True if elements are a character array.	198
isinf	True if elements are infinite.	198
isempty	True if matrix is empty.	198
isnan	True if elements are undefined.	198
isreal	True if all elements are real.	198
isnumeric	True if elements have numeric values.	198
logical	Converts a numeric array to a logical array.	198
xor	Exclusive OR.	196, 198

Miscellaneous mathematical functions

Item	Description	Pages
cross	Computes cross products.	107
dot	Computes dot products.	107
function	Creates a user-defined function.	148
nargin	Number of function input arguments.	208
nargout	Number of function output arguments.	208

Cell array functions

Item	Description	Pages
cell	Creates cell array.	112, 116
celldisp	Displays cell array.	112, 114
cellplot	Displays graphical representation of cell array.	112, 114
num2cell	Converts numeric array to cell array.	114
deal	Matches input and output lists.	112, 115
iscell	Identifies cell array.	112

Structure functions

Item	Description	Pages
fieldnames	Returns field names in a structure array.	120
getfield	Returns field contents of a structure array.	120, 121
isfield	Identifies a structure array field.	120
isstruct	Identifies a structure array.	120, 122
rmfield	Removes a field from a structure array.	24
setfield	Sets contents of field.	120, 121
struct	Creates structure array.	120

Basic *xy* plotting commands

Item	Description	Pages
axis	Sets axis limits and other axis properties.	264, 269
cla	Clears the axes.	662
fplot	Intelligent plotting of functions.	266, 269
ginput	Reads coordinates of the cursor position.	26, 27
grid	Displays gridlines.	27, 269
plot	Generates *xy* plot.	27, 269, 280, 663
print	Prints plot or saves plot to a file.	26, 269
title	Puts text at top of plot.	27, 269
xlabel	Adds text label to x-axis.	25, 269
ylabel	Adds text label to y-axis.	25, 269

Plot-enhancement commands

Item	Description	Pages
axes	Creates axes objects.	276, 280
colormap	Sets the color map of the current figure.	662
gtext	Enables label placement by mouse.	27, 280
hold	Freezes current plot.	279, 280
legend	Legend placement by mouse.	275, 280
refresh	Redraws current figure window.	276, 280
set	Specifies properties of objects such as axes.	288, 289, 664
subplot	Creates plots in subwindows.	271, 280
text	Places string in figure.	276, 280, 668

Specialized plot functions

Item	Description	Pages
bar	Creates bar chart.	284, 424
loglog	Creates log-log plot.	284
plotyy	Enables plotting on left and right axes.	284, 290
polar	Creates polar plot.	284, 290
semilogx	Creates semilog plot (logarithmic abscissa).	284
semilogy	Creates semilog plot (logarithmic ordinate).	284
stairs	Creates stairs plot.	284, 289
stem	Creates stem plot.	284, 289

Three-dimensional plotting functions

Item	Description	Pages
`contour`	Creates contour plot.	337, 338
`mesh`	Creates three-dimensional mesh surface plot.	336, 338
`meshc`	Same as `mesh` with contour plot underneath.	338
`meshgrid`	Creates rectangular grid.	336, 338
`meshz`	Same as `mesh` with vertical lines underneath.	338
`plot3`	Creates three-dimensional plots from lines and points.	335, 338
`shading`	Specifies type of shading.	662
`surf`	Creates shaded three-dimensional mesh surface plot.	338
`surfc`	Same as `surf` with contour plot underneath.	338
`surfl`	Same as `surf` with lighting.	662
`view`	Sets the angle of the view.	662
`waterfall`	Same as `mesh` with mesh lines in one direction.	338
`zlabel`	Adds text label to z-axis.	335

Program flow control

Item	Description	Pages
`break`	Terminates execution of a loop.	214
`case`	Provides alternate execution paths within `switch` structure.	225
`continue`	Passes control to the next iteration of a `for` or `while` loop.	214
`else`	Delineates alternate block of statements.	47, 52, 202
`elseif`	Conditionally executes statements.	47, 52, 205
`end`	Terminates `for`, `while`, and `if` statements.	47, 52, 202
`for`	Repeats statements a specific number of times.	48, 52, 211
`if`	Executes statements conditionally.	47, 52, 201
`otherwise`	Provides optional control within a `switch` structure.	225
`switch`	Directs program execution by comparing input with `case` expressions.	225
`while`	Repeats statements an indefinite number of times.	48, 52, 224

Optimization and root-finding functions

Item	Description	Pages
`fminbnd`	Finds the minimum of a function of one variable.	157, 160
`fminsearch`	Finds the minimum of a multivariable function.	159, 160
`fzero`	Finds the zero of a function.	156, 160

Histogram functions

Item	Description	Pages
`bar`	Creates a bar chart.	284, 424
`hist`	Aggregates the data into bins.	424

Statistical functions

Item	Description	Pages
cumsum	Computes the cumulative sum across a row.	429, 433
erf(x)	Computes the error function erf(x).	435
mean	Calculates the mean.	433, 434
median	Calculates the median.	433, 434
std	Calculates the standard deviation.	433, 434

Random number functions

Item	Description	Pages
rand	Generates uniformly distributed random numbers between 0 and 1; sets and retrieves the state.	437, 438
randn	Generates normally distributed random numbers; sets and retrieves the state.	438, 443
randperm	Generates random permutation of integers.	438, 439

Polynomial functions

Item	Description	Pages
poly	Computes the coefficients of a polynomial from its roots.	107, 108
polyfit	Fits a polynomial to data.	302, 303
polyval	Evaluates a polynomial and generates error estimates.	108, 109, 315
roots	Computes the roots of a polynomial from its coefficients.	20, 107, 108

Interpolation functions

Item	Description	Pages
interp1	Linear and cubic-spline interpolation of a function of one variable.	446, 449, 452
interp2	Linear interpolation of a function of two variables.	449, 452
spline	Cubic-spline interpolation.	451, 452
unmkpp	Computes the coefficients of cubic-spline polynomials.	451, 452

Numerical integration functions

Item	Description	Pages
quad	Numerical integration with adaptive Simpson's rule.	472, 475
quadl	Numerical integration with Lobatto quadrature.	472, 475
trapz	Numerical integration with the trapezoidal rule.	472

Numerical differentiation functions

Item	Description	Pages
`diff(x)`	Computes the differences between adjacent elements in the vector `x`.	479, 482
`polyder`	Differentiates a polynomial, a polynomial product, or a polynomial quotient.	482

ODE solvers

Item	Description	Pages
`ode23`	Nonstiff, low-order solver.	498, 499, 527
`ode45`	Nonstiff, medium-order solver.	498, 499, 527
`ode113`	Nonstiff, variable-order solver.	498, 499, 527
`ode23s`	Stiff, low-order solver.	498, 499, 527
`ode23t`	Moderately stiff, trapezoidal rule solver.	498, 499, 527
`ode23tb`	Stiff, low-order solver.	498, 499, 527
`ode15s`	Stiff, variable-order solver.	498, 499, 527
`odeset`	Creates integrator options structure for ODE solvers.	527, 528

LTI object functions

Item	Description	Pages
`ss`	Creates an LTI object in state-space form.	518, 520
`ssdata`	Extracts state-space matrices from an LTI object.	518, 520
`tf`	Creates an LTI object in transfer-function form.	518, 519
`tfdata`	Extracts equation coefficients from an LTI object.	518, 521

LTI ODE solvers

Item	Description	Pages
`impulse`	Computes and plots the impulse response of an LTI object.	521, 522
`initial`	Computes and plots the free response of an LTI object.	521, 522
`lsim`	Computes and plots the response of an LTI object to a general input.	521, 524
`step`	Computes and plots the step response of an LTI object.	521, 523

Predefined input functions

Item	Description	Pages
`gensig`	Generates a periodic sine, square, or pulse input.	525, 526
`sawtooth`	Generates a periodic sawtooth input.	525
`square`	Generates a square wave input.	525, 526
`stepfun`	Generates a step function input.	525

Functions for creating and evaluating symbolic expressions

Item	Description	Pages
`class`	Returns the class of an expression.	593, 595
`digits`	Sets the number of decimal digits used to do variable precision arithmetic.	595
`double`	Converts an expression to numeric form.	240, 591, 595
`ezplot`	Generates a plot of a symbolic expression.	594, 595, 618
`findsym`	Finds the symbolic variables in a symbolic expression.	589, 595
`numden`	Returns the numerator and denominator of an expression.	591, 595
`sym`	Creates a symbolic variable.	587, 595
`syms`	Creates one or more symbolic variables.	587, 595
`vpa`	Sets the number of digits used to evaluate expressions.	595

Functions for manipulating symbolic expressions

Item	Description	Pages
`collect`	Collects coefficients of like powers in an expression.	589, 596
`expand`	Expands an expression by carrying out powers.	590, 596
`factor`	Factors an expression.	590, 596
`poly2sym`	Converts a polynomial coefficient vector to a symbolic polynomial.	591, 596
`pretty`	Displays an expression in a form that resembles typeset mathematics.	592, 596
`simple`	Searches for the shortest form of an expression.	590, 596
`simplify`	Simplifies an expression using Maple's simplification rules.	590, 596
`subs`	Substitutes variables or expressions.	592, 596
`sym2poly`	Converts an expression to a polynomial coefficient vector.	596

Symbolic solution of algebraic and transcendental equations

Item	Description	Pages
`solve`	Solves symbolic equations.	596, 603

Symbolic calculus functions

Item	Description	Pages
`diff`	Returns the derivative of an expression.	604, 606
`Dirac`	Dirac delta function (unit impulse).	629, 636
`Heaviside`	Heaviside function (unit step).	623, 636
`int`	Returns the integral of an expression.	606, 608
`limit`	Returns the limit of an expression.	606, 613
`symsum`	Returns the symbolic summation of an expression.	606
`taylor`	Returns the Taylor series of a function.	606, 612

Symbolic solution of differential equations

Item	Description	Pages
`dsolve`	Returns a symbolic solution of a differential equation or set of equations.	616, 622

Laplace transform functions

Item	Description	Pages
`ilaplace`	Returns the inverse Laplace transform.	624
`laplace`	Returns the Laplace transform.	623

Symbolic linear algebra functions

Item	Description	Pages
`det`	Returns the determinant of a matrix.	635
`eig`	Returns the eigenvalues (characteristic roots) of a matrix.	635
`inv`	Returns the inverse of a matrix.	635
`poly`	Returns the characteristic polynomial of a matrix.	633, 635

Animation functions

Item	Description	Pages
`drawnow`	Initiates immediate plotting.	665
`getframe`	Captures current figure in a frame.	661
`movie`	Plays back frames.	661
`moviein`	Initializes movie frame memory.	663
`pause`	Pauses the display.	665

Sound functions

Item	Description	Pages
`sound`	Plays a vector as sound.	669
`soundsc`	Scales data and plays as sound.	670
`wavplay`	Plays recorded sound.	670
`wavread`	Reads Microsoft WAVE file.	670
`wavrecord`	Records sound from input device.	671
`wavwrite`	Writes Microsoft WAVE file.	671

APPENDIX B

Animation and Sound in MATLAB

B.1 Animation

Animation can be used to display the behavior of an object over time. Some of the MATLAB demos are M-files that perform animation. After completing this section, which has simple examples, you may study the demo files, which are more advanced. Two methods can be used to create animations in MATLAB. The first method uses the `movie` function. The second method uses the EraseMode property.

Creating Movies in MATLAB

The `getframe` command captures, or takes a snapshot of, the current figure to create a single frame for the movie. The `getframe` function is usually used in a `for` loop to assemble an array of movie frames. The `movie` function plays back the frames after they have been captured.

To create a movie, use a script file of the following form.

```
for k = 1:n
   plotting expressions
   M(k) = getframe;   % Saves current figure in array M
end
movie(M)
```

For example, the following script file creates 20 frames of the function $te^{-t/b}$ for $0 \le t \le 100$ for each of 20 values of the parameter b from $b = 1$ to $b = 20$.

```
% Program movie1.m
% Animates the function t*exp(-t/b).
```

```
t = [0:0.05:100];
for b = 1:20
  plot(t,t.*exp(-t/b)),axis([0 100 0 10]),xlabel('t');
  M(:,b) = getframe;
end
```

The line `M(:,b) = getframe;` acquires and saves the current figure as a column of the matrix `M`. Once this file is run, the frames can be replayed as a movie by typing `movie(M)`. The animation shows how the location and height of the function peak changes as the parameter b is increased.

Rotating a 3D Surface

The following example rotates a three-dimensional surface by changing the viewpoint. The data is created using the built-in function, `peaks`.

```
% Program movie2.m
% Rotates a 3D surface.
[X,Y,Z] = peaks(50);      % Create data.
surfl(X,Y,Z)      % Plot the surface.
axis([-3 3 -3 3 -5 5]) % Retain same scaling for each frame.
axis vis3d off % Set the axes to 3D and turn off tick marks,
               % and so forth.
shading interp      % Use interpolated shading.
colormap(winter)    % Specify a colormap.
for k = 1:30 % Rotate the viewpoint and capture each frame.
        view(-37.5+0.5*(k-1),30)
        M(k) = getframe;
end
cla      % Clear the axes.
movie(M)      % Play the movie.
```

The `colormap(map)` function sets the current figure's color map to `map`. Type `help graph3d` to see a number of colormaps to choose for `map`. The choice `winter` provides blue and green shading. The `view` function specifies the 3D graph viewpoint. The syntax `view(az,el)` sets the angle of the view from which an observer sees the current 3D plot, where `az` is the azimuth or horizontal rotation and `el` is the vertical elevation (both in degrees). Azimuth revolves about the z-axis, with positive values indicating counterclockwise rotation of the viewpoint. Positive values of elevation correspond to moving above the object; negative values move below. The choice `az = -37.5,el = 30` is the default 3D view.

Extended Syntax of the movie Function

The function `movie(M)` plays the movie in array `M` once, where `M` must be an array of movie frames (usually acquired with `getframe`). The function `movie(M,n)` plays the movie `n` times. If `n` is negative, each "play" is once forward and once backward. If n is a vector, the first element is the number of times to play the movie, and the remaining elements are a list of frames to play in the movie. For example, if M has four frames, then n = [10 4 4 2 1] plays the movie 10 times, and the movie consists of frame 4 followed by frame 4 again, followed by frame 2 and, finally, frame 1.

The function movie(M,n,fps) plays the movie at fps frames per second. If fps is omitted, the default is 12 frames per second. Computers that cannot achieve the specified fps will play the movie as fast as they can. The function movie(h,...) plays the movie in object h, where h is a handle to a figure or an axis. Handles are discussed in Section 2.2.

The function movie(h,M,n,fps,loc) specifies the location to play the movie, relative to the lower-left corner of object h and in pixels, regardless of the value of the object's Units property, where loc = [x y unused unused] is a four-element position vector, of which only the x and y coordinates are used, but all four elements are required. The movie plays back using the width and height in which it was recorded.

Note that for code to be compatible with versions of MATLAB prior to Release 11 (5.3), the moviein(n) function must be used to initialize movie frame memory for n frames. To do this, place the line M = moviein(n); before the for loop that generates the plots.

The disadvantage of the movie function is that it might require too much memory if many frames or complex images are stored.

Animation with the EraseMode Property

One form of extended syntax for the plot function is

```
plot(...,'PropertyName','PropertyValue',...)
```

This form sets the plot property specified by PropertyName to the values specified by PropertyValue for all line objects created by the plot function. One such property name is EraseMode. This property controls the technique MATLAB uses to draw and erase line objects and is useful for creating animated sequences. The allowable values for the EraseMode property are the following.

- normal This is the default value for the EraseMode property. By typing

  ```
  plot(...,'EraseMode','normal')
  ```

  the entire figure, including axes, labels, and titles, is erased and redrawn using only the new set of points. In redrawing the display, a three-dimensional analysis is performed to ensure that all objects are rendered correctly. Thus, this mode produces the most accurate picture but is the slowest. The other modes are faster but do not perform a complete redraw and are therefore less accurate. This method may cause blinking between each frame because everything is erased and redrawn. This method is therefore undesirable for animation.

- none When the EraseMode property value is set to none, objects in the existing figure are not erased, and the new plot is superimposed on the existing figure. This mode is therefore fast because it does not remove existing points, and it is useful for creating a "trail" on the screen.
- xor When the EraseMode property value is set to xor, objects are drawn and erased by performing an exclusive OR with the background color. This produces a smooth animation. This mode does not destroy other graphics objects beneath the ones being erased and does not change the color of the objects beneath. However, the object's color depends on the background color.
- background When the EraseMode property value is set to background, the result is the same as with xor except that objects behind the erased objects are destroyed. Objects are erased by drawing them in the axes' background color or in the figure background color if the axes Color property is set to none. This damages objects that are behind the erased line, but lines are always properly colored.

The drawnow command causes the previous graphics command to be executed immediately. If the drawnow command were not used, MATLAB would complete all other operations before performing any graphics operations and would display only the last frame of the animation.

Animation speed depends of the intrinsic speed of the computer and on what and how much is being plotted. Symbols such as o, *, or + will be plotted slower than a line. The number of points being plotted also affects the animation speed. The animation can be slowed by using the pause(n) function, which pauses the program execution for n seconds.

Using Object Handles

An expression of the form

```
p = plot(...)
```

assigns the results of the plot function to the variable p, which is a figure identifier called a "figure handle." This stores the figure and makes it available for future use. Any valid variable name may be assigned to a handle. A figure handle is a specific type of "object handle." Handles may be assigned to other types of objects. For example, later we will create a handle with the text function.

The set function can be used with the handle to change the object properties. This function has the general format

```
set(object handle, 'PropertyName', 'PropertyValue', ...)
```

If the object is an entire figure, its handle also contains the specifications for line color and type, marker size, and the value of the EraseMode property. Two of the properties of the figure specify the data to be plotted. Their property names are XData and YData. The following example shows how to use these properties.

Animating a Function

Consider the function $te^{-t/b}$, which was used in the first movie example. This function can be animated as the parameter b changes with the following program.

```
% Program animate1.m
% Animates the function t*exp(-t/b).
t = [0:0.05:100];
b = 1;
p = plot(t,t.*exp(-t/b),'EraseMode','xor');...
  axis([0 100 0 10]),xlabel('t');
for b = 2:20
  set(p,'XData',t,'YData',t.*exp(-t/b)),...
  axis([0 100 0 10]),xlabel('t');
  drawnow
  pause(0.1)
end
```

In this program the function $te^{-t/b}$ is first evaluated and plotted over the range $0 \leq t \leq 100$ for $b = 1$, and the figure handle is assigned to the variable `p`. This establishes the plot format for all following operations, for example, line type and color, labeling, and axis scaling. The function $te^{-t/b}$ is then evaluated and plotted over the range $0 \leq t \leq 100$ for $b = 2, 3, 4, \ldots$ in the `for` loop, and the previous plot is erased. Each call to `set` in the `for` loop causes the next set of points to be plotted. The EraseMode property value specifies how to plot the existing points on the figure (i.e., how to refresh the screen), as each new set of points is added. You should investigate what happens if the EraseMode property value is set to `none` instead of `xor`.

Animating Projectile Motion

This following program illustrates how user-defined functions and subplots can be used in animations. The following are the equations of motion for a projectile launched with a speed s_0 at an angle θ above the horizontal, where x and y are the horizontal and vertical coordinates, g is the acceleration due to gravity, and t is time.

$$x(t) = (s_0 \cos\theta)\, t \qquad y(t) = -\frac{gt^2}{2} + (s_0 \sin\theta)\, t$$

By setting $y = 0$ in the second expression, we can solve for t and obtain the following expression for the maximum time the projectile is in flight, t_{max}.

$$t_{max} = \frac{2s_0}{g}\sin\theta$$

The expression for $y(t)$ may be differentiated to obtain the expression for the vertical velocity:

$$v_{\text{vert}} = \frac{dy}{dt} = -gt + s_0 \sin\theta$$

The maximum distance $x_{\max}$ may be computed from $x(t_{\max})$, the maximum height $y_{\max}$ may be computed from $y(t_{\max}/2)$, and the maximum vertical velocity occurs at $t = 0$.

The following functions are based on these expressions, where `s0` is the launch speed s_0, and `th` is the launch angle θ.

```
function x = xcoord(t,s0,th);
% Computes projectile horizontal coordinate.
x = s0*cos(th)*t;

function y = ycoord(t,s0,th,g);
% Computes projectile vertical coordinate.
y = -g*t.^2/2+s0*sin(th)*t;

function v = vertvel(t,s0,th,g);
% Computes projectile vertical velocity.
v = -g*t+s0*sin(th);
```

The following program uses these functions to animate the projectile motion in the first subplot, while simultaneously displaying the vertical velocity in the second subplot, for the values $\theta = 45°$, $s_0 = 105$ ft/sec, and $g = 32.2$ ft/sec^2. Note that the values of `xmax`, `ymax`, and `vmax` are computed and used to set the axes scales. The figure handles are `h1` and `h2`.

```
% Program animate2.m
% Animates projectile motion.
% Uses functions xcoord, ycoord, and vertvel.
th = 45*(pi/180);
g = 32.2; s0 = 105;
%
tmax = 2*s0*sin(th)/g;
xmax = xcoord(tmax,s0,th);
ymax = ycoord(tmax/2,s0,th,g);
vmax = vertvel(0,s0,th,g);
w = linspace(0,tmax,500);
%
subplot(2,1,1)
plot(xcoord(w,s0,th),ycoord(w,s0,th,g)),hold,
h1 = plot(xcoord(w,s0,th),ycoord(w,s0,th,g),'o','EraseMode','xor');
axis([0 xmax 0 1.1*ymax]),xlabel('x'),ylabel('y')
subplot(2,1,2)
plot(xcoord(w,s0,th),vertvel(w,s0,th,g)),hold,
```

```
h2 = plot(xcoord(w,s0,th),vertvel(w,s0,th,g),'s','EraseMode','xor');
  axis([0 xmax 0 1.1*vmax]),xlabel('x'),...
  ylabel('Vertical Velocity')
for t = [0:0.01:tmax]
  set(h1,'XData',xcoord(t,s0,th),'YData',ycoord(t,s0,th,g))
  set(h2,'XData',xcoord(t,s0,th),'YData',vertvel(t,s0,th,g))
  drawnow
  pause(0.005)
end
hold
```

You should experiment with different values of the `pause` function argument.

Animation with Arrays

Thus far we have seen how the function to be animated may be evaluated in the `set` function with an expression or with a function. A third method is to compute the points to be plotted ahead of time and store them in arrays. The following program shows how this is done, using the projectile application. The plotted points are stored in the arrays `x` and `y`.

```
% Program animate3.m
% Animation of a projectile using arrays.
th = 70*(pi/180);
g = 32.2; s0=100;
tmax = 2*s0*sin(th)/g;
xmax = xcoord(tmax,s0,th);
ymax = ycoord(tmax/2,s0,th,g);
%
w = linspace(0,tmax,500);
x = xcoord(w,s0,th);y = ycoord(w,s0,th,g);
plot(x,y),hold,
h1 = plot(x,y,'o','EraseMode','xor');
axis([0 xmax 0 1.1*ymax]),xlabel('x'),ylabel('y')
%
kmax = length(w);
for k =1:kmax
  set(h1,'XData',x(k),'YData',y(k))
  drawnow
  pause(0.001)
end
hold
```

Displaying Elapsed Time

It may be helpful to display the elapsed time during an animation. To do this, modify the program `animate3.m` as shown in the following. The new lines are indicated in bold; the line formerly below the line `h1 = plot(...` has been deleted.

```
% Program animate4.m
% Like animate3.m but displays elapsed time.
th = 70*(pi/180);
g = 32.2; s0 = 100;
%
tmax = 2*s0*sin(th)/g;
xmax = xcoord(tmax,s0,th);
ymax = ycoord(tmax/2,s0,th,g);
%
t = linspace(0,tmax,500);
x = xcoord(t,s0,th);y = ycoord(t,s0,th,g);
plot(x,y),hold,
h1 = plot(x,y,'o','EraseMode','xor');
text(10,10,'Time = ')
time = text(30,10,'0','EraseMode','background')
  axis([0 xmax 0 1.1*ymax]),xlabel('x'),ylabel('y')
%
kmax = length(t);
for k = 1:kmax
  set(h1,'XData',x(k),'YData',y(k))
  t_string = num2str(t(k));
  set(time,'String',t_string)
  drawnow
  pause(0.001)
end
hold
```

The first new line creates a label for the time display using the `text` statement, which writes the label once. The program must not write to that location again. The second new statement creates the handle `time` for the text label and creates the string for the first time value, which is 0. By using the `background` value for `EraseMode`, the statement specifies that the existing display of the time variable will be erased when the next value is displayed. Note that the numerical value of time `t(k)` must be converted to a string, by using the function `num2str`, before it can be displayed. In the last new line, in which the `set` function uses the `time` handle, the property name is `'String'`, which is not a variable but a property associated with text objects. The variable being updated is `t_string`.

B.2 Sound

MATLAB provides a number a functions for creating, recording, and playing sound on the computer. This section gives a brief introduction to these functions.

A Model of Sound

Sound is the fluctuation of air pressure as a function of time t. If the sound is a *pure tone*, the pressure $p(t)$ oscillates sinusoidally at a single frequency; that is,

$$p(t) = A \sin(2\pi f t + \phi)$$

where A is the pressure amplitude (the "loudness"), f is the sound frequency in cycles per second (Hz), and ϕ is the phase shift in radians. The *period* of the sound wave is $P = 1/f$.

Because sound is an analog variable (one having an infinite number of values), it must be converted into a finite set of numbers before it can be stored and used in a digital computer. This conversion process involves *sampling* the sound signal into discrete values and *quantizing* the numbers so that they can be represented in binary form. Quantization is an issue when using a microphone and analog-to-digital converter to capture real sound, but we will not discuss it here because we will produce only simulated sounds in software.

You use a process similar to sampling whenever you plot a function in MATLAB. To plot the function you should evaluate it at enough points to produce a smooth plot. So, to plot a sine wave, we should "sample" or evaluate it many times over the period. The frequency at which we evaluate it is the *sampling frequency*. So, if we use a time step of 0.1 s, our sampling frequency is 10 Hz. If the sine wave has a period of 1 s, then we are "sampling" the function 10 times every period. So we see that the higher the sampling frequency, the better is our representation of the function.

Creating Sound in MATLAB

The MATLAB function `sound(sound_vector,sf)` plays the signal in the vector `sound_vector`, created with the sampling frequency `sf`, on the computer's speaker. Its use is demonstrated with the following user-defined function, which plays a simple tone.

```
function playtone(freq,sf,amplitude,duration)
% Plays a simple tone.
% freq = frequency of the tone (in Hz).
% sf = sampling frequency (in Hz).
% amplitude = sound amplitude (dimensionless).
% duration = sound duration (in seconds).
t = [0:1/sf:duration];
sound_vector = amplitude*sin(2*pi*freq*t);
sound(sound_vector,sf)
```

Try this function with the following values: `freq = 1000`, `sf = 10000`, `amplitude = 1`, and `duration = 10`. The `sound` function truncates or "clips" any values in `sound_vector` that lie outside the range -1 to $+1$. Try using `amplitude = 0.1` and `amplitude = 5` to see the effect on the loudness of the sound.

Of course, real sound contains more than one tone. You can create a sound having two tones by adding two vectors created from sine functions having different frequencies and amplitudes. Just make sure that they are sampled with the same frequency, have the same number of samples, and their sum lies in the range −1 to +1. You can play two different sounds in sequence by concatenating them in a row vector, as `sound([sound_vector_1, sound_vector_2], sf)`. You can play two different sounds simultaneously in stereo by concatenating them in a column vector, as `sound([sound_vector_1', sound_vector_2'], sf)`.

MATLAB includes some sound files. For example, load the MAT-file `chirp.mat` and play the sound as follows:

```
>>load chirp
>>sound(y,Fs)
```

Note that the sound vector has been stored in the MAT-file as the array `y` and the sampling frequency has been stored as the variable `Fs`. You can also try the file `gong.mat`.

A related function is `soundsc(sound_vector,sf)`. This function scales the signal in `sound_vector` to the range −1 to +1 so that the sound is played as loudly as possible without clipping.

Reading and Playing Sound Files

The MATLAB function `wavread('filename')` reads a Microsoft WAVE file having the extension `.wav`. The syntax is

```
[sound_vector, sf, bits] = wavread('filename')
```

where `sf` is the sampling frequency used to create the file, and `bits` is the number of bits per sample used to encode the data. To play the file, use the `wavplay` function as follows:

```
>>wavplay(sound_vector, sf)
```

Most computers have WAVE files to play bells, beeps, chimes, etc., to signal you when certain actions occur. For example, to load and play the WAVE file chimes.wav located in C:\windows\media on some PC systems, you type

```
>>[sound_vector, sf] = wavread('c:\windows\media\chimes.wav');
>>wavplay(sound_vector, sf)
```

You can also play this sound using the sound command, as `sound(y,sf)`, but the `wavplay` function has more capabilities than the `sound` function. See the MATLAB help for information about the extended syntax of the `wavplay` function.

Recording and Writing Sound Files

You can use MATLAB to record sound and write sound data to a WAVE file. The `wavrecord` function records sound from a PC-based audio input device. Its basic syntax is

```
sound_vector = wavrecord(n,sf)
```

where `n` is the number of samples, sampled at the rate `sf`. The default value is 11,025 Hz. For example, to record 5 sec of audio from channel 1 sampled at 11,025 Hz, speak into the audio device while the following program runs.

```
>>sf = 11025;
>>sound_vector = wavrecord(5*sf, sf);
```

Play back the sound by typing `wavplay(sound_vector,sf)`.

You can use the `wavwrite` function to write sound stored in the vector `sound_vector` to a Microsoft WAVE file. One syntax is `wavwrite (sound_vector, sf, 'filename')`, where the sampling frequency is `sf` Hz and the data is assumed to be 16-bit data. The function clips any amplitude values outside the range -1 to $+1$.

APPENDIX C

Formatted Output in MATLAB

The disp and format commands provide simple ways to control the screen output. However, some users might require more control over the screen display. In addition, some users might want to write formatted output to a data file. The fprintf function provides this capability. Its syntax is count = fprintf(fid,format,A,...), which formats the data in the real part of matrix A (and in any additional matrix arguments) under control of the specified format string format, and writes it to the file associated with file identifier fid. A count of the number of bytes written is returned in the variable count. The argument fid is an integer file identifier obtained from fopen. (It may also be 1 for standard output—the screen—or 2 for standard error. See fopen for more information.) Omitting fid from the argument list causes output to appear on the screen, and is the same as writing to standard output (fid = 1). The string format specifies notation, alignment, significant digits, field width, and other aspects of output format. It can contain ordinary alphanumeric characters, along with escape characters, conversion specifiers, and other characters, organized as shown in the following examples. Table C.1 summarizes the basic syntax of fprintf. Consult MATLAB help for more details.

Suppose the variable Speed has the value 63.2. To display its value using three digits with one digit to the right of the decimal point, along with a message, the session is

```
>>fprintf('The speed is: %3.1f\n',Speed)
The speed is:   63.2
```

Here the "field width" is 3, because there are three digits in 63.2. You may want to specify a wide enough field to provide blank spaces or to accommodate an unexpectedly large numerical value. The % sign tells MATLAB to interpret the

Table C.1 Display formats with the `fprintf` function

Syntax	Description
`fprintf('format',A, ...)`	Displays the elements of the array `A`, and any additional array arguments, according to the format specified in the string `'format'`.
`'format'` structure	`%[-][number1.number2]C`, where `number1` specifies the minimum field width, `number2` specifies the number of digits to the right of the decimal point, and `C` contains control codes and format codes. Items in brackets are optional. `[-]` specifies left justified.

Control codes		Format codes	
Code	**Description**	**Code**	**Description**
`\n`	Start new line.	`%e`	Scientific format with lowercase e.
`\r`	Beginning of new line.	`%E`	Scientific format with uppercase E.
`\b`	Backspace.	`%f`	Decimal format.
`\t`	Tab.	`%g`	`%e` or `%f`, whichever is shorter.
`''`	Apostrophe.		
`\\`	Backslash.		

following text as codes. The code `\n` tells MATLAB to start a new line after displaying the number.

The output can have more than one column, and each column can have its own format. For example,

```
>>r = [2.25:20:42.25];
>>circum = 2*pi*r;
>>y = [r;circum];
>>fprintf('%5.2f %11.5g\n',y)
  2.25       14.137
 22.25        139.8
 42.25       265.46
```

Note that the `fprintf` function displays the *transpose* of the matrix `y`.

Format code can be placed within text. For example, note how the period after the code `%6.3f` appears in the output at the end of the displayed text.

```
>>fprintf('The first circumference is %6.3f.\n',circum(1))
The first circumference is 14.137
```

An apostrophe in displayed text requires two single quotes. For example:

```
>>fprintf('The second circle''s radius %15.3e is large.\n',r(2))
The second circle's radius      2.225e+001 is large.
```

A minus sign in the format code causes the output to be left justified within its field. Compare the following output with the preceding example:

```
>>fprintf('The second circle''s radius %-15.3e is large.\n',r(2))
The second circle's radius 2.225e+001       is large.
```

Control codes can be placed within the format string. The following example uses the tab code (`\t`).

```
>>fprintf('The radii are:%4.2f \t %4.2f \t %4.2f\n',r)
The radii are:   2.25      22.25   42.25
```

The `disp` function sometimes displays more digits than necessary. We can improve the display by using the `fprintf` function instead of `disp`. Consider the program:

```
p = 8.85; A = 20/100^2;
d = 4/1000; n = [2:5];
C = ((n - 1).*p*A/d);
table (:,1) = n';
table (:,2) = C';
disp (table)
```

The `disp` function displays the number of decimal places specified by the `format` command (4 is the default value).

If we replace the line `disp(table)` with the following three lines,

```
E='';
fprintf('No.Plates Capacitance (F) X e12 %s\n',E)
fprintf('%2.0f \t \t \t %4.2f\n',table')
```

we obtain the following display:

```
2          4.42
3          8.85
4          13.27
5          17.70
```

The empty matrix `E` is used because the syntax of the `fprintf` statement requires that a variable be specified. Because the first `fprintf` is needed to display the table title only, we need to fool MATLAB by supplying it with a variable whose value will not display.

Note that the `fprintf` command truncates the results, instead of rounding them. Note also that we must use the transpose operation to interchange the rows and columns of the `table` matrix in order to display it properly.

Only the real part of complex numbers will be displayed with the `fprintf` command. For example:

```
>>z = -4+9i;
>>fprintf('Complex number:   %2.2f \n',z)
Complex number:   -4.00
```

Instead you can display a complex number as a row vector. For example, if $w = -4 + 9i$:

```
>>w = [-4,9];
>>fprintf('Real part is %2.0f. Imaginary part is %2.0f. \n',w)
Real part is -4. Imaginary part is 9.
```

APPENDIX D

References

[Brown, 1994] Brown, T. L.; H. E. LeMay, Jr.; and B. E. Bursten. *Chemistry: The Central Science*. 6th ed. Upper Saddle River, NJ: Prentice Hall, 1994.

[Eide, 1998] Eide, A. R.; R. D. Jenison; L. H. Mashaw; and L. L. Northup. *Introduction to Engineering Problem Solving.* New York: McGraw-Hill, 1998.

[Felder, 1986] Felder, R. M. and R. W. Rousseau. *Elementary Principles of Chemical Processes*. New York: John Wiley & Sons, 1986.

[Garber, 1999] Garber, N.J. and L. A. Hoel. *Traffic and Highway Engineering.* 2nd ed. Pacific Grove, CA: PWS Publishing, 1999.

[Jayaraman, 1991] Jayaraman, S. *Computer-Aided Problem Solving for Scientists and Engineers.* New York: McGraw-Hill, 1991.

[Kreyzig, 1999] Kreyzig, E. *Advanced Engineering Mathematics.* 8th ed. New York: John Wiley & Sons, 1999.

[Kutz, 1999] Kutz, M., editor. *Mechanical Engineers' Handbook.* 2nd ed. New York: John Wiley & Sons, 1999.

[Palm, 2005] Palm, W. *System Dynamics.* New York: McGraw-Hill, 2005.

[Rizzoni, 1996] Rizzoni, G. *Principles and Applications of Electrical Engineering.* 2nd ed. Homewood, IL: Irwin, 1996.

[Starfield, 1990] Starfield, A. M.; K. A. Smith; and A. L. Bleloch. *How to Model It: Problem Solving for the Computer Age.* New York: McGraw-Hill, 1990.

Answers to Selected Problems

Chapter 1

2. (*a*) -13.3333; (*b*) 0.6; (*c*) 15; (*d*) 1.0323

8. (*a*) $x + y = -3 - 2i$; (*b*) $xy = -13 - 41i$; (*c*) $x/y = -1.72 + 0.04i$

18. $-15.685, 0.8425 \pm 3.4008i$

27. $x = -3, y = 10, z = 4$

28. $L = 12.58$ m, perimeter $= 39.65$ m

Chapter 2

3.

$$\mathbf{A} = \begin{bmatrix} 0 & 6 & 12 & 18 & 24 & 30 \\ -20 & -10 & 0 & 10 & 20 & 30 \end{bmatrix}$$

7. (*a*) Length = 3, absolute value = `[2, 4, 7]`; (*b*) Same as (*a*); (*c*) Length = 3, absolute value = `[5.831, 5, 7.2801]`

12. (*a*)

$$\mathbf{A} + \mathbf{B} + \mathbf{C} = \begin{bmatrix} -4 & 2 \\ 22 & 15 \end{bmatrix}$$

(*b*)

$$\mathbf{A} - \mathbf{B} + \mathbf{C} = \begin{bmatrix} -16 & 12 \\ -2 & 19 \end{bmatrix}$$

13. (*a*) `[1024, -128; 144, 32]`; (*b*) `[4, -8; 4, 8]`; (*c*) `[4096, -64; 216, -8]`

14. (*a*) Work done on each segment, in joules (1 J = 1 N · m) is 800, 275, 525, 750, 1800; (*b*) Total work done = 4150 J.

27.

$$\mathbf{AB} = \begin{bmatrix} -47 & -78 \\ 39 & 64 \end{bmatrix}$$

$$\mathbf{BA} = \begin{bmatrix} -5 & -3 \\ 48 & 22 \end{bmatrix}$$

30. 60 tons of copper, 67 tons of magnesium, 6 tons of manganese, 76 tons of silicon, and 101 tons of zinc

33. $M = 869$ N · m if **F** is in newtons and r is in meters.

40. $2.8x - 5.12$ with a remainder of $50.04x - 11.48$

41. 0.5676

Chapter 3

1. (*a*) 3, 3.1623, 3.6056; (*b*) $1.7321i, 0.2848 + 1.7553i, 0.5503 + 1.8174i$; (*c*) $15 + 21i, 22 + 16i, 29 + 11i$; (*d*) $-0.4 - 0.2i, -0.4667 - 0.0667i, -0.5333 + 0.0667i$

2. (*a*) $|xy| = 105, \angle xy = -2.6$ rad; (*b*) $|x/y| = 0.84, \angle x/y = -1.67$ rad

3. (*a*) 1.01 rad (58°); (*b*) 2.13 rad (122°); (*c*) -1.01 rad ($-58°$); (*d*) -2.13 rad ($-122°$)

7. $F_1 = 198$ N if $\mu = 0.3$, $F_2 = 100$ N, and $\beta = 130°$.

10. For the test values, $t = 7.46$ and 2.73 sec.

Chapter 4

4. (*a*) `z = 1`; (*b*) `z = 0`; (*c*) `z = 1`; (*d*) `z = 1`

5. (*a*) `z = 0`; (*b*) `z = 1`; (*c*) `z = 0`; (*d*) `z = 4`; (*e*) `z = 1`; (*f*) `z = 5`; (*g*) `z = 1`; (*h*) `z = 0`

6. (*a*) `z = [0, 1, 0, 1, 1];`
(*b*) `z = [0, 0, 0, 1, 1];`
(*c*) `z = [0, 0, 0, 1, 0];`
(*d*) `z = [1, 1, 1, 0, 1]`

11. (*a*) `z = [1, 1, 1, 0, 0, 0];`
(*b*) `z = [1, 0, 0, 1, 1, 1];`
(*c*) `z = [1, 1, 0, 1, 1, 1];`
(*d*) `z = [0, 1, 0, 0, 0, 0]`

13. (*a*) \$7300; (*b*) \$5600; (*c*) 1200 shares; (*d*) \$15,800

25. (*a*) $x = 9$, $y = 16$ m

28. 33 years

30. $W = 300$ N. If $W = 300$, the wire tensions are $T_i = 429, 471, 267, 233, 200$, and 100 N, respectively.

42. Weekly inventory for cases (*a*) and (*b*):

Week	1	2	3	4	5
Inventory (*a*)	50	50	45	40	30
Inventory (*b*)	30	25	20	20	10
Week	6	7	8	9	10
Inventory (*a*)	30	30	25	20	10
Inventory (*b*)	10	5	0	0	(<0)

Chapter 5

1. Production is profitable for $Q \geq 10^8$ gallons per year. The profit increases linearly with Q, so there is no upper limit on the profit.

3. To two significant digits, the two roots are $x = 1$ and $x = 4.5$.

5. The left end is 47 m above the reference line. The right end is 110 m above the reference line.

10. 0.54 rad (31°).

14. The steady-state value of y is $y = 1$. $y = 0.98$ at $t = 4/b$.

18. (*a*) The ball will rise 1.68 m and will travel 9.58 m horizontally before striking the ground after 1.17 s.

30. (*a*) $y = 53.5x - 1354.5$;
(*b*) $y = 3.58 \times 10^3 x^{-0.976}$;
(*c*) $y = 2.06 \times 10^5 (10)^{-0.0067x}$

32. (*a*) $b = 1.2603 \times 10^{-4}$; (*b*) 836 years; (*c*) between 760 and 928 years ago

36. If unconstrained to pass through the origin, $f = 0.1999x - 0.0147$. If constrained to pass through the origin, $f = 0.1977x$.

38. $d = 0.0509v^2 + 1.1054v + 2.3571$, $J = 10.1786$, $S = 57{,}550$, $r^2 = 0.9998$

40. $y = 40 + 9.6x_1 - 6.75x_2$. Maximum percent error is 7.125 percent.

Chapter 6

3. $x = 2$, $y = -3$, $z = 5$

12. $T_1 = 19.8°\text{C}$, $T_2 = -7.0°\text{C}$, $T_3 = -9.7°\text{C}$. Heat loss rate is 66.8 W.

19. (*a*) $\mathbf{C} = \mathbf{B}^{-1}(\mathbf{A}^{-1}\mathbf{B} - \mathbf{A})$
(*b*)

$$\mathbf{C} = \begin{bmatrix} -0.6212 & -2.3636 \\ 1.197 & 2.1576 \end{bmatrix}$$

20. $x = 3c$, $y = -2c$, $z = c$

25. The nonunique solution is $x = 1.38z + 4.92$, $y = -0.077z - 1.38$, where z can have any value.

28. The exact and unique solution is $x = 8$, $y = 2$.

29. There is no exact solution. The least squares solution is $x = 6.09$, $y = 2.26$.

Chapter 7

4. (Answers rounded to integers.) You would expect to obtain a sum of 8 forty-two times; a sum of either 3, 4, or 5 seventy-five times, and a sum less than 9 two hundred and seventeen times.

8. (*a*) 99%; (*b*) 68%

12. (*a*) Mean pallet weight is 3000 lb, standard deviation is 10.95 lb; (*b*) 9 percent

19. Mean yearly profit = \$64,609. Minimum expected profit = \$51,340. Maximum expected profit = \$79,440. Standard deviation of yearly profit = \$5967.

23. The value at 5 P.M. is 22.5, the value at 9 P.M. is 16.5.

Chapter 8

1. 2360

5. Work = $0.8k$

7. 6 m/s

10. 13.65 ft

13. 1363.4 m/s

21. (*a*) $v(t) = (f/500)(1 - e^{-t/2})$; (*b*) Steady-state speed is $f/500$ m/s. The speed is within 2 percent of this value after $t = 8$ s.

22. (*a*) $y(t) = C_1 e^{-3t} \sin 5t + C_2 e^{-3t} \cos 5t$;
(*b*) $y(t) = C_1 e^{-8t} + C_2 e^{-5t}$

32. $\dot{x}_1 = -0.3x_1 - 0.7x_2 + 0.25f$, $\dot{x}_2 = x_1$, $y = 0.4x_2$

34. $\ddot{x}_1 + 7\dot{x}_1 + 14x_1 = 2u$

Chapter 10

3. (*a*) $60x^5 - 10x^4 + 108x^3 - 49x^2 + 71x - 24$; (*b*) 2546

4. $A = 1$, $B = -2a$, $C = 0$, $D = -2b$, $E = 1$, $F = r^2 - a^2 - b^2$

6. (*a*) $b = c\cos A \pm \sqrt{a^2 - c^2\sin^2 A}$; (*b*) $b = 5.69$

8. (*a*) $x = \pm 10\sqrt{(4b^2 - 1)/(400b^2 - 1)}$, $y = \pm b\sqrt{99/(400b^2 - 1)}$; (*b*) $x = 0.9685$, $y = 0.4976$

12. Critical points: $x = 0$ and $x = 2$. Local minimum at $x = 0$. Inflection points at $x = 2$ and $x = 2/3$

18. $h = 15\sqrt{2}$

19. 49.68 m/s

29. (*a*) 2; (*b*) 0; (*c*) 0

36. (*a*) $y(t) = [0.6y(0) + 0.2v(0)]\, e^{-3t}\sin 5t + y(0)e^{-3t}\cos 5t$; (*b*) $y(t) = (1/3)[v(0) + 8y(0)]\, e^{-5t} - (1/3)[v(0) + 5y(0)]e^{-8t}$

45. (*a*) $m(t) = (K^2/4)te^{-Kt/2}$; (*b*) $m_{peak} = K/5.4366$

48. $s^2 + 13s + 42 - 6k$, $s = \left(-13 \pm \sqrt{1 + 24k}\right)/2$

49. $x = 62/(16c + 15)$, $y = (129 + 88c)/(16c + 15)$

INDEX

Symbols

MATLAB Commands

Simulink Blocks

Topics

A